Edited by
Lutz F. Tietze

Domino Reactions

Related Titles

Hanessian, S., Giroux, S., Merner, B.L.

Design and Strategy in Organic Synthesis

From the Chiron Approach to Catalysis

2013
Print ISBN: 978-3-527-33391-2

Pignataro, B. (ed.)

New Strategies in Chemical Synthesis and Catalysis

2012
Print ISBN: 978-3-527-33090-4,
also available in digital formats

Majumdar, K.C., Chattopadhyay, S.K. (eds.)

Heterocycles in Natural Product Synthesis

2011
Print ISBN: 978-3-527-32706-5,
also available in digital formats

Majumdar, K.C., Chattopadhyay, S.K. (eds.)

Heterocycles in Natural Product Synthesis

2011
Print ISBN: 978-3-527-32706-5,
also available in digital formats

Nicolaou, K.C., Chen, J.S.

Classics in Total Synthesis III

Further Targets, Strategies, Methods

2011
Print ISBN: 978-3-527-32958-8

Carreira, E.M., Kvaerno, L.

Classics in Stereoselective Synthesis

2009
Print ISBN: 978-3-527-32452-1

Xu, P., Wang, W. (eds.)

Catalytic Cascade Reactions

2013
Print ISBN: 978-1-118-01602-2,
also available in digital formats

Zhang, W., Cue, B.W. (eds.)

Green Techniques for Organic Synthesis and Medicinal Chemistry

2012
Print ISBN: 978-0-470-71151-4,
also available in digital formats

Starkey, L.

Introduction to Strategies for Organic Synthesis

2012
Print ISBN: 978-1-118-18083-9

Oestreich, M.M. (ed.)

The Mizoroki-Heck Reaction - The Heck Reaction

2009
Print ISBN: 978-0-470-03394-4,
also available in digital formats

Edited by Lutz F. Tietze

Domino Reactions

Concepts for Efficient Organic Synthesis

WILEY-VCH
Verlag GmbH & Co. KGaA

Editor

Prof. Dr. Lutz F. Tietze
Georg-August University
Institute of Organic and Biomolecular Chemistry
Tammannstr. 2
37077 Göttingen
Germany

Cover
Jungle Photo. Source: Fotolia © Chakka

Library of Congress Card No.: applied for

British Library Cataloguing-in-Publication Data
A catalogue record for this book is available from the British Library.

Bibliographic information published by the Deutsche Nationalbibliothek
The Deutsche Nationalbibliothek lists this publication in the Deutsche Nationalbibliografie; detailed bibliographic data are available on the Internet at <http://dnb.d-nb.de>.

Print ISBN: 978-3-527-33432-2
ePDF ISBN: 978-3-527-67133-5
ePub ISBN: 978-3-527-67132-8
Mobi ISBN: 978-3-527-67131-1
oBook ISBN: 978-3-527-67130-4

Cover Design Adam-Design, Weinheim, Germany
Typesetting Laserwords Private Ltd., Chennai, India
Printing and Binding Markono Print Media Pte Ltd, Singapore

Printed on acid-free paper
Printed in Singapore

Contents

Preface

The synthesis of chemical compounds is a key issue in chemistry, both in academia and industry. A simple statement of general relevance is the saying "you cannot investigate a compound which you do not have in your hands and you cannot sell a substance which you did not make." However, the aspects of synthesis have changed over the years. At the beginning, the development of synthetic methods such as the electrophilic aromatic substitution, the aldol reaction or the Diels–Alder cycloaddition was in the focus. Then the selectivity as the chemo-, regio-, diastereo-, and enantioselectivity was the main concern. Now, new aspects in synthesis have arisen, which are part of green chemistry: efficiency, reduction of waste, saving our resources, protecting our environment, and, finally, also economic advantages by reducing the transformation time and the amount of chemicals needed. To meet all these requirements, the domino concept was introduced by me, which, since its presentation and the first reviews, has grown immensely in the last years. In this book, experts in the different fields of domino reactions have put together their knowledge, and I am very grateful to all of them for their excellent contributions. Moreover, I would like to thank Martina Pretor for her fabulous help in preparing the book. I am also grateful to the publishers Wiley/VCH, especially Dr. Elke Maase and Dr. Bernadette Gmeiner, for their support.

Finally, I would like to express my deep thanks to the University of Göttingen, the State of Lower Saxony, the German Research Foundation (DFG), the Volkswagen Foundation, the German Ministry of Education and Research (BMBF), the European Community and the Fonds der Chemischen Industrie as well as the Alexander von Humbold Foundation, the Konrad–Adenauer–Foundation and the German National Academic Foundation for their continuous support of our work on domino reactions and other topics. I am also very thankful to many Chemical Companies worldwide, in particular the BASF and the Bayer AG.

Göttingen, June 6th, 2013 *Lutz F. Tietze*

List of Contributors

Anand Acharya
New Chemistry Unit
Jawaharlal Nehru Centre for Advanced Scientific Research
Jakkur
Bangalore 560 064
Karnataka
India

Guanghui An
Texas Tech University
Department of Chemistry and Biochemistry
Street Boston and Broadway
Lubbock
TX, 79409-1061
USA

Matthias Beller
Leibniz Institute for Catalysis
Albert-Einstein-Str. 29a
18059 Rostock
Germany

Michael Boomhoff
University of Leipzig
Institute of Organic Chemistry
Johannisallee 29
04103 Leipzig
Germany

Kamal M. Dawood
Technische Universität Dresden
Department of Chemistry
Bergstrasse 66
01069 Dresden
Germany

Svenia-C. Düfert
Georg-August University
Institute of Organic and Biomolecular Chemistry
Tammannstr. 2
37077 Göttingen
Germany

Vincent Eschenbrenner-Lux
Max Planck Institute of Molecular Physiology
Otto-Hahn-Str. 11
44227 Dortmund
Germany

Dhandapani Ganapathy
Indian Institute of Technology Madras
Department of Chemistry
Chennai 600 036
Tamil Nadu
India

Judith Hierold
Georg-August University
Institute of Organic and
Biomolecular Chemistry
Tammannstr. 2
37077 Göttingen
Germany

Hiriyakkanavar Ila
New Chemistry Unit Jawaharlal
Nehru Centre for Advanced
Scientific Research
Jakkur
Bangalore 560 064
Karnataka
India

Bo Jiang
Jiangsu Normal University
School of Chemistry and
Chemical Engineering
Shanghai Road 101
New District of Tongshan
Xuzhou, 221116
P. R. China

Iyyanar Karthikeyan
Indian Institute of Technology
Madras
Department of Chemistry
Chennai 600 036
Tamil Nadu
India

Kamal Kumar
Max Planck Institute of Molecular
Physiology
Otto-Hahn-Str. 11
44227 Dortmund
Germany

Mark Lautens
Davenport Research Laboratories
University of Toronto
Department of Chemistry
80 St. George St.
Toronto
ON M5S 3H6
Canada
USA

Guigen Li
Texas Tech University
Department of Chemistry and
Biochemistry
Street Boston and Broadway
Lubbock
TX, 79409-1061
USA

and

Nanjing University
Institute of Chemistry &
BioMedical Sciences
22 Hankou Road
Nanjing 210093
P. R. China

Peter Metz
Technische Universität Dresden
Department of Chemistry
Bergstrasse 66
01069 Dresden
Germany

Helfried Neumann
Leibniz Institute for Catalysis
Albert-Einstein-Str. 29a
18059 Rostock
Germany

Lukas J. Patalag
Technical University of Braunschweig
Institute of Organic Chemistry
Hagenring 30
38106 Braunschweig
Germany

Helène Pellissier
Aix Marseille Universitè
CNRS, iSm2 UMR 7313
13397 Marseille
France

Saravanan Peruncheralathan
National Institute of Science Education and Research
Institute of Physics Campus
School of Chemical Sciences
Bhubaneswar 751 005
Orissa
India

Hiroaki Sasai
Osaka University
The Institute of Scientific and Industrial Research (ISIR)
Mihogaoka Ibaraki-shi
Osaka 567-0047
Japan

Christoph Schneider
University of Leipzig
Institute of Organic Chemistry
Johannisallee 29
04103 Leipzig
Germany

Govindasamy Sekar
Indian Institute of Technology Madras
Department of Chemistry
Chennai 600 036
Tamil Nadu
India

Scott G. Stewart
The University of Western Australia
School of Chemistry and Biochemistry
35 Stirling Highway
Crawley
WA 6009
Australia

Shinobu Takizawa
Osaka University
The Institute of Scientific and Industrial Research (ISIR)
Mihogaoka Ibaraki-shi
Osaka 567-0047
Japan

Lutz F. Tietze
Georg-August University
Institute of Organic and Biomolecular Chemistry
Tammannstr. 2
37077 Göttingen
Germany

Gavin Chit Tsui
Max-Planck-Institut für Kohlenforschung
Kaiser-Wilhelm-Platz 1
45470 Mülheim an der Ruhr
Germany

Shu-Jiang Tu
Jiangsu Normal University
School of Chemistry and
Chemical Engineering
Shanghai Road 101
New District of Tongshan
Xuzhou, 221116
P. R. China

Herbert Waldmann
Max Planck Institute of Molecular
Physiology
Otto-Hahn-Str. 11
44227 Dortmund
Germany

Qian Wang
Ecole Polytechnique Fédérale de
Lausanne
Institute of Chemical Sciences
and Engineering
1015 Lausanne
Switzerland

Daniel B. Werz
Technical University of
Braunschweig
Institute of Organic Chemistry
Hagenring 30
38106 Braunschweig
Germany

Xiao-Feng Wu
Leibniz Institute for Catalysis
Albert-Einstein-Str. 29a
18059 Rostock
Germany

Jieping Zhu
Ecole Polytechnique Fédérale de
Lausanne
Institute of Chemical Sciences
and Engineering
1015 Lausanne
Switzerland

List of Abbreviations

(*S*,*S*)-MeDuPhos	(+)-1,2-bis[(2*S*,5*S*)-2,5-dimethylphospholano]benzene
$(TMS)_2NH$	hexamethyldisilazane or bis(trimethylsilyl)amine
[Bmim]	1-butyl-3-methylimidazolium
Ac	acetyl
acac	acetylacetone
ACCN	1,1′-azobis(cyclohexanecarbonitrile)
Ac_2O	acetic anhydride
AcOH	acetic acid
AIBN	2,2′-azobisisobutyronitrile
All	allyl
Ar	aryl
ARC	anionic relay chemistry
ASG	anion stabilizing group
ATBT	allyltri-*n*-butyltin
atm	standard atmosphere
BAIB	(diacetoxyiodo)benzene
BER	borohydride exchange resin
$BF_3 \cdot OEt_2$	boron trifluoride–diethyl ether complex
BHT	butylhydroxytoluene
BINAP	2,2′-bis(diphenylphosphino)-1,1′-binaphthalene
BINAPO	2-diphenylphosphino-2′-diphenylphosphinyl-1,1′-binaphthalene
BINOL	1,1′-bi-2-naphthol
Biphep	1,1′-biphenyl-2,2′-diphenylphosphine
Bn	benzyl
Boc	*tert*-butoxycarbonyl
borsm	based on recovered starting material
bpz	2,2′-bipyrazine
Bu	butyl
Bz	benzoyl
CA	cycloaddition
CAN	ceric ammonium nitrate

Cbz	carbonylbenzyloxy
CD	circular dichroism
cf.	compare (lat. *confer*)
CM	cross-metathesis
cod	1,5-cyclooctadiene
coe	cyclooctene
Cp	cyclopentadienyl
CR	cycloreversion
CSA	camphorsulfonic acid
Cy	cyclohexyl
d	day
DA	Diels–Alder reactions
DABCO	1,4-diazabicyclo[2.2.2]octane
DAIB	(diacetoxyiodo)benzene
dba	dibenzylidenacetone
DBU	1,8-diazabicyclo[5.4.0]undec-7-ene
DCB	1,2-dichloroisobutane
DCE	1,2-dichloroethane
DCM	dichloromethane
DDQ	2,3-dichloro-5,6-dicyano-1,4-benzoquinone
de	diastereomeric excess
DFT	density functional theory
DHQ	hydroquinine
DHQD	dihydroquinidine
DIBAL	diisobutylaluminum hydride
DIOP	4,5-bis(diphenylphosphinomethyl)-2,2-dimethyl-1,3-dioxolane
DIPEA	diisopropylethylamine
DKP	diketopiperazine
DLP	1,2-dichloroethane with lauroyl peroxide
DMA	*N,N*-dimethylacetamide
DMAD	dimethyl acetylenedicarboxylate
DME	dimethoxyethane
DMF	*N,N*-dimethylformamide
DMP	Dess–Martin-periodinane
DMPU	1,3-dimethyl-3,4,5,6-tetrahydro-2(1*H*)-pyrimidinone, *N,N*-dimethyl propylene urea
DMSO	dimethyl sulfoxide
DOS	diversity-oriented synthesis
dpm	dipivaloylmethane
dppe	1,2-bis(diphenylphosphino)ethane
dppf	1,2-bis(diphenylphosphino)ferrocene
dppp	1,3-bis(diphenylphosphino)propane
dr	diastereomeric ratio
DTBP	2,6-di-*tert*-butylpyridine
E	electrophile

EC	electrocyclization
ee	enantiomeric excess
equiv	equivalent
ERO	electrocyclic ring-opening
et al.	and others (lat. *et alii*)
Et	ethyl
EWG	electron-withdrawing group
Fmoc	9-fluorenylmethoxycarbonyl
fod	(6,6,7,7,8,8,8-heptafluoro-2,2-dimethyloctane-3,5-dionate
GAP	group-assisted purification
h	hour
HAT	hydrogen atom transfer
HFIP	hexafluoroisopropanol
HIV	human immunodeficiency virus
HMPA	hexamethylphosphortriamide
HOMO	highest occupied molecular orbital
i.e.	that means (lat. *id est*)
IBX	2-iodoxybenzoic acid
IMDA	intramolecular Diels–Alder reaction
L	ligand
LDA	lithium diisopropylamide
LiHMDS	lithium hexamethyldisilazide
LUMO	lowest unoccupied molecular orbital
MAOS	microwave-assisted organic synthesis
MBH	Morita–Baylis–Hillman
MDRs	multicomponent domino reactions
Me	methyl
MeCN	acetonitrile
MEK	methyl ethyl ketone
MEM	(2-methoxyethoxy)methyl
Mes	mesityl
MOM	methoxymethyl
MTM	methylthiomethyl
MW	microwave
NADH	nicotinamide adenine dinucleotide
NBS	*N*-bromosuccinimide
NCS	*N*-chlorosuccinimide
NMM	*N*-methyl morpholine
NMO	*N*-methylmorpholine-*N*-oxide
NMP	*N*-methyl-2-pyrrolidinone
Ns	*p*-nitrobenzenesulfonyl
Nu	nucleophile
Oct	octyl
o-DCB	*ortho*-dichlorobenzene
PCC	pyridinium chlorochromate

PET	photochemical electron transfer
PEG	polyethylene glycol
PFBA	pentafluorobenzoic acid
PG	protecting group
Ph	phenyl
Phen	9,10-phenanthroline
PhMe	toluene
PIDA	phenyliodine diacetate
Piv	pivalate
PMB	*p*-methoxybenzyl
PNO	pyridine-*N*-oxide
PPh_3	triphenylphosphine
PPTS	pyridinium *p*-toluenesulfonate
Pr	propyl
PS–BEMP	polystyrene–(2-*tert*-butylimino-2-diethylamino-1,-dimethyl-perhydro-1,3,2-diazaphosphorine)
PS–DMAP	polystyrene–dimethylaminopyridine
p-TsOH or *p*-TSA	*p*-toluenesulfonic acid
PVE	propargyl vinyl ether
Py	pyridine
R	rest
rac	racemic
RCM	ring-closing metathesis
ROM	ring-opening metathesis
RRM	ring-rearrangement metathesis
rt	room temperature
SEM	2-trimethylsilylethoxymethyl
SET	single electron transfer
sigR	sigmatropic rearrangement
S_N	nucleophilic substitution
S_N1	substitution nucleophilic unimolecular
S_N2	substitution nucleophilic bimolecular
SolFC	solvent free condition
SOMO	singly occupied molecular orbital
SPPS	solid-phase peptide synthesis
t	*tert*
TADDOL	(−)-(4*R*,5*R*)- or (+)(4*S*,5*S*)-2,2-dimethyl-α,α,α′,α′-tetraphenyl-1,3-dioxolane-4,5-dimethanol
TBA	tetra-*n*-butylammonium
TBA	tribromoacetic acid
TBAB	tetra-*n*-butylammonium bromide
TBAF	tetra-*n*-butylammonium fluoride
TBAI	tetra-*n*-butylammonium iodide
TBCHD	2,4,4,6-tetrabromo-2,5-cyclohexadienone
TBD	1,5,7-triazabicyclo[4.4.0]dec-5-ene

TBDMS or TBS	*tert*-butyldimethylsilyl
TBDPS or TBPS	*tert*-butyldiphenylsilyl
t-Bu	*tert*-butyl
t-BuOH	*tert*-butyl alcohol
t-BuOK	*tert*-butylate potassium
TC	thiophene-2-carboxylate
TEA	triethylamine
TEBA	benzyltriethylammonium chloride
TEMPO	(2,2,6,6-tetramethylpiperidin-1-yl)oxy
TES	triethylsilyl
TESOTf	triethylsilyltrifluoromethanesulfonate
Tf	trifluoromethanesulfonyl
TFA	trifluoroacetic acid
TFE	2,2,2-trifluorethanol
TfO	trifluoromethanesulfonate
TFP	tri-(2-furyl)phosphine
THF	tetrahydrofuran
TMSOTf	trimethylsilyl triflurомethanesulfonate
Thio	thiophene
TIPS	triisopropylsilyl
TMEDA	tetramethylethylendiamine
TMS	trimethylsilyl
TMSI	trimethylsilyl iodide or iodotrimethylsilane
Tol	tolyl
Ts	4-toluenesulfonyl (tosyl)
TS	transition state
TsOH	*p*-toluenesulfonic acid
TTMSS	tris(trimethylsilyl)silane
VAPOL	2,2′-diphenyl-(4-biphenanthrol)
vs.	as opposed to (lat. *versus*)
XPhos	2-dicyclohexylphosphino-2′,4′,6′-triisopropylbiphenyl

Introduction

The beginning of organic synthesis can be dated back to the year 1824 when Wöhler, later professor of chemistry at the Georg-August University in Göttingen, showed that inorganic matter could be transformed into organic matter without the *vis vitalis*, the so-called power of life. At that time, he prepared the natural product oxalic acid from dicyan by simple hydrolysis. Better known is the transformation of ammonium cyanate into urea by simple heating, in 1828 (Scheme 1) [1].

A second milestone in organic synthesis is the total synthesis of the indole alkaloid reserpine by Woodward in 1956 [2] using a Diels–Alder reaction as the key step (Scheme 2), and finally with the total synthesis of palytoxin in 1994, the toxin of dinoflagellate *Ostreopsis siamensis*, with 64 stereogenic centers and several (*E*)- and (*Z*)-double bonds, Kishi [3] has shown that chemists can prepare any organic compound (Scheme 3).

However, the synthesis of such a big molecule as palytoxin using a conventional stepwise approach with more than 100 steps is a singular great feat and can almost not be repeated. Thus, a 100-step synthesis with 80% yield per step would lead to only 0.00 000 002% as the total yield.

In contrast, a much better efficiency could be accomplished using domino reactions, which have been defined by us as processes of two or more bond forming reactions under identical reaction conditions, in which the latter transformations take place at the functionalities obtained in the former bond forming reactions [4]. In the processes one, two, three, or more substrates can be involved. Thus, multicomponent transformations are domino reactions per definition. In the meantime, several excellent reviews have also been published by other authors on this topic [5].

The quality and the usefulness of domino reactions are related to the increase of complexity and diversity in the final product compared to the starting material. Thus, the more steps a domino-process includes the greater is the probability to transform simple substrates to huge compounds. A further great advantage of the domino concept is its benefit to our environment and our natural resources, as it allows reducing the waste produced compared to normal procedures and minimize the amount of chemicals required for the preparation of a product. This also makes them economically favorable; moreover, they grant a decrease of the production time, which altogether would reduce furthermore the costs of any product.

Domino Reactions: Concepts for Efficient Organic Synthesis, First Edition. Edited by Lutz F. Tietze.

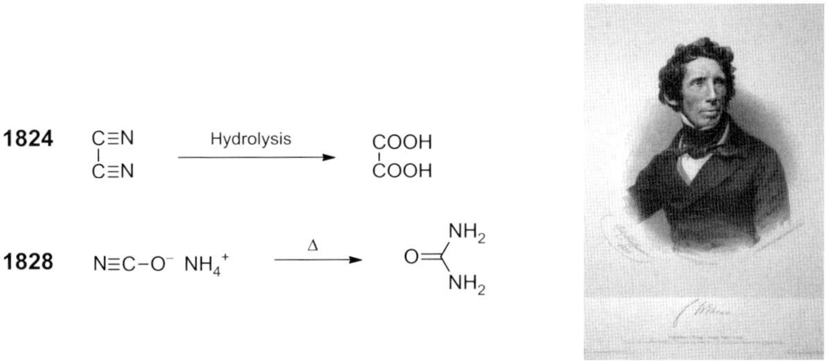

Scheme 1 Synthesis of oxalic acid and urea without a *vis vitalis* by Wöhler.

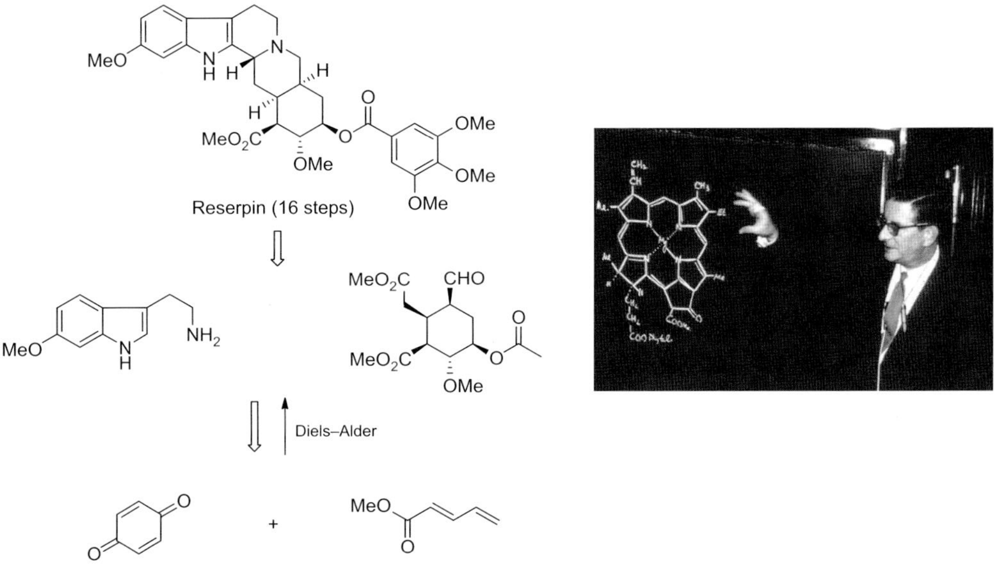

Scheme 2 Synthesis of the indole alkaloid reserpine by Woodward 1956.

Domino reactions usually show a good stereocontrol and good overall yields. Also very important is the fact that novel pathways can be developed, which cannot be followed in a stepwise approach, as in domino reactions intermediates can be unstable compounds, which are consumed as they are formed in a further step.

In our previous book on domino reactions [4h], we have classified domino reactions according to the mechanism of the different steps. This organizing principal will also be used in this book, and you will find chapters about transition metal catalysis including carbonylation, metathesis and CH-activation, nucleophilic substitutions, radical reactions, pericyclic reactions, Michael reactions, aldol reactions, oxidations, and reductions.

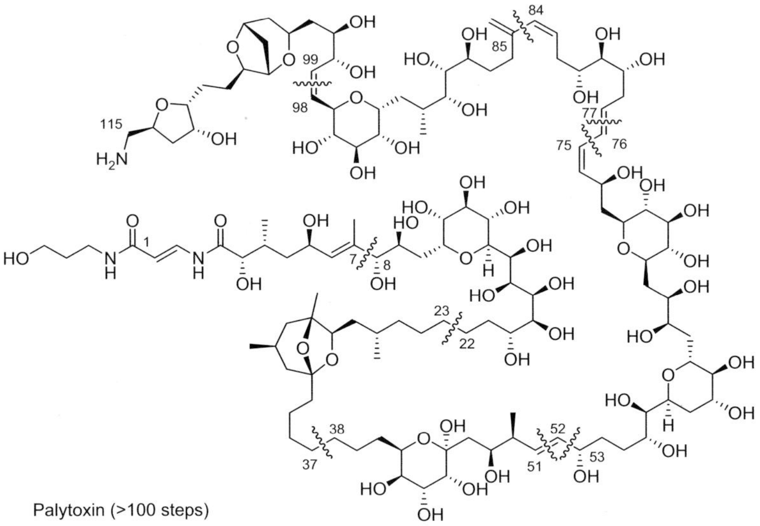

Palytoxin (>100 steps)

Scheme 3 Synthesis of palytoxin by Kishi 1994.

In addition, we have also included chapters that are related to the type of process as organocatalysis, enantio- and diastereoselective reactions, and multicomponent reactions as well as domino processes under microwave irradiation, high pressure, and in water. Finally, two chapters that are more product oriented have been included on the synthesis of compound collections and the synthesis of natural products and analogs.

This arrangement clearly leads to some overlap, which we have tried to minimize by discussing related subjects in-depth only in one chapter. However, to allow a correlation, some domino-processes are mentioned in more than one chapter.

Besides giving information to the reader about the development of domino reactions in the past years, the main purpose of this book is also to stimulate the design of novel domino reactions and use them in the synthesis of natural products and analogs, pharmaceuticals, agrochemicals, polymers, and materials not only in academic institutions but also in industry.

Per definition, all domino reactions take place in one reaction vessel without isolating any intermediates; however, they are much more than the so-called one-pot reactions, where you just put together different substrates and reagents after each other. The planning of domino reactions is like playing chess, where to be a reasonable player you will have to analyze four to five steps in advance. Thus, you have to predict the reaction pathways of all substrates and intermediates in your reaction mixture and in contrast to chess, where the movement of the different chess pieces is fixed, the reactivity of the chemical compounds can even be altered, for instance, by changing the pH-value or using different catalysts.

For the use and design of domino reactions in the synthesis of natural products, it is sometimes useful to look at the biosynthesis of these compounds. Thus, Nature is also using the concept of domino reactions and one of the most impressive examples is the biosynthesis of lanosterol from (*S*)-2,3-oxidosqualene, in which four new rings and six new stereogenic centers are formed [6]. This concept has later been exploited by developing a biomimetic synthesis of steroids [7] (Scheme 4).

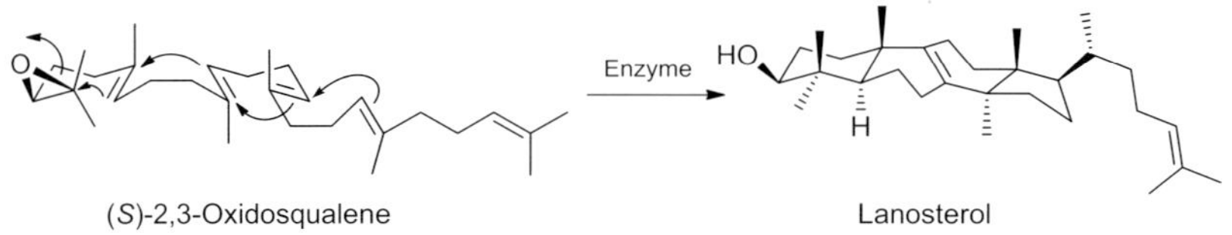

Scheme 4 Biosynthesis of lanosterol from (*S*)-2,3-oxidosqalene.

Another well-known example is the biosynthesis of atropine within the formation of the central skeleton tropinone. Using a twofold Mannich reaction, tropinone has been prepared in a single process [8] (Scheme 5).

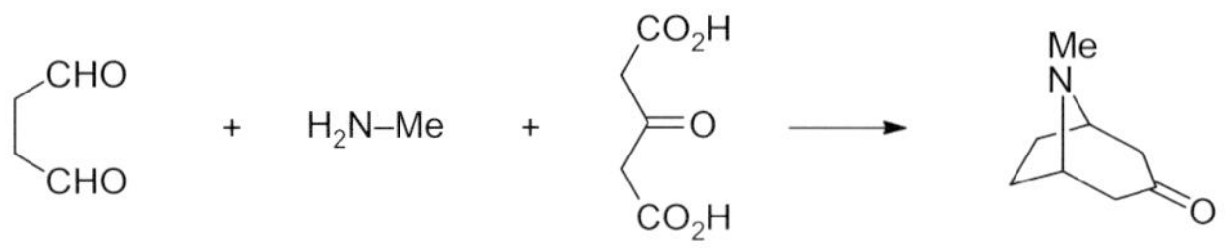

Scheme 5 Biomimetic synthesis of tropinone.

It should be stated that the book does not aim at comprehensiveness but the authors of the different chapters have looked for the most impressive examples and for clarifying the concept.

References

1. (a) Wöhler, F. (1828) *Ann. Phys. Chem.*, **88**, 253–256; (b) Wöhler, F. (1824) *Z. Physiol.*, **1**, S. 125–290.
2. Woodward, R.B. (1958) *Tetrahedron*, **2**, 1–57.
3. Suh, E.M. and Kishi, Y. (1994) *J. Am. Chem. Soc.*, **116**, 11205–11206.
4. For domino reactions, see: (a) Tietze, L.F. and Beifuss, U. (1993) *Angew. Chem. Int. Ed.*, **105**, 137–170 ; *Angew. Chem., Int. Ed. Engl.* 1993, **32**, 131–163; (b) Tietze, L.F. (1996) *Chem. Rev.*, **96**, 115–136; (c) Tietze, L.F. (1997) *Nachr. Chem. Tech. Lab.*, **45**, 1181–1187; (d) Tietze, L.F. and Lieb, M. (1998) *Curr. Opin. Chem. Biol.*, **2**, 363–37; (e) Tietze, L.F. and Haunert, F. (2000) in *Stimulating Concepts in Chemistry* (eds M. Shibasaki, J.F. Stoddart, and F. Vögtle), Wiley-VCH Verlag GmbH, Weinheim, pp. 39–64; (f) Tietze, L.F. and Modi, A. (2000) *Med. Res. Rev.*, **20**, 304–322; (g) Tietze, L.F. and Rackelmann, N. (2004) *Pure Appl. Chem.*, **76**, 1967–1983; (h) Tietze, L.F. and Rackelmann, N. (2005) in *Multicomponent Reactions* (eds J. Zhu and H. Bienaymé), Wiley-VCH Verlag GmbH, Weinheim, pp. 121–168; (i) Tietze, L.F., Brasche, G., and Gericke, K.M. (2006) *Domino Reactions in Organic Synthesis*, Wiley-VCH Verlag GmbH, Weinheim. (j) Tietze, L.F. and Levy, L. (2009) in *The Mizoroki–Heck Reaction* (ed. M. Oestreich), Wiley-VCH Verlag GmbH, Weinheim, pp. 281–344; (k) Tietze, L.F., Spiegl, D.A., and Brazel, C.C. (2009) in *Experiments in Green and Sustainable Chemistry* (eds H.W. Roesky and D.K. Kennepohl), Wiley-VCH Verlag GmbH, Weinheim, pp. 158–167; (l) Tietze, L.F.

and Düfert, A. (2010) in *Catalytic Asymmetric Conjugate Reactions* (ed. A. Cordova), Wiley-VCH Verlag GmbH, Weinheim, pp. 321–350; (m) Tietze, L.F. and Düfert, A. (2010) *Pure Appl. Chem.*, **82**, 1375–1392; (n) Tietze, L.F., Stewart, S., and Düfert, A. (2012) in *Modern Tools for the Synthesis of Complex Bioactive Molecules* (eds J. Cossy and S. Arseniyades), John Wiley & Sons, Inc, Hoboken, NJ, pp. 271–334; (o) Tietze, L.F., Düfert, M.A., and Schild, S.-C. (2012) in *Comprehensive Chirality*, Vol. 2 (eds E.M. Carreira and H. Yamamoto), Elsevier, Amsterdam, pp. 97–121.

5. (a) Pellissier, H. (2013) *Chem. Rev.*, **113**, 442–524; (b) Tejedor, D., Mendez-Abt, G., Cotos, L., and Garcia-Tellado, F. (2013) *Chem. Soc. Rev.*, **42**, 458–471; (c) Hussain, M., Van Sung, T., and Langer, P. (2012) *Synlett*, **23**, 2735–2744; (d) Aversa, M.C., Bonaccorsi, P., Madec, D., Prestat, G., and Poli, G. (2012) in *Innovative Catalysis in Organic Synthesis* (ed. P.G. Andersson), Wiley-VCH Verlag GmbH, Weinheim, pp. 47–76; (e) Galestokova, Z. and Sebesta, R. (2012) *Eur. J. Org. Chem.*, **2012**, 6688–6695; (f) Rousseaux, S., Vrancken, E., and Campagne, J.-M. (2012) *Angew. Chem. Int. Ed.*, **51**, 10934–10935; (g) Majumdar, K.C., Taher, A., and Nandi, R.K. (2012) *Tetrahedron*, **68**, 5693–5718; (h) Mueller, T.J.J. (2012) *Synthesis*, **44**, 159–174; (i) Wende, R.C. and Schreiner, P.R. (2012) *Green Chem.*, **14**, 1821–1849; (j) Grossmann, A. and Enders, D. (2012) *Angew. Chem. Int. Ed.*, **51**, 314–325; (k) Perumal, S. and Menendez, J.C. (2011) in *Targets in Heterocyclic Systems*, Vol. 15 (eds O.A. Attanasi and D. Spinelli), Royal Society of Chemistry, pp. 402–422; (l) Hummel, S. and Kirsch, S.F. (2011) *Beilstein J. Org. Chem.*, 847–859; (m) Ruiz, M., Lopez-Alvarado, P., Giorgi, G., and Menendez, J.C. (2011) *Chem. Soc. Rev.*, **40**, 3445–3454; (n) Nicolaou, K.C., Edmonds, D.J., and Bulger, P.G. (2006) *Angew. Chem. Int. Ed.*, **45**, 7134–7186.
6. (a) Corey, E.J., Russey, W.E., and Ortiz de Montellano, P.R. (1966) *J. Am. Chem. Soc.*, **88**, 4750. (b) Corey, E.J. and Virgil, S.C. (1991) *J. Am. Chem. Soc.*, **113**, 4025–4026; (c) Corey, E.J., Virgil, S.C., and Sashar, S. (1991) *J. Am. Chem. Soc.*, **113**, 8171–8172; (d) Corey, E.J., Virgil, S.C., Liu, D.R., and Sashar, S. (1992) *J. Am. Chem. Soc.*, **114**, 1524–1525.
7. Johnson, W.S. (1976) *Angew. Chem., Int. Ed. Engl.*, **15**, 9–17.
8. (a) Robinson, R. (1917) *J. Chem. Soc.*, **111**, 762–768; *J. Chem. Soc.* **111**, 876–899; (b) Schöpf, C., Lehmann, G., and Arnold, W. (1937) *Angew. Chem. Int. Ed.*, **50**, 779–787.

1
Transition-Metal-Catalyzed Carbonylative Domino Reactions

Xiao-Feng Wu, Helfried Neumann, and Matthias Beller

1.1
Introduction

"Sustainable development" has become one of the hottest terms in the twenty-first century. Of particular relevance in organic chemistry are the reaction efficiency and the avoidance of waste generation. With regard to sustainability, methodologies based on domino reactions, including multicomponent reactions, are a highly efficient strategy to synthesize complicated organic compounds. *Domino reactions* are defined as processes of two or more bond-forming reactions under identical conditions, in which the subsequent transformations take place at the functionalities that are obtained in the previous bond-forming transformations. Using domino reactions, complicated compounds can be relatively easily prepared from simple substrates. No tedious preparation of intermediates and purification processes are needed, which are the most energy-consuming and waste-generating steps in organic synthesis [1]. One prominent example is the domino Knoevenagel/hetero-Diels–Alder reaction, in which dihydropyrans could be straightforwardly synthesized from readily available starting materials [2].

Transition-metal catalysts play an ever-increasing and important role in modern chemistry [3]. Numerous transition-metal-catalyzed coupling reactions have been developed and applied in the total synthesis of natural products, such as the Suzuki reaction, the Negishi reaction, the Heck reaction, and many others [4]. Interestingly, the power of transition-metal catalysts is even more visible in the area of domino reactions, where terms such as *palladium walking* show the value of transition metals in bond formations.

Carbonylation reactions are interesting and important with regard to both industrial and academic research [5]. In these reactions, carbon monoxide (CO) can be used as one of the cheapest C1 sources. By introducing one or even more CO units into the parent molecules, carbonyl-containing products are easily prepared, which can be further modified to yield important chemicals for organic synthesis. Following the definition of domino reactions, we realized to our surprise that all the carbonylative coupling reactions belong to domino reactions, in which at least two C–C bonds were formed under the same reaction conditions.

Domino Reactions: Concepts for Efficient Organic Synthesis, First Edition. Edited by Lutz F. Tietze.

In order to assess the value of domino and carbonylation reactions, and also to differentiate them from normal transition-metal-catalyzed carbonylation reactions, in this chapter we will only describe the carbonylation reactions that produce at least three bonds under the same conditions.

1.2
Transition-Metal-Catalyzed Carbonylative Domino Reactions

Transition-metal-catalyzed carbonylation reactions have shown impressive progress during past few decades; especially, the use of ruthenium, rhodium, and palladium as catalysts is widespread. More recently, iron and copper catalysts have also been attracting the attention of synthetic chemists because of their low cost and environmentally benign properties.

1.2.1
Ruthenium-Catalyzed Carbonylative Domino Reactions

Compared with metathesis [6], the ability of ruthenium catalysts in carbonylation is also impressive.

In 1998, the first ruthenium-catalyzed cyclocarbonylation of yne-aldehydes was studied by the group of Murai [7]. Bicyclic α,β-unsaturated γ-butyrolactones were synthesized in good to excellent yields (Scheme 1.1a), and two proposed reaction mechanisms were discussed for this transformation. One involved a five-membered metalacycle formed via a [2+2+1] cycloaddition, and the other proceeded through a ruthenium acyl intermediate that was generated from the oxidative addition of an aldehyde C–H bond to ruthenium. Later on, Kang and coworkers [8] developed a ruthenium-catalyzed cyclocarbonylation of allenyl aldehydes and allenyl ketones to synthesize various α-methylene-γ-butyrolactones in 48–85% yields (Scheme 1.1b). More recently, Snapper and Finnegan prepared polycyclic lactones in moderate to good yields through ruthenium catalysis, in which a ring-closing metathesis/hetero-Pauson–Khand reaction mechanism was proposed (Scheme 1.1c) [9]. This strategy was also adopted for the preparation of cyclopentenones [10].

Some intermolecular carbonylative cycloaddition reactions were developed as well. In 1999, the group of Murai published a ruthenium-catalyzed intermolecular cyclocoupling of ketones, ethylene, and CO, producing lactones in good yields (Scheme 1.2) [11a]. This reaction showed the catalytic synthesis of heterocycles via an intermolecular carbonylative [2+2+1] cycloaddition for the first time. Many different ketones, such as α-dicarbonyl compounds and N-heterocyclic ketones, are used in this cycloaddition, and the addition of phosphines promotes the reactions of α-dicarbonyl compounds. Among the tested phosphines, $P(4\text{-}CF_3C_6H_4)_3$ has proved to be the ligand of choice. Beside ethylene cyclic olefins, unpolarized terminal olefins and internal alkynes could also be employed successfully in the synthesis, yielding highly functionalized lactones. An aromatic keto ester substituted with a CF_3 group accelerated the reaction of the keto ester with ethylene. On the other hand, by using

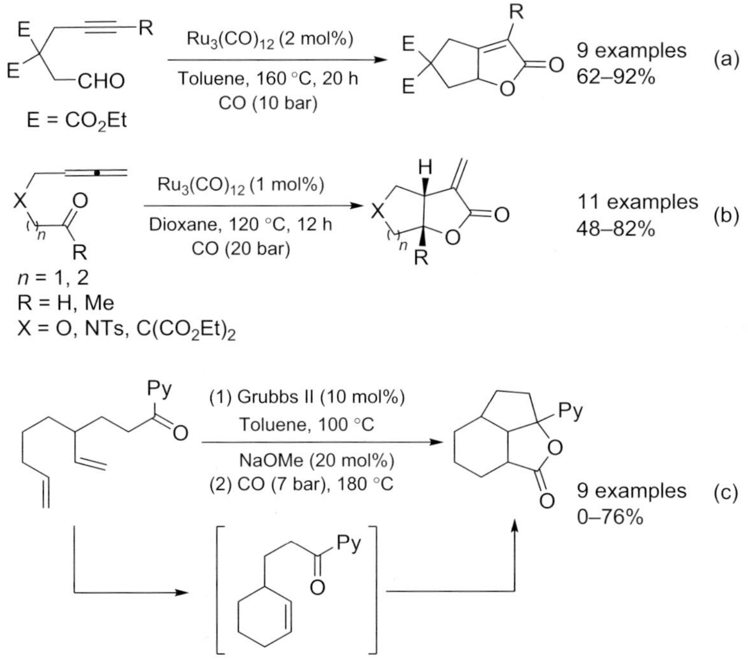

Scheme 1.1 (a–c) Ruthenium-catalyzed hetero-Pauson–Khand reactions.

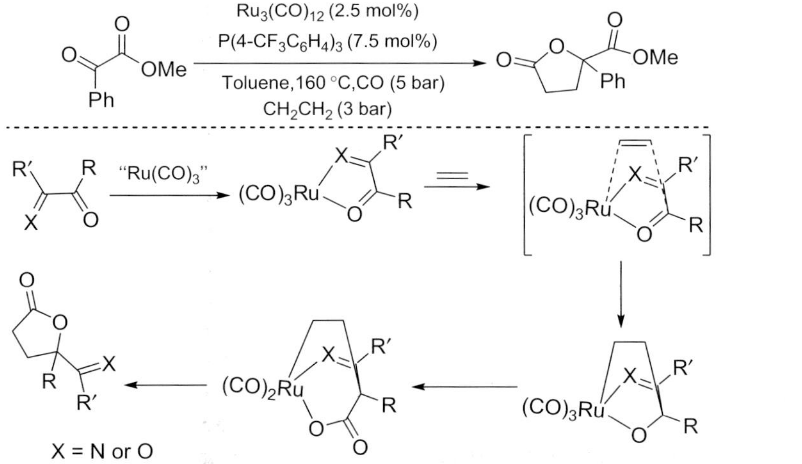

Scheme 1.2 Ruthenium-catalyzed intermolecular carbonylative cycloaddition reaction.

aromatic N-heterocyclic ketones, the rate of the reaction with ethylene slowed down when the phenyl ring contained a CF_3 and went up when CF_3 was replaced by a methoxy group [11b]. An increase in the pressure of ethylene or a lowering of the pressure of CO had a positive influence on the rate of the reaction in the case of the keto ester. Interestingly, a reversed behavior of the pressure rate was observed with N-heterocyclic ketones when using ethene.

In 2002, a novel and rapid ruthenium-catalyzed synthesis of pyranopyrandiones was developed by the group of Mitsudo [12]. Single cyclopropenones and cyclopropenones in combination with internal alkynes could be converted, in the presence of $Ru_3(CO)_{12}$ as catalyst, to pyranopyrandiones by cross-carbonylation of 2 equiv of CO in good yields (Scheme 1.3a,b). Interestingly, this reaction was successful when simple NEt_3 was used as an efficient ligand. In contrast to other amine ligands (NBu_3, *N*-methylpiperidine, pyridine, and *N,N*-diethylaniline), phosphorus ligands (PCy_3 and PBu_3) gave only moderate catalytic activity in this transformation. The right choice of the ruthenium precursor is very important, since the use of $RuCl_2(PPh_3)_3$, $RuH_2(CO)(PPh_3)_3$, and $RuCl_3{\cdot}3H_2O$ was ineffective even in the presence of NEt_3. In addition, Ryu and coworkers [13] developed a synthesis of α-pyrones based on a ruthenium-catalyzed intermolecular carbonylative cycloaddition of α,β-unsaturated ketones with silylacetylenes and CO. Moderate yields were achieved by this new transformation (Scheme 1.3c).

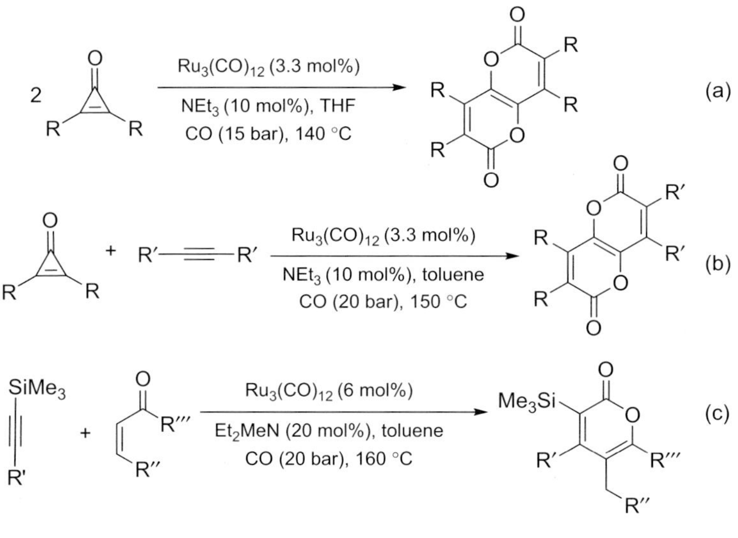

Scheme 1.3 (a–c) Ruthenium-catalyzed carbonylative synthesis of pyrones.

The group of Murai [14] could demonstrate that ruthenium-catalyzed cyclocarbonylation of yne-imines resulted in formation of lactams (Scheme 1.4a). Catalytic amounts of $Ru_3(CO)_{12}$ promote this cyclocarbonylation of 1,6- and 1,7-yne-imines, giving bicyclic α,β-unsaturated lactams. Similar to the Pauson–Khand reaction, the

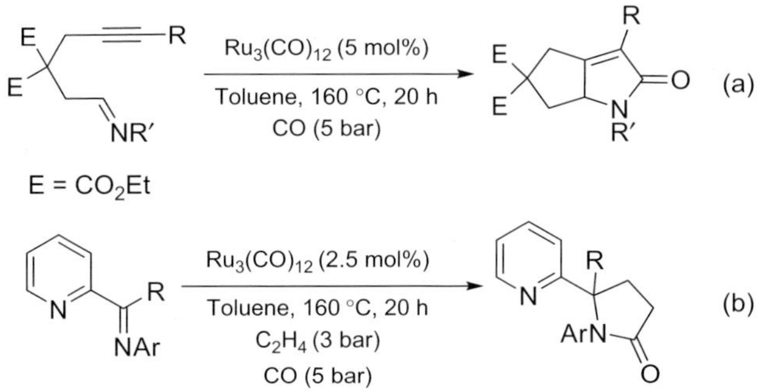

Scheme 1.4 (a, b) Ruthenium-catalyzed carbonylative synthesis of lactams.

lactam is formed in a [2+2+1] cycloaddition in which the acetylene π-bond, the imine π-bond, and the carbon atom of CO are involved. The acetylenic terminal carbon has to consist of an alkyl, an aryl, or silyl groups in order to give bicyclic α,β-unsaturated lactams via cyclocarbonylation of yne-imines. If the acetylenic terminal carbon has no substituents instead of the corresponding lactam, a dihydropyridine derivative will be generated without the incorporation of CO. Later on, the authors also showed that the cyclocarbonylation of imines, alkenes, or alkynes and CO gives γ-butyrolactams in good yields (Scheme 1.4b) [15].

In 2000, a selective cycloaddition of cyclopropyl imines, derived from cyclopropyl phenyl ketone and *tert*-butylamine and CO (2 bar), was developed by the same group (Scheme 1.5) [16]. The reaction was allowed to proceed in toluene (3 ml) in presence of a catalytic amount of $Ru_3(CO)_{12}$ (0.02 mmol) at 160 °C for 60 h, giving the pyridinone derivative in 76% isolated yield.

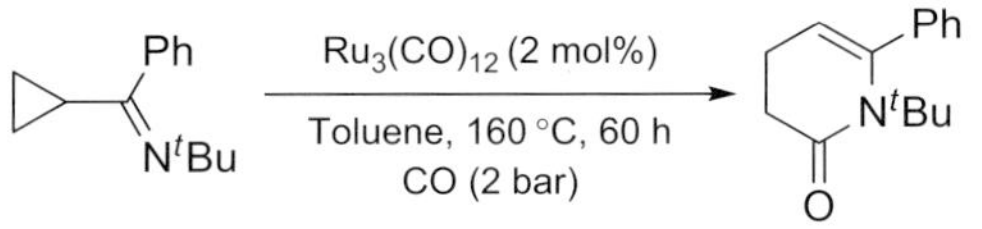

Scheme 1.5 Ruthenium-catalyzed carbonylative reaction of cyclopropyl imines.

Moreover, in 2008, a novel ruthenium-catalyzed cyclization based on a combination of isocyanates, alkynes, and CO was developed by the group of Kondo *et al.* [17]. Polysubstituted maleimides could be obtained in excellent yields under CO at atmospheric pressure with low catalyst loading (Scheme 1.6).

In 1997, Murai's [18] group developed the first ruthenium-catalyzed Pauson–Khand reaction, which originally was carried out by a cobalt catalyst. They showed that good yields of cyclopentenones could be achieved in an intramolecular, ruthenium-catalyzed cyclocarbonylation of 1,6-enynes (Scheme 1.7a). Later on, the cyclocarbonylation was extended to an intermolecular version in which alkynes, CO, and alkenes were converted into many cyclopentenones with excellent regioselectivity (Scheme 1.7b) [19]. Different substituted alkynes could be employed, showing the tolerance of the reaction. Notably, when a 2-$PyMe_2Si$-substituted alkyne was

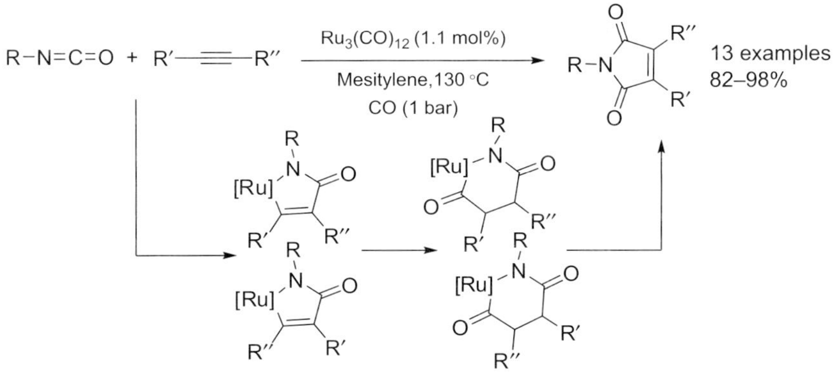

Scheme 1.6 Ruthenium-catalyzed carbonylative synthesis of maleimides.

used, the leaving group could be cleaved after the reaction. In 2000, Mitsudo and coworkers [20] reported another synthesis of cyclopentenones, which relied on a combination of allylic carbonates, alkenes, and CO. Here, the cyclocarbonylation was performed with either $[RuCl_2(CO)_3]_2/NEt_3$ or $(\eta^3\text{-}C_3H_5)RuBr(CO)_3/NEt_3$, which constitutes a highly effective catalyst systems (Scheme 1.7c).

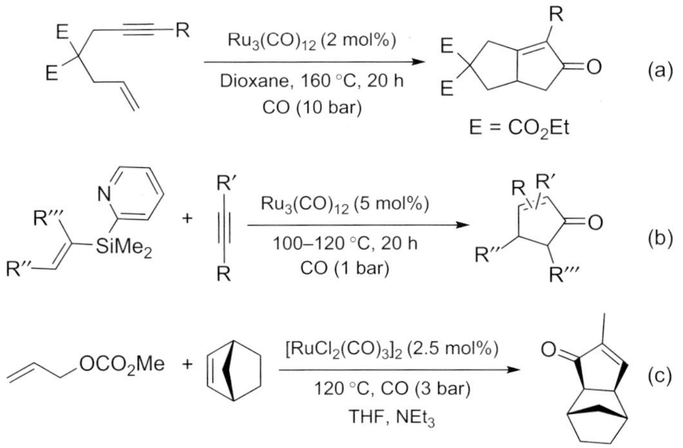

Scheme 1.7 (a–c) Ruthenium-catalyzed carbonylative synthesis of cyclopentenones.

Starting from the same substrates, even hydroquinones can be prepared by insertion of two molecules of CO. In 1998, Mitsudo and coworkers [21a] demonstrated that hydroquinones could be achieved in a ruthenium-catalyzed cyclocarbonylation by using alkynes and 2-norbornenes. Unsymmetrically substituted hydroquinones were obtained in high yields by this novel ruthenium-catalyzed transformation. For the preparation of higher substituted hydroquinones, functionalized alkenes could also be employed as starting material. Here, $[Cp^{*}RuCl_2]_2$ was used as the catalyst (Scheme 1.8) [21b].

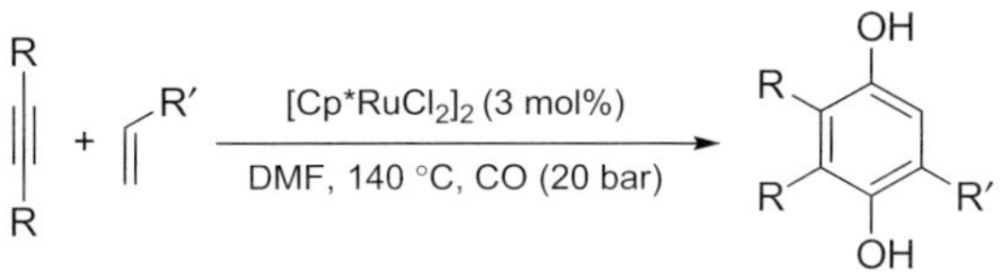

Scheme 1.8 Ruthenium-catalyzed carbonylative synthesis of hydroquinones.

1.2.2
Rhodium-Catalyzed Carbonylative Domino Reactions

In 2006, the group of Artok showed that 5-aryl-2(5*H*)-furanones could be prepared in moderate to good yields by a rhodium-catalyzed carbonylative arylation of internal alkynes with aryl boronic acids (Scheme 1.9a) [22]. α,β-Unsaturated ketones (chalcone derivatives) were formed as the major product when some TFA (trifluoroacetic acid) was added under the same reaction conditions [23a]. By varying the catalytic system, indanones could be produced as the main product [23b]. The chemical behavior of terminal alkynes is different, and either α,β-unsaturated ketones or furans starting from propargylic alcohols can be achieved (Scheme 1.9b) [24, 25]. In the case of vinyl ketones, 1,4-diketones were obtained by rhodium-catalyzed coupling of arylboronic acids in the presence of 20–40 bar of CO [26]. In 2007, Chatani demonstrated that indenones could be accessed by a carbonylative rhodium-catalyzed cyclization of alkynes with 2-bromophenylboronic acids (Scheme 1.9c) [27]. Here, the key intermediate is a vinylrhodium(I) species that is formed by transmetallation of RhCl with 2-bromophenylboronic acid followed by insertion of

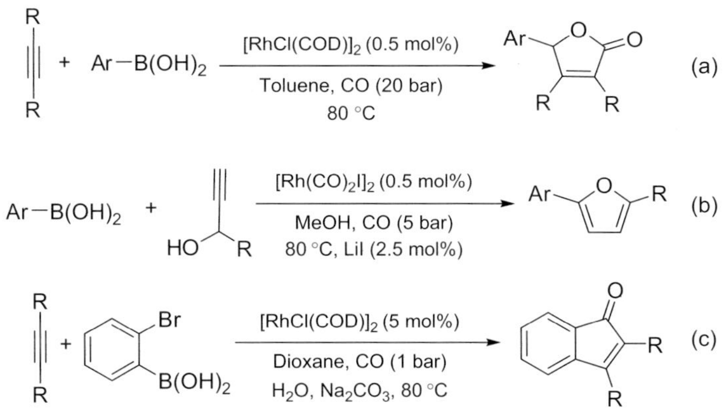

Scheme 1.9 (a–c) Rhodium-catalyzed carbonylation of alkynes and boronic acids.

an alkyne. Next, the C–Br bonds on the adjacent phenyl ring were oxidatively added to provide a benzorhodacyclopentene species. After CO insertion and reductive elimination, the desired indenone was obtained. With regard to the regioselectivity, an alkyne substituted with a bulky and electron-withdrawing group favors the α-position of indenones. The highest regioselectivity was achieved in the case of silyl- or ester-substituted alkynes in the order $SiMe_3 > COOR \gg aryl \gg alkyl$. Similarly, also indanone derivatives could be obtained when 2-bromophenylboronic acid was reacted with norbornene under 1 bar of CO. On conducting the reaction without CO, two molecules of alkynes were incorporated during the reaction sequence with 2-bromophenylboronic acid to give naphthalene derivatives. With the aid of carbonylative rhodium-catalyzed cyclization of 1-(2-bromophenyl)-hept-2-yn-1-one and $PhB(OH)_2$, indan-1,3-dione derivatives were obtained.

In 2001, a novel rhodium-catalyzed cyclohydrocarbonylation of imino alkynes was developed by Alper and Van den Hoven [28]. The reaction was catalyzed by a zwitterionic rhodium complex and $P(OPh)_3$, giving aldehyde-substituted pyrrolinones in 67–82% yield (Scheme 1.10a). Imino alkynes with alkyl, alkoxyl, vinyl, and aryl substituents can be used in this unique transformation. This synthetic approach is a convenient way for the synthesis of highly functionalized pyrrolinones, which constitute often biologically active compounds. In 2001, the group of Saito reported on the intramolecular carbonylation of alkyne-carbodiimides, giving 4,5-dihydro-1*H*-pyrrolo[2,3-*b*]pyrrolin-2-ones and 1*H*-pyrrolo[2,3-*b*]indol-2-ones in reasonably good yields (Scheme 1.10b) [29]. Later on, they could apply their methodology on *N*-[2-(2-alkyn-1-yl)phenyl]carbodiimides to synthesize 2,3-dihydro-1*H*-pyrrolo[2,3-*b*]quinolin-2-ones in good yields (Scheme 1.10c) [30].

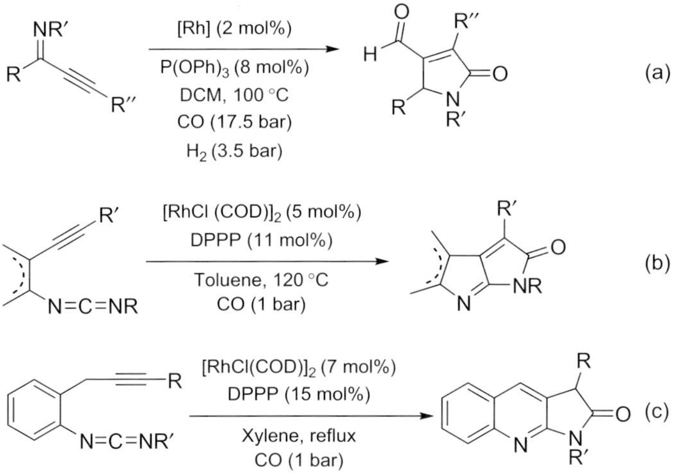

Scheme 1.10 (a–c) Rhodium-catalyzed cyclocarbonylation reactions.

In 2004, the synthesis of indazolo[2,1-a]indazole-6,12-diones was carried out by a rhodium-catalyzed cyclocarbonylation of azobenzenes by the group of

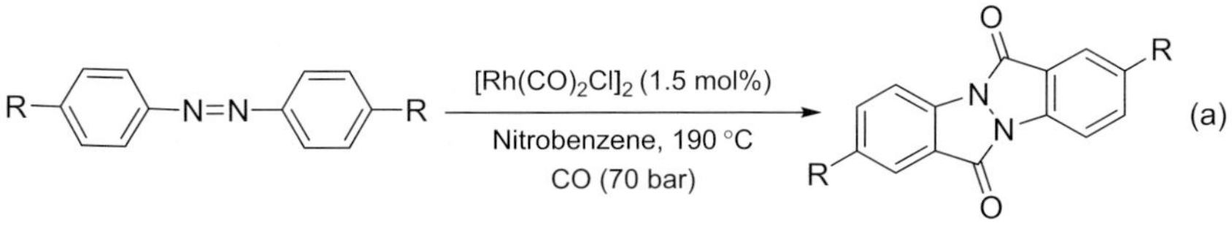

Scheme 1.11 (a, b) Rhodium-catalyzed carbonylative C–H activation.

Takahashi [31]. To get good yields, nitrobenzene was added as a hydrogen acceptor (Scheme 1.11a). But on running the carbonylation of azobenzene via cobalt catalysis, quinazoline was obtained as the terminal product. Furthermore, Chatani and coworkers described a rhodium-catalyzed synthesis of maleimides starting from a combination of alkynes and pyridine-2-ylmethylamine in the presence of CO (Scheme 1.11b) [53].

Lautens and coworkers investigated an asymmetrical ring-opening reaction in which *meso*-diazabicycles were opened by acyl anion nucleophiles in a catalytic manner to give functionalized *trans*-1,2-hydrazinoacyl cyclopentenes stereoselectively [32]. Under very mild conditions, an acyl anion is generated *in situ*, starting from readily available organoboron precursors (Scheme 1.12).

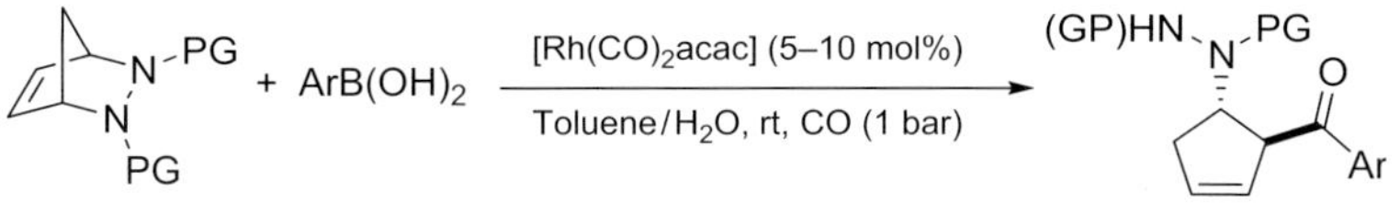

Scheme 1.12 Rhodium-catalyzed carbonylative ring-opening of diazabicycles.

In 2007, a series of 3-methylcyclopent-2-enones were synthesized by a rhodium-catalyzed carbonylation of spiropentanes [33]. Here, two different types of carbon–carbon bond cleavage processes were involved to get the product in good yield (Scheme 1.13)

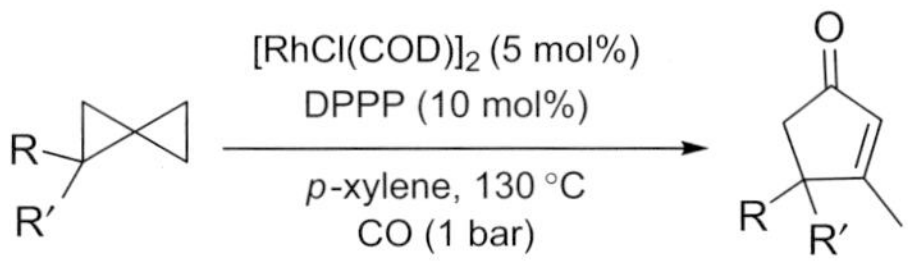

Scheme 1.13 Rhodium-catalyzed carbonylation of spiropentanes.

1.2.3
Palladium-Catalyzed Carbonylative Domino Reactions

The outstanding ability of palladium catalysts was demonstrated in the area of carbonylative coupling reactions mainly with activated arenes. Nevertheless, palladium catalysts can also be used in oxidative cyclization chemistry. The group of Gabriele succeeded in producing substituted furans from the corresponding alkynols under oxidative conditions (Scheme 1.14a) [34]. Here, in the presence of catalytic amounts of $[PdI_4]^{2-}$ in conjunction with an excess of KI, 4-yn-1-ols containing a terminal triple bond undergo oxidative cyclization/alkoxycarbonylation in methanol at 70 °C and 100 bar of a 9 : 1 mixture of CO and air to give 2*E*-[(methoxycarbonyl)methylene]tetrahydrofurans in good yield. A side reaction, producing 2-methoxy-2-methyltetrahydrofurans via a cycloisomerization/hydromethoxylation sequence, could be easily prevented by increasing the KI excess. Without KI excess and in the absence of carbon monoxide, the latter product can be formed from 4-yn-1-ols and methanol in high yields using the same catalytic system. Another system that needs no KI and high pressure leading to different products was developed by Akita and coworkers [35] (Scheme 1.14b). Following this procedure, they were able to perform the reaction in an asymmetric manner by applying chiral bisoxazolines as ligands.

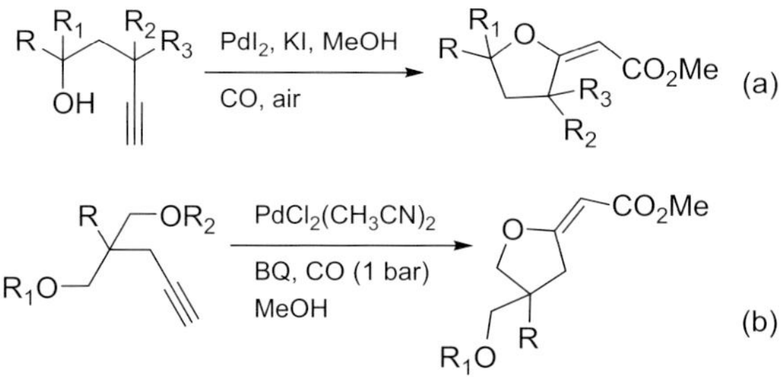

Scheme 1.14 (a, b) Palladium-catalyzed carbonylative synthesis of tetrahydrofurans.

Gabriele and coworkers [36] showed that, besides lactones, furans could also be prepared by a similar process starting from different substrates. Here, a variety of (*Z*)-2-en-4-yn-1-ols have been carbonylated under oxidative conditions to give substituted furan-2-acetic esters in good yields (Scheme 1.15a). The cyclization/alkoxycarbonylation sequence was carried out in alcoholic media at 50–70 °C under 100 bar pressure of a 9 : 1 mixture of CO and air. As catalyst system, PdI_2 in combination with KI was used. The proposed reaction pathway involves the *in situ* isomerization of the initially formed (*E*)-2-[(alkoxycarbonyl)methylene]-2,5-dihydrofuran species, which in some cases have been isolated and proved to be the intermediates. Under similar reaction conditions, 3-yne-1,2-diols were transformed into the corresponding furan-3-carboxylic esters in good yield (Scheme 1.15b).

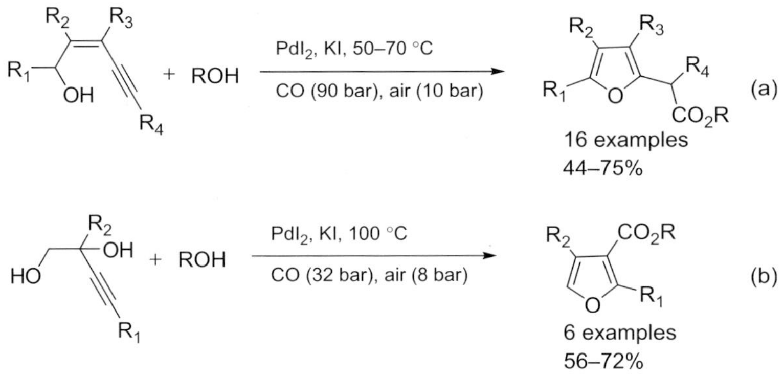

Scheme 1.15 (a, b) Palladium-catalyzed carbonylative synthesis of furans.

The palladium-catalyzed carbonylation of alkynols resulted in the formation of synthetically interesting lactones and furans. In 1994, Sakamoto and coworkers showed that the palladium-catalyzed carbonylation reaction of 2-alkynylanilines and 2-alkynylphenols in methanol could give the corresponding indoles and benzofurans in moderate yields. Starting from 2-alkynylbenzamides, 3-alkylidenisoindoles were obtained (Scheme 1.16) [37a]. A similar methodology was applied by Scammells for the synthesis of XH-14 and its derivatives, which contain a benzofuran as the main skeleton [37b].

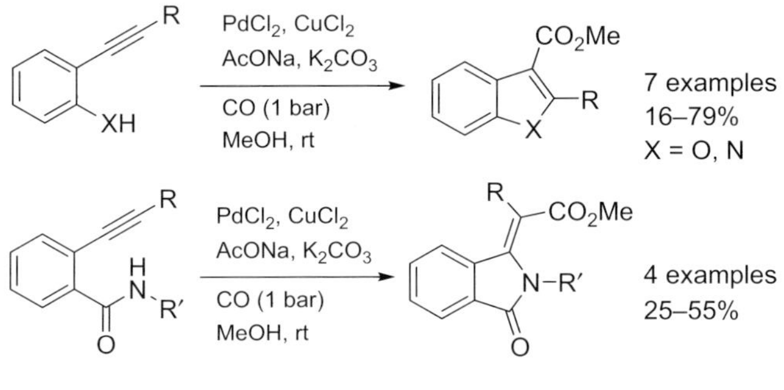

Scheme 1.16 Palladium-catalyzed carbonylative synthesis of benzofurans and indoles.

Costa and coworkers [38] used an oxidative Pd-catalyzed cyclization/alkoxycarbonylation sequence for the synthesis of 1-(alkoxycarbonyl)methylene-1,3-dihydroisobenzofurans and 4-(alkoxycarbonyl)benzo[*c*]-pyrans starting from 2-alkynylbenzyl alcohols and 2-alkynylbenzaldehydes or 2-alkynylphenyl ketones. The reactions were run in ROH or CH_3CN/ROH (R = Me, *i*-Pr) mixtures as solvent at 70–105 °C in the presence of catalytic amounts of PdI_2 in combination with KI under a CO/air mixture in the ratio 4 : 1 or 3 : 1 (20 or 32 bar total pressure at 25 °C). The reaction proceeds via an intramolecular attack of nucleophilic oxygen atom (either already present in the starting material or generated in *situ* by ROH attack on carbonyl group) directed to the triple bond which is coordinated to

Pd(II). The reaction sequence is closed by a subsequent alkoxycarbonylation. The presence of substituents at the alkyne terminal position and at the carbon atom α to the hydroxy group control the selectivity of the process by forming a five- or six-membered ring (Scheme 1.17). Alternatively, the reaction of alkynyloxiranes could also lead to 1,3-dihydroisobenzofurans and tetrahydrofurans. Moderate to good yields of the products were obtained under similar reaction conditions ($PdI_2/KI/CO/O_2$).

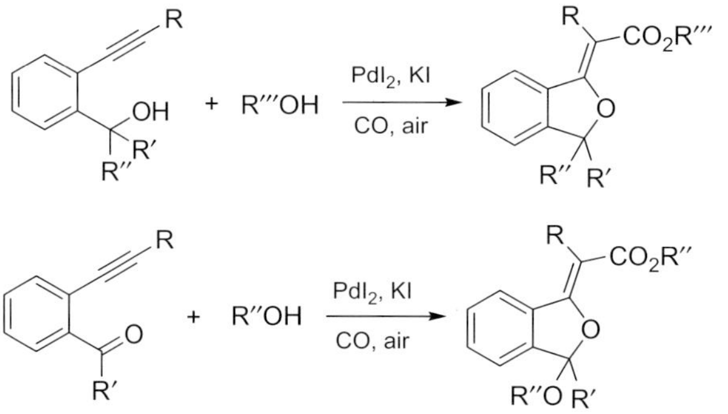

Scheme 1.17 Palladium-catalyzed carbonylative synthesis of furans and pyrans.

Because of the interesting biological properties of 3(2*H*)-furanone derivatives, many methodologies have been developed for their syntheses [39a]. In 1988, Inoue and coworkers described the coupling of α-ethynyl tertiary alcohols and acyl chlorides to give 3(2*H*)-furanones in the presence of a palladium catalyst and CO_2. Next, they started with the same reaction conditions with CO and CO_2 under pressure but using aryl halides instead of acyl chlorides [39b]. They showed that acetylenic ketone was formed as an intermediate by a combination of acetylenic alcohol, CO, and the aryl halides. Subsequent reaction with CO_2 resulted in the formation of a cyclic carbonate, which was decarboxylated to 3(2*H*)-furanones with the release of CO_2 (Scheme 1.18a). Alternatively, Kiji and coworkers [39c] showed that, in the absence of CO_2, 3-isopropylidene-5-phenyl-2(2*H*)-furanone could be achieved as the main product. Carbonylative coupling of iodobenzene and 2-methyl-3-butyn-2-ol in aqueous biphasic NaOH/benzene system was carried out by using $Pd(OAc)_2/PPh_3/Bu_4PBr$ as catalyst. This biphasic solvent system gave, in sharp contrast to a homogeneous Et_3N solution, 3-isopropylidene-5-phenyl-2(2*H*)-furanone in moderate yield accompanied by 2,2-dimethyl-5-phenyl-3(2*H*)-furanone and benzoic acid as side products. The formation of the main product was explained by a carbonylative coupling of iodobenzene with 2-methyl-3-butyn-2-ol, forming 4-hydroxy-4-methyl-1-phenyl-2-pentyn-1-one, which underwent hydrogenolysis to yield 4-methyl-1-phenyl-2,3-pentadien-1-one. Subsequent cyclocarbonylation yielded 3-isopropylidene-5-phenyl-2(2*H*)-furanone as the final product (Scheme 1.18b). Concerning the formation of 3-alkylidenefuran-2-ones, the group of Alper [39d] established a palladium catalyst system for the carbonylative coupling of aryl

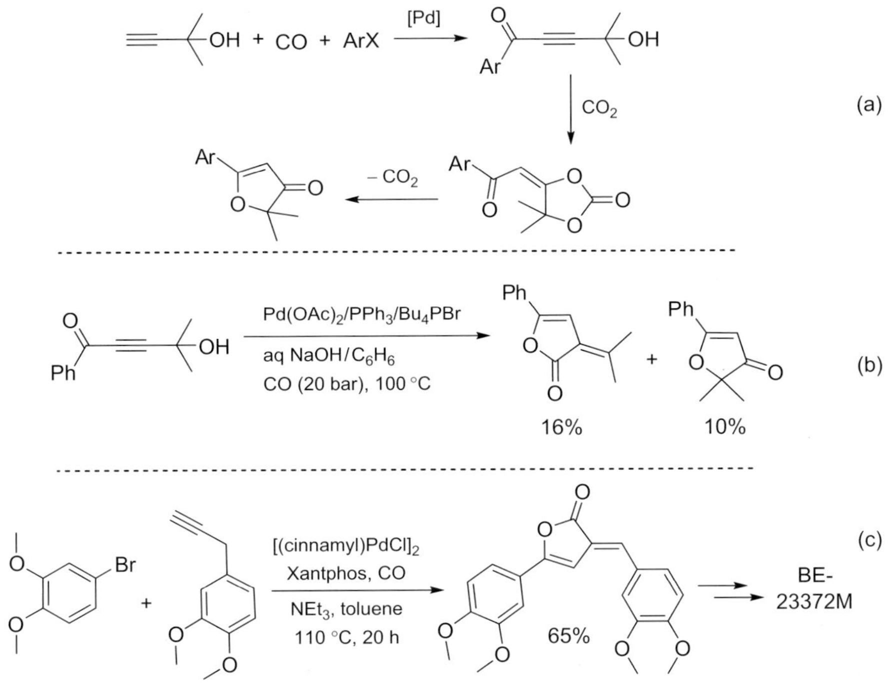

Scheme 1.18 (a–c) Palladium-catalyzed carbonylative synthesis of furanones.

iodides with benzyl acetylenes. More recently, our group developed a general and efficient method for the synthesis of furanones starting from aryl bromides and aryl triflates [39e]. After double carbonylation with benzyl acetylenes, furanones were produced in good yields. Methylated BE-23372M, a kinase inhibitor, was also produced in a one-pot sequence with 65% yield (Scheme 1.18c).

In 2005, Chatani and coworkers [40a] reported on the carbonylation of yne esters giving lactones in good yields under 1 bar of CO (Scheme 1.19a). It turned out that the 2-pyridinyloxy moiety was a good leaving group among the groups tested. Similarly, the cyclocarbonylation of 2-propynyl-1,3-dicarbonyls with organo halides or triflates gave rise to the formation of furans in good yields (Scheme 1.19b) [40b]. Kato and coworkers [40c] developed a palladium-mediated oxidative cyclocarbonylation of 2-alkyl-2-propargylcyclohexane-1,3-diones, generating bicyclic-β-alkoxyacrylates in 51–74% yield with 72–82% ee (Scheme 1.19c). The authors were able to extend their palladium-catalyzed cyclocarbonylation to propargylic esters, propargylic acetates, 4-yn-1-ones, and allenyl ketones. Mukai and coworkers [40d] were able to apply this methodology in the total synthesis of naturally occurring diacetylenic spiroacetal enol ethers. A related mechanistic study supported by both experiment and DFT (density functional theory) study was carried out by Carfagna and coworkers [40e]. They proposed that, under the carbonylative conditions of Gabriele *et al.*, the concatenation occurs between a Pd(0)-promoted deallylation and

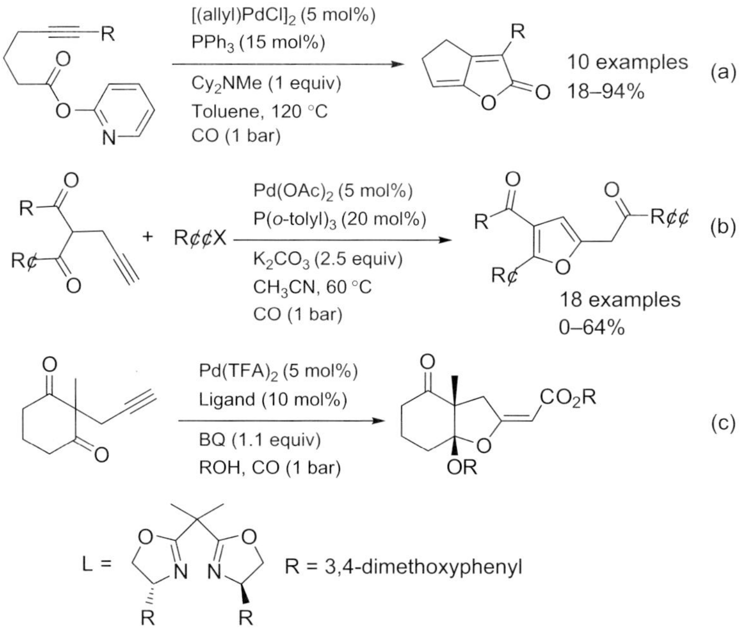

Scheme 1.19 (a–c) Palladium-catalyzed cyclocarbonylation of alkynones.

a Pd(II)-promoted heterocyclization catalytic cycle to convert 1-(2-allyloxyphenyl)-2-yn-1-ols to 2-benzofuran-2-ylacetic esters and β,γ-unsaturated esters in high yields. This reaction sequence is named *sequential homobimetallic catalysis* [40f]. Owing to the theoretical and synthetic importance of the process, a closer look at the mechanism and scope of the reaction revealed that the experimental results fit to the sequential homobimetallic mechanism. In place of the esters, under the same conditions, amides could also be produced in the presence of amines [40g]. The methodology could be extended to the synthesis of coumarins by using similar reaction conditions [40h]. Here, 3-[(methoxycarbonyl)-methyl]coumarins were prepared starting from readily available 2-(1-hydroxyprop-2-ynyl)phenols. In the presence of catalytic amounts of PdI_2 and an excess of KI in MeOH at room temperature and under 90 bar of CO, the product was obtained in good to high isolated yields (62–87%).

Moreover, Shim and coworkers [41a] studied the cyclocarbonylation of 2-(2-bromophenyl)-2-oxazolines to the corresponding isoindolinones. By using a palladium–nickel catalyst under 3 bar of CO, the products were produced in high yields (Scheme 1.20a). Later on, isoindolinones could also be achieved by coupling 2-iodobenzoyl chloride with imines in moderate yields using NEt_3 and $Pd(PPh_3)_2Cl_2$/PPh_3 (Scheme 1.20b) [41b]. The same group could synthesize even more complex isoindolinones through a palladium-catalyzed carbonylative coupling of 2-bromobenzaldehydes with aminoalcohols or diamines [41c]. At lower temperature

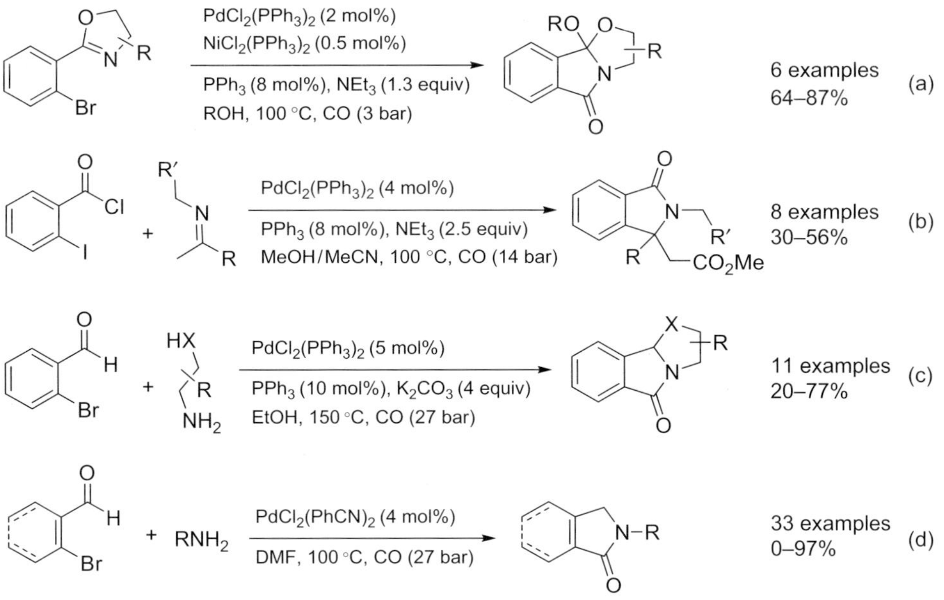

Scheme 1.20 (a–d) Palladium-catalyzed carbonylative synthesis of isoindolin-1-ones.

and lower catalyst loading, the corresponding isoindolinones were achieved in good isolated yields (Scheme 1.20c). Interestingly, when primary amines were used for the palladium-catalyzed coupling of 2-bromobenzaldehydes or 2-bromocyclohex-1-enecarbaldehydes, no base was needed (Scheme 1.20d) [41d]. The mechanism is believed to start with the condensation of the aldehyde and the primary amine, forming an imine. After the oxidative addition of the carbon–bromide bond of the imine to the active palladium(0) catalyst and subsequent CO insertion, an aroylpalladium(II) intermediate is formed. Next, an intramolecular acylpalladation to the imine gives the alkylpalladium(II) intermediate, which is decomposed to isoindolin-1-one by hydrogenolysis with molecular hydrogen. It is assumed that hydrogen is produced by the water-gas shift reaction of CO and H_2O which comes from the initial condensation stage.

The group of Arndtsen developed a number of elegant multicomponent reactions that introduce one or two CO groups into the parent molecules [42]. A combination of alkynes, imines, acid chlorides, and CO gives pyrroles as the terminal products in the presence of a palladium catalyst (Scheme 1.21a). By using α-amidoesters and alkynes, the reaction proceeds to give the same products (Scheme 1.21b). Even imidazoles could be formed when the reaction was carried out with imines and acid chlorides. Interestingly, by simply changing the reaction sequence of adding the substrates, imidazolinium salts or imidazolines could be produced. In general, these methods offer convenient pathways for the production of heterocycles from easily available substrates.

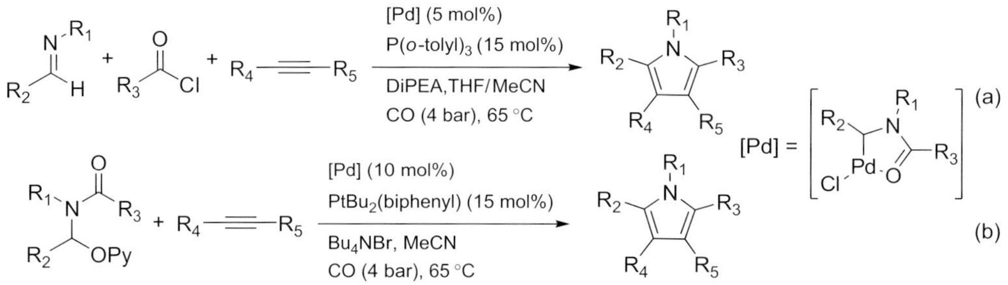

Scheme 1.21 (a, b) Palladium-catalyzed carbonylation of imines.

Recently, Alper and coworkers [43] reported on novel processes for the synthesis of carbonylated indole derivatives via a palladium-catalyzed N–C coupling/carbonylation sequence. 2-Carboxyindoles with a variety of functional groups were achieved in good yields (Scheme 1.22a). Similarly, 2-aroylindoles could also be obtained from the same substrates in moderate yields (Scheme 1.22b). In 2011, Alper and Zeng [43c] published a facile and selective palladium-catalyzed domino synthesis of carbonylated benzothiophenes. By a carbonylative intramolecular C–S coupling/intermolecular cascade sequence, 2-carbonylbenzo[*b*]thiophene derivatives were produced from 2-*gem*-dihalovinylthiophenols in 24–73% yield (Scheme 1.22c). This protocol allows access to various highly functionalized benzo[*b*]thiophenes.

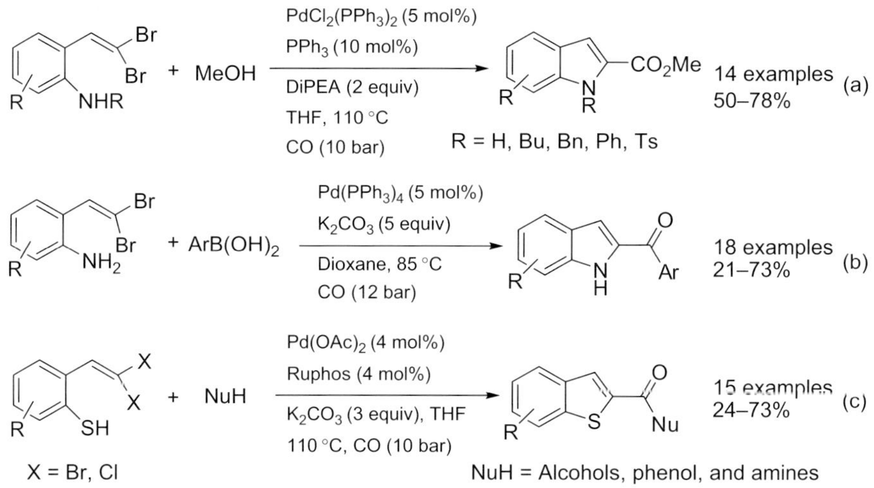

Scheme 1.22 (a–c) Palladium-catalyzed carbonylative synthesis of indoles and thiophenes.

In 2000, a palladium-catalyzed decarboxylative carbonylation of 5-vinyloxazolidin-2-ones was studied by Knight and coworkers [44]. By a palladium-catalyzed decarboxylative carbonylation process, 5-vinyloxazolidin-2-ones, which are prepared from amino acids, reacted to form 3,6-dihydro-1*H*-pyridin-2-ones in good yields (Scheme 1.23).

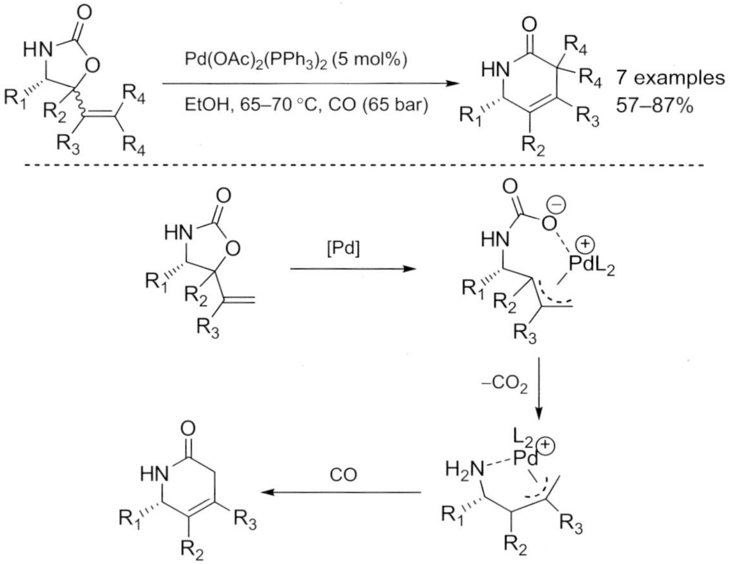

Scheme 1.23 Palladium-catalyzed carbonylative synthesis of pyridinones.

Alper and Xiao [45] synthesized thiochromanones by palladium-catalyzed carbonylative ring-forming reactions of 2-iodothiophenol derivatives, allenes, and CO. The thiochroman-4-ones were achieved in good to excellent isolated yields with high regioselectivity, which was probably caused by electronic effects (Scheme 1.24). This catalytic heteroannulation comprises the regioselective addition of the sulfur moiety on the more electrophilic carbon center of the allene, arylpalladium formation, CO insertion, subsequent intramolecular cyclization, and, finally, the reductive elimination.

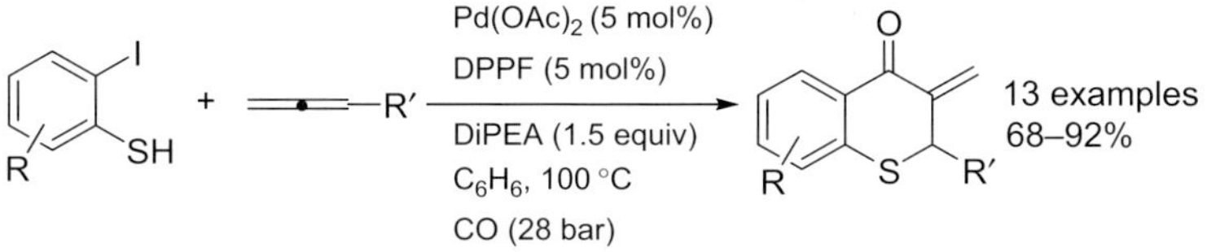

Scheme 1.24 Palladium-catalyzed carbonylative synthesis of thiochromanones.

Recently, an efficient method for the synthesis of 1,4-benzo- and pyrido-oxazepinones was also disclosed [46]. This reaction proceeds via a domino process through one-pot ring-opening/carboxamidation reaction sequences of *N*-tosylaziridines with 2-halophenols/pyridinol under phase-transfer conditions (benzyltriethylammonium chloride, TEBA). A variety of 1,4-benzo- and pyrido-oxazepinones could be easily synthesized by using a range of *N*-tosylaziridines and 2-halophenols/pyridinol (Scheme 1.25a). Analogously, when 2-iodothiophenols were employed, 1,4-benzothiazepin-5-ones were obtained in good yields (Scheme 1.25b).

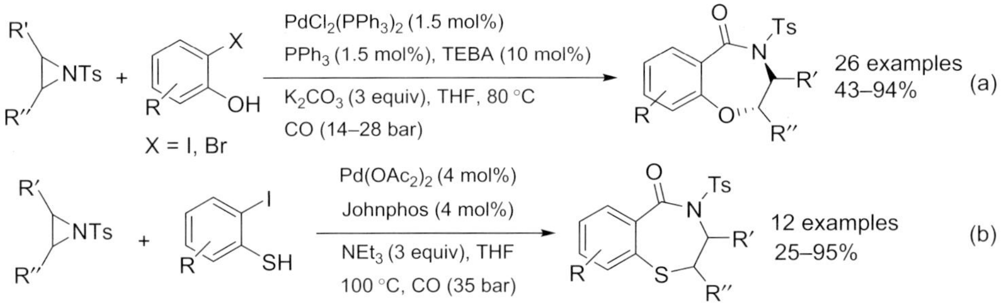

Scheme 1.25 (a,b) Palladium-catalyzed carbonylation of *N*-tosyl aziridines.

1.2.4
Iron-, Copper-, Nickel-, and Cobalt-Catalyzed Carbonylative Domino Reactions

Compared to palladium, rhodium, and ruthenium, iron and copper are less developed in carbonylation reactions. But the advantages of iron and copper are attracting more and more chemists to work in this area. $Fe(CO)_5$ as a more easily available iron–carbonyl complex has found an important place in the stoichiometric and catalytic carbonylation reactions [47]. The group of Periasamy applied $Fe(CO)_5$ as precursor for the *in situ* generation of $NaHFe(CO)_4$ for double carbonylation of alkynes to cyclobutenediones. In their procedures, $CuCl_2$ was needed as the oxidant reagent. The active species was $[Fe(CO)_4]$, which could be generated from various reagents, such as MeI, $NaBH_4$, amines, Me_3NO, and NaH. Additionally, using these methods, α,ß-unsaturated acids, benzoquinones, and cyclic anhydrides could also be produced as unexpected products (Scheme 1.26). In the mentioned reactions, stoichiometric amount of iron salts were still needed.

Notably, Beller's group developed a series of iron-catalyzed aminocarbonylation of alkynes in 2009 [48a–d]. Starting from alkynes and amines, succinimides were prepared in good yields in the presence of carbon monoxide (Scheme 1.27). This methodology was also applied for the synthesis of himanimides A and B. Under the same conditions, cinnamides were also synthesized by adding 1,4-diazabutadiene as ligand or by using microwave irradiation. NEt_3 was the ligand of choice [48e]. A combination of $Fe(CO)_5$ and irradiation was also used for producing vinylesters and lactones from alkynes at 0 °C [48f].

Additionally, the reactions with alkynes and iron catalyst were also applied to the carbonylative homocoupling of aryl iodides to give benzophenones [49]. As catalyst system, $Fe(CO)_5$-$Co_2(CO)_8$ was used under phase-transfer conditions to give carbonylate aryl iodides in moderate yields.

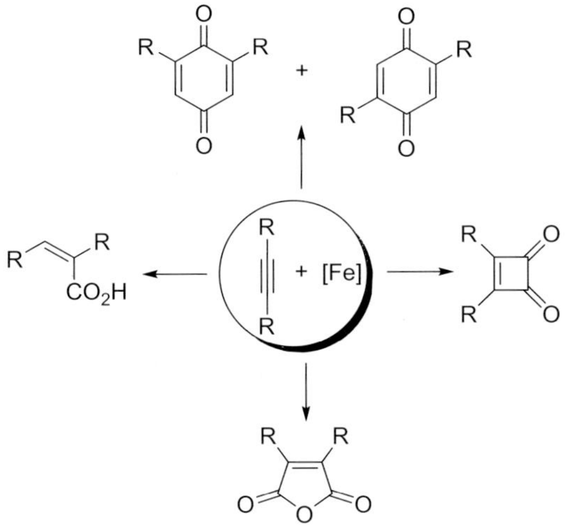

Scheme 1.26 Double carbonylation of alkynes.

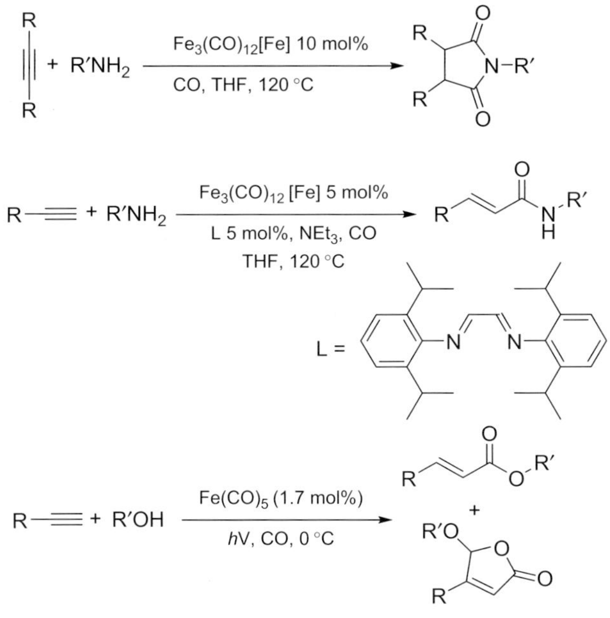

Scheme 1.27 Iron-catalyzed carbonylative synthesis of amides and esters.

In 2008, Bhanage and coworkers [50a] reported on a copper-catalyzed carbonylative Sonogashira reaction of aryl iodides. In this procedure, copper bis(2,2,6,6-tetramethyl-3,5-heptanedionate) [Cu(TMHD)$_2$] was used as the catalyst for this transformation and NEt$_3$ as base. Alkynones were produced in good yields. Recently, Xia and coworkers [50b] described a general and efficient copper-catalyzed double aminocarbonylation of aryl iodides (Scheme 1.28). Aryl iodides were double-carbonylated with amines in good yields by using the NHC–Cu catalyst (72–93%).

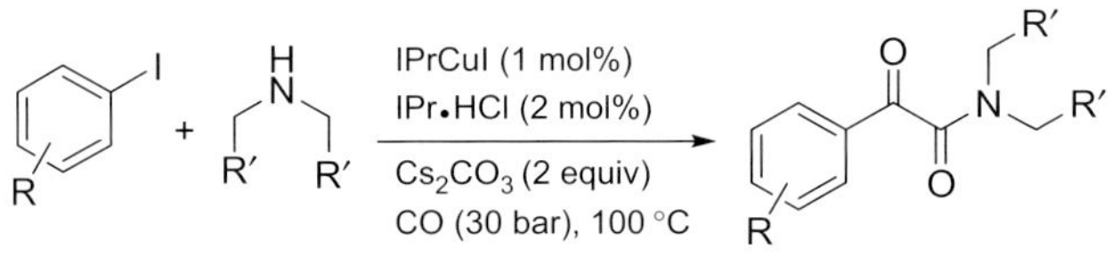

Scheme 1.28 Copper-catalyzed double carbonylation of aryl iodides.

Skoda-Földes and coworkers [51] investigated the domino reaction of ethyl diazoacetate, CO, and ferrocenylimines in the presence of $Co_2(CO)_8$ as catalyst (Scheme 1.29). In most cases, the main products were 2-(1-ferrocenylmethylidene) malonates formed by an N(1)–C(4) cleavage of the primarily derived β-lactams. The latter compounds could only be isolated when the reaction was carried out at relatively low CO pressure, using an excess of ethyl diazoacetate. Among these compounds, *trans-N-(tert*-butyl)-3-ethoxycarbonyl-4-ferrocenyl-b-lactam proved to be the most stable one and could be isolated in 55% yield. *N*-Alkyl β-lactams were shown to undergo an acidic cleavage, leading to the (*E*)-isomers of 2-(1-ferrocenylmethylidene)malonates as the main products. The structures of the two new compounds, namely (*E*)-2-ethoxycarbonyl-3-ferrocenyl-*N*-((*R*)-1-phenylethyl)-2-propenamide and *trans-N-(tert*-butyl)-3-ethoxycarbonyl-4-ferrocenyl-b-lactam, were confirmed by X-ray crystallography. The relative thermodynamic stability of the products as well as the energetics of the acid-mediated cleavage of the β-lactam ring was elucidated with DFT calculations.

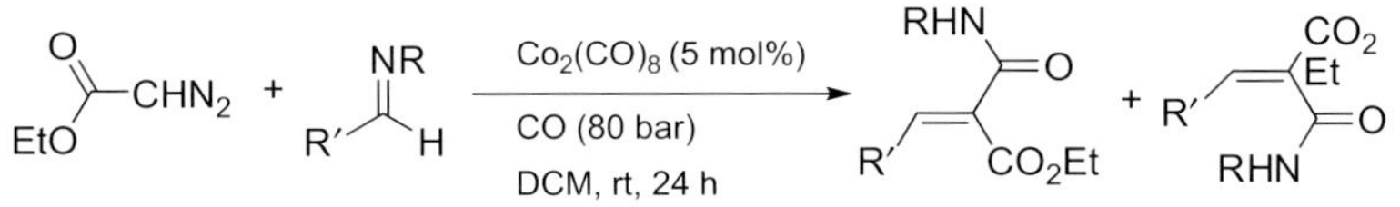

Scheme 1.29 Cobalt-catalyzed carbonylative synthesis of malonic acid derivatives.

Even though $Ni(CO)_4$ is called *liquid death*, this nickel catalyst has been applied in carbonylation reactions [52]. The group of Ricart reported a nickel-catalyzed carbonylative cycloaddition of alkynes and allyl halides to cyclopentanes. The desired products were obtained in high yields and with controlled stereoselectivity. Iron was used as a reductant. An extension of the reaction to new substrates led to the conclusion that, although the steric and electronic effects of the alkyne substituents are generally irrelevant in relation to the adducts and their yields, those of the allylic counterpart may have a significant influence on the outcome of the reaction. However, the presence of the amine moiety in the alkyne completely inhibited the reaction. The feasibility of a multicentered reaction was verified with a triacetylene, in which up to 12 bonds were created simultaneously and in good yield (Scheme 1.30).

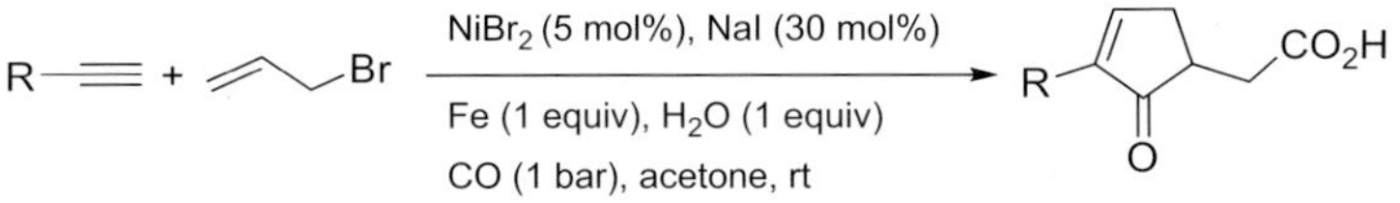

Scheme 1.30 Nickel-catalyzed carbonylative synthesis of cyclopentanes.

1.3 Outlook

In summary, we have summarized representative examples of transition-metal-catalyzed carbonylative domino reactions. In the area of carbonylations, palladium, rhodium, and cobalt are still the main actors. The ability of palladium catalysts in carbonylative cross-coupling, rhodium catalysts in carbonylative C–H activation, and cobalt catalyst in carbonylative reactions with unsaturated bonds is impressive.

In the future, cheap catalysts such as iron and copper are expected to be explored and applied. In the case of noble metals, their reaction efficiency and selectivity should be improved. The use of nickel catalysts in carbonylation is potentially accompanied with the formation of $Ni(CO)_4$, which is highly dangerous for the operators. Therefore, methods for stabilizing Ni must be developed before Ni can be used in catalytic reactions.

With regard to oxidative carbonylations, green oxidants, such as air or oxygen, are much more interesting than equal amounts of $Cu(OAc)_2$ or BQ.

In conclusion, the main direction for methodology development in the future is looking at "sustainable development."

References

1. For selected reviews and book on domino reactions, see: (a) Tietze, L.F. and Düfert, A. (2010) *Pure Appl. Chem.*, **82**, 1375–1392. (b) Arndtsen, B.A. (2009) *Chem. Eur. J.*, **15**, 302–313. (c) Tietze, L.F., Brasche, G., and Gericke, K. (2006) *Domino Reactions in Organic Synthesis*, Wiley-VCH Verlag GmbH. (d) Tietze, L.F. and Rackelmann, N. (2004) *Pure Appl. Chem.*, **76**, 1967–1983. (e) Neuschütz, K., Velker, J., and Neier, R. (1998) *Synthesis*, **3**, 227–255. (f) Tietze, L.F. (1996) *Chem. Rev.*, **96**, 115–136. (g) Tietze, L.F. and Beifuss, U. (1993) *Angew. Chem., Int. Ed. Engl.*, **32**, 131–163.
2. (a) Gallos, J.K. and Koumbis, A.E. (2003) *Arkivoc*, **VI**, 135–144. (b) Tietze, L.F. and Modi, A. (2000) *Med. Res. Rev.*, **20**, 304–322. (c) Tietze, L.F. and Lieb, M. (1998) *Curr. Opin. Chem. Biol.*, **2**, 363–371.
3. For selected reviews on transition metal-catalyzed coupling reactions, see: (a) Wu, X.F., Anbarasan, P., Neumann, H., and Beller, M. (2010) *Angew. Chem. Int. Ed.*, **49**, 9047–9050. (b) Nakao, Y. and Hiyama, T. (2011) *Chem. Soc. Rev.*, **40**, 4893–4901. (c) Baudoin, O. (2011) *Chem. Soc. Rev.*, **40**, 4902–4911. (d) Littke, A.F. and Fu, G.C. (2002) *Angew. Chem. Int. Ed.*, **41**, 4176–4211. (e) Rosen, B.M., Quasdorf, K.W., Wilson, D.A., Zhang, N., Resmerita, A.M., Garg, N.K., and Percec, V. (2011) *Chem. Rev.*, **111**, 1346–1416. (f) Jana, R., Pathak, T.P., and Sigman, M.S. (2011) *Chem. Rev.*, **111**, 1417–1492.
4. For selected reviews, see: (a) Nicolau, K.C., Vourloumis, D., Winssinger, N., and Baran, P.S. (2000) *Angew. Chem.*

Int. Ed., **39**, 44–122. (b) Nicolaou, K.C., Edmonds, D.J., and Bulger, P.G. (2006) *Angew. Chem. Int. Ed.*, **45**, 7134–7186. (c) Nicolaou, K.C., Bulger, P.G., and Sarlah, D. (2005) *Angew. Chem. Int. Ed.*, **44**, 4442–4489. (d) Nicolaou, K.C., Bulger, P.G., and Sarlah, D. (2005) *Angew. Chem. Int. Ed.*, **44**, 4490–4527.

5. For reviews on palladium-catalyzed carbonylations, see: (a) Brennführer, A., Neumann, H., and Beller, M. (2009) *Angew. Chem. Int. Ed.*, **48**, 4114–4133. (b) Brennführer, A., Neumann, H., and Beller, M. (2009) *ChemCatChem*, **1**, 28–41. (c) Beller, M. (2002) in *Applied Homogeneous Catalysis with Organometallic Compounds*, 2nd edn (eds B. Cornils and W.A. Herrmann), Wiley-VCH Verlag GmbH, Weinheim, pp. 145–156. (d) Skoda-Foldes, R. and Kollör, L. (2002) *Curr. Org. Chem.*, **6**, 1097–1119. (e) Beller, M., Cornils, B., Frohning, C.D., and Kohlpaintner, C.W. (1995) *J. Mol. Catal. A: Chem.*, **104**, 17–85. (f) Grigg, R. and Mutton, S.P. (2010) *Tetrahedron*, **66**, 5515–5548. (g) Wu, X.F., Neumann, H., and Beller, M. (2011) *Chem. Soc. Rev.*, **40**, 4986–5009. and one review on Ru- and Rh-catalyzed carbonylation reactions, see: (h) Wu, X.F. and Neumann, H. (2012) *ChemCatChem*, **4**, 447–458.
6. For reviews on metathesis, see: (a) Lozano-Vila, A.M., Monsaert, S., Bajek, A., and Verpoort, F. (2010) *Chem. Rev.*, **110**, 4865–4909. (b) Vougioukalakis, G.C. and Grubbs, R.H. (2010) *Chem. Rev.*, **110**, 1746–1787. (c) Samojtowicz, C., Bieniek, M., and Grela, K. (2009) *Chem. Rev.*, **109**, 3708–3742.
7. Chatani, N., Morimoto, T., Fukumoto, Y., and Murai, S. (1998) *J. Am. Chem. Soc.*, **120**, 5335–5336.
8. Kang, S., Kim, K., and Hong, Y. (2002) *Angew. Chem. Int. Ed.*, **41**, 1584–1586.
9. Finnegan, D.F. and Snapper, M.L. (2011) *J. Org. Chem.*, **76**, 3644–3653.
10. (a) Rosillo, M., Casarrubios, L., Dominguez, G., and Pérez-Castells, J. (2003) *Org. Biomol. Chem.*, **1**, 1450–1451. (b) Rosillo, M., Arnáiz, E., Abdi, D., Blanco-Urgoiti, J., Dominguez, G., and Pérez-Castells, J. (2008) *Eur. J. Org. Chem.*, **2008**, 3917–3927.
11. (a) Chatani, N., Tobisu, M., Asaumi, T., Fukumoto, Y., and Murai, S. (1999) *J. Am. Chem. Soc.*, **121**, 7160–7161. (b) Tobisu, M., Chatani, N., Asaumi, T., Amako, K., Ie, Y., Fukumoto, Y., and Murai, S. (2000) *J. Am. Chem. Soc.*, **122**, 12663–12674. (c) Chatani, N., Amako, K., Tobisu, M., Asaumi, T., Fukumoto, Y., and Murai, S. (2003) *J. Org. Chem.*, **68**, 1591–1593.
12. Kondo, T., Kaneko, Y., Taguchi, Y., Nakamura, A., Okada, T., Shiotsuki, M., Ura, Y., Wada, K., and Mitsudo, T. (2002) *J. Am. Chem. Soc.*, **124**, 6824–6825.
13. Fukuyama, T., Higashibeppu, Y., Yamaura, R., and Ryu, I. (2007) *Org. Lett.*, **9**, 587–589.
14. Chatani, N., Morimoto, T., Kamitani, A., Fukumoto, Y., and Murai, S. (1999) *J. Organomet. Chem.*, **579**, 177–181.
15. Chatani, N., Tobisu, M., Asaumi, T., and Murai, S. (2000) *Synthesis*, 925–928.
16. Kamitani, A., Chatani, N., Morimoto, T., and Murai, S. (2000) *J. Org. Chem.*, **65**, 9230–9233.
17. Kondo, T., Nomura, M., Ura, Y., Wada, K., and Mitsudo, T. (2006) *J. Am. Chem. Soc.*, **128**, 14816–14817.
18. Morimoto, T., Chatani, N., Fukumoto, Y., and Murai, S. (1997) *J. Org. Chem.*, **62**, 3762–3765.
19. Itami, K., Mitsudo, K., and Yoshida, J. (2002) *Angew. Chem. Int. Ed.*, **41**, 3481–3484.
20. Morisaki, Y., Kondo, T., and Mitsudo, T. (2000) *Org. Lett.*, **2**, 949–952.
21. (a) Suzuki, N., Kondo, T., and Mitsudo, T. (1998) *Organometallics*, **17**, 766–769. (b) Fukuyama, T., Yamaura, R., Higashibeppu, Y., Okamura, T., Ryu, I., Kondo, T., and Mitsudo, T. (2005) *Org. Lett.*, **7**, 5781–5783.
22. Aksin, Ö., Dege, N., Artok, L., Türkmen, H., and Cetinkaya, B. (2006) *Chem. Commun.*, 3187–3189.
23. (a) Kus, M., Artok, O.A., Ziyanak, F., and Artok, L. (2008) *Synlett*, 2587–2592. (b) Artok, L., Kus, M., Aksin-Artok, Ö., Dege, F.N., and Özkilinc, F.Y. (2009) *Tetrahedron*, **65**, 9125–9133.

24. Dheur, J., Sauthier, M., Castanet, Y., and Mortreux, A. (2007) *Adv. Synth. Catal.*, **349**, 2499–2506.

25. Dheur, J., Sauthier, M., Castanet, Y., and Mortreux, A. (2010) *Adv. Synth. Catal.*, **352**, 557–561.

26. (a) Sauthier, M., Castanet, Y., and Mortreux, A. (2004) *Chem. Commun.*, 1520–1521. (b) Chochois, H., Sauthier, M., Maerten, E., Castanet, Y., and Mortreux, A. (2006) *Tetrahedron*, **62**, 11740–11746. (c) Sauthier, M., Lamotte, N., Dheur, J., Castanet, Y., and Mortreux, A. (2009) *New J. Chem.*, **33**, 969–971.

27. Harada, Y., Nakanishi, J., Fujihara, H., Tobisu, M., Fukumoto, Y., and Chatani, N. (2007) *J. Am. Chem. Soc.*, **129**, 5766–5771.

28. Van den Hoven, B.G. and Alper, H. (2001) *J. Am. Chem. Soc.*, **123**, 10214–10220.

29. Saito, T., Sugizaki, K., Otani, T., and Suyama, T. (2007) *Org. Lett.*, **9**, 1239–1241.

30. Saito, T., Furukawa, N., and Otani, T. (2010) *Org. Biomol. Chem.*, **8**, 1126–1132.

31. Zhou, D.Y., Kioke, T., Suetsugu, S., Onitsuka, K., and Takahashi, S. (2004) *Inorg. Chim. Acta*, **357**, 3057–3063.

32. Menard, F., Weise, C.F., and Lautens, M. (2007) *Org. Lett.*, **9**, 5365–5367.

33. Matsuda, T., Tsuboi, T., and Murakami, M. (2007) *J. Am. Chem. Soc.*, **129**, 12596–12597.

34. Gabriele, B., Salerno, G., De Pascali, F., Costa, M., and Chiusoli, G.P. (2000) *J. Organomet. Chem.*, **593–594**, 409–415.

35. (a) Kato, K., Nishimura, A., Yamamoto, Y., and Akita, H. (2001) *Tetrahedron Lett.*, **42**, 4203–4205. (b) Kato, K., Tanaka, M., Yamamoto, Y., and Akita, H. (2002) *Tetrahedron Lett.*, **43**, 1511–1513. (c) Kato, K., Matsuba, C., Kusakabe, T., Takayama, H., Yamamura, S., Mochida, T., Akita, H., Peganova, T.A., Vologdin, N.V., and Gusev, O.V. (2006) *Tetrahedron*, **62**, 9988–9999. (d) Motodate, S., Kobayashi, T., Fujii, M., Mochida, T., Kusakabe, T., Katoh, S., Akita, H., and Kato, K. (2010) *Chem. Asian J.*, **5**, 2221–2230.

36. (a) Gabriele, B., Salerno, G., De Pascali, F., Sciano, G.T., Costa, M., and Chiusoli, G.P. (1997) *Tetrahedron Lett.*, **38**, 6877–6880. (b) Gabriele, B., Salerno, G., De Pascali, F., Sciano, G.T., Costa, M., and Chiusoli, G.P. (1999) *J. Org. Chem.*, **64**, 7693–7699. (c) Gabriele, B., Veltri, L., Mancuso, R., Plastina, P., Salerno, G., and Costa, M. (2010) *Tetrahedron Lett.*, **51**, 1663–1665.

37. (a) Kondo, Y., Shiga, F., Murata, N., Sakamoto, T., and Yamanaka, H. (1994) *Tetrahedron*, **50**, 11803–11812. (b) Lütjens, H. and Scammells, P.J. (1998) *Tetrahedron Lett.*, **39**, 6581–6584.

38. (a) Bacchi, A., Costa, M., Della Cà, N., Fabbricatore, M., Fazio, A., Gabriele, B., Nasi, C., and Salerno, G. (2004) *Eur. J. Org. Chem.*, **2004**, 574–585. (b) Della Cá, N., Campanini, F., Gabriele, B., Salerno, G., Massera, C., and Costa, M. (2009) *Adv. Synth. Catal.*, **351**, 2423–2432.

39. (a) Inoue, Y., Ohuchi, K., and Imaizumi, S. (1988) *Tetrahedron Lett.*, **46**, 5941–5942. (b) Inoue, Y., Ohuchi, K., Yen, I.F., and Imaizumi, S. (1989) *Bull. Chem. Soc. Jpn.*, **62**, 3518–3522. (c) Kiji, J., Okano, T., Kimura, H., and Saiki, K. (1998) *J. Mol. Catal. A: Chem.*, **130**, 95–100. (d) Huang, Y. and Alper, H. (1991) *J. Org. Chem.*, **56**, 4534–4536. (e) Wu, X.F., Sundararaju, B., Anbarasan, P., Neumann, H., Dixneuf, P.H., and Beller, M. (2011) *Chem. Eur. J.*, **17**, 8014–8017.

40. (a) Harada, Y., Fukumoto, Y., and Chatani, N. (2005) *Org. Lett.*, **7**, 4385–4387. (b) Arcadi, A., Cacchi, S., Fabrizi, G., Marinelli, F., and Parisi, L.M. (2003) *Tetrahedron*, **59**, 4661–4671. (c) Kato, K., Yamamoto, Y., and Akita, H. (2002) *Tetrahedron Lett.*, **43**, 6587–6590. (d) Miyakoshi, N., Aburano, D., and Mukai, C. (2005) *J. Org. Chem.*, **70**, 6045–6052. (e) Carfagna, C., Gatti, G., Mosca, L., Paoli, P., and Guerri, A. (2005) *Chem. Eur. J.*, **11**, 3268–3278. (f) Gabriele, B., Mancuso, R., Lupinacci, E., Salerno, G., and Veltri, L. (2010) *Tetrahedron*, **66**, 6156–6161. (g) Gabriele, B., Mancuso, R., Salerno, G., and Costa, M. (2007) *J. Org. Chem.*, **72**, 9278–9282.

(h) Gabriele, B., Mancuso, R., Salerno, G., and Plastina, P. (2008) *J. Org. Chem.*, **73**, 756–759.

41. (a) Cho, C.S., Lee, J.W., Lee, D.Y., Shim, S.C., and Kim, T.J. (1996) *Chem. Commun.*, 2115–2116. (b) Cho, C.S., Shim, H.S., Choi, H., Kim, T., Shim, S.C., and Kim, M.C. (2000) *Tetrahedron Lett.*, **41**, 3891–3893. (c) Cho, C.S., Jiang, L.H., and Shim, S.C. (1998) *Synth. Commun.*, **28**, 849–857. (d) Cho, C.S., Kim, H.B., and Lee, S.Y. (2010) *J. Organomet. Chem.*, **695**, 1744–1748.
42. (a) Dhawan, R. and Arndtsen, B.A. (2004) *J. Am. Chem. Soc.*, **126**, 468–469. (b) Lu, Y. and Arndtsen, B.A. (2008) *Angew. Chem. Int. Ed.*, **47**, 5430–5433. (c) Siamaki, A.R. and Arndtsen, B.A. (2006) *J. Am. Chem. Soc.*, **128**, 6050–6051. (d) Worrall, K., Xu, B., Bontemps, S., and Arndtsen, B.A. (2011) *J. Org. Chem.*, **76**, 170–180. (e) Bontemps, S., Quesnel, J.S., Worrall, K., and Arndtsen, B.A. (2011) *Angew. Chem. Int. Ed.*, **50**, 8948–8951. (f) Dhawan, R., Dghaym, R.D., and Arndtsen, B.A. (2003) *J. Am. Chem. Soc.*, **125**, 1474–1475.
43. (a) Vieira, T.O., Meaney, L.A., Shi, Y., and Alper, H. (2008) *Org. Lett.*, **10**, 4899–4901. (b) Arthuis, M., Pontikis, R., and Florent, J.C. (2009) *Org. Lett.*, **11**, 4608–4611. (c) Zeng, F. and Alper, H. (2011) *Org. Lett.*, **13**, 2868–2871.
44. (a) Knight, J.G., Ainge, S.W., Harm, A.M., Harwood, S.J., Maughan, H.I., Armour, D.R., Hollinshead, D.M., and Jaxa-Chamiec, A.A. (2000) *J. Am. Chem. Soc.*, **122**, 2944–2945. (b) Knight, J.G., Lawson, I.M., and Johnson, C.N. (2006) *Synthesis*, 227–230.
45. Xiao, W. and Alper, H. (1999) *J. Org. Chem.*, **64**, 9646–9652.
46. (a) Chouhan, G. and Alper, H. (2010) *Org. Lett.*, **12**, 192–195. (b) Zeng, F. and Alper, H. (2010) *Org. Lett.*, **12**, 5567–5569.
47. (a) Periasamy, M., Radhakrishnan, U., Brunet, J.J., Chauvin, R., and Zaizi, A.W.E. (1996) *Chem. Commun.*, 1499–1500. (b) Periasamy, M., Rameshkumar, C., and Radhakrishnan, U. (1997) *Tetrahedron Lett.*, **38**, 7229–7232. (c) Periasamy, M., Rameshkumar, U., Radhakrishnan, U., and Brunet, J.J. (1998) *J. Org. Chem.*, **63**, 4930–4935. (d) Rameshkumar, C. and Periasamy, M. (2000) *Organometallics*, **19**, 2400–2402. (e) Rameshkumar, C. and Periasamy, M. (2000) *Tetrahedron Lett.*, **41**, 2719–2722. (f) Periasamy, M., Mukkanti, A., and Raj, D.S. (2004) *Organometallics*, **23**, 6323–6326. (g) Periasamy, M., Beesu, M., and Raj, D.S. (2008) *J. Organomet. Chem.*, **693**, 2843–2846. (h) Beesu, M. and Periasamy, M. (2011) *J. Org. Chem.*, **76**, 543–549.
48. (a) Driller, K.M., Klein, H., Jackstell, R., and Beller, M. (2009) *Angew. Chem. Int. Ed.*, **48**, 6041–6044. (b) Prateeptongkum, S., Driller, K.D., Jackstell, R., Spannenberg, A., and Beller, M. (2010) *Chem. Eur. J.*, **16**, 9606–9615. (c) Prateeptongkum, S., Driller, K.D., Jackstell, R., and Beller, M. (2010) *Chem. Asian J.*, **5**, 2173–2176. (d) Driller, K.D., Prateeptongkum, S., Jackstell, R., and Beller, M. (2011) *Angew. Chem. Int. Ed.*, **50**, 537–541. (e) Pizzetti, M., Russo, A., and Petricci, E. (2011) *Chem. Eur. J.*, **17**, 4523–4528. (f) Mathur, P., Joshi, R.K., Jha, B., Singh, A.K., and Mobin, S.M. (2010) *J. Organomet. Chem.*, **695**, 2687–2694.
49. (a) Brunet, J.J. and Taillefer, M. (1990) *J. Organomet. Chem.*, **384**, 193–197. (b) Erunet, J.J. and Zaizi, A.E. (1995) *J. Organomet. Chem.*, **486**, 275–277.
50. (a) Tambade, P.J., Patil, Y.P., Nandurkar, N.S., and Bhanage, B.M. (2008) *Synlett*, 886–888. (b) liu, J., Zhang, R., Wang, S., Sun, W., and Xia, C. (2009) *Org. Lett.*, **11**, 1321–1324.
51. (a) Balogh, J., Kégl, T., Ungváry, F., and Skoda-Földes, R. (2009) *Tetrahedron Lett.*, **50**, 4727–4730. (b) Balogh, J., Kégl, T., Párkányi, L., Kollár, L., Ungváry, F., and Skoda-Földes, R. (2011) *J. Organomet. Chem.*, **696**, 1394–1403.
52. del Moral, D., Ricart, S., and Moretó, J.M. (2010) *Chem. Eur. J.*, **16**, 9193–9202.
53. Inoue, S., Fukumoto, Y., and Chatani, N. (2007) *J. Org. Chem.*, **72**, 6588–6590.

2
Metathesis Reactions in Domino Processes

Kamal M. Dawood and Peter Metz

Recent years have witnessed an extensive use of domino metathesis reactions for the rapid generation of complex molecular structures including a range of synthetically challenging natural products [1]. The aim of this chapter is to provide an overview on the developments in this area that have been published between 2005 and 2012. In the first part of this survey, domino processes featuring solely metathesis events are covered, while the second part will describe domino processes featuring both metathesis and non-metathesis events. Throughout this chapter, the classification of a reaction as a domino process will follow the stringent definition provided in the 2006 book on domino reactions in organic synthesis [2]. As a consequence, many useful one-pot transformations embedding a metathesis step are outside the scope of this review, if they are sequential in nature [3].

Scheme 2.1 depicts the structure of some of the catalysts that have been used most often for metathesis reactions in domino processes.

2.1
Domino Processes Featuring Solely Metathesis Events

This section has been divided into reactions involving only alkenes as the reactive functional groups, and those involving both alkene and alkyne functionalities.

2.1.1
Reactions Involving Only Alkenes

The C_2-symmetric trienes **7** and **9** readily derived from D-mannitol underwent a symmetrical cleavage via a domino process featuring two ring-closing metathesis (RCM) events to afford the enantiopure lactones **8** and **10**, respectively (Scheme 2.2) [4]. While the second-generation Grubbs catalyst **2** led to an efficient conversion of the β,γ-unsaturated diester **9**, the more reactive phosphine-free catalyst **4** was required for a clean transformation of the α,β-unsaturated diester **7**.

Domino reactions consisting of a cross-metathesis (CM) and an RCM step were used to construct macrocycles in a one-pot manner [5]. Thus, the macrocyclic

Domino Reactions: Concepts for Efficient Organic Synthesis, First Edition. Edited by Lutz F. Tietze.

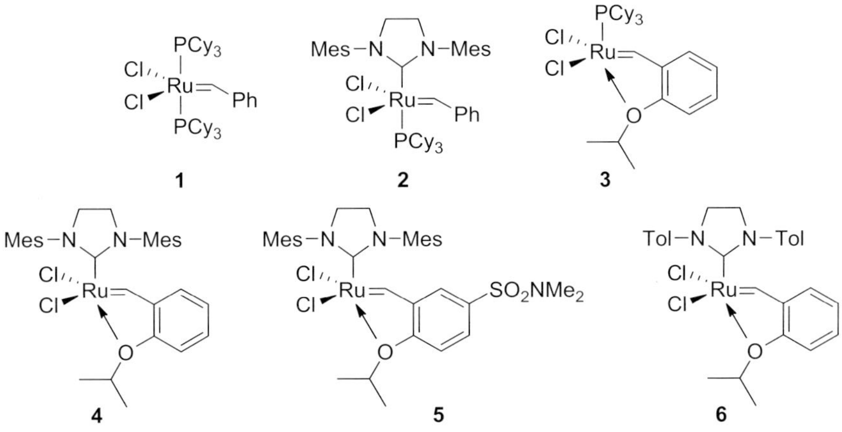

Cy = cyclohexyl, Mes = mesityl, Tol = *o*-tolyl

Scheme 2.1 Catalysts for domino metathesis reactions.

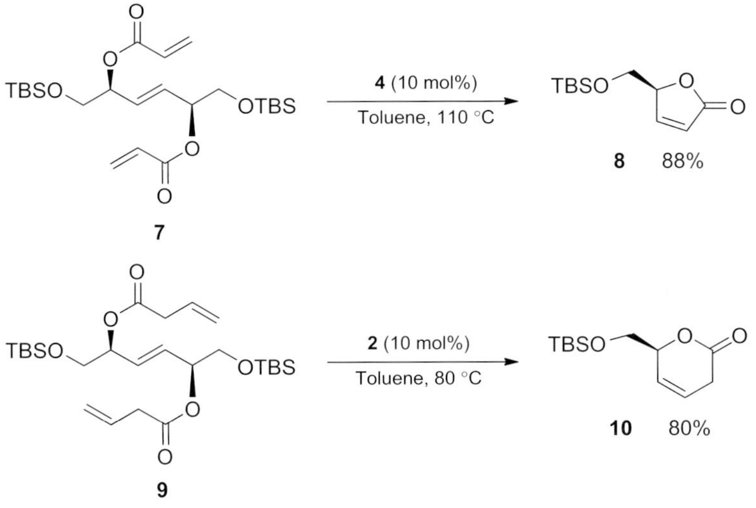

Scheme 2.2 Domino RCM/RCM of C_2-symmetric trienes.

cyclophane **12** was readily assembled from diene **11** with the Grubbs first-generation catalyst **1** in the presence of titanium(IV) isopropoxide through an (*E*)-selective self-metathesis CM event followed by an (*E*)-selective RCM (Scheme 2.3) [5a].

Similarly, the 18-membered macrocycle **14**, which served as a precursor for the synthesis of a bioactive analog of the anticancer agent antimycin A_3, was quickly produced by domino self-metathesis/RCM from the threonine derivative **13** (Scheme 2.4) [5b]. The Zhan catalyst **5** was shown to be more effective for this key step than the Grubbs catalysts **1** or **2**.

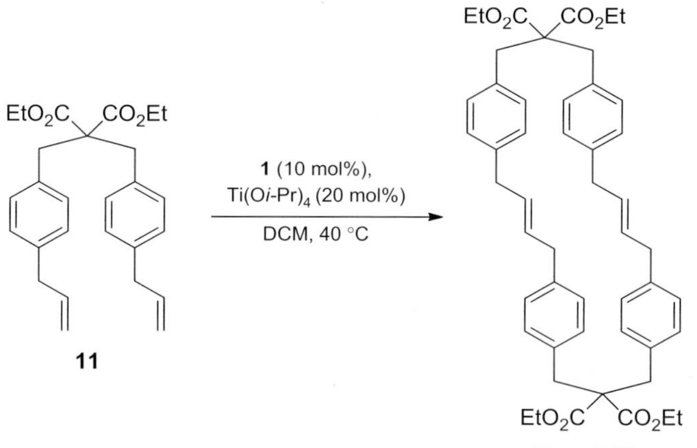

11

12 54%

Scheme 2.3 Domino CM/RCM for cyclophane synthesis.

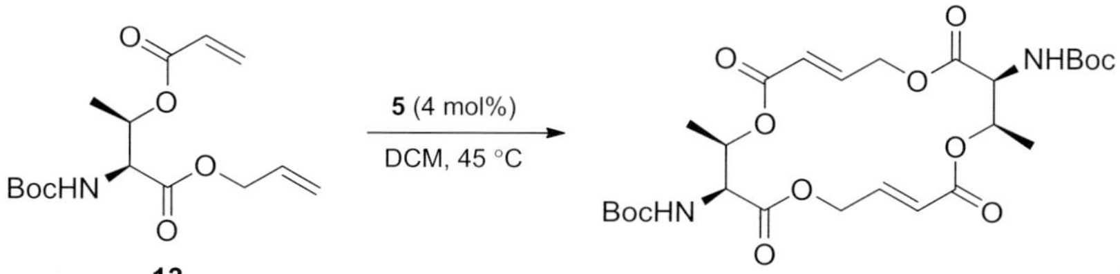

13

14 73%

Scheme 2.4 Domino CM/RCM for macrotetrolide generation.

Domino processes involving a ring-opening metathesis (ROM) and an RCM event have been extensively used in recent years. These transformations, also termed *ring-rearrangement metathesis* (RRM), often rely on the smooth opening of the strained alkene embedded in norbornene [6]. A rapid access to the basic out–out bicyclo[4.4.1]undecene framework of the bioactive tetracyclic diterpene ingenol was accomplished by treatment of the easily available norbornene derivative **15** with the first-generation Grubbs catalyst **1** in the presence of ethylene (Scheme 2.5) [6a]. Next to the RRM process involving the norbornene moiety and one allyl unit, an

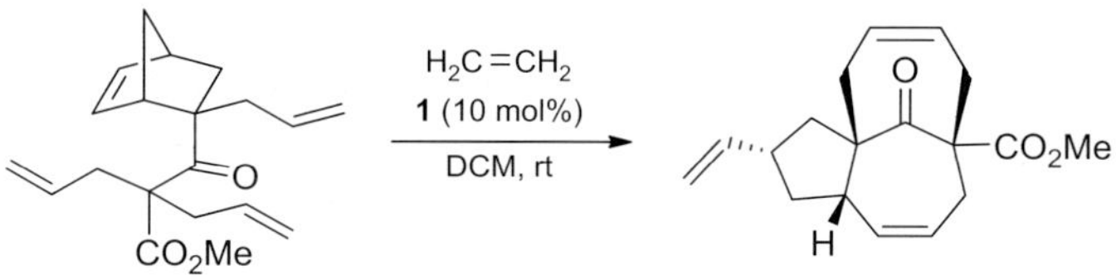

15

16 59%

Scheme 2.5 Domino ROM/RCM of norbornene **15**.

independent RCM of the remaining terminal olefins allowed the direct formation of the desired tricycle **16**.

Under similar conditions, the norbornene derivative **17** was readily transformed into the enantiopure hydroazulene **18** with a densely functionalized lactarane ring system (Scheme 2.6) [6b].

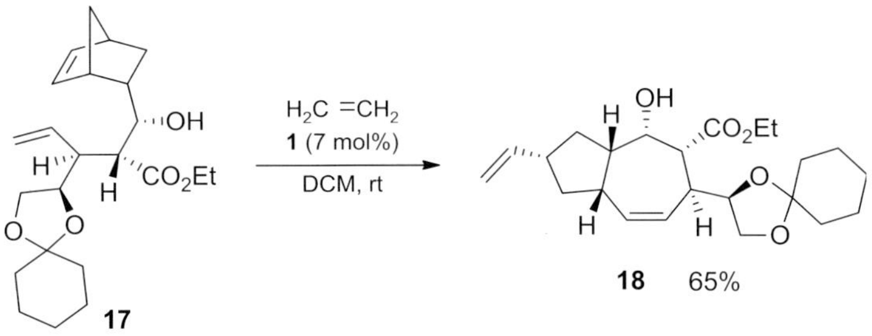

Scheme 2.6 Domino ROM/RCM of norbornene **17**.

A smooth RRM reaction of norbornene **19**, featuring a vinyl ketone as the internal olefinic partner, was the key step in the construction of the enantiomerically pure bicyclo[3.3.0]octane subunit of the bioactive macrolactam aburatubolactam A (Scheme 2.7) [6c]. The resultant bicyclic enone **20** was then elaborated to the target natural product in a concise manner.

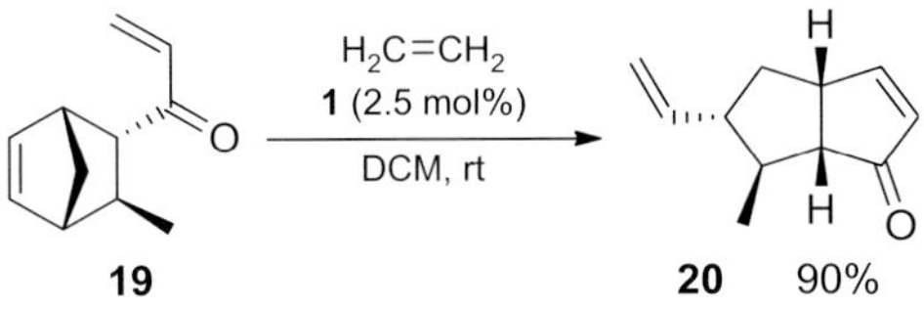

Scheme 2.7 Domino ROM/RCM of norbornene **19**.

The advanced intermediate **22** for the anticancer diterpene umbellactal was efficiently prepared by the RRM of the highly functionalized norbornene derivative **21** under standard conditions (Scheme 2.8) [6d,e].

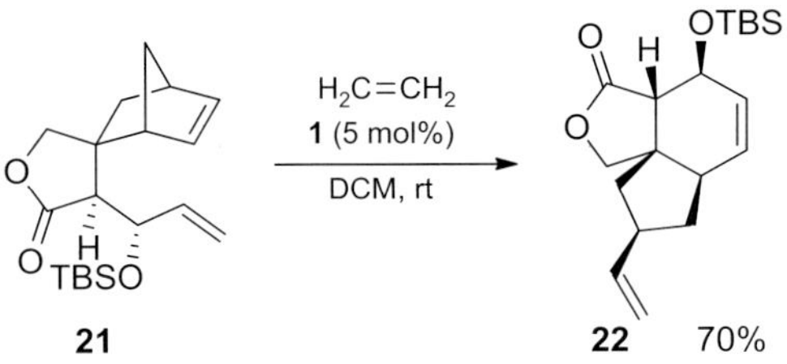

Scheme 2.8 Domino ROM/RCM of norbornene **21**.

While the sequence of events in RRM transformations is not always known, a first attack of the ruthenium carbene on the exocyclic olefin followed by RCM with the norbornene unit and CM with ethylene was established for the conversion of **23** to **24** (Scheme 2.9) [6f]. This highly efficient generation of the linear triquinane **24** also constitutes a formal synthesis of the sesquiterpene $\Delta^{9(12)}$-capnellene in racemic form.

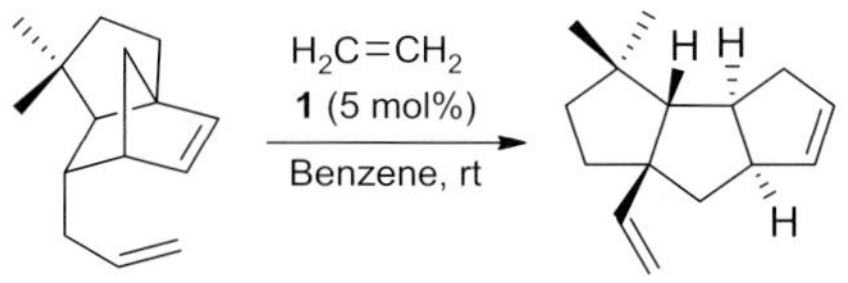

23 **24** 98%

Scheme 2.9 Domino RRM of norbornene **23**.

Using the second-generation Grubbs catalyst **2**, norbornene **25** was converted to the tricyclic compound **26** representing the CDE ring system of the nortriterpenoid schintrilactones A and B in high yield (Scheme 2.10) [6g]. In this case, scission of the norbornene double bond is coupled with two RCM events.

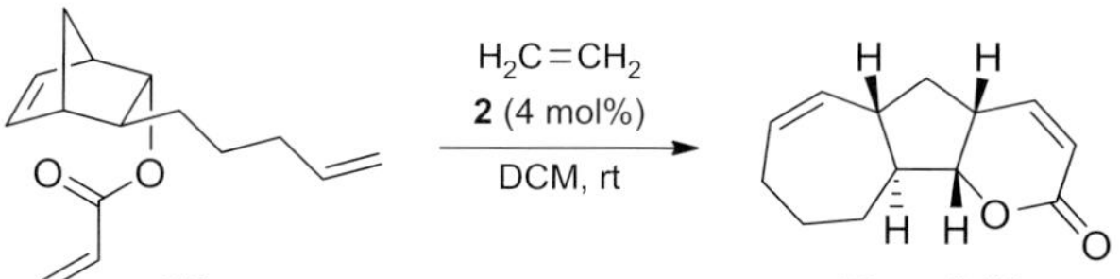

25 **26** 94%

Scheme 2.10 Domino ROM/double RCM of norbornene **25**.

Similarly, the bicyclo[2.2.2]octene double bond in **28** was readily opened in a domino ROM/double RCM process, leading to tricyclic bisenone **29** with catalyst **2** under ethylene atmosphere in good overall yield from dialdehyde **27** (Scheme 2.11) [7]. Compound **29** was successfully utilized as the key intermediate in a total synthesis of the bioactive diterpenoid (+)-cyanthiwigin U.

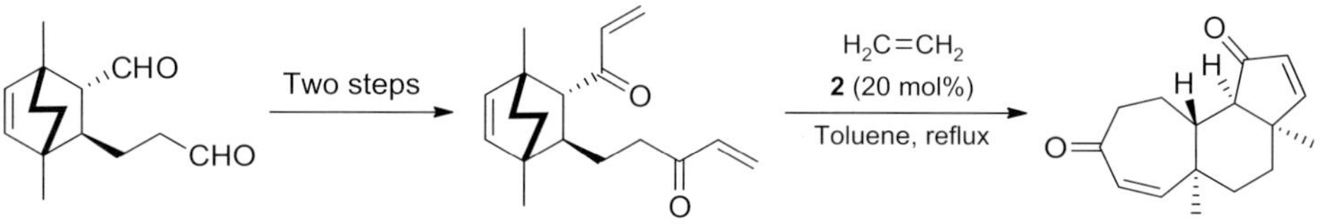

27 **28** **29** 43% (three steps)

Scheme 2.11 Domino ROM/double RCM of bicyclo[2.2.2]octene **28**.

Oxanorbornenes are valuable substrates for RRM processes as well [8]. Thus, a smooth rearrangement of the 7-oxabicyclo[2.2.1]heptene **30** via domino ROM/double RCM occurred on treatment with the second-generation Grubbs catalyst **2** under ethylene-free conditions (Scheme 2.12) [8b]. The spiro compound **31** produced with high efficacy represents the DEF ring fragment of steroidal alkaloid cyclopamine.

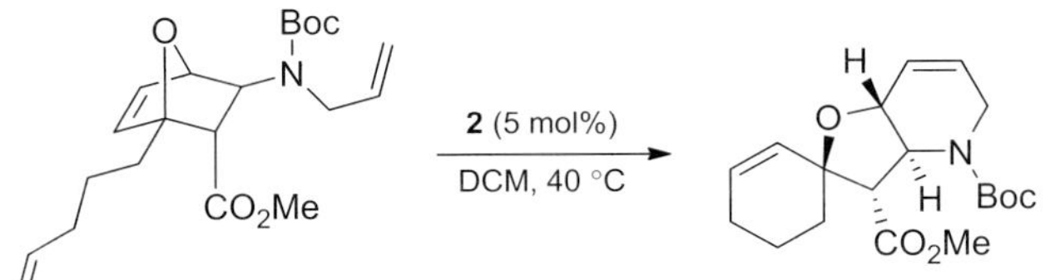

30 **31** 96%

Scheme 2.12 Domino ROM/double RCM of 7-oxabicyclo[2.2.1]heptene **30**.

In the presence of ethylene, the 8-oxabicyclo[3.2.1]octane **32** underwent a clean rearrangement to give the pyranopyran **33** with the Grubbs catalyst **2** (Scheme 2.13) [9]. The enantiopure heterobicycle **33** was then applied as a building block for a total synthesis of the structurally complex marine polyether norhalichondrin B.

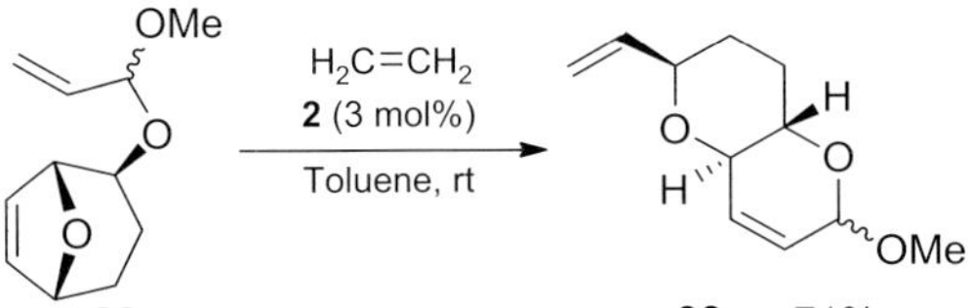

32 **33** 71%

Scheme 2.13 Domino ROM/RCM of 8-oxabicyclo[3.2.1]octane **32**.

Norbornene scaffolds incorporating nitrogen atoms at different positions have been used successfully for RRM reactions, too [10]. In an approach to bicyclic heterocycles as versatile intermediates for dipeptide mimetics, 2-azanorbornene **34** was rearranged to give azabicyclo[5.3.1]alkene **35** on exposure to the Grubbs catalyst **2** under ethylene atmosphere (Scheme 2.14) [10a].

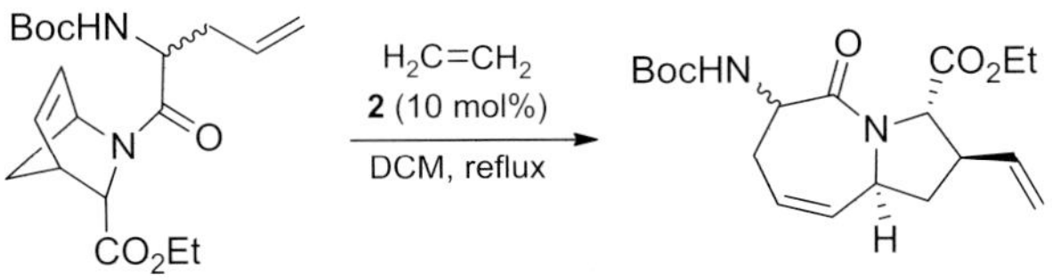

34 **35** 57%

Scheme 2.14 Domino ROM/RCM of azanorbornene **34**.

Shifting the nitrogen atom to the 7-position as in azanorbornene **36** enabled the synthesis of bicyclic lactam **37** in good yield under the conditions already used for **34** (Scheme 2.15) [10b].

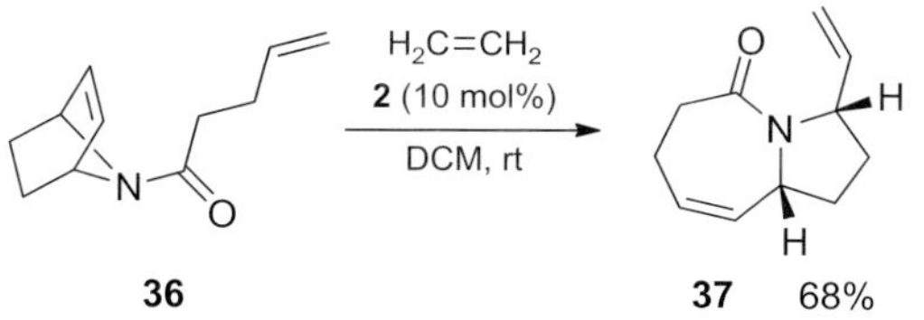

Scheme 2.15 Domino ROM/RCM of azanorbornene **36**.

Nonstrained heterocyclic systems can also serve as suitable substrates for domino ROM/RCM processes [11]. Scheme 2.16 illustrates extended RRM transformations of the tetraenes **38** and **40** embedding two dihydropyran moieties, which proceed with remarkable efficiency using the ruthenium complex **2** to give the rearranged products **39** and **41**, respectively, without unwanted side reactions such as dimer or macrocycle formation [11a].

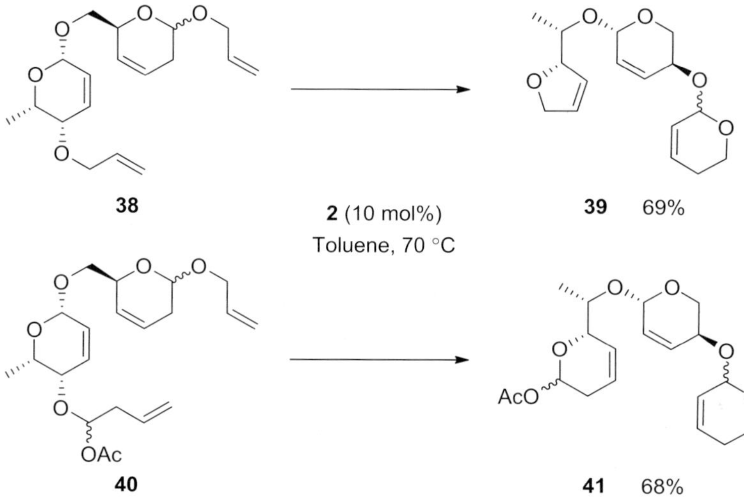

Scheme 2.16 Extended domino RRM processes with bisdihydropyrans **38** and **40**.

The bisdihydrofuran substrate **43** likewise led to the extended RRM product **44** in a preparatively useful yield (Scheme 2.17) [11b]. However, it is currently not easy to predict the feasibility of such a domino process, because a compound similar to **43** lacking the two methyl groups and featuring a 2,5-cis substitution of the heterocycles failed to give the desired RRM product and only produced the corresponding macrocyclic dimer instead.

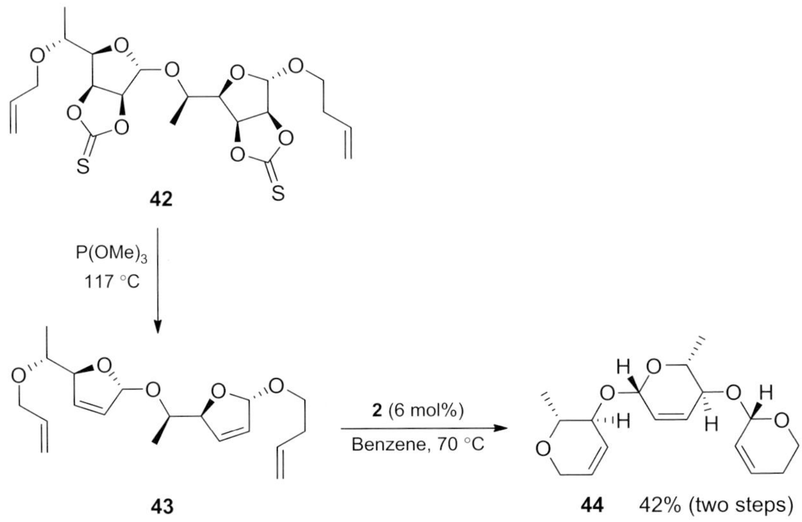

Scheme 2.17 Extended domino RRM transformation of bisdihydrofuran **43**.

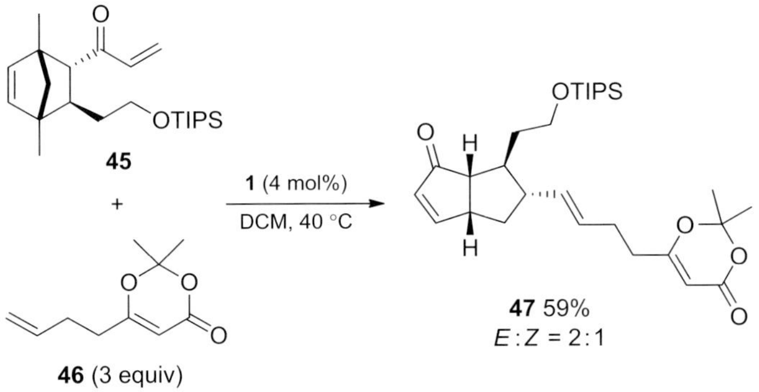

Scheme 2.18 Domino ROM/RCM/CM reaction with norbornene **45**.

RRM processes can also be coupled with an additional CM event. Such a scenario is depicted for a norbornene derivative in Scheme 2.18 [12a]. Domino ROM/RCM of **45** with the first-generation Grubbs catalyst **1** in the presence of terminal olefin **46** yielded the bicyclic enone **47** as a 2 : 1 mixture of separable (*E*)- and (*Z*)-diastereomers. Subsequently, (*E*)-**47** was successfully advanced to the cytotoxic macrolactam (+)-cylindramide A in a straightforward manner. The enantiopure norbornene **45** was also used in a domino RRM/CM approach to the bicyclo[3.3.0]octene core of the macrolactam geodin A [12b].

Oxanorbornenes have also been widely applied for domino RRM/CM transformations [13]. A time-controlled process of this kind is illustrated in Scheme 2.19 [13c]. Reaction of oxanorbornene **48** with 5 mol% of the Grubbs first-generation

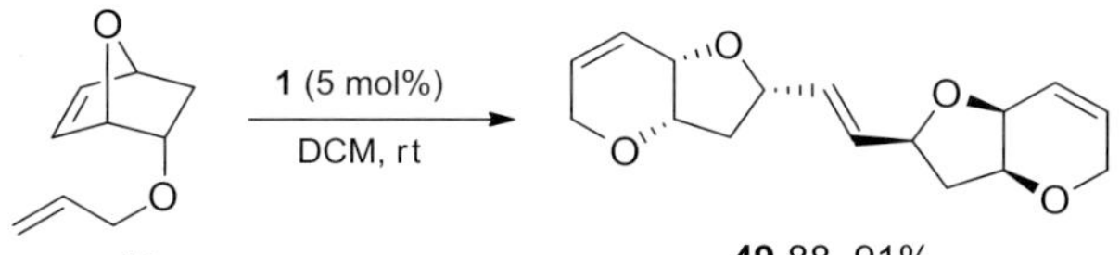

Scheme 2.19 Domino ROM/RCM/CM reaction with oxanorbornene **48**.

catalyst **1** at room temperature for 1 h afforded the bicyclic RRM product (not depicted) in 87% isolated yield but the RRM/CM product **49**, formed by self-metathesis of the RRM product, in only 11% isolated yield. Simply by extending the reaction time from 1 to 12 h, the yield of the RRM/CM product **49** could be increased to 88–91%.

In a synthetic approach to artificial glutamate analogs, a domino ROM/RCM/CM reaction of oxanorbornene **50** with vinyl acetate as the co-olefin was studied (Scheme 2.20) [13d,e]. Using only 0.5 mol% of the phosphine-free catalyst **4**, the tricyclic product **51** was generated quantitatively and with complete control over the geometry of the exocyclic olefin.

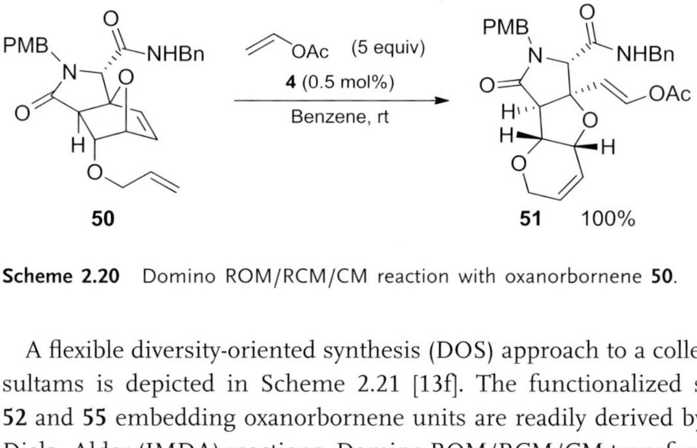

Scheme 2.20 Domino ROM/RCM/CM reaction with oxanorbornene **50**.

A flexible diversity-oriented synthesis (DOS) approach to a collection of tricyclic sultams is depicted in Scheme 2.21 [13f]. The functionalized sultam scaffolds **52** and **55** embedding oxanorbornene units are readily derived by intramolecular Diels–Alder (IMDA) reactions. Domino ROM/RCM/CM transformations of these substrates using the Grubbs catalyst **2** with a set of co-olefins **53** gave rise to the desired heterocyclic systems **54** and **56**, respectively, in good to very good yields. Interestingly, only one of the two epimers **55** participated in the RRM process.

In a formal synthesis of the excitatory amino acids dysiherbaine and neodysiherbaine, oxanorbornene **57** was subjected to a domino RRM/CM process with vinyl acetate using the phosphine-free catalyst **4** (Scheme 2.22) [13g]. The heterobicyclic product **58** was obtained in excellent yield with high (*E*)-selectivity in the CM event.

A domino RRM/CM transformation has been applied to azanorbornene **59**, too (Scheme 2.23) [14]. Reaction of this *N*-Boc substrate with methyl acrylate in the

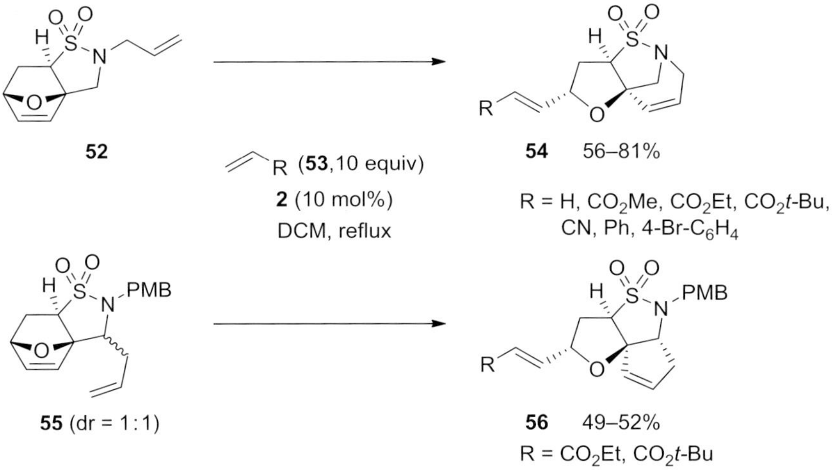

Scheme 2.21 Domino ROM/RCM/CM reactions with oxanorbornenes **52** and **55**.

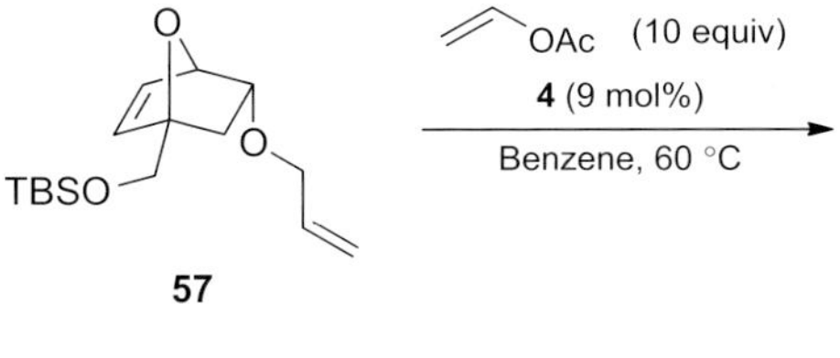

Scheme 2.22 Domino ROM/RCM/CM reaction with oxanorbornene **57**.

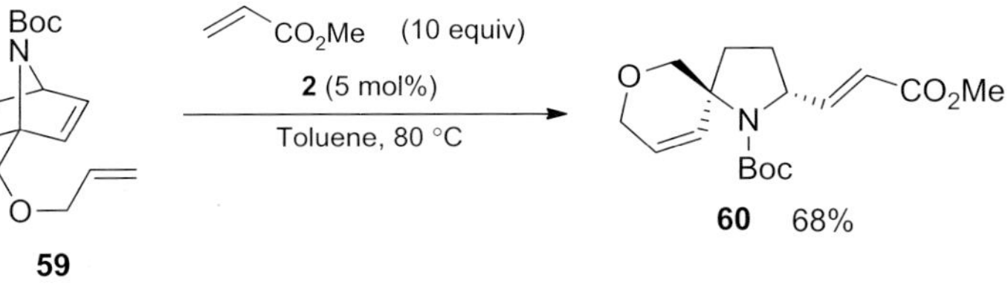

Scheme 2.23 Domino ROM/RCM/CM reaction with azaanorbornene **59**.

presence of Grubbs catalyst **2** gave rise to the spirocyclic product **60**, which was subsequently elaborated to a range of novel γ-amino acids.

Efficient RRM/CM processes were also achieved with the bicyclic dihydro-1,2-oxazine **61** (Scheme 2.24) [15]. While allyl alcohol as the co-olefin did not allow reproducible results, probably due to competitive double-bond isomerization, high yields of the RRM/CM products **63** were obtained next to only small amounts of the RRM products **64** with the unsaturated alcohols **62** and the Grubbs catalyst

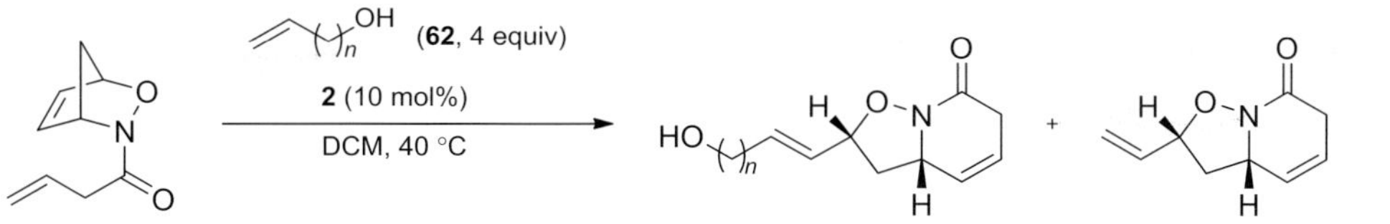

61 63 64

	63		64	
$n = 2$	90	:	10	(87%)
$n = 3$	96	:	4	(91%)

Scheme 2.24 Domino ROM/RCM/CM reaction with bicyclic dihydro-1,2-oxazine **61**.

2. Compound **63** ($n = 2$) was used for the rapid synthesis of a multifunctional quinolizinone.

2.1.2
Reactions Involving Alkenes and Alkynes

Enyne RCM reactions have been successfully coupled with a CM event [16]. Thus, a domino enyne RCM/CM process with enyne **65** and methyl vinyl ketone was used as the final step in a concise total synthesis of the bioactive sesquiterpene lactone (+)-8-*epi*-xanthatin (**66**) (Scheme 2.25) [16a,b]. Using the phosphine-free ruthenium catalyst **4**, an efficient transformation was achieved with the required electron-deficient co-olefin.

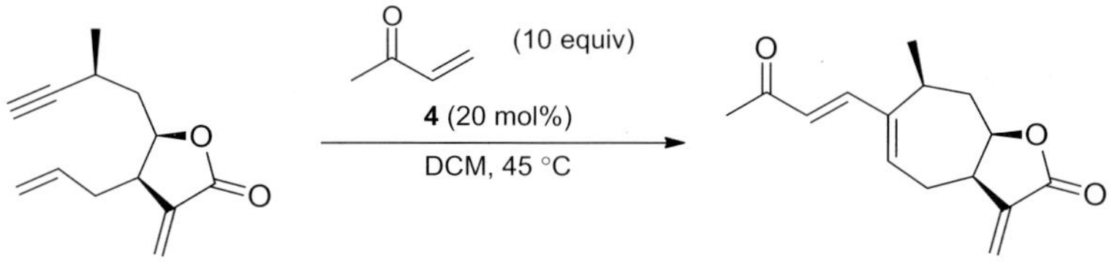

65 **66** 83%

Scheme 2.25 Domino enyne RCM/CM reaction to give (+)-8-*epi*-xanthatin (**66**).

The second-generation Grubbs catalyst **2** already performed very well in the domino enyne RCM/CM process with enyne **67** and 1-tetradecene to afford the semicyclic 1,3-diene **68** with full control over the exocyclic olefin geometry (Scheme 2.26) [16d]. Compound **68** was then advanced to several conformationally restricted piperidine-containing sphingoid base analogs with significant cytotoxic activities.

A relay enyne RCM/CM reaction of dienyne **69** and 2-methyl-3-buten-2-ol was the key step in a total synthesis of the epoxyquinoid natural product (+)-panepophenanthrin (Scheme 2.27) [16e]. This domino transformation, which is initiated by attack of the ruthenium carbene at the terminal alkene followed by an RCM with release of 2,5-dihydrofuran, worked best with the phosphine-free

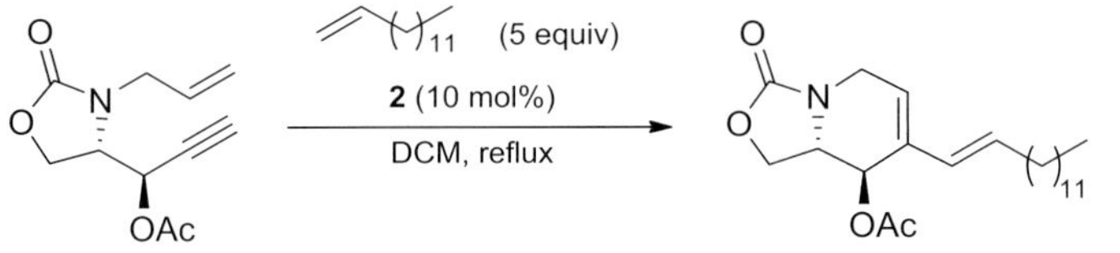

67 **68** 76%

Scheme 2.26 Domino enyne RCM/CM reaction to furnish the heterobicyclic system **68**.

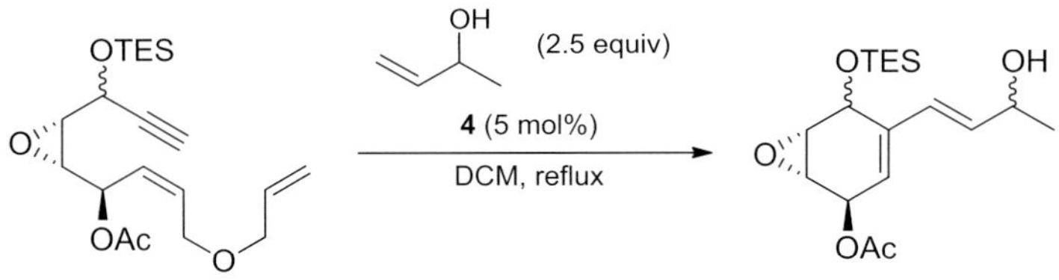

69 **70** 51%

Scheme 2.27 Relay enyne RCM/CM reaction to give the semicyclic 1,3-diene **70**.

ruthenium catalyst **4**, whereas no CM was observed when the second-generation Grubbs catalyst **2** was used.

Domino processes involving an enyne CM followed by an RCM have served for the regioselective formation of heterocycles and carbocycles [17]. The reaction of siloxy-tethered enyne **71** featuring a sterically hindered alkene unit with *cis*-2-butene in the presence of Grubbs catalyst **2** afforded the cyclic siloxane **72** in high yield (Scheme 2.28) [17a]. An initial regio- and stereoselective enyne CM reaction between a propagating species derived from the external olefin and catalyst **2** and the alkyne generates a new alkylidene intermediate that undergoes a subsequent RCM to give **72**. This six-membered heterocycle is a formal endo-mode RCM product, which is unachievable by a direct enyne RCM reaction.

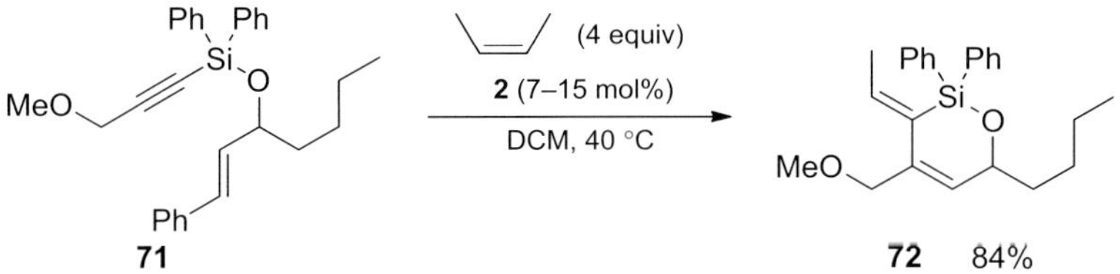

71 **72** 84%

Scheme 2.28 Domino enyne CM/RCM reaction to afford the cyclic siloxane **72**.

In the reaction of the α-amino acid derivative **73** with 1,5-cyclooctadiene catalyzed by ruthenium complex **2** depicted in Scheme 2.29, an enyne CM followed by an RCM was coupled in a domino manner with an initial ROM of the cyclic diene [17b,c]. The resulting enantiomerically pure 1,3-cyclohexadiene **74** was then used to construct the diketopiperazine core of the scabrosin epidithiodiketopiperazine antibiotics [17c].

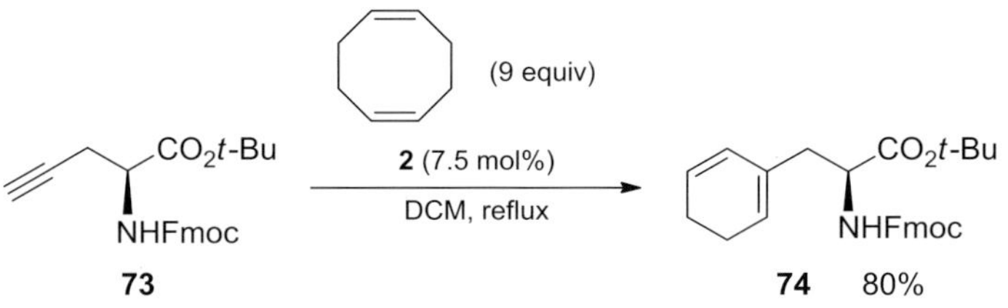

Scheme 2.29 Domino ROM/enyne CM/RCM reaction to give the 1,3-cyclohexadiene **74**.

This domino ROM/enyne CM/RCM process has been exploited for a regio- and chemoselective ring expansion of substituted cyclopentenes to provide 1,3-cycloheptadienes (Scheme 2.30) [17d]. Thus, reaction of the cyclopentenes **75**, **77**, and **79** with different alkynes and Grubbs catalyst **2** readily afforded the depicted products **76**, **78**, and **80**, respectively.

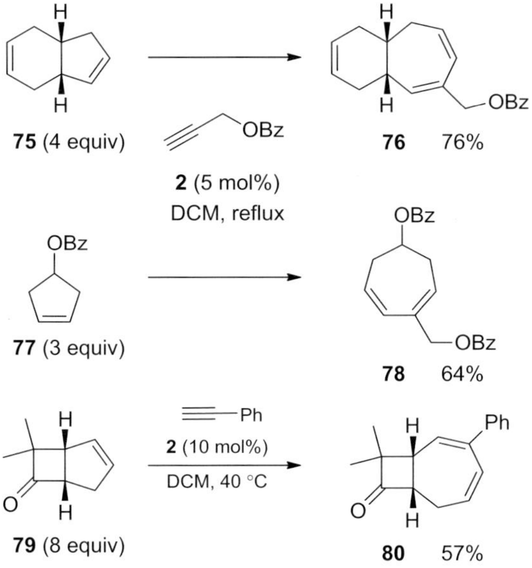

Scheme 2.30 Ring expansion to 1,3-cycloheptadienes via domino ROM/enyne CM/RCM.

On the other hand, treatment of enyne **81** with 1,5-hexadiene in the presence of Grubbs catalyst **2** led to the 12-membered lactone **82** via a domino enyne CM/RCM process (Scheme 2.31) [17f]. In addition to **82**, 54% of the simple enyne CM product from the alkyne moiety of **81** and 1,5-hexadiene with an (*E*)-configured 1,2-disubstituted alkene was isolated.

An efficient approach to the 4-methylene-2-cyclohexenone and 4-methylene-2-cyclohexenol substructures, respectively, of the bioactive ottelione and loloanolide natural products has been developed using domino enyne CM/RCM transformations (Scheme 2.32) [17g]. To this end, the highly functionalized enynes **83**

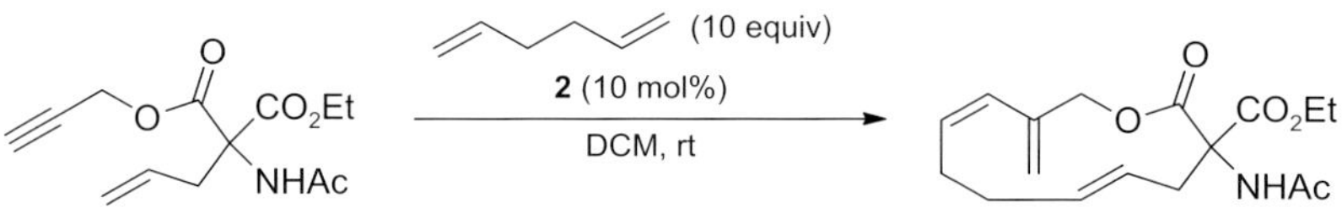

Scheme 2.31 Domino enyne CM/RCM process leading to 12-membered lactone **82**.

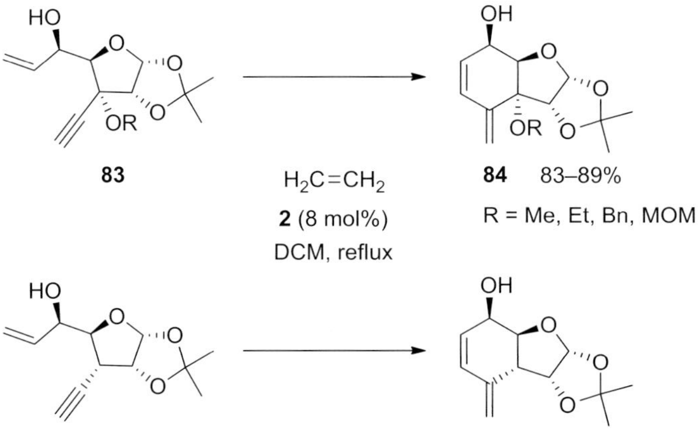

Scheme 2.32 Domino enyne CM/RCM transformations to afford 4-methylene-2-cyclohexenols.

and **85** were treated with the second-generation Grubbs catalyst **2** under ethylene atmosphere to give the *cis*- or *trans*-fused 4-methylene-2-cyclohexenols **84** and **86**, respectively, in high yields. Subsequent Dess–Martin periodinane oxidation then gave rise to the 4-methylene-2-cyclohexenone moieties as well.

A domino metathesis process that was broadly applied to the synthesis of structurally complex target molecules in recent years involves an enyne RCM coupled with another RCM event. The majority of such dienyne metatheses was applied to the construction of fused bicyclic product structures [18]. Scheme 2.33 illustrates the generation of a fused 5/6-bicyclic substructure in the final step of an enantioselective total synthesis of the *Erythrina* alkaloid (+)-β-erythroidine (**89**) [18c]. Treatment of the β,γ-unsaturated substrate **88** containing small amounts of the α,β-unsaturated isomer **87** with the first-generation Grubbs catalyst **1** at room temperature in dichloromethane yielded the desired natural product **89** after separation of the minor by-product **90** resulting from a simple diene RCM reaction. Formation of **89** was enabled by selective initiation at the sterically less hindered N-allyl unit. Interestingly, catalyst **1** was superior to the second-generation catalyst **2** in this particular case.

The corresponding assembly of the fused 7/7-bicyclic substructure present in **93** by domino dienyne metathesis of silyl ether **92** with Grubbs catalyst **2** was

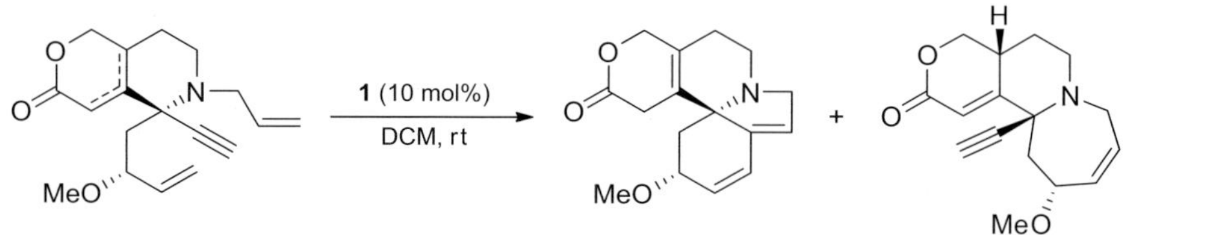

87: α,β-unsaturated
88: β,γ-unsaturated
(**87**:**88** = 6:94)

89 : **90**
91 : 9 (66% borsm)

Scheme 2.33 Domino dienyne metathesis in the total synthesis of (+)-β-erythroidine (**89**).

again made possible by a chemoselective initiation at the sterically more accessible monosubstituted olefinic site (Scheme 2.34) [18e]. As **93** could be elaborated to a known advanced colchicine intermediate, a formal synthesis of this alkaloid was accomplished.

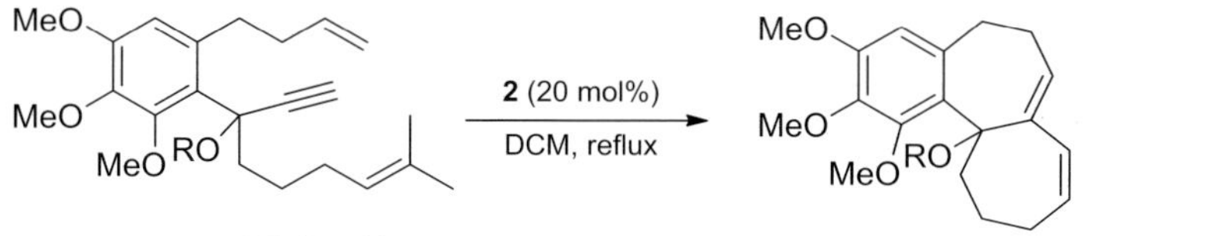

TMS-Im **91**:R = H → **92**:R = TMS

93:R = TMS 74% (two steps)

Scheme 2.34 Domino dienyne metathesis of **92** in a formal synthesis of colchicine.

The selective domino dienyne metathesis of **94** with Grubbs catalyst **2** to give the fused 5/9-bicyclic substructure embedded in **95** controlled by olefin substitution was coupled in a sequential manner with a chemoselective ruthenium-catalyzed hydrogenation of the disubstituted alkene of **95** to yield the tricyclic protected amino ketone **96** in a one-pot process (Scheme 2.35) [18h]. Only two further steps were required to convert **96** into the *Lycopodium* alkaloid (+)-lycoflexine.

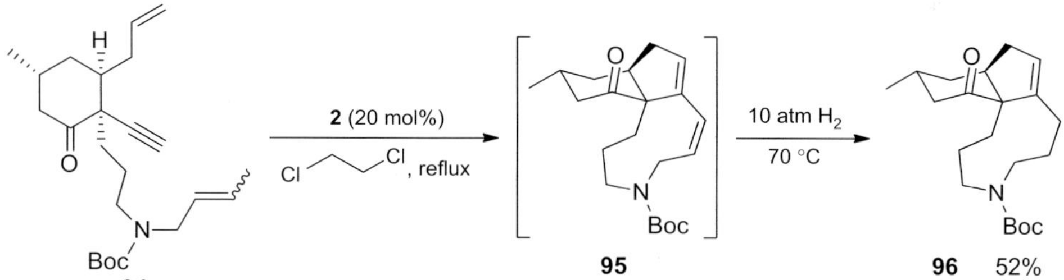

94 **95** **96** 52%

Scheme 2.35 Domino dienyne metathesis of **94** in a total synthesis of (+)-lycoflexine.

Domino metathesis reactions of the dienynes **97** and **99** catalyzed by the phosphine-free ruthenium catalyst **4** in the presence of ethylene allowed a rapid access to the naturally occurring marine trisnorsesquiterpenes (−)-clavukerin A (**98**) and (−)-isoclavukerin A (**100**), respectively, by preferential initiation at the disubstituted alkene (Scheme 2.36) [18i]. As the dienyne substrates **97** and **99** were easily available from (*S*)- and (*R*)-citronellal, respectively, in only three high yielding operations, the 1,3-dienes **98** and **100** now offer themselves as enantiopure building blocks for the preparation of structurally more complex hydroazulene targets.

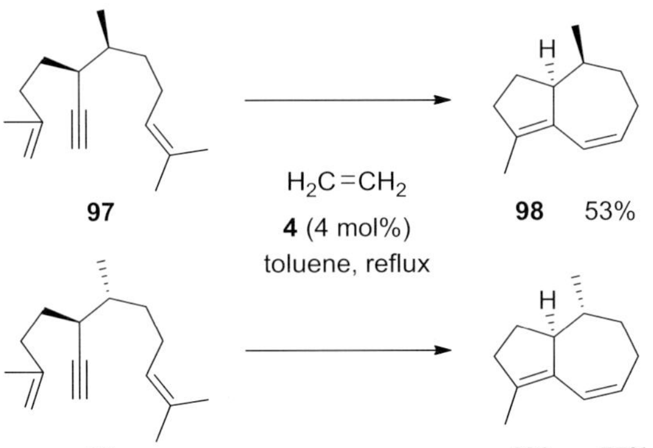

97 **98** 53%

99 **100** 55%

Scheme 2.36 Domino dienyne metathesis in the total synthesis of (−)-clavukerin A (**98**) and (−)-isoclavukerin A (**100**).

The tetracyclic kempane skeleton **102** incorporating a fused 5/7-bicyclic substructure was obtained with high efficacy from a domino metathesis of the dienyne **101** with the second-generation Grubbs catalyst **2** through selective initiation at the monosubstituted olefinic site (Scheme 2.37) [18j]. In a few steps, compound **102** could be converted into the kempane diterpenes (+)-kempene-2, (+)-kempene-1, and (+)-3-*epi*-kempene-1, which have been isolated from the defense secretion of higher termites.

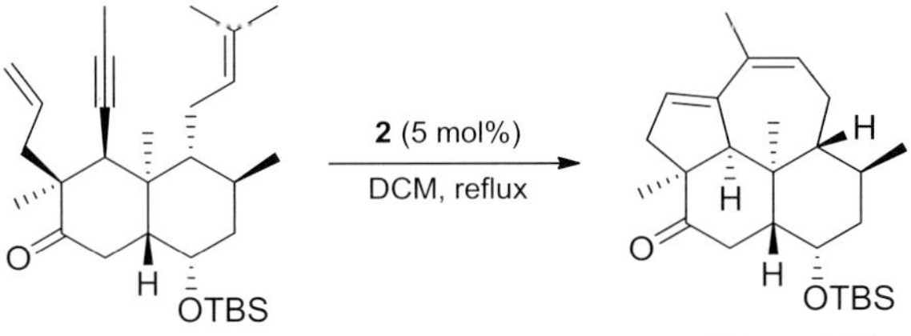

101 **102** 92%

Scheme 2.37 Domino metathesis of dienyne **101** in the total synthesis of kempane diterpenes.

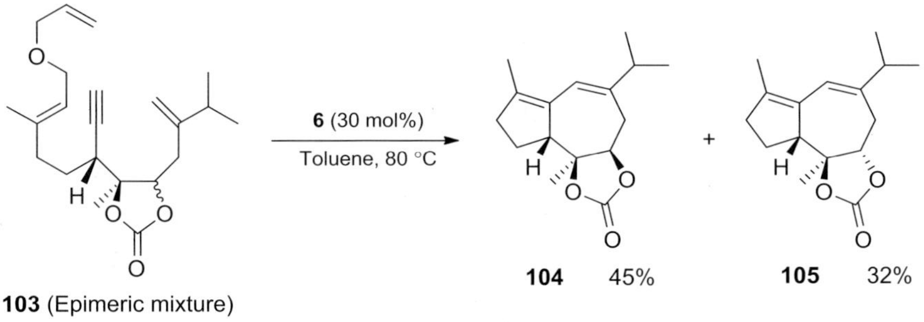

Scheme 2.38 Domino relay dienyne metathesis in a formal synthesis of (−)-englerin A.

Scheme 2.38 depicts a relay dienyne metathesis transformation to afford the hydroazulene ring system [18m]. In the presence of the phosphine-free catalyst **6**, substrate **103** used as an epimeric mixture underwent a completely regioselective domino cyclization due to exclusive attack of the ruthenium carbene at the terminal olefin followed by RCM with release of 2,5-dihydrofuran. Diastereomer **104** was then further elaborated to achieve a formal synthesis of the guaiane (−)-englerin A, which shows a highly selective action against renal cancer cell lines.

Another domino relay dienyne metathesis with initial generation of cyclopentene to position the ruthenium carbene selectively adjacent to the ester carbonyl group is illustrated in Scheme 2.39 [18n]. Subjecting the proline-derived substrates **106** and **108** to the Zhan catalyst **5** in refluxing toluene turned out to be more efficient than catalysis with ruthenium complexes **2** or **4** in different solvents and provided the fused 5/6-bicyclic substructure of the unsaturated lactones **107** and **109**, respectively, in good yields. Compounds **107** and **109** were then successfully utilized for the total synthesis of the *Securinega* alkaloids (−)-flueggine A and (+)-virosaine B, respectively.

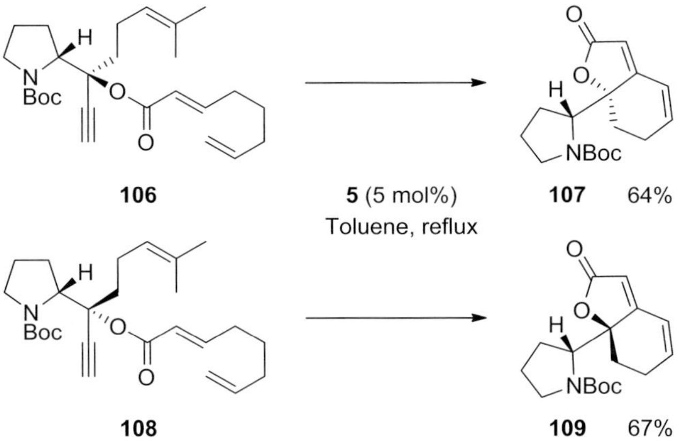

Scheme 2.39 Domino relay dienyne metathesis in the total synthesis of *Securinega* alkaloids.

Next to fused bicyclic systems, the domino dienyne metathesis process can also be applied to the construction of bridged bicyclic dienes. Thus, dienyne **110** bearing differently substituted alkene units to ensure a regiocontrolled cyclization readily afforded the taxosteroid hybrid compound **111** without competing cyclopentene formation in the presence of the Grubbs second-generation catalyst **2** (Scheme 2.40) [19].

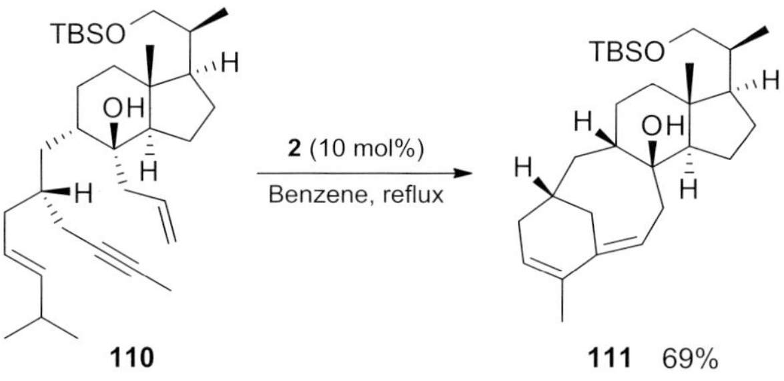

Scheme 2.40 Domino dienyne metathesis in the construction of a bridged taxosteroid hybrid.

Using a disubstituted alkyne with appropriately placed alkenes at either side, the domino dienyne metathesis also enables the formation of bicyclic 1,3-dienes where two cyclic olefins are connected by a C–C single bond [20]. Scheme 2.41 depicts such a scenario for the double cyclization of diether **112** to give a virtually quantitative yield of the bisdihydrofuran **113** with Grubbs catalyst **1** [20a,c].

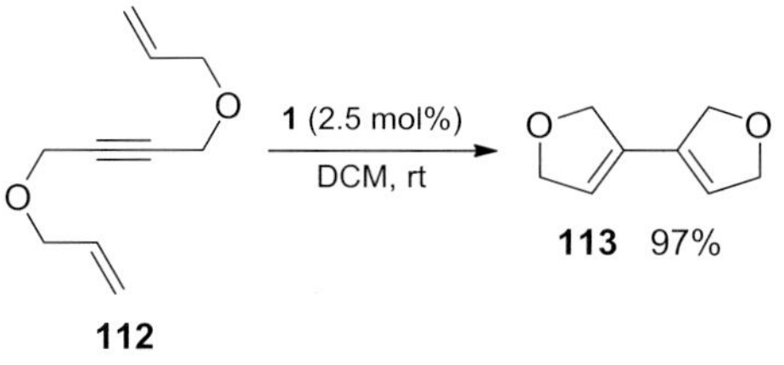

Scheme 2.41 Domino dienyne metathesis to give the bisdihydrofuran **113**.

Similarly, the dienynes **114** and **115** underwent a domino cyclization catalyzed by ruthenium complex **1** in good to excellent yields with generation of two cyclohexene rings embedded into the steroid-like derivatives **116** and **117**, respectively (Scheme 2.42) [20b].

Domino metathesis of siloxy-tethered dienynes offers attractive opportunities for the synthesis of stereochemically defined 1,3-dienes after desilylation [21]. These reactions also allow the efficient preparation of highly substituted cycloolefins, which is illustrated in Scheme 2.43 [21b,c]. Selective double cyclization of silaketal **118** via initiation at the monosubstituted alkene led to a fused 6/7-bicyclic product which was subsequently desilylated to afford the triol **119** in high overall yield.

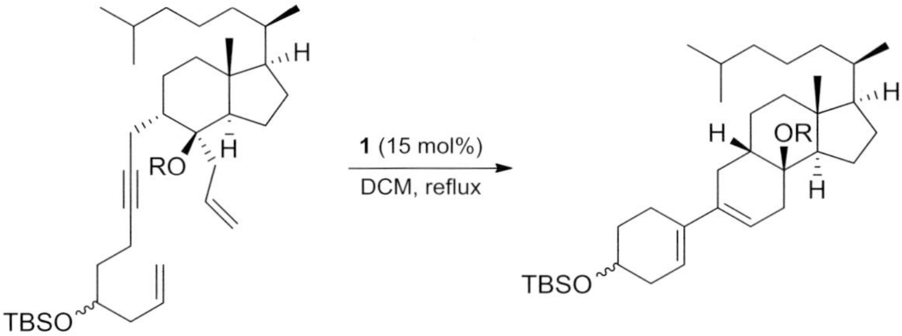

114 R = H
115 R = Me

116 66%
117 93%

Scheme 2.42 Domino dienyne metathesis in the construction of steroid-like polycyclic systems.

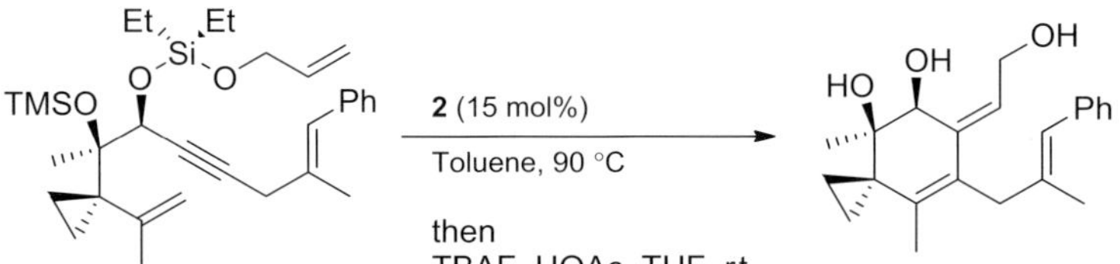

118

119 79% (two steps)

Scheme 2.43 Domino dienyne metathesis in the total synthesis of illudin sesquiterpenes.

Compound **119** was then used straightforwardly for the total synthesis of the antitumor agents (−)-acylfulvene and (−)-irofulvene belonging to the illudin family of sesquiterpenes.

Dendralenes can be rapidly synthesized by consecutive enyne RCM of allylsilyl propargyl ethers followed by rhenium oxide-catalyzed 1,4-elimination of the resultant siloxenes [21f]. Hence, an efficient preparation of [4]dendralene **122** was accomplished via domino dienyne metathesis of the bis(silyl ether) **120** to give the bis-siloxene **121**, which was subjected to catalytic amounts of rhenium oxide to induce a double 1,4-elimination with the formation of **122** (Scheme 2.44).

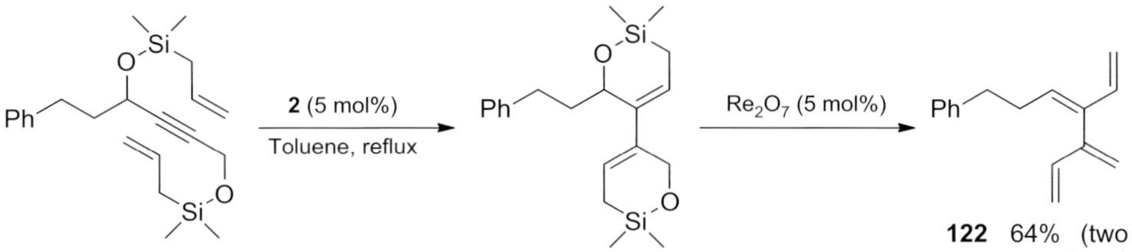

120

121

122 64% (two steps)

Scheme 2.44 Domino dienyne metathesis in the synthesis of [4]dendralene **122**.

Coupling an ROM event with enyne metathesis opens up further possibilities for RRM. As already discussed for reactions involving only olefins, norbornenes are valuable strained cycloolefins for the ring-opening step [22]. Treatment of the norbornene derivative **123** with the first-generation Grubbs catalyst **1** under ethylene atmosphere induced a smooth domino ROM/enyne RCM/diene RCM process with quantitative formation of the tricyclic compound **124** (Scheme 2.45) [22b].

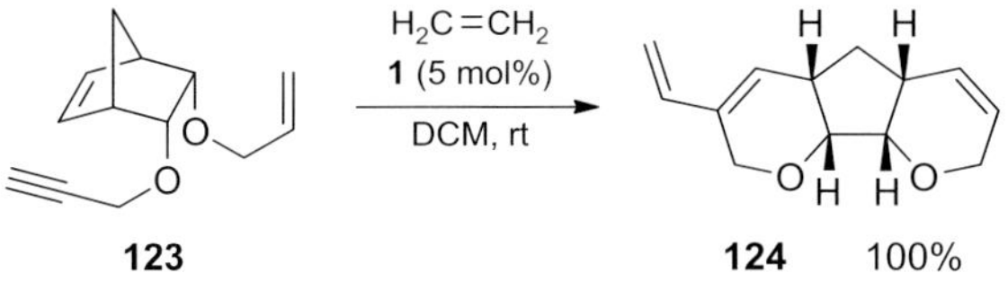

Scheme 2.45 Domino ROM/enyne RCM/diene RCM of norbornene **123**.

Already in the absence of ethylene, the oxanorbornene propargyl ethers **125** and **126** bearing an internal or terminal alkyne readily participated in a domino ROM/enyne RCM process catalyzed by ruthenium complex **1** as well to give the heterobicycles **127** and **128**, respectively, in high yields (Scheme 2.46) [23]. An efficient domino ROM/double enyne RCM transformation of the bis-propargyl ethers **129** and **130** likewise furnished the bis-dienes **131** and **132**, respectively, under similar reaction conditions.

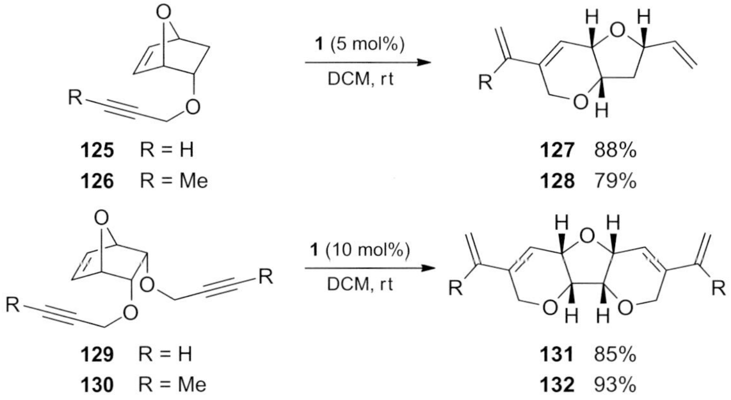

Scheme 2.46 Domino ROM/enyne RCM of oaxanorbornene propargyl ethers.

Strained small ring olefins are also highly suitable for ROM/enyne RCM reactions. An extended domino ROM/enyne RCM/diene RCM process leading to a ring expansion of cyclobutenes **133** to afford the bicyclic products **134** incorporating a 1,3-cyclohexadiene moiety was achieved with the second-generation Grubbs catalyst **2** under ethylene atmosphere (Scheme 2.47) [24].

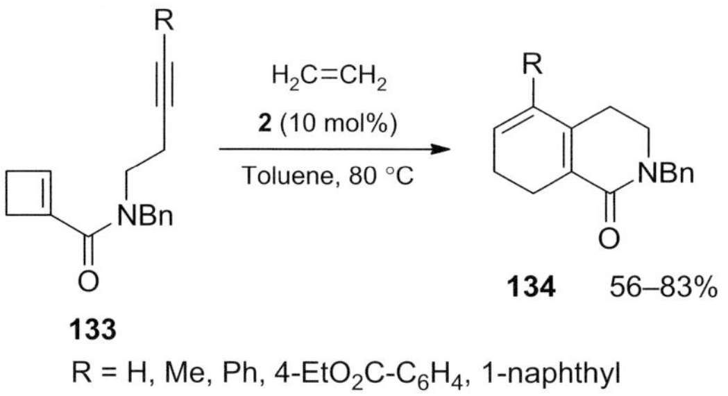

Scheme 2.47 Domino ROM/enyne RCM/diene RCM of cyclobutenes **133**.

The cyclopropene-derived propargyl ether **135** and *p*-nitrobenzenesulfonamide-linked enyne **136** took part in a smooth ROM/enyne RCM/CM domino reaction at room temperature on stirring in styrene as the solvent with the Grubbs first-generation catalyst **1** (Scheme 2.48) [25]. However, there was almost no control over the geometry of the disubstituted alkene in the resulting trienes **137** and **138**.

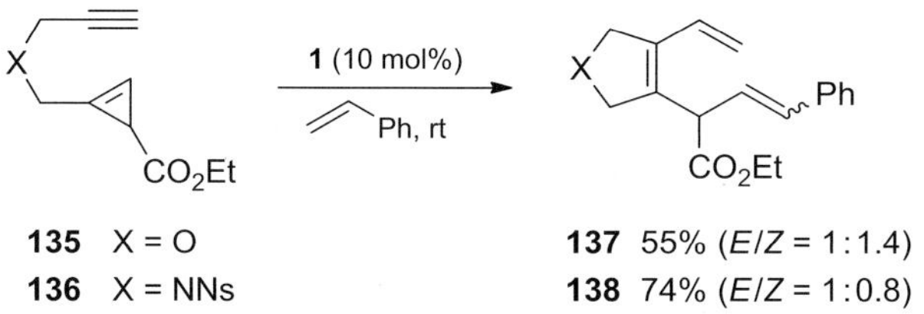

Scheme 2.48 Domino ROM/enyne RCM/CM of cyclopropenes **135** and **136**.

In the context of a DOS approach to natural product-like compound libraries, extended domino RRM transformations of polyunsaturated substrates involving also monocyclic olefins in the ROM step have been successfully applied [26]. As exemplified in Scheme 2.49, the trienyne **139** featuring a cyclohexene building block is rearranged with good yield into the heterocyclic triene **140** with the phosphine-free catalyst **4**. A similar "zipping-up" of the trienyne **141** embedding a cyclopentene unit, again with initiation at the monosubstituted olefin by the Grubbs first-generation catalyst **1**, was followed by desilylation to give the heterocyclic scaffold **142** with two free primary alcohols.

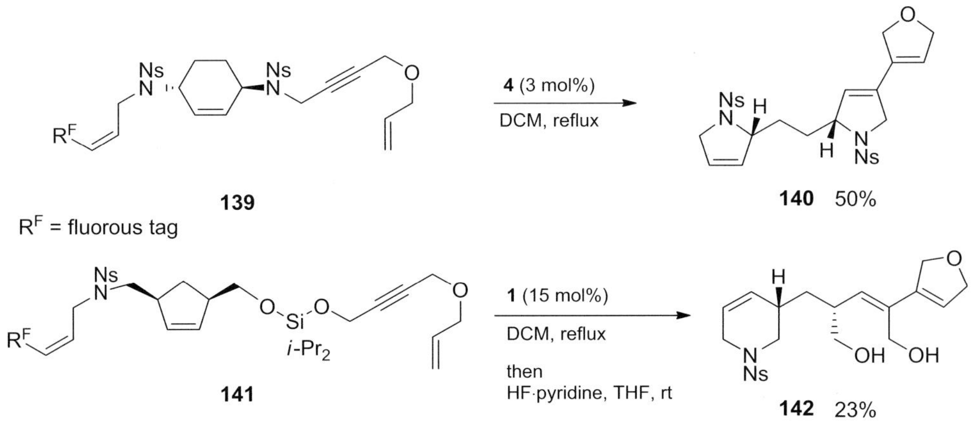

Scheme 2.49 Extended domino RRM transformations of trienynes **139** and **141**.

2.2 Domino Processes Featuring Metathesis and Non-metathesis Events

In this section, a division has been made according to the nature of the non-metathesis event spanning a wide range of reaction types from redox transformations to isomerization, cycloaddition, and substitution to conjugate addition and carbonyl olefination.

2.2.1 Metathesis/Redox Transformation

Treatment of dienes **143** with the Grubbs second-generation catalyst **2** at 80–110 °C in toluene induced a domino RCM/dehydrogenation process to give the indenones **144** in moderate to good yields (Scheme 2.50) [27]. Separately subjecting the indenol intermediate obtained from **143** ($R^1 = R^2 = H$) at room temperature in dichloromethane to catalyst **2** under anaerobic conditions at 60 °C promoted the formation of the corresponding indenone **144**, suggesting that the oxidation step is indeed ruthenium-mediated.

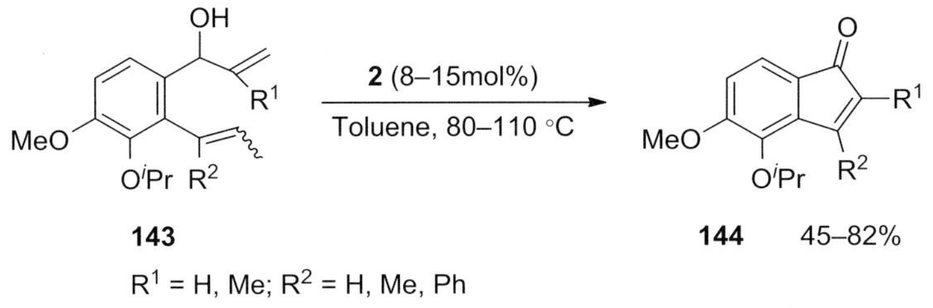

Scheme 2.50 Domino RCM/dehydrogenation of dienes **143**.

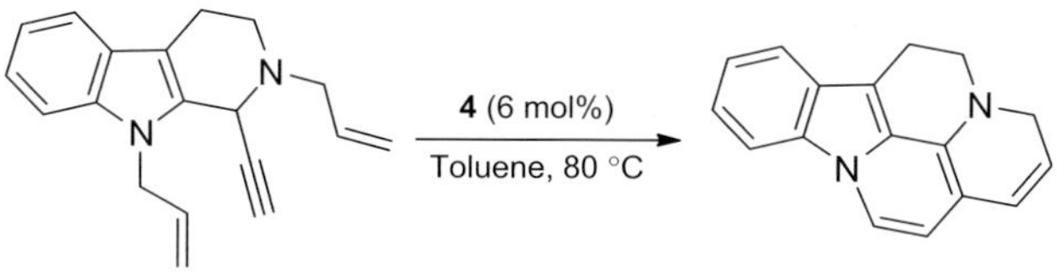

145 **146** 67%

Scheme 2.51 Domino dienyne metathesis/dehydrogenation of dienyne **145**.

A dehydrogenation was also observed when the β-carboline dienyne **145** was subjected to the phosphine-free catalyst **4** at 80 °C in toluene (Scheme 2.51) [28]. Formation of **146** implies initial attack of the ruthenium carbene on the allyl group attached to the nitrogen atom embedded in the five-membered ring.

A domino RCM/reduction transformation of dienes **147** and **149** to furnish the saturated heterocycles **148** and **150**, respectively, was realized by simultaneous treatment of the substrates with Grubbs catalyst **1** and triethylsilane (Scheme 2.52) [29]. Possibly, the hydrogenation step was catalyzed by the metathesis-inactive catalyst after completion of the RCM event.

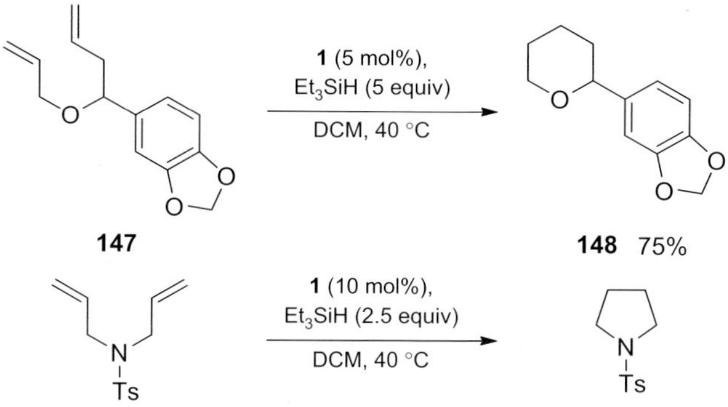

147 **148** 75%

149 **150** 76%

Scheme 2.52 Domino RCM/reduction of dienes **147** and **149**.

2.2.2
Metathesis/Isomerization

A ruthenium-catalyzed double-bond isomerization may occur before the desired metathesis event, giving rise to the formation of an unexpected constitutional isomer. Thus, reaction of dienes **151** with the second-generation Grubbs catalyst **2** under ethylene atmosphere was intended to construct a fused 6/7-carbobicyclic substructure through a domino ROM/RCM process involving a cleavage of the bicyclo[2.2.2]octene (Scheme 2.53) [30]. However, only the RRM products **152** were isolated in high yields, implying that the terminal olefin in **151** was isomerized to

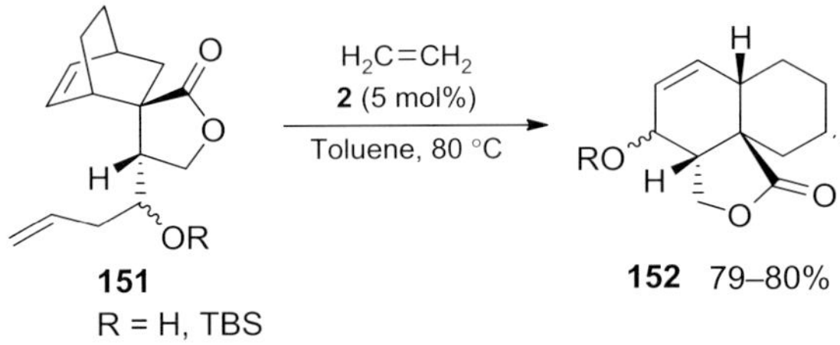

Scheme 2.53 Domino isomerization/RRM of dienes **151**.

give the allyl alcohol or allyl silyl ether, respectively, before the RCM step took place. In line with this rationale, treatment of the homoallyl silyl ether **151** (R = TBS) with Grubbs catalyst **1** in toluene under ethylene atmosphere provided the more stable allyl silyl ether in 80% yield.

Similarly, the synthesis of a fused 7/7-bicyclic scaffold was attempted by dienyne metathesis of **153** with Grubbs catalyst **2** under ethylene atmosphere (Scheme 2.54) [31]. Surprisingly, under these conditions only the 6/7- and 6/6-bicyclic compounds **154** and **155**, respectively, were isolated, which resulted from an initial isomerization of one or two double bonds before ring closure.

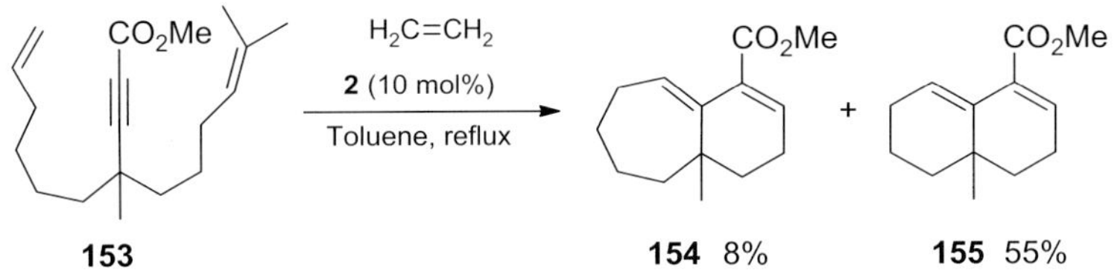

Scheme 2.54 Domino isomerization/dienyne metathesis of **153**.

However, a double-bond isomerization coupled with a metathesis event in a domino manner can also be highly beneficial. As depicted in Scheme 2.55, an RCM step leading to an allyl amine derivative followed by isomerization to give an iminium ion and subsequent intramolecular nucleophilic attack onto this function can generate a range of interesting heterocycles [32]. Subjecting substrate **156** to the ruthenium catalyst **3** at elevated temperature provided compound **157** with a tetracyclic indolizinoindol core via a diastereoselective trapping of an *N*-acyliminium intermediate by the nucleophilic indol moiety. Under similar conditions, the hydroxyl group present in **158** acts as a heteroatom nucleophile to furnish the *N*,*O*-acetal **159** in a diastereoselective manner, as well. Furthermore, an *N*-alkyliminium ion can also be produced from **160** through RCM/isomerization with Grubbs catalyst **2** in refluxing toluene to eventually afford the bicyclic compound **161**.

Another synthetically useful isomerization that can be combined with a metathesis event in a domino transformation is a metallotropic [1,3]-shift of a transient ruthenium alkinyl carbene complex [33]. Treatment of the polyunsaturated substrate **162** with Grubbs catalyst **2** afforded the oligoenyne **163** in 86% yield

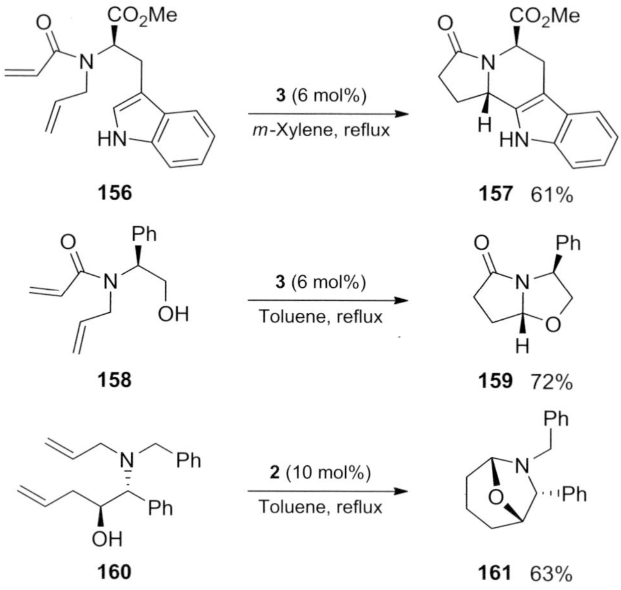

Scheme 2.55 Domino RCM/isomerization/iminium cyclization to give nitrogen heterocycles.

(Scheme 2.56) [33a]. Interestingly, the yield of this process dropped to 67% when the reaction was performed under ethylene atmosphere.

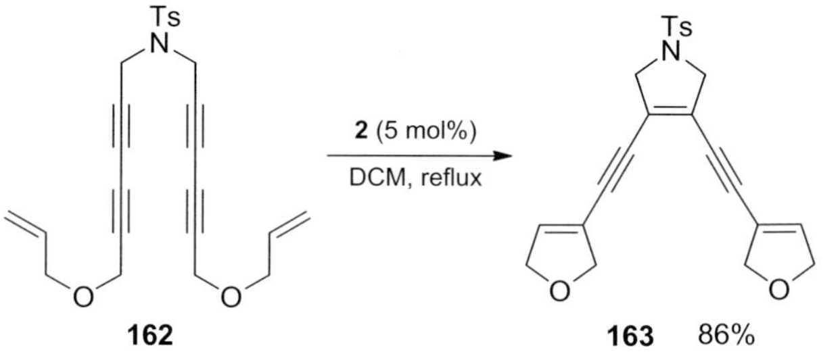

Scheme 2.56 Extended enyne RCM/metallotropic [1,3]-shift domino process for the synthesis of oligoenynes.

The two examples shown in Scheme 2.57 illustrate that the metallotropic shift depends on the nature of the substituent on the 1,3-diyne [33b]. When the substrates **164** and **166** were subjected to 1-octene in the presence of catalyst **2**, an initial enyne CM occurred followed by cyclization to give a ruthenium alkinyl carbene intermediate. Termination by CM with the co-olefin after metallotropic [1,3]-shift was observed for the benzyloxymethyl substituent with production of **165**, whereas termination at the sterically less substituted carbon delivered the complementary regioisomer **167** in case of triisopropylsilyl substitution.

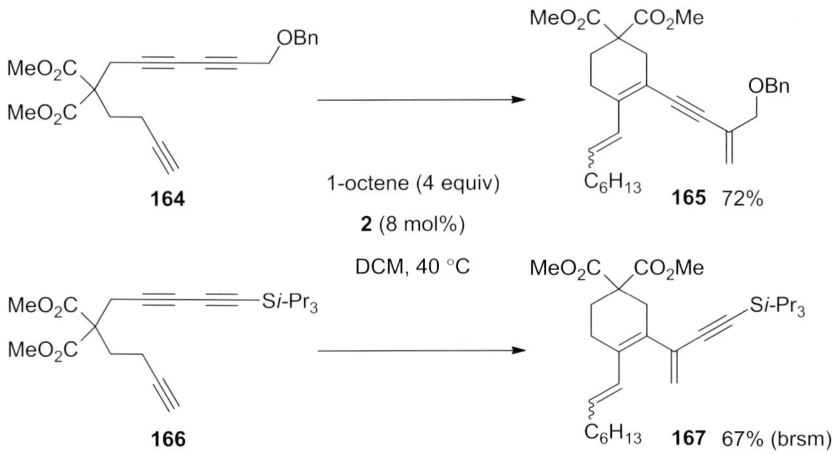

Scheme 2.57 Dependence of the metallotropic shift on the nature of the 1,3-diyne substituent.

Enyne metathesis/metallotropic [1,3]-shift domino processes are also valuable for natural product synthesis [33c,d]. Reaction of substrate **168** with *cis*-1,4-diacetoxy-2-butene in the presence of Grubbs catalyst **2** generated the intermediate ruthenium alkinyl carbene through a relay RCM with the liberation of 2,5-dihydrofuran followed by metallotropic [1,3]-shift and terminating (*Z*)-selective CM with the co-olefin to yield the conjugated enediyne **169** (Scheme 2.58) [33c]. The antitumor active *Panax ginseng* constituent (3*R*,9*R*,10*R*)-panaxytriol was readily synthesized from **169** in six steps.

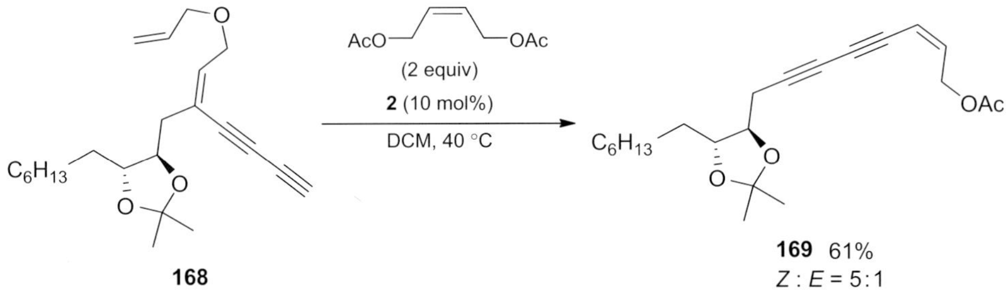

Scheme 2.58 Domino relay RCM-initiated metallotropic [1,3]-shift/enyne CM in the synthesis of (3*R*,9*R*,10*R*)-panaxytriol.

2.2.3 Metathesis/Cycloaddition

The 1,3-dienes resulting from an enyne metathesis can often be coupled with Diels–Alder reactions in a domino transformation. Heating of benzyldimethylsilylacetylene with the styrenes **170** and *N*-phenylmaleimide in the presence of the phosphine-free catalyst **4** led to the cyclohexenes **171** directly via regioselective enyne CM and subsequent cycloaddition (Scheme 2.59) [34]. Remarkably, the endo-selectivity of the Diels–Alder reaction was significantly higher under these conditions compared to a sequential enyne CM of the silylalkyne and **170** (R = H) to furnish the 1,3-diene and its separate [4 + 2]-cycloaddition with *N*-phenylmaleimide in the absence of ruthenium catalyst **4**.

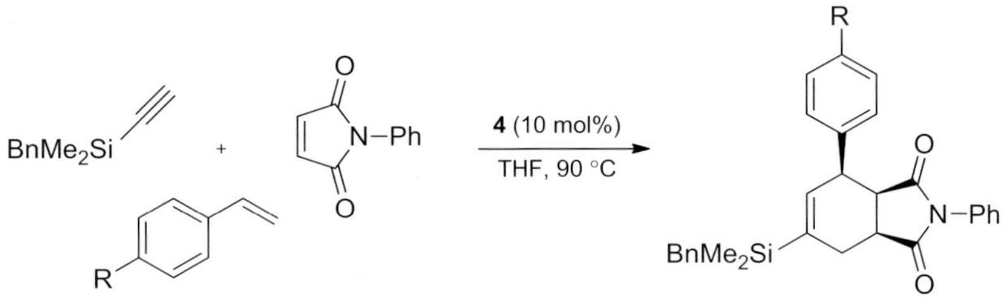

170 R = H, OMe, Cl

171 53–69%

Scheme 2.59 Domino enyne CM/intermolecular Diels–Alder reaction to give the cyclohexene **171**.

Enyne metathesis can also be combined with an intramolecular Diels–Alder reaction (IMDA) in a domino manner [35]. Reaction of the dienes **172** with the terminal alkynes **173** in the presence of the phosphine-free catalyst **4** at reflux temperature yielded the carbobicyclic products **174** preferentially via an exo-(*E*)-anti or endo-(*Z*)-anti transition state for the [4 + 2]-cycloaddition step, respectively (Scheme 2.60) [35a].

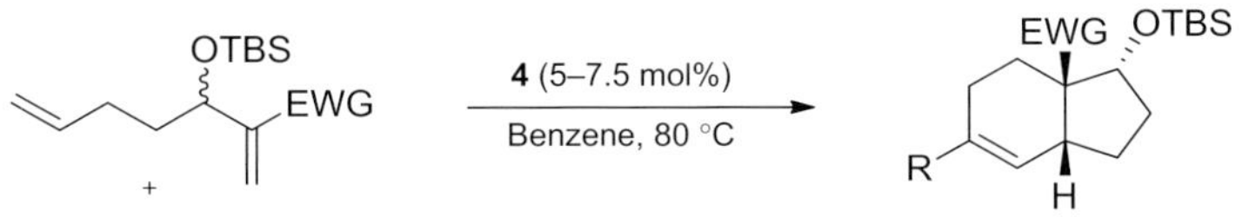

172 **173** **174**

EWG/R	Yield	ds
CO_2Me/Ph	34%	(83% ds)
CN/Ph	90%	(83% ds)
CO_2Me/4-(MeO)C_6H_4	64%	(90% ds)
CN/4-(MeO)C_6H_4	62%	(80% ds)

Scheme 2.60 Domino enyne CM/IMDA to afford the carbobicycles **174**.

Treatment of substrate **175** with Grubbs catalyst **2** under ethylene atmosphere in refluxing toluene likewise triggered a domino enyne CM/IMDA process with the generation of the bicyclo[5.3.1]undecenone **176** as a single diastereomer (Scheme 2.61) [35b,c]. Framework **176** corresponds to the AB ring fragment of taxol lacking the geminal dimethyl group.

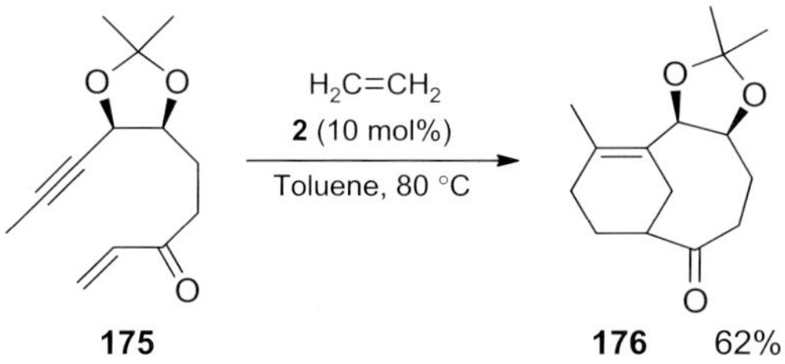

Scheme 2.61 Domino enyne CM/IMDA to give the bridged bicycle **176**.

Under the same conditions, a domino enyne CM/IMDA transformation of **177** yielded the oxa-taxane–sugar hybrid **178** as a single diastereomer (Scheme 2.62) [35d].

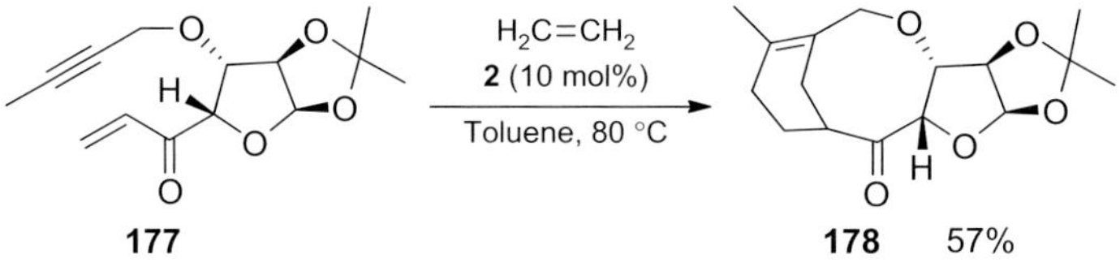

Scheme 2.62 Domino enyne CM/IMDA to give the bridged product **178**.

Heating of the alkynes **179** with ethyl vinyl ether and ethyl glyoxylate under microwave irradiation in the presence of the second-generation Grubbs catalyst **2** directly furnished a 2 : 1 mixture of the dihydropyrans **180** and **181** in a domino enyne CM/hetero-Diels–Alder process (Scheme 2.63) [36]. These mixtures could be readily equilibrated to give only the cis isomers **181** on treatment with zinc chloride in dichloromethane, subsequently.

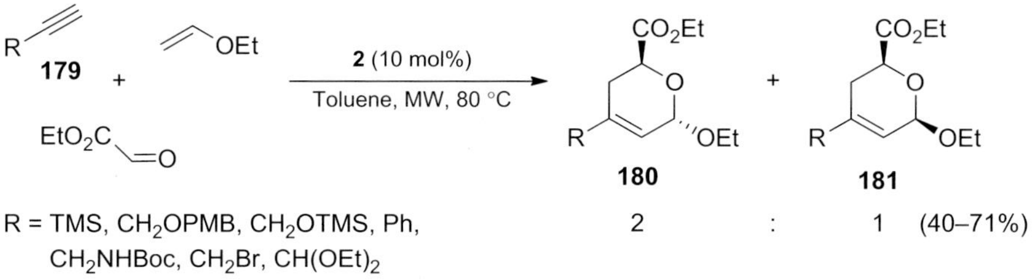

Scheme 2.63 Domino enyne CM/hetero-Diels–Alder reaction to give the dihydropyrans **180** and **181**.

2.2.4 Metathesis/Substitution

Metathesis events and substitution reactions can be readily combined in attractive domino processes. Thus, a dual catalysis with palladium and ruthenium complexes allowed an efficient preparation of the cycloalkenes **183** from substituted malonates **182** and allyl acetate (Scheme 2.64) [37]. Addition of the sodium enolate of **182** to Grubbs catalyst **2** and tetrakis(triphenylphosphine)palladium(0) in refluxing dichloromethane induced a palladium-catalyzed allylic alkylation to give an α,ω-diene which underwent a smooth RCM to give the cyclic olefins **183** in good yields.

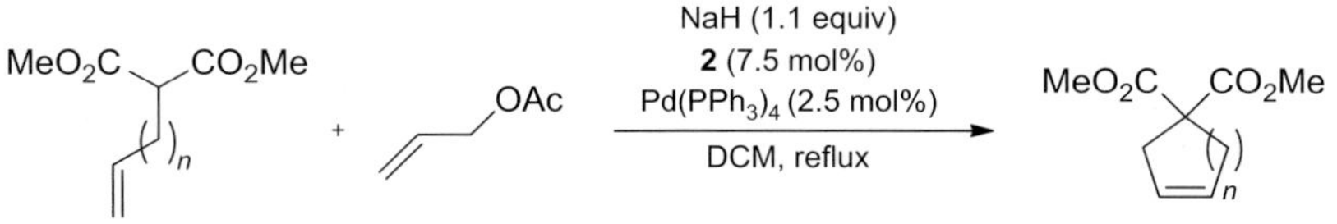

182 **183** 67–74%

n = 1, 2, 3

Scheme 2.64 Domino allylic substitution/RCM for the synthesis of cycloalkenes **183**.

Coupling of CM and intramolecular S_N2' substitution to domino processes leading to tetrahydrofurans is feasible as well [38]. However, a corresponding combination to give tetrahydropyrans is better carried out as a sequential process [38d]. Reaction of the terminal olefin **184** with allyl chloride catalyzed by ruthenium complex **2** led to tetrahydrofuran **185** by a completely diastereoselective cyclization of the intermediate CM product (Scheme 2.65) [38a,b]. Compound **185** was then successfully advanced to the immunosuppressive diterpenoid pyrone subglutinol B.

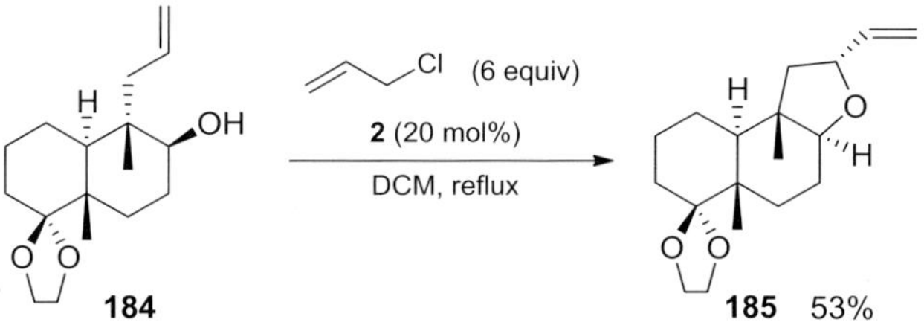

184 **185** 53%

Scheme 2.65 Domino CM/intramolecular S_N2' substitution to give tetrahydrofuran **185**.

Under similar conditions, oxazolidinone **186** was readily converted to a diastereomeric mixture of tetrahydrofurans **187** and **188** (Scheme 2.66) [38c]. The enantiopure major diastereomer **187** was subsequently utilized for the total synthesis of an epimer of the antitumor marine natural product pachastrissamine.

2.2.5 Metathesis/Conjugate Addition

Domino reactions consisting of a CM and an oxa-Michael addition provide an efficient access to 2,6-cis disubstituted tetrahydropyrans [39]. Coupling of the

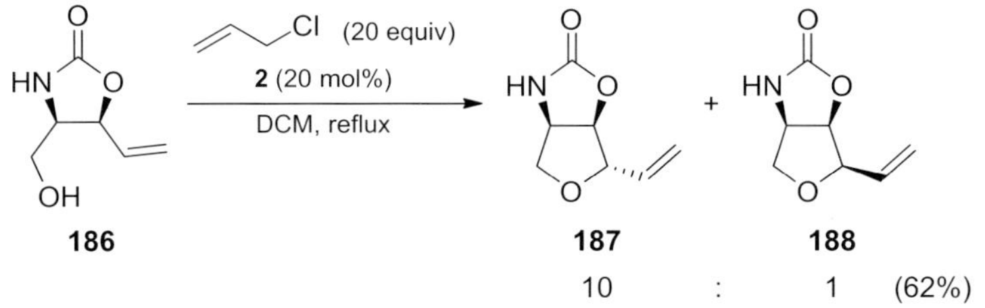

Scheme 2.66 Domino CM/intramolecular S_N2' substitution to give tetrahydrofurans **187** and **188**.

unsaturated alcohol **189** with enones **190** in the presence of the phosphine-free catalyst **4** under microwave irradiation gave rise to the tetrahydropyrans **191** as single diastereomers in high yields (Scheme 2.67) [39a].

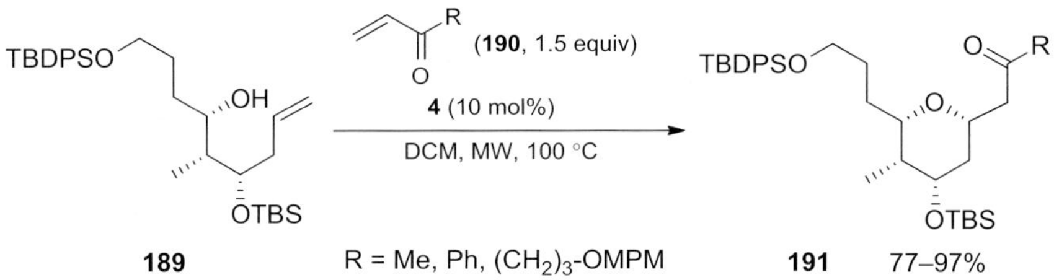

Scheme 2.67 Domino CM/oxa-Michael addition for the synthesis of tetrahydropyrans **191**.

Application of this methodology to the two-directional conversion of the C_2-symmetric diol **192** with ethyl vinyl ketone yielded the mono oxa-Michael adduct of the intermediate double CM product with a high diastereomeric ratio (Scheme 2.68) [39e]. The depicted major isomer **193** was then used for a concise total synthesis of the molluscicidal macrodiolide cyanolide A.

Next to an oxa-Michael addition, an aza-Michael reaction is a suitable partner for a CM event in a domino process, too [40]. On heating the unsaturated carbamates

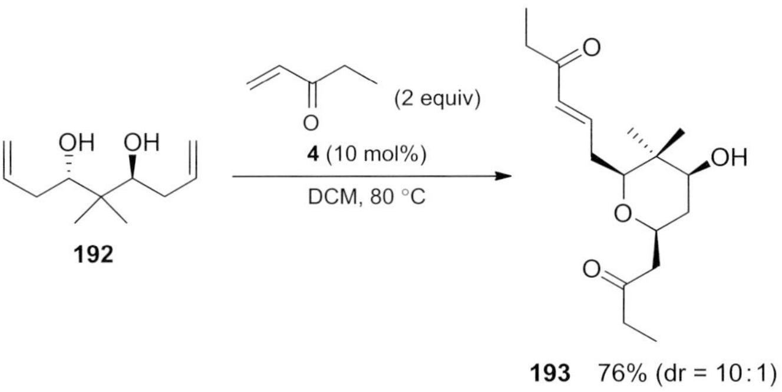

Scheme 2.68 Domino CM/oxa-Michael addition in a total synthesis of cyanolide A.

194 with methyl vinyl ketone in the presence of the phosphine-free ruthenium catalyst **4** and small amounts of boron trifluoride etherate to activate the cyclization step, both five- and six-membered rings **195** were formed in high yield (Scheme 2.69) [40a]. Furthermore, the cyclization of the α-substituted carbamate **196** proceeded with a useful level of diastereoselectivity to give the 2,5-trans disubstituted pyrrolidine **197** as the major product.

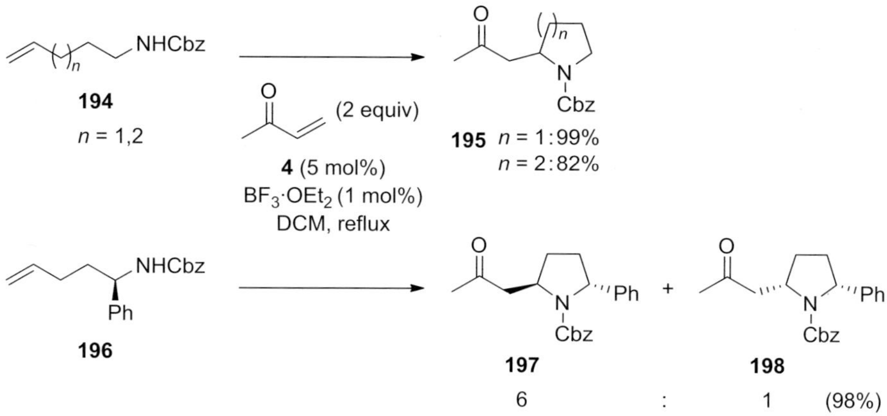

Scheme 2.69 Domino CM/aza-Michael addition for the synthesis of pyrrolidines and piperidines.

After exchanging boron trifluoride etherate against titanium(IV) isopropoxide to promote the aza-Michael step, the enantiomerically pure *N*-sulfinyl amine **199** could be coupled to methyl vinyl ketone as well to give a mixture of the pyrrolidines **200** and **201** in nearly quantitative yield with remarkable substrate-induced diastereoselectivity (Scheme 2.70) [40b].

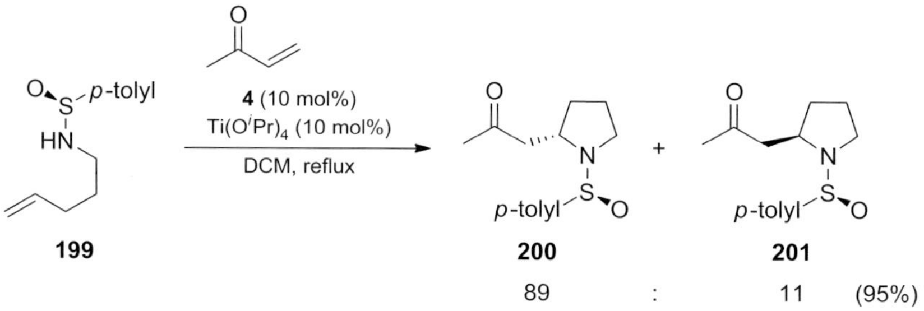

Scheme 2.70 Domino CM/aza-Michael addition with *N*-sulfinyl amine **199**.

Again, using the phosphine-free catalyst **4**, the pyrrole derivative **202** could be used in a diastereoselective domino CM/aza-Michael reaction with acrolein, resulting in the formation of the heterocyclic products **203** and **204** without adding

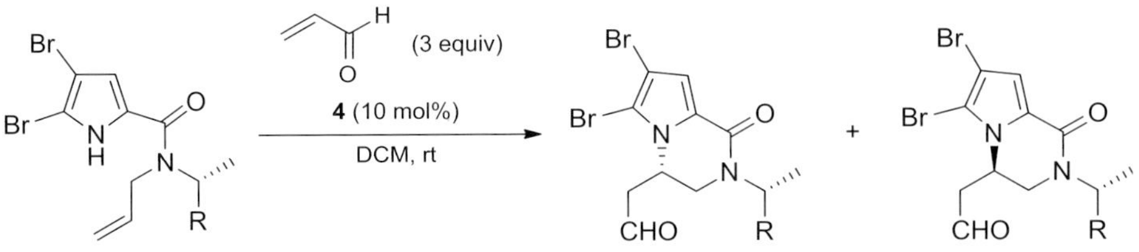

202 R = 1-naphthyl **203** **204**

3.3 : 1 (56%)

Scheme 2.71 Domino CM/aza-Michael addition to furnish pyrrolopiperazinones **203** and **204**.

an external Lewis acid (Scheme 2.71) [40d]. Stereoisomer **203** was then efficiently converted in two steps to a known key intermediate for the total synthesis of a range of marine pyrrolopiperazinone natural products with potent biological activities.

2.2.6 Metathesis/Carbonyl Olefination

Titanium alkylidenes are involved in synthetically valuable olefinic ester cyclizations using the Takai–Utimoto reagent [41]. These transformations probably proceed by initial olefin metathesis to generate a titanium carbene followed by intramolecular carbonyl olefination. Owing to the diminished Lewis acid activity of the Takai–Utimoto reagent, the highly oxygenated substrate **205** successfully underwent a domino process, affording the cyclic enol ether **206** in good yield (Scheme 2.72) [41a].

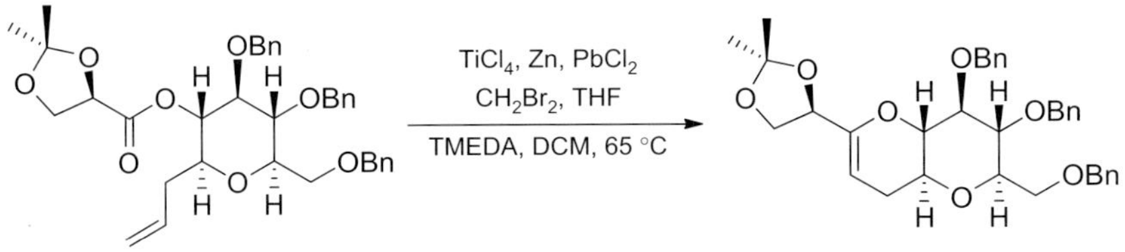

205 **206** 71%

Scheme 2.72 Domino olefin metathesis/carbonyl olefination to give the cyclic enol ether **206**.

Analogous treatment of the unsaturated ester **207** using 1,1-dibromoethane in this case cleanly produced the dihydropyran **208** in high yield on a gram scale (Scheme 2.73) [41b]. Cyclic enol ether **208** was then efficiently advanced to a spiroacetal, which corresponds to the core structure of the bioactive polyketide spirofungin A.

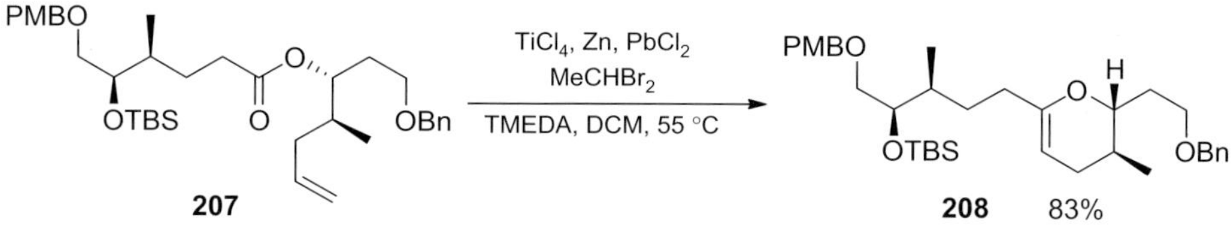

Scheme 2.73 Domino olefin metathesis/carbonyl olefination furnishing the cyclic enol ether **208**.

2.3 Conclusion and Outlook

As illustrated by the schemes above, metathesis reactions in domino processes have swiftly become an established tool in modern organic synthesis. Especially, RRM and domino dienyne metathesis have found extensive uses and will surely continue to catch the attention of synthetic chemists as elegant and atom-economic transformations that allow a quick enhancement of molecular complexity. Next to combinations of metathesis events only, the incorporation of non-metathesis steps into true domino processes has advanced to a remarkable level and will certainly provide even more opportunities in the years to come.

Acknowledgments

Financial support of our own work in this area by the Deutsche Forschungsgemeinschaft, the Fonds der Chemischen Industrie, and the Alexander von Humboldt-Stiftung is gratefully acknowledged.

References

1. Reviews: (a) Pellissier, H. (2013) *Chem. Rev.*, **113**, 442–524; (b) Li, J. and Lee, D. (2011) *Eur. J. Org. Chem.*, 4269–4287; (c) López, J.C. and Plumet, J. (2011) *Eur. J. Org. Chem.*, 1803–1825; (d) Porta, M. and Blechert, S. (2010) in *Metathesis in Natural Product Synthesis* (eds J. Cossy, S. Arseniyadis, and C. Meyer), Wiley-VCH Verlag GmbH, Weinheim, pp. 313–341; (e) Nolan, S.P. and Clavier, H. (2010) *Chem. Soc. Rev.*, **39**, 3305–3316; (f) Mori, M. (2010) *Materials*, **3**, 2087–2140; (g) Holub, N. and Blechert, S. (2007) *Chem. Asian J.*, **2**, 1064–1082; (h) Dragutan, V., Dragutan, I., Delaude, L., and Demonceau, A. (2007) *Coord. Chem. Rev.*, **251**, 765–794; (i) Diver, S.T. (2007) *Coord. Chem. Rev.*, **251**, 671–701; (j) Mori, M. (2007) *Adv. Synth. Catal.*, **349**, 121–135; (k) Schmidt, B. and Hermanns, J. (2006) *Curr. Org. Chem.*, **10**, 1363–1396; (l) Dragutan, V. and Dragutan, I. (2006) *J. Organomet. Chem.*, **691**, 5129–5147; (m) Hansen, E.C. and Lee, D. (2006) *Acc. Chem. Res.*, **39**, 509–519; (n) Maifeld, S.V. and Lee, D. (2005) *Chem. Eur. J.*, **11**, 6118–6126; (o) Nicolaou, K.C., Bulger, P.G., and Sarlah, D. (2005) *Angew. Chem. Int. Ed.*, **44**, 4490–4527; (p) Kaliappan, K.P. (2005) *Lett. Org. Chem.*, **2**, 678–686; (q) Aitken, S.G. and Abell, A.D. (2005) *Aust. J. Chem.*, **58**, 3–13.
2. Tietze, L.F., Brasche, G., and Gericke, K.M. (2006) *Domino Reactions in Organic*

Synthesis, Wiley-VCH Verlag GmbH, Weinheim.

3. Schmidt, B. (2006) *Pure Appl. Chem.*, **78**, 469–476.
4. (a) Riache, N., Blond, A., and Nay, B. (2008) *Tetrahedron*, **64**, 10853–10859; (b) Nay, B., Gaboriaud-Kolar, N., and Bodo, B. (2005) *Tetrahedron Lett.*, **46**, 3867–3870.
5. (a) Kotha, S., Chavan, A.S., and Shaikh, M. (2012) *J. Org. Chem.*, **77**, 482–489; (b) Arsianti, A., Tanimoto, H., Morimoto, T., Bahtiar, A., Takeya, T., and Kakiuchi, K. (2012) *Tetrahedron*, **68**, 2884–2891; (c) Lejkowski, M., Banerjee, P., Schüller, S., Münch, A., Runsink, J., Vermeeren, C., and Gais, H.J. (2012) *Chem. Eur. J.*, **18**, 3529–3548.
6. (a) Malik, C.K. and Ghosh, S. (2007) *Org. Lett.*, **9**, 2537–2540; (b) Mondal, S., Malik, C.K., and Ghosh, S. (2008) *Tetrahedron Lett.*, **49**, 5649–5651; (c) Henderson, J.A. and Phillips, A.J. (2008) *Angew. Chem. Int. Ed.*, **47**, 8499–8501; (d) Maity, S. and Ghosh, S. (2009) *Tetrahedron*, **65**, 9202–9210; (e) Maity, S. and Ghosh, S. (2008) *Tetrahedron Lett.*, **49**, 1133–1136; (f) Nguyen, N.N.M., Leclère, M., Stogaitis, N., and Fallis, A.G. (2010) *Org. Lett.*, **12**, 1684–1687; (g) Matcha, K., Maity, S., Malik, C.K., and Ghosh, S. (2010) *Tetrahedron Lett.*, **51**, 2754–2757.
7. Pfeiffer, M.W.B. and Phillips, A.J. (2005) *J. Am. Chem. Soc.*, **127**, 5334–5335.
8. (a) Clavier, H., Broggi, J., and Nolan, S.P. (2010) *Eur. J. Org. Chem.*, 937–943; (b) Oblak, E.Z., Dayanandan, N.G., and Wright, D.L. (2011) *Org. Lett.*, **13**, 2433–2435.
9. Jackson, K.L., Henderson, J.A., Motoyoshi, H., and Philipps, A.J. (2009) *Angew. Chem. Int. Ed.*, **48**, 2346–2350.
10. (a) Büchert, M., Meinke, S., Prenzel, A.H.G.P., Deppermann, N., and Maison, W. (2006) *Org. Lett.*, **8**, 5553–5556; (b) Maison, W., Büchert, M., and Deppermann, N. (2007) *Beilstein J. Org. Chem.*, **3**, No. 48.
11. (a) Donnard, M., Tschamber, T., and Eustache, J. (2008) *Tetrahedron Lett.*, **49**, 7325–7327; (b) Donnard, M., Tschamber, T., Le Nouën, D., Desrat, S., Hinsinger, K., and Eustache, J. (2011) *Tetrahedron*, **67**, 339–357; (c) Donnard, M., Tschamber, T., Desrat, S., Hinsinger, K., and Eustache, J. (2008) *Tetrahedron Lett.*, **49**, 1192–1195.
12. (a) Hart, A.C. and Phillips, A.J. (2006) *J. Am. Chem. Soc.*, **128**, 1094–1095; (b) Phillips, A.J., Hart, A.C., and Henderson, J.A. (2006) *Tetrahedron Lett.*, **47**, 3743–3745.
13. (a) Ikoma, M., Oikawa, M., and Sasaki, M. (2008) *Tetrahedron*, **64**, 2740–2749; (b) Ikoma, M., Oikawa, M., and Sasaki, M. (2009) *Eur. J. Org. Chem.*, 72–84; (c) Aljarilla, A. and Plumet, J. (2009) *Heterocycles*, **77**, 1333–1340; (d) Oikawa, M., Ikoma, M., Sasaki, M., Gill, M.B., Swanson, G.T., Shimamoto, K., and Sakai, R. (2009) *Eur. J. Org. Chem.*, 5531–5548; (e) Ikoma, M., Oikawa, M., Gill, M.B., Swanson, G.T., Sakai, R., Shimamoto, K., and Sasaki, M. (2008) *Eur. J. Org. Chem.*, 5215–5220; (f) Jeon, K.O., Rayabarapu, D., Rolfe, A., Volp, K., Omar, I., and Hanson, P.R. (2009) *Tetrahedron*, **65**, 4992–5000; (g) Lee, H.Y., Lee, S.S., Kim, H.S., and Lee, K.M. (2012) *Eur. J. Org. Chem.*, 4192–4199.
14. Carreras, J., Avenoza, A., Busto, J.H., and Peregrina, J.M. (2011) *J. Org. Chem.*, **76**, 3381–3391.
15. Calvet, G., Blanchard, N., and Kouklovsky, C. (2007) *Org. Lett.*, **9**, 1485–1488.
16. (a) Kummer, D.A., Brenneman, J.B., and Martin, S.F. (2006) *Tetrahedron*, **62**, 11437–11449; (b) Kummer, D.A., Brenneman, J.B., and Martin, S.F. (2005) *Org. Lett.*, **7**, 4621–4623; (c) Dieltiens, N. and Stevens, C.V. (2006) *Synlett*, 2771–2776; (d) Cho, J., Lee, Y.M., Kim, D., and Kim, S. (2009) *J. Org. Chem.*, **74**, 3900–3904; (e) Li, J. and Lee, D. (2010) *Chem. Asian J.*, **5**, 1298–1302.
17. (a) Park, S., Kim, M., and Lee, D. (2005) *J. Am. Chem. Soc.*, **127**, 9410–9415; (b) Peppers, B.P., Kulkarni, A.A., and Diver, S.T. (2006) *Org. Lett.*, **8**, 2539–2542; (c) Middleton, M.D., Peppers, B.P., and Diver, S.T. (2006) *Tetrahedron*, **62**, 10528–10540; (d) Diver, S.T., Clark, D.A., and Kulkarni, A.A. (2008) *Tetrahedron*, **64**, 6909–6919; (e) Kotha, S. and Singh, K. (2007) *Eur.*

J. Org. Chem., 5909–5916; (f) Kotha, S., Bansal, D., Singh, K., and Banerjee, S. (2011) *J. Organomet. Chem.*, **696**, 1856–1860; (g) Betkekar, V.V., Panda, S., and Kaliappan, K.P. (2012) *Org. Lett.*, **14**, 198–201.

18. (a) Kaliappan, K.P. and Nandurdikar, R.S. (2005) *Org. Biomol. Chem.*, **3**, 3613–3614; (b) Boyer, F.D. and Hanna, I. (2006) *Eur. J. Org. Chem.*, 471–482; (c) Fukumoto, H., Takahashi, K., Ishihara, J., and Hatakeyama, S. (2006) *Angew. Chem. Int. Ed.*, **45**, 2731–2734; (d) Kaliappan, K.P., Nandurdikar, R.S., and Shaikh, M.M. (2006) *Tetrahedron*, **62**, 5064–5073; (e) Boyer, F.D. and Hanna, I. (2007) *Org. Lett.*, **9**, 2293–2295; (f) Krishna, P.R. and Narsingam, M. (2007) *Tetrahedron Lett.*, **48**, 8721–8724; (g) Park, H., Hong, Y.L., Kim, Y.B., and Choi, T.L. (2010) *Org. Lett.*, **12**, 3442–3445; (h) Ramharter, J., Weinstabl, H., and Mulzer, J. (2010) *J. Am. Chem. Soc.*, **132**, 14338–14339; (i) Knüppel, S., Rogachev, V.O., and Metz, P. (2010) *Eur. J. Org. Chem.*, 6145–6148; (j) Schubert, M. and Metz, P. (2011) *Angew. Chem. Int. Ed.*, **50**, 2954–2956; (k) Oguri, H., Hiruma, T., Yamagishi, Y., Oikawa, H., Ishiyama, A., Otoguro, K., Yamada, H., and Ōmura, S. (2011) *J. Am. Chem. Soc.*, **133**, 7096–7105; (l) Chang, L., Jiang, H., Fu, J., Liu, B., Li, C., and Yang, Z. (2012) *J. Org. Chem.*, **77**, 3609–3614; (m) Lee, J. and Parker, K.A. (2012) *Org. Lett.*, **14**, 2682–2685; (n) Wei, H., Qiao, C., Liu, G., Yang, Z., and Li, C. (2013) *Angew. Chem. Int. Ed.*, **52**, 620–624.

19. Aldegunde, M.J., García-Fandiño, R., Castedo, L., and Granja, J.R. (2007) *Chem. Eur. J.*, **13**, 5135–5150.

20. (a) Sashuk, V. and Grela, K. (2006) *J. Mol. Catal. A: Chem.*, **257**, 59–66; (b) Aldegunde, M.J., Codesido, E.M., Castedo, L., and Granja, J.R. (2006) *Tetrahedron Lett.*, **47**, 6587–6589; (c) Clavier, H., Correa, A., Escudero-Adán, E.C., Benet-Buchholz, J., Cavallo, L., and Nolan, S.P. (2009) *Chem. Eur. J.*, **15**, 10244–10254.

21. (a) Grimm, J.B., Otte, R.D., and Lee, D. (2005) *J. Organomet. Chem.*, **690**, 5508–5516. (b) Siegel, D.S., Piizzi, G., Piersanti, G., and Movassaghi, M. (2009) *J. Org. Chem.*, **74**, 9292–9304; (c) Movassaghi, M., Piizzi, G., Siegel, D.S., and Piersanti, G. (2006) *Angew. Chem. Int. Ed.*, **45**, 5859–5863; (d) Kim, Y.J., Grimm, J.B., and Lee, D. (2007) *Tetrahedron Lett.*, **48**, 7961–7964; (e) Movassaghi, M., Piizzi, G., Siegel, D.S., and Piersanti, G. (2009) *Tetrahedron Lett.*, **50**, 5489–5492; (f) Park, S. and Lee, D. (2007) *Synthesis*, 2313–2316.

22. (a) Spandl, R.J., Rudyk, H., and Spring, D.R. (2008) *Chem. Commun.*, 3001–3003; (b) Groaz, E., Banti, D., and North, M. (2007) *Eur. J. Org. Chem.*, 3727–3745; (c) Groaz, E., Banti, D., and North, M. (2008) *Tetrahedron*, **64**, 204–218.

23. Aljarilla, A., Murcia, M.C., Csákÿ, A.G., and Plumet, J. (2009) *Eur. J. Org. Chem.*, 822–832.

24. Mori, M., Wakamatsu, H., Tonogaki, K., Fujita, R., Kitamura, T., and Sato, Y. (2005) *J. Org. Chem.*, **70**, 1066–1069.

25. Zhu, Z.B. and Shi, M. (2010) *Org. Lett.*, **12**, 4462–4465.

26. (a) Morton, D., Leach, S., Cordier, C., Warriner, S., and Nelson, A. (2009) *Angew. Chem. Int. Ed.*, **48**, 104–109; (b) O'Leary-Steele, C., Pedersen, P.J., James, T., Lanyon-Hogg, T., Leach, S., Hayes, J., and Nelson, A. (2010) *Chem. Eur. J.*, **16**, 9563–9571.

27. van Otterlo, W.A.L., Coyanis, E.M., Panayides, J.L., de Koning, C.B., and Fernandes, M.A. (2005) *Synlett*, 501–505.

28. (a) González-Gómez, Á., Domínguez, G., and Pérez-Castells, J. (2009) *Tetrahedron*, **65**, 3378–3391; (b) González-Gómez, Á., Domínguez, G., and Pérez-Castells, J. (2005) *Tetrahedron Lett.*, **46**, 7267–7270.

29. Menozzi, C., Dalko, P.I., and Cossy, J. (2005) *Synlett*, 2449–2452.

30. Bose, S., Ghosh, M., and Ghosh, S. (2012) *J. Org. Chem.*, **77**, 6345–6350.

31. Boyer, F.D. and Hanna, I. (2006) *J. Organomet. Chem.*, **691**, 5181–5188.

32. Ascic, E., Jensen, J.F., and Nielsen, T.E. (2011) *Angew. Chem. Int. Ed.*, **50**, 5188–5191.

33. (a) Kim, M. and Lee, D. (2005) *J. Am. Chem. Soc.*, **127**, 18024–18025; (b) Yun,

S.Y., Wang, K.P., Kim, M., and Lee, D. (2010) *J. Am. Chem. Soc.*, **132**, 8840–8841; (c) Cho, E.J. and Lee, D. (2008) *Org. Lett.*, **10**, 257–259; (d) Li, J., Park, S., Miller, R.L., and Lee, D. (2009) *Org. Lett.*, **11**, 571–574.

34. Junker, C.S. and Welker, M.E. (2012) *Tetrahedron*, **68**, 5341–5345.

35. (a) Mix, S. and Blechert, S. (2005) *Org. Lett.*, **7**, 2015–2018; (b) Kaliappan, K.P., Ravikumar, V., and Pujari, S.A. (2008) *J. Chem. Sci.*, **120**, 205–216; (c) Kaliappan, K.P., Ravikumar, V., and Pujari, S.A. (2006) *Tetrahedron Lett.*, **47**, 981–984; (d) Nandurdikar, R.S., Subrahmanyam, A.V., and Kaliappan, K.P. (2010) *Eur. J. Org. Chem.*, 2788–2799.

36. (a) Castagnolo, D., Botta, L., and Botta, M. (2009) *Carbohydr. Res.*, **344**, 1285–1288; (b) Castagnolo, D., Botta, L., and Botta, M. (2009) *Tetrahedron Lett.*, **50**, 1526–1528.

37. Kammerer, C., Prestat, G., Gaillard, T., Madec, D., and Poli, G. (2008) *Org. Lett.*, **10**, 405–408.

38. (a) Kim, H., Baker, J.B., Park, Y., Park, H.B., DeArmond, P.D., Kim, S.H., Fitzgerald, M.C., Lee, D.S., and Hong, J. (2010) *Chem. Asian J.*, **5**, 1902–1910; (b) Kim, H., Baker, J.B., Lee, S.U., Park, Y., Bolduc, K.L., Park, H.B., Dickens, M.G., Lee, D.S., Kim, Y., Kim, S.H., and Hong, J. (2009) *J. Am. Chem. Soc.*, **131**, 3192–3194; (c) Lee, D. (2012) *Synlett*, **23**, 2840–2844; (d) Lee, K., Kim, H., and Hong, J. (2009) *Org. Lett.*, **11**, 5202–5205.

39. (a) Fuwa, H., Noto, K., and Sasaki, M. (2010) *Org. Lett.*, **12**, 1636–1639; (b) Fuwa, H., Noto, K., and Sasaki, M. (2010) *Heterocycles*, **82**, 641–647; (c) Park, H., Kim, H., and Hong, J. (2011) *Org. Lett.*, **13**, 3742–3745; (d) Fuwa, H., Noguchi, T., Noto, K., and Sasaki, M. (2012) *Org. Biomol. Chem.*, **10**, 8108–8112; (e) Waldeck, A.R. and Krische, M.J. (2013) *Angew. Chem. Int. Ed.*, **52**, 4470–4473.

40. (a) Fustero, S., Jiménez, D., Sánchez-Roselló, M., and del Pozo, C. (2007) *J. Am. Chem. Soc.*, **129**, 6700–6701; (b) Fustero, S., Monteagudo, S., Sánchez-Roselló, M., Flores, S., Barrio, P., and del Pozo, C. (2010) *Chem. Eur. J.*, **16**, 9835–9845; (c) Fustero, S., Báez, C., Sánchez-Roselló, M., Asensio, A., Miro, J., and del Pozo, C. (2012) *Synthesis*, **44**, 1863–1873; (d) Kwon, S.H., Lee, H.J., and Cho, C.W. (2011) *Bull. Korean Chem. Soc.*, **32**, 315–318.

41. (a) Majumder, U. and Rainier, J.D. (2005) *Tetrahedron Lett.*, **46**, 7209–7211; (b) Neumaier, J. and Maier, M.E. (2011) *Synlett*, 187–190.

3
C–H Activation Reactions in Domino Processes

Gavin Chit Tsui and Mark Lautens

Functionalization of C–H bonds has become one of the most intensively studied fields in organic chemistry since the turn of the century. The applications of C–H activation reactions [1] in domino processes have seen a dramatic increase in recent years. By incorporating C–H activation as a key step in the synthetic strategies with subsequent carbon–carbon or carbon–heteroatom bond formations, complex molecules can be constructed in a highly efficient manner without the need for prefunctionalization of the reacting partners. Creative methods utilizing C–H activation in domino reactions have been developed to synthesize pharmaceutically important heterocycles and novel molecular motors possessing interesting material properties. These remarkable transformations are often accompanied by multiple bond formations all in a single operation. While extensive reviews are available on transition-metal-catalyzed C–H activations focusing on both the mechanistic and synthetic aspects, this chapter discusses some of the most recent synthetic applications of domino reactions where two or more new bonds are formed, at least one of which arises from a C–H activation process. This review is not intended to be comprehensive, and the contents are divided into four sections based on the reaction pathways in the domino sequence: Heck reactions/C–H activations, carbopalladations and aminopalladations of alkynes/C–H activations, palladium-catalyzed/norbornene-mediated ortho C–H activations, and heteroatom-directed C–H activations in domino reactions.

3.1
Heck Reactions/C–H Activations

A series of recent reports from three research groups have described the synthesis of spirocyclic N-heterocycles utilizing domino Heck reaction/C–H activation sequences (Scheme 3.1). Spirocycles are key structural units in many biologically active natural products. These methods offered new strategies for constructing the quaternary carbon center and the spirocyclic scaffold.

Ruck and coworkers [2] developed an efficient method for synthesizing spirocyclic oxindoles **2** from acrylamides **1** (Scheme 3.1a). The key event in

Domino Reactions: Concepts for Efficient Organic Synthesis, First Edition. Edited by Lutz F. Tietze.

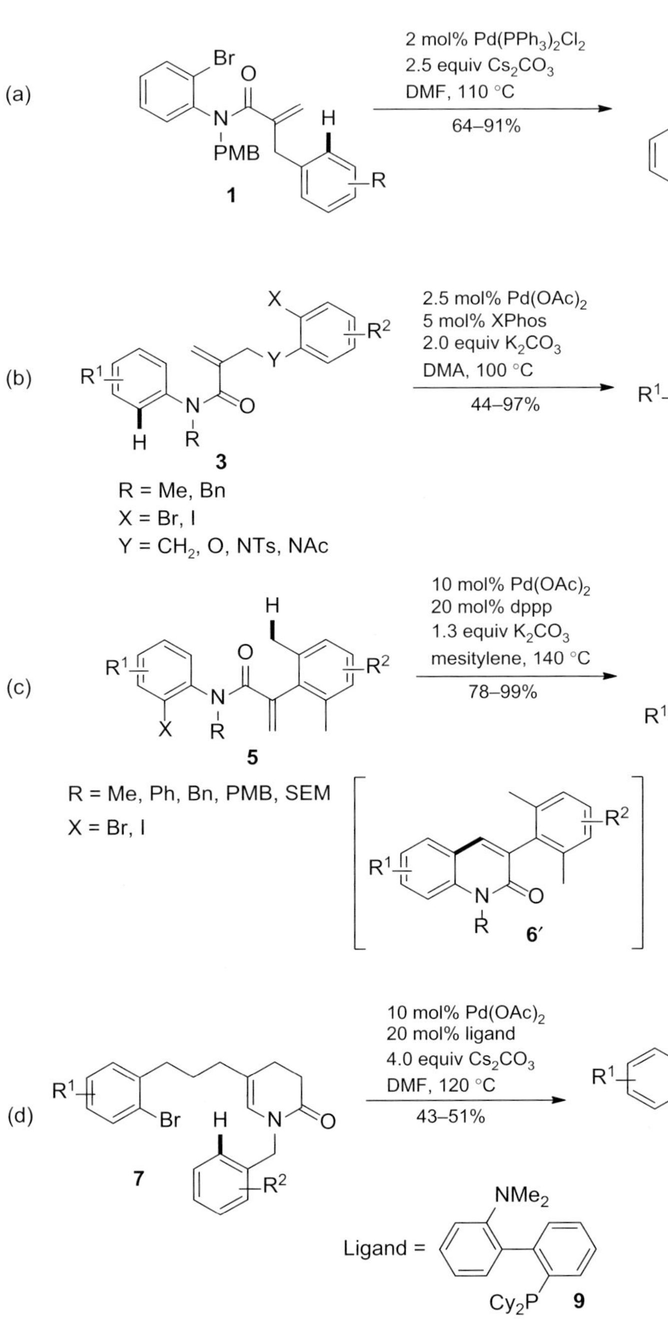

Scheme 3.1 (a–d) Synthesis of spirocyclic N-heterocycles.

the proposed catalytic cycle is the intramolecular Heck reaction in a 5-exo-trig manner, forming ring A and a primary alkylpalladium species. Subsequent C–H activation/cyclization constructs the five-membered ring B.

Zhu and coworkers [3] synthesized spirocyclic dihydroquinolinones **4** from acrylamides **3** utilizing a similar strategy (Scheme 3.1b). In this case, ring A was accessed by an analogous 5-exo-trig Heck cyclization followed by C–H activation of the anilide ring to furnish the six-membered ring B. The authors identified XPhos as a superior ligand compared to PPh_3. Nevertheless, the reaction was only slightly dependent on the ligand, as high yield was obtained even in the absence of the phosphine ligand. The use of tertiary amide was found to be necessary. Heteroatom linker Y was also tolerated, affording a broad scope of spirocyclic heterocycles. Unfortunately, both the above-mentioned methods were limited to racemic product formation. Considerable screening of chiral ligands failed to yield enantiomerically enriched products.

The same group [4] has extended this methodology to activate a more challenging $C(sp^3)$–H bond in the synthesis of spirocyclic oxindoles **6** (Scheme 3.1c). Optimization studies revealed that significant amounts of undesirable quinolinone products **6′** were formed by 6-endo-trig Heck cyclization. The regioselectivity was influenced by the ligand, base, temperature, and the substituent group (R^1) on the anilide ring. The authors argued that the strong chelating effect of 1,3-bis(diphenylphosphino)propane (dppp) on Pd favors the less sterically hindered neopentyl-type alkylpalladium intermediate (via 5-exo cyclization) over the tertiary alkylpalladium intermediate (via 6-endo cyclization). The method also suffered from low enantioselectivities when using chiral bidentate ligands such as (*R*,*R*)-BINAP (2,2′-bis(diphenylphosphanyl)-1,1′-binaphthyl) or (*S*,*S*)-4,5-bis(diphenylphosphinomethyl)-2,2-dimethyl-1,3-dioxolane (DIOP).

Structurally distinct enamides **7** were employed by Maier and coworkers [5] to synthesize spirocyclic amides **8** (Scheme 3.1d). The key steps involved 6-exo-trig Heck cyclization (forming ring A) and C–H activation of the *N*-benzyl group (forming ring B). The Buchwald aminophosphine ligand **9** provided products in higher yields, but the reaction was generally low yielding because of significant debromination of **7** (25–34%).

A reaction sequence involving double Heck reactions/C–H activations was adopted by Hu and coworkers [6] to synthesize tri- and tetracyclic [7] fused N-heterocycles **13** and **14** from dienes **10** and **11**, respectively (Scheme 3.2).

The reaction is initiated by oxidative addition of Pd(0) to aryl bromide followed by intermolecular Heck reaction, generating the intermediate **15** (Scheme 3.3). The subsequent intramolecular Heck reaction takes place to form the intermediate **16**, which then undergoes C–H activation/cyclization, releasing the product **14** and regenerating the Pd(0) catalyst. Overall, three C–C bonds are constructed in this sequence. A wide range of functional groups were tolerated in the aryl bromide. Compounds **14** containing five- to eight-membered rings ($n = 1–4$) were all successfully synthesized.

While intramolecular C–H activations have been successfully incorporated in the domino Heck reactions, the intermolecular C–H activations with external

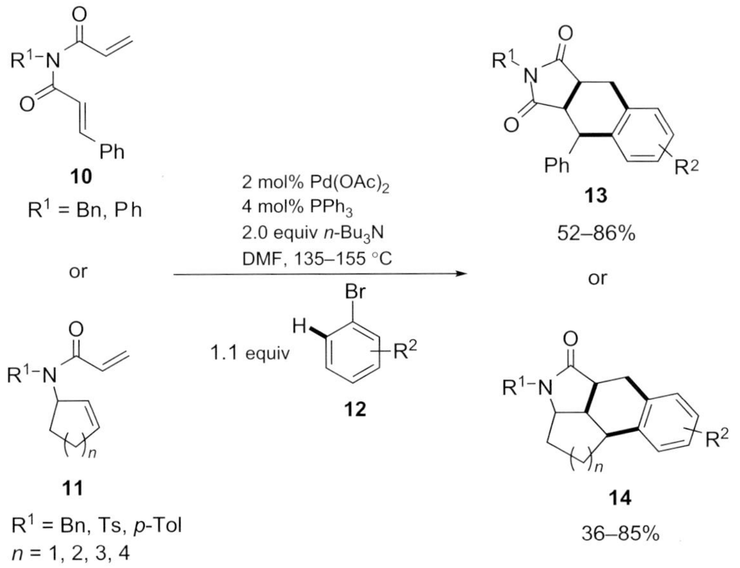

Scheme 3.2 Synthesis of fused N-heterocycles.

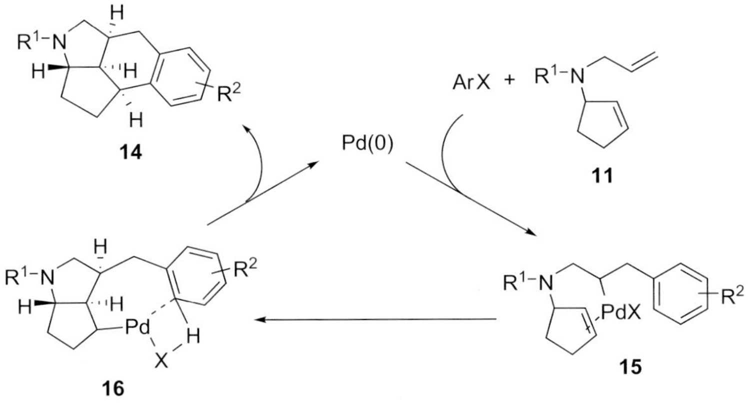

Scheme 3.3 Proposed catalytic cycle for the formation of tetracyclic fused N-heterocycles.

coupling partners are arguably more challenging because of competing reaction pathways. Fagnou and coworkers [8] developed a domino reaction involving an intramolecular Heck reaction followed by intermolecular C–H activation with an external heterocycle as the termination step (Scheme 3.4). Therefore, reactions of aryl bromides **17–19** with sulfur-containing heterocycles afforded a variety of dihydrobenzofurans **20**, indolines **21**, and oxindoles **22**, respectively. Initial optimization studies showed that the yield and regioselectivity were strongly influenced by the base and the ligand. Adding pivalic acid increased the yields and reproducibility.

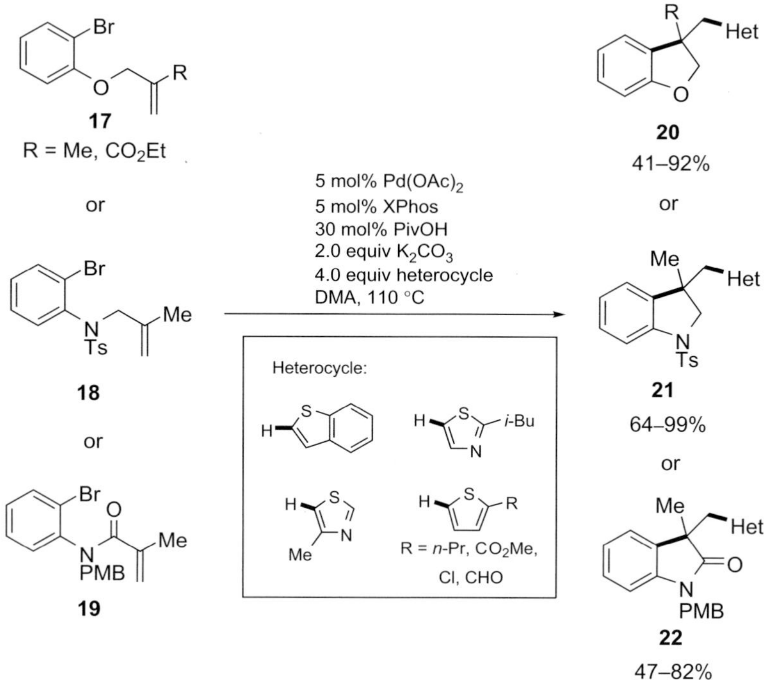

Scheme 3.4 Synthesis of dihydrobenzofurans, indolines, and oxindoles.

Interestingly, when the aryl bromide homolog **23** was used, two regioisomeric isochroman products **24** and **25** were obtained in 1.7 : 1 ratio (Scheme 3.5). The desired product **24** is presumably formed via an alkylpalladium intermediate **24′**. On the other hand, an arylpalladium intermediate **25′**, generated by 1,4-palladium alkyl to aryl migration, would give product **25**. This type of palladium migration has also been observed by Larock and coworkers [9] in the synthesis of fused polycycles.

By changing the reaction conditions, Jia and coworkers [10] were able to control the regioselectivity and trap the palladium intermediates with external nucleophiles. For example, when aryl iodide **26** was subjected to the domino Heck/cyanation sequence, it was found that the presence of water completely shut down the palladium migration, affording product **27** exclusively (Scheme 3.6). In the absence of water, product **28** was obtained in good yield via a Heck/C–H activation/Pd migration/cyanation sequence.

The reaction was not only limited to cyanation, but other Pd-catalyzed reactions such as Heck and Suzuki couplings were also used as the termination step to afford a variety of functionalized carbo- and heterocycles **30** via the C–H activation/Pd migration pathway (Scheme 3.7a). Under these conditions, isochroman **25** was obtained as a single regioisomer [11] by the aforementioned intermolecular C–H activation of heterocycle as the termination step (Scheme 3.7b).

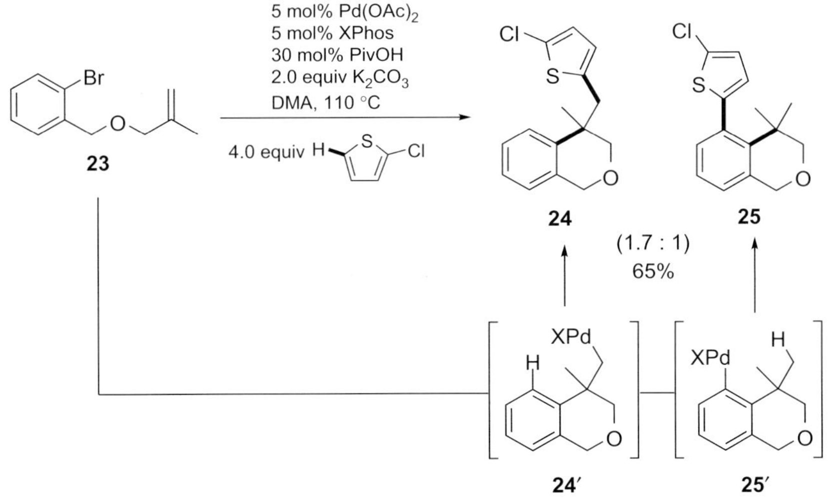

Scheme 3.5 Formation of regioisomeric mixture of isochromans.

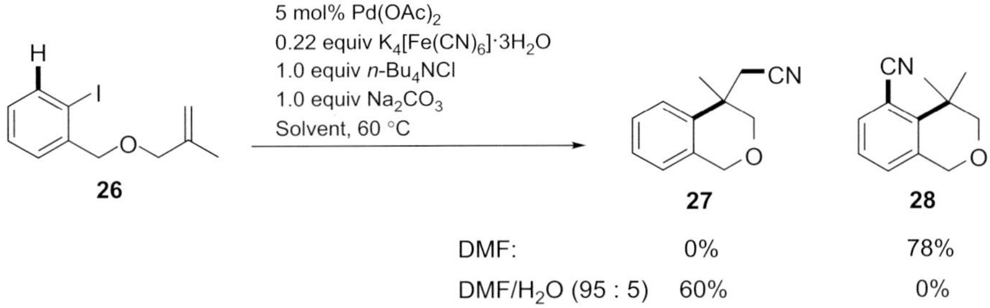

Scheme 3.6 Controlling the regioselectivity in the presence/absence of water.

3.2
Carbopalladations and Aminopalladations of Alkynes/C–H Activations

Zhu and coworkers [12] reported the synthesis of unsymmetrically substituted 3-(diarylmethylenyl)indolinones via a palladium-catalyzed domino reaction involving intermolecular carbopalladation and C–H activation. For example, the reaction between propiolamide **31** and aryl iodide **32** afforded the oxindole product **33** in 82% yield as a single stereoisomer (Scheme 3.8). The *E*-configuration of the tetrasubstituted double bond was confirmed by X-ray analysis.

The proposed mechanism involves initial oxidative addition of aryl iodide to Pd(0), forming an arylpalladium(II) species. The regioselective syn carbopalladation to the triple bond of the propiolamide leads to the vinylpalladium species **34** (Scheme 3.9). The syn insertion would also account for the observed double-bond geometry. Interestingly, in the presence of two benzenoid C–Hs (ring A and ring B), only the C–H of ring A was activated (Path a) to give the six-membered palladacycle **35**.

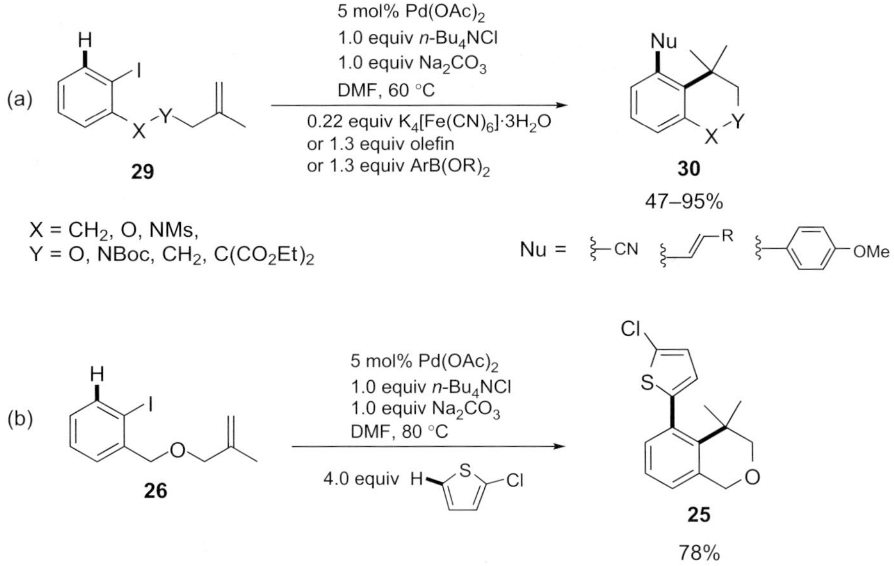

Scheme 3.7 (a, b) Synthesis of carbo- and heterocycles via C–H activation/Pd migration pathway.

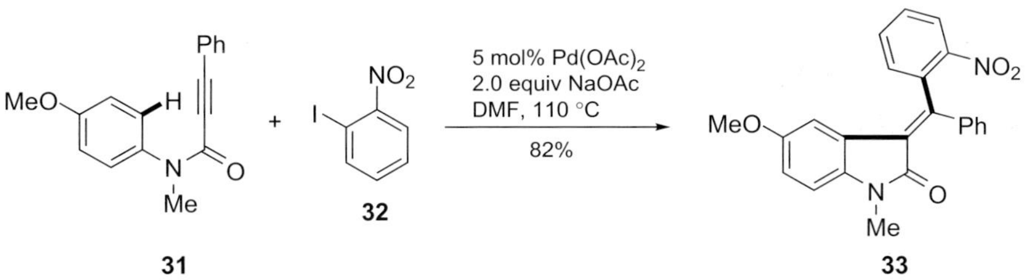

Scheme 3.8 Synthesis of oxindoles.

The activation of the C–H of ring B (Path b) to five-membered palladacycle **36** was not observed, which would lead to the fluorene derivative **37** via a vinyl to aryl Pd migration and cyclization [13].

The same group extended this methodology to incorporate a Sonogashira reaction into the domino sequence [14]. As a result, oxindole **40** could be synthesized from propiolamide **38** and 2 equiv of aryl iodide **39** in the presence of catalytic $Pd(PPh_3)_4$/CuI via a domino Sonogashira reaction/carbopalladation/C–H activation sequence (Scheme 3.10). Overall three C–C bonds were formed from distinct catalytic cycles using only one catalyst. Unsymmetrically substituted oxindoles were also synthesized by a one-pot/two-step protocol where the second aryl iodide was simply added to the same reaction vessel after the Sonogashira reaction of the first aryl iodide.

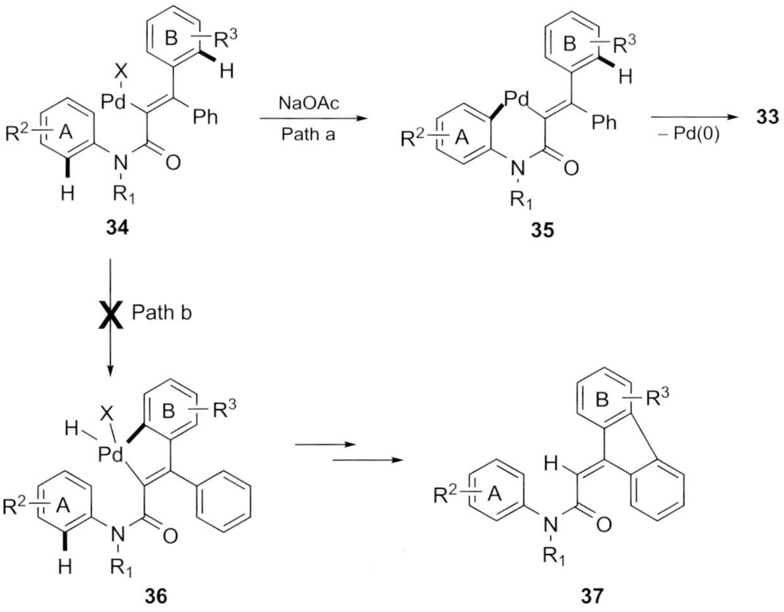

Scheme 3.9 Proposed mechanism for the formation of oxindoles via intermolecular carbopalladation/C–H activation.

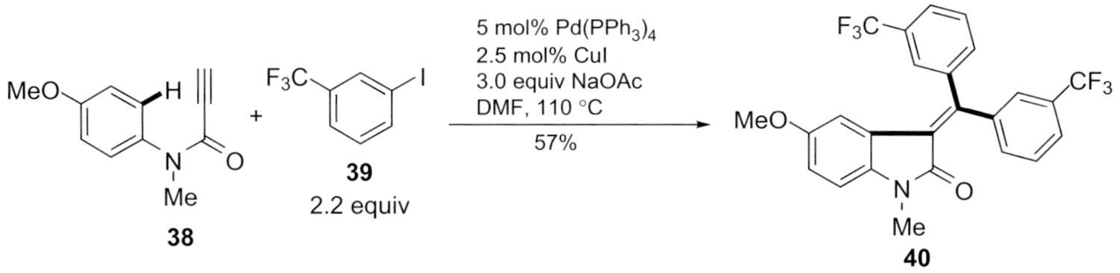

Scheme 3.10 Synthesis of oxindoles via domino Sonogashira reaction/carbopalladation/C–H activation.

Gong and coworkers [15] also showed that *N*-arylalkynylimines **41** were suitable substrates for the analogous carbopalladation/C–H activation sequence to synthesize substituted 3-methylene-3*H*-indoles **42** (Scheme 3.11). The method was largely limited to symmetrically substituted products (Ar = R^3) since the (*E*/*Z*)-isomers of **42** could easily undergo isomerization at room temperature.

A double inter-/intramolecular carbopalladation of 1,6-diynes **43** with aryl bromide **44** followed by C–H activation and cyclization was reported by Hu and coworkers [16] to synthesize polyaromatic compounds **45** (Scheme 3.12). Under the same catalytic conditions, biologically important fused phenanthridines **46** and **47** were synthesized using 3-bromoquinoline and 3-bromoisoquinoline as the coupling partners, respectively [17].

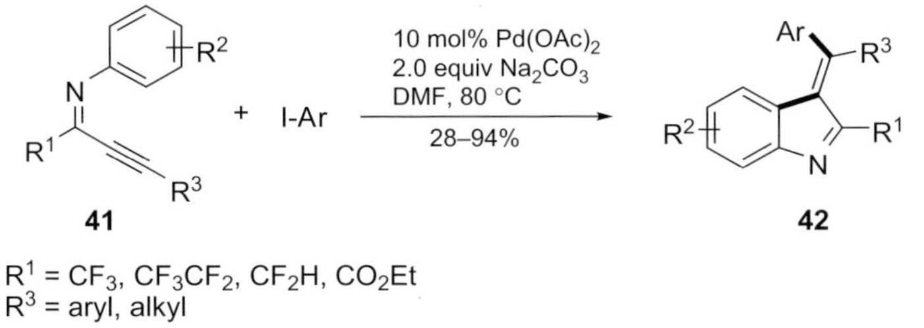

Scheme 3.11 Synthesis of 3-methylene-3*H*-indoles.

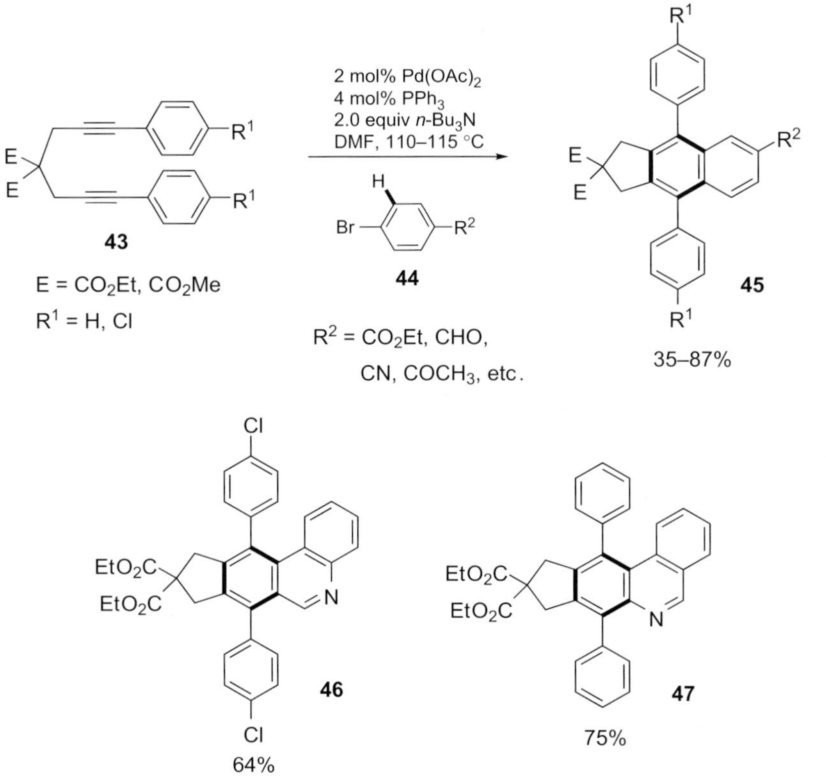

Scheme 3.12 Synthesis of polyaromatic hydrocarbons and fused phenanthridines.

Tietze and coworkers [18] had previously demonstrated that helical tetrasubstituted alkenes such as **49** could be successfully synthesized by a domino carbopalladation/Stille coupling. Related compounds have shown interesting applications as molecular switches. When incorporating C–H activations [19] in the synthetic strategy using propargylic alcohol **48** as the starting material, the attempted reaction failed to give product **49** (activation of C–H^a bond) but afforded acenaphthylene **50** (activation of C–H^b bond) exclusively (Scheme 3.13).

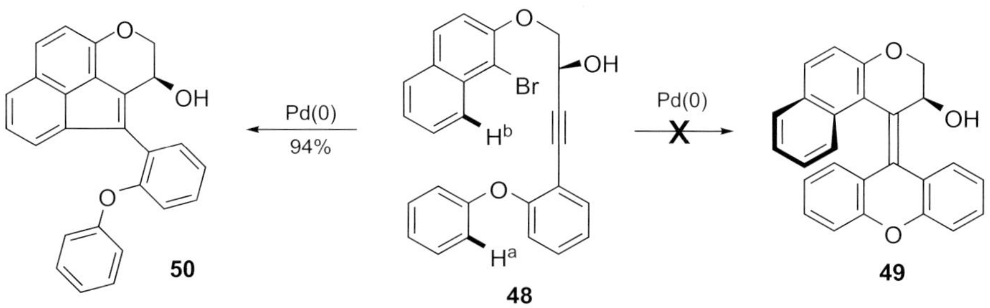

Scheme 3.13 Synthesis of acenaphthylenes.

By omitting the second aromatic ring in the naphthalene moiety of compound **48**, helical tetrasubstituted alkenes **53** were successfully obtained from substrate **51** in a highly efficient and diastereoselective manner [20a] via an intramolecular carbopalladation/C–H activation sequence through the intermediate **52** (Scheme 3.14). Although most studies were performed using racemic **51**, highly enantiomerically pure product **53** (R = H, X = O, 95% ee) could be synthesized from the corresponding chiral propargylic alcohol. The method has also been applied to synthesize tetrasubstituted alkenes containing five- and seven-membered ring systems [20b].

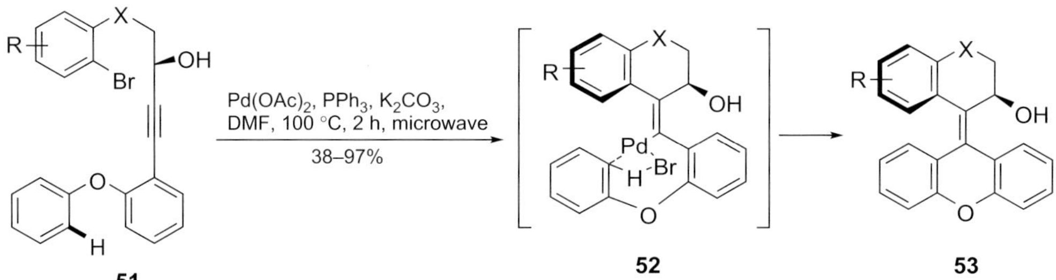

Scheme 3.14 Synthesis of helical tetrasubstituted alkenes.

Quite recently, they also succeeded in preparing naphthalene containing helical tetrasubstituted alkenes of type **49** by a two-component Pd-catalyzed threefold domino process starting from a mixture of **54** and **55** (Scheme 3.15) [21]. Reaction of this mixture with $Pd(OAc)_2$ in the presence of PPh_3 and tetrabutyl ammonium acetate led to the desired product **58** via the Sonogashira intermediate **56** and the vinylpalladium species **57** in 67–96% yield.

A domino reaction involving biscarbocyclization of alkynes and benzynes via C–H activation was developed by Cheng and coworkers [22] to synthesize biologically important isochromenone and oxepine derivatives (Scheme 3.16).

By employing a benzyne precursor **61**, isochromenone **62** and oxepine **63** were synthesized in good yields from alkynes **59** and **60**, respectively. The proposed

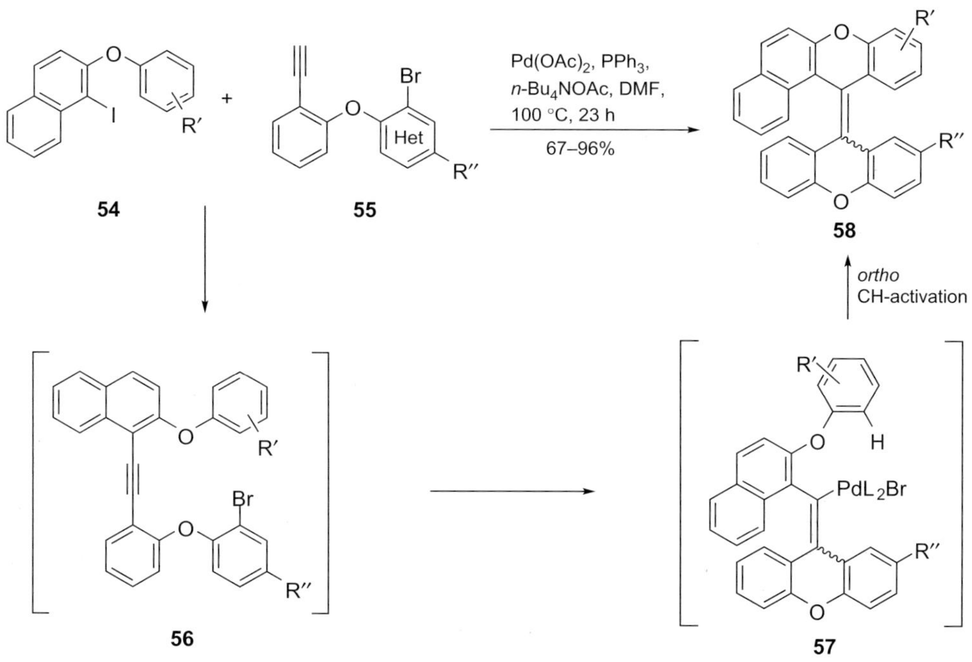

Scheme 3.15 Two-component Pd-catalyzed triple domino process for the synthesis of **58**.

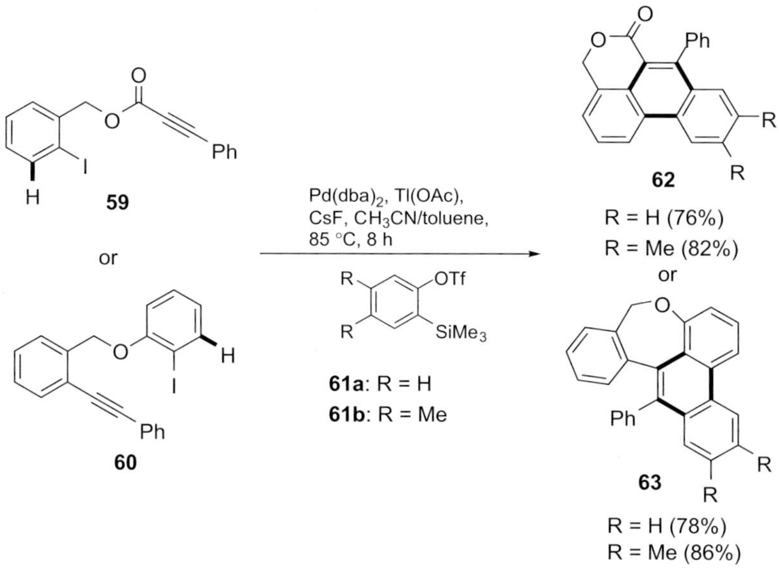

Scheme 3.16 Synthesis of isochromenone and oxepine derivatives.

mechanism begins with oxidative addition of the aryl iodide **59** to Pd(0) followed by intramolecular carbopalladation to the alkyne moiety, leading to the intermediate **64** (Scheme 3.17).

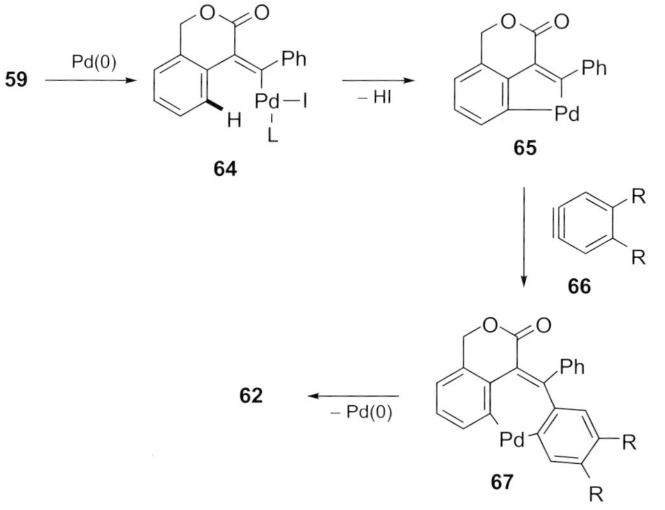

Scheme 3.17 Proposed mechanism for the formation of isochromenones.

Subsequent C–H activation generates the five-membered palladacycle **65**, which undergoes intermolecular carbopalladation with aryne **66** to give seven-membered palladacycle **67**. Finally, reductive elimination furnishes the product **62** and regenerates the palladium catalyst. Overall, three C–C bonds are constructed in this reaction sequence. In an earlier report, Tanaka and coworkers [23] had observed similar domino reactions with bromoenynes to synthesize benzoisoindoles in the absence of external benzyne acceptors.

The potential of using acyloximes in Pd-catalyzed annulation reactions with arynes and alkynes was explored by Zhu and coworkers [24]. The underlying principle involves the oxidative addition of acyloxime **68** with Pd(0), forming the species **69** (Scheme 3.18).

Intermolecular aminopalladation with arynes or alkynes leads to the vinylpalladium species **70**. A key C–H activation followed by reductive elimination furnishes the heterocycle **71**. This concept was successfully applied in the synthesis of phenanthridines **73** and isoquinolines **75** using acyloxime **72** in the reactions with benzyne precursor **61a** and alkyne **74**, respectively (Scheme 3.19).

The same group also recently [25] developed a Pd(II)-catalyzed intramolecular aminopalladation/C–H activation sequence to synthesize pyrrolo[1,2-*a*]indoles of the type **77** from alkyne **76** (Scheme 3.20).

It was found that the presence of an electron-withdrawing group at the terminal position of the alkyne was necessary. The proposed mechanism involves the

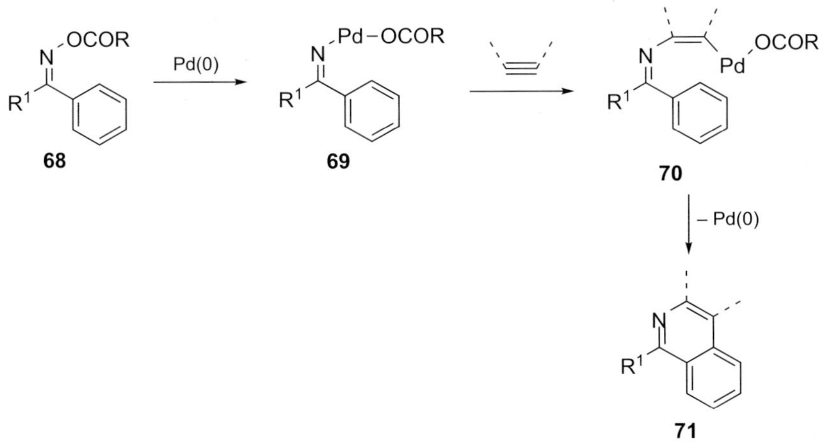

68 69 70 71

Scheme 3.18 Palladium-catalyzed annulation of acyloximes with arynes and alkynes via C–H activation.

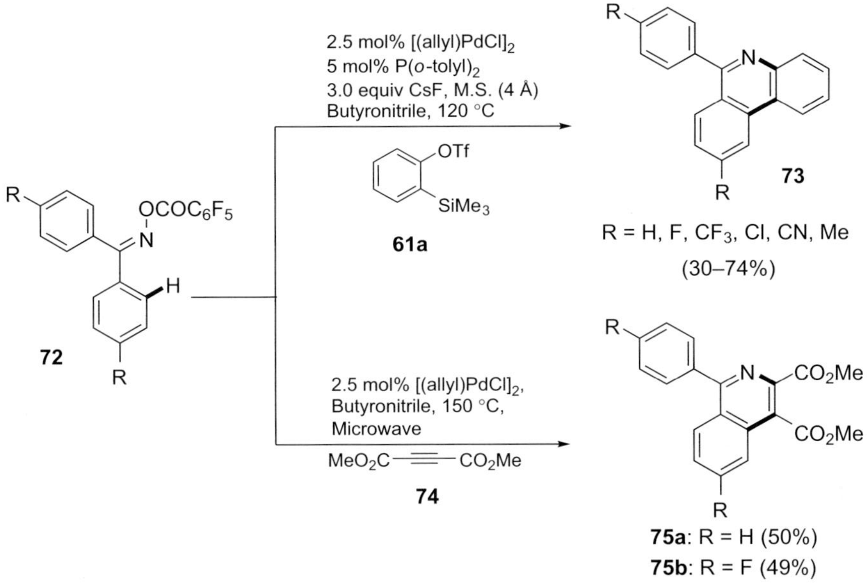

72 61a 73 74 75

Scheme 3.19 Synthesis of phenanthridines and isoquinolines.

initial formation of the vinylpalladium intermediate **78** by syn-aminopalladation. Alternatively, an intramolecular aza-Michael addition was also possible, leading to the intermediate **79**. This pathway was ruled out, as **79** gave only a small amount of product **77** together with pyrrole **80**, and recovered starting material under identical catalytic conditions.

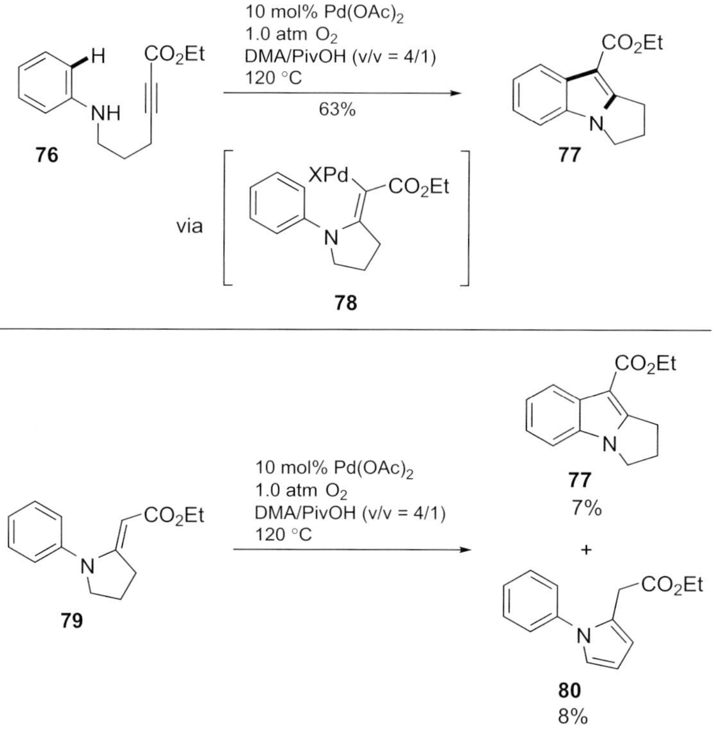

Scheme 3.20 Synthesis of pyrrolo[1,2-*a*]indoles.

3.3 Palladium-Catalyzed/Norbornene-Mediated ortho C–H Activations

Catellani and coworkers [26] reported a remarkable palladium-catalyzed/norbornene-mediated domino reaction where aryl iodides were alkylated at the ortho positions followed by a Heck reaction to afford the *o,o*-dialkylvinylarenes **81** (Scheme 3.21). Multiple bonds are formed in a single reaction from simple starting materials.

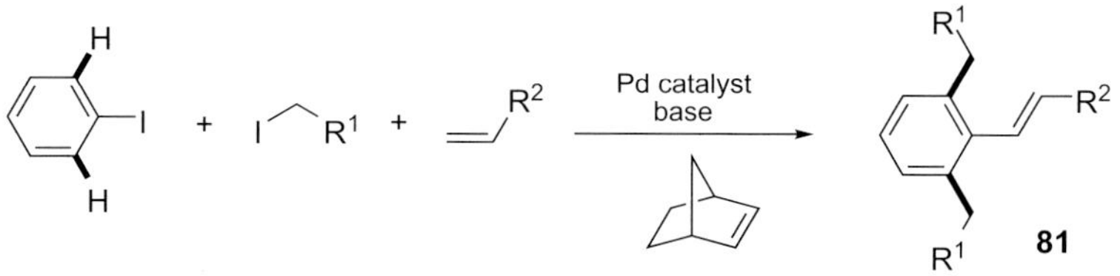

Scheme 3.21 Initial report of the Catellani reaction.

The proposed catalytic cycle involves the transition of multiple oxidation states including Pd(0), (II), and (IV) (Scheme 3.22). Initial oxidative addition of Pd(0) to mono-ortho-substituted aryl iodide **82** leads to the Pd(II) intermediate **83**. Carbopalladation with norbornene **84** gives norbornylpalladium(II) **85**. As *syn*-β-hydride elimination is not possible, C–H activation at the ortho position takes place with the assistance of a base to generate the palladacycle **86**. A second oxidative addition with an alkyl halide (R^1-X) leads to the Pd(IV) intermediate **87**. Reductive elimination forms the ortho C–C bond and generates the norbornylpalladium(II) **88**. Extrusion of norbornene by β-carbon elimination is favorable because of the increased steric strain of **88**, affording arylpalladium(II) **89**. Conventional Pd-catalyzed coupling reactions of **89** with a coupling partner (R^2-Y) furnishes the final product **90** where the second C–C bond is constructed and the Pd(0) catalyst is regenerated. If there is no preexisting ortho substituent in the substrate **82**, the cycle repeats to alkylate both ortho positions.

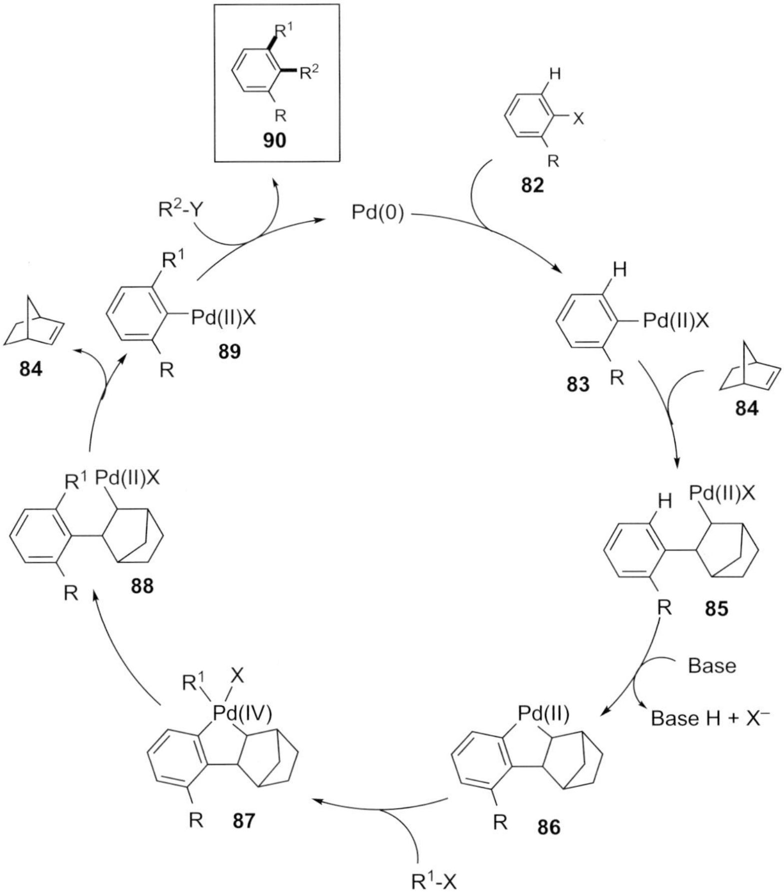

Scheme 3.22 Proposed catalytic cycle of the Catellani reaction.

As the initial discovery of this reaction, tremendous progress has been made in unlocking its synthetic utility. Reviews on the mechanistic and synthetic aspects of the Catellani reaction have been written by the Lautens group [27, 28] and the Catellani group [29, 30]. This section focuses on the recent developments in this area since 2009.

Catellani and coworkers [31] observed dramatic effects of added olefins on the product selectivity in Pd-catalyzed/norbornene-mediated reactions of aryl iodides, aryl bromides, and aryl boronic acids in the synthesis of *o*-teraryls **91** (Scheme 3.23).

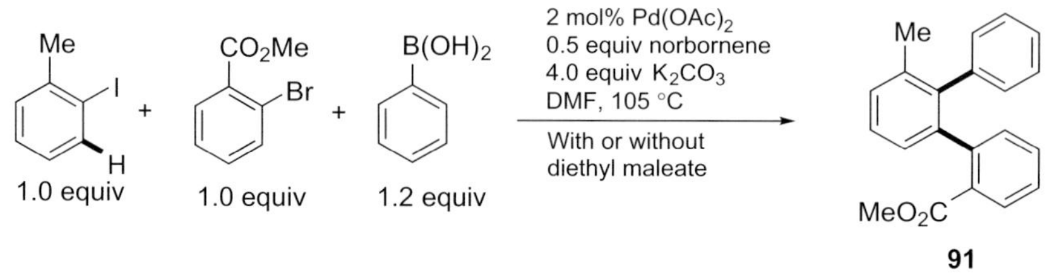

	Yield (%)		
	91	**92**	**93**
Without diethyl maleate:	30	50	50
With diethyl maleate (1.6 equiv):	87	3	3

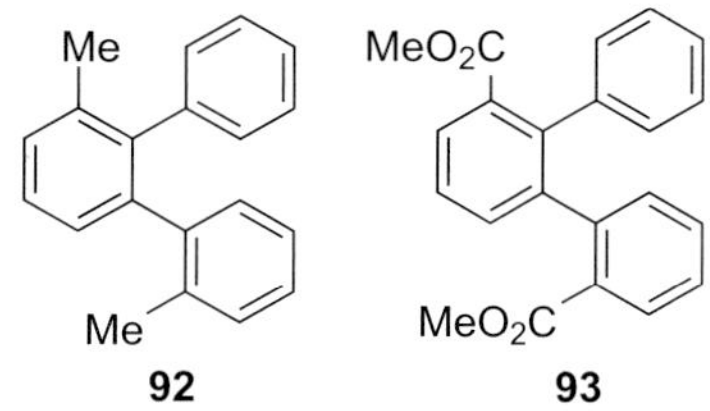

Scheme 3.23 Effects of added olefin on product selectivity in the synthesis of *o*-teraryls.

It was found that adding diethyl maleate drastically increased the product ratio, favoring the desired cross-coupled product **91** over the homo-coupled products **92** and **93**. Under these conditions, a variety of multisubstituted *o*-teraryls **94** were synthesized with good selectivities in which the Suzuki coupling with aryl boronic acids served as the termination step (Scheme 3.24). The authors argued that the

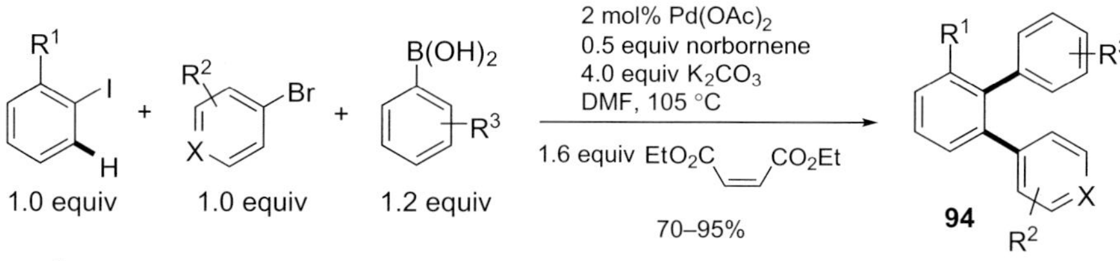

R^1 = Me, naphthyl

R^2 = H, 4-CO_2Me, 3-CF_3, 4-CN, etc.

R^3 = H, 4-Me, 4-F

X = CH, N

Scheme 3.24 Synthesis of *o*-teraryls.

effects of the added olefin arose from its ability to coordinate to and stabilize the Pd species by acting as a ligand. As a result, the oxidative addition was depressed and the reductive elimination was accelerated.

Building on the previously discovered synthesis of tetracyclic fused pyrroles via a threefold domino reaction [32], Lautens and coworkers [33] reported the synthesis of seven-membered ring-fused indoles **97** (Scheme 3.25). Three C–C bonds were constructed in the reaction between iodobenzylindoles **95** and bromoalkylalkynes **96**, two of which were formed by C–H activation. Tri-(2-furyl)phosphine (TFP) was used as the ligand. The termination step was the direct arylation of the pendant indole. Seven-membered ring-fused pyrroles were also synthesized by this method, albeit in lower yields because of the formation of by-products.

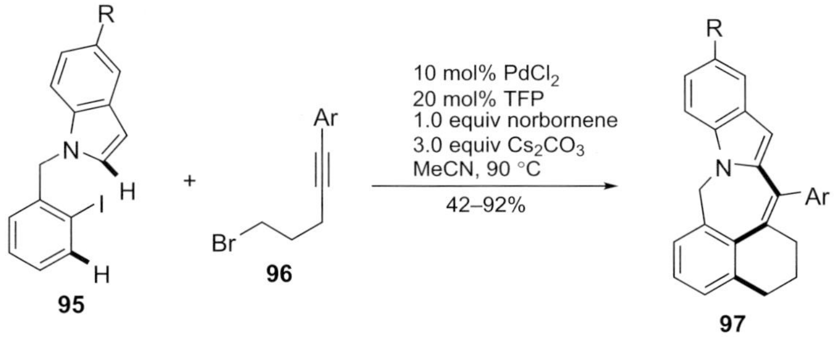

Scheme 3.25 Synthesis of seven-membered ring-fused indoles.

By applying an analogous direct arylation of indole as the termination step, Jafarpour and Hazrati [34] developed a domino annulation of indoles (Scheme 3.26). Intermolecular ortho alkylation of aryl iodides **98** with bromoalkylindoles **99** followed by intramolecular indole C–H activation afforded the dihydrobenzo[*a*]carbazoles **100**, which are structural motifs in biologically active compounds.

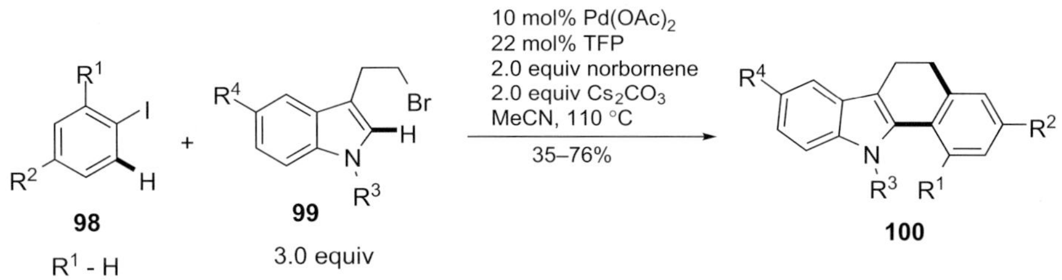

Scheme 3.26 Synthesis of annulated indoles.

A series of reports from the Lautens group and others have described the synthesis of phenanthridines, an important class of compounds with a broad range of biological activities, utilizing palladium-catalyzed/norbornene-mediated domino reactions. Lautens and Candito [35] identified that *N*-silylimines **102**, easily accessible from the corresponding aldehydes, reacted efficiently with ortho-substituted aryl iodide **101** to form phenanthridines **103** (Scheme 3.27). The use of *N*-silylimines was advantageous because of the nucleophilic nature of the nitrogen atom and the easy cleavage of the silyl group under the reaction conditions. The reaction tolerated a variety of substituent groups on the aryl iodide and the imines. Both *N*-silylaldimines and *N*-silylketimines were successfully employed, the latter offering the possibility of substitution (R^3) at the 6-position of the phenanthridine ring. Even unsubstituted (*N*-H) ketimines participated in the reaction effectively. The *N*-TBDMS (*tert*-butyldimethylsilyl) ketimine gave the product in lower yields possibly because of the increased stability of the TBDMS group. The reaction presumably proceeds through the intermediate **104** after ortho-arylation. Cleavage

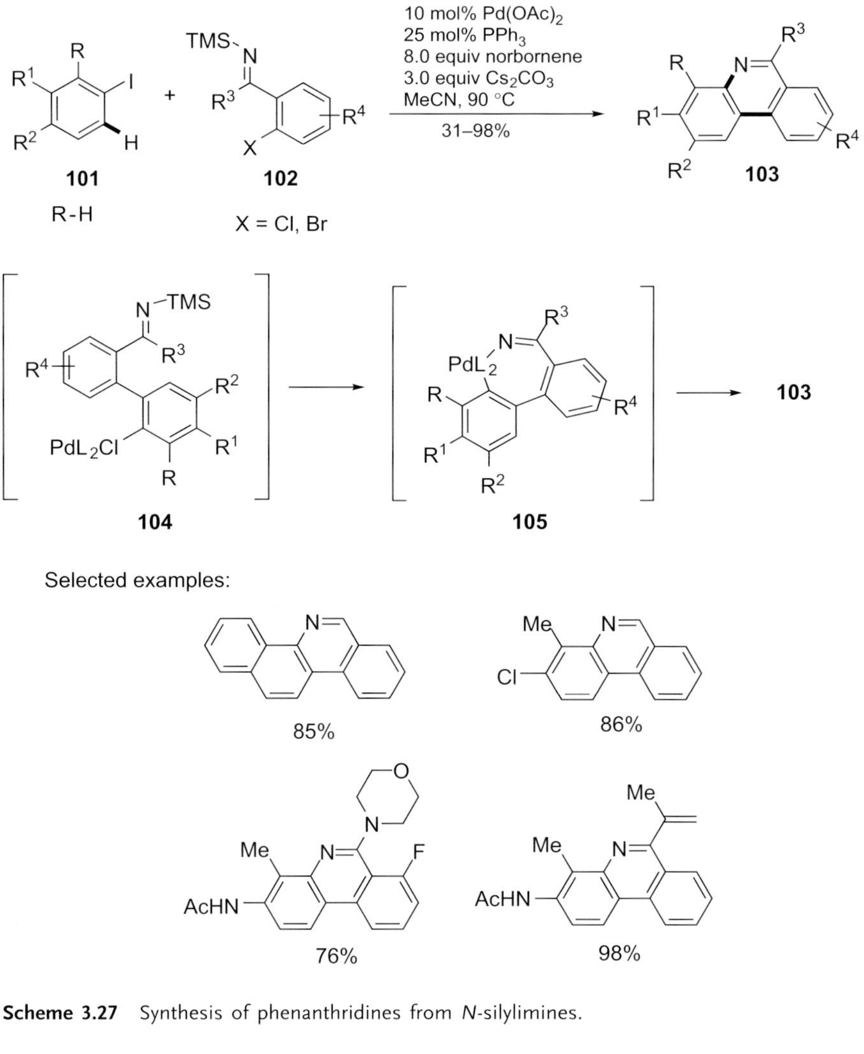

Scheme 3.27 Synthesis of phenanthridines from *N*-silylimines.

of the N–Si bond leads to the formation of the Pd-imino intermediate **105**, which releases the product on reductive elimination to form the final C–N bond.

This methodology was successfully applied in the formal synthesis of biologically relevant benzo[*c*]phenanthridine alkaloids nitidine and NK109 (Scheme 3.28) [36]. The aryl triflate **107**, which is more easily accessible than the corresponding aryl iodide, was synthesized in three steps from commercially available dibromide **106** in high yields. The use of aryl triflates in norbornene-mediated C–H activations had not been demonstrated before this study. In the investigations on the scope of the

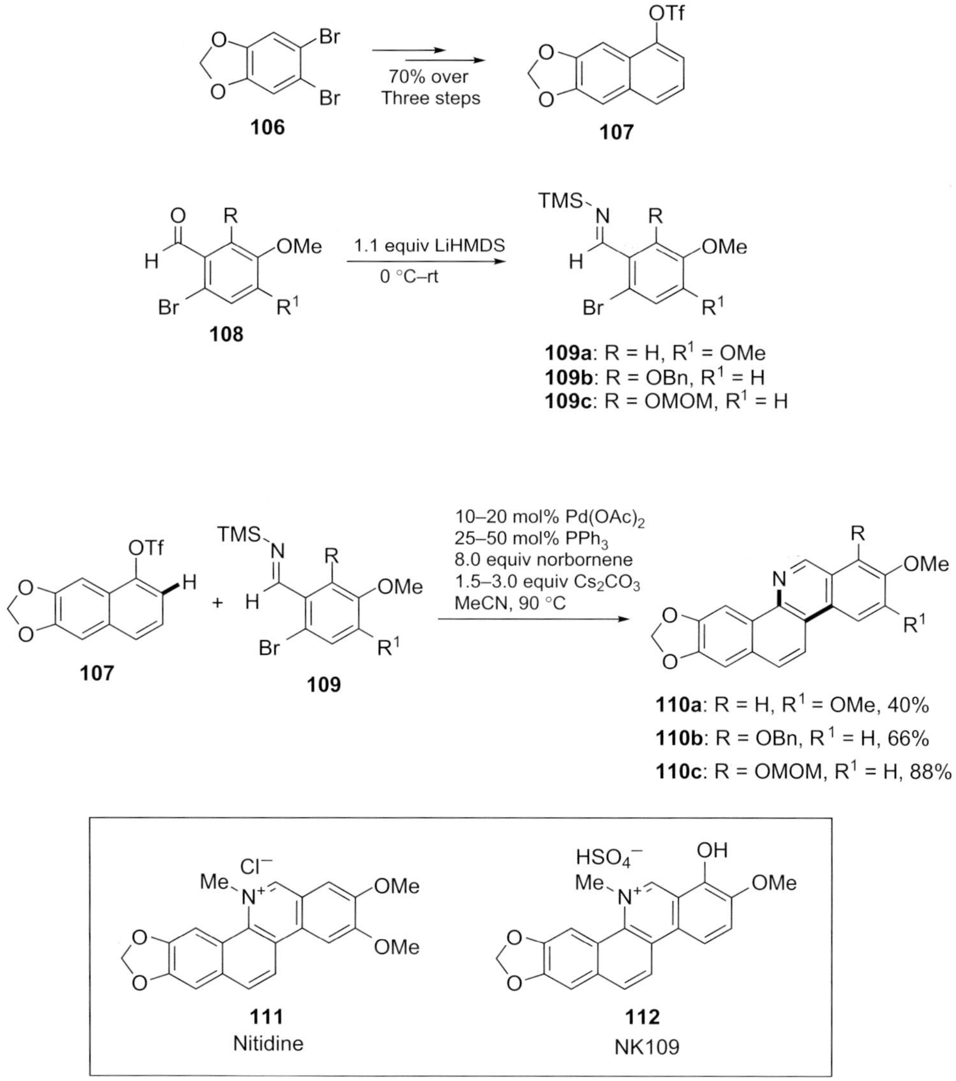

Scheme 3.28 Formal synthesis of nitidine and NK109.

reaction, the authors showed that aryl triflates reacted effectively with *N*-silylimines to afford phenanthridine products with good functional group tolerability. The imines **109** were prepared from aldehydes **108** with lithium hexamethyldisilazide (LiHMDS) in one step, which were used as crude products in the subsequent domino reactions because of the difficulty in their purification. The key domino ortho-arylation/N-arylation reaction sequence furnished the structural cores **110** of Nitidine **111** and NK109 **112**. The reaction was performed on a gram scale in the synthesis of **110c**.

A different strategy to synthesize phenanthridines was shown by Malacria and coworkers [37] using unprotected 2-bromobenzylamines **113** as the reaction partner (Scheme 3.29). The reaction follows a similar pathway of ortho-arylation/N-arylation, leading to dihydrophenanthridines **115**. Introduction of an oxygen balloon at the end of the reaction (visualized by precipitation of Pd black) completely converted **115** to the phenanthridines **114** by a Pd(II)-catalyzed oxidative dehydrogenation. In the absence of oxygen, the reaction still produced the phenanthridine product in a 4 : 1 ratio (**114** : **115**) using excess norbornene (3.5 equiv). The authors reasoned that dioxygen or a sacrificial olefin such as norbornene acted as the hydrogen scavenger in the dehydrogenation step. Overall, a dual Pd catalysis was developed in a simple one-pot/two-step setting using the same Pd precatalyst.

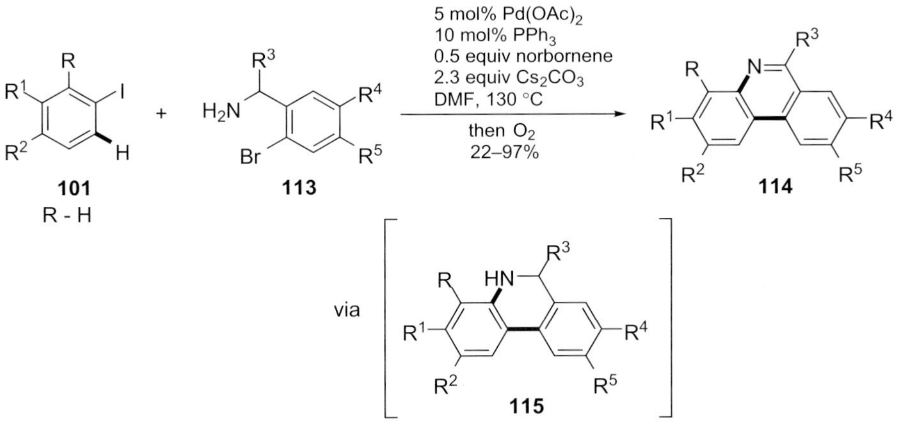

Scheme 3.29 Synthesis of phenanthridines from 2-bromobenzylamines.

Catellani and coworkers [38] discovered another interesting domino reaction to synthesize phenanthridines **118** from aryl iodide **101**, *o*-bromo-*N*-trifluoroacetanilides **116**, and methyl vinyl ketone **107** (Scheme 3.30). This multicomponent/domino reaction consists of four distinct steps: (i) ortho-arylation of the aryl iodide via C–H activation leading to **119**, (ii) Heck coupling with methyl vinyl ketone giving **120**, (iii) facile cleavage of the *N*-trifluoroacetyl group and aza-Michael addition to the enone moiety furnishing dihydrophenanthridine **121**, and (iv) retro-Mannich reaction ejecting acetone and affording the phenanthridine product **118**. The authors reasoned that the formation of **118** must occur very fast since the intermediates **120** and **121** could not be isolated. However, the use of methyl acrylate instead of methyl vinyl ketone led to the isolation of the analog of **121**, which supported it being along the reaction path. Despite the complex reaction mechanism, the reaction setup was fairly simple, allowing the generation of various substituted phenanthridines from easily available aryl iodides, 2-bromoanilides, and olefins.

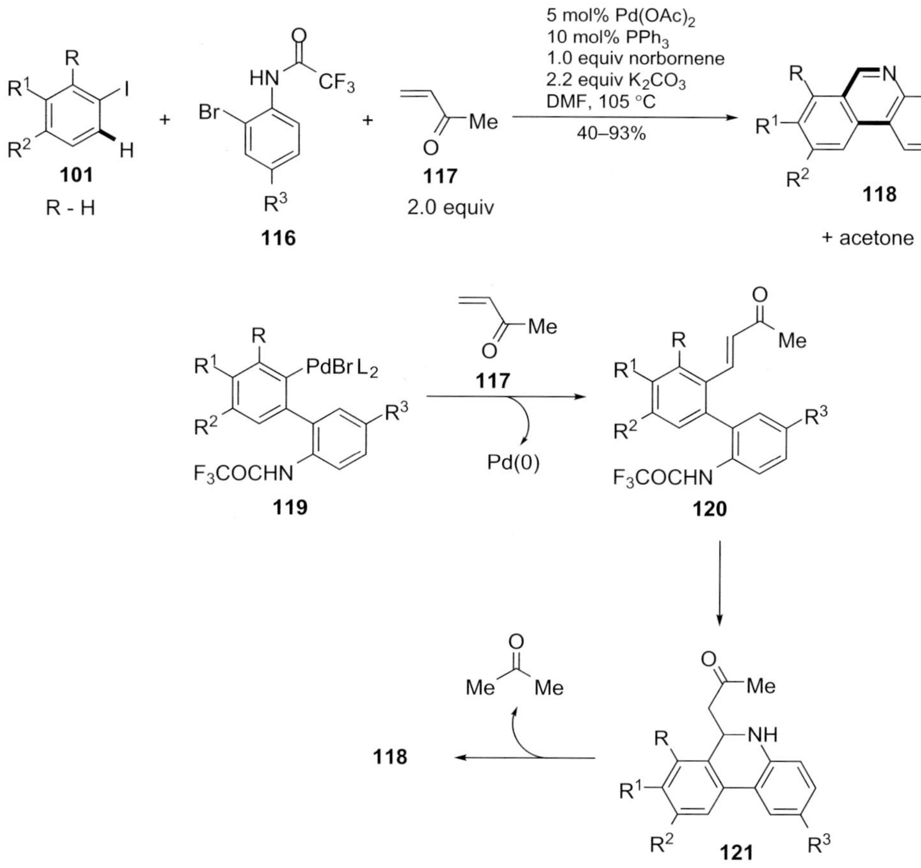

Scheme 3.30 Synthesis of phenanthridines from *o*-bromo-*N*-trifluoroacetanilides.

As an extension of the direct arylation/N-arylation methodology in the synthesis of phenanthridines, Lautens and Candito [39] attempted the synthesis of indole **124** from iodonaphthalene **122** and α-chloroimine **123** (Scheme 3.31). While the concept was successfully demonstrated, the availability of the α-haloimines was rather limited because of their challenging syntheses which are often accompanied by decomposition of the imine. To circumvent this problem, the authors decided to employ the highly strained 2*H*-aziridines **125** as 1,3-dipole equivalents. Optimization studies identified the phosphine ligand P(*m*-Cl-C_6H_4), which possessed

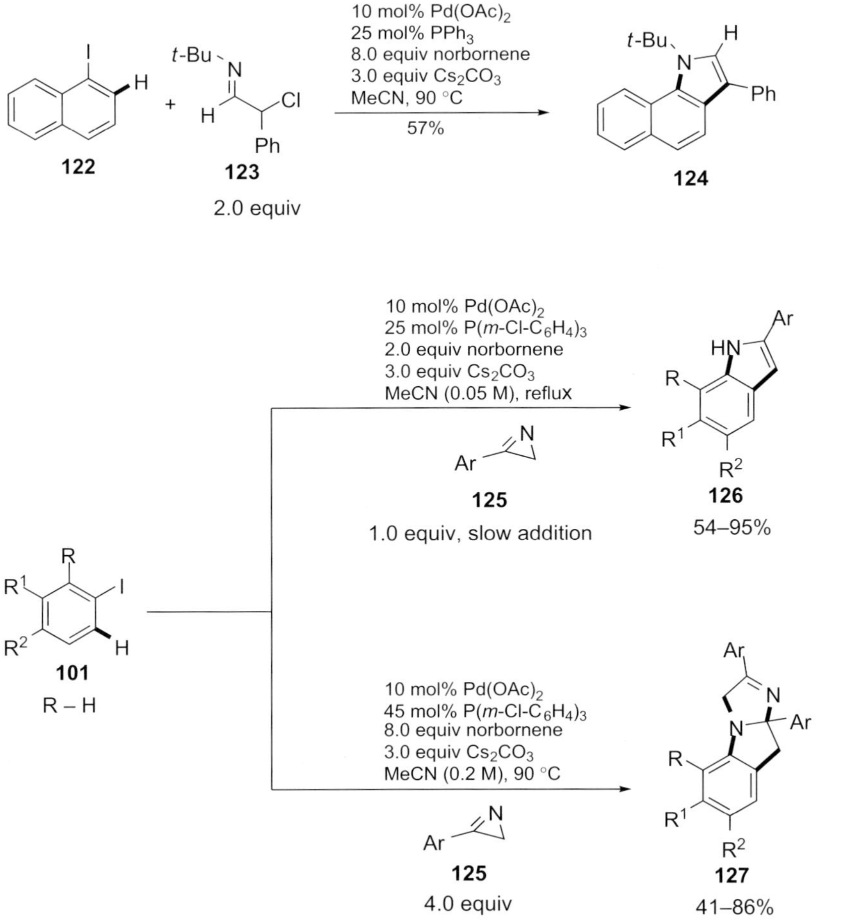

Scheme 3.31 Synthesis of indoles and polycyclic dihydroimidazoles.

the optimal steric and electronic properties for the reaction. Depending on the reaction conditions, two products could be obtained: the desired indoles **126**, and the unusual polycyclic dihydroimidazoles **127**. It was found that the slow addition of aziridine **125** was the key to the selectivity toward the indole product **126**. On the other hand, increasing the concentration and using large excess of **125** favored the formation of dihydroimidazole product **127**.

The proposed mechanism involves the initial formation of the palladacycle **118** via the Catellani reaction sequence between the aryl iodide and norbornene (Scheme 3.32). Coordination of the aziridine **125** followed by oxidative addition of the N–C single bond by Pd leads to the intermediate **129**. The fission of N–C single bond is commonly observed in transition-metal-catalyzed reactions with aziridines. Chemoselective C–C bond formation gives **130**, which on decarbopalladation affords palladacycle **131** and extrudes norbornene. Reductive elimination furnishes

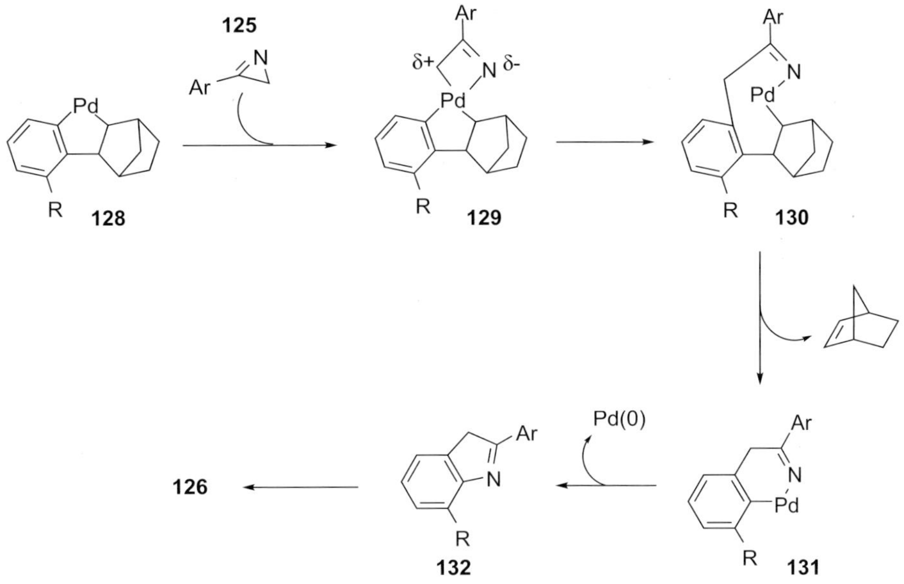

Scheme 3.32 Proposed mechanism for the synthesis of indoles.

132, which tautomerizes to give the final indole product **126**. The formation of the dihydroimidazoles **127** can be reasoned by a Pd(0)-catalyzed formal [3+2] cycloaddition of an equivalent of the aziridine with intermediate **132**, but the exact mechanism remains unclear. The unique feature of this reaction sequence is the exploitation of the strain energy of two small molecules, namely norbornene and 2*H*-aziridine, to facilitate complex transformations in the context of heterocycle synthesis.

Catellani and coworkers [40] observed that the reactions between aryl iodides **101** and 2-bromobenzyl alcohols **133** could lead to *o*-biaryl carbaldehydes **134** and dibenzopyrans **135** (Scheme 3.33).

Both products are presumably generated via a common intermediate, the seven-membered oxapalladacycle **136**, from the domino Pd-catalyzed/norbornene-mediated reaction sequence. In the presence of a benzylic C–H, an intramolecular redox process takes place by hydrogen migration from the benzylic position to the aromatic position, affording the aldehyde products **134**. When using tertiary benzyl alcohols, C–O coupling is the predominant pathway leading to the dibenzopyran products **135**. It was found that cesium pivalate was the base of choice in the former process and potassium carbonate was suitable for the latter.

Interest in chiral tetrasubstituted helical alkenes has grown rapidly in recent years, owing to their unique switching properties under UV irradiation [41]. In particular, the helical compounds **137–139**, developed by Feringa and coworkers [42], have shown potential applications as molecular switches and motors (Scheme 3.34).

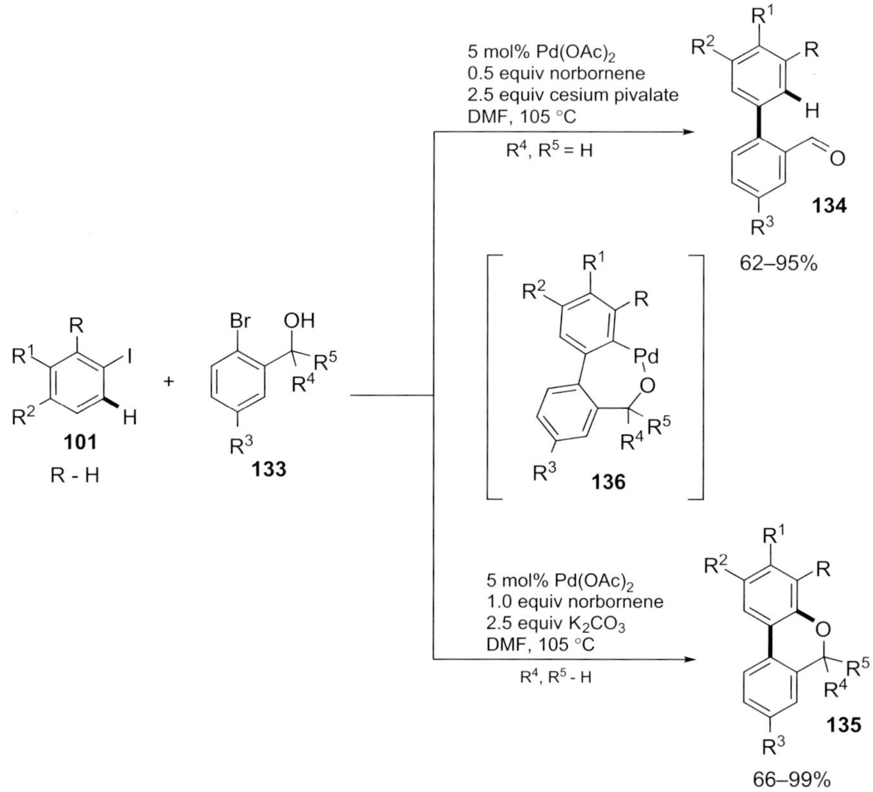

Scheme 3.33 Synthesis of *o*-biaryl carbaldehydes and dibenzopyrans.

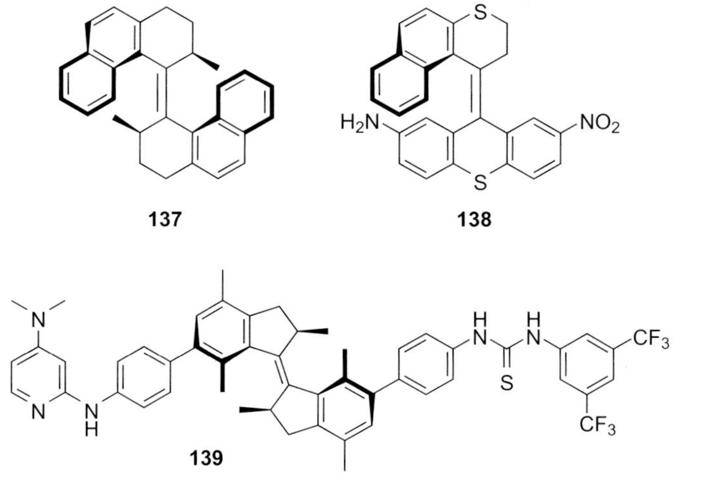

Scheme 3.34 Representative examples of light-driven molecular switches and motors.

Lautens [32, 33] and Tietze (Section 3.2) have developed approaches utilizing Pd-catalyzed domino reactions to construct sterically crowded tetrasubstituted alkene structures in a highly efficient manner. A remarkable Pd-catalyzed domino reaction of aryl iodide **101** and alkyne **96** was reported by Lautens and coworkers [43], where a norbornene molecule is incorporated in the product to afford a new class of tetrasubstituted helical alkenes **140** (Scheme 3.35). A total of four C–C bonds were formed in a single operation, two of which were from C–H activations. TFP was an

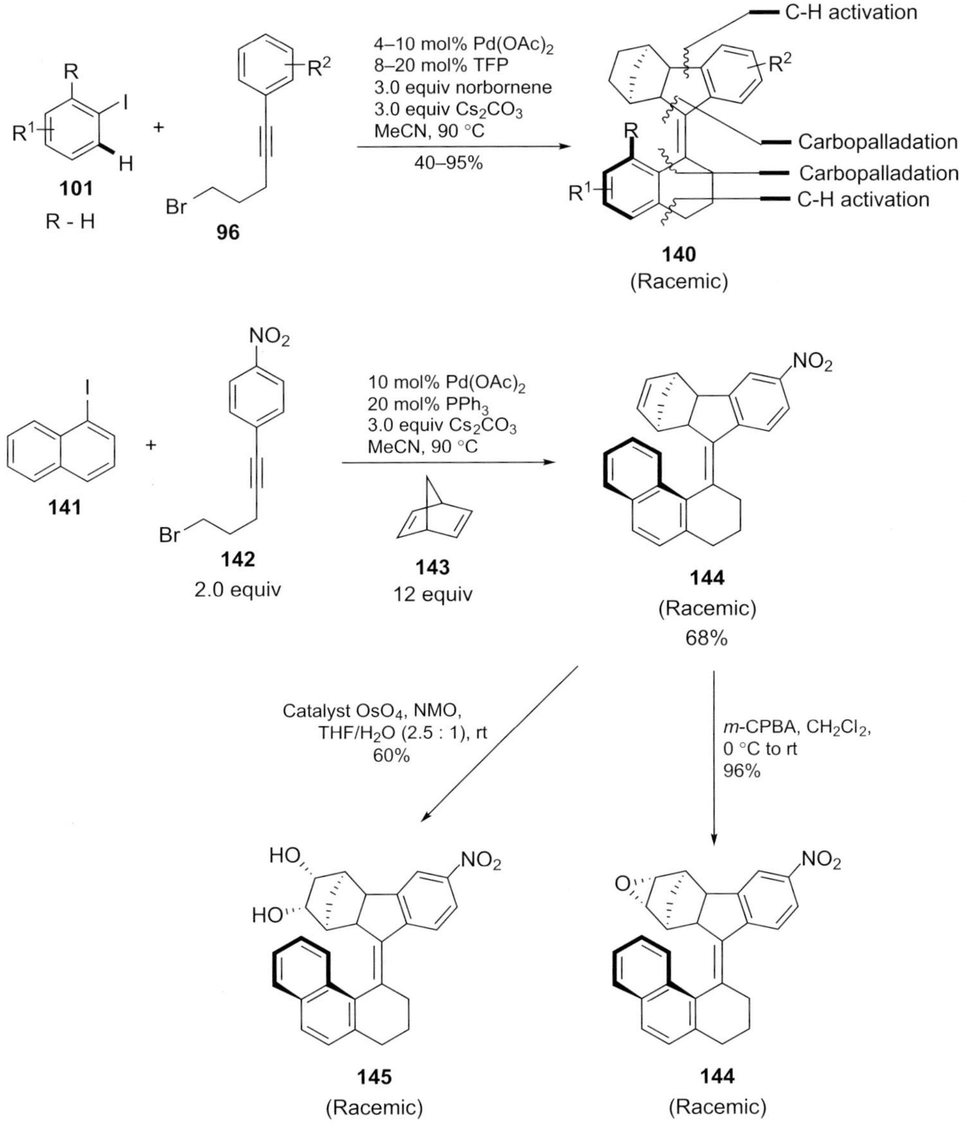

Scheme 3.35 Synthesis of tetrasubstituted helical alkenes.

effective ligand in this reaction. The use of norbornadiene **143** in the reaction with 1-iodonaphthalene **141** and alkyne **142** gave tetrasubstituted alkene **144**. In this case, triphenylphosphine was the ligand of choice. The unreacted double bond in the norbornene unit provided an extra handle for further derivatization. Therefore, epoxidation and dihydroxylation from the less hindered side furnished the epoxide **144** and the diol **145**, respectively.

In this novel domino sequence, norbornene (or norbornadiene) acts as a promoter as well as a reaction partner. It enters and exits the catalytic cycle and is eventually incorporated in the product, and therefore the reaction requires an excess of this reagent. The standard Pd-catalyzed/norbornene-mediated ortho alkylation of the aryl iodide with the alkyne would form the arylpalladium species **146** (Scheme 3.36). Intramolecular carbopalladation onto the tethered alkyne unit gives vinylpalladium **147**. Intermolecular carbopalladation with a molecule of norbornene

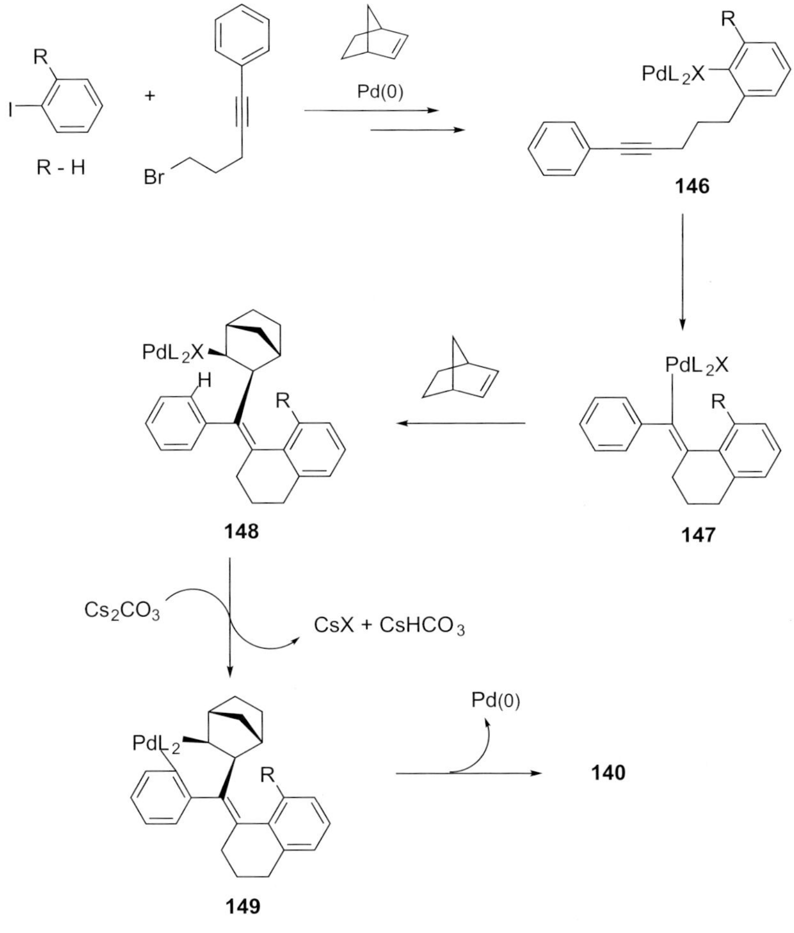

Scheme 3.36 Proposed mechanism for the formation of tetrasubstituted helical alkenes.

affords the intermediate **148** where the second C–H activation takes place leading to palladacycle **149**. The final reductive elimination releases the product **140** and regenerates the Pd(0) catalyst.

An extension of this methodology by the same group [44] led to the synthesis of enantiomerically pure tetrasubstituted helical alkenes. The key substrates were the chiral bromoalkyl aryl alkynes **152** (Scheme 3.37), which were synthesized in four steps from easily accessible chiral α-lactones **150**. Therefore, reduction of the lactones **150** to lactols with diisobutylaluminum hydride (DIBAL-H) followed by homologative rearrangement with trimethylsilyldiazomethane ($TMSCHN_2$) and lithium diisopropylamide (LDA) gave alcohols **151**. Sonogashira coupling with an aryl iodide installed the aryl group, and an Appel reaction converted the alcohols to the enantiomerically enriched bromides **152** (98.5 to >99% ee). Both enantiomers of **152** could be obtained from the corresponding enantiomers of the lactones **150**.

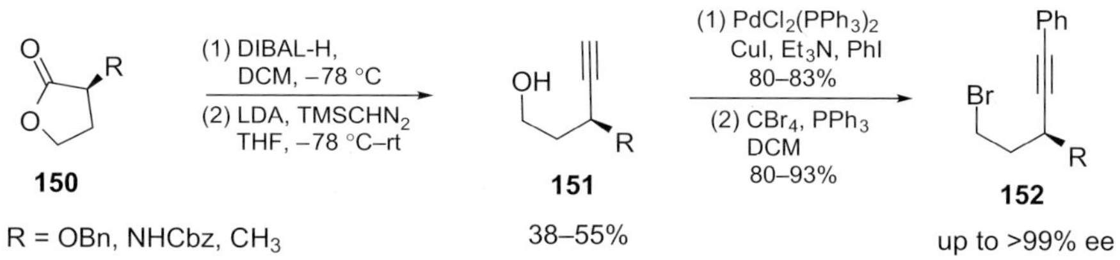

Scheme 3.37 Synthesis of chiral bromoalkyl aryl alkynes.

The Pd-catalyzed domino reactions between alkynes **152** and 2-iodotoluene afforded the tetrasubstituted helical alkenes **153** in good yields. Excellent enantioselectivities (up to >99% ee) were obtained, suggesting retention of stereochemistry in the alkyne substrates under the reaction conditions. The diastereoselectivities were moderate (5 : 1–6.6 : 1 dr). When comparing **153f** with previously obtained **153g** [43], the former gave only 5 : 1 dr but the latter was obtained as a single diastereomer, clearly indicating the effect of the substituent group (R) on the diastereoselective outcome of the reaction. The authors reasoned that the diastereoselectivity results from the helical chirality induced in the first intramolecular carbopalladation step. The 1,3-allylic strain between the R group and the phenyl group (top part of the molecule) may have played a part in controlling the conformations of the intermediates leading to the observed diastereomeric ratios (Scheme 3.38).

Very recently, Lautens and coworkers [45] reported the synthesis of a new class of tetrasubstituted helical alkenes **155** without the incorporation of the norbornene unit (Scheme 3.39). The authors envisioned that, by using the alkyne **154** containing a pendant aryl/heteroaryl group, an intramolecular ortho C–H activation would take place instead of intermolecular carbopalladation with norbornene to form the last C–C bond. This concept was successfully demonstrated in the formation of a wide range of tetrasubstituted helical alkenes **155** containing pyrrole, indole, furan, thiophene, and benzene moieties. In the pyrrole cases, PPh_3 was found to give higher yields than TFP as a ligand. The products were prone to acid-catalyzed

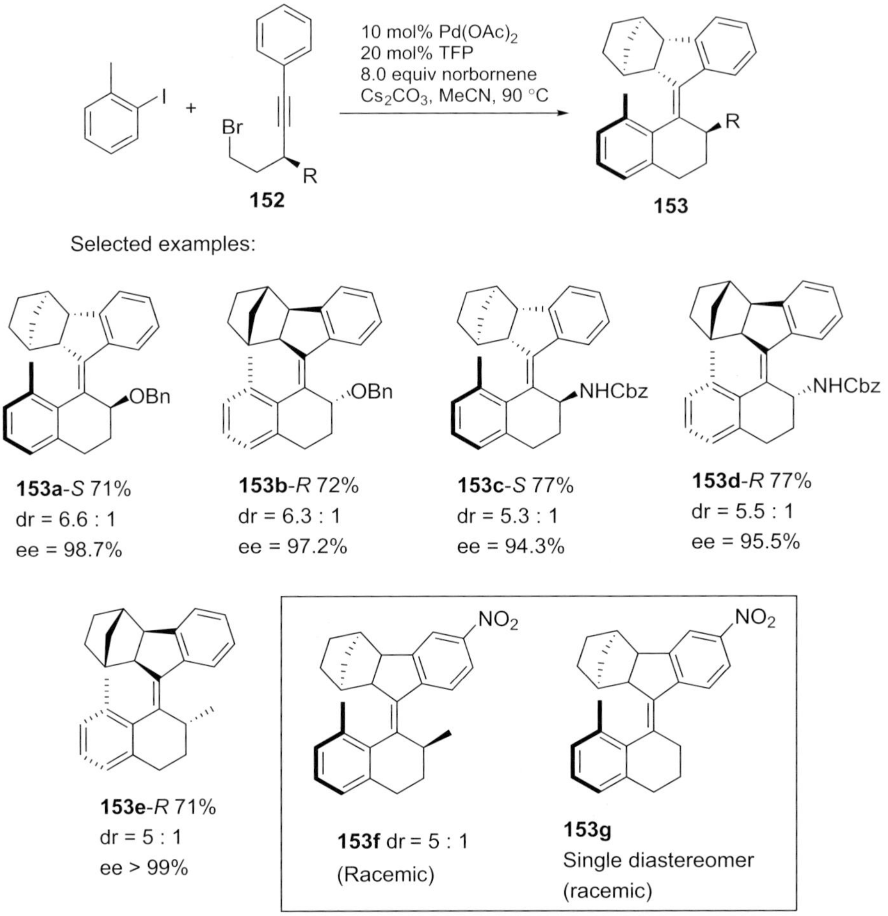

Scheme 3.38 Synthesis of chiral tetrasubstituted helical alkenes.

or light-promoted E/Z isomerization, which could be avoided with precautions, except in the case of the very light-sensitive **155i**.

The reaction was also studied using enantiomerically pure alkyne precursors **156** (Scheme 3.40). The reactions proceeded with retention of stereochemistry of the precursors to afford the chiral products **157** in excellent enantioselectivites (97.4 to >99% ee) and moderate diastereoselectivities.

The proposed mechanism follows a similar pathway of ortho alkylation of the aryl iodide with pyrrole-containing alkyne **154a** affording arylpalladium **158**, which undergoes intramolecular carbopalladation to form vinylpalladium **159** (Scheme 3.41). At this point, the intermediate can induce an intramolecular C–H activation on the adjacent pyrrole ring to form the palladacycle **160**, which undergoes reductive elimination to form the desired product **155a**. Alternatively, vinylpalladium **159** can also undergo intermolecular carbopalladation with norbornene to

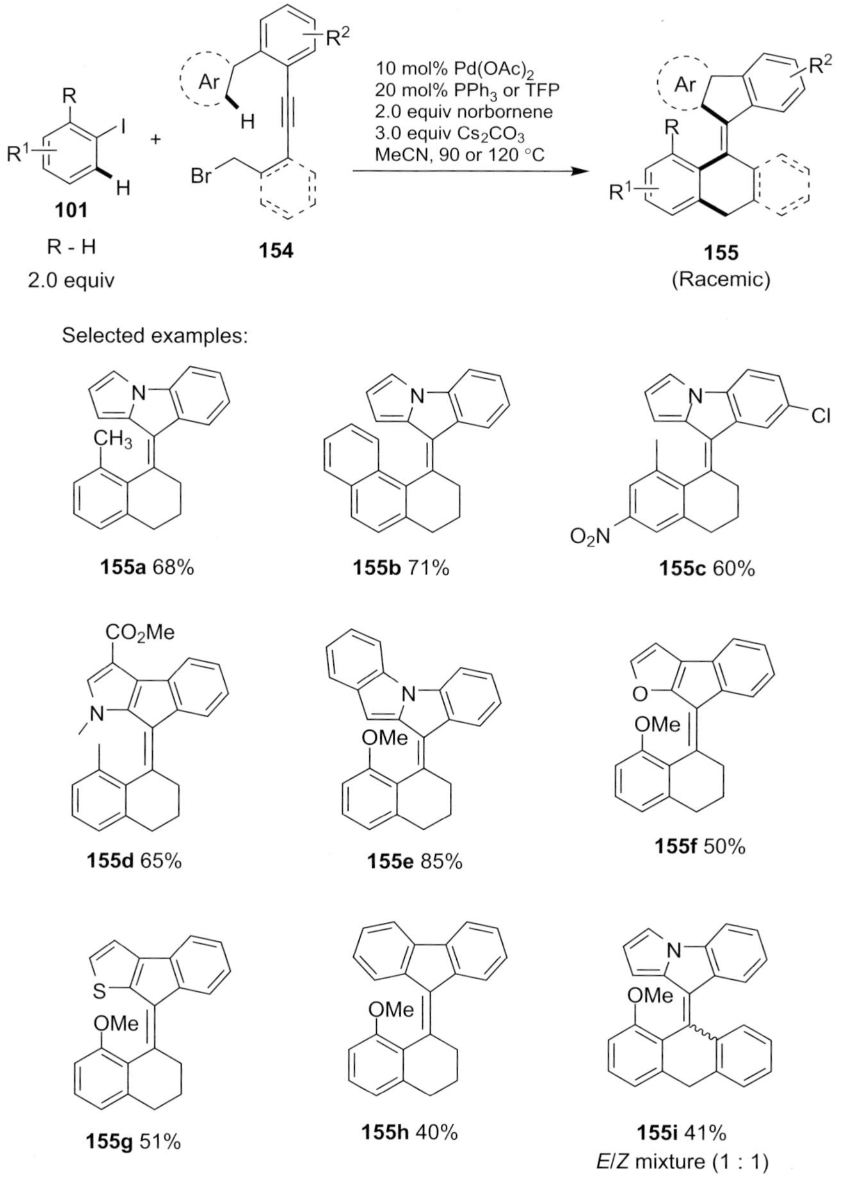

Scheme 3.39 Synthesis of aryl-/heteroaryl-containing tetrasubstituted helical alkenes.

form **161**, which would eventually give the norbornene-incorporated product **162**. The latter pathway was not followed, presumably because the intramolecular C–H activation to form **160** is much faster than the intermolecular carbopalladation to form **161**. Another possible reason is that the increased steric congestion of intermediate **161**, due to the presence of the ortho-substituted heteroaryl group,

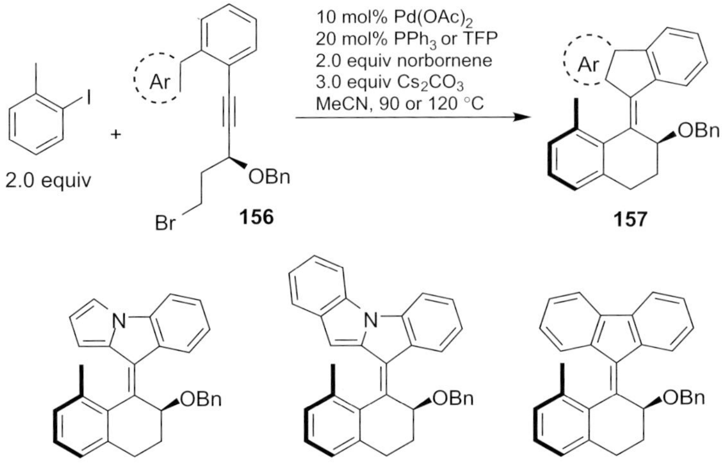

(*S*)-**157a** 90%
dr = 10 : 1, ee = 97.4%

(*S*)-**157c**, 92%
dr = 10 : 1, ee = 98.2%

(*S*)-**157e**, 89%
dr = 9 : 1, ee > 99%

(*R*)-**157b** 88%
dr = 10 : 1, ee = 97.4%

(*R*)-**157d**, 90%
dr = 10 : 1, ee > 99%

(*R*)-**157f**, 92%
dr = 9.3 : 1, ee > 99%

Scheme 3.40 Synthesis of chiral aryl-/heteroaryl-containing tetrasubstituted helical alkenes.

would favor retro-carbopalladation to regenerate **159**, despite the use of a large excess of norbornene.

3.4
Domino Reactions Involving Heteroatom-Directed C–H Activations

Transition-metal-catalyzed (Pd, Rh, and Ru) heteroatom-directed C–H activations have been successfully employed in domino strategies for constructing multiple carbon–carbon/carbon–heteroatom bonds. This area has already been extensively reviewed [46]. Selected recent examples of domino reactions involving heteroatom-directed C–H activations are chosen in this section to highlight this concept.

Ellman and coworkers [47] reported an Rh(III)-catalyzed synthesis of biologically important phthalides **165** (Scheme 3.42). The key step of the reaction sequence was the imidate **163**-directed ortho C–H activation, followed by addition to aldehydes **164** generating alcohols **166** and intramolecular cyclization via intermediates **167**. The initial studies showed that the use of methyl benzoate was ineffective presumably because of the poor directing ability of the ester group for C–H activation. The previously less explored *N*-alkoxy imidates **163a** containing an oxime directing group was found to be effective. A cationic rhodium catalyst generated *in situ* from $[Cp^{*}RhCl_2]_2$ and $AgSbF_6$ (or $AgBF_4$) was necessary. Other halide abstractors gave significantly decreased yields. A variety of substituted benzimidates

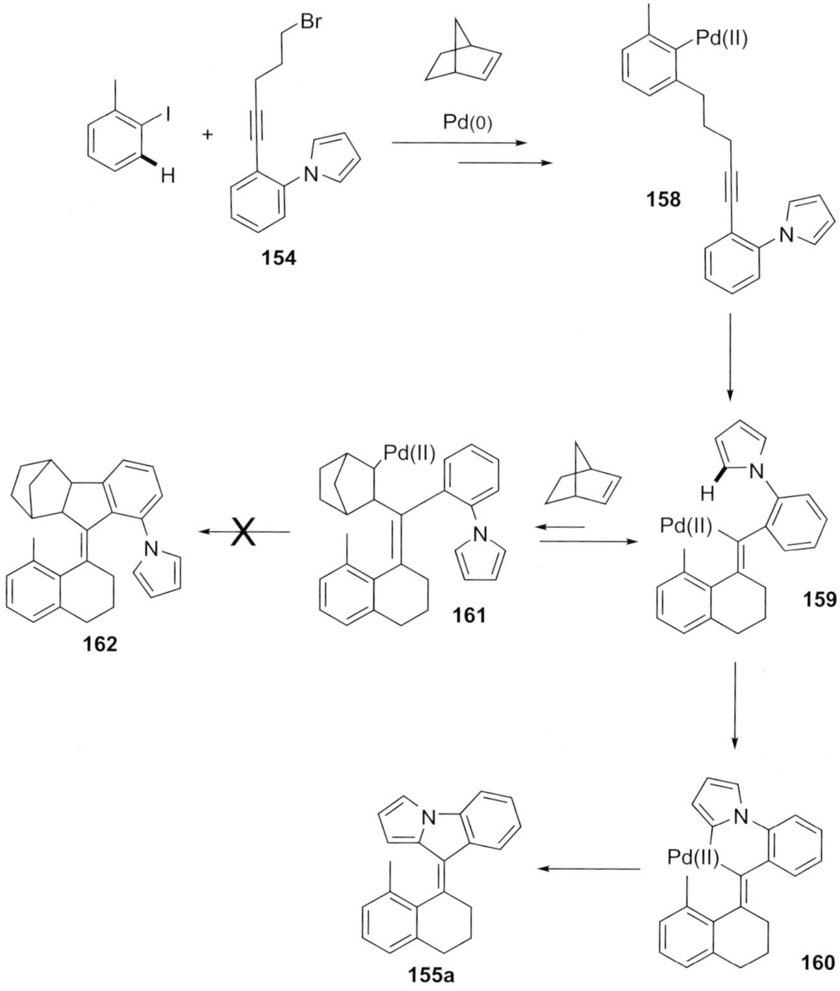

Scheme 3.41 Proposed mechanism for the formation of heteroaryl-containing tetrasubstituted helical alkenes.

163a and aromatic aldehydes were employed and showed good functional group tolerance. The *meta*-methyl substituted benzimidate provided the product arising from selective C–H activation at the less hindered C–H site. When reacting with unactivated aliphatic aldehydes, *N*-3,5-bistrifluoromethylphenyl benzimidate **163b** ($R^1 = H$) generally gave higher yields than the *N*-methoxy benzimidate **163a**.

Li and Shi [48] also reported a cationic Rh(III)-catalyzed synthesis of phthalides **165** utilizing C–H activation as a key step (Scheme 3.43). In this case, benzoic acids **168** were the substrates, which could undergo carboxylic acid-directed ortho C–H

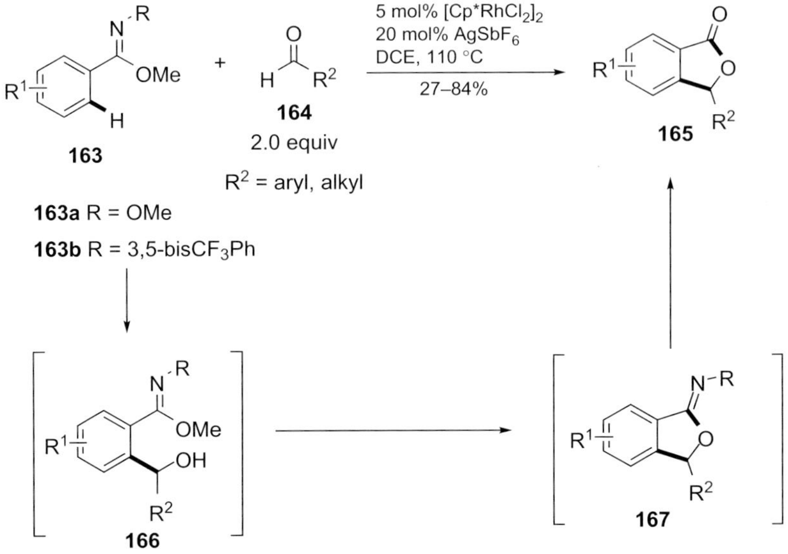

Scheme 3.42 Synthesis of phthalides from benzimidates and aldehydes.

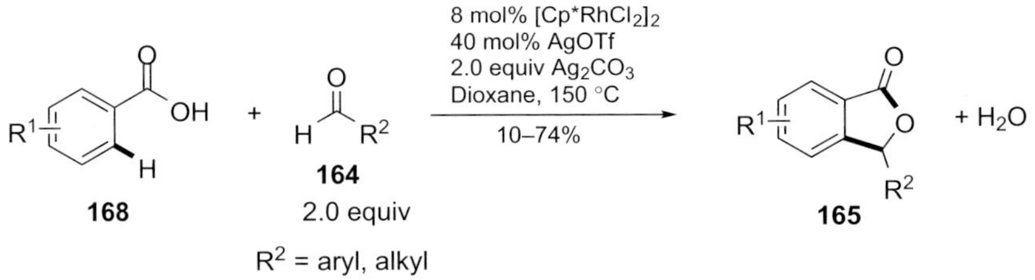

Scheme 3.43 Synthesis of phthalides from benzoic acids and aldehydes.

activation, followed by addition to aldehydes **164** and intramolecular lactonization. While a variety of substituent groups were tolerated on the benzoic acids, the reaction was largely limited to aromatic aldehydes. An attractive feature of this transformation is that the only by-product generated is water.

Besides carboxylic acids, commonly used directing groups for Rh(III)-catalyzed ortho C–H activations also include ketones, imines, benzamides, and acetanilides. The use of pharmaceutically relevant aryl sulfonamides as directing groups has seen limited study. Cramer and coworkers [49] recently reported the synthesis of benzosultams **171** utilizing the directing ability of acylated sulfonamides **169** in a key C–H activation step (Scheme 3.44). The domino sequence presumably involves the initial formation of the five-membered rhodacycle **172** via sulfonamide-directed ortho C–H activation. It was found that the acetyl group on nitrogen was crucial for the reaction. Primary or N-alkylated sulfonamides were completely unreactive. The trifluoroacetyl group underwent facile unproductive deacylation,

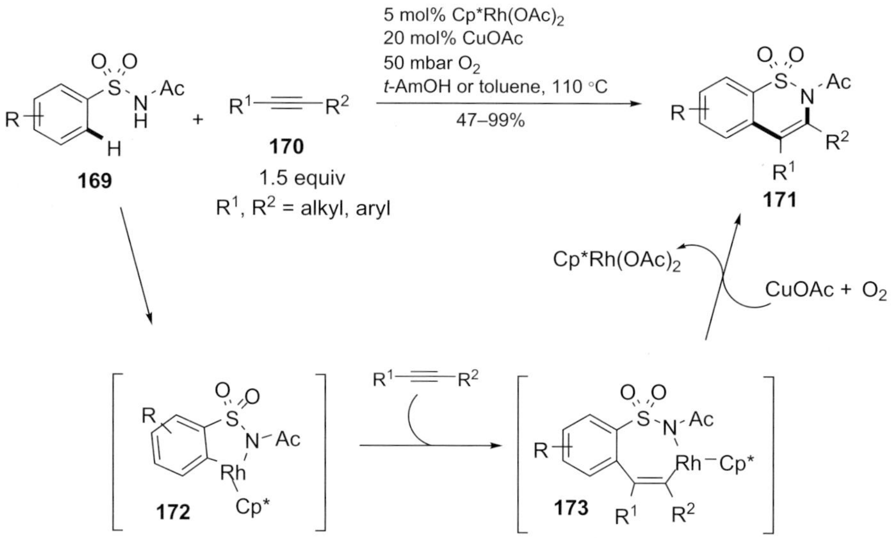

Scheme 3.44 Synthesis of sultams.

and the bulky pivaloyl group shut down the reactivity. The use of defined and highly soluble Cp*Rh$(OAc)_2$ was superior to the commonly used [Cp*Rh$Cl_2]_2$ precatalyst. Subsequent migratory insertion with alkynes **170** would give the seven-membered rhodacycle **173**. The C–N bond is formed by reductive elimination, leading to products **171** and generating Rh(I) species which is oxidized to Rh(III) to reenter the catalytic cycle via the classical Cu(I)/Cu(II) redox couple in the presence of molecular oxygen.

The reaction tolerated a wide range of substituents on the sulfonamides. Both symmetrical diaryl and dialkyl alkynes participated but the latter required

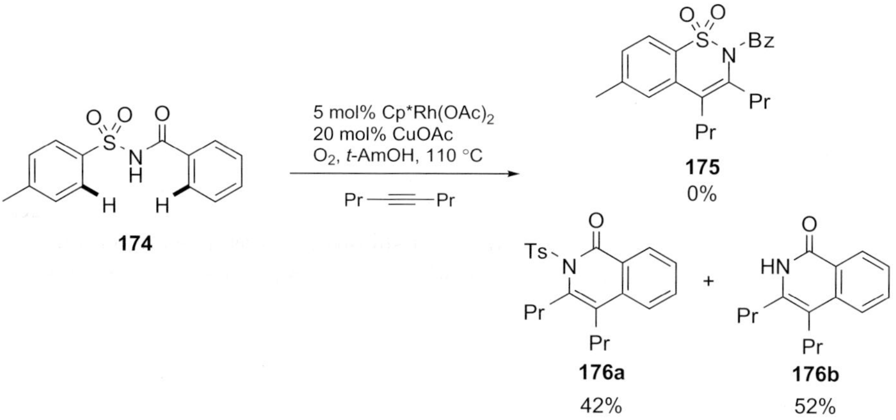

Scheme 3.45 C–H activation competition between sulfonamide and amide.

tert-amyl alcohol as the solvent. Unsymmetrical alkynes gave moderate to good regioselectivities. An interesting competition experiment using tosyl benzamide **174** revealed exclusive ortho C–H activation of the benzamide fragment, forming products **176a** and **176b** in high yields (Scheme 3.45). The results showed that sulfonamides were weaker directing groups than carbonyl derivatives, possibly because of a less favorable alignment for the cyclometalation step (sp^3-hybridized sulfur vs sp^2-hybridized carbon).

The same group [50] developed a remarkable enantioselective synthesis of chiral indenylamines **179** via domino Rh(I)-catalyzed C–H activation/cyclization of aryl ketimines **177** and internal alkynes **170** (a formal [3+2] cycloaddition) (Scheme 3.46). A major challenge of this transformation was the requirement of a catalyst that could accommodate the many different steps in the reaction sequence and show selectivity: for instance, imine-directed ortho C–H activation/

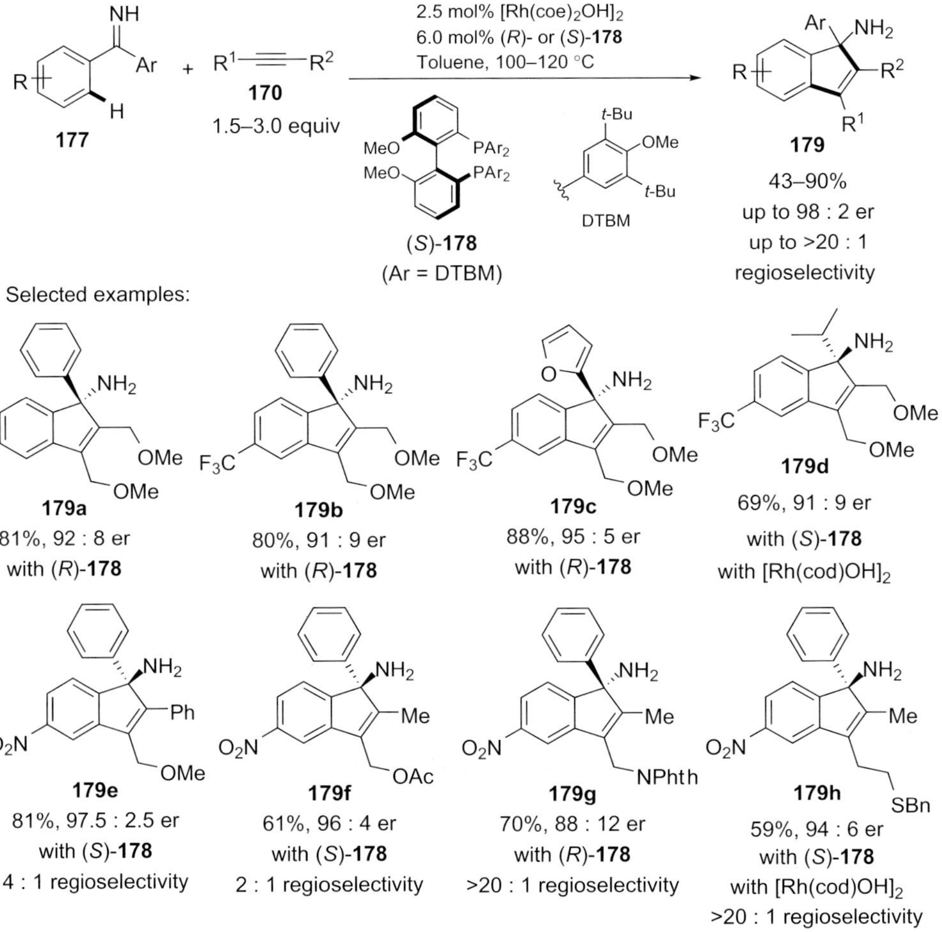

Scheme 3.46 Synthesis of chiral indenylamines.

metallation, regioselective carborhodation to unsymmetrical alkynes, and enantioselective addition to the imine moiety. Rhodium(I) hydroxides showed superior activities compared to cationic rhodium or halide-containing rhodium complexes. The complex $[Rh(coe)_2OH]_2$, which contains a labile cyclooctene ligand, was a more suitable precatalyst than $[Rh(cod)OH]_2$ because of the convenient premixing with the chiral ligand (at ambient temperature for only a few minutes). The steric bulk of the chiral ligand influenced the reaction yield and enantioselectivity. DTBM-MeOBiphep **178** showed the overall best performance.

Symmetrical diphenyl ketimine afforded **179a** in good yield and high enantiomeric ratio. When unsymmetrical diaryl ketimines that have electronically different aryl rings were employed, the C–H activation generally took place on the more electron-poor ring, affording **179b** and **179c** in good positional selectivities (20 : 1). An aryl alkyl ketimine was also employed to give **179d** in a slightly lower enantiomeric ratio as a result of a higher reaction temperature (120 °C). When unsymmetrical internal alkynes were employed, the aryl alkyl alkynes gave higher regioselectivities than the more challenging dialkyl-substituted alkynes, as illustrated in products **179e** and **179f**. However, the use of alkynes containing coordinating functional groups such as phthalimide and thioether dramatically enhanced the regioselectivities. In these cases, the carbon–carbon bond formation occurred preferentially at the position proximal to the directing group, leading to **179g** and **179h** as single regioisomers.

3.5 Conclusions

As illustrated in the examples, a wide range of domino reactions utilizing C–H activations have been achieved. Complex targets become readily available using simple starting materials. The domino nature of the reaction sequence allows multiple bond formations in one reaction setup. By selectively activating diverse C–H bonds in the reaction partners, more atom-economical and scalable processes that avoid wasteful by-products can be developed toward the synthesis of valuable products. It is truly remarkable that a plethora of structurally distinct heterocycles can be selectively synthesized by only a few combinations of C–H activations in the domino reactions: Heck reactions, carbopalladations and aminopalladations of alkynes, norbornene-mediated ortho C–H activations, and heteroatom-directed C–H activations. Therefore, the potential of this new synthetic strategy is limitless, and the combinations of newly discovered C–H activations with other types of reactions will certainly allow access to an even greater variety of complex molecules that would be otherwise be challenging and time consuming to obtain.

References

1. For recent reviews on C-H activation, see: (a) White, M.C. (2012) *Synlett*, **23**, 2746–2748. (b) Arockiam, P.B., Bruneau, C., and Dixneuf, P.H. (2012) *Chem. Rev.*, **112**, 5879–5918. (c) Kuhl, N., Hopkinson, M.N., Wencel-Delord, J., and Glorius, F. (2012) *Angew. Chem. Int. Ed.*, **51**, 10236–10254. (d) Yamaguchi, J.,

Yamaguchi, A.D., and Itami, K. (2012) *Angew. Chem. Int. Ed.*, **51**, 8960–9009. (e) Chen, D.Y.K. and Youn, S.W. (2012) *Chem. Eur. J.*, **18**, 9452–9474. (f) Song, G., Wang, F., and Li, X. (2012) *Chem. Soc. Rev.*, **41**, 3651–3678. (g) Cho, S.H., Kim, J.Y., Kwak, J., and Chang, S. (2011) *Chem. Soc. Rev.*, **40**, 5068–5083. (h) Zhang, S.Y., Zhang, F.M., and Tu, Y.Q. (2011) *Chem. Soc. Rev.*, **40**, 1937–1949. (i) Wencel-Delord, J., Dröge, T., Liu, F., and Glorius, F. (2011) *Chem. Soc. Rev.*, **40**, 4740–4761. (j) Yu, J.Q. and Shi, Z. (eds) (2010) *C-H Activation*, Topics in Current Chemistry, Vol. 292, Springer, Heidelberg. (k) Mkhalid, I.A.I., Barnard, J.H., Marder, T.B., Murphy, J.M., and Hartwig, J.F. (2010) *Chem. Rev.*, **110**, 890–931. (l) Giri, R., Shi, B.F., Engle, K.M., Maugel, N., and Yu, J.Q. (2009) *Chem. Soc. Rev.*, **38**, 3242–3272. (m) Alberico, D., Scott, M.E., and Lautens, M. (2007) *Chem. Rev.*, **107**, 174–238.

2. Ruck, R.T., Huffman, M.A., Kim, M.M., Shevlin, M., Kandur, W.V., and Davies, I.W. (2008) *Angew. Chem. Int. Ed.*, **47**, 4711–4714.
3. Piou, T., Neuville, L., and Zhu, J. (2012) *Org. Lett.*, **14**, 3760–3763.
4. Piou, T., Neuville, L., and Zhu, J. (2012) *Angew. Chem. Int. Ed.*, **51**, 11561–11565.
5. Satyanarayana, G., Maichle-Mössmer, C., and Maier, M.E. (2009) *Chem. Commun.*, 1571–1573.
6. Hu, Y., Song, F., Wu, F., Cheng, D., and Wang, S. (2008) *Chem. Eur. J.*, **14**, 3110–3117.
7. (a) Hu, Y., Yu, C., Ren, D., Hu, Q., Zhang, L., and Cheng, D. (2009) *Angew. Chem. Int. Ed.*, **48**, 5448–5451. (b) Hu, Y., Ren, D., Zhang, L., Lin, X., and Wan, J. (2010) *Eur. J. Org. Chem.*, **2010**, 4454–4459.
8. René, O., Lapointe, D., and Fagnou, K. (2009) *Org. Lett.*, **11**, 4560–4563.
9. Huang, Q., Fazio, A., Dai, G., Campo, M.A., and Larock, R.C. (2004) *J. Am. Chem. Soc.*, **126**, 7460–7461.
10. Lu, Z., Hu, C., Guo, J., Li, J., Cui, Y., and Jia, Y. (2010) *Org. Lett.*, **12**, 480–483.
11. Lu, Z., Cui, Y., and Jia, Y. (2011) *Synthesis*, **2011**, 2595–2599.
12. Pinto, A., Neuville, L., Retailleau, P., and Zhu, J. (2006) *Org. Lett.*, **8**, 4927–4930.
13. Zhao, J. and Larock, R.C. (2005) *Org. Lett.*, **7**, 701–704.
14. Pinto, A., Neuville, L., and Zhu, J. (2007) *Angew. Chem. Int. Ed.*, **46**, 3291–3295.
15. Chen, Z., Zhu, J., Xie, H., Li, S., Wu, Y., and Gong, Y. (2011) *Adv. Synth. Catal.*, **353**, 325–330.
16. Hu, Y., Yao, H., Sun, Y., Wan, J., Lin, X., and Zhu, T. (2010) *Chem. Eur. J.*, **16**, 7635–7641.
17. Hu, Y., Sun, Y., Hu, J., Zhu, T., Yu, T., and Zhao, Q. (2011) *Chem. Asian J.*, **6**, 797–800.
18. (a) Tietze, L.F., Düfert, A., Lotz, F., Sölter, L., Oum, K., Lenzer, T., Beck, T., and Herbst-Irmer, R. (2009) *J. Am. Chem. Soc.*, **131**, 17879–17884. (b) Tietze, L.F., Düfert, M.A., Hungerland, T., Oum, K., and Lenzer, T. (2011) *Chem. Eur. J.*, **17**, 8452–8461.
19. Tietze, L.F. and Lotz, F. (2006) *Eur. J. Org. Chem.*, **2006**, 4676–4684.
20. (a) Tietze, L.F., Hungerland, T., Düfert, A., Objartel, I., and Stalke, D. (2012) *Chem. Eur. J.*, **18**, 3286–3291. (b) Tietze, L., Hungerland, T., Depken, C., Maaß, C., and Stalke, D. (2012) *Synlett*, **23**, 2516–2520.
21. Tietze, L.F., Hungerland, T., Eichhorst, C., Düfert, A., Maaß, C., and Stalke, D. (2013) *Angew. Chem. Int. Ed.*, **52**, 3668–3671.
22. Parthasarathy, K., Han, H., Prakash, C., and Cheng, C.H. (2012) *Chem. Commun.*, **48**, 6580–6582.
23. Ohno, H., Yamamoto, M., Iuchi, M., and Tanaka, T. (2005) *Angew. Chem. Int. Ed.*, **44**, 5103–5106.
24. Gerfaud, T., Neuville, L., and Zhu, J. (2009) *Angew. Chem. Int. Ed.*, **48**, 572–577.
25. Piou, T., Neuville, L., and Zhu, J. (2013) *Tetrahedron*. doi: 10.1016/j.tet.2013.01.003
26. Catellani, M., Frignani, F., and Rangoni, A. (1997) *Angew. Chem., Int. Ed. Engl.*, **36**, 119–122.
27. Martins, A., Mariampillai, B., and Lautens, M. (2010) *Top. Curr. Chem.*, **292**, 1–33.

28. Thansandote, P. and Lautens, M. (2009) *Chem. Eur. J.*, **15**, 5874–5883.

29. Catellani, M., Motti, E., and Della Ca, N. (2008) *Acc. Chem. Res.*, **41**, 1512–1522.

30. Chiusoli, G.P., Catellani, M., Costa, M., Motti, E., Della Ca, N., and Maestri, G. (2010) *Coord. Chem. Rev.*, **254**, 456–469.

31. Motti, E., Della Ca, N., Deledda, S., Fava, E., Panciroli, F., and Catellani, M. (2010) *Chem. Commun.*, **46**, 4291–4293.

32. Gericke, K.M., Chai, D.I., and Lautens, M. (2008) *Tetrahedron*, **64**, 6002–6014.

33. Aureggi, V., Davoust, M., Gericke, K., and Lautens, M. (2009) *Synlett*, 1004–1008.

34. Jafarpour, F. and Hazrati, H. (2010) *Adv. Synth. Catal.*, **352**, 363–367.

35. Candito, D.A. and Lautens, M. (2009) *Angew. Chem. Int. Ed.*, **48**, 6713–6716.

36. Blanchot, M., Candito, D.A., Larnaud, F., and Lautens, M. (2011) *Org. Lett.*, **13**, 1486–1489.

37. Maestri, G., Larraufie, M.H., Derat, E., Ollivier, C., Fensterbank, L., Lacôte, E., and Malacria, M. (2010) *Org. Lett.*, **12**, 5692–5695.

38. Ca, N.D., Motti, E., Mega, A., and Catellani, M. (2010) *Adv. Synth. Catal.*, **352**, 1451–1454.

39. Candito, D.A. and Lautens, M. (2010) *Org. Lett.*, **12**, 3312–3315.

40. Motti, E., Ca, N.D., Xu, D., Piersimoni, A., Bedogni, E., Zhou, Z.M., and Catellani, M. (2012) *Org. Lett.*, **14**, 5792–5795.

41. (a) Feringa, B.L. (2007) *J. Org. Chem.*, **72**, 6635–6652. (b) Feringa, B.L., van Delden, R.A., Koumura, N., and Geertsema, E.M. (2000) *Chem. Rev.*, **100**, 1789–1816. (c) Ruangsupapichat, N., Pollard, M.M., Harutyunyan, S.R., and Feringa, B.L. (2011) *Nat. Chem.*, **3**, 53–60.

42. (a) Feringa, B.L. and Wynberg, H. (1977) *J. Am. Chem. Soc.*, **99**, 602–603. (b) Jager, W.F., de Jong, J.C., de Lange, B., Huck, N.P.M., Meetsma, A., and Feringa, B.L. (1995) *Angew. Chem. Int. Ed Engl.*, **34**, 348–350. (c) Feringa, B.L., Jager, W.F., de Lange, B., and Meijer, E.W. (1991) *J. Am. Chem. Soc.*, **113**, 5468. (d) Wang, J. and Feringa, B.L. (2011) *Science*, **331**, 1429–1432.

43. Gericke, K.M., Chai, D.I., Bieler, N., and Lautens, M. (2009) *Angew. Chem. Int. Ed.*, **48**, 1447–1451.

44. Liu, H., El-Salfiti, M., Chai, D.I., Auffret, J., and Lautens, M. (2012) *Org. Lett.*, **14**, 3648–3651.

45. Liu, H., El-Salfiti, M., and Lautens, M. (2012) *Angew. Chem. Int. Ed.*, **51**, 9846–9850.

46. (a) Ackermann, L. (2013) *Acc. Chem. Res.* doi: 10.1021/ar3002798 (b) Mei, T.S., Kou, L., Ma, S., Engle, K., and Yu, J.Q. (2012) *Synthesis*, **44**, 1778–1791. (c) Colby, D.A., Tsai, A.S., Bergman, R.G., and Ellman, J.A. (2012) *Acc. Chem. Res.*, **45**, 814–825. (d) Zhu, C., Wang, R., and Falck, J.R. (2012) *Chem. Asian J.*, **7**, 1502–1514. (e) Colby, D.A., Bergman, R.G., and Ellman, J.A. (2010) *Chem. Rev.*, **110**, 624–655. (f) Lyons, T.W. and Sanford, M.S. (2010) *Chem. Rev.*, **110**, 1147–1169. (g) Chen, X., Engle, K.M., Wang, D.H., and Yu, J.Q. (2009) *Angew. Chem. Int. Ed.*, **48**, 5094–5115.

47. Lian, Y., Bergman, R.G., and Ellman, J.A. (2012) *Chem. Sci.*, **3**, 3088–3092.

48. Shi, X. and Li, C.J. (2012) *Adv. Synth. Catal.*, **354**, 2933–2938.

49. Pham, M.V., Ye, B., and Cramer, N. (2012) *Angew. Chem. Int. Ed.*, **51**, 10610–10614.

50. Tran, D.N. and Cramer, N. (2011) *Angew. Chem. Int. Ed.*, **50**, 11098–11102.

4
Domino Reactions Initiated by Nucleophilic Substitution

Hiriyakkanavar Ila, Anand Acharya, and Saravanan Peruncheralathan

Domino reactions are defined as two or more bond forming transformations under identical reaction conditions, in which the latter transformations take place at the functionalities formed in the former bond forming reactions. Introduction of the domino concept in organic chemistry has greatly improved the efficiency of syntheses, since it not only allows the synthesis of complex compounds from simple substrates in a short way, but has also ecological and economic advantages, minimizing waste and saving labor, as well as our resources [1].

In this chapter, the class of domino reactions that are initiated by a nucleophilic substitution, as the first step, are highlighted. These reactions are classified as anionic processes, independently of whether a carbocation is the intermediate as the second substrate (e.g., S_N1 reaction) [1b]. The primary step in this process is the attack of a nucleophilic species, either an anion (e.g., a carbanion, an enolate, alkoxide, or thiolate) or a "pseudo" anion, as an uncharged nucleophile (e.g., an amine or an alcohol) on to an electrophilic center bearing a leaving group. A bond formation takes place with the creation of a new "real" or "pseudo-anionic" functionality, which can undergo further transformations, by being trapped by an electrophilic center present in the molecule (conjugated enone, aldehyde, epoxide, etc.). The sequence can then be terminated either by the addition of a proton or by the elimination of an X group.

In the first part of the chapter, we have covered mostly those reactions, initiated by S_N2 type nucleophilic substitution, followed by a Michael reaction on a α,β-unsaturated ketone moiety, present in the intermediate adduct (S_N/Michael). A few examples of the reactions with carbonyl compounds and alkyl halides (twofold S_N reactions) in the second step are also included.

The large numbers of domino reactions, which are initiated by nucleophilic substitution, are those, involving nucleophilic ring opening of three membered rings, for example, epoxides, aziridines, and a few activated cyclopropanes. These strained three membered compounds undergo facile ring opening by various nucleophiles, resulting in the generation of another nucleophilic species (e.g., alkoxide or amine), which can react further inter- or intramolecularly, undergoing a variety of reactions, including transition metal catalyzed cross-coupling reactions. In this context, it should be noted that it is of no relevance, whether the process

Domino Reactions: Concepts for Efficient Organic Synthesis, First Edition. Edited by Lutz F. Tietze.

involves a nucleophilic or electrophilic ring opening of an epoxide or aziridine and whether a carbocationic intermediate is of formal or real nature. Thus, the second part of the chapter deals with those domino reactions involving nucleophilic ring opening of epoxides, aziridines, and cyclopropyl ketones as the first step.

The third section of the chapter describes recently developed "anion relay chemistry" (ARC) involving threefold domino S_N-Brook rearrangement/S_N reactions, discovered by Tietze and extensively explored by Amos Smith III and his group. These reactions are also initiated by nucleophilic ring opening of epoxides, usually by silyl-substituted dithiane anions, and find useful application in natural product synthesis.

4.1 Domino S_N/Michael Addition and Related Reactions

The earlier domino reactions initiated by nucleophilic substitution, mostly involve substitution on an alkyl halide or its variants by carbon- or hetero-nucleophiles (amines, alcohols, phenols, thiols, etc.) and subsequent Michael addition of regenerated carbon or hetero-nucleophiles on an enone functionality present in the alkyl halide precursor, thus resulting in the formation of cyclic, bicyclic, and sometimes polycyclic compounds [2].

Gharpure and coworkers [3] have earlier reported a stereoselective synthesis of 1,2,2-trisubstituted indane derivatives employing a domino S_N2/Michael sequence of *o*-halomethyl cinnamates with various active methylene compounds. Following the similar strategy, Chiba and coworkers [4] have developed a stereocontrolled synthesis of highly substituted cyclopentanes as shown in Scheme 4.1. Thus ethyl (*E*)-4-substituted 6-bromo-2-hexenoates of the type **1** are reacted with a series of active methylene compounds in the presence of K_2CO_3 at room temperature in *N*,*N*-dimethylformamide (DMF), to yield a variety of 1,2,3-trisubstituted cyclopentanes, such as **2** bearing three successive stereogenic centers with varying diastereoselectivities, whereas in the reactions with (phenylsulfonyl)acetonitrile ($X = SO_2Ph$), only one diastereomer is obtained (**2d–f**). Similar reactions of (*E*)-4,5-disubstituted 6-bromo-2-hexenoate **3** with substituted malononitriles, provide tetrasubstituted cyclopentanes **4a,b** in high yields with high diastereoselectivity through an S_N/Michael domino reaction (Scheme 4.1). A few of these polysubstituted cyclopentanes were further derivatized to bicyclic lactones such as **5** through a series of transformations.

Desmaele and d'Angelo have reported a new spiroannulation reaction based on a domino alkylation/Michael addition of 2-tetralones **6** with iodoalkenoate **7**, allowing the efficient access to spiroketoesters **8a,b** in a highly stereoselective manner (Scheme 4.2) [5, 6]. Thus treatment of 2-tetralone **6a** with **7** in the presence of cesium carbonate in DMF at 0 °C furnishes the spiroketoester **8a** in 66% yield with high diastereoselection. The reaction is shown to proceed through an initial alkylation of tetralone enolate followed by intramolecular Michael addition of resulting ketoester **9**. In a similar manner, condensation of 6,7-dimethoxy-2-tetralone

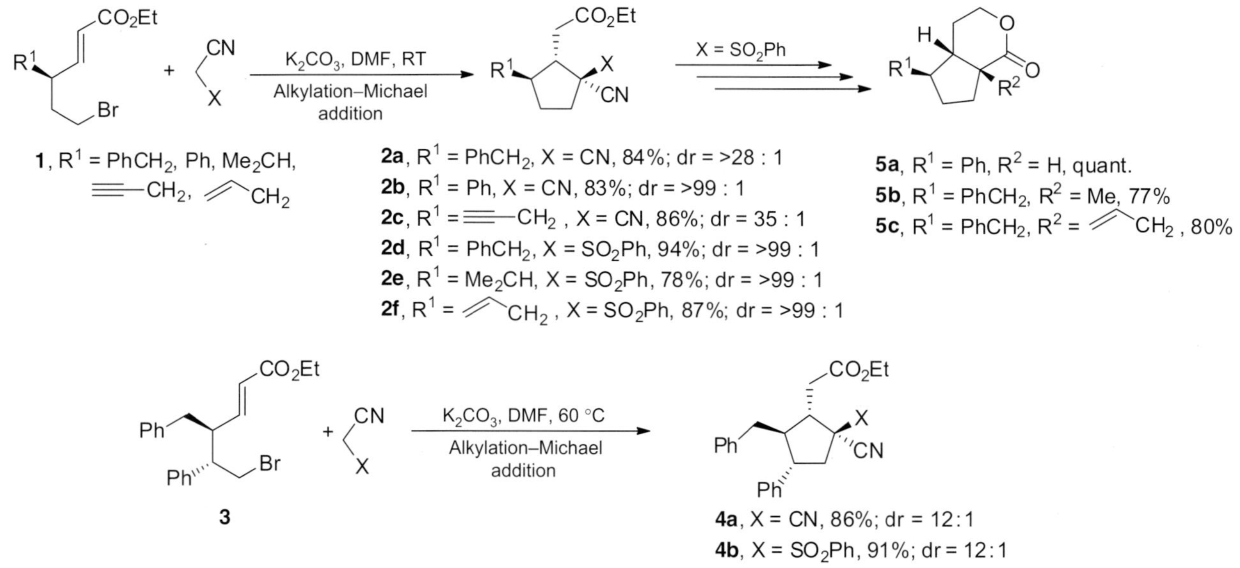

Scheme 4.1 Stereoselective syntheses of highly substituted cyclopentanes through domino S_N2/Michael addition.

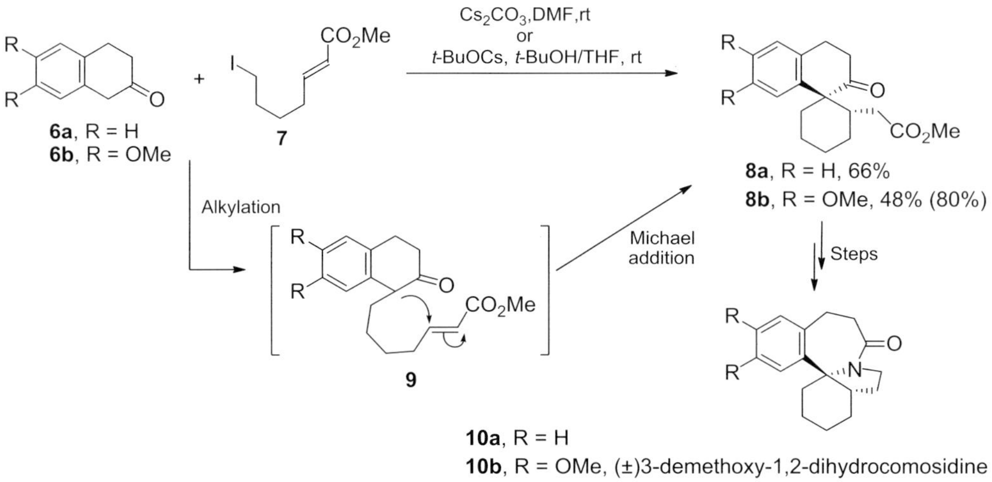

Scheme 4.2 Syntheses of homoerythrina alkaloids by domino alkylation/Michael addition.

6b with iodide **7** gave spiro compound **8b** in 48% yield, which could be subsequently increased to 80% by using cesium *t*-butoxide as a base in *t*-BuOH/THF (tetrahydrofuran) solvent [5]. The ketoesters **8a,b** were found to be useful precursors for Homoerythrina alkaloids such as **10**. Thus, (±)-3-demethoxy-1,2-dihydrocomosidine **10b** was obtained in good yield in four steps form ketoesters **8b**, involving a Schmidt rearrangement as the final step (Scheme 4.2) [6].

Watanabe and coworkers have developed an efficient synthesis of *trans*-decalin core **13** of the natural product (−)-tanabalin starting from enantiopure iodoalkane **11**, involving a domino alkylation/intramolecular Robinson annulation sequence (Scheme 4.3) [7a]. Thus, the alkylation of enolate of ketoester **12** with alkyl iodide **11** initially forms the intermediate **14**, which then undergoes an intramolecular Michael addition furnishing the intermediate cyclohexane derivative **15** containing four contiguous stereocenters. Subsequent cyclization of **15** through an aldol condensation affords the *trans*-decalin core **13** in excellent yield. This highly efficient domino process culminating in the formation of three new C–C bonds, is used by the same workers as part of a stereoselective total synthesis of (−)-tanabalin (**16**) [7a], displaying potent activity against pink bollworm *Pectinophora gossypiella*, a severe cotton pest (Scheme 4.3) [7b].

Gharpure and coworkers have recently utilized the domino S_N/Michael protocol for the construction of 2,3,3,6-tetrasubstituted tetrahydropyrans **18** in a highly diastereoselective manner, by reaction of various active methylene compounds with vinylogous carbonates **17** in the presence of cesium carbonate as base (Scheme 4.4) [8]. Symmetrical active methylene compounds such as diethyl malonate or malononitrile as well as unsymmetrical active methylene compounds such as sulfone nitrile, sulfone ester, or ethyl cyanoacetate gave good results and interestingly, better diastereoselectivities were obtained, when using unsymmetrical nucleophiles rather than symmetrical ones (de > 90%). The scope of this domino process was

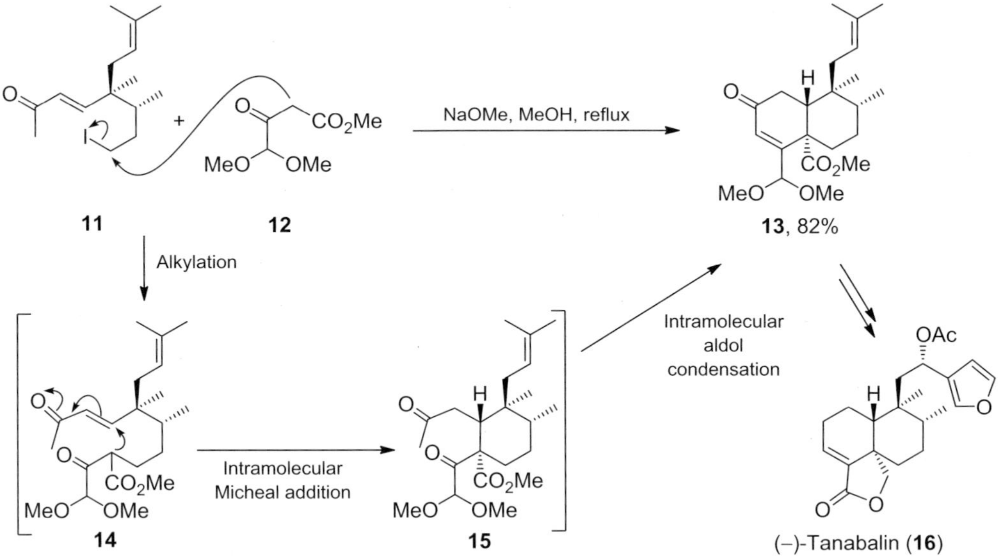

Scheme 4.3 Watanabe's synthesis of *trans*-decalin core of (−)-tanabalin by domino intermolecular alkylation/intramolecular Robinson annulation.

extended further for the synthesis of fused bicyclic bistetrahydropyrans such as **21**, starting from tetrahydropyranyl iodide **20** (derived from tri-*O*-acetyl-(D)-glucal), which on reaction with sulfone nitrile gave the corresponding **21** in 64% yield and with excellent relative diastereoselectivity of >90% de (Scheme 4.4).

These workers have further extended the domino S_N2/Michael approach for the synthesis of 2,3,3,5-tetrasubstituted tetrahydrofurans such as **23**, although with moderate diastereoselectivities, by reacting vinylogous carbonate **22** with active methylene compounds in the presence of potassium carbonate as base (Scheme 4.5) [9]. The method was further elaborated for the synthesis of diversely substituted bis(tetrahydrofuran) derivative such as **25**, starting from iodovinylogous carbonate **24** (prepared from D-glucose) (Scheme 4.5). It should be noted that adjacent bis(tetrahydrofuran) motif is present in a variety of natural products belonging to acetogenin family [10].

Srikrishna and coworkers have synthesized bicyclo[3.3.1]nonane ring system, as useful synthons for the construction of various natural products, by domino S_N/Michael process using allyl bromide derived from (*R*)-carvone [11]. Bunce has employed S_N/Michael domino reactions for the synthesis of N-protected pyrrolidine and piperidine derivatives bearing functionalized side chain at C-2 [12]. By employing domino S_N/Michael strategy, Goff and coworkers have developed a solid phase synthesis of peptoid based 2-oxopiperazine library with a possibility of introducing a variety of substituents at 1,3,4,5-positions of piperazine ring [13].

In addition to the presented S_N/Michael transformations, a twofold domino S_N/S_N' cycloalkylation of stabilized carbanions from cyclic β-ketoesters with 1,4-dibromobut-2-yne yielding annulated 2-vinylidenehydrofurans has been reported

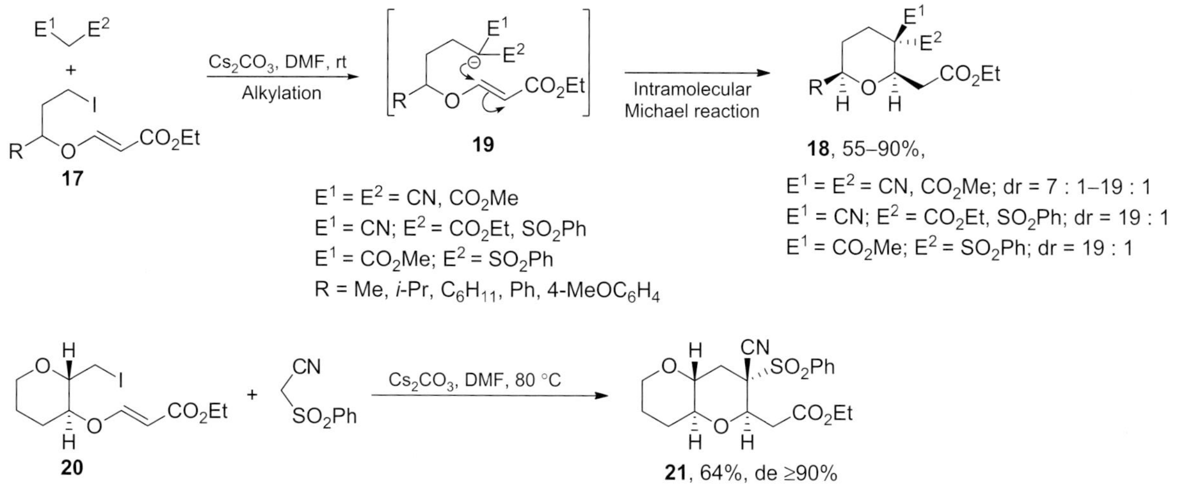

Scheme 4.4 Synthesis of substituted tetrahydropyrans by domino S_N2/Michael reaction.

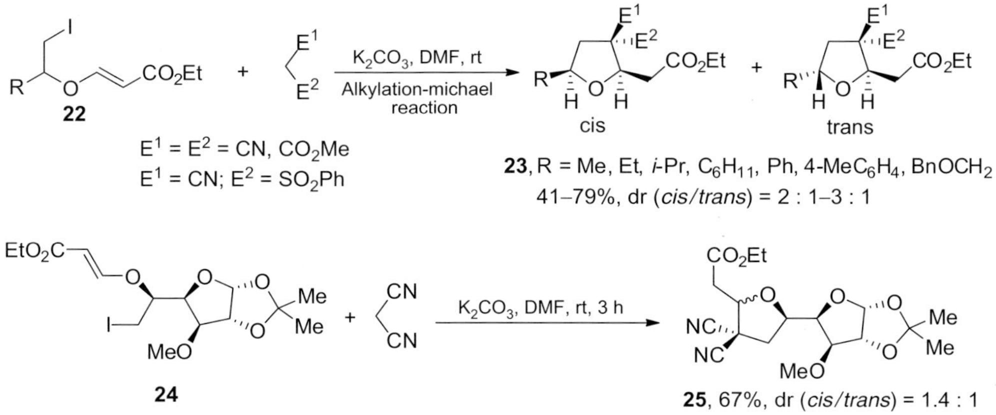

Scheme 4.5 Stereoselective synthesis of 2,3,3,5-tetrasubstituted tetrahydrofurans by domino S_N2/Michel reaction.

by Rodriguez and coworkers [14]. The same group has also described a domino S_N/retro-Dieckmann condensation, applicable for the construction of substituted cycloheptanes as well as cyclooctanes [15]. A number of S_N-domino reactions have been employed in the synthesis of natural products by Holton [16], Takikawa *et al.* [17], Barrero *et al.* [18a] as well as by Pandey and coworkers [18b].

A twofold S_N' type domino organocatalytic reaction for the synthesis of γ-butenolides **28** has been reported by Krische and coworkers [19], by treatment of Morita–Baylis–Hillman acetates **26** with trimethylsilyloxyfuran (**27**) in the presence of triphenylphosphine to afford **28** in yields up to 94% and diastereoselectivities of >95 : 5 dr (Scheme 4.6).

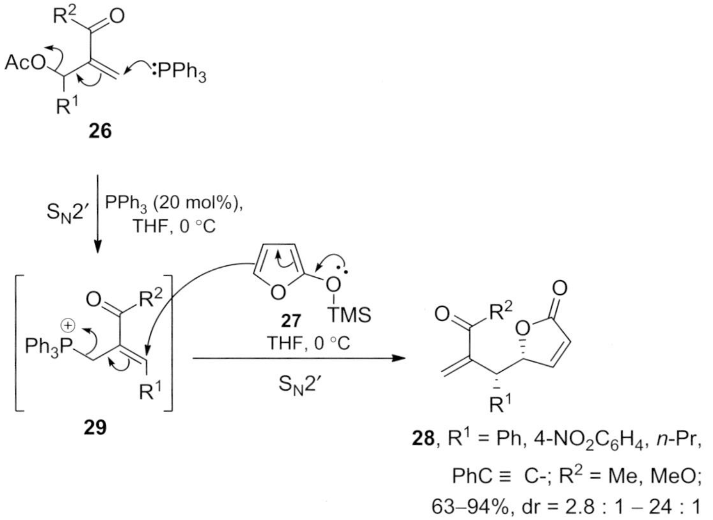

Scheme 4.6 Synthesis of γ-butenolide through a twofold S_N' domino process.

Langer and coworkers have employed a twofold domino alkylation process for the synthesis of functionalized 2-alkylidene-5-hydroxymethyltetrahydrofurans **32** involving base-induced reaction of dianions of β-ketoesters or 1,3-diketones **30** with epibromohydrin **31** (Scheme 4.7) [20]. The overall reaction proceeds by initial attack of dianion **33** at bromomethyl group of **31** and subsequent intramolecular 5-exo ring opening/ cyclization by nucleophilic attack of the alkoxide ion in the resulting intermediate **34**, leading to the products **32**. The reaction is found to be quite general and both acyclic and cyclic 1,3-dicarbonyl compounds could be transformed into the corresponding hydroxymethyltetrahydrofurans in moderate to good yields with only few exceptions (Scheme 4.7).

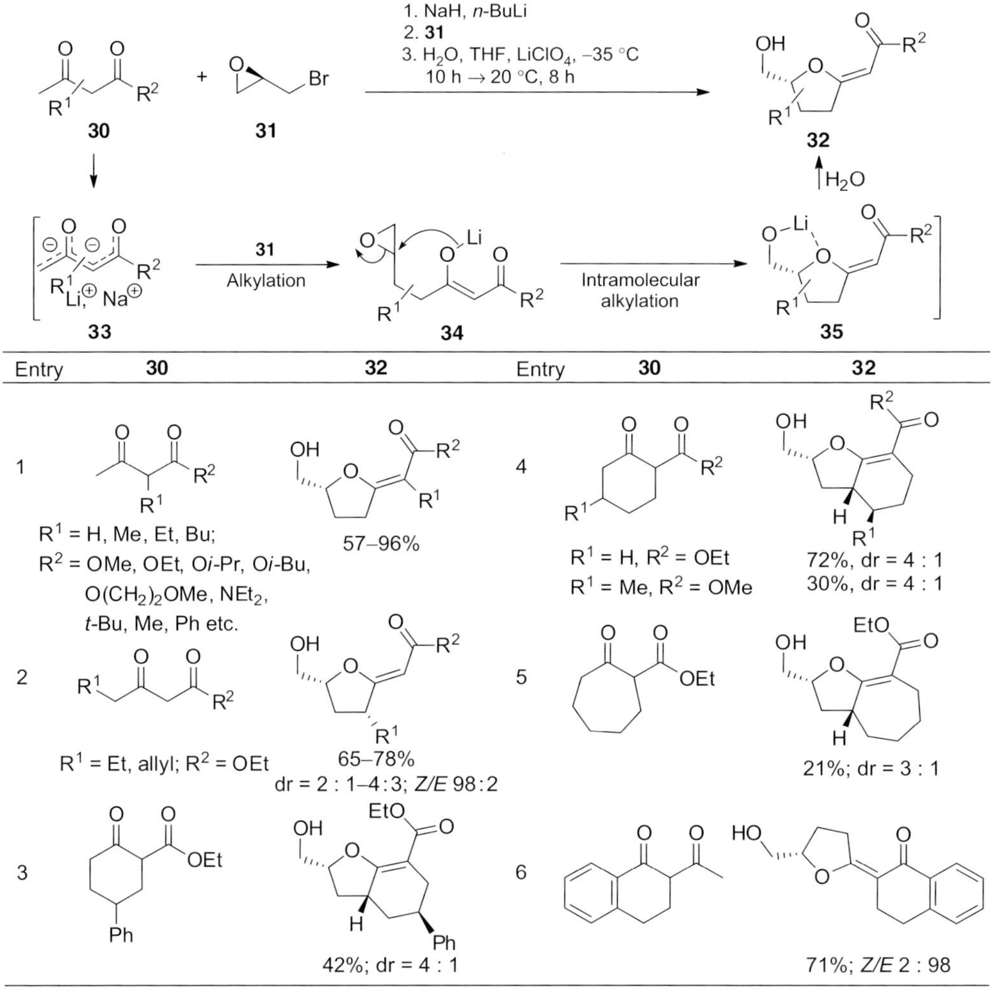

Scheme 4.7 Synthesis of 2-alkylidene-5-hydroxymethyltetrahydrofurans through a twofold domino S_N reaction of epoxides **31** with 1,3-dicarbonyl compounds.

Huang and coworkers have reported synthesis of substituted benzoxepinones **38** via a twofold S_N/Wittig reaction, as shown in the Scheme 4.8 [21]. Thus, a phenolate, generated by treatment of salicylaldehyde (**36**) with sodium ethoxide, undergoes alkylation with chloroacetonyl phosphorane **37** to give intermediate **39**, which on intramolecular Wittig reaction affords the benzoxepinone **38** in varying yields. The versatility of the above domino S_N2/Wittig reaction is further demonstrated by its application in the total synthesis of (*Z*)-pterulinic acid, a natural product isolated from fermentation of the *Pterula* sp. 82168 [22].

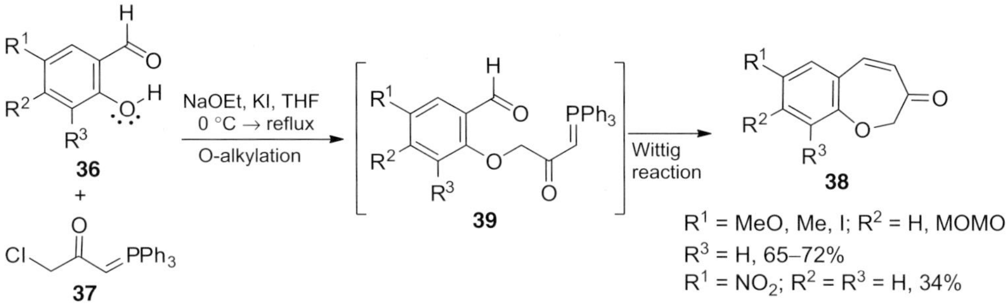

Scheme 4.8 Synthesis of benzoxipinones through a domino S_N/Wittig reaction of salicylaldehyde and phosphorane.

An interesting threefold anionic domino process initiated by a S_N2'-type reaction, has been reported by Fujimoto and coworkers [23], involving reaction of cyclic oxosulfonium ylide **40** with Morita–Baylis–Hillman adducts **41**, furnishing cycloheptene oxide derivatives **42** in varying yields (Scheme 4.9). With unsubstituted α-methylene-β-acetoxy ketone **41** ($R^1 = Me$; $R^2 = H$), in the presence of 2 equiv of

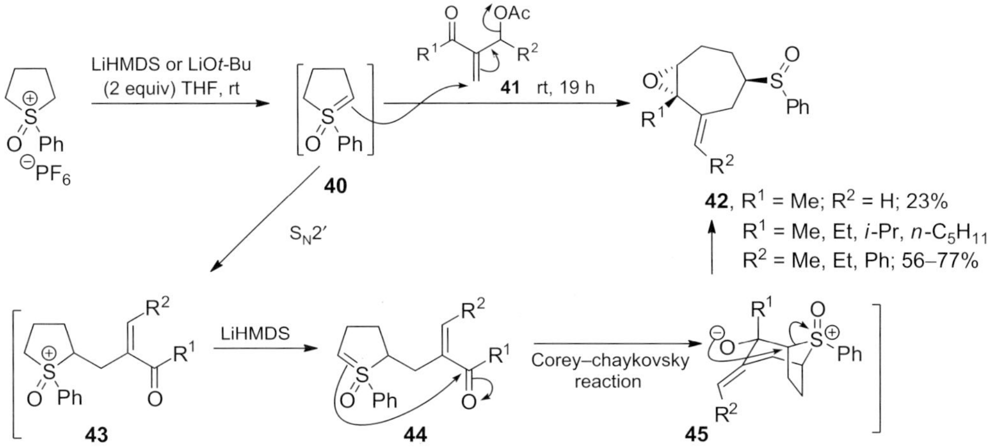

Scheme 4.9 Domino S_N'/intramolecular Corey–Chaykovsky reaction of cyclic oxosulfonium ylide with Baylis–Hillman adducts.

LiHMDS, the cycloheptene oxide derivative **42** was obtained in lower yield (23%), whereas with substituted **41** and LiO*t*-Bu as base, yields of **42** could be greatly enhanced up to 77% ($R^1 = i$-Pr, $R^2 = $ Ph) [23b]. The cycloheptene oxides **42** are considered to be formed through S_N2' substitution of ylide **40** on **41** and subsequent intramolecular Corey–Chaykovsky reaction of the resulting ylide **44** through intermediate **45** (Scheme 4.9).

Chandrasekaran and coworkers [24] have demonstrated in a series of papers, that benzyltriethylammonium tetrathiomolybdate ($[PhCH_2NEt_3]_2MoS_4$) (**46**) is a useful sulfur transfer reagent in domino reactions (Schemes 4.10 and 4.11). It has been suggested that the initially formed disulfide, by treatment of an alkyl halide with tetrathiomolybdate **46**, undergoes a reductive cleavage at higher temperature to produce alkyl thiolate and $Mo_2S_8^{2-}$, involving an induced redox reaction (Eqs. (1) and (2) in Scheme 4.10) [25, 26]. Using this strategy, Chandrasekaran's [25] group has synthesized thiabicyclo[3.3.1]nonane skeleton through a threefold domino nucleophilic sulfur transfer/reduction/Michael addition by treatment of allylic halide **47** with 2.2 equiv of molybdate reagent **46** in acetonitrile at room temperature, thus furnishing the thiabicyclic ketone **48** in 98% yield (Scheme 4.10).

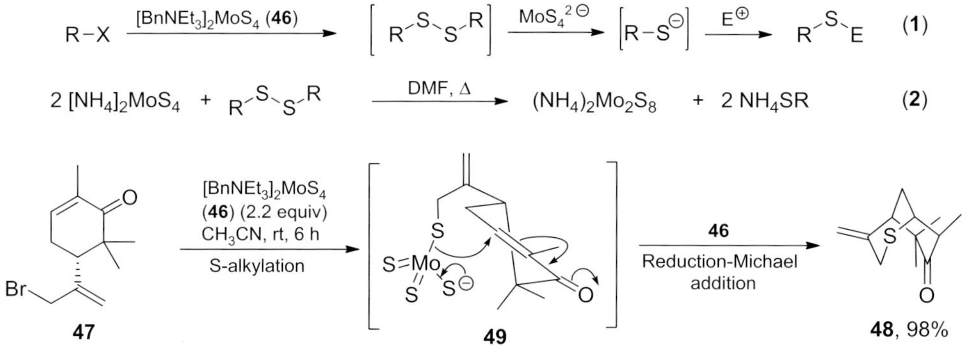

Scheme 4.10 Domino S_N'/reduction/Michael addition mediated by tetrathiomolybdate **46**.

These workers have further employed an analogous strategy for the efficient synthesis of 1-deoxythioglyconojirimycins such as **52** (Scheme 4.11) [27]. Thus, 2,6-dibromo-2,6-dideoxy-D-idono-1,4-lactone (**50**), on reaction with benzyltriethylammonium tetrathiomolybdate (**46**) in dimethyl sulfoxide (DMSO) at room temperature, smoothly furnishes the axially enriched bicyclic 1-deoxy-5-thio-D-glucopyrano-3,6-lactone (**51**) in 56% yield. Subsequent reduction of the lactone **51** with borohydride exchange resin (BER) in methanol gives 1-deoxythionojirimycin **52** in good yield (Scheme 4.11). The intermediate lactone **51** is supposed to be formed by reductive cleavage of initially formed disulfide intermediate **53**, followed by an intramolecular nucleophilic displacement in the resulting thiolate intermediate **54** (Scheme 4.11) [27].

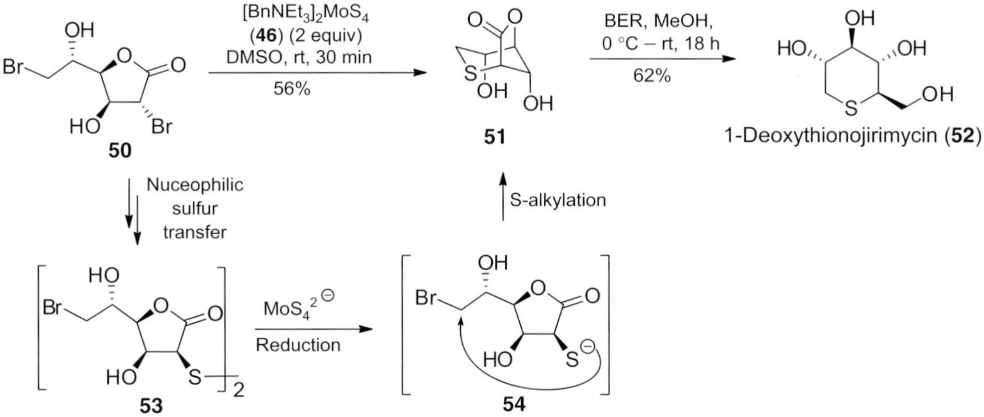

Scheme 4.11 Synthesis of 1-deoxythionojirimycin by domino nucleophilic substitution mediated by tetrathiomolybdate.

Bao and coworkers [28] have reported efficient synthesis of 2*H*-1,4-benzoxazin-3-(4*H*)-ones and quinoxalin-2-(1*H*)-ones via domino S_NAr/copper-catalyzed coupling of *o*-halophenols or *N*-(2-halophenyl)methylsulfonamides with 2-haloamides. Dai and coworkers [29] have also developed a general base mediated synthesis of 2-alkyl-3,4-dihydro-3-oxo-2*H*-1,4-benzoxazine by domino S_N condensation of 2-aminophenols with 2-bromoalkanoates under controlled microwave heating.

4.2 Domino Reactions Initiated by Nucleophilic Ring Opening of Aziridines, Epoxides, and Activated Cyclopropanes

Copper-catalyzed domino ring opening/cyclization of cyclic aziridines with *o*-halophenols leading to benzoxazines has been described by Sekar [30] and independently by Ranu [31] and coworkers (Scheme 4.12). Thus, 2,3-annulated 2,3-dihydro-2*H*-1,4-benzoxazines **57** are obtained in high yields by stereoselective ring opening of aziridines **55** with substituted *o*-iodophenol **56** followed by copper-catalyzed coupling cyclization of the resulting intermediate **58** (Scheme 4.12) [30, 31]. Ranu and coworkers have also demonstrated efficient domino ring opening/cyclization of acyclic aziridines with a number of substituted *o*-iodophenols in the presence of alumina-supported copper(II) catalyst, which proceeds in highly regioselective manner and a sterically congested 2,5,8-trisubstituted benzoxazine such as **60** carrying a sensitive aldehyde functionality could also be obtained in 88% yield, following this procedure [31]. Formation of tetracyclic benzoxazine **62** in high yield, as a single regioisomer, from the reaction of *o*-iodophenol and indene-fused aziridine **61** is also noteworthy (Scheme 4.12) [31].

A domino ring opening/carboxamidation of *N*-tosylaziridines leading to the synthesis of 1,4-benzo- and pyridoxazepinones has been recently described by Alper

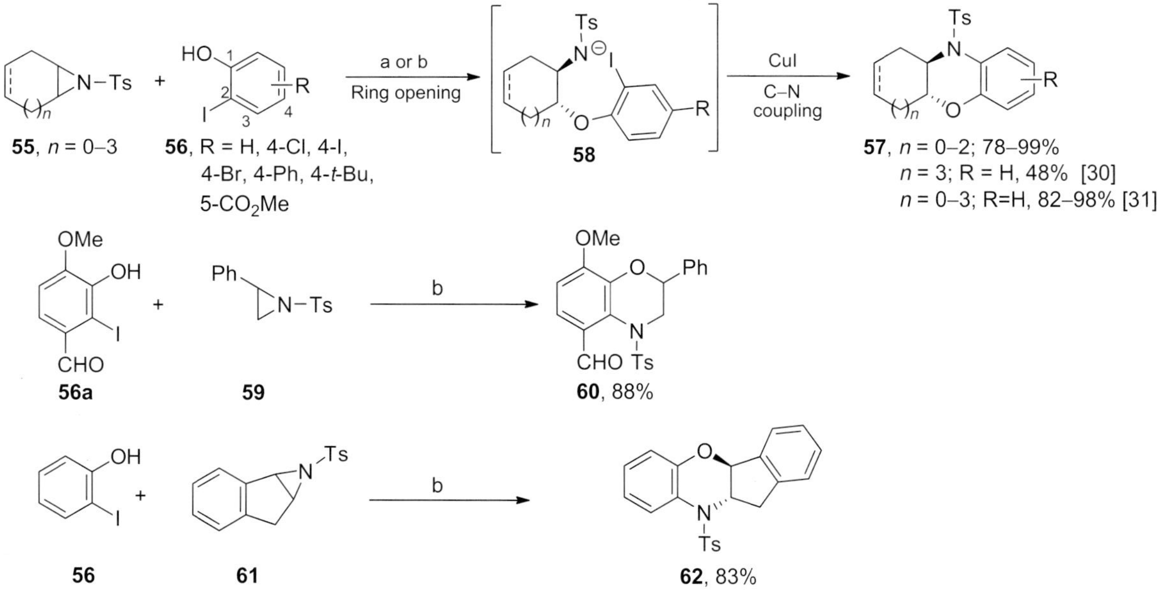

a = CuI (2.5 mol%), $H_2NCH_2CH_2NH_2$ (5 mol%), K_2CO_3 (2 equiv), DMF, 110 °C [30]

b = Cu/Al_2O_3 (4 mol%), K_2CO_3, DMF, 100 °C [31]

Scheme 4.12 Copper-catalyzed domino ring opening/cyclization of cyclic and acyclic aziridines with *o*-iodophenols.

and coworkers [32] (Scheme 4.13). Thus, various five- to seven-membered cyclic *N*-tosylaziridines **55** were reacted with substituted *o*-halophenols and pyridinols **63** in the presence of $PdCl_2/PPh_3$ catalyst and carbon monoxide at 200 psi, to yield fused 1,4-benzoxazepinones (X = CH) and the corresponding 1,4-pyridoxazepinone (X = N) **64** in good yields. Carboxamidation of seven-membered cyclic aziridine **55** ($n = 2$) required higher temperature (110 °C) and 400 psi carbon monoxide pressure for the formation of cyclic benzoxazepinone **64** ($n = 2$, X = CH, Y = Ph) in 72% yield. The protocol could also be extended to acyclic *N*-tosylaziridines **65**, which on reaction with **63** afforded the corresponding 1,4-benzo- and pyridoxazepinones **66** in 64–92% yield under similar conditions, whereas the disubstituted aziridine **65** ($R^1 = R^2$ = Et) required higher temperature and pressure for its conversion to the corresponding benzoxazepinone in moderate yield (Scheme 4.13).

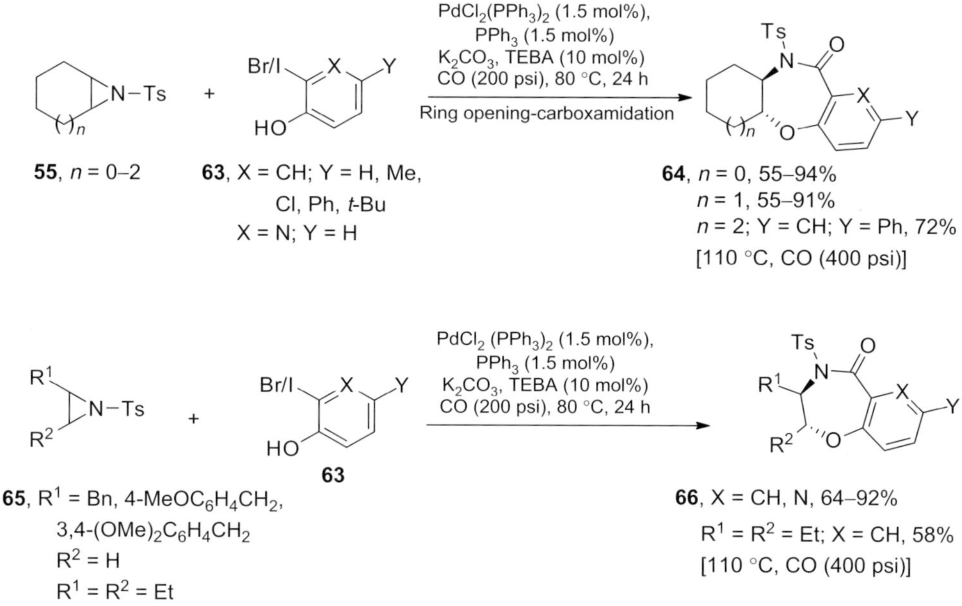

Scheme 4.13 Domino ring opening/carboxamidation of *N*-tosylaziridines with 2-halophenols/pyridinols.

An interesting threefold S_N-domino reaction for the synthesis of biologically important 1,4-benzodiazepine derivatives **69** has been developed by Wang and coworkers [33] from a one-pot reaction of methyl 1-arylaziridine-2-carboxylates **68** with *N*-[2-bromomethyl(aryl)]trifluoroacetamides **67** (Scheme 4.14). The reaction proceeds through initial N-benzylation of the aziridine and a regioselective ring opening of the resulting aziridinium ion **70** by bromide anion to give α-bromo-β-benzylaminopropionate intermediate **71** as the major product (route a), which on Et_3N mediated intramolecular nucleophilic displacement of bromide by amide nitrogen affords substituted 1,4-benzodiazepines **69** in moderate to good yields,

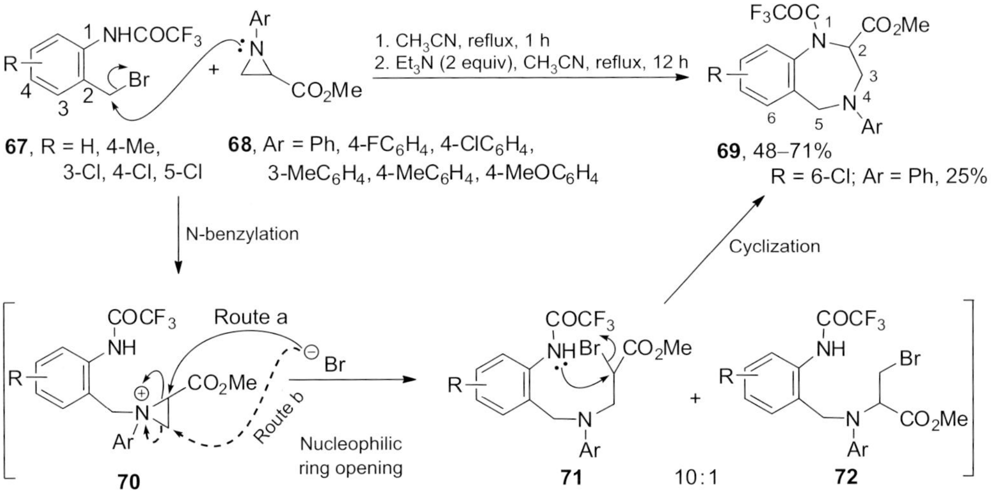

Scheme 4.14 Synthesis of 1,4-benzodiazepines through domino N-alkylation/ring opening/cyclization reaction.

whereas with 3-chloro derivative **67** (R = 3-Cl), the corresponding benzodiazepine was obtained only in 25% yield, probably due to unfavorable steric effects. Interestingly, the primary bromide **72** formed as minor product (through route b) does not undergo intramolecular cyclization under these conditions, which is due to lower reactivity of nonactivated primary alkyl bromide (Scheme 4.14).

A highly regioselective domino nucleophilic ring opening/cyclization of 2-(bromomethyl)aziridine **73** with *o*-aminothiophenol leading to 3-sulfonamido-2,3,4,5-tetrahydro-1,5-benzothiazepines **75** has been recently described by Kimpe and coworkers [34] (Scheme 4.15). The product benzothiazepines **75** are apparently formed by attack of thiophenolate anion (from **74**) on the less hindered carbon atom of the aziridine followed by intramolecular nucleophilic displacement of bromide in the intermediate **76**, which was confirmed by NMR spectroscopy. Interestingly, the reaction of 2-aminothiophenol with trans-2,3-substituted aziridine tosylate **77a** (R = 4-MeC_6H_4) in the presence of K_2CO_3 affords 4-substituted 3-aminobenzothiazepine **78**, whereas the 2-phenylaziridine **77b** with a mesylate leaving group (R = Me) was transformed into 2-phenyl-3-aminobenzothiazepine **79** under similar conditions (Scheme 4.15). Although the yields of products **78** and **79** are moderate, the authors claim that no traces of the other regioisomers were detected in either of the reaction mixtures. It has been suggested that tosylate group being more reactive in comparison to mesylate, results in initial nucleophilic displacement of tosyloxy group in **77a** by thiophenolate anion to give intermediate **80**, which cyclizes spontaneously via intramolecular ring opening of the aziridine to yield thiazepine **78** (pathway a). On the other hand, 2-(mesyloxymethyl)aziridine **77b** suffers from initial ring opening by 2-aminothiophenol at the benzylic position to give intermediate **81** followed by cyclization through intramolecular nucleophilic substitution of mesyloxy moiety to afford benzothiazepine **79** (Scheme 4.15).

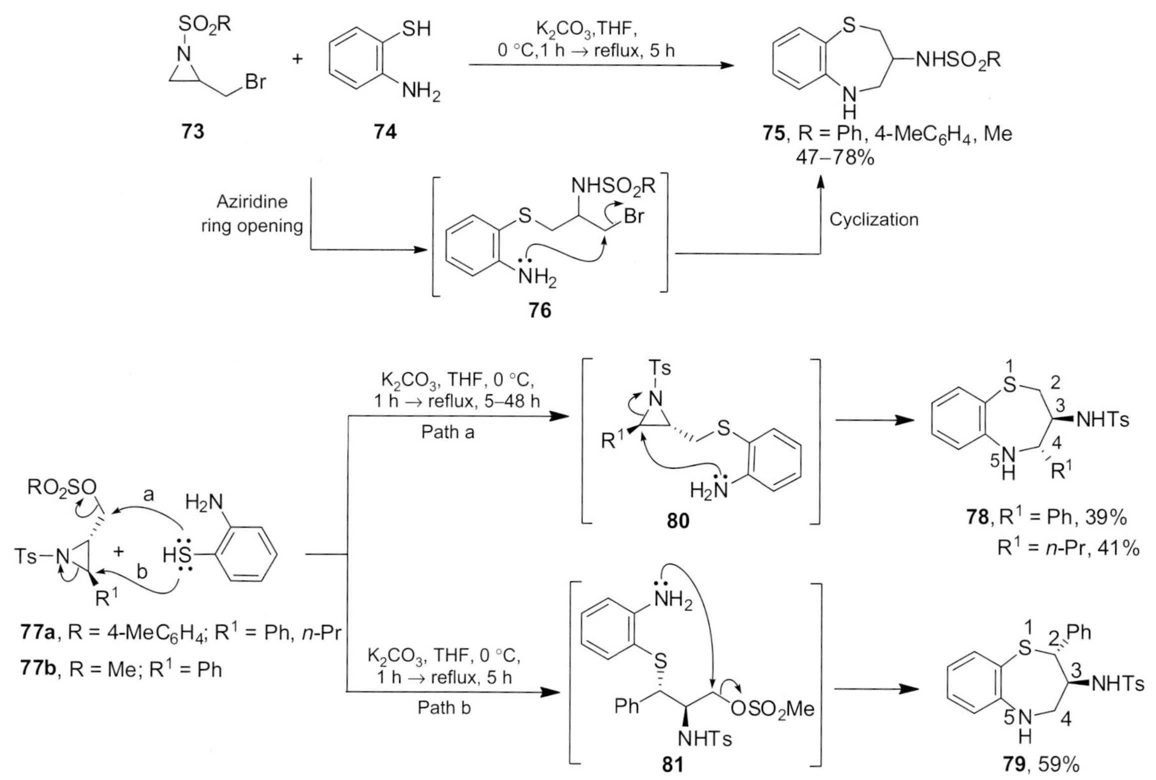

Scheme 4.15 Regio and stereocontrolled synthesis of substituted 2,3,4,5-tetrahydro-1,5-benzothiazepines through domino S_N reaction.

Substituted *N*-tosylaziridines **82** have been shown to undergo domino ring opening/cyclization with arylpropargyl alcohols **83** in the presence of potassium *t*-butoxide as base, affording dihydrooxazine derivatives **84** in moderate to good yields (Scheme 4.16) [35]. A plausible reaction mechanism involving isomerization of the ring-opened intermediate **85** to allene **86** and its subsequent intramolecular cyclization to give dihydrooxazines **84** has been proposed on the basis of experimental evidences (Scheme 4.16).

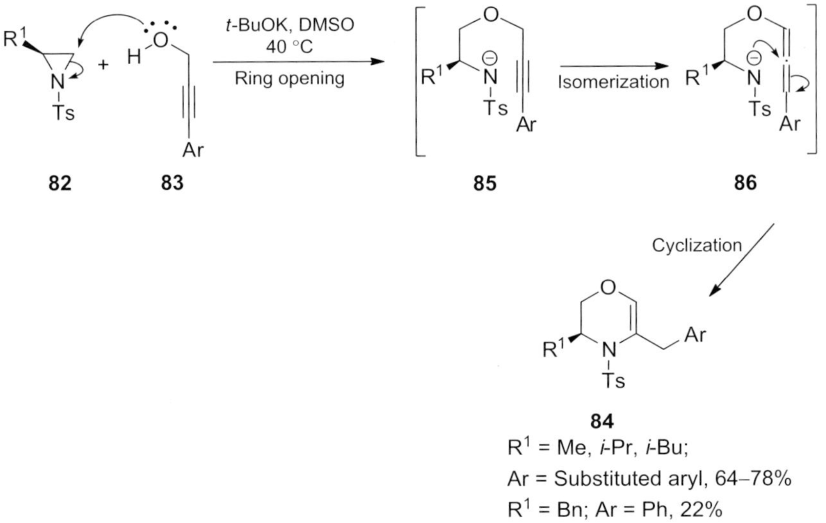

Scheme 4.16 Domino base-induced ring opening/cyclization of *N*-tosylaziridines with propargyl alcohols.

Wang and coworkers [36] have developed an efficient and general approach to substituted tetrahydrotriazines **89** via copper-catalyzed domino nucleophilic ring opening/cyclization of *N*-tosylaziridines **87** with aryl and alkyl hydrazones **88** (Scheme 4.17). A wide range of hydrazones bearing both electron-donating and - withdrawing aryl and alkyl groups could be tolerated, yielding triazines **89** in moderate to good yields. A mechanism involving copper-catalyzed nucleophilic ring opening of the aziridine and intramolecular oxidative C–H amination of the intermediate **90** has been suggested for the formation of **89** (Scheme 4.17).

A domino organophosphine-catalyzed nucleophilic ring opening/cyclization of cyclic and acyclic aziridines with carbon disulfide and isothiocyanates providing a simple and convenient route to thiazolidinones **91** (from CS_2) has been reported by Hou and coworkers [37] (Scheme 4.18). Thus, the reaction of cyclic aziridine **55** from cyclohexene ($n = 1$) with carbon disulfide in the presence of 10 mol% of Bu_3P in *t*-butyl alcohol affords *trans*-1,3-thiazolidine **91a** in 92% yield. On the other hand, no product was obtained from aziridine derived from cyclopentene ($n = 0$), whereas the reaction of aziridine from cycloheptene ($n = 2$) afforded the corresponding thiazolidine **91c** in 41% yield. Similarly, the acyclic aziridines **65**

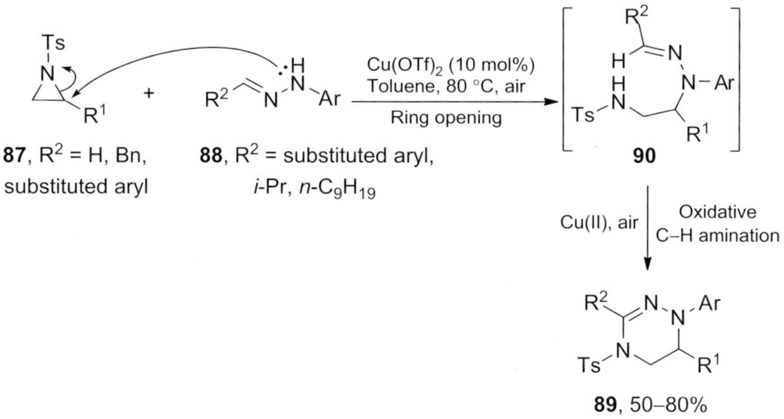

Scheme 4.17 Copper-catalyzed domino ring opening/intramolecular oxidative amidation of *N*-tosylaziridines and hydrazones.

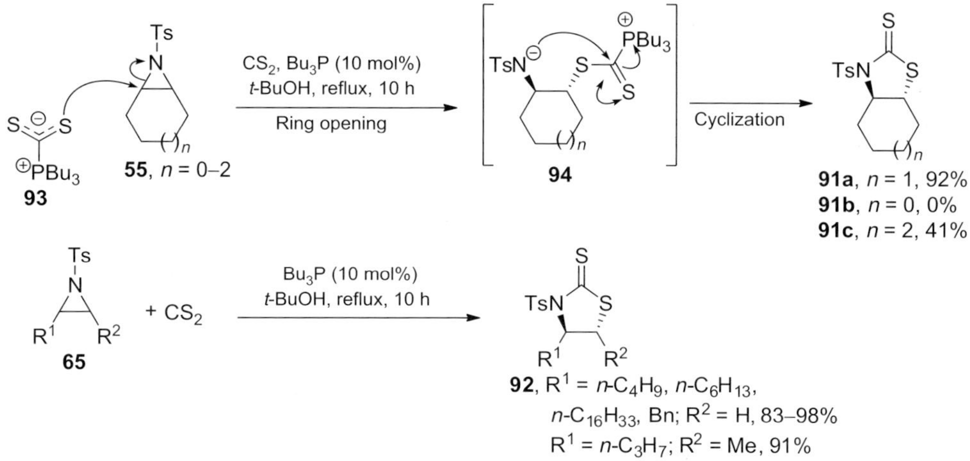

Scheme 4.18 Tributylphosphine-catalyzed domino ring opening/cyclization of *N*-tosylaziridines with carbon disulfide.

also underwent facile ring opening/cyclization with CS_2 under these conditions to yield the substituted thiazolidines **92** in high yields (Scheme 4.18). A plausible mechanism involving ring opening of aziridine by zwitterionic intermediate **93** (formed by reaction of Bu_3P with CS_2) to give intermediate **94** followed by ring closure, furnishing **91** and release of Bu_3P in a catalytic cycle, has been suggested (Scheme 4.18).

Chandrasekaran and coworkers have recently described multifold domino reactions initiated by nucleophilic ring opening of aziridines by benzyltrimethylammonium tetrathiomolybdate **46**, a useful sulfur transfer reagent (see previous Schemes 4.10 and 4.11) [38]. Thus, treatment of *N*-tosylaziridinemethyl tosylate

95 with 2 equiv of molybdate **46**, in acetonitrile at room temperature, affords *trans*-thiirane **96** as the major product, along with cyclic disulfide **97** as the minor product with excellent regio- and stereocontrol (Scheme 4.19). The overall reaction proceeds by regioselective ring opening of **95** by tetrathiomolybdate to give the intermediate **98**, which affords the thiirane **96** through a thiaaza-Payne-type rearrangement, with migration of nitrogen occurring from C-3, C-2 to C-1 position (path a). Alternatively, intramolecular nucleophilic substitution in thiomolybdate **98** and subsequent cleavage of the resulting intermediate **100** through internal redox process affords cyclic thiaperoxide **97** as the minor product (path b) (Scheme 4.19).

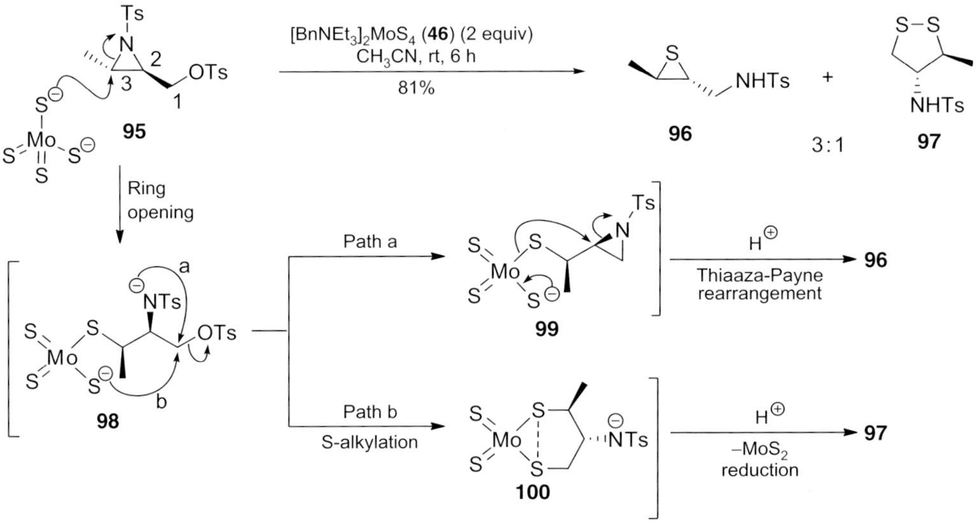

Scheme 4.19 Domino aziridine ring opening–nucleophilic substitution with tetrathiomolybdate.

A variety of substrates carrying aziridine and other reactive functionalities like α,β-unsaturated ketone, lactone, and epoxide groups are shown to undergo multifold domino reactions in the presence of tetrathiomolybdate **46** yielding diverse products [38]. Interestingly, the aziridine lactone **101** underwent regio- and chemoselective aziridine ring opening and subsequent reductive intramolecular nucleophilic substitution of the tosylate in the intermediate **103** to afford thiabicyclic lactone derivative **102**, which is a precursor for sulfur containing unnatural amino acids (Scheme 4.20). Similarly the aziridine epoxide **104** derived from (*R*)-(−)-carvone, yielded the bicyclic derivative **105** bearing an amino, hydroxy, and sulfide functionality, on treatment with 2 equiv of **46**, involving sequential ring opening of aziridine, as well as epoxide via intermediate **106** (Scheme 4.20).

Ghorai and Tiwari [39] have recently described synthesis of enantiopure, functionalized γ-lactams on the basis of a highly stereoselective copper-catalyzed domino aziridine ring opening/cyclization reactions (Scheme 4.21). Thus, chiral monosubstituted aziridines **107** react with active methylene carbon nucleophiles

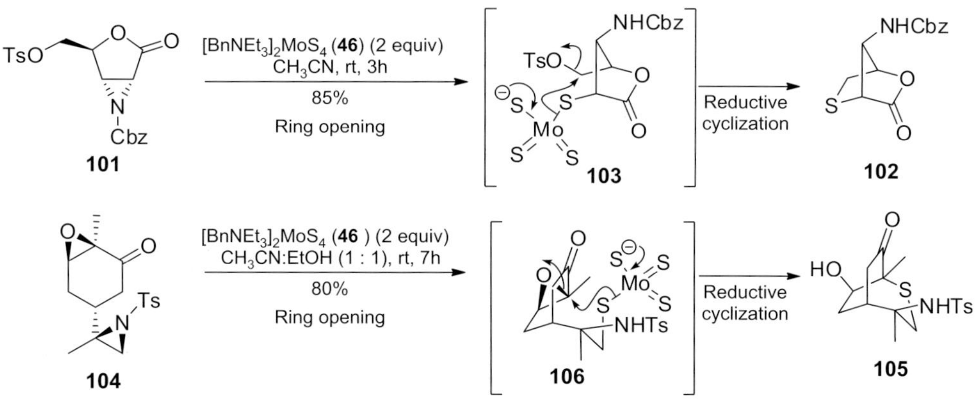

Scheme 4.20 Domino nucleophilic ring opening of aziridines mediated by tetrathiomolybdate.

such as diethyl malonate and ethyl acetoacetate in the presence of $Cu(OTf)_2$ as catalyst and NaH as base, to afford the corresponding γ-lactams **108** in good to excellent yields (64–99%) with very high diastereo- and enantioselectivities (Scheme 4.21). The scope of this methodology could also be extended to *trans-N*-tosyl-2,3-disubstituted aziridines **110**, which led to the corresponding tetrasubstituted γ-lactams **111** as a single diastereomer in high yields. These authors have suggested a Cu-catalyzed S_N2-type ring opening of aziridines with enolates followed by intramolecular cyclization of the intermediate **109** for the formation of these lactams in highly stereo and diastereoselective manner (Scheme 4.21).

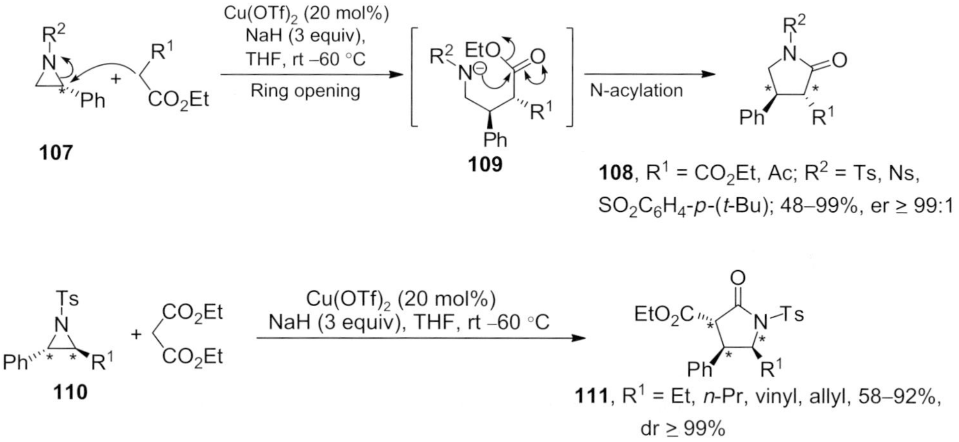

Scheme 4.21 Synthesis of enantiopure-functionalized γ-lactams by domino ring opening/cyclization of aziridines.

The same workers have recently extended this protocol for an efficient synthesis of highly functionalized racemic and nonracemic 4,5-dihydropyrroles in excellent yields and stereoselectivities via domino ring opening/cyclization of N-activated aziridines with malononitrile as shown in Scheme 4.22 [40]. Thus, treatment of enantiomerically pure alkyl (**112**) and phenyl aziridines (**113**) with malononitrile in the presence of $Sc(OTf)_3$ as LA and *t*-BuOK as base furnishes enantiomerically pure dihydropyrroles **115** and **116** as single regioisomers. Similarly chiral penta-substituted dihydropyrroles **117** could be synthesized in quantitative yields and as single diastereomers from the ring opening/cyclization of chiral 2,3-disubstituted aziridines **114**.

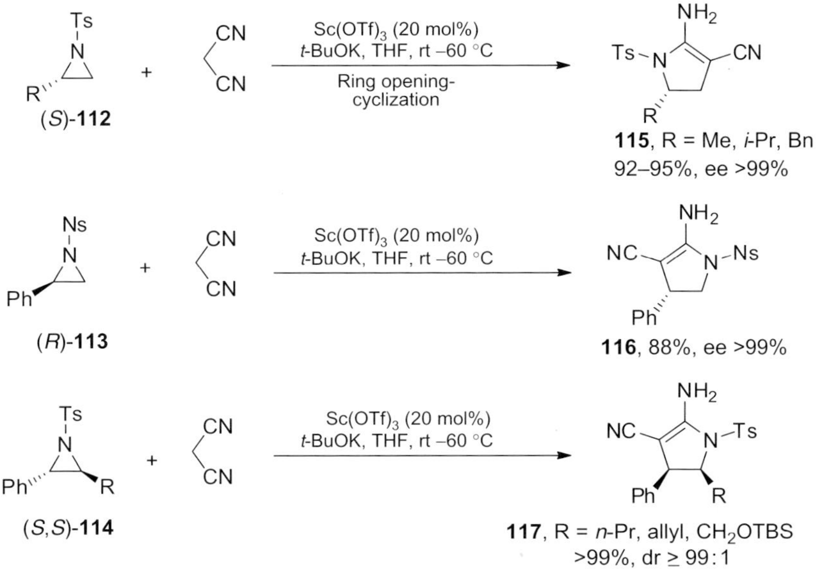

Scheme 4.22 Enantioselective synthesis of 4,5-dihydropyrroles via domino ring opening/cyclization of N-activated aziridines with malononitrile.

Bao and coworkers [41] have reported synthesis of 2,3-dihydro-1,4-benzodioxins **119** in moderate to good yields by domino ring opening/coupling cyclization of substituted *o*-iodophenols with epoxides catalyzed by Cu_2O/1,10-phenanthroline/Cs_2CO_3 system. However the reaction of unsymmetrically substituted *o*-iodophenols **56** with substituted epoxides affords regioisomeric mixture of benzodioxins **119a,b** in varying ratios (Scheme 4.23). On the other hand, group of Ranu has described highly regioselective ring opening of epoxides with unsymmetrically substituted *o*-iodophenols in the presence of alumina-supported Cu(II) catalyst, yielding substituted 1,4-benzodioxins **119a** in high yields by initial nucleophilic attack of phenolate on more substituted carbon of epoxides (Scheme 4.23) [31].

A few examples of base promoted domino ring opening lactonization of epoxides with active methylene compounds (malonate, cyanoacetate, acetoacetates) are known earlier in the literature [42]. Pizzo's group has recently described domino nucleophilic ring opening of epoxides by active methylenes, catalyzed by polymer-supported bases, under solvent-free conditions (SolFCs) (Scheme 4.24)

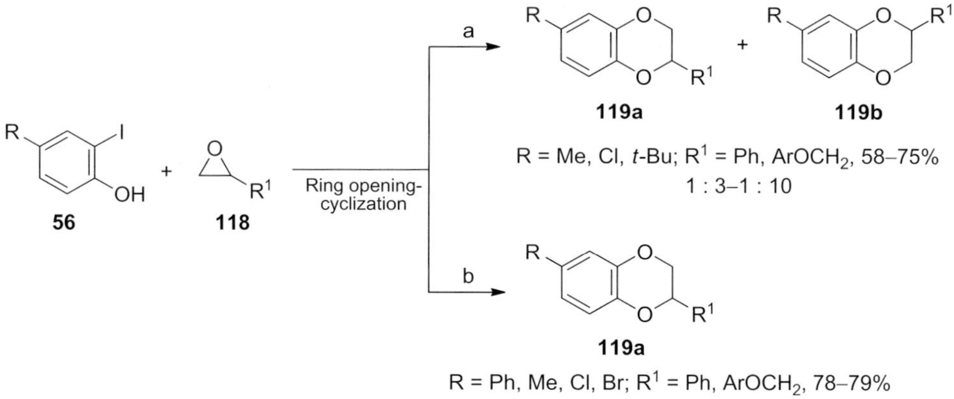

a: Cu_2O (15 mol%), 1, 10-phenanthroline, Cs_2CO_3, DMF, 110 °C, 48 h [41]

b: Cu/Al_2O_3 (4 mol%), K_2CO_3, DMF, 100 °C [31]

Scheme 4.23 Cu-catalyzed domino ring opening/coupling cyclization of *o*-iodophenols with epoxides: Synthesis of 2,3-dihydro-1,4-benzodioxins.

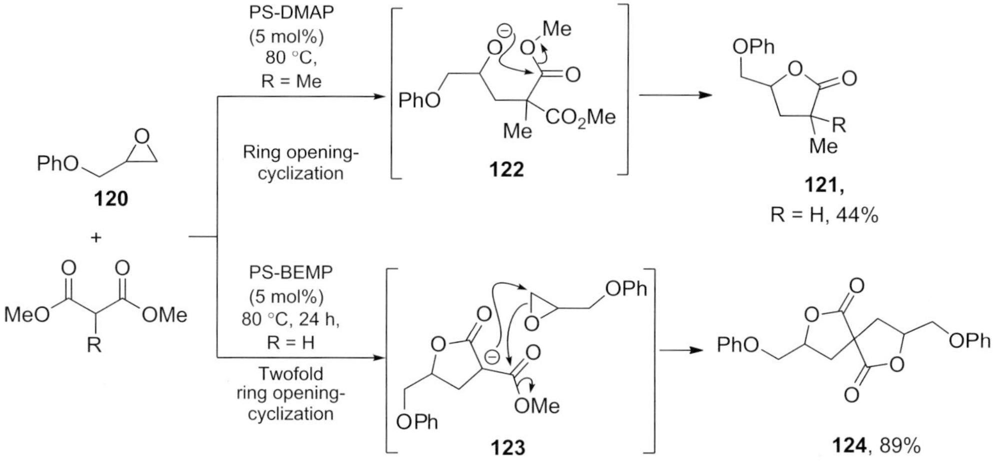

BEMP = 2-*tert*-Butylimino-2-diethylamino-1,3-dimethyl-perhydro-1,3,2-diazaphosphorine

Scheme 4.24 Polymer-supported base-catalyzed domino ring opening/cyclization of epoxide **120**.

[43]. Thus ring opening/cyclization of epoxide **120** with methyl-substituted malonate in the presence of polystyrene-dimethylaminopyridine (PS-DMAP) at 80 °C under SolFC gave **121** in moderate yields. On the other hand, dimethyl malonate undergoes twofold domino ring opening/cyclization with 2 equiv of *O*-phenylgycidol **120** in the presence of supported base, polystyrene-(2-*tert*-butylimino-2-diethylamino-1,-dimethyl-perhydro-1,3,2-diazaphosphorine) (PS-BEMP) under

SolFC, furnishing the spirolactone **124** in 89% yield, through the intermediate lactone **123** (Scheme 4.24).

As part of the investigation toward novel synthesis of taxane framework, a domino epoxide opening/retro-aldol/semiacetalization process (through intermediates **127–129**) for the formation of A/B ring part of taxane **126**, involving acid induced ring expansion of tricyclic epoxide derivative **125**, has been reported by Blechert's group [44] (Scheme 4.25).

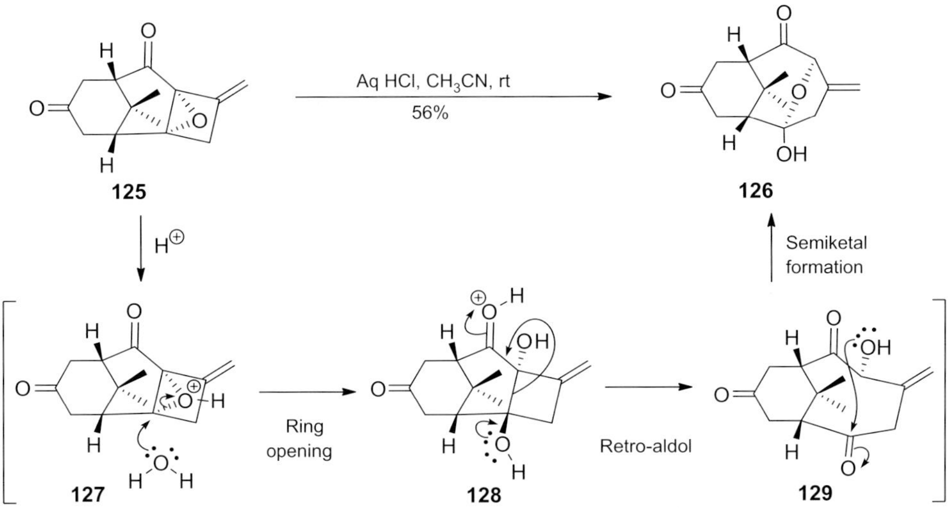

Scheme 4.25 Ring enlargement of the assembly of the taxane skeleton via domino ring opening of epoxide/retro-aldol and intramolecular semiketal formation.

Fukumoto and coworkers [45, 46] have observed a threefold domino S_N/Michael/aldol condensation reaction on treatment of an acetylcyclopropane **130** with iodotrimethylsilane (TMSI) and hexamethyldisilazane, furnishing the bicyclo[3.2.0]heptane **131** in 67% yield as a single stereoisomer (Scheme 4.26). It is proposed that silyl enol ether intermediate **133** initially formed by nucleophilic ring opening of cyclopropane with TMSI through the transition state **132**, undergoes intramolecular domino Michael/aldol reaction, providing the bicyclo[3.2.0]heptane **131** in highly stereoselective manner [45] (Scheme 4.26).

Similarly the bicyclic cyclopropyl ketone **134** is readily converted to the tricyclo[5.3.0.0^{3,7}]decane **135** under similar conditions via a domino ring opening/Michael/aldol reaction sequence as shown in the Scheme 4.27 [45]. The selective attack of the iodide ion at the 6-position of **136** leading to the intermediate **137** may be due to the effective overlap between cleaved bond and the π orbital of the carbonyl group. This novel domino reaction producing polycyclic cyclobutanes was exploited for the synthesis of natural products like (±)-anthoplalone and (±)-lepidozene by the same workers [45].

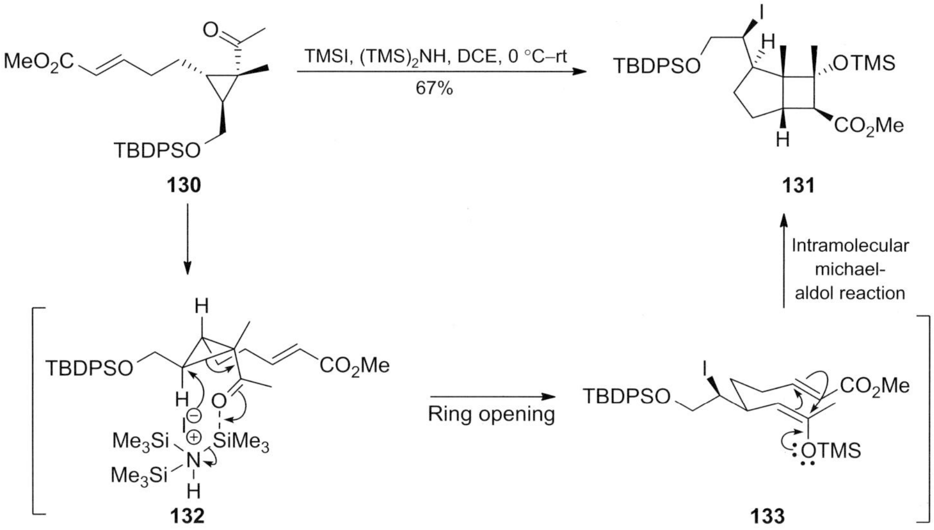

Scheme 4.26 Threefold domino ring opening/Michael/aldol reaction of cyclopropyl ketone.

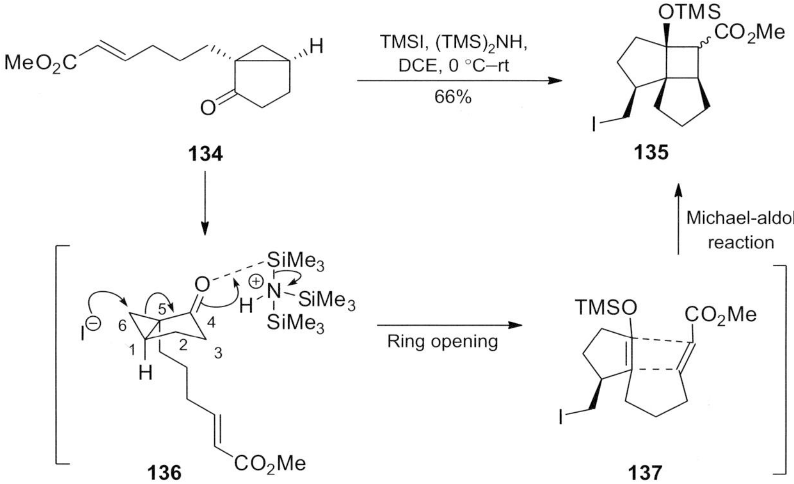

Scheme 4.27 Synthesis of complex polycylic cyclobutane skeleton through domino reaction of cyclopropyl ketone.

4.3
Domino S_N/Brook Rearrangements

A highly useful threefold domino S_N/Brook rearrangement/S_N reaction of silyl-1,3-dithiane with epoxides was first reported by Tietze and coworkers [47] (Scheme 4.28). Thus, treatment of 2.2 equiv of enantiopure epoxides (*R*)-**139a**

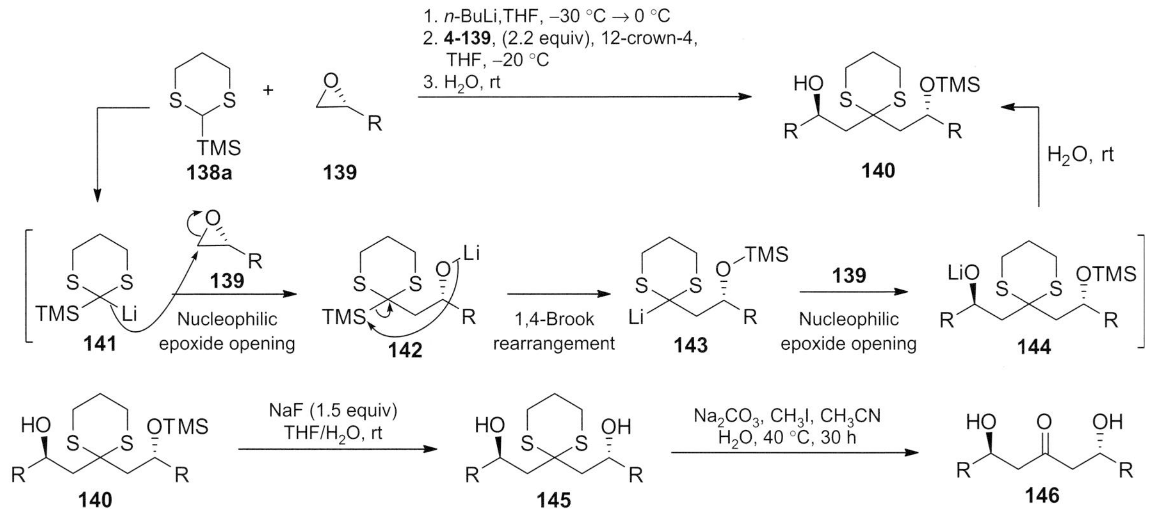

Entry	Substrate, **139**	R	Product, **140**	Yield (%)
1	(*R*)-**139a**	Ph	(*S,S*)-**140a**	65
2	(*R*)-**139b**	4-ClC_6H_4	(*S,S*)-**140b**	41
3	(*R*)-**139c**	4-MeC_6H_4	(*S,S*)-**140c**	54
4	(*R*)-**139d**	4-$MeOC_6H_4$	(*S,S*)-**140d**	63
5	(*R*)-**139e**	Me	(*S,S*)-**140e**	89

Scheme 4.28 Synthesis of enantiopure 1,5-diols via domino epoxide ring opening/1,4-Brook rearrangement/S_N reaction.

with lithiated silyldithiane **141** in the presence of crown ether furnishes the three component coupling adduct (*S,S*)-**140a**, after workup, in 65% yield. The other substituted enantiopure diol adducts **140b–e** were similarly obtained in good yields through this protocol. The overall transformation proceeds by initial attack of lithiated 1,3-dithiane **141** at less hindered position of epoxide **139** and subsequent 1,4-Brook rearrangement of the resulting lithiated alkoxide **142** to furnish the lithium dithioacetal **143**, which on further reaction with another molecule of epoxide **139** yields after workup, the enantiopure carbinol adducts **140**. Treatment of **140** with NaF leads to the diols **145**, which could be converted into dihydroxyketones **146** by cleavage of the 1,3-dithiane moiety (Scheme 4.28).

Using a similar strategy, Schaumann and coworkers [48] have reported a synthesis of functionalized cyclopentanols of type **150**, by treatment of lithiated silyldithioacetals **147** to epoxyhomoallyl tosylates **148** (Scheme 4.29). The domino reaction is initiated by the chemoselective nucleophilic attack of the lithiated carbanion **147** on the terminal carbon atom of epoxyhomoallyl tosylate **148** to give the lithiated adduct **149**, after 1,4-Brook rearrangement. Intramolecular nucleophilic substitution in the intermediate **149** affords the cyclopentane derivatives **150** in moderate to good yields [48a,c] (Scheme 4.29). In a further refinement of this cyclopentane forming process, the epoxyhomoallyl tosylate unit was incorporated into a cyclohexane ring as in **151**, which provides the bicyclic carbinol **152** in 83% yield on reaction with lithiated dithioacetal species **147a**, following similar sequence of reactions [48a] (Scheme 4.29). By using a similar strategy, groups of Takeda *et al.* [49], Block *et al.* [50], and Schaumann [48] have independently reported synthesis of substituted cyclobutyl silyl ethers in good yields by reaction of anion of bis-silylated sulfur species and epichlorohydrin or the corresponding tosylates.

Based on the pioneering work by Tietze *et al.* [47], Smith III and coworkers [51] have developed a series of multicomponent "one flask" domino protocols, where

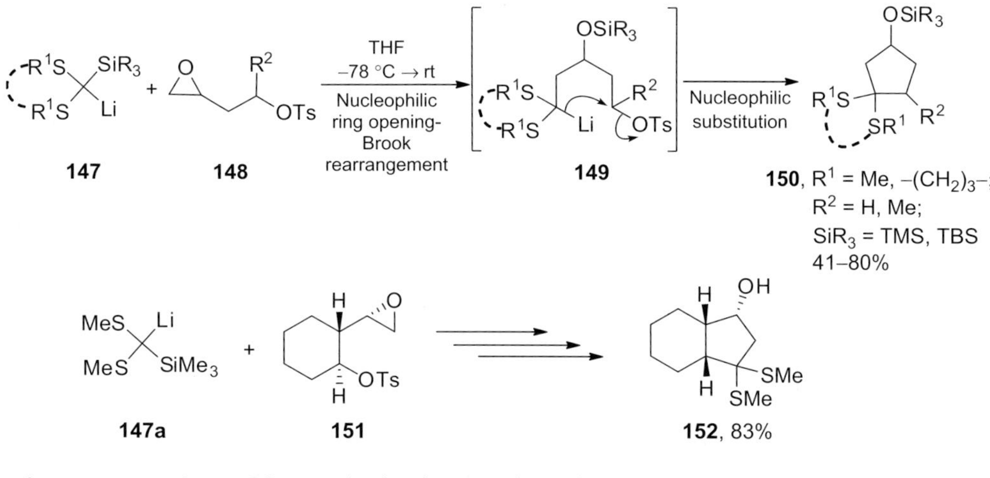

Scheme 4.29 Synthesis of functionalized cyclopentanols via domino epoxide ring opening/1,4-Brook rearrangement/S_N reaction.

an epoxide and a second different electrophile (epoxide, aziridine, alkyl bromide, aldehyde) is reacted with lithiated silyldithianes. This protocol is labeled as "linchpin coupling" by these workers and has found elegant applications, particularly in the synthesis of mycoticins A and B [52], subtargets of spongistatin precursors [53] and also in second generation approach toward synthesis of (−)-indolizidine alkaloids 223AB and (−)-205B (Schemes 4.30 and 4.31) [51, 53, 54].

The natural antibiotics mycoticin A and B (**159** and **160**) belong to a large family of poly-1,3-hydroxyated macrocyclic polyene-lactones, many of which exhibit antifungal properties [52]. Schreiber's group [55] has previously reported the synthesis of (+)-mycoticin A and B involving the pseudo C2-symmetric trisacetonide intermediate **158** (Scheme 4.30). Smith's group [52] has developed a domino approach for the synthesis of (+)-diol intermediate **157**, as precursor for trisacetonide **158**, starting from silyl-1,3-dithiane **138**, epoxide **153**, and bisepoxide **156** (Scheme 4.30). Thus, **157** containing four stereogenic centers could be prepared in a highly efficient way and with excellent stereoselectivity by this procedure involving five component coupling. Treatment of dithiane **138b** with *t*-BuLi and subsequent reaction of lithiated silyldithiane **147b** with enantiopure epoxide **153** gives the lithiated dithioacetal **155** via a 1,4-Brook rearrangement. Treatment of **155** with an equivalent of bisepoxide **156** affords the (+)-diol **157** in 59% yield. This intermediate was in turn converted to the trisacetonide **158**, which is required as intermediate in Schreiber's [55] synthesis of (+)-mycoticin A (**159**) (Scheme 4.30).

Following a similar protocol, using combination of an epoxide **161** and aziridine derivative **162**, group of Smith III has synthesized useful intermediate **163** for the synthesis of indolizidine alkaloid 223AB (**164**) (Scheme 4.31) [54a]. Moreover by this strategy, the synthesis of alkaloid (−)-205B (**165**) could also be accomplished [54b,c]. These natural alkaloids were isolated from neotropical dart-poison frogs, belonging to genus *Dendrobates*, displaying interesting biological activity [54a]. Smith III and coworkers [53] have also applied these multicomponent domino reactions of 2-silyl-1,3-dithiane with epoxides in a number of other synthetic ventures, including spogistatin 1 and 2, (+)-rimocidin [56] and the Hale bryostatin [57].

The success and considerable utility of this domino "linchpin coupling," led group of Smith III to develop additional synthetic tactic, termed as *anion relay chemistry*, which would allow rapid access to high level of molecular diversity and complexity. Early ARC studies focused on the use of the dithiane moieties both as terminal nucleophile and as anion stabilizing group (ASG). Based on the resulting locus of transported anion in this "through space" anion relay domino reactions, Smith III termed them as Type I ARC and Type II ARC processes (Scheme 4.32) [58]. Type I ARC is defined as a multicomponent coupling, wherein a linchpin nucleophile (**166**) reacts with an electrophile (**167**) to generate an oxyanion (**168A**), which subsequently relays the negative charge back to initiating site through Brook rearrangement (**166** → **168A** → **168B**). The domino reactions described in Schemes 4.28–4.31 utilizing lithiated silyl-1,3-dithianes as nucleophiles belong to Type I ARC reactions. In 2004, these workers further extended the Type I ARC process, which allows anion migration along the linchpin chain to generate a new reactive anion at a distal site (Scheme 4.32). Termed as *Type II ARC*, the process

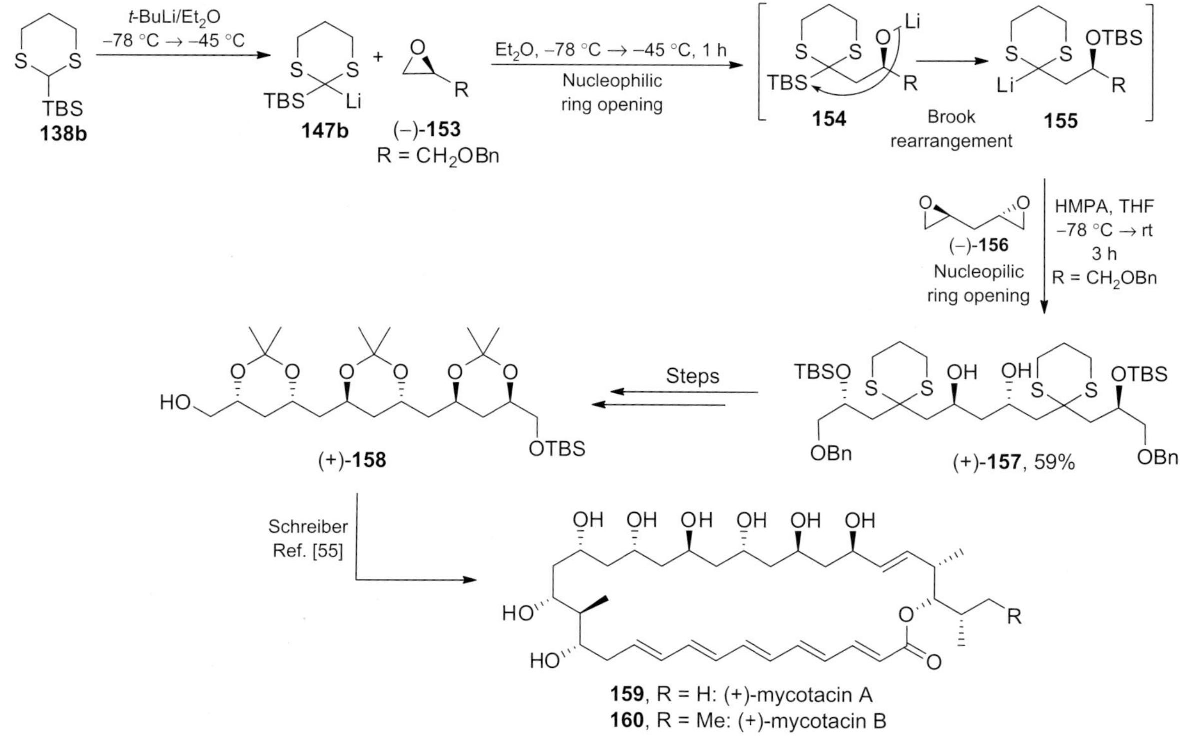

Scheme 4.30 Smith III's synthesis of triacetonide precursor in the total synthesis of (+)-mycoticin A.

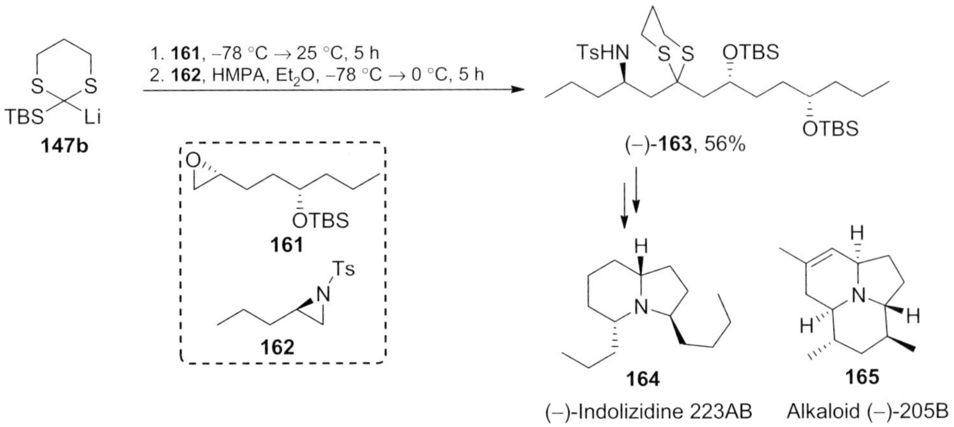

Scheme 4.31 Smith III's synthesis of indolizidine alkaloids 223AB and 205B.

involves initial attack of an anion (nucleophile) at electrophilic site of the linchpins (**170**) (epoxide or aldehyde), bearing on a distal carbon, a trialkylsilyl group and an ASG to furnish an oxyanion **171**. Addition of hexamethylphosphoramide (HMPA) or other polar solvents triggers a solvent controlled Brook rearrangement of **171** leading to a new distal anion **172**, which can be captured by a variety of electrophiles to generate in a "single flask," diverse class of three-component adducts such as **173** (Scheme 4.32).

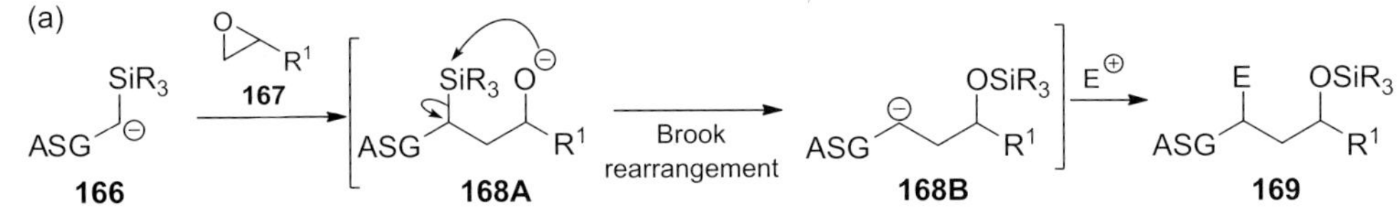

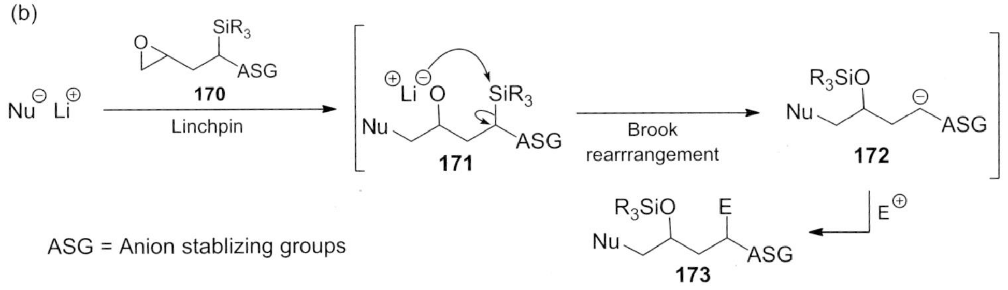

Scheme 4.32 Domino reactions involving (a) Type I and (b) Type II anion relay chemistry (ARC).

The viability of this new Type II ARC involving threefold domino reactions was demonstrated by developing a one-flask three component coupling protocol using 1,3-dithiane epoxy compound **176** as "linchpin" (Scheme 4.33) [59]. Thus,

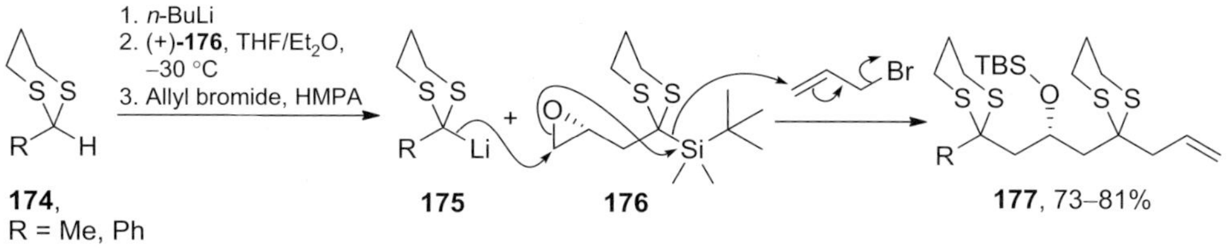

Scheme 4.33 Domino three-component ARC II coupling of "linchpin" **176**.

the reaction of 2-lithiated 2-substituted 1,3-dithiane **175** with linchpin **176** and subsequent alkylation with allyl bromide affords the three-component adduct **177** in 81% yield (Scheme 4.33). The synthetic utility of this Type II ARC domino protocol was further demonstrated by Smith's [60] group in the construction of an advanced intermediate **179**, in their perspective synthesis of (+)-spirastrellolides A and B, a class of architecturally complex sponge metabolites (Scheme 4.34). Thus exploiting the Schlosser base (*n*-BuLi/KO*t*-Bu), the first alkylation of **180** (with linchpin **176**) was achieved with concomitant Brook rearrangement and subsequent reaction with the (−)-epoxide **181** furnishes the intermediate **179** in 77% yield in a single operation [60].

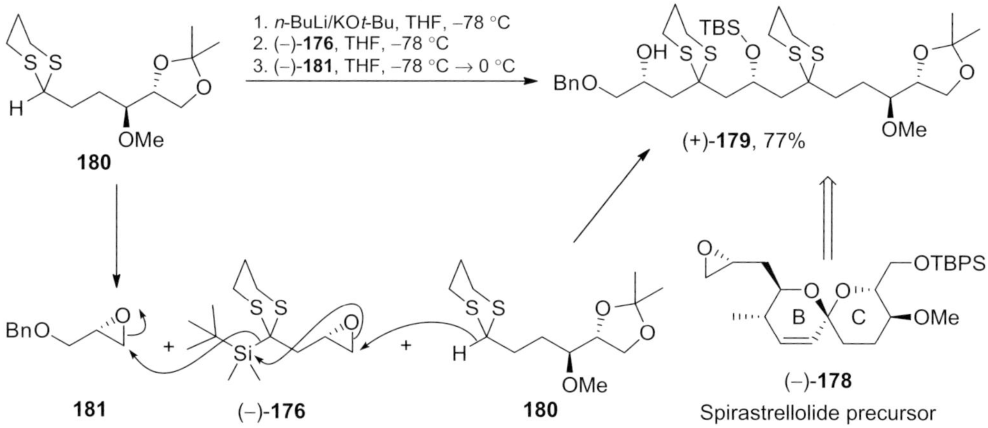

Scheme 4.34 Three-component domino "ARC tactic" for construction of spirastrellolide A precursor with linchpin **176**.

By exploiting this efficient three-component domino Type II ARC protocol, utilizing linchpin **176** and aziridines **182** as electrophiles, group of Smith has recently designed and validated an efficient general approach for diversity oriented synthesis of all possible stereoisomers of the selected 2,4,6-trisubstituted piperidine scaffolds **183**, through construction of an acyclic precursor **184** [61]. Subsequent intramolecular S_N2 cyclization, removal of the dithiane moieties and reduction of carbonyl groups in the intermediate **184** affords the substituted piperidines in good yields (Scheme 4.35) [61].

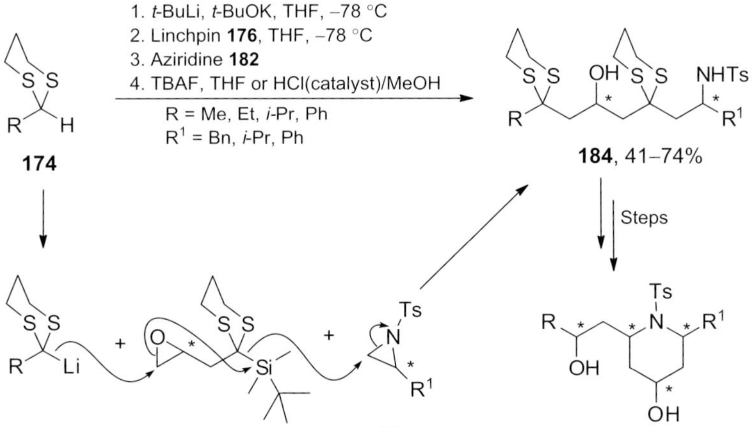

Scheme 4.35 Diversity oriented synthesis of 2,4,6-trisubstituted piperidines via Type II anion relay chemistry.

Smith and coworkers next extended Type II ARC domino protocol beyond dithiane arena by designing a variety of effective ARC linchpins with different ASGs capable of three and four component couplings. In their first effort toward this goal, they developed readily available 2-bromoallylsilane [62a], and subsequently, allyltrimethylsilane **185** [63], and recently *o*-TMS (trimethylsilyl) benzaldehyde **188** as promising linchpins for Type II ARC process [64, 65]. However, domino anionic relay reactions with these linchpins **185** and **188** are not discussed in this chapter, since they are not initiated by nucleophilic substitution as the first step.

Subsequently, Smith and coworkers [63] demonstrated the possibility of a four-component union employing the consecutive addition of two different linchpins such as **176** and **185** (Scheme 4.36). Thus the dithiane (+)-**180**, an advanced intermediate in the spirastrellolide venture, was treated with the Schlosser base (*n*-BuLi/KO*t*-Bu) and the resulting anion was reacted in sequence with linchpins (+)-**176** and **185** at −78 °C and a final electrophile, allyl bromide, affording the four-component adduct **186** as diastereomeric mixture, in 44% yields. Treatment of the **186** with tetra-*n*-butylammonium fluoride (TBAF), yielded the diol **187** as a separable mixture (1.3 : 1) of diastereomers in 37% overall yield (Scheme 4.36) [63].

In a further study, comprising of a "proof of concept" sequence to "natural product like" libraries, Smith and coworkers [64] have developed another Type II ARC, four-component domino process in an iterative manner employing linchpins (−)-**176**, **188**, and 1,3-dithiane **180** as initial nucleophile and allyl bromide as final electrophile (Scheme 4.37). The four-component adduct **189**, thus obtained in 59% yield as diastereomeric mixture (1.25 : 1 at carbinol site) was converted to natural product like macrolactone **190** in five steps in overall yields of 12 and 13% (Scheme 4.37).

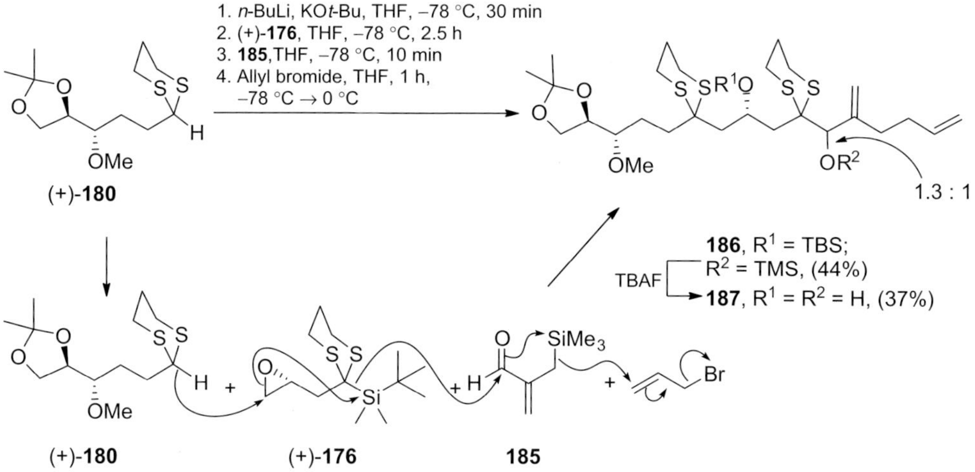

Scheme 4.36 Four-component domino ARC coupling employing linchpins **176** and **185**.

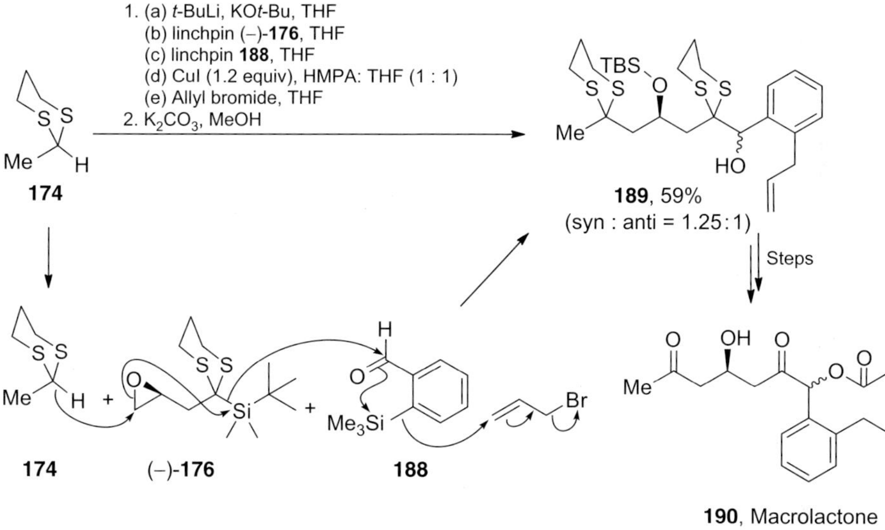

Scheme 4.37 Synthesis of macrolactone through four-component domino ARC coupling employing linchpins **176** and **188**.

In a further effort for search of other viable ASGs that would extend the scope and general utility of the ARC tactic, Smith and coworkers have developed synthesis and reactions of novel achiral and chiral linchpins, with other ASGs such as nitrile, phenyl, and phenylthio functionalities (**191–193**) (Scheme 4.38) [63, 66–68]. An evaluation of these newly synthesized compounds as linchpins for Type II ARC tactics, reveals that both phenyl and phenylthio moieties are viable as ASGs for

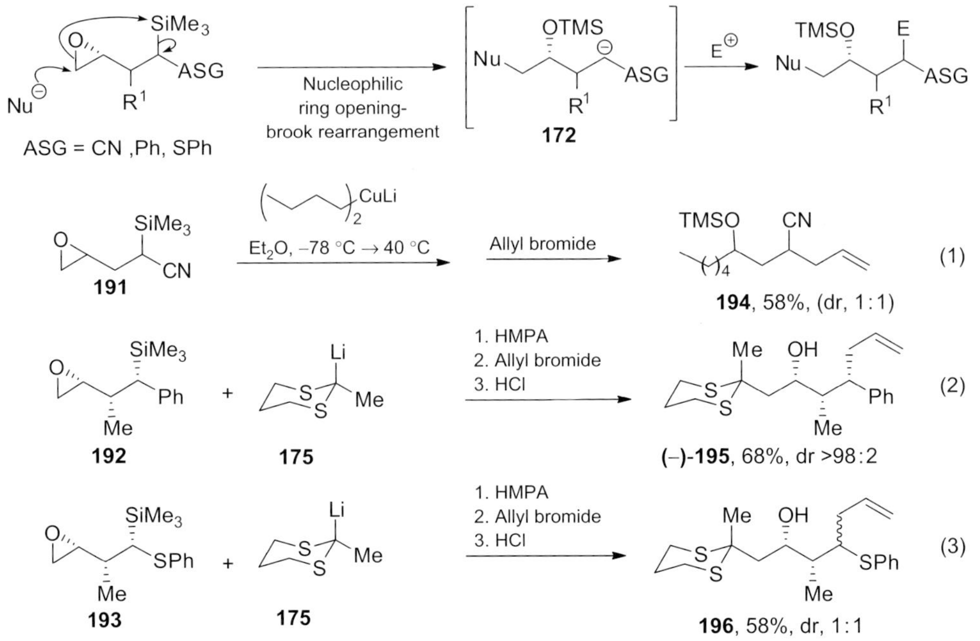

Scheme 4.38 New bifunctional linchpins for Type II ARC domino reactions.

ARC process [67]. The initiating nucleophiles included either lithiated 2-methyl-1,3-dithiane or lithium di-*n*-butylcuprate, which on reaction with these linchpins (**191–193**), followed by trapping of the rearranged anion **172** by allyl bromide as electrophile, provide three-component adducts such as **194–196** in 58–68% overall yields (Scheme 4.38) [63, 67]. Of particular note is the stereochemical outcome of these reactions. High diastereoselectivity was observed with linchpins possessing a methyl substituent α to the silyl group, when a phenyl moiety serve as ASG (Scheme 4.38, Eq. (2)), while poor diastereoselectivity is observed when a phenylthio moiety is employed as the ASG irrespective of the presence of α-substituent (Scheme 4.38, Eq. (3)) [67, 68].

Finally in a recent paper, Smith and coworkers [69] have demonstrated an elongated long-range version of domino Type II ARC protocol through newly designed linchpins **197** and **202** (Schemes 4.39 and 4.40) thus allowing "long range" relocation of negative charge across the space and to generate reactive distal nucleophile center for alkylation with electrophiles. It was reasoned that an appropriately placed silyl ether might provide a vehicle to augment charge location and thereby overcome the current distance constraints of higher 1,*n*-Brook rearrangements ($n \leq 5$) [70]. Thus, as anticipated, transfer of silyl group (i.e., *tert*-butyldimethylsilyl (TBS)) present on benzylic oxygen in the compound **197**, occurs upon addition of lithium dimethylcuprate as nucleophile, involving epoxide ring opening and a 1,5-O–O silyl migration (**199** → **200**). A 1,4-Brook rearrangement

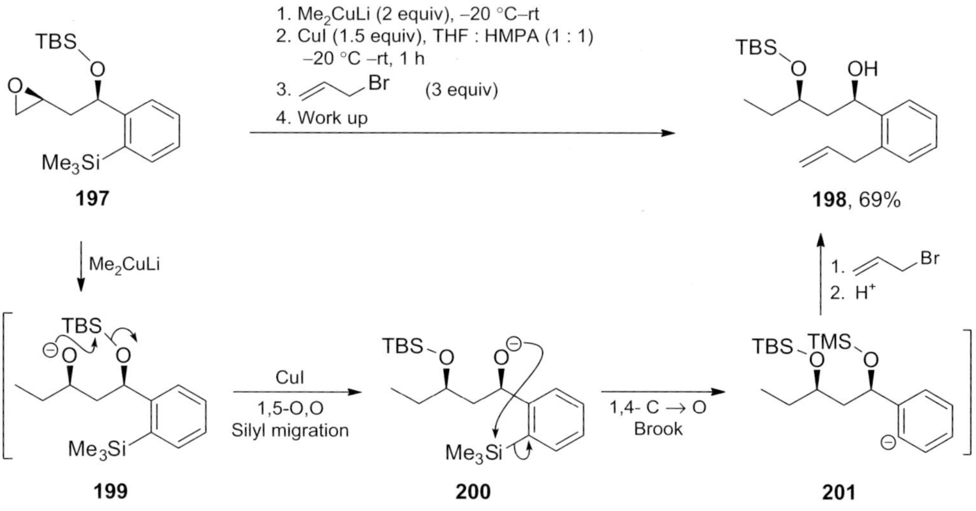

Scheme 4.39 Long-range domino ARC II coupling with linchpin **197**.

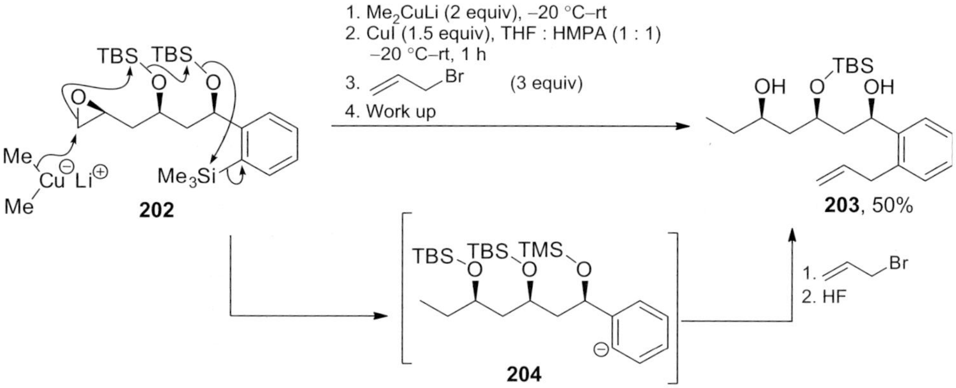

Scheme 4.40 Long-range domino ARC II three-component coupling by threefold silyl group migration in linchpin **202**.

in the intermediate **200** generates a reactive anion **201** in which negative charge is relocated six atom distal to the origin. Subsequent trapping of the anion **201** with allyl bromide as electrophile affords the product **198** in 69% yield (Scheme 4.39).

Interestingly under same long range anionic relay chemistry (LR-ARC) conditions, a newly designed and constructed epoxybis-silyloxy linchpin **202** furnished the corresponding three-component allylation adduct **203** in 50% yield (Scheme 4.40) [69], thereby validating the unprecedented possibility of threefold silyl group migration in a sequential manner and subsequent transfer of negative charge to a position eight atoms removed from origin in the intermediate **204** (Scheme 4.40).

References

1. Reviews: (a) Tietze, L.F. (1996) *Chem. Rev.*, **96**, 115–136. (b) Tietze, L.F., Brasche, G., and Gericke, K. (2006) *Domino Reactions in Organic Synthesis*, Wiley-VCH Verlag GmbH, Weinheim. (c) Tietze, L.F., Kinzel, T., and Brazel, C. (2009) *Acc. Chem. Res.*, **42**, 367–378 and references therein. (d) Tietze, L.F. and Rackelmann, N. (2005) in *Multicomponent Reactions* (eds J. Zhu and H. Bienaymé), Wiley-VCH Verlag GmbH, Weinheim, pp. 121–168. (e) Tietze, L.F., Spiegl, D.A., and Brazel, C.C. (2009) in *Experiments in Green and Sustainable Chemistry* (eds H.W. Roesky and D.K. Kennepoh), Wiley-VCH Verlag GmbH, Weinheim, pp. 158–167. (f) (2009) *Chem. Soc. Rev.*, **38**, 2969–3276Special issue on *Rapid Formation of Molecular Complexity in Organic Synthesis*. (g) Pellissier, H. (2013) *Chem. Rev.*, **113**, 442–524 and references therein.
2. Liu, H.J. and Han, Y. (1991) *J. Chem. Soc., Chem. Commun.*, 1484–1485.
3. Gharpure, S.J., Reddy, S.R.B., and Sanyal, U. (2007) *Synlett*, 1889–1892.
4. Tong, B.M.K., Chen, H., Chong, S.Y., Heng, Y.L., and Chiba, S. (2012) *Org. Lett.*, **14**, 2826–2829.
5. Desmaele, D. and Louvet, J.M. (1994) *Tetrahedron Lett.*, **35**, 2549–2552.
6. Le Dreau, M.A., Desmaele, D., Dumas, F., and d'Angelo, J. (1993) *J. Org. Chem.*, **58**, 2933–2935.
7. (a) Watanabe, H., Onoda, T., and Kitahara, T. (1999) *Tetrahedron Lett.*, **40**, 2545–2548. (b) Kubo, I., Jamalamadaka, V., Kamikawa, T., Takahashi, K., Tabata, K., and Kusumi, T. (1996) *Chem. Lett.*, 441–442.
8. Gharpure, S.J. and Reddy, S.R.B. (2009) *Org. Lett.*, **11**, 2519–2522.
9. Gharpure, S.J. and Reddy, S.R.B. (2010) *Tetrahedron Lett.*, **51**, 6093–6097.
10. Bermejo, A., Figadere, B., Zafra-Polo, M.C., Barrachina, I., Estornell, E., and Cortes, D. (2005) *Nat. Prod. Rep.*, **22**, 269–303.
11. Srikrishna, A., Reddy, T.J., and Kumar, P.P. (1997) *Synlett*, 663–664.
12. (a) Bunce, R.A., Peeples, C.J., and Jones, P.B. (1992) *J. Org. Chem.*, **57**, 1727–1733. (b) Bunce, R.A. and Allison, J.C. (1999) *Synth. Commun.*, **29**, 2175–2186. (c) Bunce, R.A., Kotturi, S.V., Peeples, C.J., and Holt, E.M. (2002) *J. Heterocycl. Chem.*, **39**, 1049–1054.
13. Goff, D.A. (1998) *Tetrahedron Lett.*, **39**, 1473–1476.
14. Lavoisier-Gallo, T. and Rodriguez, J. (1997) *J. Org. Chem.*, **62**, 3787–3788.
15. Lavoisier-Gallo, T., Charonnet, E., and Rodriguez, J. (1998) *J. Org. Chem.*, **63**, 900–902.
16. Zakarian, A., Batch, A., and Holton, R.A. (2003) *J. Am. Chem. Soc.*, **125**, 7822–7824.
17. Takikawa, H., Hirooka, M., and Sasaki, M. (2003) *Tetrahedron Lett.*, **44**, 5235–5238.
18. (a) Barrero, A.F., Oltra, J.E., and Alvarez, M. (2000) *Tetrahedron Lett.*, **41**, 7639–7643. (b) Pandey, G. and Prasanna Kumara, C. (2011) *Org. Lett.*, **13**, 4672–4675.
19. Cho, C.W. and Krische, M.J. (2004) *Angew. Chem. Int. Ed.*, **43**, 6689–6691.
20. Langer, P. and Freifeld, I. (2001) *Chem. Eur. J.*, **7**, 565–572.
21. Lin, Y.L., Kuo, H.S., Wang, Y.W., and Huang, S.T. (2003) *Tetrahedron*, **59**, 1277–1281.
22. Engler, M., Anke, T., Sterner, O., and Brandt, U. (1996) *J. Antibiot.*, **50**, 325–329.
23. (a) Akiyama, H., Fujimoto, T., Ohshima, K., Hoshino, K., and Yamamoto, I. (1999) *Org. Lett.*, **1**, 427–430. (b) Akiyama, H., Fujimoto, T., Ohshima, K., Hoshino, K., Saito, Y., Okamoto, A., Yamamoto, I., Kakehi, A., and Iriye, R. (2001) *Eur. J. Org. Chem.*, **2001**, 2265–2272.
24. (a) Prabhu, K.R., Devan, N., and Chandrasekaran, S. (2002) *Synlett*, 1762–1778. (b) Sureshkumar, D., Gunasundari, T., Ganesh, V., and Chandrasekaran, S. (2007) *J. Org. Chem.*, **72**, 2106–2117. (c) Devan, N., Sridhar, P.R., Prabhu, K.R., and Chandrasekaran, S. (2002) *J. Org. Chem.*, **67**, 9417–9420.
25. Prabhu, K.R., Sivanand, P.S., and Chandrsekaran, S. (2000) *Angew. Chem. Int. Ed.*, **39**, 4316–4319.

26. Pan, W.H., Harmer, M.A., Halbert, T.R., and Stiefel, E.I. (1984) *J. Am. Chem. Soc.*, **106**, 459–460.
27. (a) Gunasundari, T. and Chandrsekaran, S. (2010) *J. Org. Chem.*, **75**, 6685–6688. (b) Gunasundari, T. and Chandrasekaran, S. (2012) *Eur. J. Org. Chem.*, **2012**, 6986–6995.
28. (a) Chen, D., Shen, G., and Bao, W. (2009) *Org. Biomol. Chem.*, **7**, 4067–4073. (b) Chen, D. and Bao, W. (2010) *Adv. Synth. Catal.*, **352**, 955–960.
29. (a) Dai, W.M., Wang, X., and Ma, C. (2005) *Tetrahedron*, **61**, 6879–6885. (b) Feng, G., Wu, J., and Dai, W.M. (2006) *Tetrahedron*, **62**, 4635–4642.
30. Rao, R.K., Naidu, A.B., and Sekar, G. (2009) *Org. Lett.*, **11**, 1923–1926.
31. Bhadra, S., Adak, L., Samanta, S., Islam, A.K.M.M., Mukherjee, M., and Ranu, B.C. (2010) *J. Org. Chem.*, **75**, 8533–8541.
32. Chouhan, G. and Alper, H. (2010) *Org. Lett.*, **12**, 192–195.
33. Wang, J.Y., Guo, X.F., Wang, D.X., Huang, Z.T., and Wang, M.X. (2008) *J. Org. Chem.*, **73**, 1979–1982.
34. Karikomi, M., D'hooghe, M., Verniest, G., and De Kimpe, N. (2008) *Org. Biomol. Chem.*, **6**, 1902–1904.
35. Wang, L., Liu, Q.B., Wang, D.S., Li, X., Han, X.W., Xiao, W.J., and Zhou, Y.G. (2009) *Org. Lett.*, **11**, 1119–1122.
36. Hong, D., Lin, X., Zhu, Y., Lei, M., and Wang, Y. (2009) *Org. Lett.*, **11**, 5678–5681.
37. Wu, J.Y., Luo, Z.B., Dai, L.X., and Hou, X.L. (2008) *J. Org. Chem.*, **73**, 9137–9139.
38. Sureshkumar, D., Koutha, S.M., and Chandrasekaran, S. (2005) *J. Am. Chem. Soc.*, **127**, 12760–12761.
39. Ghorai, M.K. and Tiwari, D.P. (2010) *J. Org. Chem.*, **75**, 6173–6181.
40. Ghorai, M.K. and Tiwari, D.P. (2013) *J. Org. Chem.*, **78**, 2617–2625.
41. Bao, W., Liu, Y., Lv, X., and Qian, W. (2008) *Org. Lett.*, **10**, 3899–3902.
42. (a) Dyker, G. and Thone, A. (2000) *Tetrahedron*, **56**, 8669–8672. (b) DePuy, C.H., Breitbeil, F.W., and Eilers, K.L. (1964) *J. Org. Chem.*, **29**, 2810. (c) Zuidema, G.D., Cook, P.L., and Van Zyl, G. (1953) *J. Am. Chem. Soc.*, **75**, 294–296.
43. Angelini, T., Fringuelli, F., Lanari, D., Pizzo, F., and Vaccaro, L. (2010) *Tetrahedron Lett.*, **51**, 1566–1569.
44. Jansen, R., Velder, J., and Blechert, S. (1995) *Tetrahedron*, **51**, 8997–9004.
45. (a) Ihara, M., Taniguchi, T., Tokunaga, Y., and Fukumoto, K. (1994) *J. Org. Chem.*, **59**, 8092–8100. (b) Ihara, M., Taniguchi, T., Tokunaga, Y., and Fukumoto, K. (1995) *Synthesis*, **1995**, 1405–1410.
46. Ihara, M., Taniguchi, T., Makita, K., Takano, M., Ohnishi, M., Taniguchi, N., Fukumoto, K., and Kabuto, C. (1993) *J. Am. Chem. Soc.*, **115**, 8107–8115.
47. Tietze, L.F., Geissler, H., Gewert, J.A., and Jakobi, U. (1994) *Synlett*, 511–512.
48. (a) Scaumann, E. and Kirschning, A. (2007) *Synlett*, 177–190. (b) Fischer, M.R., Kirschning, A., Michel, T., and Schaumann, E. (1994) *Angew. Chem., Int. Ed. Engl.*, **33**, 217–218. (c) Michel, T., Kirschning, A., Beier, C., Brauer, N., Schumann, E., and Adiwidjaja, G. (1996) *Liebigs Ann.*, **11**, 1811–1821. (d) Brauer, N., Michel, T., and Schaumann, E. (1998) *Tetrahedron*, **54**, 11481–11488. (e) Tries, F. and Schaumann, E. (2003) *Eur. J. Org. Chem.*, **2003**, 1085–1090.
49. Takeda, T., Naito, S., Ando, K., and Fujiwara, T. (1983) *Bull. Chem. Soc. Jpn.*, **56**, 967–968.
50. Block, E., Laffitte, J.A., and Eswarakrishnan, V. (1986) *J. Org. Chem.*, **51**, 3428–3435.
51. (a) Smith, A.B. III, Pitram, S.M., Boldi, A.M., Gaunt, M.J., Sfouggatakis, C., and Moser, W.H. (2003) *J. Am. Chem. Soc.*, **125**, 14435–14445. (b) Smith, A.B. III, and Adams, C.M. (2004) *Acc. Chem. Res.*, **37**, 365–377.
52. Smith, A.B. III, and Pitram, S.M. (1999) *Org. Lett.*, **1**, 2001–2004.
53. (a) Smith, A.B. III, Doughty, V.A., Lin, Q., Zhuang, L., McBriar, M.D., Boldi, A.M., Moser, W.H., Murase, N., Nakayama, K., and Sobukawa, M. (2001) *Angew. Chem. Int. Ed.*, **40**, 191–195. (b) Smith, A.B. III, Lin, Q., Doughty, V.A., Zhuang, L., McBriar, M.D., Kerns, J.K., Brook, C.S., Murase, N., and Nakayama, K. (2001) *Angew. Chem.*

Int. Ed., **40**, 196–199. (c) Smith, A.B. III, Zhu, W., Shirakami, S., Sfouggatakis, C., Doughty, V.A., Bennett, C.S., and Sakamoto, Y. (2003) *Org. Lett.*, **5**, 761–764.
54. (a) Smith, A.B. III, and Kim, D.S. (2004) *Org. Lett.*, **6**, 1493–1495. (b) Smith, A.B. III, and Kim, D.S. (2005) *Org. Lett.*, **7**, 3247–3250. (c) Smith, A.B. III, and Kim, D.S. (2006) *J. Org. Chem.*, **71**, 2547–2557.
55. Poss, C.S., Rychnovsky, S.D., and Schreiber, S.L. (1993) *J. Am. Chem. Soc.*, **115**, 3360–3361.
56. Smith, A.B. III, Pitram, S.M., and Fuertes, M.J. (2003) *Org. Lett.*, **5**, 2751–2754.
57. (a) Hale, K.J., Hummersone, M.G., and Bhatia, G.S. (2000) *Org. Lett.*, **2**, 2189–2192. (b) Manaviazar, S., Frigerio, M., Bhatia, G.S., Hummersone, M.G., Aliev, A.E., and Hale, K.J. (2006) *Org. Lett.*, **8**, 4477–4480.
58. Smith, A.B. III, and Wuest, W.M. (2008) *Chem. Commun.*, 5883–5895.
59. Smith, A.B. III, and Xian, M. (2006) *J. Am. Chem. Soc.*, **126**, 66–67.
60. Smith, A.B. III, and Kim, D.S. (2007) *Org. Lett.*, **9**, 3311–3314.
61. Smith, A.B. III, Han, H., and Kim, W.S. (2011) *Org. Lett.*, **13**, 3328–3331.
62. For allyltrimethylsilane based linchpins, See also: (a) Smith, A.B. III, and Duffey, M.O. (2004) *Synlett*, 1363–1366. (b) Takaku, K., Shinokubo, H., and Oshima, K. (1998) *Tetrahedron Lett.*, **39**, 2575–2578.
63. Smith, A.B. III, Kim, D.S., and Xian, M. (2007) *Org. Lett.*, **9**, 3307–3309.
64. Smith, A.B. III, Kim, W.S., and Wuest, W.M. (2008) *Angew. Chem. Int. Ed.*, **47**, 7082–7086.
65. For aldehyde based aromatic bifunctional linchpins, see also: (a) Moser, W.H., Endsley, K.E., and Colyer, J.T. (2000) *Org. Lett.*, **2**, 717–719. (b) Moser, W.H., Zhang, J., Lecher, C.S., Frazier, T.L., and Pink, M. (2002) *Org. Lett.*, **4**, 1981–1984. (c) Taguchi, H., Takami, K., Tsubouchi, A., and Takeda, T. (2004) *Tetrahedron Lett.*, **45**, 429–432. (d) Deavrie-Baez, N.O., Shuhler, B.J., Wang, H., and Xian, M. (2007) *Org. Lett.*, **9**, 4655–4658. (e) Deavrie-Baez, N.O., Kim, W.S., Smith, A.B. III, and Xian, M. (2009) *Org. Lett.*, **11**, 1861–1864. (f) Smith, A.B. III, and Kim, W.S. (2011) *Proc. Natl. Acad. Sci. U.S.A.*, **108**, 6787–6792.
66. Matsuda, I., Murata, S., and Ishii, Y. (1979) *J. Chem. Soc., Perkin Trans. 1*, 26–30.
67. Smith, A.B. III, and Tong, R. (2010) *Org. Lett.*, **12**, 1260–1263.
68. Smith, A.B. III, Kim, W.S., and Tong, R. (2010) *Org. Lett.*, **12**, 588–591.
69. Sanchez, L. and Smith, A.B. III, (2012) *Org. Lett.*, **14**, 6314–6317.
70. Smith, A.B. III, Xian, M., Kim, W.S., and Kim, D.S. (2006) *J. Am. Chem. Soc.*, **128**, 12368–12369 and references therein.

5
Radical Reactions in Domino Processes

Guanghui An and Guigen Li

5.1
Introduction

Increasing demand for chemical and pharmaceutical products has been raising serious global questions of how to achieve high efficiencies of chemical synthesis and production and, concurrently, how to reduce hazardous waste generation and pollution. The concept of domino reaction was established for this purpose [1, 2]. Since 1993, when Tietze [2] formally introduced this concept to chemistry, impressive progress has taken place in chemical sciences, from which both academia and industry benefited substantially, as revealed by the citations in the literature. By taking advantage of the domino strategy, multiple chemical bonds and rings of complex products can be readily assembled, and the use of starting material, energy, and manpower for making products can be reduced [3]. Besides the attractive characteristics above, domino reactions can substantially accelerate the progress of drug discovery, development, and synthesis because of the shortened synthesis times and fast reaction rates under domino conditions. It is believed that the domino impact on chemistry is comparable to that of solid-phase peptide synthesis (SPPS), which was invented by the Nobel Laureate in chemistry, Bruce Merrifield [4], three decades ago.

While several excellent comprehensive review articles and books have been published on domino reactions and synthesis [5], this chapter is mostly focused on the latest development on radical reactions in domino processes in the past 5 years. So far, radical reactions have been not appreciated sufficiently because of the high reactivity of radicals and the difficulty in controlling the reactions. The extremely reactive radicals can react with each other at very high rates with reaction constants of $10^9-10^{10}\ M^{-1}\ S^{-1}$ to give dimeric side products. Fortunately, domino reactions help to minimize this disadvantage by directing the rapid reactions toward forming wanted products, which is particularly suitable for intramolecular cyclizing processes. In the meanwhile, the updated physical organic and stereochemistry enabled a deeper understanding on the behavior of a series of radicals with different stabilities and reactivities so as to control stereoselectivities under concise domino conditions.

Domino Reactions: Concepts for Efficient Organic Synthesis, First Edition. Edited by Lutz F. Tietze.

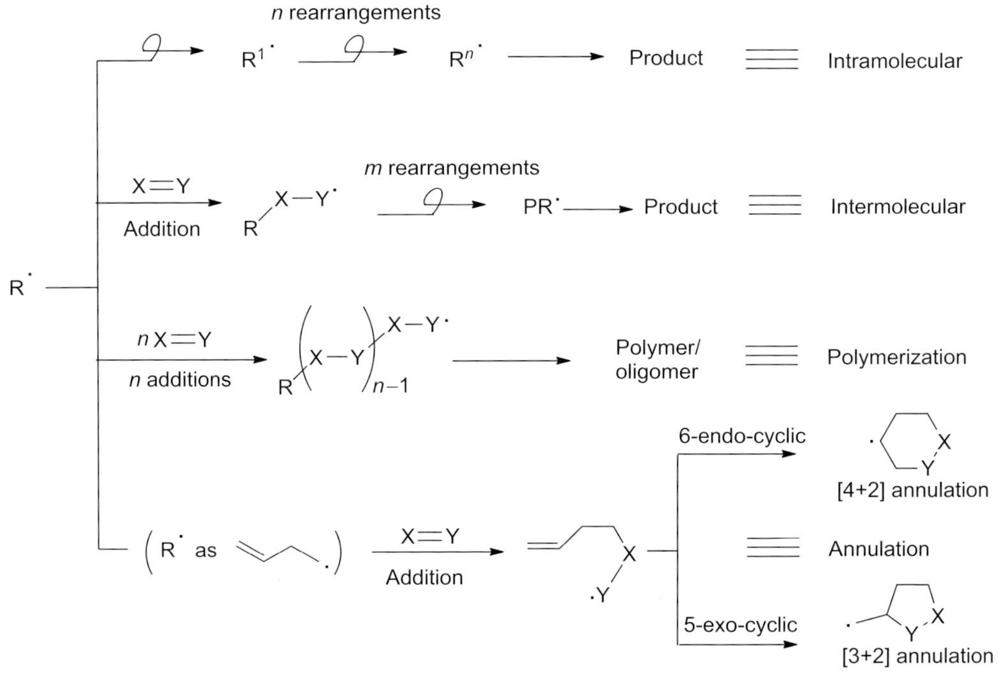

Scheme 5.1 General scheme of domino radical reactions.

As described by Tietze, Walton, and McCarroll [1b, 6], radical domino reactions are divided into the four types, as shown in a slightly modified Scheme 5.1. Among them, the unimolecular radical domino reactions are the most common in general organic synthesis and can be divided into six types [1, 6]. The domino annulation belongs to unimolecular reactions after the first addition step of generating radicals. Usually, intramolecular domino processes have an inbuilt entropy advantage, and this can result in numerous invaluable cyclic addition products including heterocycles. The domino cyclizations are usually furnished through addition and substitution cyclizations, which will occupy most of the material for this chapter.

A general intramolecular free-radical domino process was illustrated in an excellent review by Walton and McCarroll [6] (Scheme 5.2). In this scheme, each capital letter is assigned as C (cyclizations), S (substitutions), H (abstractions), M (1,2-migrations), and F (fragmentations). In the literature, these symbols can be combined with suffixes (exo = x, endo = n) to clarify the mechanistic details: for example, $\mathbf{C}^{5x}$ = 5-exo-cyclization and $\mathbf{C}^{6n}$ = 6-endo-cyclization. As shown in Scheme 5.2, the initial step involves the conversion of the configured precursor (S^b) into a free-radical intermediate ($R^{1\bullet}$) initiated by radical initiators (In) through abstractions or substitutions.

This initiation step is not included in the central chain process. However, the radical intermediate $R^{1\bullet}$ then undergoes unimolecular rearrangement and forms $R^{2\bullet}$ with or without the production of the side product B^1. Next, the domino reactions

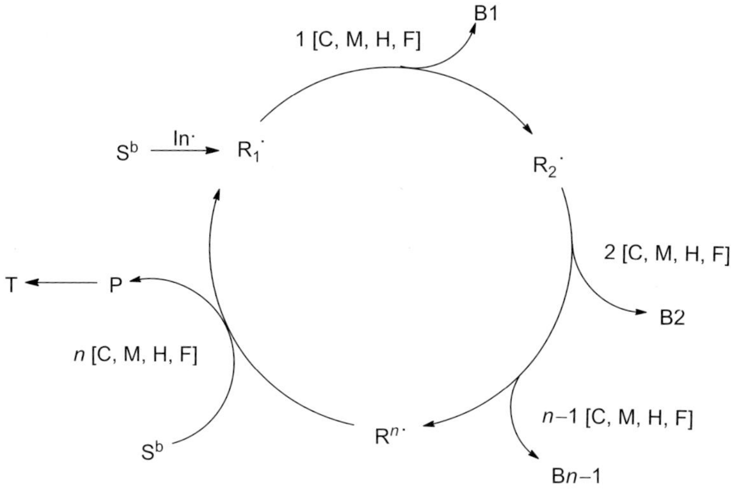

Scheme 5.2 Walton and McCarroll's description of radical domino sequence.

proceed through $(n-1)$ unimolecular rearrangements until the final intermediate $R^{n\bullet}$ is added onto the desired product (P); P will then be converted into the target molecule (T) via different functional manipulations. For most domino reactions, the final step regenerates the first radical intermediate, which is necessary for catalytic cycles.

This whole domino sequence becomes a chain reaction, which is not absolutely required for a domino reaction. For efficiently functionalizing domino products, individual rearrangements must be faster than termination of combination, disproportionation, and redox reactions of $R^{x\bullet}$, and faster than their reactions with the solvent, precursor, and initiator molecules. Also, the final intermediate $R^{n\bullet}$ should react selectively with S^b while the rest of the radical species do not. Hence, if the final rearrangement results in changes in polarity or reactivity of the propagating radical species, the generation of versatile O-centered radicals or vinyl-type radicals for efficient domino synthesis can be made.

5.2
Radical/Cation Domino Processes

A reaction sequence involving an initiating radical process and terminating cationic transformation is defined as radical/cation domino process. This type of reaction has been used in the synthesis of several heterocycles such as oxazines, indoles, quinolines, and isoquinolines. 5,6-Dihydro-4*H*-1,3-oxazines are present in several bioactive natural products and drugs. Also, they have been used in the syntheses of the Taxol side chain, unsaturated amino acids, alkylhomoserines and homocysteines, sphingosines, jasmonic acid derivatives, and amino alcohols [7]. Because of their high value, a number of synthetic routes have been developed toward the synthesis of these oxazines. Recently, Boto and coworkers reported a synthesis of 6-aryl-5,6-dihydro-4*H*-1,3-oxazines through a domino radical/cation reaction starting from the serine derivative *rac*-**1** [8]. This domino process consists of a sequential radical scission/oxidation/nucleophilic addition/intramolecular cyclization (Scheme 5.3).

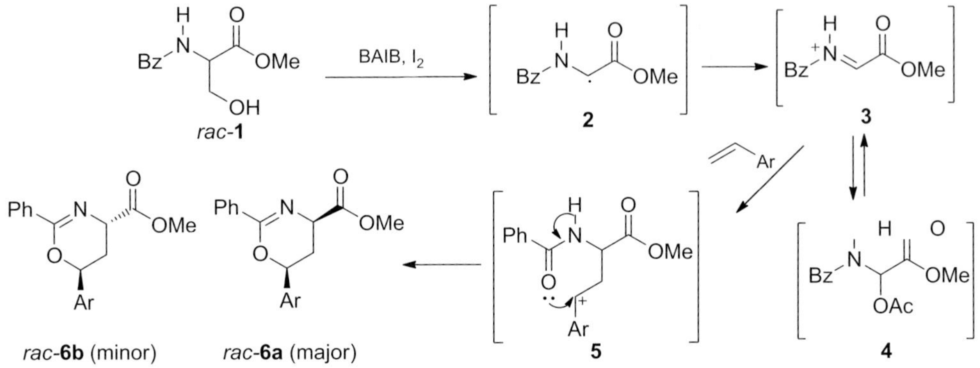

Scheme 5.3 One-pot domino conversion of serine derivatives into oxazines.

Methyl ester of serine *rac*-**1** under treatment with (diacetoxyiodo)benzene (BAIB) and iodine gives radical **2**, which is oxidized to give acyliminium ion **3**. This intermediate is then trapped with acetate to yield *N,O*-acetal **4**. The acyliminium ion **3** can regenerate from the acetal and react with vinyl arenes to produce the intermediate **5**. In the final step, an intramolecular cyclization furnishes the 6-aryl-5,6-dihydro-4*H*-1,3-oxazine **6**, with the cis isomer *rac*-**6a** as the major product (Scheme 5.3) [8]. In a similar domino process, aminosugars provide substituted oxazines, which are analogs of C-nucleosides [8].

Indoles are N-heterocyclic compounds present in several bioactive compounds and pharmaceuticals [9]. An interesting domino process to prepare *N*-arylindoles was developed by Zhang and coworkers through a visible-light-mediated oxidative C–N bond formation/1,2-carbon shift/aromatization sequence starting from styryl aniline **7**. First, the nitrogen-centered radical cation **8** is generated, which undergoes an electrophilic addition to the tethered alkene, thus initiating the domino reaction

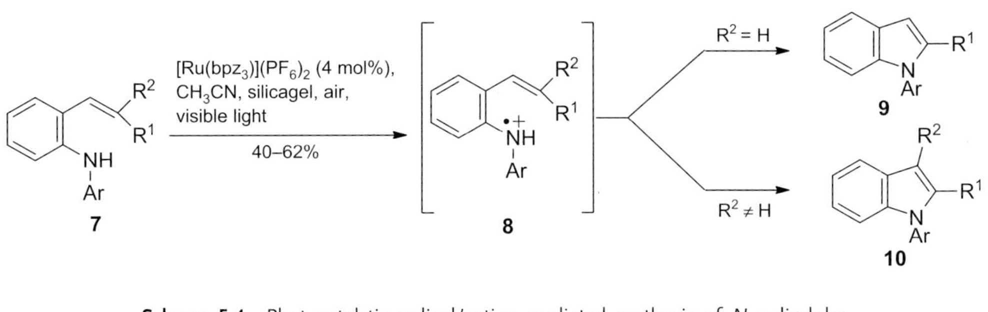

Scheme 5.4 Photocatalytic radical/cation-mediated synthesis of *N*-arylindoles.

involving either aromatization to give indole **9** (when $R_2 = H$) or C–C bond migration followed by aromatization to yield indole **10** (when $R_2 \neq H$) (Scheme 5.4) [10].

Masson and coworkers reported α-alkylation of imines through a three-component radical/cation domino process using bromomalonate **11**, an enamide **13**, and an alcohol as starting materials. First, an iridium species forms the alkyl radical **12** from **11** by single electron transfer (SET). Then enamide **13** reacts with **12** to produce the amidoalkyl radical **14**. An oxidative radical–polar crossover reaction then furnishes an *N*-acyliminium cation **15**, which finally reacts with nucleophilic alcohol to produce α-alkylated α-amido ether **16** (Scheme 5.5) [11].

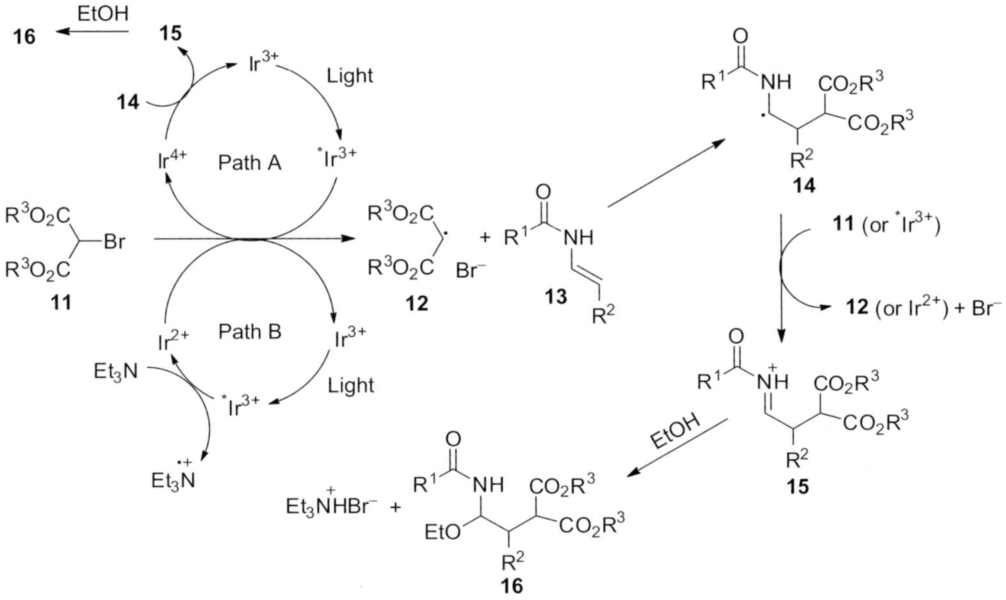

Scheme 5.5 Domino radical/cationic alkylation of enamides.

Wang and coworkers developed an interesting methodology for the synthesis of quinoline-2-carboxylates via domino Povarov reaction/aromatization of iminoethyl glyoxylates **17** with *N*-vinylamides **18**, induced by a catalytic amount of the radical cation salt $TBPA^{+\cdot}$ ([tris(4-bromophenyl)aminium hexachloroantimonate]) [12]. $TBPA^{+\cdot}$ oxidizes *N*-vinylamide **18** to its radical cation, which adds to imine **17** to produce a radical cation intermediate **A**. This undergoes an intramolecuar cyclization to generate the radical cation **B**. Then 1,7-H shift and a second electron transfer from **18** afford a tetrahydroquinoline **C**. Removal of the amido group to give intermediate **D** is followed by an oxidation, providing quinoline-2-carboxylate **19** in 45–90% yield (Scheme 5.6). In addition, they have also developed a domino process in which imines undergo a domino cyclization with *N*-vinyllactams catalyzed by a radical cation salt to give 2-methyl-4-anilino-1,2,3,4-tetrahydroquinolines [13].

In 2011, Hayashi and coworkers [14] reported an iron-catalyzed oxidative coupling of alkylamides **20** and arenes **21** to give benzylamides **25**. This domino process

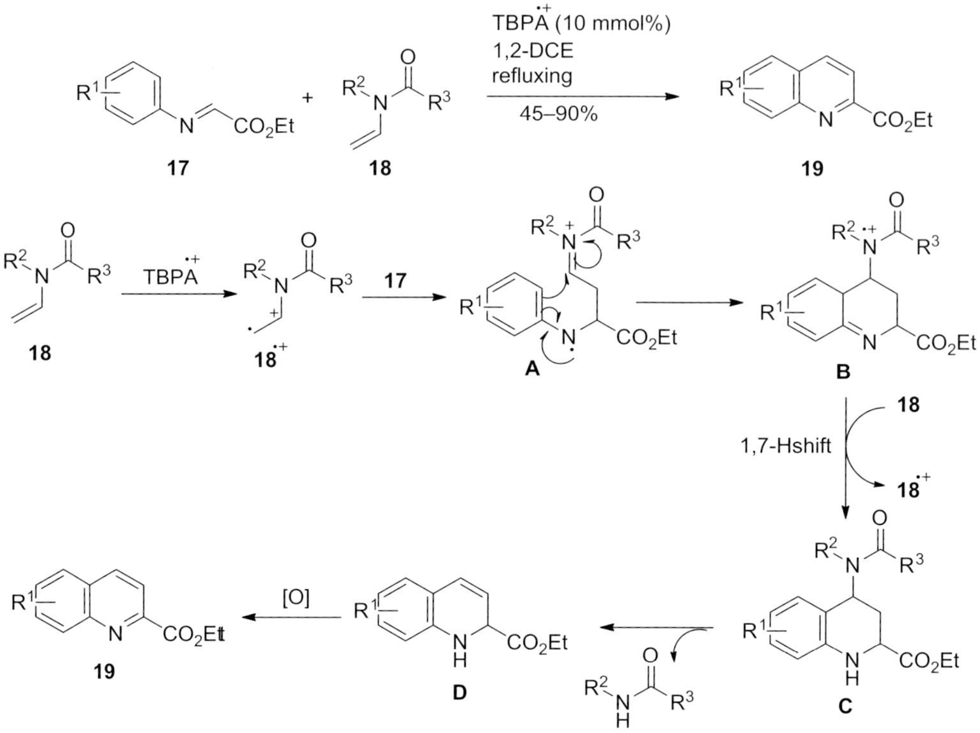

Scheme 5.6 Domino Povarov reaction/aromatization.

consists of an alkylamide oxidation (Cycle A) and a Friedel–Crafts reaction (Cycle B). The relatively stable α-(*tert*-butoxy)alkylamide **23** is produced from **20** through oxidation with *t*-Bu-OO-*t*-Bu/$FeCl_3$ in Cycle A and gets into Cycle B. Here, by Lewis acid-catalyzed elimination of *t*-BuOH, acyliminium salt **22** is formed, which reacts with the highly nucleophilic arene **21** to give the coupling product **25** (Scheme 5.7).

Furthermore, in an intramolecular oxidative coupling reaction starting from the lactam **26**, isoquinoline alkaloids have been prepared. Thus, reaction of **26** with *t*-Bu-OO-*t*-Bu/$FeCl_3$ led to **27**, which was converted into *rac*-crispine A (**28**) by reduction with $LiAlH_4$ and *rac*-trolline (**29**) by demethylation of the two methyl ethers with BBr_3 (Scheme 5.8).

Cha and coworkers [15] described an interesting example of a radical/cation-mediated domino reaction for the stereoselective preparation of bicyclo[5.3.0]decan-3-ones **31** using aminocyclopropane **30** as substrate which was prepared by employing an intramolecular Kulinkovich cyclopropanation of olefin-tethered amide.

Aerobic oxidation of **30** leads to the radical cation **A**, which undergoes a ring expansion/cyclization sequence to give **31** via the intermediates **B–D** (Scheme 5.9). In this process, **C** is trapped by oxygen to furnish **D**, which by electron transfer with **30** yields the corresponding hydroperoxide **31**. It is important to note that the overall ring expansion/cyclization process is accelerated using silica gel and fluorocarbon

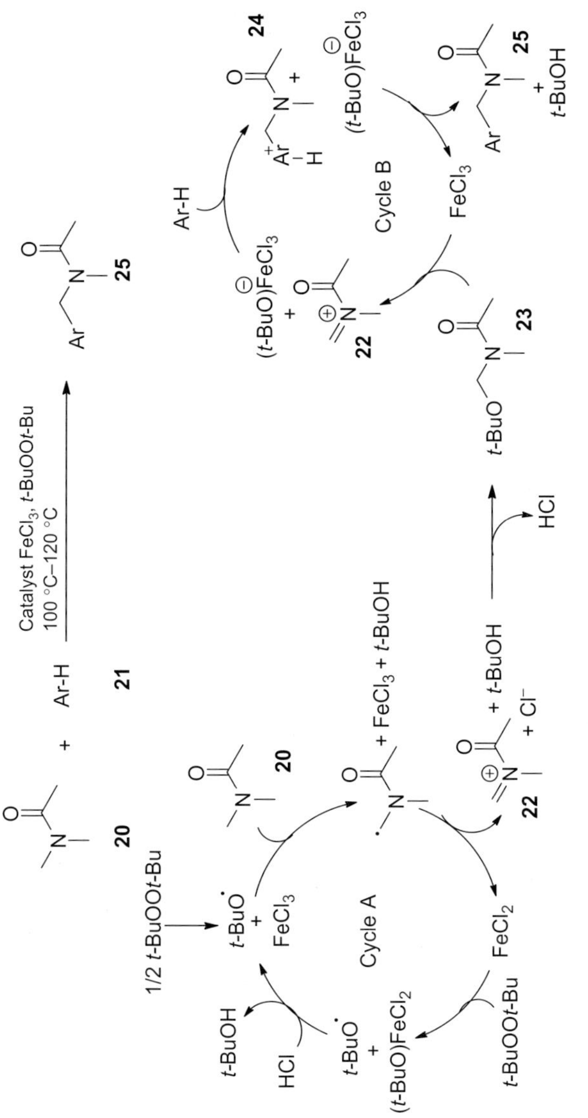

Scheme 5.7 Radical/cation-mediated domino oxidative coupling.

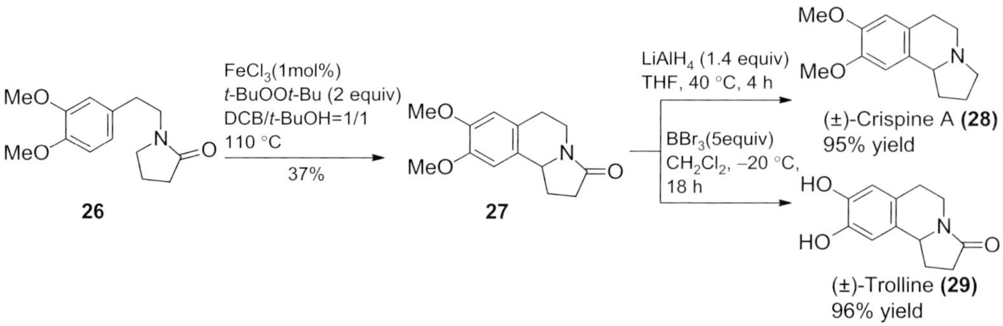

Scheme 5.8 Domino intramolecular oxidative coupling reaction in isoquioline alkaloids synthesis.

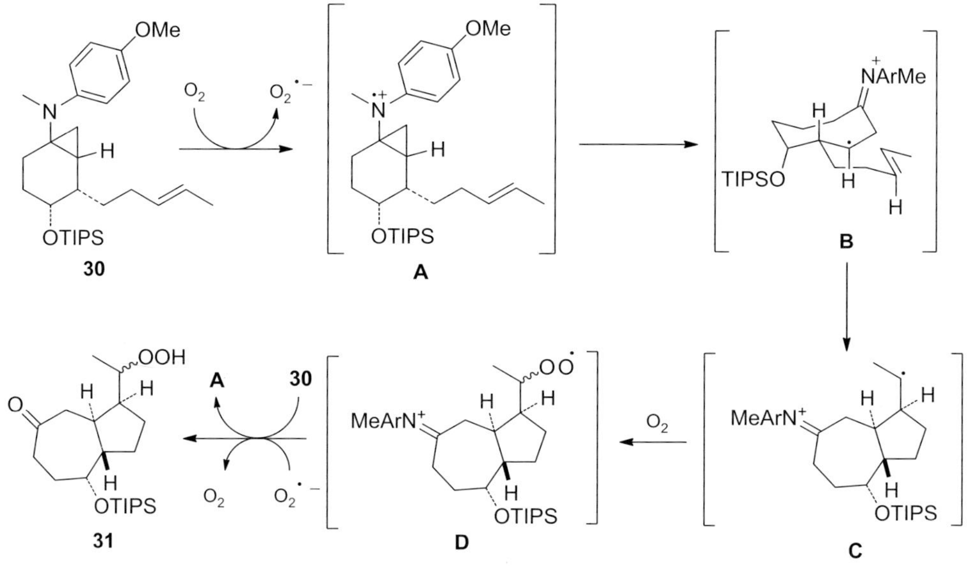

Scheme 5.9 Radical/cation-mediated domino ring expansion/cyclization.

solvents, such as 2,2,2-trifluro ethanol. Thus, the reaction rate increases with enhancement of the solubility of molecular oxygen in the solvent.

5.3
Radical/Anionic Domino Processes

Samarium(II) iodide has been proven to be an efficient reagent to initiate radical processes that can be utilized in domino reactions. In 2001, Schwartz and coworkers [16] reported a domino reductive fragmentation and aldol reaction with 1,4-diketones promoted by SmI_2 to afford the ring contracted products. The Reissig group then found that ketyl radicals formed from **32** by SmI_2 could react with an indole moiety intramolecularly to form a cyclic intermediate **33**, which was further transformed via reactions with a phenol or a halide [17]. Recently, they have also found that the cyclic intermediate **33** could also undergo an intramolecular acylation to form a tetracyclic product, which was utilized to synthesize the natural alkaloid strychnine [18]. When precursor **34** was subjected to 2.4 equiv of samarium diiodide in the presence of HMPA (hexamethylphosphoramide), the cyclization and subsequent acylation readily occurred and the desired tetracyclic **35** was isolated as the major product in diastereomerically pure form (Scheme 5.10). The domino reaction apparently proceeded within seconds as indicated by the decolorization of SmI_2. As byproduct, **34** was also isolated in a low amount, which was probably formed by a reductive fragmentation of compound **35**. Addition of bromoacetonitrile to the reaction mixture improved the overall yield to 75–80% because of the *in situ* reconversion of **36** into product **35**.

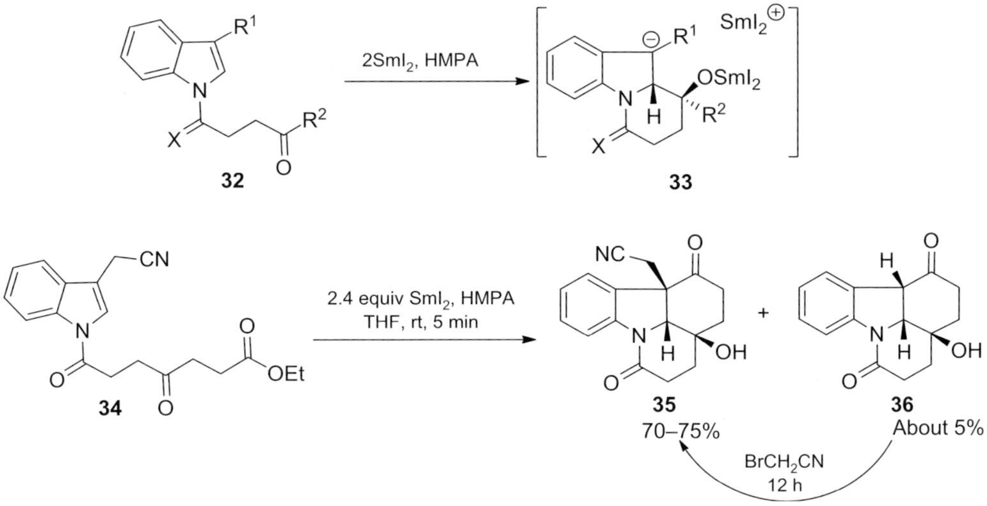

Scheme 5.10 Samarium(II) iodide-induced domino radical cyclization/acylation reaction.

In general, samarium(II) iodide-mediated radical/anionic domino processes display an important role in total syntheses [19]. Thus, Naito and coworkers [20] used a samarium(II) iodide-mediated domino process for the preparation of martinellic acid (**40**). In this approach SmI_2 proved to be the preferred radical initiator, as it gave the desired precursor **39** starting from **37** in 41% isolated yield. The proposed formation of a α-aza radical species **A** might explain the enhanced diastereoselectivity in the reaction because of a chelation of Sm(III) through the ester moiety and the nitrogen atom. As a result, the desired cis-fused **39** was obtained as the main product, which was formed from the intermediate radical **38** by subsequent reductive cleavage of the nitrogen–oxygen bond and spontaneous lactamization (Scheme 5.11).

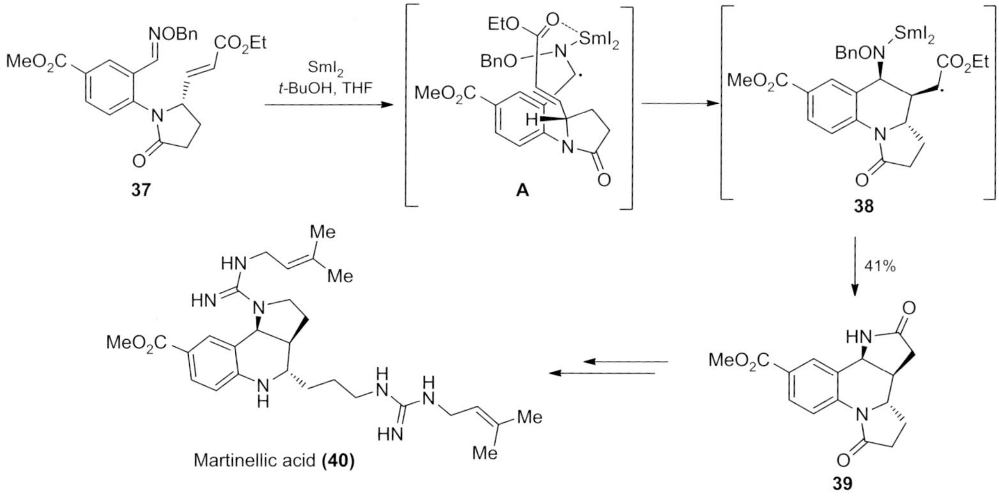

Scheme 5.11 SmI_2-promoted cyclization protocol in the total synthesis of martinellic acid (**40**).

A cyclization/Barbier-type reaction was reported by Curran and coworkers [21] in 2004 in the total synthesis of penitrem D **46**. In this work, the aryl radical generated from the iodoarene **41** and SmI_2 proceeded to attack the tethered cyclobutene to from a cyclobutyl radical **42**. Subsequently, reaction with SmI_2 led the organo-samarium species **43**, which underwent a Barbier-type reaction with acetone to give the tertiary alcohol **44** in 40% yield. The product contains the BCD ring system of penitrem D **46** (Scheme 5.12).

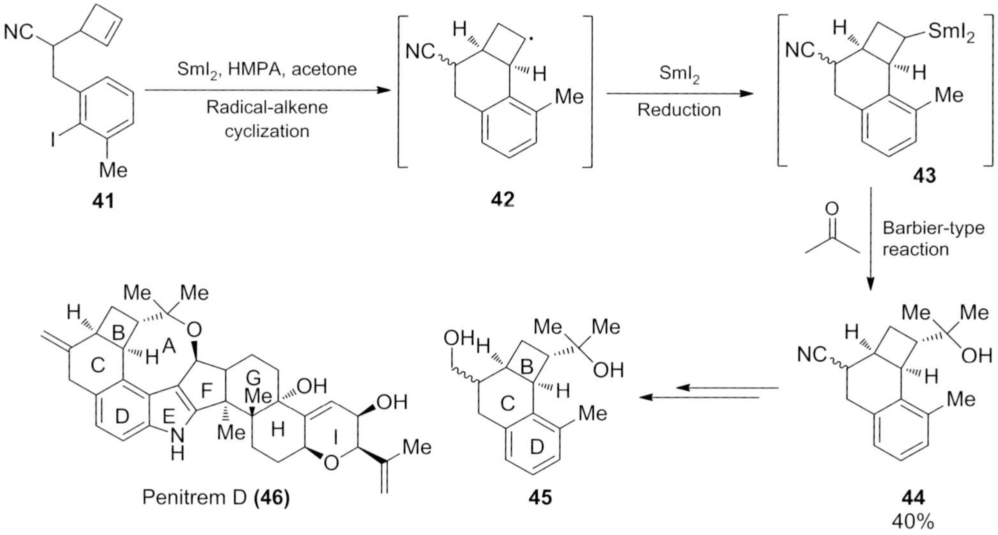

Scheme 5.12 Synthesis of the BCD system of penitrem D **46** with SmI_2.

Another SmI_2-mediated domino process was reported by the Baran group in the synthesis of (+)-Cortistatin A **51** [22]. When the bromoketone **47** was treated with SmI_2 in the presence of the additive DMPU (1,3-dimethyl-3,4,5,6-tetrahydro-2(1*H*)-pyrimidinone), a cyclopropane fragmentation occurred to form an α-bromo radical **48**, which underwent subsequent reduction by SmI_2 and then bromide elimination to give **49** with a newly formed C=C double bond and a Sm(III) enolate. After a few minutes, the addition of 2,4,4,6-tetrabromo-2,5-cyclohexadienone (TBCHD) resulted in the α-brominated product **50**, which was converted into cortistatin A **51** (Scheme 5.13).

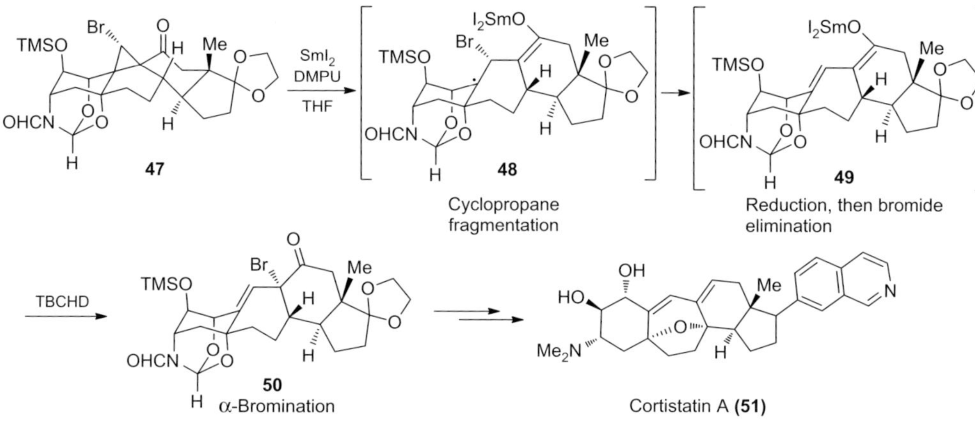

Scheme 5.13 SmI_2-promoted domino reaction in the total synthesis of cortistatin A.

In 2012, the Procter group reported several compelling SmI_2-promoted domino reactions [23]. For example, they described a "lactone radical cyclization and cyclization process mediated by SmI_2–H_2O" [23e]. The radical anion **53** generated by SmI_2 from lactone **52** would attack the tethered C=C double bond to form the cyclic hemiacetal **54** after the reduction with SmI_2. The equilibrium between **54** and the ketone **55** leads to further reduction of **54** to form cycloheptan-1,4-diols **56**. Through oxidation with DMP (Dess–Martin Periodinane), **56** can be transformed to the hemiacetal **57** (Scheme 5.14). This cyclization process allows access to complex molecular architectures in a single reaction using simple and readily available reagents.

It is important to note that other Sm species (such as $SmBr_2$, $SmCl_2$) were also used in radical/anionic domino reactions. In addition, some other lanthanides, such as TmI_2, DyI_2, NdI_2, YbI_2, CeI_2, $Eu(ClO_4)_2$, and so on, can also serve as potential substitutes for SmI_2 [24].

The Renaud group has developed a wide range of C–C bond-forming radical reactions such as conjugate addition, allylation, alkynylation, and vinylation using organoboranes as radical precursors [25]. In 2009, they reported a domino procedure involving radical conjugate addition of alkylcatecholboranes **59** to enones **58**, followed by intramolecular aldol reaction (Scheme 5.15).

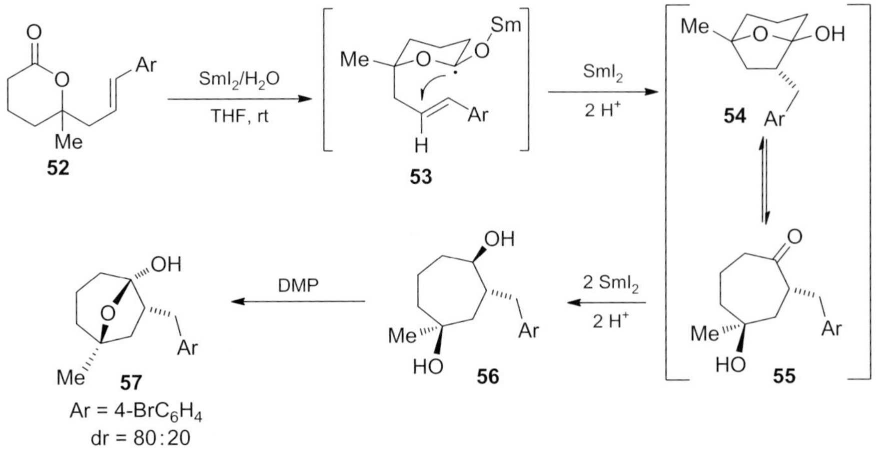

Scheme 5.14 SmI_2-promoted reductive cyclizations of alkenyllactones.

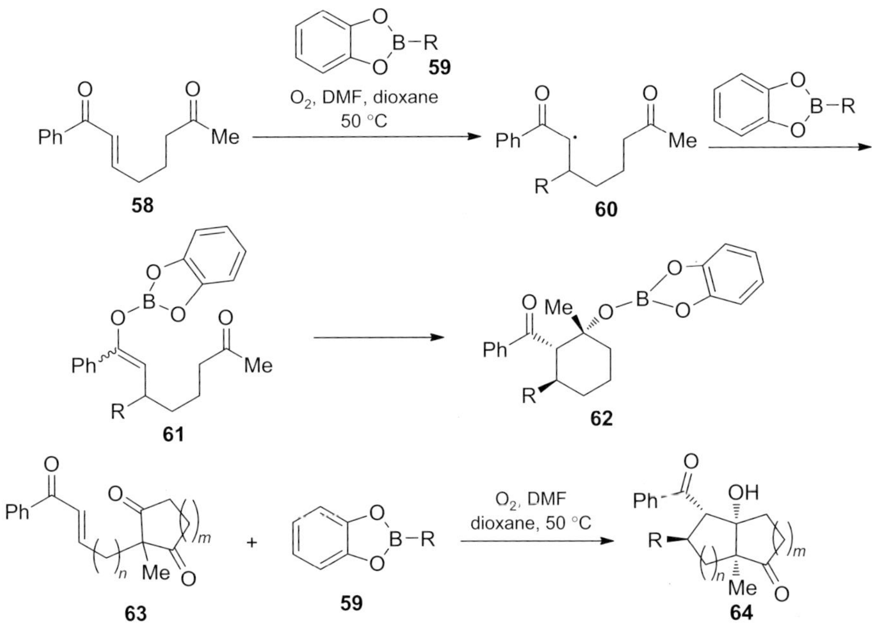

Scheme 5.15 Organoboranes-mediated domino radical conjugated addition/aldol cyclization.

The applications in stereoselective synthesis of monocyclic and bicyclic products with up to four contiguous stereogenic centers have been reported, as shown in Scheme 5.15 in which **63** reacts with the borane **59** to give **64** [26]. In these processes, the addition of the radical to the enone is followed by trapping of the enolate radical **60** with an organoborane and the formation of an intermediate boron enolate **61**, which can undergo an intramolecular aldol reaction to form the cyclic products **62**. The proposed mechanism is supported by the stereochemical outcome of the reaction, where the OH and the acyl group are cis configured in accordance with similar intramolecular aldol reactions reported by Krische [27].

In 2010, Maury and coworkers reported a radical/anionic domino processes initiated by $ZnMe_2$ and air. Dimethyl zinc-mediated addition of acyloxymethyl radicals **66**, formed from **65**, to diethyl fumarate **67** led via the proposed intermediates **68–70** to the highly stereoselective formation of disubstituted γ-lactones **71** in medium to good yields. Five equivalents of the starting iodide **65** and 3 equiv of dimethylzinc are necessary to achieve high yields (up to 98% yield). As shown in Scheme 5.16,

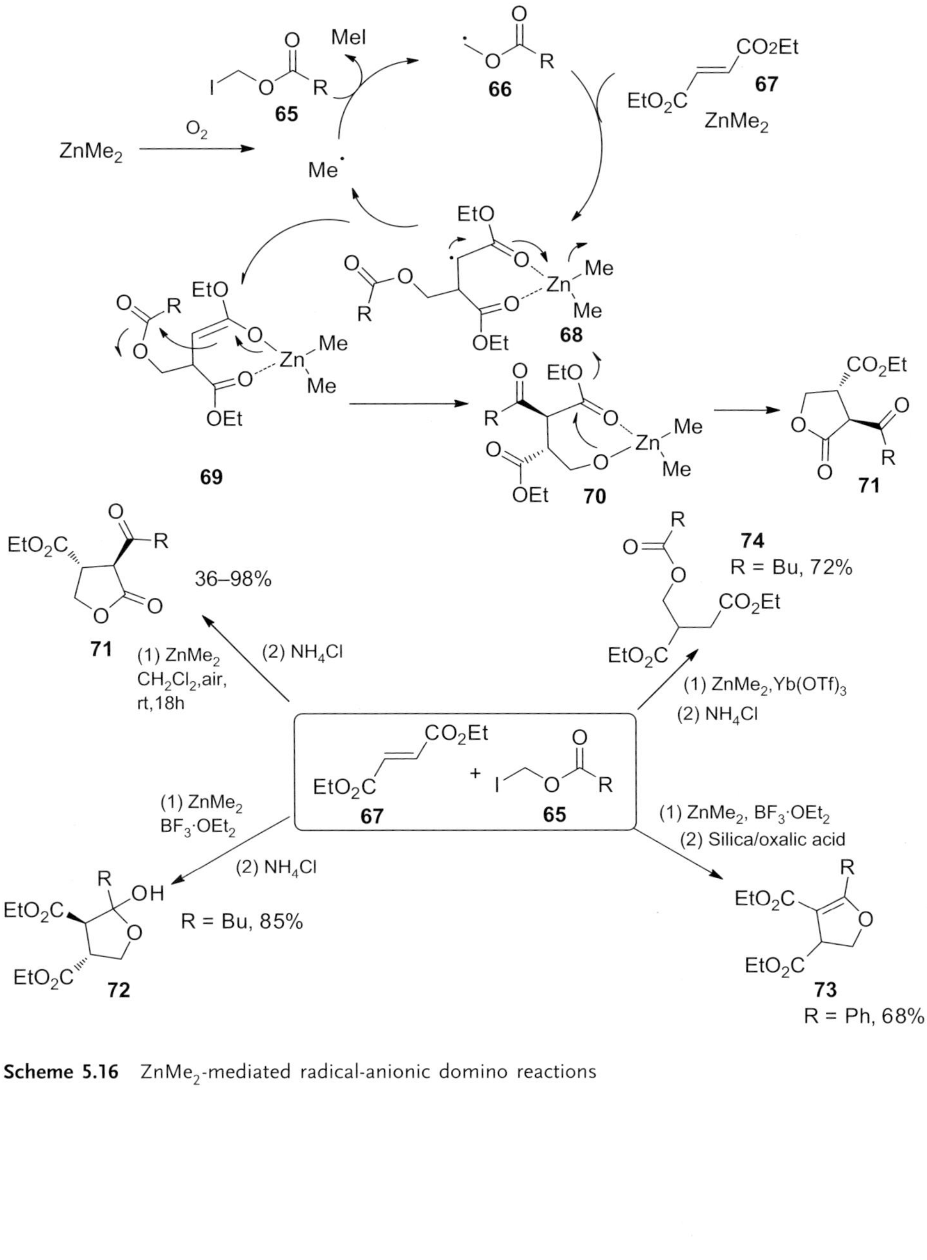

Scheme 5.16 $ZnMe_2$-mediated radical-anionic domino reactions

the acyloxymethyl radical **66** formed through iodine atom transfer adds to the activated double bond. The resulting α-acyloxy radical undergoes homolytic substitution at the zinc ion leading to a zinc enolate **68**, which evolves via intramolecular acyl transfer. The resulting zinc alkoxide affords the disubstituted γ-lactone **71** through intramolecular nucleophilic substitution at the ester group. This protocol provides a new example of a one-pot process mediated by a dimethylzinc involving a domino radical/anionic reaction. Interestingly, the addition of monodentate or polydentate Lewis acids to the reaction mixture allows the preparation of totally modified products such as **74**, **73**, and **72** (Scheme 5.16) [28]. When the strong Lewis acid $BF_3{\cdot}OEt_2$ was used, only lactol **72** was detected in the crude mixture. If treated with silica/oxalic acid, the dehydration product **73** was isolated in 68% yield. The outcome of the reaction changed upon addition of 1.2 equiv of $Yb(OTf)_3$. In this case, the acyl group transfer was suppressed, and only the adduct **74** of the acyloxymethyl radical was isolated in 72% yield.

5.4 Domino Radical/Radical Process

Radical/radical domino processes represent the majority of radical domino reactions. The synthesis of natural products and the methodology development associated with this topic still remain one of the most attractive fields in organic chemistry [29–31].

A subcategory of the Melodinus alkaloids, such as epimeloscine **77** or meloscine **78**, has a unique pentacyclic scaffold with four stereogenic centers. After the first total synthesis of meloscine by Overman in 1989 [32], a few other syntheses of these alkaloids have been reported [33, 34]. An effective domino radical annulation of divinylcyclopropane **75** has been developed by the Curran group in the course of the first total synthesis of (±)-epimeloscine **77**, which readily epimerized to (±)-meloscine **78** (Scheme 5.17) [30, 31]. In Curran's strategy, substrate **75** was subjected to classical radical conditions (tributyltin hydride/AIBN, 2,2′-azobisisobutyronitrile, refluxing toluene) to generate a radical which was consumed by annulation with a cyclopropane motif in an intramolecular approach to afford **76** with rings B and C concurrently in 38% yield. Compound **76** was then converted into epimeloscine **77**.

The Stachybotrys metabolites bisabosquals A–D (**79–82**), isolated in a screen for inhibition of microsomal squalene synthase [35], have disclosed activity against squalene syntheses from rat liver and HepG2 cells, and are of interest as antihypercholesteremic agents (Scheme 5.18) [36]. Owing to their hexahydrobenzofurobenzopyran ring in the cis, cis-fused tetracyclic ring system bearing five stereogenic centers, their synthesis is a difficult task [37]. The first total synthesis of *rac*-bisabosqual A **79** was carried out by Parker and coworkers [38] employing a domino 5-exo/6-exo radical cyclization as the key step (Scheme 5.18). Treatment of aryl iodide **83** with s-Bu_3B and $(TMS)_3SiH$ (TMS, trimethylsilyl) in air provided the aryl radical **84**, which underwent a 5-exo-trig cyclization forming the intermediate **85**. The following 6-exo-trig cyclization afforded a 3 : 2 mixture of desired product **87**

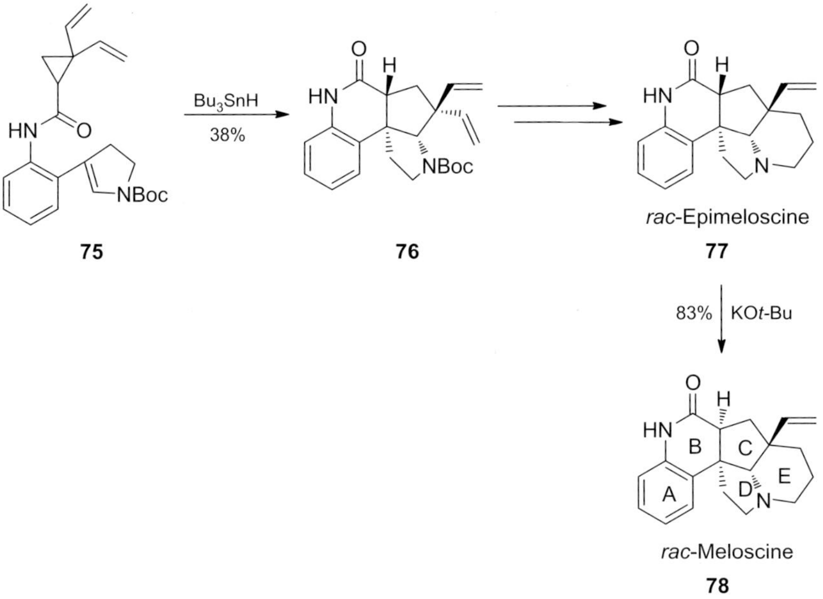

Scheme 5.17 Synthesis of *rac*-epimeloscine **77** and *rac*-meloscine **78**.

and its epimer **86** in 73% yield. The unwanted epimer could be partially transformed into the desired product by treatment with TBD (1,5,7-triazabicyclo[4.4.0]dec-5-ene) and $NaHCO_3$, resulting in a 2 : 1 mixture of **87** and **86** in 94% yield.

Another domino radical cyclization approach, which allows the construction of the B and D rings of the alkaloid (−)-dihydrocodeinone **92**, was also described by the same group (Scheme 5.19) [39]. In this reaction, the β-alcohol **88** was transformed into the tetracycle **91** in 30% yield as a single diastereoisomer by attack at the carbon–bromide bond under common radical conditions (tributyltin hydride/AIBN, refluxing toluene). The resulting aryl radical **89** initiated a 5-exo-trig/6-endo-trig cyclization sequence to give **91** via **90** with the diastereoselective formation of two stereogenic centers.

During their study on 6-exo-trig radical cyclizations of axially chiral α-halo-ortho-alkenyl anilides, Curran and coworkers [40] prepared the tricyclic amide **93/94** in moderate yield (49–51%) and excellent chirality transfer (98–99%) via a 6-exo-trig/5-exo-trig sequence (Scheme 5.20). The reaction was proposed to be initiated by an attack at the carbon–bromide bond with Bu_3SnH, providing α-amidyl radical **97** (Scheme 5.21).

The subsequent 6-exo-trig cyclization proceeded via a constrained transition state (TS) **97**, and led to the desired product **100** in a 5-exo-trig manner via **98** and **99**.

The fungal metabolite (+)-fusarisetin A **108**, produced by the soil fungus *Fusarium* sp. FN080326, holds promise as a valuable anticancer agent (Scheme 5.22) [41]. To investigate the structure–activity relationship of (+)-fusarisetin A, Gao's

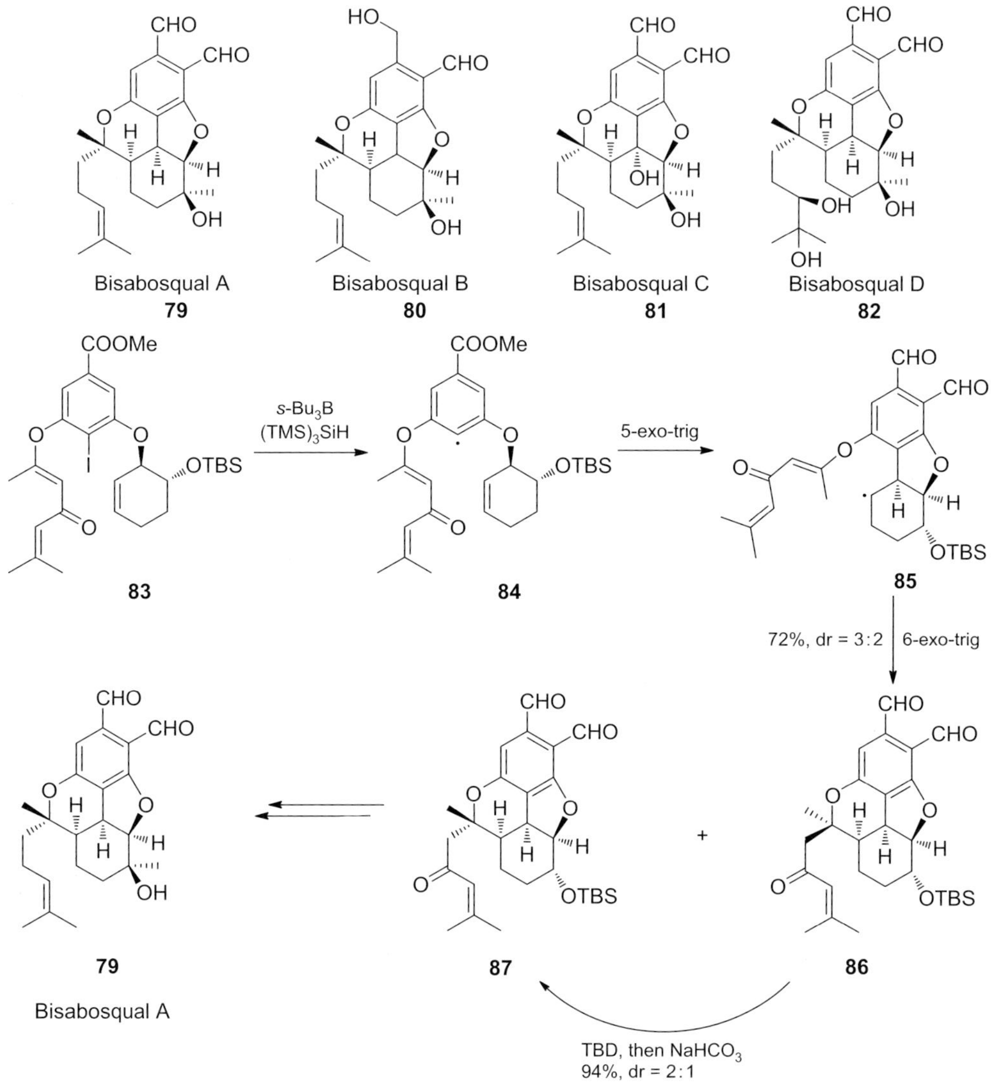

Scheme 5.18 Bisabosquals A–D and synthesis of (±)-bisabosquals A **79**.

group developed the first asymmetric synthesis of (+)-fusarisetin A with a 5-exo-trig/oxidation/6-exo-trig sequence as the key step [42]. Equisetin **101** was subjected to $Mn(OAc)_3$ and air in acetic acid to generate Mn-enolate **102**, which was transformed into the spiro tetracyclic compound **103**. It followed an oxidation in which a hydrogen abstraction took place after the 6-exo-trig cyclization of **104** to give **106/107** via **105**. (+)-Fusarisetin A (**108**) was then obtained by the reduction of the peroxide **106/107** with Zn in acetic acid in 44% yield over two steps from equisetin.

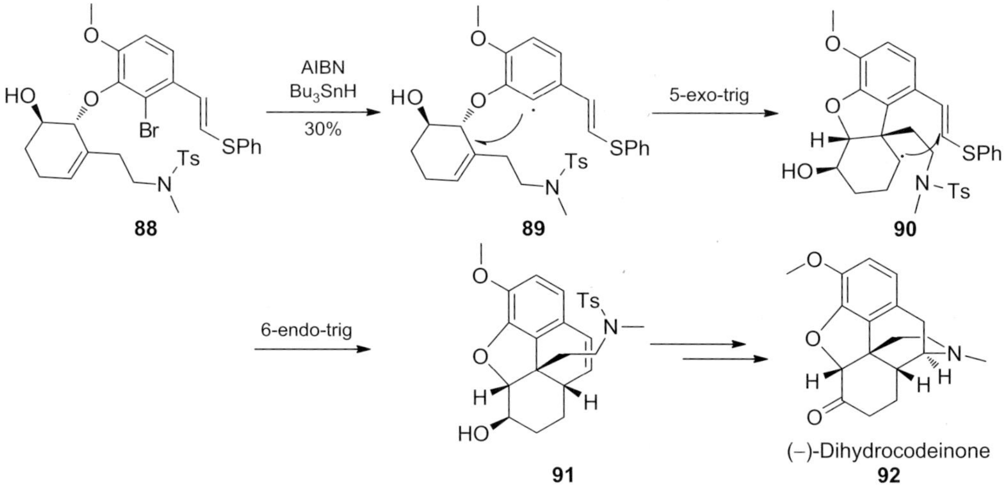

Scheme 5.19 Synthesis of (−)-dihydrocodeinone **92**.

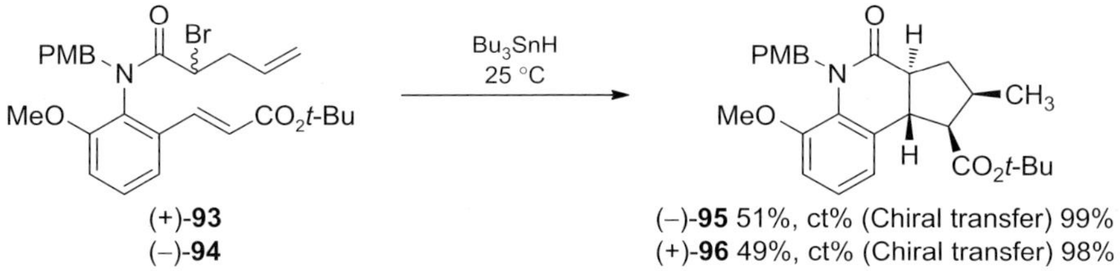

Scheme 5.20 Domino 6-exo-trig/5-exo-trig sequence.

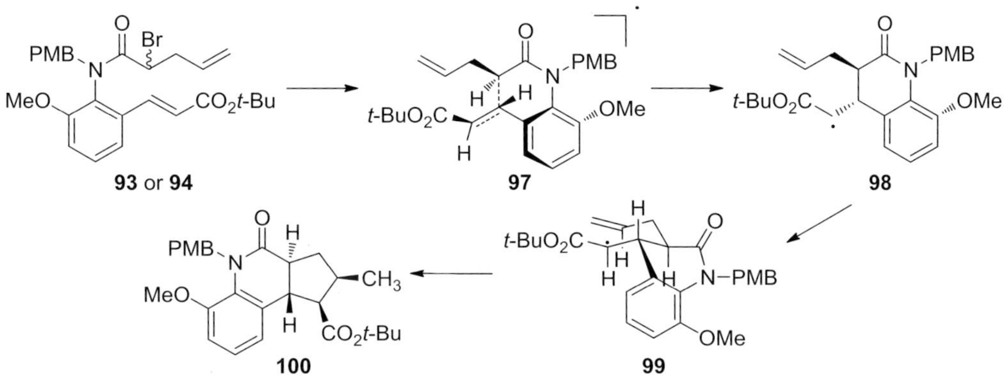

Scheme 5.21 Mechanism and constraint transition state of domino 6-exo-trig/5-exo-trig sequence.

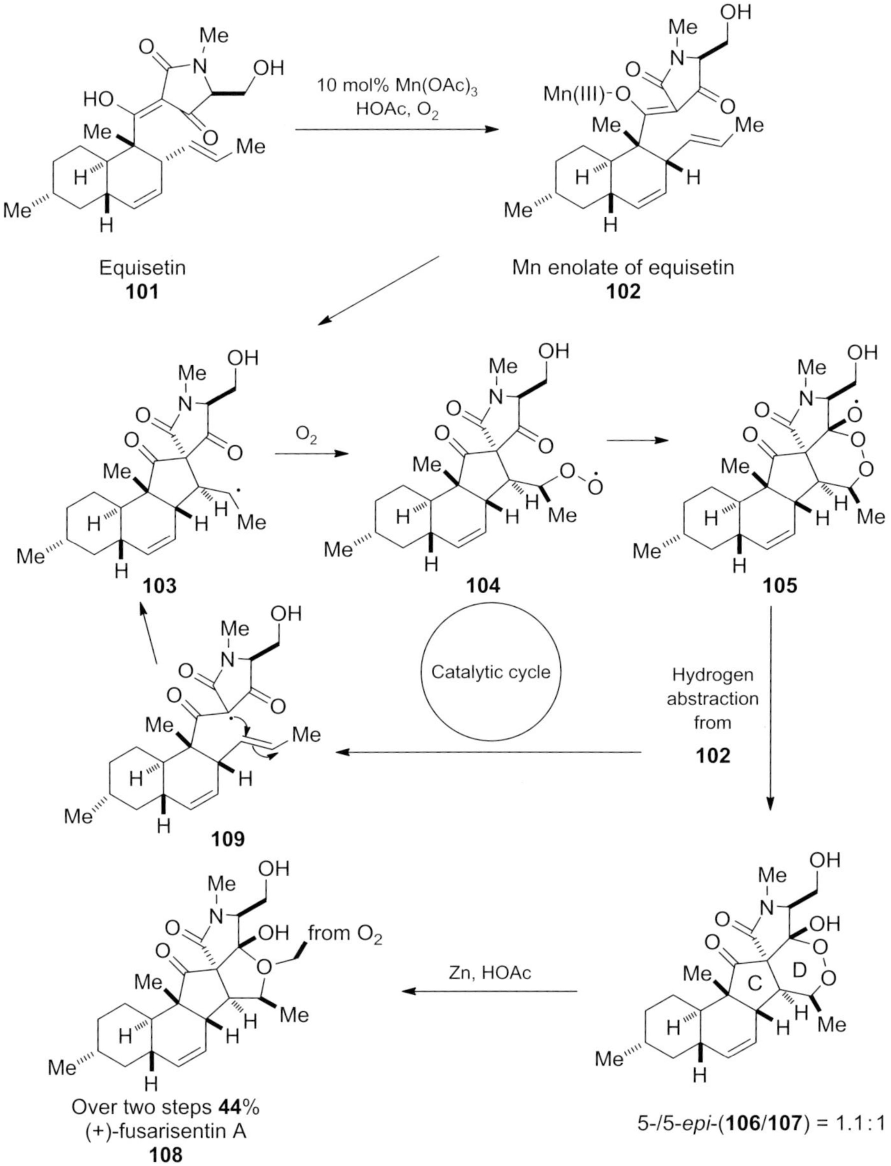

Scheme 5.22 Synthesis of (+)-fusarisetin A **108**.

Fortucine **112**, first isolated from the Fortune variety of narcissus by Tokhtabaeva *et al.* [43] in 1987, is a Lycorine alkaloid. To clarify the structure, Zard and coworkers [44] carried out the first total synthesis of (±)-fortucine using a radical domino process as the key step (Scheme 5.23). It was suggested that an amidyl-nitrogen radical is generated from **110**, followed by a 5-exo-trig cyclization in sequence with

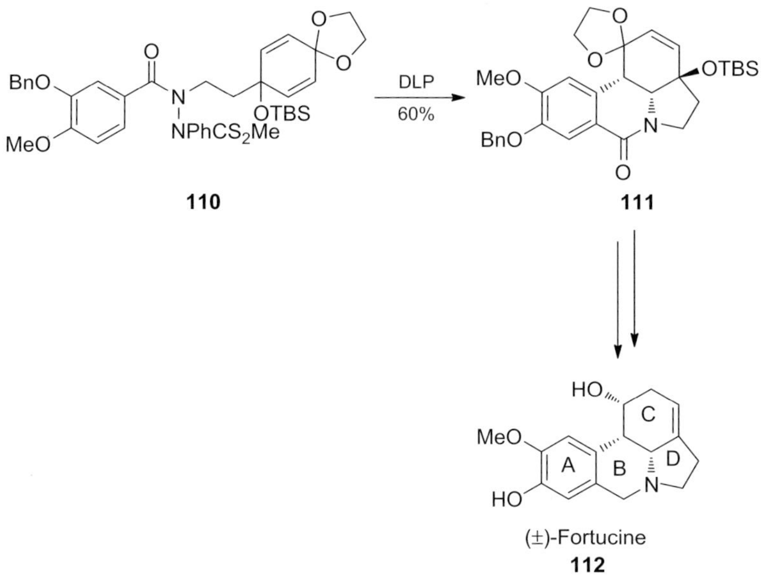

Scheme 5.23 Synthesis of (±)-fortucine **112**.

radical addition to the phenyl ring. The tetracyclic compound **111** was obtained, both stereoselectively and regioselectively, as the major product in 60% yield. Isomers arising from the trans B/C ring junction were not observed, and the desired regioisomer was formed in a 14 : 1 ratio (para/ortho with respect to the benzyloxy group).

Naturally occurring oroidin alkaloids possess interesting biological properties, such as antibiotic, antiproliferative, and immunosuppressive activities, along with an attractive molecular skeleton [45]. There are two types of domino radical processes that were devised by Chen and coworkers [46] in order to access their core skeleton (Scheme 5.24). (*E*)-Allylic β-ketoesters **113** were converted into lactones **117** and **118**, respectively, via the radical intermediate **114** using $Mn(OAc)_3$ in HOAc depending on the controlling element X in **113**. Moreover, lactone **117** and its decarboxylated compound **119** could readily be converted into oroidin dimers of type **118** and **120** by oxidation. In the radical cyclization of **113** toward **117** with a stereogenic center in R′, only two out of eight possible diastereomers (2.1 : 1) were obtained. The stereogenic center in the side chain controls the diastereoselectivity of this transformation through a moderate $A^{1,3}$ strain. Using the corresponding compound with a *Z*-double bond gives the opposite selectivity (1 : 1.5), and constraining the *Z*-double bond in a cyclic system affords the corresponding lactone as a single stereoisomer.

Garcibracteatone **126**, isolated from the bark of *Garcinia bracteata*, is a polycyclic polyprenylated acylphloroglucinol (PPAP) natural product [47]. It is assumed that biosynthesis of **126** and of nemorosonol **124** takes place by a 7-endo-trig/5-exo-trig radical cyclization of Weddellianone A via the radical intermediates **121–123**

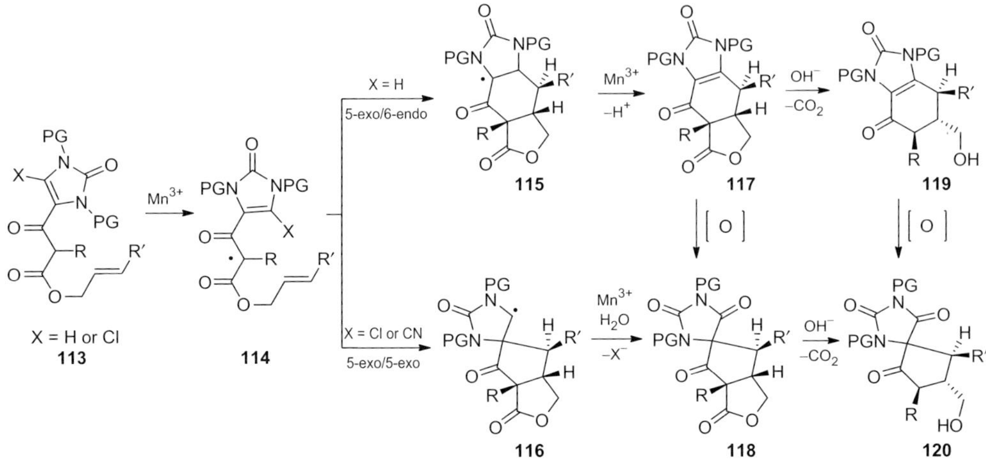

Scheme 5.24 Proposed mechanism of the 5-exo/6-endo and 5-exo/5-exo cyclization domino process.

(Scheme 5.25). Accordingly, a biomimetic total synthesis of garcibracteatone **126** was achieved through a domino radical reaction starting from **127** by treatment with $Mn(OAc)_3$ to give the diketo radical intermediate **121**, which underwent a 7-endo-trig/5-exo-trig/5-exo-trig/aromatic radical substitution sequence to afford **126** in 14% yield (Scheme 5.26) [48]. In addition, 8% of 5-*epi*-garcibracteatone **128** was obtained.

Compounds possessing a bicyclo[2.2.2]diazaoctane core with a keto or diketopiperazine (DKP) moiety can be found in many natural products and pharmaceutical compounds [49–51]. Their interesting bioactivity has drawn the attention of organic chemists, resulting in the first total synthesis of marcfortine B **135** by Trost and coworkers [52] in 2007 (Scheme 5.27). They created a radical domino process in forming the bicyclo[2.2.2]diazaoctane core structure **134**. Under classical radical conditions (tributyltin hydride/AIBN, refluxing toluene), the substrate **130** was transformed into the radical intermediate **131**, creating a quaternary carbon via a 6-exo ring closure. This was further reacted with AIBN, inducing a 1,4-hydrogen abstraction to give **134** via **132** and **133** in 61% yield as the sole product with the evolution of nitrogen.

Triquinanes **137** serve as good examples of a challenging framework in natural product synthesis [53]. In 2012, a domino radical cyclization to give vinylogous carbonates and carbamates was developed by Gharpure and coworkers [54], which involves an unprecedented, highly stereoselective formation of heterocyclic rings (Scheme 5.28).

The strategy is also useful to gain access to oxa- and azatriquinanes and lactone-bearing as well as uracil-fused angular triquinanes. Exposure of the domino precursor **138** to AIBN and Bu_3SnH provided the aryl radical **139**, which led to a 5-exo-trig/5-exo-trig sequence terminated by hydrogen abstraction (Scheme 5.29).

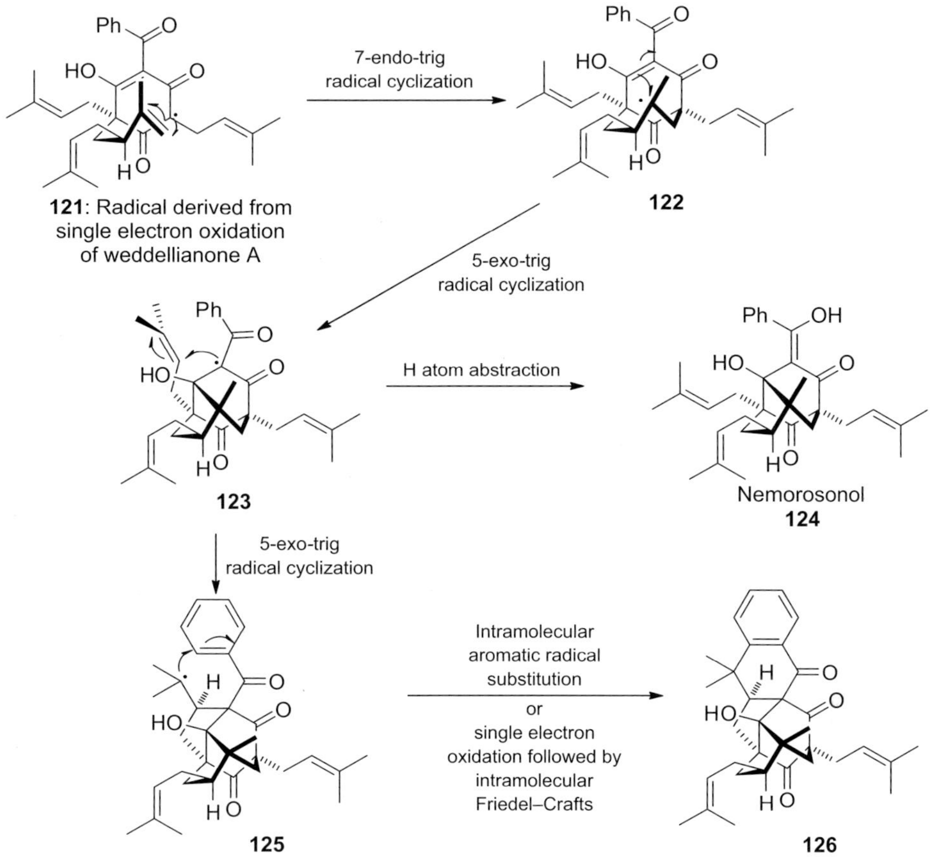

Scheme 5.25 Biosynthesis of garcibracteatone and nemorosonol from weddellianone A.

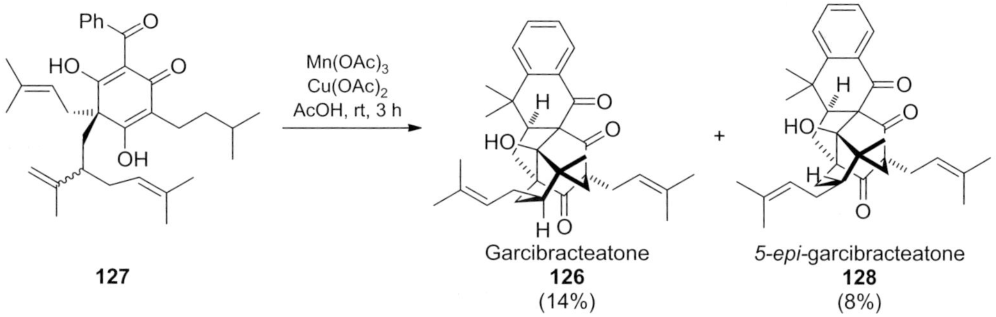

Scheme 5.26 Synthesis of garcibracteatone **126**.

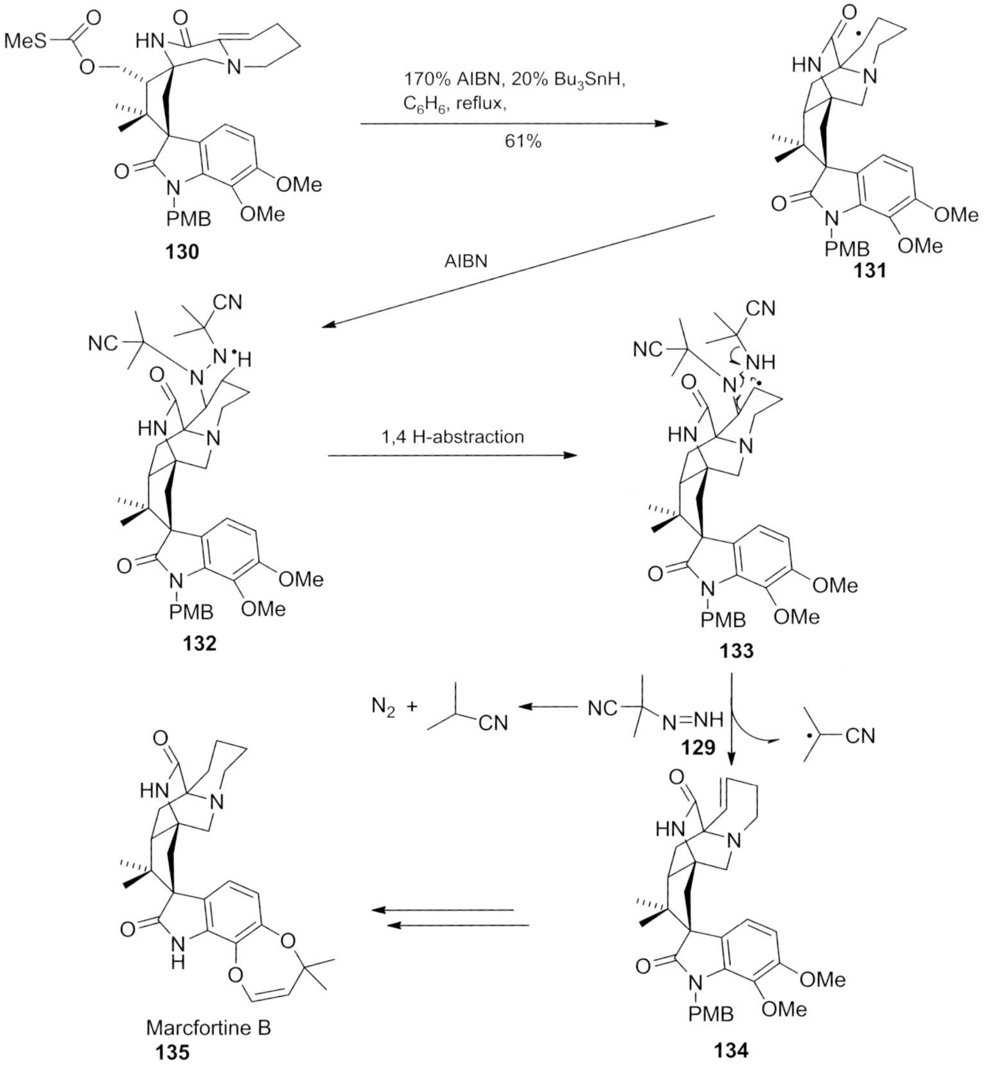

Scheme 5.27 Synthesis of marcfortine B **135**.

The resulting tetracyclic product **137** was obtained in moderate to very good yields (45–82%) and excellent diastereoselectivity ($\geq$19 : 1).

Davies and coworkers [55] disclosed a radical domino approach for the facile construction of bicyclic ring systems during their investigation on bis-vinyl ethers (Scheme 5.30). Thus, submitting the bis-vinyl ethers **142** to radical conditions (triphenyltin hydride/AIBN, refluxing benzene) led to the desired product **143** in moderate to good yields (33–81%) with up to 8 : 1 diastereoselectivity.

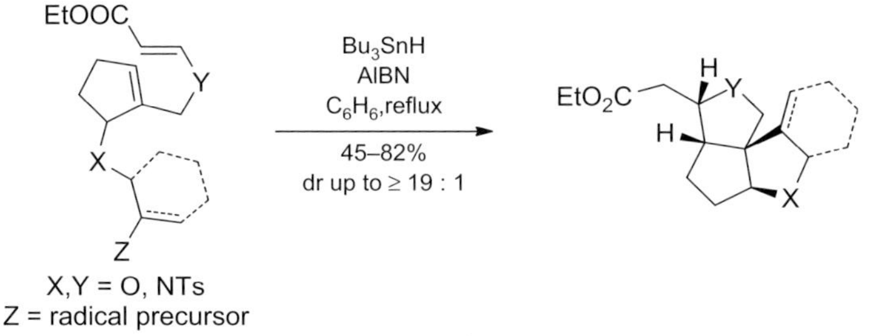

136 **137**

Scheme 5.28 Synthesis of triquinane **137**.

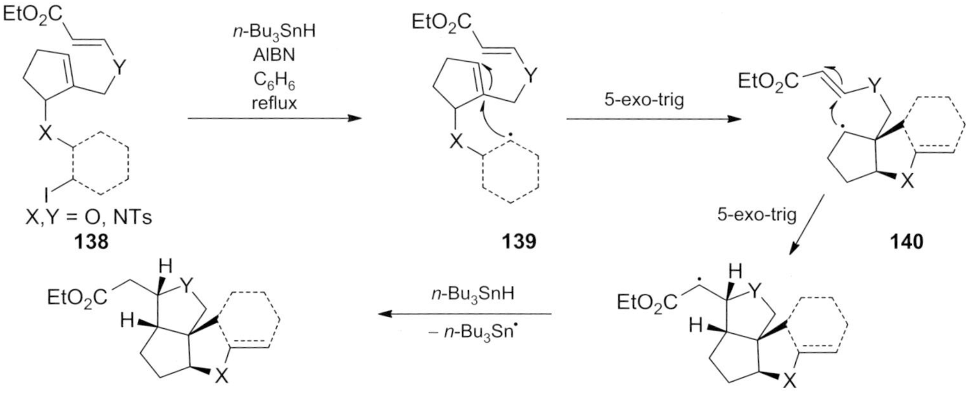

138 **139** **140** **137** **141**

Scheme 5.29 Mechanism of the domino radical 5-exo-trig/5-exo-trig cyclization.

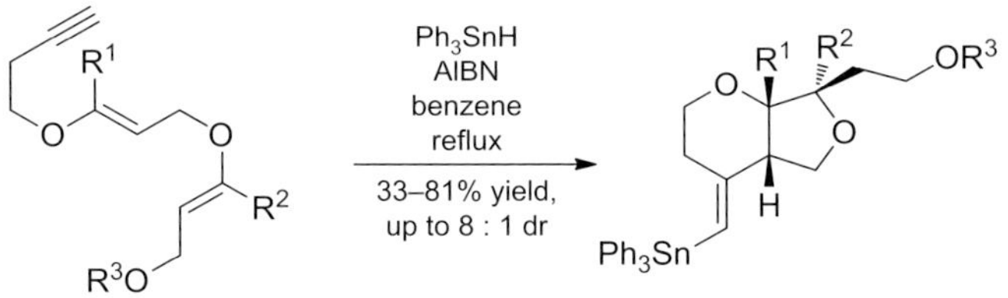

142 **143**

Scheme 5.30 Domino 6-endo-trig/5-exo-trig sequence.

Initially, a 6-endo-trig cyclization took place to afford a tetrahydro-2*H*-pyran, which was followed by a 5-exo-trig cyclization.

Baran *et al.* developed a radical cross-coupling reaction between arylboronic acids and trifluoroborates, in 2011 (Scheme 5.31) [56]. This methodology was applied to achieve a domino cyclization/coupling reaction. When the aryl boronic acid **144** was subjected to $Ag^{+}/S_2O_8^{2-}$, a benzofuran radical was generated, which underwent a radical trap with 1,4-benzoquinone as the terminating radicophile.

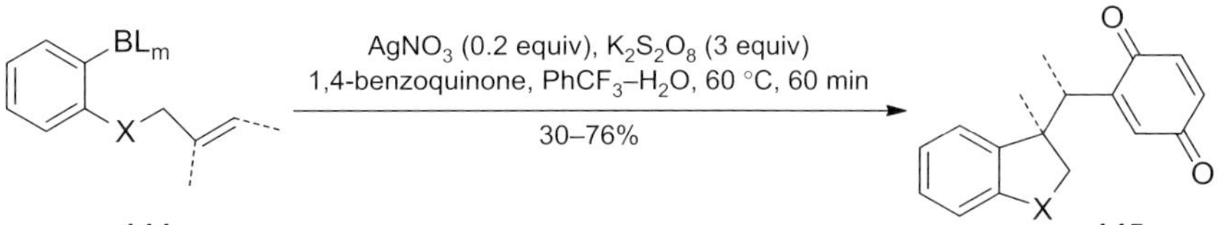

Scheme 5.31 Domino 5-exo-trig/benzoquinone-trap sequence.

As products, the substituted benzoquinones **145** were obtained in moderate to good yields (23–76%). The overall process can be characterized as a vicinal olefin difunctionalization. In the reaction, 5-exo, 6-exo, and 6-endo radical cyclizations are combined with an intermolecular radical capture, and increasing degrees of olefin substitution attenuate the efficiency of the process.

Radical reactions of allenes are largely unknown [57]. Recently, Ma and coworkers [58] developed the first example of a radical addition/cyclization reaction of allene–enes in an alkene-to-allene manner (Scheme 5.32). The reaction of the allene **146** and polyfluoro alkyl iodide **147** with zinc powder as a cheap, readily available, efficient, and mild initiator led to **148** in moderate to good yields (53–86%), which on treatment with TBAF (tetra-*n*-butylammonium fluoride) gave the allene **149**.

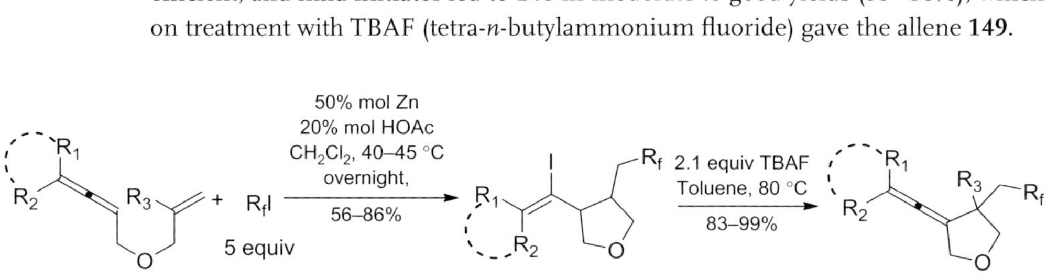

Scheme 5.32 Radical addition/cyclization reactions of allene–enes.

As shown in Scheme 5.33, the first step of the domino process is the generation of the polyfluoro alkyl radical **154**, which gives **152** via radical addition/5-exo cyclization of **150**. In the final step, the product **155** is formed via an atomic transformation process (Scheme 5.33).

Similarly, a polyfluoroalkyl radical-mediated radical addition/cyclization of the polarity inverted ene–ene acceptor **156** was carried out by Yoshioka *et al.* [59] (Scheme 5.34). The reaction was induced by the polyfluoroalkyl radical **154** through attacking at the electron-deficient alkene moiety in **156**. The radical intermediate followed an electronically matched pathway (path a/path c through **157**) through a 5-exo-cyclization to afford product **159** in a highly regioselective manner.

In Cramer's [60] heteroatom-nucleophile-induced C–C fragmentations, an allenic hydroxamic acid **162** was employed for a domino oxidative radical cyclization to afford the densely functionalized 1,2-oxazine **166** in 52% yield

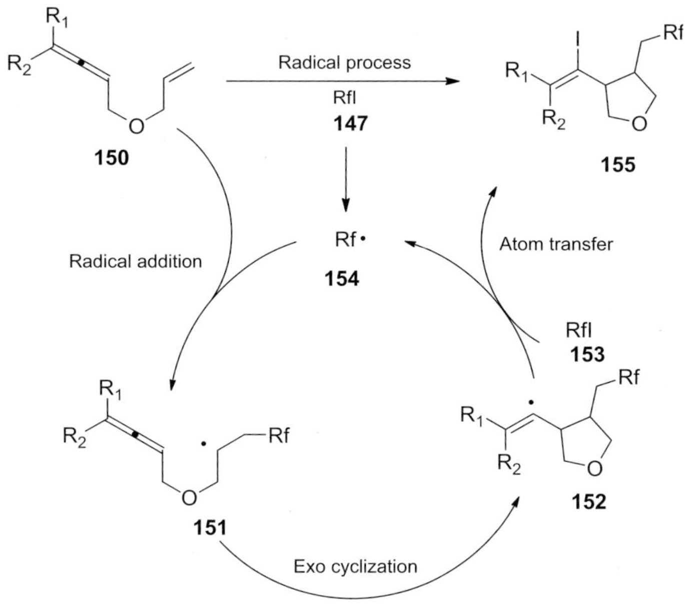

Scheme 5.33 Mechanism of radical addition/cyclization reactions of allenenes.

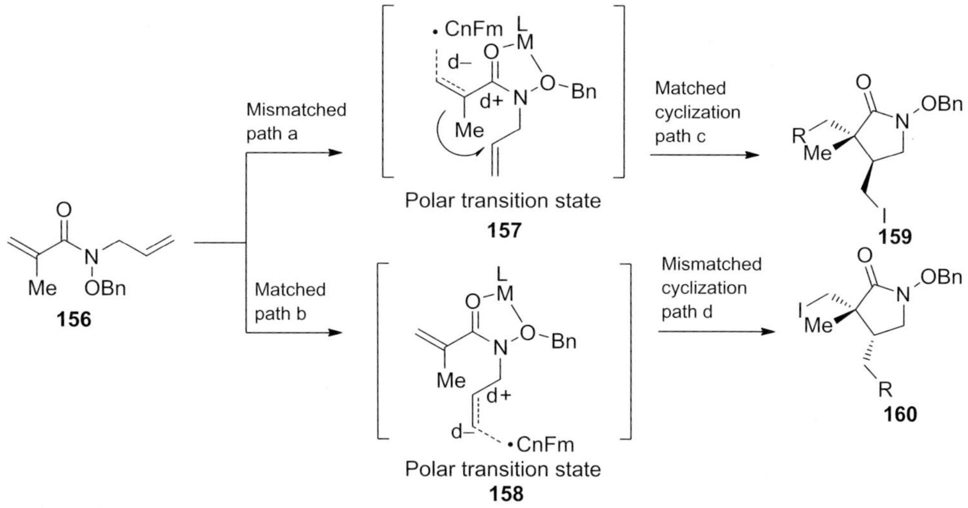

Scheme 5.34 Radical addition/cyclization of polarity inverted ene–ene reaction.

(Scheme 5.35). The hydroperoxide intermediate **165** was formed via radical cyclization of the intermediates **163** and **164**.

Two radical polar crossover domino reactions involving alkenes and alkynes were reported by Chemla and coworkers [61] in 2006 and 2008, respectively (Scheme 5.36). The radical **167** was generated from a zinc reagent anion and

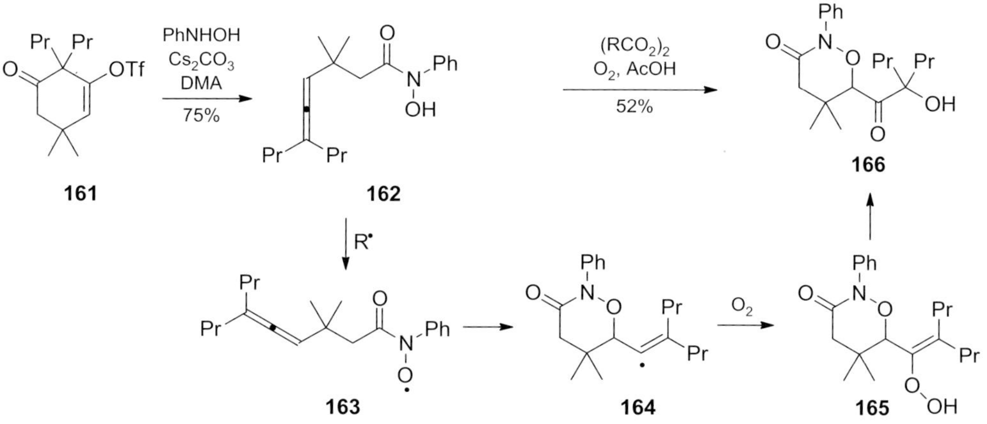

Scheme 5.35 Domino oxidative radical cyclization.

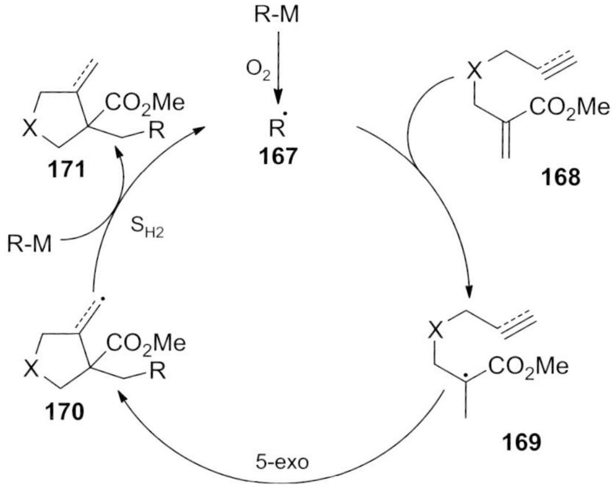

Scheme 5.36 Radical–polar crossover domino reactions.

underwent a radical addition/cyclization process onto β-*N*-allyl amino enoate **168** (X = N, alkene). The final products **171** were formed in moderate yield (56–64%) and up to ≥95 : 5 dr by hydrolysis of the corresponding carbon–metal bond. For substrates **168** with an alkyne moiety, the final products were obtained in moderate to excellent yields (42–91%).

A Co(III)–carbene radical-initiated reaction was recently reported by Zhang and coworkers [62] (Scheme 5.37). It allows the regioselective synthesis of multisubstituted furans through cyclization of alkynes with α-diazocarbonyls via a metallo radical. The tertiary radical intermediate generated from the initial sequence of consecutive radical additions undergoes radical fragmentation via consecutive radical scission of two β-C–O bonds to give the key acyl radical intermediate. The release of acetone, further radical addition/substitution, and rearrangement affords

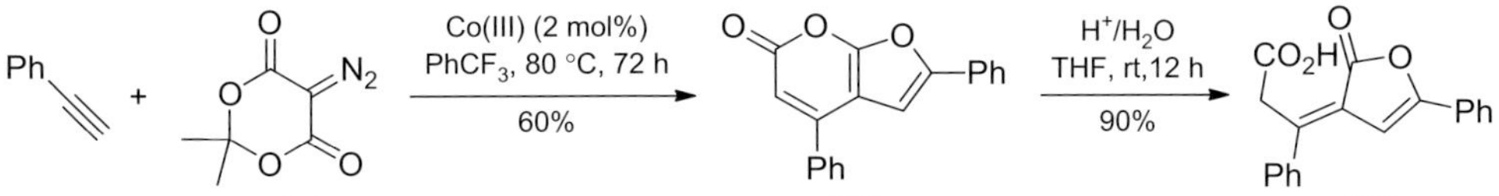

Scheme 5.37 Co(III)–carbene radical-initiated domino reactions.

the stable product **173** in 60% yield, which is readily converted into product **174** by hydrolysis and isomerization.

An unprecedented domino radical cyclization/Smiles rearrangement process of **175** to give 3-(2-aryl-*N*-methylacetamido)indolin-2-ones **178** or **180** was reported by Gérard, Sapi, *et al.* [63] (Scheme 5.38). The reaction proceeds via a 5-exo-trig cyclization followed by radical substitution. The radical intermediate **177**, formed via the radicals A and **176**, could undergo either a normal termination of a radical to form product **178** in 11–66% yields or an additional amidyl radical cyclization, giving the product **180** in up to 40% yield depending on the reaction conditions.

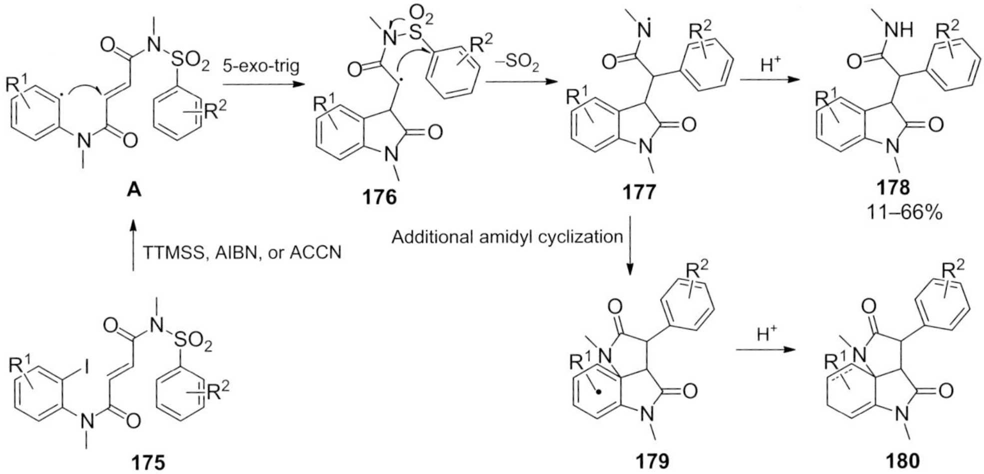

Scheme 5.38 Domino radical cyclization/Smiles rearrangement process.

A titanium-radical-initiated radical substitution was reported by Gansäuer, Flowers II, *et al.* [64], in which the radical metal complex Cp_2TiCl opened the epoxide ring in **181** followed by subsequent aromatic radical substitution to form an indoline **183** via **182** (Scheme 5.39). This was readily converted into product **184** by hydrogen abstraction in excellent yields (92–98%).

A new methodology for the synthesis of fused polycyclic vinylcyclopropanes such as **187** from **185** was developed by Malacria, Fensterbank, *et al.* [65] in 2007 (Scheme 5.40). A radical trigger leaving group in **185** induced a 5-exo-dig/5-exo-trig/3-exo-trig process, and the formed intermediate **186** underwent a

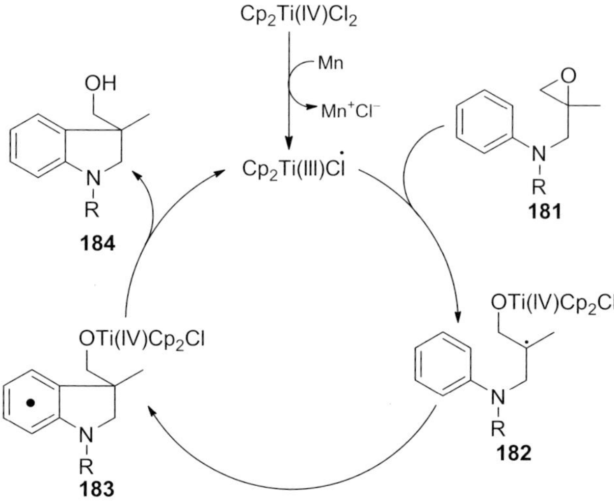

Scheme 5.39 Domino titanium-radical-initiated process.

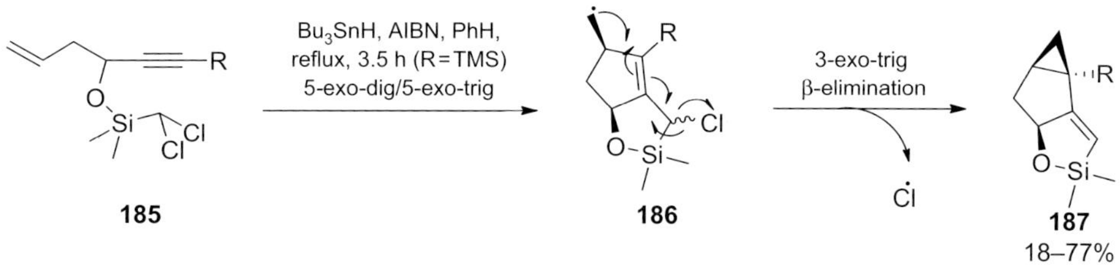

Scheme 5.40 Domino 5-exo-dig/5-exo-trig/3-exo-trig process.

3-exo-trig/β-elimination to form **187** by releasing a chlorine atom. The tricyclic products were obtained in up to 77% yield. If the reaction of **188** was performed in the presence of acrylonitrile and oxidative workup, **189** was obtained in 30% yield with 80 : 20 dr (Scheme 5.41).

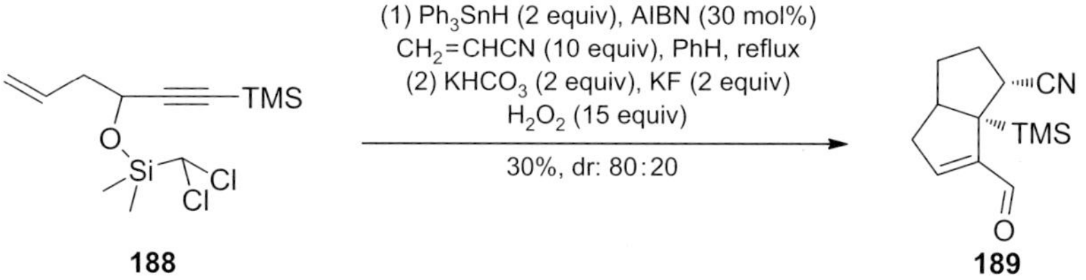

Scheme 5.41 Domino 5-exo-dig/5-exo-trig/5-endo-trig process.

A radical addition/cyclization reaction for the synthesis of indolines, indoles, and benzopyrrolizidinones was reported by Brucelle and Renaud (Scheme 5.42) [66]. The produced α-radical of ethyl acetate was added to the alkene, followed by radical

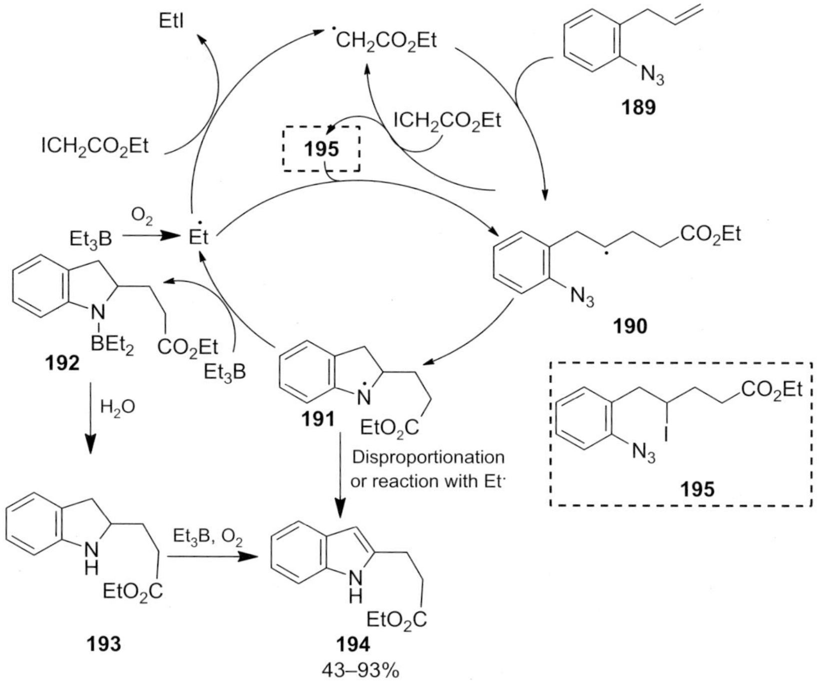

Scheme 5.42 Domino radical addition/cyclization process.

substitution cyclization to form an indoline skeleton with the release of nitrogen. The product **194** was obtained by hydrolysis or hydrogen abstraction in moderate to excellent yields (42–93%).

A radical process involving *N*-acyl cyanamides such as **196** was reported in 2010 (Scheme 5.43) [67]. A quinozaline was prepared from **196** by an initial alkenyl radical attack at the nitrile triple bond to give the radical intermediate **197**. Subsequent aromatic substitution led to alkyl radical **198**, and further *in situ* addition/reduction provided the product **200** in reasonable yields (24–69%). By involving an aryl radical in the beginning, the yields of the corresponding quinoxalines were good (71–88%).

During their studies of Ugi postcondensation reactions, Kaim *et al.* [68] discovered an intramolecular aryl transfer/oxidative cleavage/cyclization process (Scheme 5.44). Reaction of the Ugi product **201** with a radical obtained from malonate led to the radical intermediate **202**, which underwent a 1,4 aryl transfer to give radical **203**. Oxidative addition of AcOH followed by hydrolysis of the formed *N*,*O*-acetal led to **205**. The final step in the domino process was an oxidative 5-exo-trig cyclization to form the indane motif **206** in 59% yield.

For the preparation of C-glycosides, a hydrogen-atom transfer (HAT)/intermolecular radical allylation process has been devised as a new methodology (Scheme 5.45) [69]. Through the reaction of **207** with allyltri-*n*-butyltin, a radical **208**

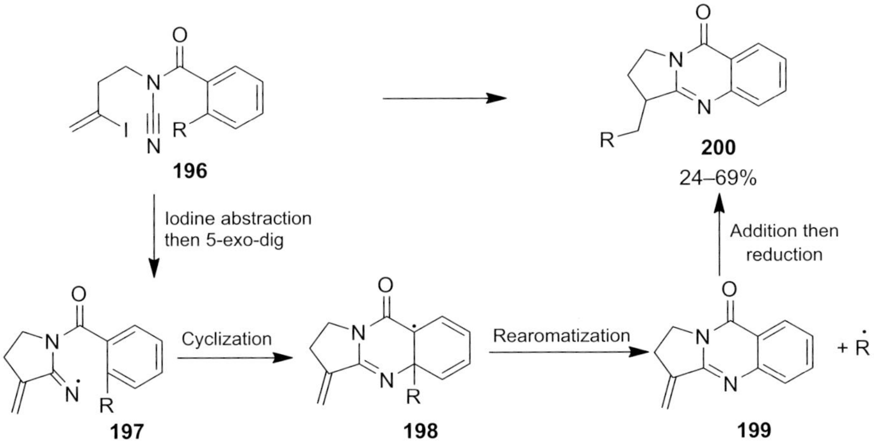

Scheme 5.43 Domino radical cyclization process.

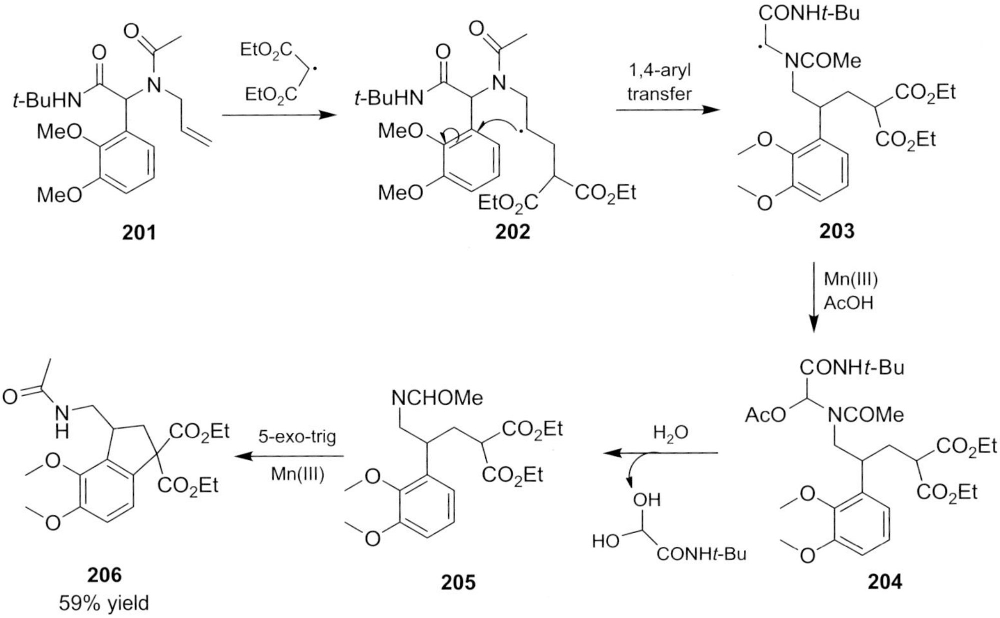

Scheme 5.44 Domino Ugi postcondensation reaction.

was generated at the oxygen atom, which underwent 1,5-, 1,6-, or 1,8-HATs depending on the substrate. The formed radical **209** was then converted into the product **210** in moderate yields (33–68%) by radical allylation with the allyltin derivative.

Spirocyclic lactones and the structures derived from them are prevalent motifs in medicinal and natural product chemistry [70]. A Beckwith–Dowd ring expansion was reported for the synthesis of γ-lactones **212** spirofused to 6- to 10-membered cycloalkanones (Scheme 5.46) [71]. The reaction involves a one-, three-, four-, and

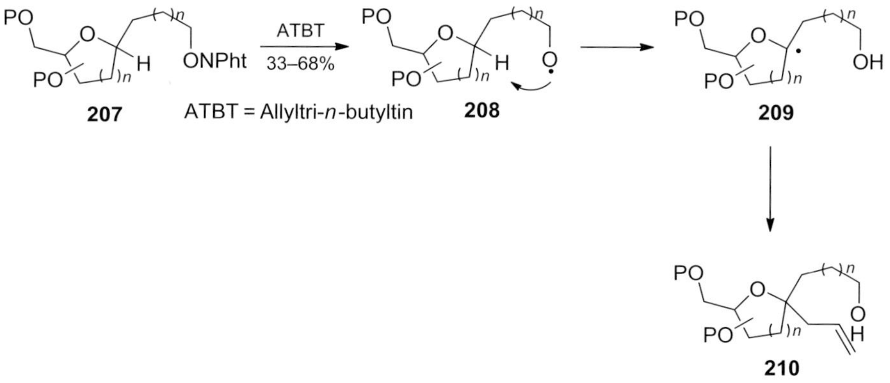

Scheme 5.45 Domino hydrogen-atom transfer (HAT)/intermolecular radical allylation.

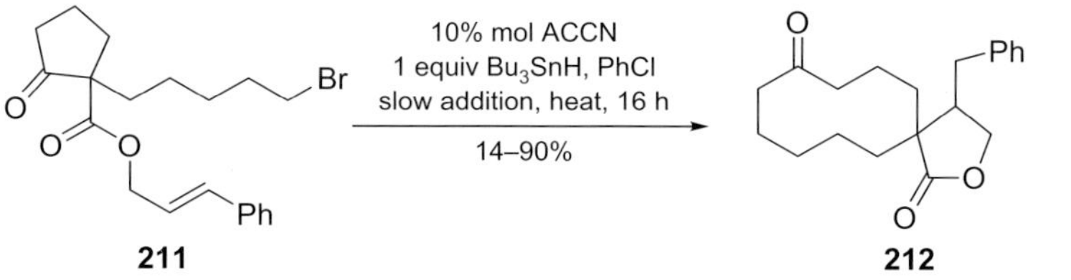

Scheme 5.46 Domino Beckwith–Dowd ring expansion/5-exo-cyclization process.

five-carbon expansion of cyclopentanone or cyclohexanone, followed by a 5-exo-trig or 5-exo-dig cyclization. The lactones **212** were obtained in moderate to excellent yield (14–90%) (Scheme 5.47).

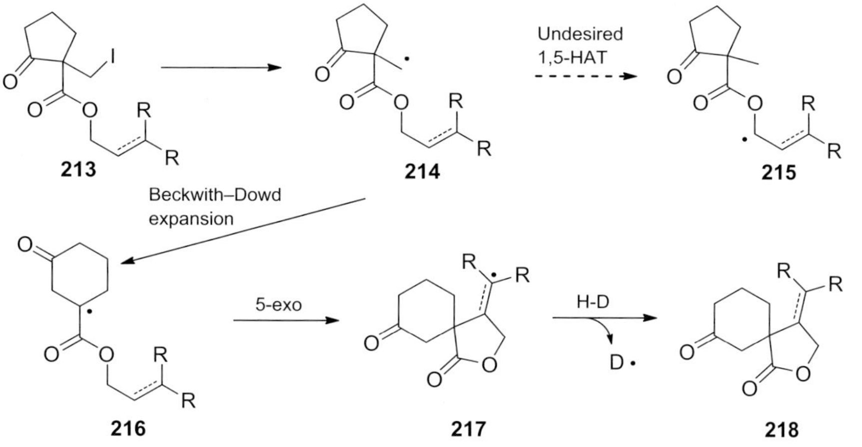

Scheme 5.47 Mechanism of domino Beckwith–Dowd ring expansion/5-exo-cyclization process.

5.5
Radical/Pericyclic Domino Processes

Gerald Pattenden's [72] group has reported a domino radical-mediated cyclization of conjugated ynone electrophores to synthesize steroids and other ring-fused polycyclic carbocycles. When a solution of **219** in dry, degassed benzene was treated with Bu_3SnH–AIBN at 80 °C for 8 h, a single diastereoisomer of the expected tricyclic enone **224** was obtained after workup and chromatography in a modest yield of 22%. The exclusive formation of the cis, syn, cis-diastereoisomer **224** from **219** is interesting, and most likely implicates a pathway involving a 13-endo-dig radical macrocyclization to give **221** via the radical **220** followed by an *in situ* structure set up for an intramolecular Diels–Alder transannulation through an "endo-like" TS **225** (Scheme 5.48). It is conceivable that a domino 6-exo-trig/5-exo-trig radical cyclization from the *E,E, E*-trienone intermediate **222** via **223** [73] would also produce **224**, but by this pathway one would expect that a mixture of diastereoisomers of the tricyclic enone would be formed.

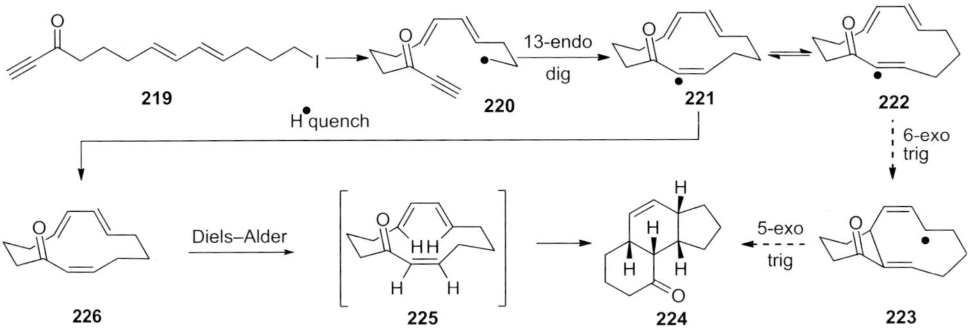

Scheme 5.48 Synthesis of tricyclic enone **224**.

In 2011, Stephenson and coworkers [74] reported a domino visible-light-mediated radical cyclization using an iridium(III) salt. In this reaction, dialkynyl cyclopropane **230** was transformed to afford tricyclic pyrrolidinone **232** via a radical-initiated cyclization followed by a Cope rearrangement/rearomarization. The reaction was initiated by a 5-exo-dig radical cyclization of **230** and subsequent quenching of the vinyl radical by hydrogen-atom abstraction to give **231**. Mild heating (40 °C) in DMF (*N,N*-dimethylformamide) in the absence of any redox reaction conditions cleanly afforded the desired fused tricycle **232** as the sole product in 69% yield.

On the basis of the above observation, a thermal [3,3] sigmatropic rearrangement, presumably driven by the release of the cyclopropyl ring strain, was proposed to explain the transformation from **230** to **232** (Scheme 5.49, path b). However, it was also stated that product **232** might also be formed by path a via **227–229** or path c via **33** and **34**.

Recently, Beifuss's group has described a selective formation of carpanones and benzopyrans using the enzyme laccase in the presence of O_2 (Scheme 5.50) [75].

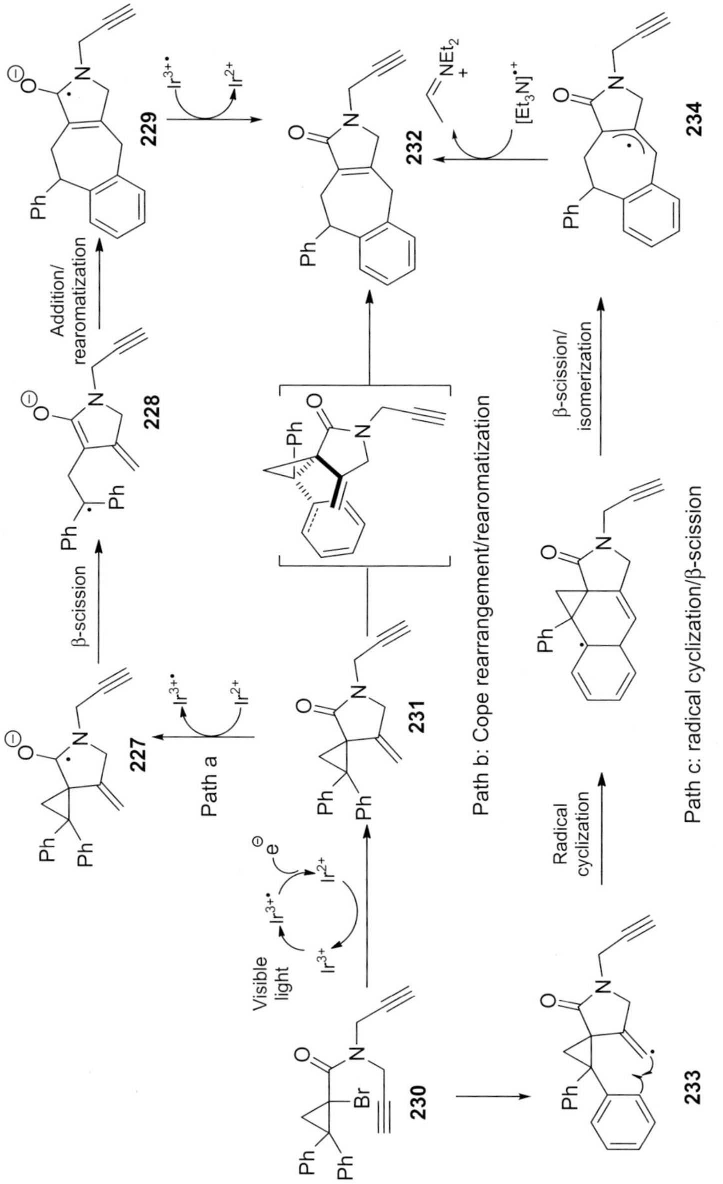

Scheme 5.49 Possible mechanism to form **232**.

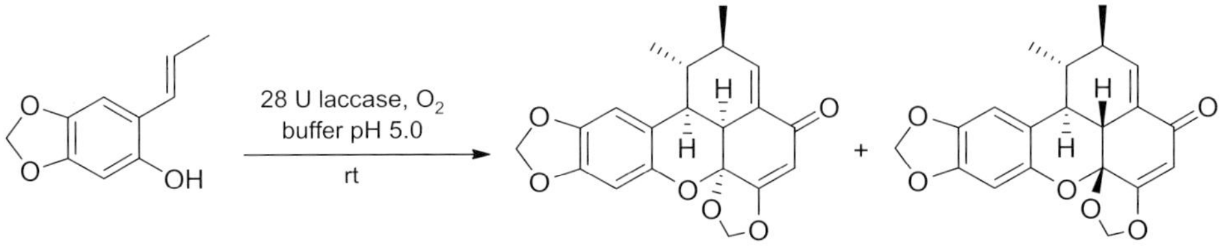

(*E*)-**235** **236** **237**

Scheme 5.50 Oxidative dimerization of (*E*)-**235**.

The key step is the dimerization of **235** to give **236** and **237**. This transformation will be described in Chapter 9.

5.6
Asymmetric Radical Domino Processes

This topic has been reviewed by Malacria, Pellissier, and their coworkers [5] in two excellent articles. The present section deals with asymmetric radical domino processes covering chiral auxiliary-directed and chiral catalyst-driven processes mainly developed in the past several years. However, chiral reagent-controlled and participated radical domino reactions have rarely been reported.

5.6.1
Chiral Auxiliary-Directed Asymmetric Radical Domino Processes

Chiral oxazolidinones are the most popular auxiliaries in the field of asymmetric synthesis. Chen and Sibi [76] employed this auxiliary for the domino asymmetric C–C/C–C single bond formation under convenient radical conditions (Scheme 5.51). During the reaction, the radical intermediates **240** was coordinated onto the Lewis acid $Yb(OTf)_3$ at −78 °C to form the desired s-cis conformation. Upon the treatment of this intermediate **240** by 2-methylallyl stannane, the product **241** was generated with a diastereoselectivity of >20 : 1. Similar radical conditions using a bromooxazolidinone imide source also resulted in allylated products with excellent diastereoselectivity (>50 : 1) and yield (>90%) even at room temperature.

A chiral menthol auxiliary has been utilized by Hoffmann and coworkers for the radical addition/cyclization reaction of *N,N*-dimethylaniline **242** with (5*R*)-menthyloxyfuran-2(5*H*)-one **245** using photochemical electron transfer (PET). The reaction led to **246** and **247** with 44% yield and 3.3 : 1 diastereoselectivity [77, 78] (Scheme 5.52). A modified menthol auxiliary was utilized for the asymmetric synthesis of a chiral precursor to the natural terpenoid (+)-triptocallol by Yang and coworkers [79]. The reaction was performed under oxidative radical conditions mediated by 1.0 equiv of $Yb(OTf)_3$ and 2.2 equiv of $Mn(OAc)_3$-$2H_2O$ $Mn(OAc)_3$. Reasonable yields of 55–72% and a diastereoselectivity of 24.2 : 1–2.8 : 1 were

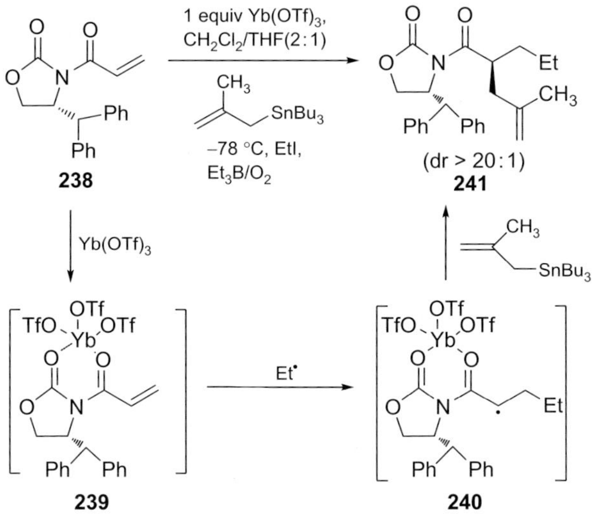

Scheme 5.51 Chiral oxazolidinone asymmetric C–C/C–C bond formations.

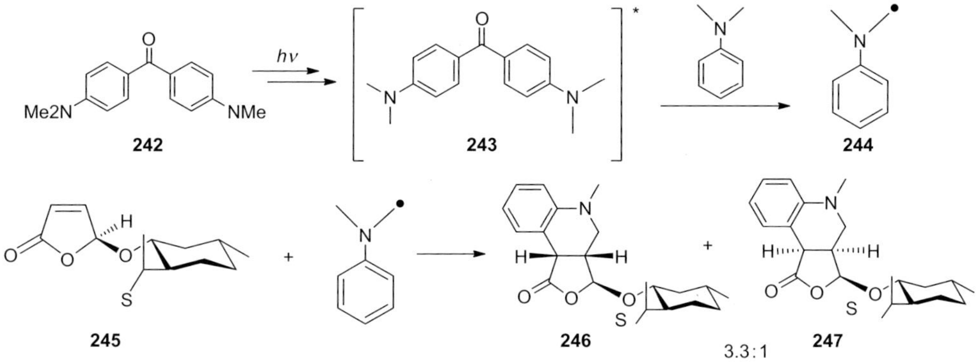

Scheme 5.52 Menthol-controlled asymmetric radical addition/cyclization reaction.

achieved. The product **249** obtained from **248** using this method could be further transformed into (+)-triptocallol (**250**) with 90% ee (Scheme 5.53).

Another auxiliary-controlled asymmetric methodology was developed by Simpkins and coworkers [80] using a chiral proline–glycine DKP template for a concise synthesis of the asperparaline bridged bicyclic unit that exists in alkaloid derivatives. The radical domino reaction included stereocontrolled 6-exo-trig and 5-exo-trig ring closures, and the product was isolated as a single diastereomer and confirmed to have the desired configuration existing in most known natural products. An efficient domino synthesis of chiral bicyclic lactams was reported by Sutherland and coworkers [81]. During the synthesis, the chiral precursor was pregenerated *in situ* by using a chiral palladium(II) complex such as [(*S*)-COP–Cl], which was subjected to a ring-closing metathesis catalyzed by the first-generation Grubbs

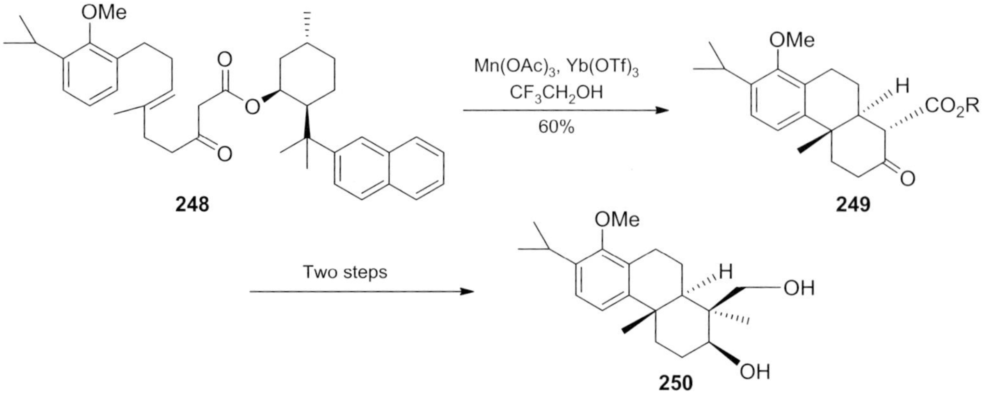

Scheme 5.53 Chiral ester-controlled domino cyclization and total synthesis application for (+)-triptocallol.

catalyst and then followed by a Kharasch cyclization. Very recently, Fensterbank, Malacria, and coworkers developed a highly diastereoselective ionic/radical domino reactions via SET-induced cyclization of bis-sulfoxides. For this, the initial domino radical was generated by treating the bis-sulfoxides with NaH, followed by SET. This is the first application of bis-sulfinyl radicals as synthetic equivalents of chiral acyl or methylene radicals to asymmetric synthesis [82].

5.6.2
Chiral Catalyst-Driven Asymmetric Radical Domino Processes

Yang's group has developed an asymmetric catalytic PhSe-group transferring radical domino cyclization by using a complex of Mg^{2+} with C_2-symmteric bisoxazoline ligands [83]. The resulting products are very useful for assembling many monocyclic and bicyclic core structures present in biologically important natural products. The authors used a model of a similar process to explain the stereochemistry of this domino reaction, in which a six-membered ring TS is assumed. The magnesium cation is chelated by the dicarbonyl moiety of the substrate, adopting an octahedral geometry so that the ClO_4^- ions occupy two axial positions (Scheme 5.54). The steric interactions with the α-*tert*-butyl group forces the olefin moiety to approach the α-radical center from its *re* face to give intermediate C. TS A is favored because of steric interactions between the olefin group and the β-*tert*-butyl group. The SePh abstraction preferentially happens from the less hindered *re* face to afford a single diastereoisomer of product **256** (Scheme 5.55).

On the basis of the general strategy of organo-SOMO (singly occupied molecular orbital) catalysis (Scheme 5.56), MacMillan's group developed a highly enantioselective domino radical-mediated (4+2) coupling reaction [84]. This process was carried out by reacting aldehydes **257** with conjugated olefins **258** in the presence of the imidazolidinone **259** to give a series of cyclic products **262** with good yields of 70–90%, a diastereoselectivity of 6 : 1 to >20 : 1, and an enatioselectivity of 70–94% ee; it

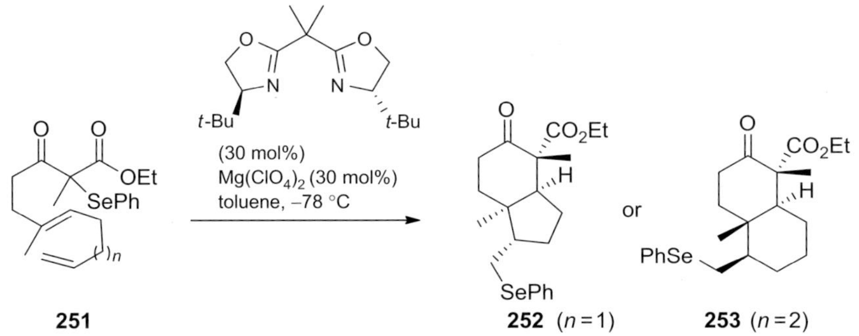

Scheme 5.54 Mg^{2+}-bisoxazoline ligand-driven asymmetric domino catalysis.

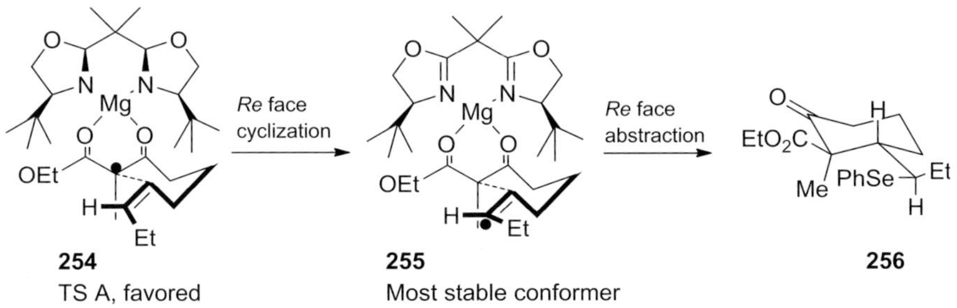

Scheme 5.55 Model for the asymmetric induction using a bisoxazoline ligand.

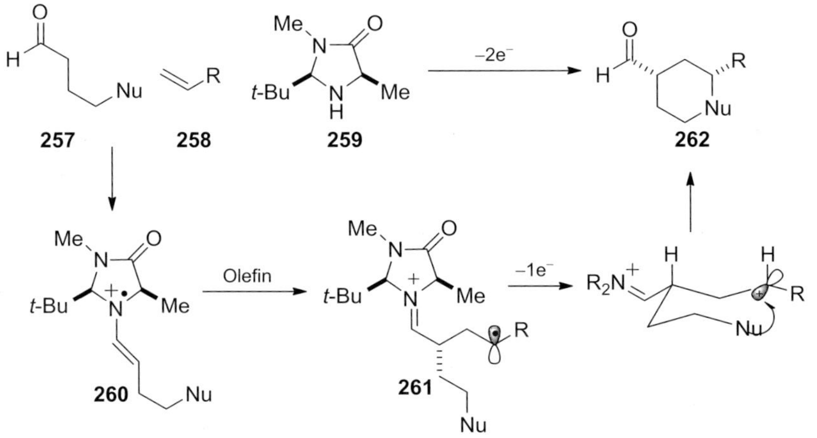

Scheme 5.56 General strategy of organo-SOMO catalysis.

should be noted that the reactions are completely regioselective. Very interestingly, carbogenic π-nucleophiles are also suitable for the cation-trapping cyclization.

The most recent highlight in enantioselective radical domino processes can be attributed again to the MacMillan group, in their development of an efficient polyene domino cyclization via the organo-SOMO catalysis strategy [85] (Scheme 5.57).

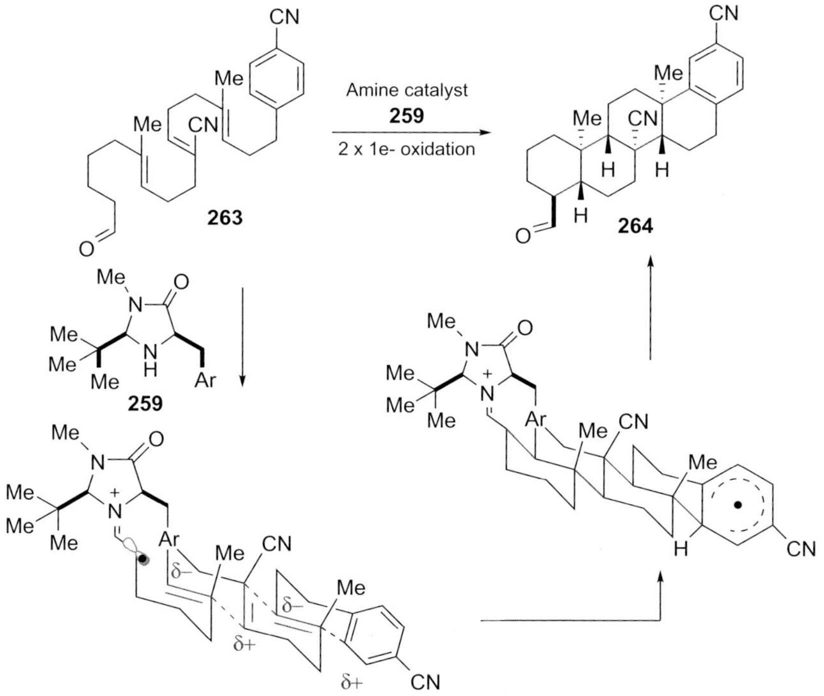

Scheme 5.57 Enantioselective domino cyclization via SOMO catalysis.

This work presents the first catalytic enantioselective cyclization strategy for accessing steroidal and terpenoidal frameworks using organocatalysis. It should be pointed out that the present domino cyclization could be conducted at room temperature. The asymmetric construction of multiple C–C bonds and contiguous stereocenters using substrate **263** in the presence of the chiral imidazolidinone **259** resulted in the formation of **264** with four new C–C bonds, seven stereogenic centers, and three quaternary carbon centers with about 90% yield for each chemical bond formation and over 90% ee.

5.7 Conclusion and Outlook

Domino radical reactions consisting of radical/cationic, radical/anionic, radical/radical, and radical/pericyclic processes have become increasingly important and useful tools in organic synthesis. As known also for all other types of domino processes, domino radical reactions will continue to offer an environmentally friendly and economically appropriate molecular assembly method of all sorts of compounds, particularly those containing multiple carbon and heterocyclic rings with tertiary and quaternary carbon centers. One can anticipate that more work on asymmetric catalysis of domino processes will be developed and applied to organic, medicinal, and material chemistry in the coming years.

Acknowledgments

We are grateful to Robert A. Welch Foundation (D-1361, USA) and Jiangsu Innovational Team Grant (P. R. China) for their generous support. We thank Dr Jianbo Xie, Mr Padmanabha V. Kattamuri, Suresh Pindi, Jianbin Wu, and Patrick McDowell for their assistance.

References

1. (a) Tietze, L.F. and Haunert, F. (2000) in *Stimulating Concepts in Chemistry* (eds M. Shibasaki, J.F. Stoddart, and F. Vögtle), Wiley-VCH Verlag GmbH, Weinheim, pp. 39–64. (b) Tietze, L.F., Brasche, G., and Gericke, K.M. (2006) *Domino Reactions in Organic Synthesis*, Wiley-VCH Verlag GmbH, Weinheim.
2. (a) Tietze, L.F. (1996) *Chem. Rev.*, **96**, 115–136. (b) Tietze, L.F. and Beifuss, U. (1993) *Angew. Chem., Int. Ed. Engl.*, **32**, 131–163. (c) Tietze, L.F., Brazel, C.C., Hoelsken, S., Magull, J., and Ringe, A. (2008) *Angew. Chem. Int. Ed.*, **47**, 5246–5249.
3. (a) Padwa, A. and Bur, S.K. (2007) *Tetrahedron*, **63**, 5341–5378. (b) Padwa, A. (2009) *Chem. Soc. Rev.*, **38**, 3072–3081. (c) Jiang, B., Feng, B.M., Wang, S.L., Tu, S.J., and Li, G. (2012) *Chem. Eur. J.*, **18**, 9823–9826.
4. Merrifield, R.B. (1963) *J. Am. Chem. Soc.*, **85**, 2149–2154.
5. (a) Albert, M., Fensterbank, L., Lacôte, E., and Malacria, M. (2006) *Top. Curr. Chem.*, **264**, 1–62. (b) Claviera, H. and Pellissiera, H. (2012) *Adv. Synth. Catal.*, **354**, 3347–3403.
6. McCarroll, A.J. and Walton, J.C. (2001) *Angew. Chem. Int. Ed.*, **40**, 2224–2248.
7. (a) Kim, C.M. and Youn, J.H. (1999) *Tetrahedron Lett.*, **40**, 3581–3582. (b) Grigorieva, L. and Jirgensons, A. (2011) *Eur. J. Org. Chem.*, **2011**, 2421–2425. (c) Kim, T.S., Lee, Y.J., Lee, K., Jeong, B.S., Park, H., and Jew, S. (2009) *Synlett*, 671–674. (d) Singh, O.V. and Han, H. (2007) *Tetrahedron Lett.*, **48**, 2345–2348. (e) Singh, O.V., Kampf, D.J., and Han, H. (2004) *Tetrahedron Lett.*, **45**, 7239–7242. (f) Balázs, Á., Szakonyi, Z., and Fülöp, F. (2007) *J. Heterocycl. Chem.*, **44**, 403–406. (g) Meyers, A.I. and Nazarenko, N. (1978) *J. Org. Chem.*, **58**, 175–176.
8. Boto, A., Gallardo, J.A., and Álvarez, E. (2012) *Eur. J. Org. Chem.*, **2012**, 391–397.
9. Cacchi, S. and Fabrizi, G. (2011) *Chem. Rev.*, **111**, 215–283.
10. Maity, S. and Zheng, N. (2012) *Angew. Chem. Int. Ed.*, **51**, 9562–9566.
11. Courant, T. and Masson, G. (2012) *Chem. Eur. J.*, **18**, 423–427.
12. Jia, X., Qing, C., Huo, C., Peng, F., and Wang, X. (2012) *Tetrahedron Lett.*, **53**, 7140–7142.
13. Jia, X., Ren, Y., Huo, C., Wang, W., Chen, X., Xu, X., and Wang, X. (2010) *Tetrahedron Lett.*, **51**, 6779–6782.
14. Shirakawa, E., Uchiyama, N., and Hayashi, T. (2011) *J. Org. Chem.*, **76**, 25–34.

15. Lee, H.B., Sung, M.J., Blackstock, S.C., and Cha, J.K. (2001) *J. Am. Chem. Soc.*, **123**, 11322–11324.
16. Schwartz, A. and Seger, C. (2001) *Monatsh. Chem.*, **132**, 855–858.
17. Gross, S. and Reissig, H.U. (2003) *Org. Lett.*, **5**, 4305–4307.
18. Beemelmanns, C. and Reissig, H.U. (2010) *Angew. Chem. Int. Ed.*, **49**, 8021–8025.
19. Nicolaou, K.C., Ellery, S.P., and Chen, J.S. (2009) *Angew. Chem. Int. Ed.*, **48**, 7140–7165.
20. Shirai, A., Miyata, O., Tohnai, N., Miyata, M., Procter, D.J., Sucunza, D., and Naito, T. (2008) *J. Org. Chem.*, **73**, 4464–4475.
21. Rivkin, A., Turiso, F.G.L., Nagashima, T., and Curran, D.P. (2004) *J. Org. Chem.*, **69**, 3719–3725.
22. Shenvi, R.A., Guerrero, C.A., Shi, J., Li, C.C., and Baran, P.S. (2008) *J. Am. Chem. Soc.*, **130**, 7241–7243.
23. (a) Helm, M.D., Silva, M.D., Sucunza, D., Findley, T.J.K., and Procter, D.J. (2009) *Angew. Chem. Int. Ed.*, **48**, 9315–9317. (b) Parmar, D., Price, K., Spain, M., Matsubara, H., Bradley, P.A., and Procter, D.J. (2011) *J. Am. Chem. Soc.*, **133**, 2418–2420. (c) Harb, Y.H. and Procter, D.J. (2012) *Synlett*, 6–20. (d) Sautier, B., Lyons, S.E., Webb, M.R., and Procter, D.J. (2012) *Org. Lett.*, **14**, 146–149. (e) Parmar, D., Matsubara, H., Price, K., Spain, M., and Procter, D.J. (2012) *J. Am. Chem. Soc.*, **134**, 12751–12757.
24. Szostak, M. and Procter, D.J. (2012) *Angew. Chem. Int. Ed.*, **51**, 9238–9256.
25. (a) Ollivier, C. and Renaud, P. (1999) *Chem. Eur. J.*, **5**, 1468–1473. (b) Brecht-Forster, A., Fitremann, J., and Renaud, P. (2002) *Helv. Chim. Acta*, **85**, 3965–3974. (c) Schaffner, A.P. and Renaud, P. (2004) *Eur. J. Org. Chem.*, 2291–2298. (d) Kumli, E., Montermini, F., and Renaud, P. (2006) *Org. Lett.*, **8**, 5861–5864.
26. Beauseigneur, A., Ericsson, C., Renaud, P., and Schenk, K. (2009) *Org. Lett.*, **11**, 3778–3781.
27. (a) Huddleston, R.R., Cauble, D.F., and Krische, M.J. (2003) *J. Org. Chem.*, **68**, 11–14. (b) Agapiou, K., Cauble, D.F., and Krische, M.J. (2004) *J. Am. Chem. Soc.*, **126**, 4528–4529.
28. Maury, J., Feray, L., Perfetti, P., and Bertrand, M.P. (2010) *Org. Lett.*, **12**, 3590–3593.
29. Wille, U. (2013) *Chem. Rev.*, **113**, 813–853.
30. Zhang, H. and Curran, D.P. (2011) *J. Am. Chem. Soc.*, **133**, 10376–10378.
31. (a) Bernauer, K., Englert, G., Vetter, W., and Weiss, E. (1969) *Helv. Chim. Acta*, **52**, 1886–1905. (b) Daudon, M., Mehri, H., Plat, M.M., Hagaman, E.W., Schell, F.M., and Wenkert, E. (1975) *J. Org. Chem.*, **40**, 2838–2839. (c) Oberhänsli, W.E. (1969) *Helv. Chim. Acta*, **52**, 1905–1911.
32. (a) Overman, L.E., Robertson, G.M., and Robichaud, A.J. (1989) *J. Org. Chem.*, **54**, 1236–1238. (b) Overman, L.E., Robertson, G.M., and Robichaud, A.J. (1991) *J. Am. Chem. Soc.*, **113**, 2598–2610.
33. (a) Selig, P. and Bach, T. (2008) *Angew. Chem. Int. Ed.*, **47**, 5082–5084. (b) Selig, P., Herdtweck, E., and Bach, T. (2009) *Chem. Eur. J.*, **15**, 3509–3525.
34. Hayashi, Y., Inagaki, F., and Mukai, C. (2011) *Org. Lett.*, **13**, 1778–1780.
35. (a) Minagawa, K., Kouzuki, S., Nomura, K., Kawamura, Y., Tani, H., Terui, Y., Nakai, H., and Kamigauchi, T. (2001) *J. Antibiot.*, **54**, 896–903. (b) Minagawa, K., Kouzuki, S., Nomura, K., Yamaguchi, T., Kawamura, Y., Matsushima, K., Tani, H., Ishii, K., Tanimoto, T., and Kamigauchi, T. (2001) *J. Antibiot.*, **54**, 890–895.
36. (a) Davidson, M.H. (2007) *Curr. Atheroscler. Rep.*, **9**, 78. (b) El Harchaoui, K., Akdim, F., Stroes, E.S.G., Trip, M.D., and Kastelein, J.J.P. (2008) *Am. J. Cardiovasc. Drugs*, **8**, 233–242. (c) Tavridou, A. and Manolopoulos, V.G. (2008) *Curr. Med. Chem.*, **15**, 792–802. (d) Tavridou, A. and Manolopoulos, V.G. (2009) *Curr. Pharm. Des.*, **15**, 3167–3178.
37. (a) Snider, B.B. and Lobera, M. (2004) *Tetrahedron Lett.*, **45**, 5015–5018. (b) Zhou, J., Lobera, M., Neubert-Langille, B.J., and Snider, B.B. (2007) *Tetrahedron*, **63**, 10018–10024.

38. am Ende, C.W., Zhou, Z., and Parker, K.A. (2012) *J. Am. Chem. Soc.*, **135**, 582–585.
39. Parker, K.A. and Fokas, D. (2005) *J. Org. Chem.*, **71**, 449–455.
40. Guthrie, D.B., Geib, S.J., and Curran, D.P. (2009) *J. Am. Chem. Soc.*, **131**, 15492–15500.
41. Jang, J.H., Asami, Y., Jang, J.P., Kim, S.O., Moon, D.O., Shin, K.S., Hashizume, D., Muroi, M., Saito, T., Oh, H., Kim, B.Y., Osada, H., and Ahn, J.S. (2011) *J. Am. Chem. Soc.*, **133**, 6865–6867.
42. Yin, J., Wang, C., Kong, L., Cai, S., and Gao, S. (2012) *Angew. Chem. Int. Ed.*, **51**, 7786–7789.
43. Tokhtabaeva, G.M., Sheichenko, V.I., Yartseva, I.V., and Tolkachev, O.N. (1987) *Khim. Prir. Soedin.*, **23**, 727.
44. Biechy, A., Hachisu, S., Quiclet-Sire, B., Ricard, L., and Zard, S.Z. (2008) *Angew. Chem. Int. Ed.*, **47**, 1436–1438.
45. For a review of the pyrrole–imidazole family of natural products, see: Hoffmann, H. and Lindel, T. (2003) *Synthesis*, **12**, 1753–1783.
46. Tan, X. and Chen, C. (2006) *Angew. Chem. Int. Ed.*, **45**, 4345–4348.
47. (a) Ciochina, R. and Grossman, R.B. (2006) *Chem. Rev.*, **106**, 3963–3986. (b) Pal Singh, I. and Bharate, S.B. (2006) *Nat. Prod. Rep.*, **23**, 558–591. (c) Thoison, O., Cuong, D.D., Gramain, A., Chiaroni, A., Hung, N.V., and Sévenet, T. (2005) *Tetrahedron*, **61**, 8529–8535.
48. Pepper, H.P., Lam, H.C., Bloch, W.M., and George, J.H. (2012) *Org. Lett.*, **14**, 5162–5164.
49. (a) Hayashi, H., Nishimoto, Y., and Nozaki, H. (1997) *Tetrahedron Lett.*, **38**, 5655–5658. (b) Banks, R.M., Blanchflower, S.E., Everett, J.R., Manger, B.R., and Reading, C. (1997) *J. Antibiot.*, **50**, 840. (c) Hayashi, H., Nishimoto, Y., Akiyama, K., and Nozaki, H. (2000) *Biosci. Biotechnol. Biochem.*, **64**, 111.
50. Hirata, K., Kataoka, S., Furutani, S., Hayashi, H., and Matsuda, K. (2011) *PLoS ONE*, 6. doi: 10.1371/journal.pone.0018354
51. (a) Lee, B.H. (2002) *Curr. Top. Med. Chem.*, **2**, 779. (b) Geary, T.G., Sangster, N.C., and Thompson, D.P. (1999) *Vet. Parasitol.*, **84**, 275.
52. Trost, B.M., Cramer, N., and Bernsmann, H. (2007) *J. Am. Chem. Soc.*, **129**, 3086–3087.
53. (a) Marchand, A.P. (1991) *Synlett*, **1991**, 73–79. (b) Mehta, G. and Srikrishna, A. (1997) *Chem. Rev.*, **97**, 671–720. (c) Singh, V. and Thomas, B. (1998) *Tetrahedron*, **54**, 3647–3692.
54. Gharpure, S.J., Niranjana, P., and Porwal, S.K. (2012) *Org. Lett.*, **14**, 5476–5479.
55. Davies, K.A. and Wulff, J.E. (2011) *Org. Lett.*, **13**, 5552–5555.
56. Lockner, J.W., Dixon, D.D., Risgaard, R., and Baran, P.S. (2011) *Org. Lett.*, **13**, 5628–5631.
57. For reviews of radical reactions of allenes, see: (a) Pan, F., Fu, C., and Ma, S. (2004) *Chin. J. Org. Chem.*, **24**, 1168. (b) Hartung, J. and Kopf, T. (2004) in *Modern Allene Chemistry*, Vol. 2 (eds N. Krause and A.S.K. Hashmi), Wiley-VCH Verlag GmbH, Weinheim, p. 701.
58. Zeng, R., Fu, C., and Ma, S. (2012) *Angew. Chem. Int. Ed.*, **51**, 3888–3891.
59. Yoshioka, E., Kohtani, S., Sawai, K., Kentefu, Tanaka, E., and Miyabe, H. (2012) *J. Org. Chem.*, **77**, 8588–8604.
60. Saget, T. and Cramer, N. (2010) *Angew. Chem. Int. Ed.*, **49**, 8962–8965.
61. (a) Denes, F., Cutri, S., Perez-Luna, A., and Chemla, F. (2006) *Chem. Eur. J.*, **12**, 6506–6513. (b) Pérez-Luna, A., Botuha, C., Ferreira, F., and Chemla, F. (2008) *Chem. Eur. J.*, **14**, 8784–8788.
62. Cui, X., Xu, X., Wojtas, L., Kim, M.M., and Zhang, X.P. (2012) *J. Am. Chem. Soc.*, **134**, 19981–19984.
63. Pudlo, M., Allart-Simon, I., Tinant, B., Gérard, S., and Sapi, J. (2012) *Chem. Commun.*, **48**, 2442–2444.
64. Gansäuer, A., Behlendorf, M., von Laufenberg, D., Fleckhaus, A., Kube, C., Sadasivam, D.V., and Flowers, R.A. (2012) *Angew. Chem. Int. Ed.*, **51**, 4739–4742.
65. Maddess, M.L., Mainetti, E., Harrak, Y., Brancour, C., Devin, P., Dhimane, A.L., Fensterbank, L., and Malacria, M. (2007) *Chem. Commun.*, 936–938.

66. Brucelle, F. and Renaud, P. (2012) *Org. Lett.*, **14**, 3048–3051.

67. Larraufie, M.H., Courillon, C., Ollivier, C., Lacote, E., Malacria, M., and Fensterbank, L. (2010) *J. Am. Chem. Soc.*, **132**, 4381–4387.

68. El Kaim, L., Grimaud, L., and Vieu, E. (2007) *Org. Lett.*, **9**, 4171–4173.

69. León, E.I., Martín, Á., Pérez-Martín, I., Quintanal, L.M., and Suárez, E. (2012) *Eur. J. Org. Chem.*, **2012**, 3818–3829.

70. (a) Brocksom, T.J., Coelho, F., Deprés, J.P., Greene, A.E., Freire de Lima, M.E., Hamelin, O., Hartmann, B., Kanazawa, A.M., and Wang, Y. (2002) *J. Am. Chem. Soc.*, **124**, 15313–15325. (b) Bunyapaiboonsri, T., Yoiprommarat, S., Intereya, K., Rachtawee, P., Hywel-Jones, N.L., and Isaka, M. (2009) *J. Nat. Prod.*, **72**, 756–759. (c) Ghantous, A., Gali-Muhtasib, H., Vuorela, H., Saliba, N.A., and Darwiche, N. (2010) *Drug Discov. Today*, **15**, 668–678. (d) Lane, J.F., Koch, W.T., Leeds, N.S., and Gorin, G. (1952) *J. Am. Chem. Soc.*, **74**, 3211–3215. (e) Mathema, V., Koh, Y.S., Thakuri, B., and Sillanpää, M. (2012) *Inflammation*, **35**, 560–565. (f) Reddy, D.M., Qazi, N.A., Sawant, S.D., Bandey, A.H., Srinivas, J., Shankar, M., Singh, S.K., Verma, M., Chashoo, G., Saxena, A., Mondhe, D., Saxena, A.K., Sethi, V.K., Taneja, S.C., Qazi, G.N., and Sampath Kumar, H.M. (2011) *Eur. J. Med. Chem.*, **46**, 3210–3217. (g) Yamada, K., Takada, S., Nakamura, S., and Hirata, Y. (1968) *Tetrahedron*, **24**, 199–229.

71. Hierold, J. and Lupton, D.W. (2012) *Org. Lett.*, **14**, 3412–3415.

72. Pattenden, G., Stoker, D.A., and Thomson, N.M. (2007) *Org. Biomol. Chem.*, **5**, 1776–1788.

73. Jenkins, P.R., Symons, M.C.R., Booth, S.E., and Swain, C.J. (1992) *Tetrahedron Lett.*, **33**, 3543–3546.

74. Tucker, J.W., Corey, R.J., and Stephenson, C.R.J. (2011) *Org. Lett.*, **13**, 5468–5471.

75. Constantin, M.A., Conrad, J., Mericcedil or, E., Koschorreck, K., Urlacher, V.B., and Beifuss, U. (2012) *J. Org. Chem.*, **77**, 4528–4543.

76. Sibi, M.P. and Chen, J. (2001) *J. Am. Chem. Soc.*, **123**, 9472–9473.

77. Bertrand, S., Hoffmann, N., Humbel, S., and Pete, J.P. (2000) *J. Org. Chem.*, **65**, 8690–8703.

78. Marinkovic, S., Brule, C., Hoffmann, N., Prost, E., Nuzillard, J.M., and Bulach, V. (2004) *J. Org. Chem.*, **69**, 1646–1651.

79. Yang, D., Xu, M., and Bian, M.Y. (2001) *Org. Lett.*, **3**, 111–114.

80. Crick, P.J., Simpkins, N.S., and Highton, A. (2011) *Org. Lett.*, **13**, 6472–6475.

81. Fiona, I.M., Lindsay, B., Cooke, A., and Sutherland, A. (2010) *Org. Biomol. Chem.*, **8**, 3418–3425.

82. Goddard, J.P., Gomez, C., Brebion, F., Beauvière, S., Fensterbank, L., and Malacria, M. (2007) *Chem. Commun.*, 2929–2931.

83. Yang, D., Zheng, B.F., Gao, Q., Gu, S., and Zhu, N.Y. (2006) *Angew. Chem. Int. Ed.*, **45**, 255–258.

84. Jui, N.T., Lee, E.C.Y., and MacMillan, D.W.C. (2010) *J. Am. Chem. Soc.*, **132**, 10015–10017.

85. Rendler, S. and MacMillan, D.W.C. (2010) *J. Am. Chem. Soc.*, **132**, 5027–5029.

6
Pericyclic Reactions in Domino Processes

Lukas J. Patalag and Daniel B. Werz

6.1
Introduction

Among the numerous reactions that have been developed to date, a certain class, namely the pericyclic reactions, especially stands out for several reasons. Originally termed *no-mechanism reactions*, these transformations have occupied one of the most prominent positions in organic synthesis. Beside ionic and radical reactions, they constitute the third distinct class of reaction mechanisms. In all kinds of pericyclic reactions – cycloadditions, sigmatropic rearrangements, electrocyclizations, and ene reactions – one can observe cyclic transition states. No intermediates are formed, and all bond-forming and bond-breaking processes take place in concert [1].

From a synthetic point of view, pericyclic reactions commonly yield a high degree of molecular complexity brought about in a very elegant and efficient way. Manifold processes can take place, such as ring closure or ring cleavage, rearrangements of fragments, or the shift of hydrogens along conjugated π systems. One of the first pericyclic reactions that have been observed – already as part of a domino process – might be the diaza-Cope rearrangement, being part of Fischer's [2] famous indole synthesis published in 1883. Of course, chemists at that time were far from understanding such a pericyclic transformation. This situation was not different when Diels and Alder [3] published their report on six-membered ring formation in 1928. Even in 1950, when both of them received the Nobel Prize for their so-called Diels–Alder reaction, chemists were not able to explain the complete stereoselectivity of this synthetically highly useful transformation. It took another 15 years before Woodward and Hoffmann presented a comprehensive theoretical treatment of all pericyclic processes [4]. The key principle consists of the *conservation of orbital symmetry* [4, 5]. If orbitals of the starting material and product correlate with each other regarding their symmetry properties, the reaction is allowed to occur; otherwise, the transformation is associated with a huge activation barrier and commonly does not take place (at least not by a pericyclic process).

This deeper understanding (together with its impact on thermal vs photochemical reactions and the respective issues of stereoselectivity) paved the way to a plethora of

Domino Reactions: Concepts for Efficient Organic Synthesis, First Edition. Edited by Lutz F. Tietze.

syntheses in which pericyclic reactions have been exploited [6]. Very soon, chemists recognized that products of pericyclic reactions are often ideal starting materials for another pericyclic process [7, 8]. Thereby, cascades of two or even more pericyclic transformations were designed without isolation of the intermediate. In other cases, ionic reactions initiated a process in which precursors for a certain pericyclic reaction were obtained. These reactions instantaneously follow the advent of the reactive intermediate [8].

In this chapter, we highlight the most interesting and fascinating domino processes in which pericyclic reactions are involved. These sequences might start with a pericyclic transformation; however, it is not a prerequisite. Also, other types of reactions (e.g., metal-catalyzed transformations) that generate an intermediate which is prone for a subsequent pericyclic reaction are discussed. Because of the large number of reports available in the literature, we have limited ourselves to the results obtained in the last couple of years. Only in those cases where recent results are less informative or can only be understood with similar domino sequences developed earlier do we deviate from this restriction. In general, we do not give a comprehensive overview of the field – other examples for pericyclic domino reactions are also found in other chapters of the book. In this chapter, we present pericyclic domino processes according to the first two types of reactions that take place. To facilitate ease of reading, the class of pericyclic reaction involved in each step is stressed in small boxes below the reaction arrow.

6.2
Cycloadditions

6.2.1
Cycloaddition/Cycloaddition

Cycloadditions form a huge and diverse group within pericyclic reactions. Diels' and Alder's discovery of a formal [4+2]-addition to shape Nature's most abundant six-membered rings nowadays appears as one of many more possible construction patterns. As soon as charge separation or photochemical excitation gets involved, the merging building blocks can theoretically be of any size, facing the synthetic chemist with a preparative challenge to prove the rules of conservation of orbital symmetry.

The conversion of furan moieties into preliminary motifs of aromaticity was used by Guitian's group in the synthesis of perylene derivatives [9]. Difurylnaphthalene **1**, which was obtained directly by cross coupling from the corresponding diiodo species, underwent two domino-type cycloaddition steps initiated by the *in situ* formed aryne out of **2** (Scheme 6.1). Considering the mechanistic options to form four different diastereoisomers, the highly diastereoselective cascade evolved only one Diels–Alder adduct, namely the *exo–exo* species **4**, in moderate to excellent yields. Treatment of these with concentrated aqueous HCl in refluxing EtOH

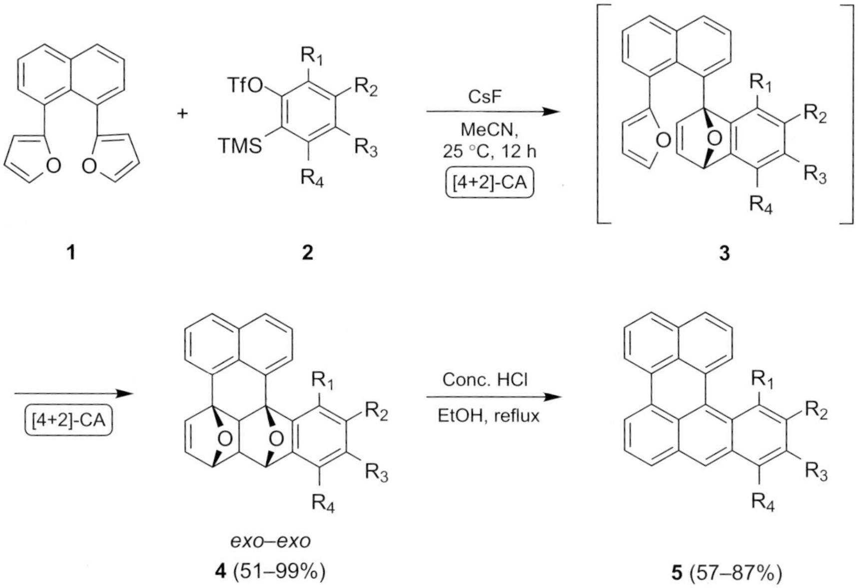

Scheme 6.1 Two subsequent Diels–Alder reactions form precursors for perylene derivatives as new tools in molecular electronics.

gave the final perylene congeners **5** in yields up to 87% by an elimination and deoxygenation procedure.

6.2.2
Cycloaddition/Cycloreversion

A domino sequence comprising a cycloaddition and subsequent cycloreversion step can often find a more general application in organic synthesis, especially in the formation of aromatic compounds such as furans or pyrroles. Oxazole moieties as electron-deficient dienes often serve as the crucial reactive centers which cycloadd to a triple bond and eliminate a nitrile upon cycloreversion. If the first step is intramolecular, the impelling enthalpy preserved in the stability of the formed CN function is additionally accompanied by a positive entropy when the nitrile, sometimes volatile, leaves the substrate. In an older example from 1984 [10], Jacobi and coworkers devised a scheme for the preparation of a highly substituted furan on their synthetic way to paniculide A. An intramolecular Diels–Alder reaction was followed by the critical extrusion of volatile acetonitrile, furnishing the bicycle **8** in 94% yield (Scheme 6.2).

An impressive and more current multicomponent reaction by Masson and Zhu [11] orchestrates the same strategy as a key step in another furanoid synthesis (Scheme 6.3). Again, a triple bond, unactivated in this example, undergoes a Diels–Alder reaction with a more electron-deficient oxazole moiety. A consecutive

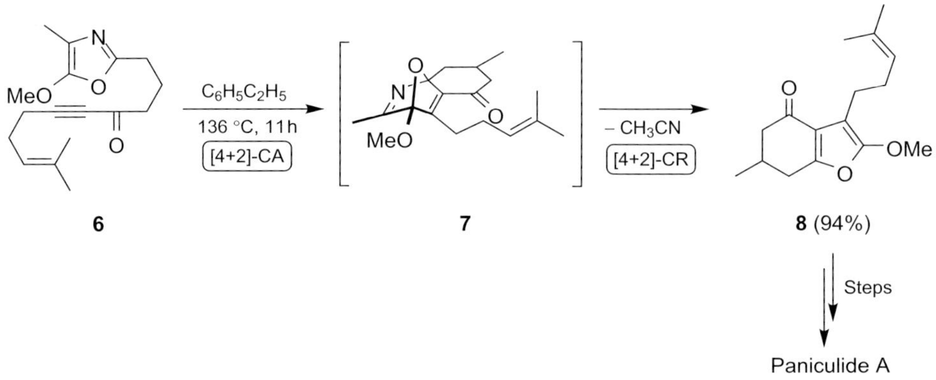

Scheme 6.2 Cycloaddition/cycloreversion protocol for the preparation of highly substituted furanoids implemented in a total synthesis by Jacobi *et al.*

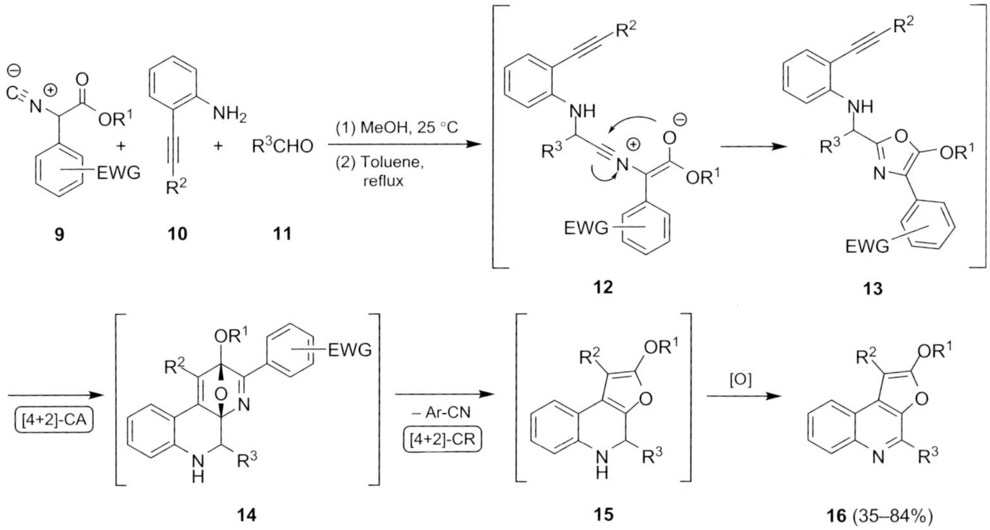

Scheme 6.3 Another cycloaddition/cycloreversion protocol with an oxazole moiety exploiting the extrusion of an aryl nitrile.

cycloreversion from the adduct **14** releases an aryl nitrile and establishes the furanoid core.

Heteroaromatic motifs are not the only accessible products of a cycloaddition/cycloreversion strategy. Cleverly shaped six-membered dienes with an intrinsic leaving group might also act as precursors for phenyl moieties. A definitely irreversible cycloreversion tailors this route for the integration of arenes into strained scaffolds. Beaudry and coworkers [12] embarked on the synthesis of the conformationally chiral macrocyclic natural product cavicularin. A vinyl sulfone moiety, acting as an alkyne equivalent, underwent a Diels–Alder reaction with a neighboring pyrone unit under perfect regiocontrol to furnish the intramolecular adduct **18** (Scheme 6.4). As soon as phenylsulfinic acid is eliminated, the system gains considerable stabilization on emitting CO_2 and establishing aromaticity, though the corresponding new "B-ring" adds severe strain and is known to be significantly distorted.

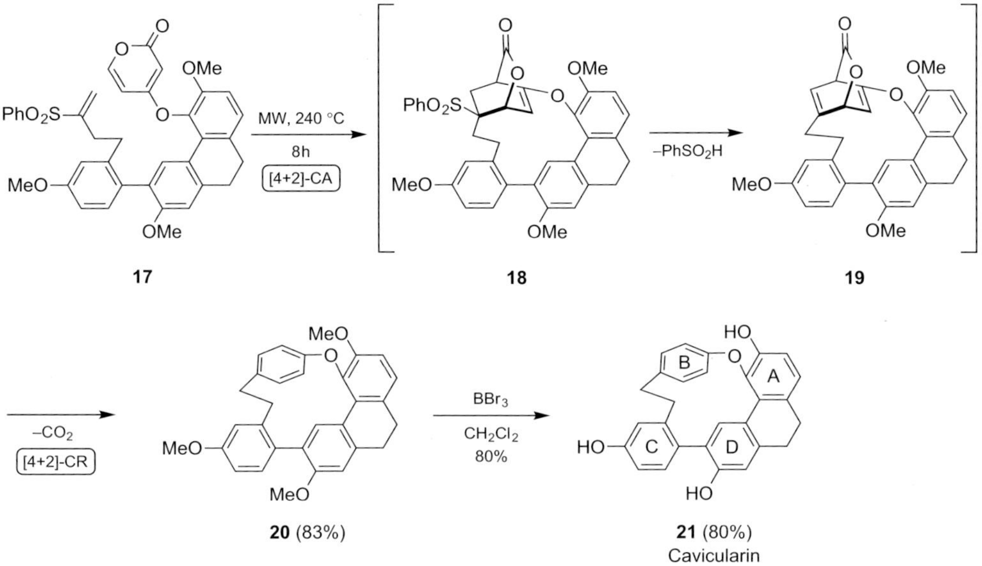

Scheme 6.4 Synthesis of cavicularin with elimination of CO_2 as key step.

If triple bonds are implemented in cycloadditions with six-membered cyclic dienes, an irreversible release of an alkene moiety claims for another type of arene synthesis. Driven by the search for novel antitumor agents, Danishefsky exploited this strategy during the synthesis of cycloproparadicicol [13]. The dimedone-derived electron-rich diene **22** cycloadds to the activated triple bond of the compound termed *ynolide* **23** at elevated temperatures, whereupon a concomitant extrusion of isobutene uncovers the aromatic core (Scheme 6.5). The trimethylsilyl (TMS) ether groups get cleaved during chromatography, furnishing the precursor **25** of cycloproparadicicol in 75% yield.

A remarkable most likely pericyclic rearrangement was observed in the laboratories of Wegner [14]. This final [3,9]-sigmatropic rearrangement is to date not fully confirmed; however, the happenings before are definitely pericyclic. The bidentate Lewis acid **28** decreases the lowest unoccupied molecular orbital (LUMO) energy of phthalazine **27** to such a level that an inverse-electron-demand Diels–Alder reaction with a furan reactant such as **26** can take place at temperatures usually higher than 115 °C (Scheme 6.6). A cycloreversion releases nitrogen and regenerates the catalyst, whereupon the loss of aromaticity associated therewith triggers another rearrangement, leading finally to the cyclopropane derivative **30**. The recovered benzene core in interplay with the ester resonance is probably the main thermodynamic

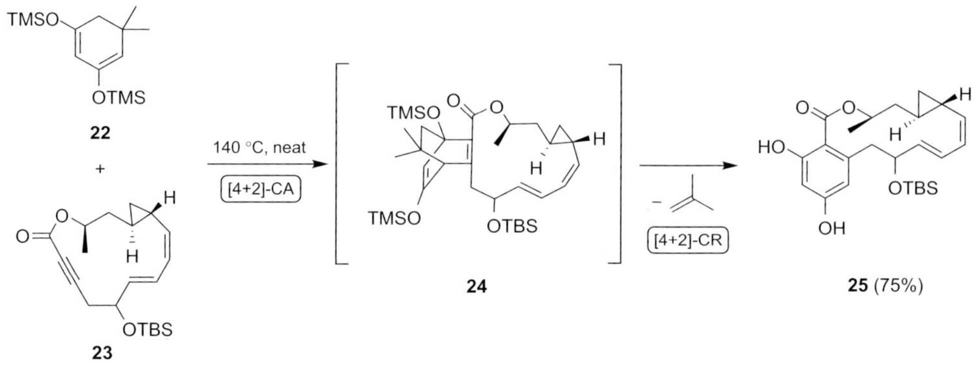

Scheme 6.5 A volatile alkene like isobutene filling the role of a pericyclic leaving group in the synthesis of cyclopropane-derivatized radicicol.

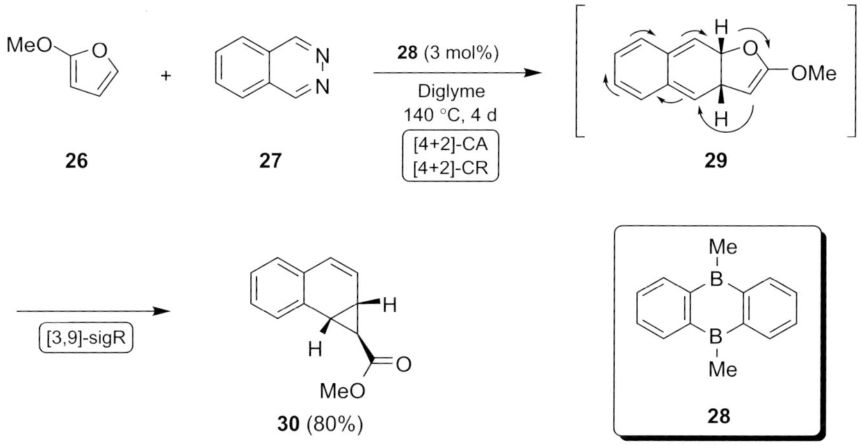

Scheme 6.6 Cycloaddition/cycloreversion cascade followed by an unusual three-membered ring formation developed by Wegner *et al.*

driving force. The compatibility with many substitution patterns on the phthalazine part provides the reaction with a reasonably wide scope.

6.2.3
Cycloaddition/Sigmatropic Rearrangement

A contribution to the synthesis of oxoindoles equipped with quaternary stereocenters of high stereochemical purity was made by Smith *et al.* [15] in 2009. Nitrones **31** derived from Garner's aldehyde cycloadd readily to ketenes **32**, a process that intensifies its asymmetric induction with increasing size of the protecting group (PG). An initially proposed ionic addition to the quaternary carbon of the ketene was subsequently ruled out through computational methods [16]. The weak N–O bond in **33** is now ideally suited to migrate and to be transformed into a C–C bond with an attached stereocenter (Scheme 6.7). The seven-membered aminal **34** is then labile enough to get cleaved, presumably after a very quick rearomatization. Upon workup, the imino acid **35** is hydrolyzed and a cyclic amide is formed. Tethering the substituents R^1 and R^2 gives rise to a wide range of spirocyclic oxoindoles (e.g., **37**) with only seldomly decreased yields [17].

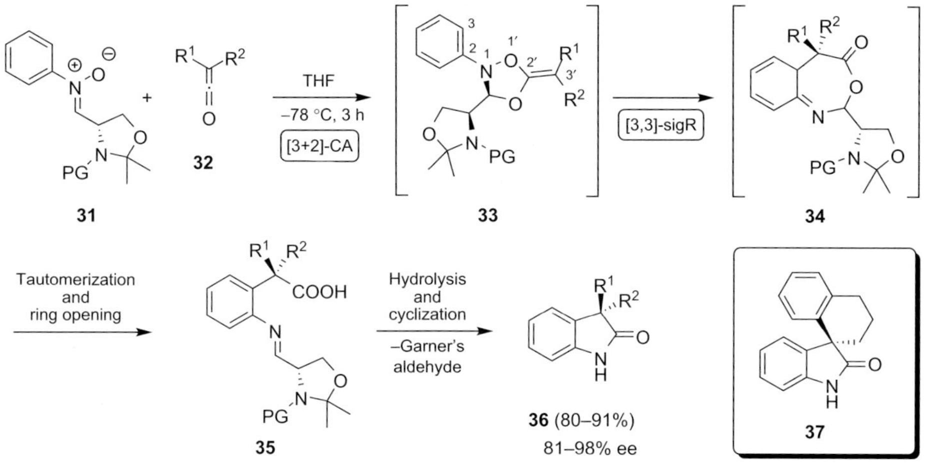

Scheme 6.7 A cycloaddition followed by a sigmatropic rearrangement equipping an arene with a quaternary stereocenter and realized in the preparation of various oxoindoles.

6.2.4
Cycloaddition/Electrocyclization

A domino pericyclic route to polysubstituted salicylic acid derivatives was found in the laboratories of Suzuki [18]. The terminal electron-deficient triple bond of enynone **38** attains perfect regioselectivity upon cycloaddition with a ketene silyl acetal such as **39** already at 60 °C. The neat reaction mixture is now diluted with mesitylene and heated up to initiate an electrocyclic ring opening. The silylated ester enolate moiety of **41** is then converted into a neighboring keto enolate in terms of a [1,5]-silyltropic rearrangement which transforms the compound into an ideal substrate for 6π-electrocyclization. This transformation either finishes the reaction progress to give **43** or might be followed by an elimination to furnish aromatic products such as **44** (Scheme 6.8). Yields range from 50 to 86% and more complex substitution patterns at the double bond of **38** do not adversely affect the efficiency.

Doyle and coworkers [19] presented an elaborate mechanism for their formation of tricyclic heterocycles, although still in a more or less speculative state. Treating both reactants with an appropriate Rh catalyst transforms a diazo compound into **46**. Presumably, the vinylogous double bond now undergoes an electronic activation, which tailors it for a cycloaddition with nitrone **45** furnishing exclusively

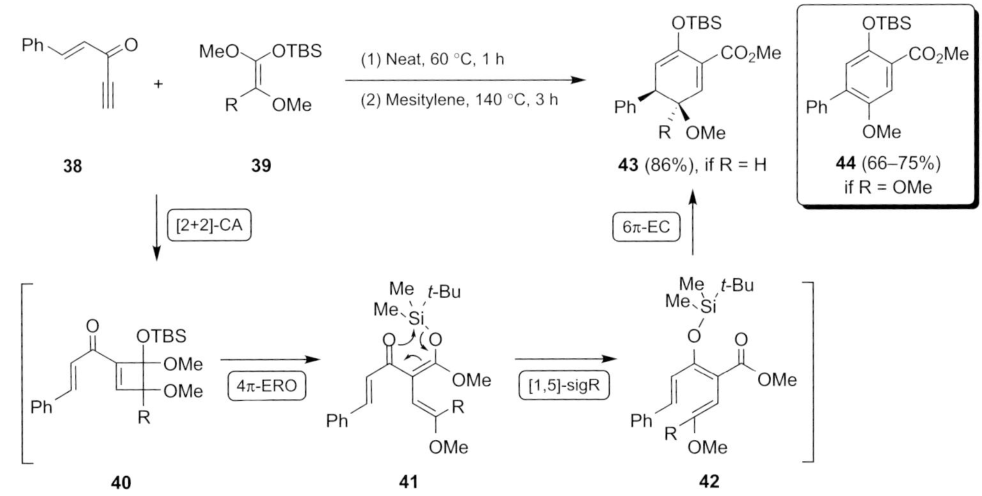

Scheme 6.8 A sequence of four pericyclic processes affording the dihydrosalicyclic acid derivatives **43** and **44**.

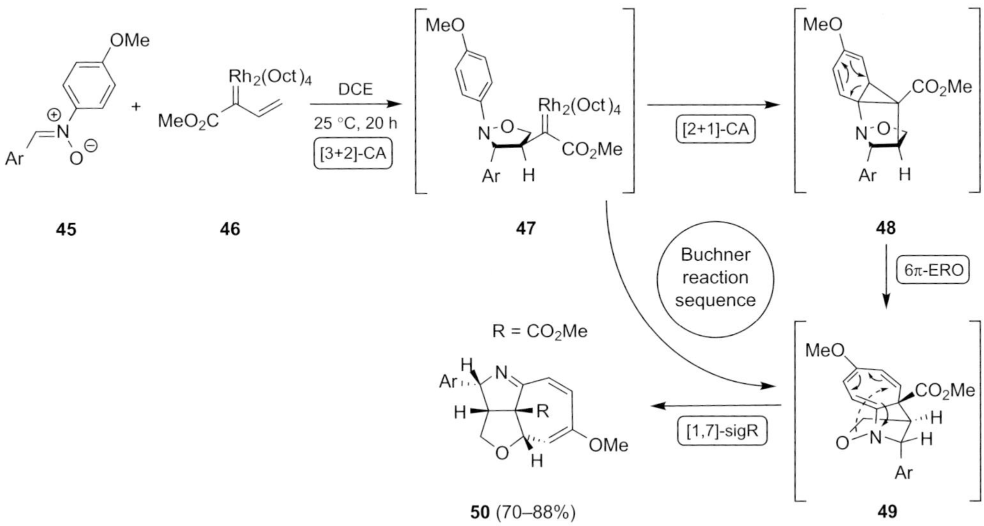

Scheme 6.9 Four pericyclic reaction steps proposed for the formation of tricyclic heterocycles, including a dipolar cycloaddition, a Buchner reaction sequence, and a final sigmatropic rearrangement of a C–O bond.

the *trans* intermediate **47** (Scheme 6.9). After cyclopropanation of an electron-rich aromatic double bond, the system is particularly suited for a 6π-electrocyclic ring opening, leading to the product of a typical Buchner reaction. An oxygen migration as part of a [1,7]-sigmatropic rearrangement finally stabilizes the reaction's progress in the formation of the sophisticated tricycles **50**.

6.2.5
Cycloaddition/Mixed Transformations

A very promising, fundamentally new synthetic technique for highly substituted benzenoid compounds was published by the group of Hoye *et al.* [20]. This technique, termed the *hexadehydro-Diels–Alder* (*HDDA*) reaction, originates with an aryne by linking a 1,3-diyne with an alkyne under metal-free conditions at elevated temperatures. An electronic activation in the π-systems of both reactants is no prerequisite and, although they were hitherto tethered, numerous linkages appear highly suitable. The mode of the consecutive trapping step can be manifold but was first studied by means of a formal intramolecular Si–O bond addition (Scheme 6.10). However, a further intramolecular cycloaddition or an ene reaction [21] (Scheme 6.11) did not compromise the compatibility and efficiency of the domino sequence, and even intermolecular aryne trapping announces itself as a wide field for future research.

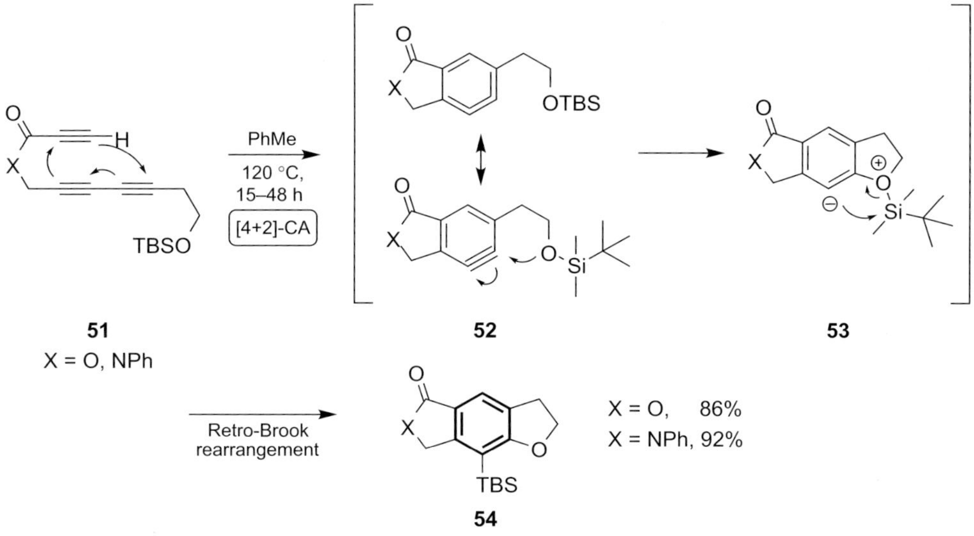

Scheme 6.10 HDDA reaction followed by an intramolecular silyl ether trapping event.

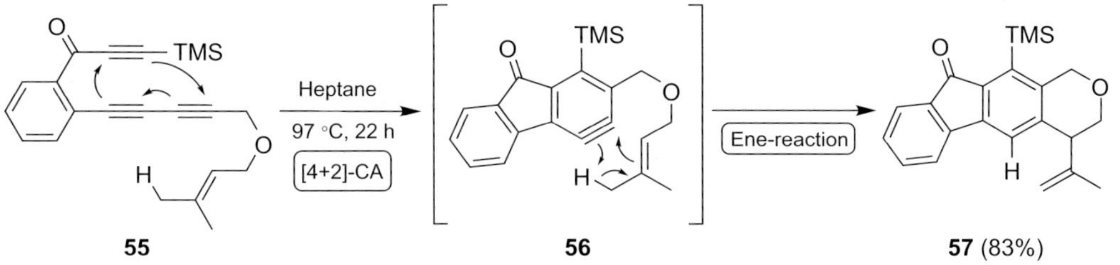

Scheme 6.11 A consecutive ene reaction as one of many trapping modes for the HDDA-generated aryne.

Toward the biomimetic synthesis of chloropupukeananin, Kobayashi *et al.* [22] successfully implemented a domino sequence of two pericyclic reactions. Masked benzoquinone **59** as the diene reactant classifies the Diels–Alder reaction as inverse-electron-demanding one to which the vinylallene **58** can contribute its extra double bond forming a caged skeleton for the subsequent carbonyl–ene reaction. Longer reaction times and higher pressures were able to direct the reaction pathway toward the desired *endo* product **60** against the backdrop of numerous conceivable side mechanisms. Choosing the pure enantiomer of **59**, as depicted in Scheme 6.12, furthermore increased the stereoselectivity and completed the synthesis of the model compound **61** in 70% yield.

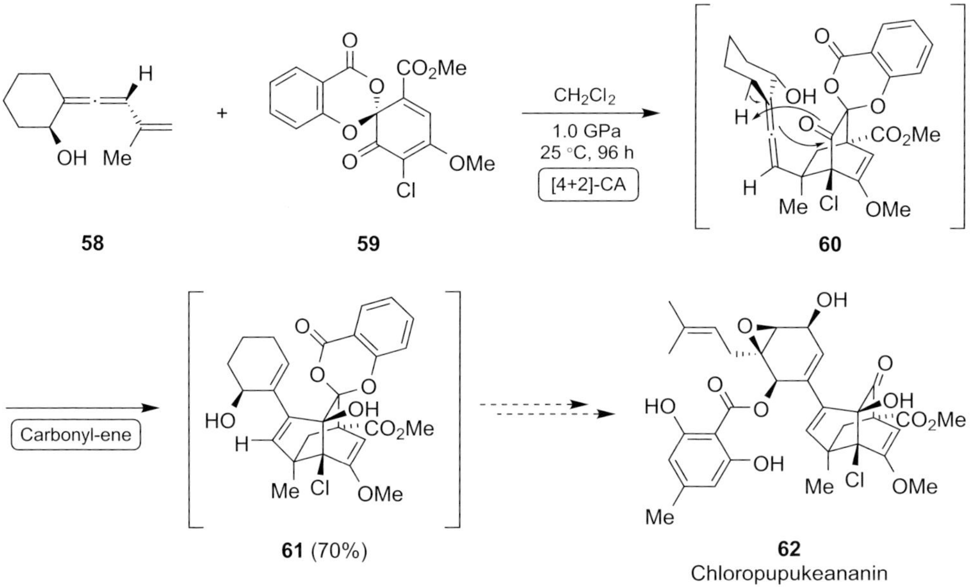

Scheme 6.12 Domino process consisting of an inverse-electron-demanding Diels–Alder reaction and a carbonyl–ene reaction to access the central core of chloropupukeananin.

6.3 Sigmatropic Rearrangements

6.3.1 Sigmatropic Rearrangement/Sigmatropic Rearrangement

The traditional pattern for a sigmatropic rearrangement is revealed in a classic Cope rearrangement; however, the often-demanded high temperatures encouraged chemists to place heteroatoms, substituents, and charge to smoothen the reaction pathway. Several named reactions, among these the numerous Claisen-type rearrangements, benefit from the σ-bond shift when the electronic reorganization creates a favored carbonyl function. But, for launching a concerted domino reaction, too much stability can sometimes hamper any progression, turning the design of cascade reactions into a walk on a fine line. One sigmatropic rearrangement involving cleavage and migration of a σ bond along a π system often sets the stage for another subsequent sigmatropic rearrangement. A domino aryl-Claisen rearrangement of cyclic and acyclic bisaryloxy-substituted alkenes **63** was performed in high yield catalyzed by lanthanide(III) salts such as $Eu(fod)_3$ [23]. In the first step, one aryl moves affording intermediate **64**; in the second step, the other aryl residue follows yielding the bisphenolic alkene **65** (Scheme 6.13). For acyclic substrates, only modest transfer of chirality is observed, whereas cyclic starting materials undergo the domino process in a highly diastereoselective manner.

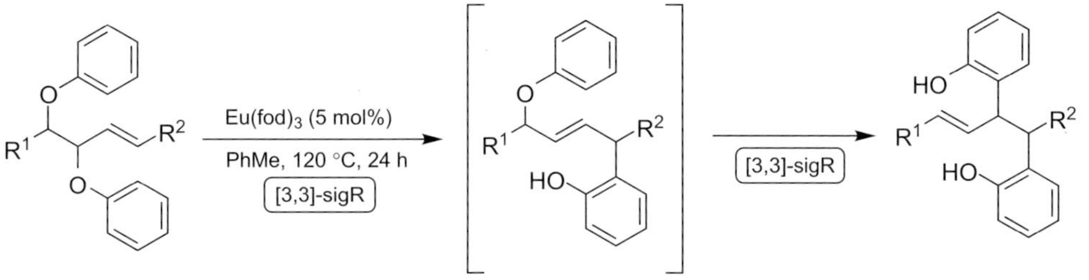

63 **64** **65**

Scheme 6.13 Two subsequent aryl-Claisen rearrangements.

An interesting cascade of two sigmatropic rearrangements followed by an ene reaction was developed for the synthesis of the *neo*-clerodane skeleton that is a part of the natural product teucrolivin A (Scheme 6.14). Promoted by microwave heating, the precursor **66**, which is easily available by the attack of a lithiumorganyl

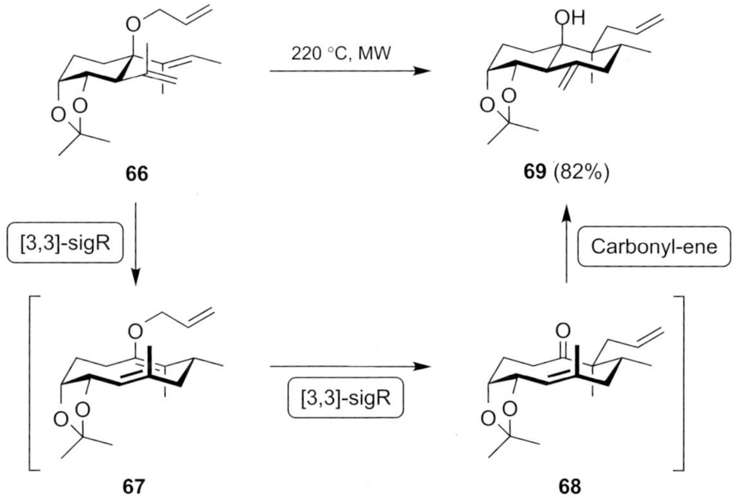

66 **69** (82%) **67** **68**

Scheme 6.14 Consecutive Cope, Claisen, and carbonyl–ene reactions to afford the *neo*-clerodane skeleton.

to the cyclohexanone derivative and allylation of the respective alcohol, yields the bicyclic compound **66** in 78% yield as the only detectable diastereomer [24]. A Cope rearrangement provides the 10-membered macrocycle **67**. This intermediate instantaneously undergoes a Claisen rearrangement, generating a second 10-membered ring with a ketone moiety **68**. The latter is prone to reacting in a transannular carbonyl–ene reaction leading to the formation of the *trans*-decalin-like structure **69**. The authors state that the diastereoselectivity of this domino process is a direct result of preferred conformations of the macrocyclic intermediates and their impact on the respective transition states [25].

Not only the common [3,3]-sigmatropic rearrangements can be part of domino cascades, but even processes involving [2,3]-sigmatropic rearrangements can be designed. The synthesis of highly substituted cyclopentanes **77** starts with the Rh-catalyzed decomposition of the diazo compound **70** and subsequent reaction with alcohol **71** to generate *O*-ylide **73** [26]. Immediately, a [2,3]-sigmatropic shift occurs, leading to a Cope system (Scheme 6.15). If the scandium triflate is omitted, this intermediate can be isolated. To accelerate the following Cope rearrangement, the temperature is increased. At this point, one might discuss whether this process is a real domino reaction according to the definition of Tietze [7], which states all the conditions have to stay the same during the complete domino process. Nevertheless, the transformation that takes place afterward,

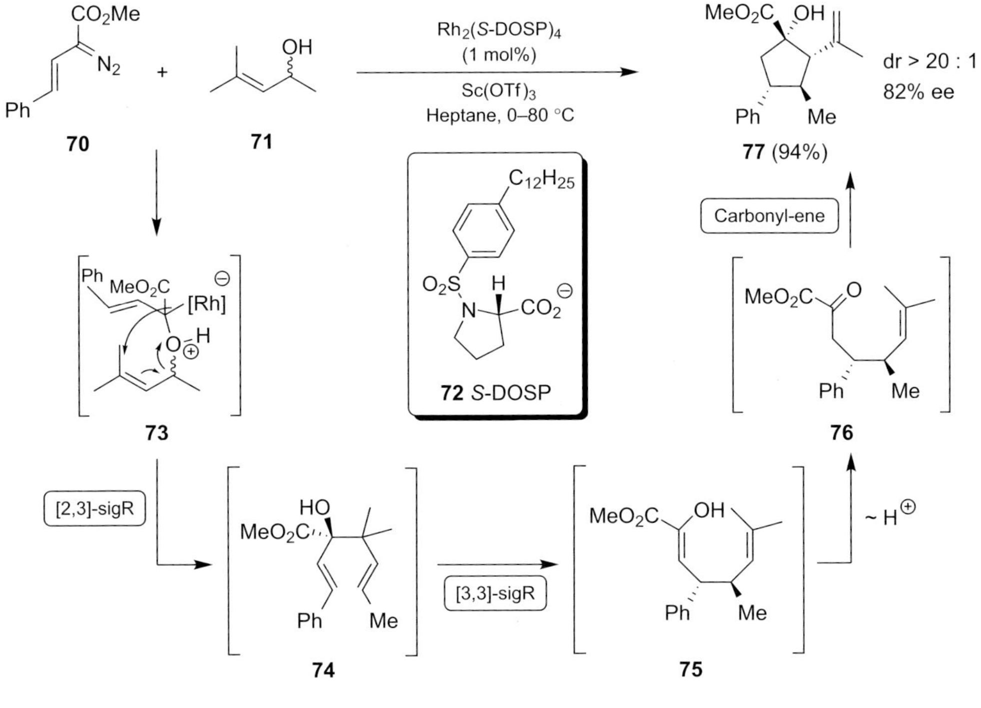

Scheme 6.15 Five-step domino sequence including a [2,3]-, [3,3]-sigmatropic rearrangement and a carbonyl–ene reaction.

consisting of [3,3]-sigmatropic rearrangement, keto–enol tautomerization, and final carbonyl–ene reaction leading to the five-membered ring, is definitely a domino reaction. Good to excellent yields have been observed, and high levels of diastereoselectivity and moderate enantiomeric excess values obtained when proline-derived DOSP was used as the chiral ligand at the rhodium dimer. The high degree of diastereoselectivity is the result of unfavorable *pseudo*-1,3-diaxial interactions in one of the two possible transition states of the carbonyl–ene reaction. In contrast, the enantioselectivity is already determined by the formation of the Cope educt **74**; however, a slight loss of enantiopurity is observed during the following steps as demonstrated by the isolation and analysis of the intermediate **76**.

6.3.2
Sigmatropic Rearrangement/Cycloaddition

Although the heart of the reaction does not meet the criteria of a domino reaction, a preluding, undisputed cascade linked with a simple but elegant strategy to proline scaffolds of broad applicability warrants its presentation in this chapter. Building on the studies of Overman [27] and Bennett [28], the group of Waters [29] succeeded in upgrading the mutual iminium formation with a final cycloaddition step. An aryl-substituted homoallyl amine such as **78** is the crucial substrate to facilitate the following pericyclic events. After achieving the imine **80**, the π system of the aryl-substituent, which can be neutral, electron-donating, or preferably electron-withdrawing, enforces the 2-aza-Cope rearrangement to the imino acid esters **81** (Scheme 6.16). The acidified saturated α-position is now the origin of

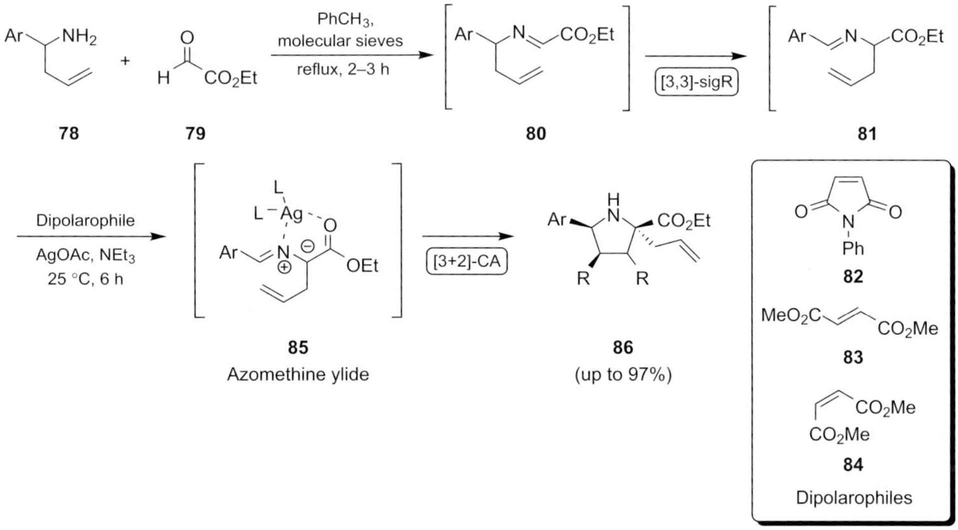

Scheme 6.16 Imine formation, sigmatropic rearrangement, and a final 1,3-dipolar cycloaddition upon changing the reaction conditions, classifying this one-pot reaction sequence to highly substituted prolines as "semi-domino" according to its key step.

the azomethine formation when $AgOAc/NEt_3$ is added, and the reaction mixture is cooled down to room temperature. The dipolarophiles **82–84** show comparable reactivity toward the 1,3-dipole cycloadding in an *endo* orientation to provide the proline derivatives with a quaternary carbon in good to excellent yields. Except for the fumaric acid ester **83**, the substituents *R* are consistently *cis*, as mechanistically predicted for a concerted process.

6.3.3
Sigmatropic Rearrangement/Electrocyclization

Extensive investigations by cleverly designed experiments and computational means suggest that transition states of sigmatropic rearrangements are very sensitive to the nature and position of substituents. As a result of the weakening of the central single bond by $n \rightarrow \sigma^*$ interaction, anionic oxy-Cope rearrangements proceed much faster than their neutral counterparts. In the case phenyl substituents are attached to the two formal allyl systems, a dissociative mechanism is also discussed. An interesting domino reaction where all these insights play in concert deals with 1,5-hexadiyn-3,4-olates which are prepared in one step from 1,2-diketones and the corresponding lithiated alkynes. Because of the various substituents (anionic oxygens as well as phenyl moieties), the Cope rearrangement step is considered to be highly asynchronous: that is, also a dissociative pathway might play a role. It turns out that the type of silyl substituent plays a crucial role in determining the product distribution. With the relatively small TMS groups, the anionic Cope rearrangement takes place, whereas larger silyl residues such as triisopropyl silyl (TIPS) favor a dissociative pathway leading mainly to **92**. The product of the oxy-Cope rearrangement is the bis-allenic structure **88**, which immediately undergoes a 4π-electrocyclic ring closure to cyclobutenes containing two enolate units (Scheme 6.17). Hydrolysis yields a mixture of 1,2-*cis*- and 1,2-*trans*-substituted cyclobutenes **90** and **91** [30]. If alkyl or aryl residues at the alkyne are utilized, an aldol-like reaction occurs as the second step of the [3,3]-sigmatropic rearrangement and cyclopentenones of type **93** are formed.

Another type of four-membered ring systems, namely highly substituted fused cyclobutane moieties, was also accessed by a one-pot domino process consisting of sigmatropic rearrangement and electrocyclization. The Yudin group prepared *N*-vinyl β-lactams **94**, which were heated under microwave conditions to 160–200 °C. As a first step, an aza-Cope rearrangement took place, leading to an eight-membered lactam with two double bonds. A proton shift mediated by cesium carbonate afforded the 1,3,5-heterotriene system which immediately underwent ring closure to the [4.2]-bicyclic system (Scheme 6.18) [31, 32]. High levels of diastereoselectivity were obtained. Interestingly, computational studies have revealed that the reaction should proceed through an eight-membered ring intermediate that contains an (*E*)-configured double bond (not depicted in Scheme 6.18) [32], which is an unusual feature in the area of small-ring strain-release Cope rearrangements.

Double bonds being part of hetero-Cope systems can be generated by Michael addition of nucleophiles to electron-poor carbon–carbon triple bonds.

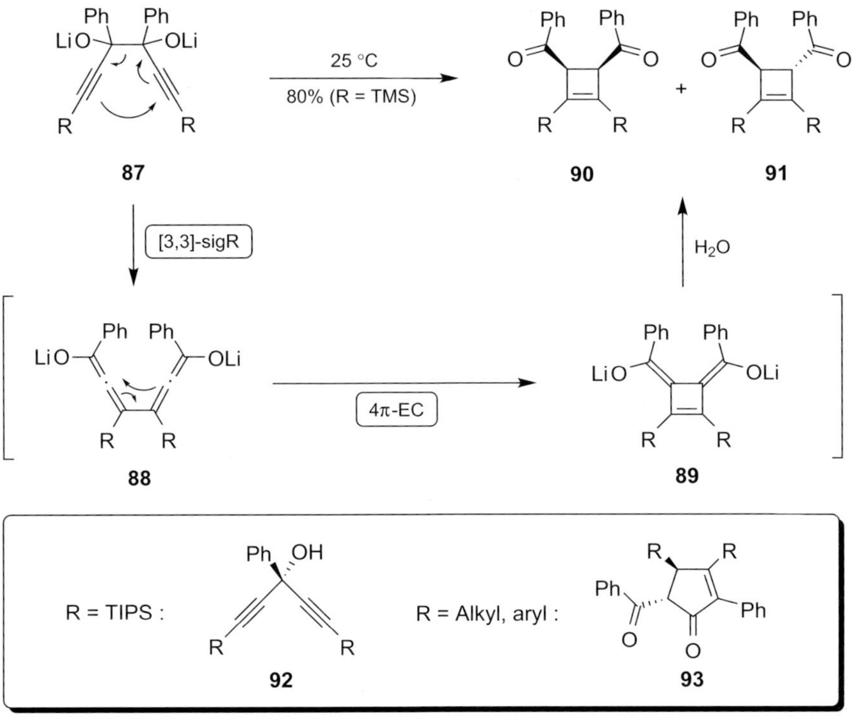

Scheme 6.17 Anionic oxy-Cope rearrangement followed by electrocyclization to provide cyclobutenes.

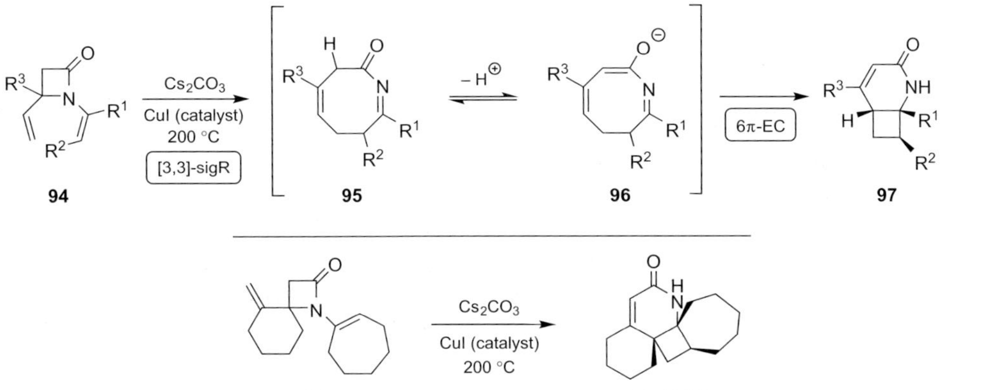

Scheme 6.18 Domino process for the synthesis of highly substituted cyclobutane fused ring systems.

Pyrrolidinyl-substituted indol derivatives **100** were brought to reaction with tosyl-substituted acetylene **101**. The conjugate addition formed a zwitterionic aza-Cope system with cationic nitrogen and a negative charge located at the CC double bond (Scheme 6.19). A [3,3]-sigmatropic rearrangement occurred as the next step, leading to the formation of a nine-membered ring annelated to the indole scaffold. The generation of this intermediate was followed by deprotonation of C-3 of the indol nucleus and by further electrocyclic ring closure of the resulting anionic 6π-electron system [33]. The latter process proceeded in a disrotatory manner when thermal conditions were applied; thus, *cis*-fused cyclopentene and piperidine rings were found in the final product **105**. It is noteworthy that the electron-withdrawing tosyl substituent at the indol is a necessary prerequisite for the success of this domino process. The use of benzyl at this position leads to other products, probably due to the much less acidic C-3 proton of the indol moiety in intermediate **103**.

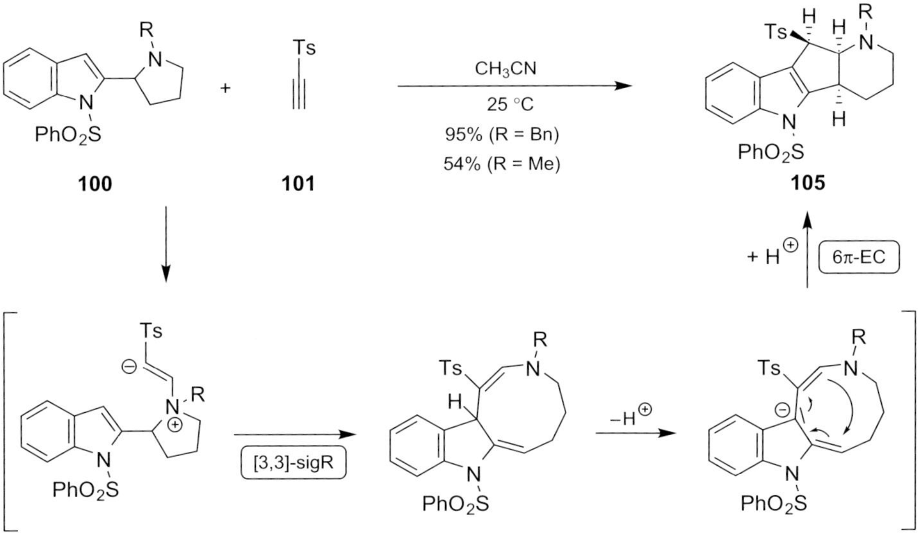

Scheme 6.19 Michael, aza-Cope, and electrocyclic ring closure to obtain **105**.

Fisher's indole synthesis is a traditional textbook example for a domino reaction with inherent pericyclic reactivity. The pyrrole annelation it adds can be further extended with a strategy developed by Mohanakrishnan [34] to obtain various carbazoles. Indole substrates such as **106** act as potent electrophiles in a Friedel–Crafts alkylation with a preferentially activated arene when $ZnBr_2$ is present at elevated temperatures. After a 1,5-hydrogen shift established continuous conjugation, an electrocyclic ring closure furnishes intermediate **109** as ideally suited for a terminating elimination step (Scheme 6.20). The malonate ester moiety is obligatory for a pericyclic pathway; electron-donating substituents on both reactants tend to give somehow better yields. Not only indole species undergo the presented process but

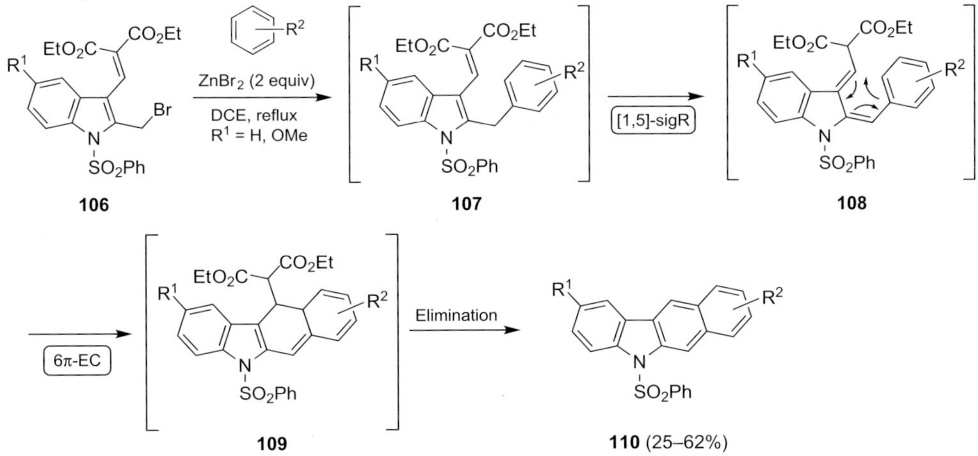

Scheme 6.20 A novel access to annelated carbazoles, an important class of compounds used in optoelectronics, semiconducting materials, and organic light emitting diodes (OLEDs).

also different thiophene derivatives can be reacted with a wide range of arenes. The authors studied several examples and comprehensively state that the number of reaction steps involved excuse the moderate yields.

6.3.4
Sigmatropic Rearrangement/Mixed Transformations

A domino process similar to the one described in Section 6.3.1 consists of an oxy-Cope/carbonyl–ene/Claisen sequence, whereas the former consists of two [3,3]-sigmatropic rearrangements terminated by a carbonyl–ene reaction. Again, a 10-membered macrocyclic intermediate is formed in the first step upon heating compound **111** at 220 °C [35]. After tautomerization, the ketone is forced to perform a transannular carbonyl–ene reaction to furnish two annelated six-membered rings. The corresponding enol ether with the terminal double bond sets the stage for the final Claisen rearrangement. The spirocyclic transition state is formed in such a way that the less hindered face of **113** is attacked. The resulting enol moiety tautomerizes to the aldehyde, forming with its tertiary hydroxyl group the hemiacetal **114** (Scheme 6.21). However, slight modifications in the starting material (e.g., the use of unsubstituted allyl systems and further substituents of the alkene units in the starting material) result in the formation of other products. Of course, the conformational flexibility of the 10-membered ring allows several transition states of the ene reaction that are similar in energy; it turns out that the Curtin–Hammett principle rules, and that the ratio of products corresponds to the relative energetic values of the different transition states.

Acetals incorporated in annelated five-membered rings are prominent examples in natural products such as, for instance, aflatoxins. A new route to this structural motif weaving the different reactivities of a furan core into a reaction pathway was

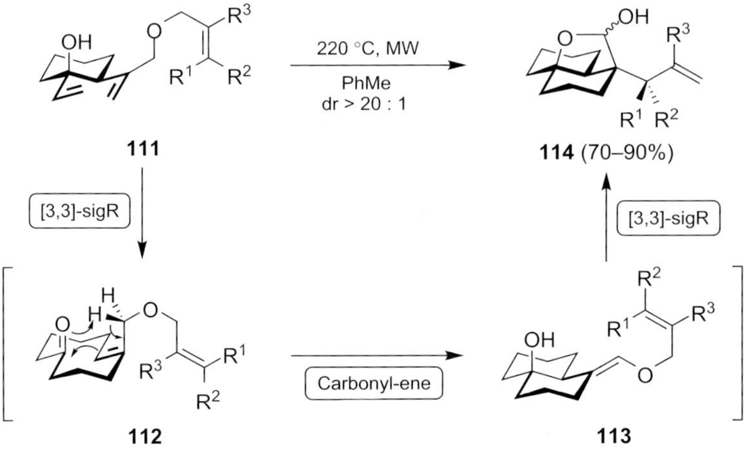

Scheme 6.21 Consecutive oxy-Cope, carbonyl–ene, and Claisen rearrangement.

disclosed by the group of Yin [36] in 2011. A Michael addition of a furfuryl alcohol to an acetylenedicarboxylate is the preluding step that furnishes precursor **115** of the following events in the same pot. If a Pd catalyst is added, its coordination to the aromatic diene system triggers a sigmatropic rearrangement and consequently appends a new substituent to the furan moiety after rearomatization (Scheme 6.22). Anyhow, **117** is not a stable compound under these conditions; by virtue of two

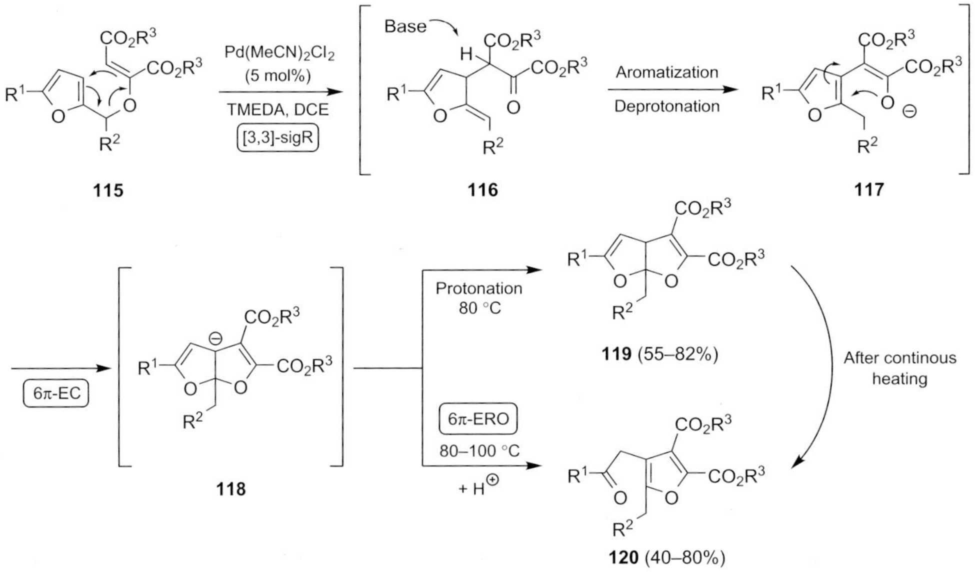

Scheme 6.22 A tunable one-pot synthesis of furan motifs developed by Yin *et al.*

electron-withdrawing ester groups, the liberated oxygen can force its charge upon the aromatic ring which takes up a proton at the annelated carbon to yield the highly substituted furofurans **119** in good yields. This process can be interpreted as electrocyclic; nonetheless, it is highly reversible as well. Hence, **118** appears as a key substrate of an equilibrium which is able to be more extensively accessed when the reaction mixture is exposed to higher temperatures. In so doing, compounds **119** rearrange in up to quantitative yields to the persubstituted furans **120**. Presumably, the electron-withdrawing ester groups profit from the electron-rich furan core by establishing a thermodynamically favored push-pull system. Substituting the furan oxygen of **115** by sulfur gives substrates of similar reactivity but with somewhat poorer yields for the proposed reaction pathway.

6.4 Electrocyclizations

6.4.1 Electrocyclization/Electrocyclization

During an electrocyclization, a pericyclic shift of electrons concertedly transforms a π-bond into a σ-bond, or vice versa. Depending on the direction, rings can thus be closed or opened strictly dictated by the rules of Woodward and Hoffmann, even though electronic and steric effects might play an important role for more exhaustive explanations of stereoselectivity as well (e.g., torquoselectivity). It is quite apparent that systems that accumulate many π-electrons in close vicinity appear very prone for an electrocyclic event to take place. On this count, conjugated oligoenes are ideal candidates and are effectively used to build up caged skeletons by electrocyclic ring closures. Oligocyclic skeletons such as fenestradienes were obtained by the Suffert group [37, 38] by employing such sequences. The group of Trauner [39] exploited this synthetic field very successfully in the construction of many natural products such as, for instance, pyrone derivatives found in several mollusks. Tetraene **123** with the (*E,Z,Z,E*)-configuration is directly derived from a Stille coupling conducted in the same pot. Slight warming is enough to bring both ends of the tetraene in sufficient proximity to initiate an electrocyclic ring closure and form the eight-membered ring of **124** (Scheme 6.23). The flexibility of the shortened π system now allows two conformations, from which the *endo* type triggers a kinetically controlled 6π-electrocyclization to cyclobutane **126**. Although the original natural product "Ocellapyrone A" **125** is only the minor product from a corresponding *exo* transition state, a separate oxidation with oxygen discloses a photochemical access to the natural congener "Ocellapyrone B" in an impressive yield of 89%. Interestingly, a first approach with an (*E,Z,E,E*)-configured pyrone similarly gave the presented product distribution as a result of a complex product equilibration after 5 days at 45 °C, but with reduced yields.

Another wonderful example of a related strategy implemented in a natural product synthesis is depicted in Scheme 6.24. The kingianins are a class of natural products which are homo- or heterochiral dimers of bicyclo[4.2.0]octadienes. These

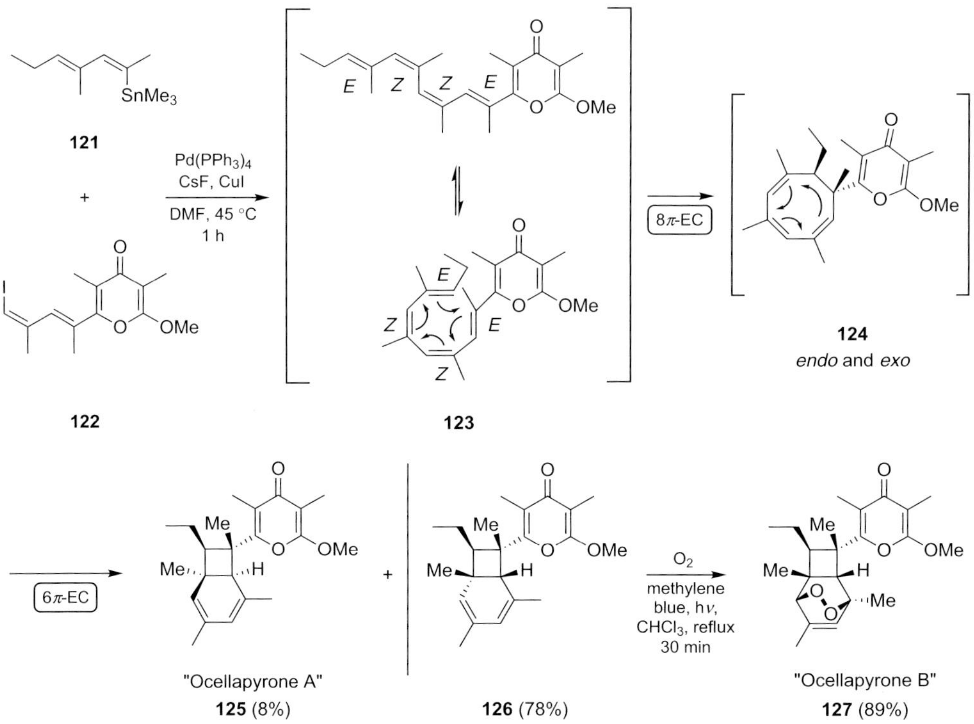

Scheme 6.23 Exploiting extended π-systems in domino-fashioned electrocyclizations for the total synthesis of complex polycyclic molecular skeletons.

latter compounds are available by an electrocyclization/electrocyclization domino process starting with tetrayne **128**. A fourfold reduction with Riecke zinc in ethanol afforded (*Z,Z,Z,Z*)-tetraene **129** in a completely chemo- and diastereoselective manner [40]. Heating a solution of this unsaturated system to 100 °C in toluene triggered the anticipated 8π/6π-electrocyclization sequence to bicyclic dienes **130** and **132**, respectively. Deprotection and some functional group interconversions set the stage for a radical-promoted Diels–Alder reaction to furnish the complex oligocyclic kingianin core.

6.4.2
Electrocyclization/Cycloaddition

Electrocyclizations often produce six-membered dienes or four-membered strained enes which can be part of a cycloaddition if appropriate reaction partners such as enes or dienes are available. Such cascades are able to quickly generate highly complex oligocyclic frameworks with excellent stereocontrol [41]. The Trauner group has made remarkable contributions to this field and has exploited this field for elegant and efficient natural product syntheses. An example yielding the tricyclic core structure of the fungal polyketide (−)-PF-1018 is presented in Scheme 6.25.

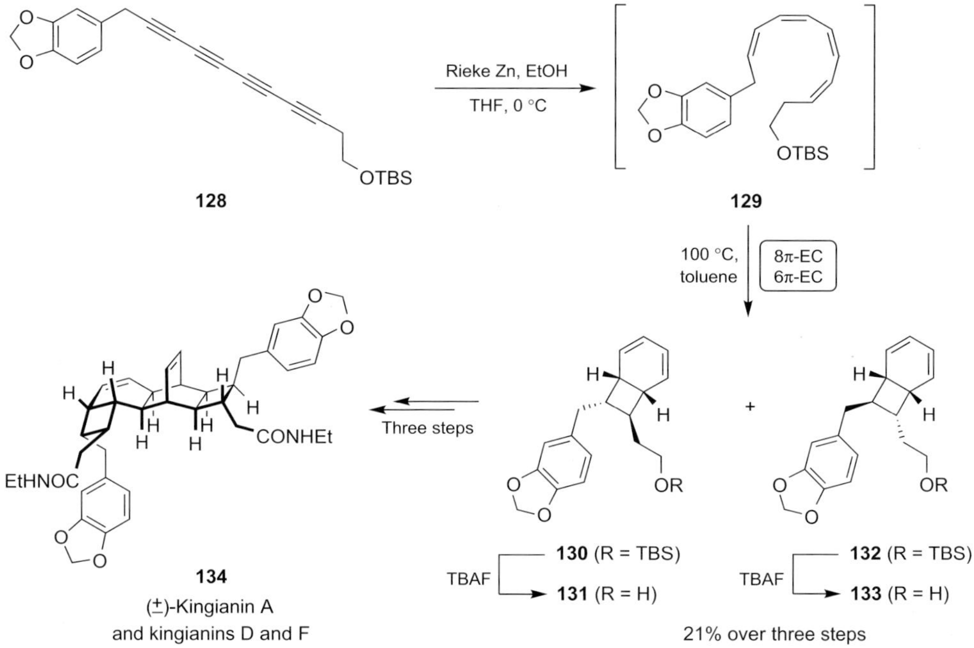

Scheme 6.24 8π-EC/6π-EC cascade of reduced tetrayne **128** as the key step in the synthesis of kingianins.

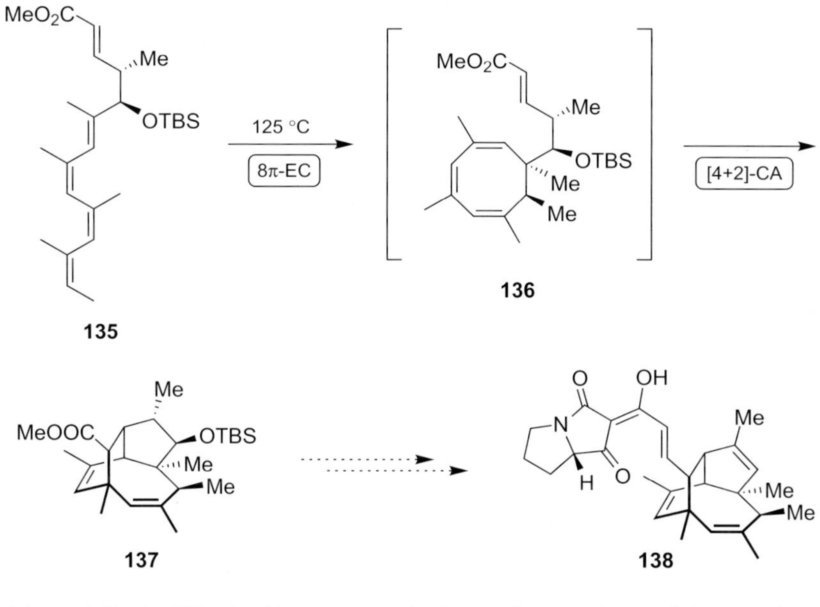

Scheme 6.25 8π-EC/cycloaddition process leading to the central core of the natural product (−)-PF-1018.

Pentaene **135** with four conjugated double bonds, on being heated to 125 °C, undergoes a thermal 8π-conrotatory electrocyclization yielding 1,3,5-cyclooctatriene **136** [42]. Instead of another disrotatory electrocyclization, the external double bond comes into play and a Diels–Alder reaction takes place.

An ynamide-based benzannulation strategy proceeding via a cascade of four consecutive pericyclic reactions was devised by Danheiser and coworkers [43]. A 4π-electrocyclic cleavage of cyclobutenone **139** is believed to form the ketene **141**. The high-lying highest occupied molecular orbital (HOMO) of the electron-rich nitrogen-substituted alkyne **140** makes a [2+2]-cycloaddition with the carbon–carbon double bond of the ketene moiety possible because the electron-withdrawing effect of the carbonyl strongly decreases the LUMO energy of this unit. During this step, again strain is formed as a result of the generation of cyclobutenone **142** (Scheme 6.26). Thus, the next step is driven by a release of strain. Cyclobutenone undergoes a 4π-electrocyclic cleavage – similar to the very first step – to afford the highly unsaturated ketene **143**. A comparison with the first ketene formed reveals that a formal insertion of ynamide into the carbon chain has

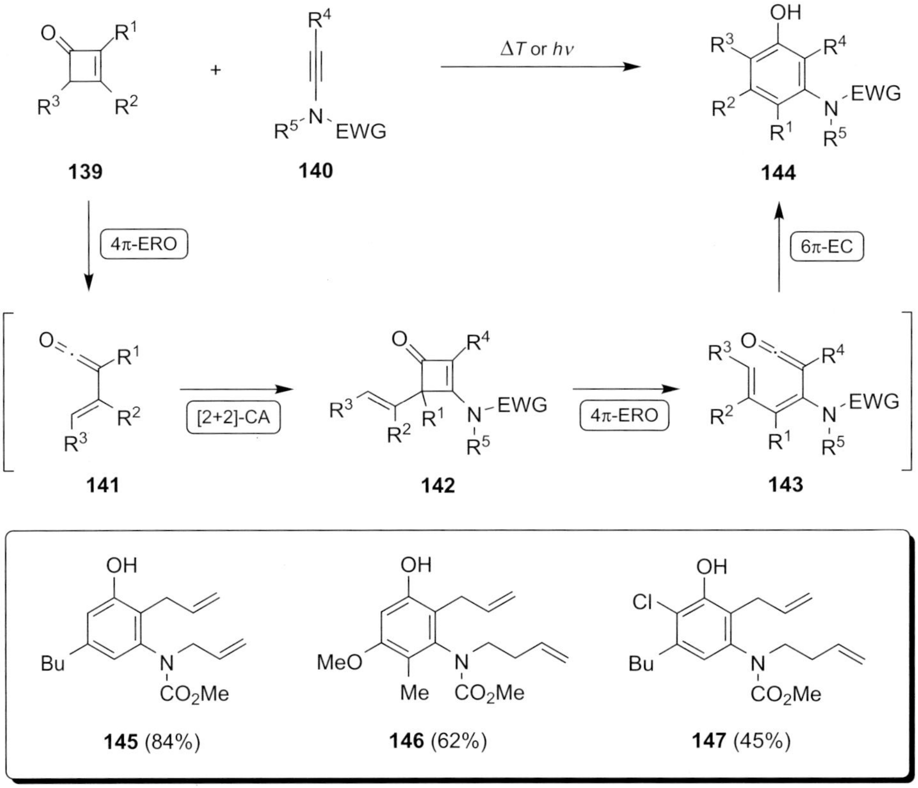

Scheme 6.26 Cascade of four pericyclic reactions to obtain phenol derivatives according to Danheiser *et al.*

taken place. This domino process is terminated by a 6π-electrocyclic ring closure and a subsequent tautomerization to yield the persubstituted phenol derivative **144**. Besides the mechanism, Scheme 6.26 also depicts some examples of phenol derivatives obtained by this sequence together with their respective yields. If ynamides are replaced by ynamines, which are much more nucleophilic, a mixture of **144** together with other regioisomers and allenic products was observed. The latter are explained by nucleophilic attack to the sp-hybridized carbon of the ketene.

Heteroatoms might be also part of systems that undergo electrocyclizations. The symmetry properties of the underlying orbitals are commonly not changed by their introduction; only the energetic levels of the molecular orbitals are varied and respective coefficients are increased or decreased in size. Of course, the latter two points are often made responsible for reactions that do not proceed since HOMO–LUMO interactions have become unfavorable by the introduction of heteroelements or the sizes of coefficients have changed dramatically. A nice example of a bisheterohexatriene-electrocyclization followed by a 1,3-dipolar cycloaddition was recently devised by Kwon (Scheme 6.27). Starting from nitrobuta-1,3-diene **148**, a 6π-electrocyclic ring closure yields a nitronate **150**. In the presence of an electron-poor alkene such as **149**, the 1,3-dipolar intermediate immediately undergoes cycloaddition to afford nitroso acetals of type **151** [44]. In total, two rings and a quaternary stereocenter are formed by this one-pot process. High diastereoselectivities and good to excellent yields were observed. The compounds obtained could easily be transformed further by hydrogenolysis in oligocyclic lactams.

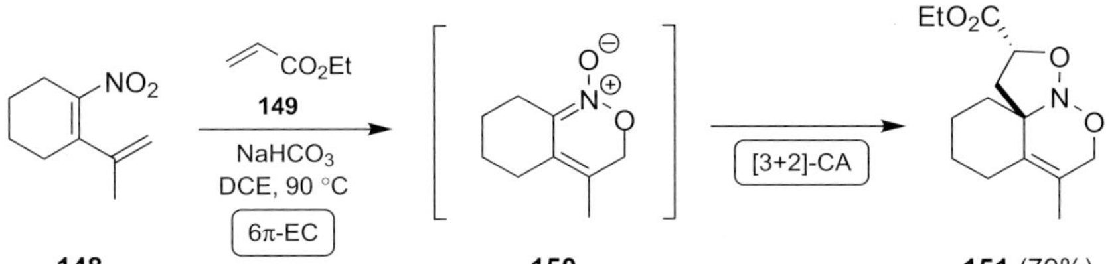

Scheme 6.27 Electrocyclization of nitrodienes to nitronates and subsequent 1,3-dipolar cycloaddition.

6.4.3
Electrocyclization/Sigmatropic Rearrangement

Fascinating transformations starting with an electrocyclization followed by a sigmatropic rearrangement were recently described by Vanderwal. Donor–acceptor dienes known as *Zincke aldehydes*, available by simple aminolysis of pyridinium salts [45, 46], rearrange cleanly at temperatures of 200–220 °C to (*Z*)-dienes under transformation of the aldehyde portion to an amide. The reaction is commonly performed in 1,2-dichlorobenzene under microwave irradiation (Scheme 6.28). The authors consider as the first step a thermally instigated (*E*)- to (*Z*)-isomerization

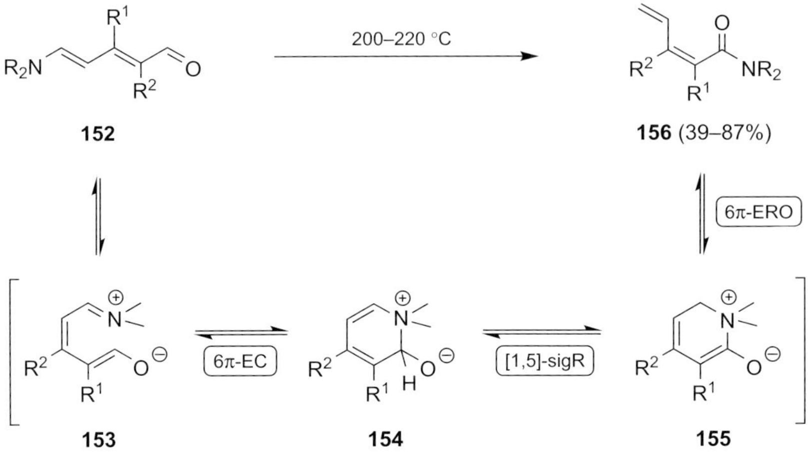

Scheme 6.28 Isomerization of Zincke aldehyde **152** to amides of type **156** by three consecutive pericyclic reactions.

from **152** to **153**, which should occur relatively easily because of the push-pull substituted nature of the highly unsaturated system. This change in geometry might enable a 6π-electrocyclization to afford a dihydropyridinium intermediate **154** [47]. A [1,5]-sigmatropic hydrogen shift paves the way for an electrocyclic ring opening. The (*Z*)-geometry of the final product is the result of the cyclic nature of **155**. An originally discussed alternative mechanism involving pyran intermediates resulting in oxygen transposition rather than amino group transposition was ruled out by regiochemical investigations.

During its mechanistic studies, the Vanderwal group realized that the cascade could even be extended when suitable substrates were employed. They posited that the zwitterionic dienolate intermediate **155** could be trapped by allyl residues attached to the amine. As a result, dihydropyridones **158** would be generated. Indeed, the anticipated product was formed in minor amount (Scheme 6.29), supporting the involvement of intermediate **155**. However, as the major product of this transformation, bicyclic lactam **160** was identified, suggesting that a Diels–Alder cycloaddition was the final step of the cascade [48]. Single diastereoisomers with a *cis*-configured ring junction were formed. This route opens a possibility to access highly diverse scaffolds for natural product synthesis. An example **161** is depicted in Scheme 6.29, where instead of the commonly used terminal double bond a tosyl-substituted indole derivative undergoes this reaction cascade; the tetracyclic lactam is generated in 30% yield in only one step starting from the respective Zincke aldehyde.

Recently, an in-depth theoretical study of this pericyclic cascade rearrangement was performed by Houk and coworkers [49]. Although the mechanistic picture illustrated in Scheme 6.28 seems to be plausible and is in accordance with experimental details, it suffers from energetic barriers that appear too high to allow

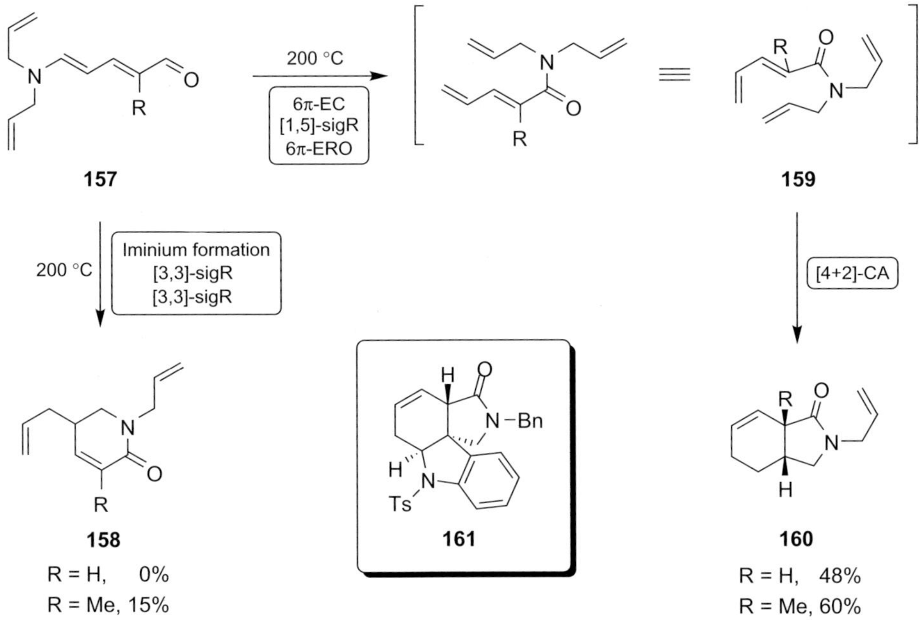

Scheme 6.29 Formation of bicyclic lactams from Zincke aldehydes (right), and side reaction to monocyclic lactams (left).

the rearrangement under the conditions used. The computations implicate the formation of vinylketenes as intermediates after the (*E*)- to (*Z*)-isomerization. These ketenes **164** would be the result of a [1,5]-sigmatropic rearrangement as the first pericyclic step (Scheme 6.30). Attack of the amine to the highly electron-deficient

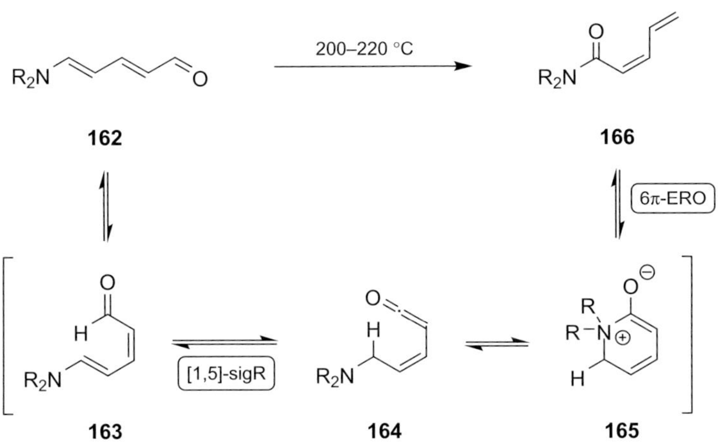

Scheme 6.30 Ketene-based mechanism for the rearrangement of Zincke aldehydes (as revealed by computation).

sp-hybridized ketene carbon is assumed to be the next step, affording six-membered rings of type **165** that undergo electrocyclic ring opening to (*Z*)-α,β,γ,δ-unsaturated amides **166**.

This example illustrates the difficulties that are associated with a proper investigation of the reaction mechanisms. In addition, it clearly demonstrates that the outcome of a pericyclic cascade can often be described by several alternative reaction mechanisms differing in the type and order of pericyclic reaction(s) involved. Thus, the classification made in this chapter might be taken with care. We (as well as the authors of the papers we rely on) commonly tried to provide plausible mechanistic scenarios; however, often only the products of a transformation are what we really know.

6.4.4
Electrocyclization/Mixed Transformations

As previously described, electrocyclic ring closures or ring openings are often followed by other pericyclic reactions. But there are also examples of polar reactions associated with the respective electrocyclization. Of course, charged intermediates are beneficial for the success of such a transformation. Treatment of 1,4-pentadien-3-ones **167**, which are ideal starting materials for the Nazarov cyclization, was used in a domino process consisting of 4π-electrocyclic ring closure and homologous Mukaiyama reaction (Scheme 6.31). By the action of borontrifluoride-etherate, a cyclopentenyl cation **168** is formed in the first step which is captured by the electron-rich silyloxyalkene. Hydrolysis yields the highly substituted carbocycle

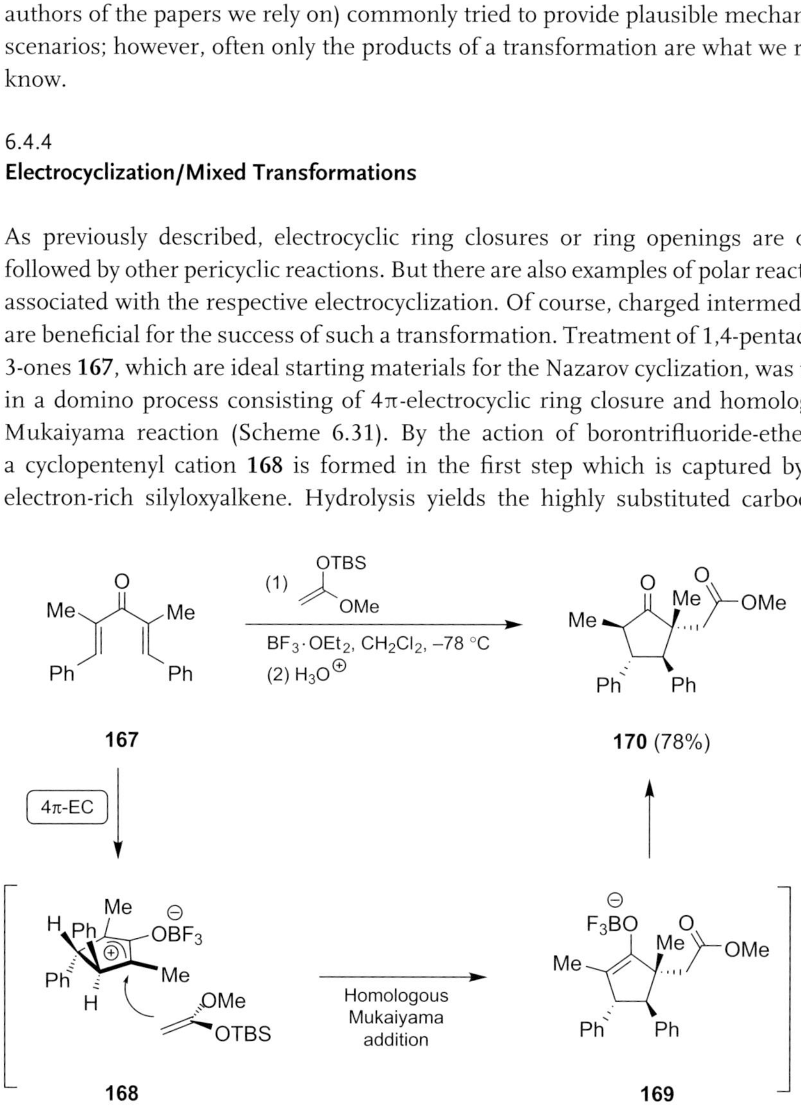

Scheme 6.31 Homologous Mukaiyama reaction to trap the Nazarov intermediate **168**.

170 in good yield. During this domino process, up to five new stereocenters are established; excellent regio- and facial diasteroselectivities are observed [50].

6.5
Mixed Transformations

6.5.1
Mixed Transformations Followed by Pericyclic Reactions

Among the various preluding and guiding reaction steps, an ene reaction can also adjust a substrate for further pericyclic transformations such as a cycloadditon. The group of Danheiser [51] encountered ene-type reactions in their elementary synthesis of highly substituted six-membered ring systems (Scheme 6.32) and derivatized pyridines (Scheme 6.33) [52]. The initial substrates share the shape of **171**: two alkyne moieties can be quite differently tethered and substituted, either with extended conjugation or with an electron-withdrawing ester group (**G**). Exposing such a molecular geometry to temperatures above 110 °C launches a first propargylic ene reaction providing vinylallenes of the type **172**. An electron-poor dienophile present in the reaction mixture readily undergoes cycloaddition and exclusively forms the *endo* product **177** in good yields. Aromatic congeners **178** are equally accessible when an activated alkyne such as **176** is operative. Anyhow, transforming linear architectures into complex scaffolds is always a challenge with respect to stereoselectivity and is, in this example especially, present in the formation of (*E*)/(*Z*)-isomers of **177**. The ratio of approximately 9 : 1 is nevertheless satisfactory, whereas the regioselectivity of the cycloaddition step appears even more effectively controlled. To decrease the required temperatures of the initial ene reaction, Danheiser successfully attached another silylated triple bond at the

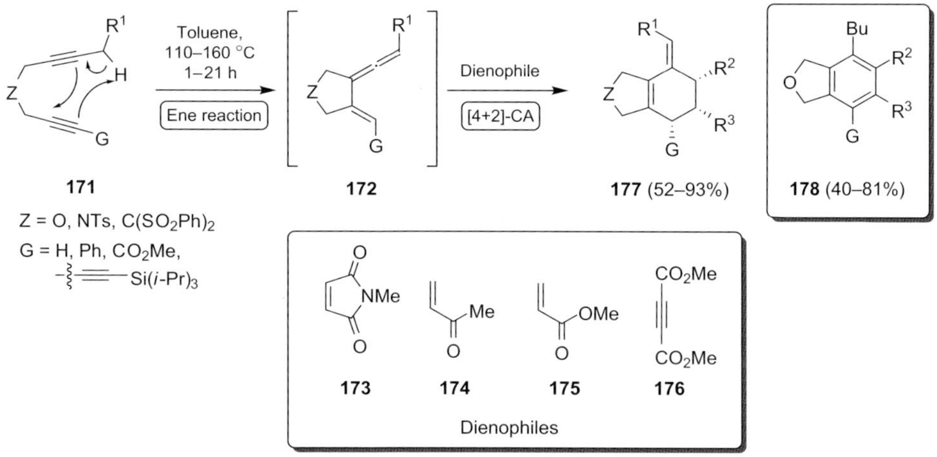

Scheme 6.32 A propargylic ene reaction forming a vinylallene as the diene part of a subsequent Diels–Alder reaction in the preparation of six-membered ring systems.

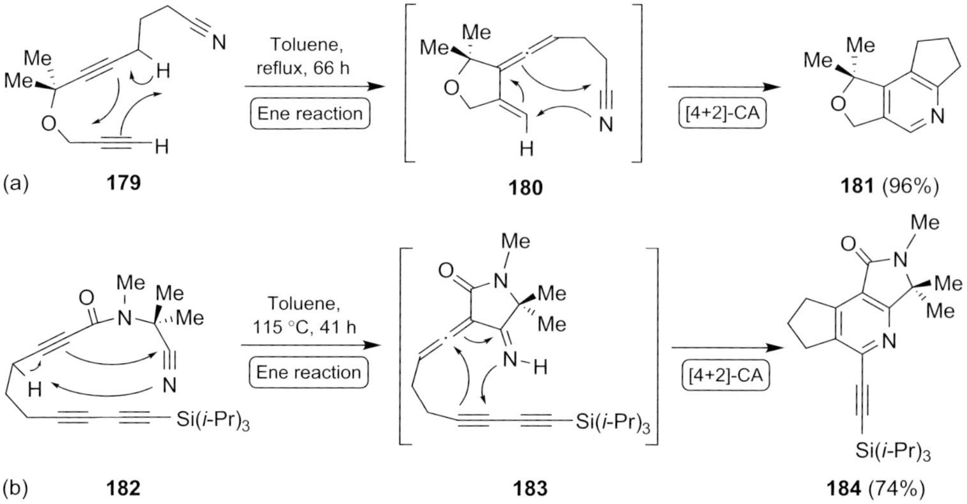

(a) 179 180 181 (96%)

(b) 182 183 184 (74%)

Scheme 6.33 Same strategy as in Scheme 6.32, but new products and insights in another similar domino reaction to access pyridines developed by Danheiser *et al.* (a) Pathway A. (b) Pathway B.

terminal position of **171**, which particularly functions as an activator but is equally useful as a synthetic equivalent for further transformations. Finally, the knowledge about an involved ene reaction was a novel achievement since Ley and coworkers proposed a radical intermediate in place of the respective species **172** in a similar system [53]. The distinct arrangement of the double bonds in **177** was eventually the critical feature for distinguishing between both pathways.

The potency of Danheiser's pericyclic cascade was further demonstrated in the construction of pyridine cores, disclosing presumably the first example in which an unactivated nitrile function participates in a [4+2]-cycloaddition. Two examples with established mechanisms are depicted in Scheme 6.33. If the required hydrogen for the anticipated ene reaction is present, pathway A dominates and follows the common domino reaction sequence to pyridine **181**. If, however, the crucial position is substituted, for instance by an amide moiety, the system is able to overcome this hurdle and utilize its nitrile group for the preluding ene reaction (pathway B). This time, the alkyne group eventually terminates the cascade in a cycloaddition to give the tricyclic pyridine **184**. Some efforts has been made to prove that pathway A is usually faster; a *gem*-dialkyl effect in the substrates **179** as well as **182** was shown to play a role in order to facilitate the reaction progress.

So far, we have shown in this subchapter ene reactions that adjust the starting material for a subsequent pericyclic step. Of course, also numerous other reaction types are able to pave the way for a pericyclic transformation. In Scheme 6.34 we present a domino three-component synthesis of α-spiro-δ-lactams. This transformation is a prime example to demonstrate how difficult it might be to really distinguish what kind of pericyclic reaction is involved or – furthermore – to distinguish whether any pericyclic reaction is involved or whether the course of

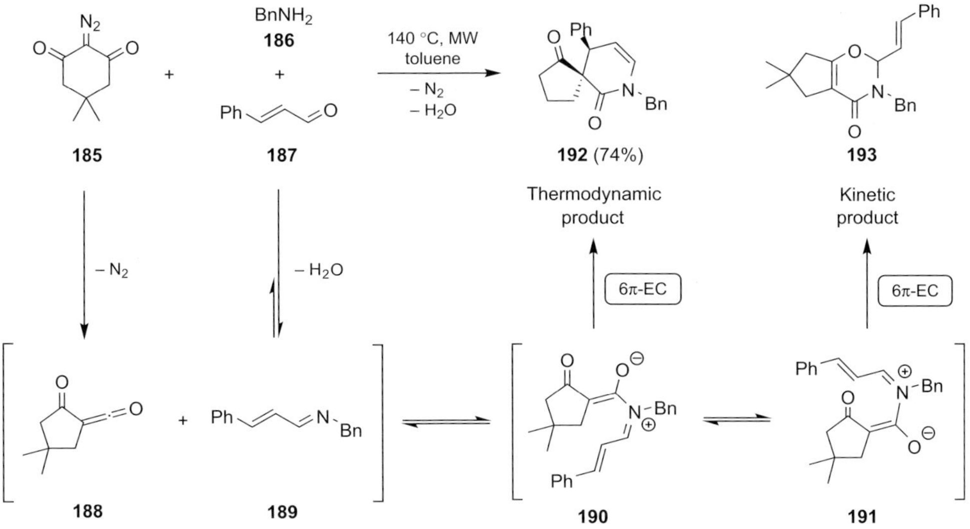

Scheme 6.34 Three-component domino reaction to yield spirolactams.

the reaction might also be explained by a series of other (polar) reactions. In a first step, cyclic 2-diazo-1,3-diketone **185** forms in a microwave-assisted Wolff rearrangement the acyl ketene **188** [54]. On the other hand, the respective amine reacts with α,β-unsaturated aldehyde **187** to afford the imine **189**. To rationalize the generation of the spiro compound, different scenarios are conceivable. Either a [4+2]-cycloaddition of the ketene with the unsaturated imine takes place (not shown), or the nucleophilic imine nitrogen attacks the sp-hybridized ketene carbon atom to generate **190** and **191** as intermediates. These zwitterionic structures might undergo 6π-electrocyclizations. One possibility (under thermodynamic control) would yield the spirolactam **192**; the other possibility involving the carbonyl moiety would annelate an oxazinone to the five-membered ring (under kinetic control) to afford **193**. However, the electrocyclizations might also be regarded as nucleophilic attacks of an enolate to a (homologous) iminium ion – in the first case via the carbon of the enolate, and in the second via the oxygen. Whether the process really involves a pericyclic step is hard to distinguish.

6.5.2
Cascades of Carbopalladations Followed by Pericyclic Reactions

Carbometalation cascades are often ideal starting points for the *in situ* preparation of conjugated π systems that might undergo pericyclic reactions. An unusual synthesis to highly substituted chromans and isochromans developed by the Werz group made use of this concept which was also used by others to access oligocyclic frameworks [55]. Commonly, the preparation of chroman-like structures starts with a benzene moiety to which a pyrane unit is annelated. If chiral chromans are in the focus, chirality has to be introduced during the annelation process.

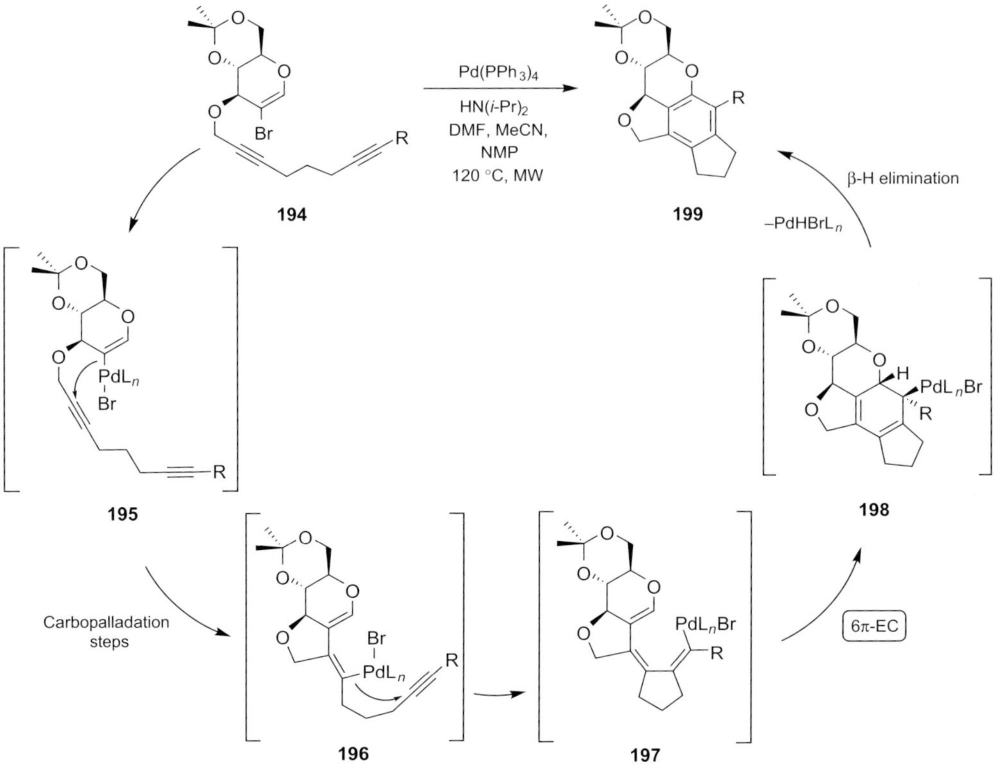

Scheme 6.35 Chroman formation starting from carbohydrates by a sequence of carbopalladations, an electrocyclic ring closure, and termination by aromatization.

In contrast, the domino route depicted in Scheme 6.35 was built on the idea to start with the pyrane subunit [56]. Stereochemistry could be obtained from the chiral pool since modified sugars serve as the pyrane moiety. A Pd-catalyzed domino reaction consisting of two carbopalladation steps affording 1,3,5-triene **197** followed by a disrotatory electrocyclic ring closure and terminated by *syn*-β-hydride elimination generated the benzene unit being the second half of chroman **199**. Complex oligocyclic structures are obtained with little effort. The facile variation of the carbohydrate part on the one hand and the diyne part on the other lead to a high degree of structural diversity. Starting materials derived from glucose, galactose, arabinose, and rhamnose were utilized. In addition, the length and the type of the tether connecting the two alkynes in the starting materials can be varied to obtain annelated five- or six-membered rings. Further diversity can be introduced by the choice of the residue R at the terminal alkyne; the reaction proceeded smoothly with aryl, alkyl, silyl groups, and even terminal hydrogen [56, 57]. Of course, it is worth discussing whether the cyclization step is a pericyclic reaction. Hypothetical alternatives might be a C–H activation process or a Heck-type reaction; however, the latter would yield a Pd species with a hydrogen in the *anti* position to the metal. But even such a course cannot be definitely excluded since there are reports dealing with an *anti*-β-hydride elimination in case highly stabilized aromatic units are formed. By a similar route, the respective isochromans could be obtained starting from bromoglycals. In these cases, the diyne tether was attached to the pseudoanomeric center and Pd-catalyzed domino process took place in an analogous manner [56].

The disadvantage of the chroman formation presented in Scheme 6.35 is the intramolecular nature of this process, that is, a diyne has to be prepared first and the tether between the triple bonds affords a further ring in the final molecule. To circumvent this structural limitation, an intermolecular process would be desirable [58]. Bromoglycal **201** and easily available propargylic halide **202** react in a nucleophilic substitution. With this product in hand, the Pd-catalyzed domino process is launched and a second alkyne **203** needed to access the ring is taken out of the reaction mixture. Chromans of type **200** were obtained (Scheme 6.36). In a similar manner, isochromans **204** were assembled. The mechanism is similar to the one depicted in Scheme 6.35. To obtain good yields, an excess of the second alkyne

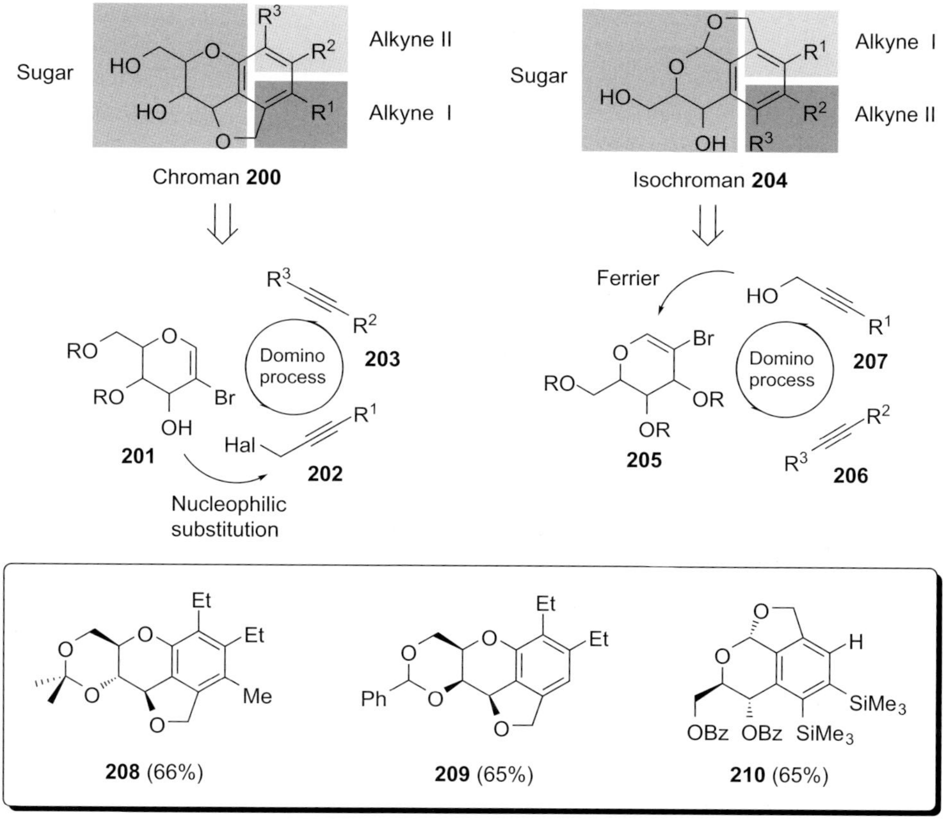

Scheme 6.36 Formation of chromans and isochromans by an intermolecular carbopalladation sequence followed by electrocyclic ring closure and aromatization.

was required. Fu's salt deliberating tris(*tert*-butyl)phosphine as ligand proved to be a crucial additive to gain moderate to good yields. Three examples with their corresponding yields are shown in Scheme 6.36.

A formal Glaser coupling of the diynes **194** (R = H) used previously yielded tetraynes of type **211/212**. The question raised was whether the same domino reaction might take place on both sides of the molecule. According to the nomenclature developed by Tietze, two domino processes at separate parts of the molecule are denominated as *tandem-domino reactions* [7, 8]. Indeed, the anticipated transformation consisting of two independent processes, each consisting of a twofold carbopalladation and cyclization, took place [59]. Six carbon–carbon bonds, six rings, and one chiral axis are formed in only one step (Scheme 6.37). If galactose residues protected with 4,6-benzylidene groups were used, the axial chirality between the both newly generated benzene units could be adjusted with complete diastereoselectivity. Such an approach might become a valuable tool for the preparation of other complex persubstituted biaryl systems.

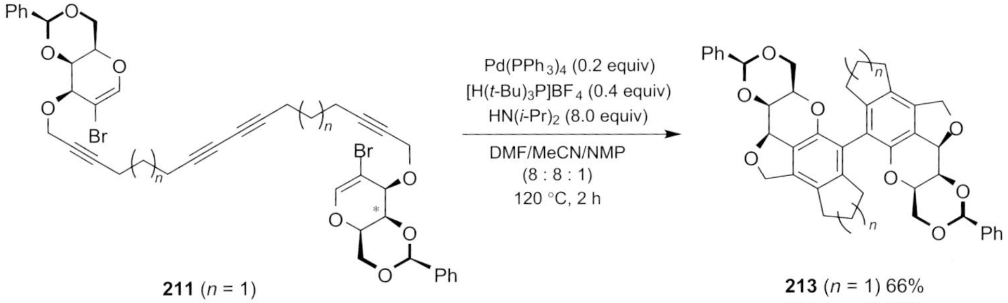

211 (n = 1)
212 (n = 2)

213 (n = 1) 66%
214 (n = 2) 60%

Scheme 6.37 Tandem-domino approach to persubstituted biphenyls according to Werz *et al.*

6.5.3
Domino Knoevenagel/Hetero Diels–Alder Reaction

One of the most versatile domino reactions involving a cycloaddition step is the domino Knoevenagel/hetero Diels–Alder reaction, which was developed by Tietze *et al.* It has been used for the efficient synthesis of innumerous heterocycles and many natural products. A novel example is the preparation of the amino sugar D-forosamin by Tietze *et al.* using a three-component reaction, which is described in detail in Chapter 14 [60].

6.6
Concluding Remarks

Pericyclic reactions are among the most versatile transformations in organic synthesis. Without the formation of any intermediates, cycloadditions, electrocyclic ring closures, and sigmatropic rearrangements are able to access – starting from simple educts – a degree of complexity that is unrivalled in most other reactions. Because of their concerted nature, pericyclic transformations commonly yield products with predictable and well-defined stereochemistry. It is obvious that a combination of pericyclic processes, in which the product of one pericyclic reaction serves as a substrate for the following one, has the potential to further increase the complexity of the final product. Two, three, or even more of these transformations can be combined, whereas also cycloreversions and electrocyclic ring openings may be involved. In addition, polar reactions might set the stage for pericyclic transformations or terminate them. In total, a plethora of such domino cascades have been implemented during the last decades. Numerous natural product syntheses have been designed involving a cascade of pericyclic reactions as the key step; by doing so, the synthetic routes have often been shortened tremendously.

The underlying theory is the *conservation of orbital symmetry* (or, in other words, the rules of Woodward and Hoffmann); they specify what might be possible with respect to the number of electrons involved, whether a ring closure or a ring opening proceeds in a conrotatory or disrotatory manner, whether supra- or antarafacial attacks in cycloadditions take place, or whether a reaction should be carried out photochemically rather than thermally. Although these rules are a mandatory prerequisite to understand the processes that take place and were even awarded with the Nobel Prize in 1981, they are not able to replace chemists' creative imagination to design novel fascinating domino cascades involving pericyclic reactions.

Acknowledgments

D.B. Werz thanks the German Research Foundation (DFG) for his Emmy Noether and Heisenberg Fellowships. The Fonds der Chemischen Industrie is gratefully acknowledged for a Dozentenstipendium. L.J. Patalag thanks the German–Israeli Foundation (G.I.F.) for financial support.

References

1. Sankararaman, S. (2005) *Pericyclic Reactions – A Textbook*, Wiley-VCH Verlag GmbH, Weinheim.
2. Fischer, E. and Jourdan, F. (1883) *Ber. Dtsch. Chem. Ges.*, **16**, 2241–2245.
3. Diels, O. and Alder, K. (1928) *Liebigs Ann. Chem.*, **460**, 98–122.
4. (a) Woodward, R.B. and Hoffmann, R. (1965) *J. Am. Chem. Soc.*, **87**, 395–397; (b) Hoffmann, R. and Woodward, R.B. (1965) *J. Am. Chem. Soc.*, **87**, 2046–2048; (c) Woodward, R.B. and Hoffmann, R. (1965) *J. Am. Chem. Soc.*, **87**, 2511–2513.
5. Woodward, R.B. and Hoffmann, R. (1969) *Angew. Chem., Int. Ed. Engl.*, **81**, 797–869.
6. Gilchrist, T.L. and Storr, R.C. (1972) *Organic Reactions and Orbital Symmetry*, Cambridge University Press, Cambridge.
7. Tietze, L.F. (1996) *Chem. Rev.*, **96**, 115–136.
8. Tietze, L.F., Brasche, G., and Gericke, K. (2006) *Domino Reactions in Organic Synthesis*, Wiley-VCH Verlag GmbH, Weinheim.

9. Criado, A., Peña, D., Cobas, A., and Guitían, E. (2010) *Chem. Eur. J.*, **16**, 9736–9740.
10. Jacobi, P.A., Kaczmarek, C.S.R., and Udodong, U.E. (1984) *Tetrahedron Lett.*, **25**, 4859–4862.
11. Bouma, M.J., Masson, G., and Zhu, J. (2012) *Eur. J. Org. Chem.*, **2012**, 475–479.
12. Zhao, P. and Beaudry, C.M. (2013) *Org. Lett.*, **15**, 402–405.
13. Yang, Z.Q. and Danishefsky, S.J. (2003) *J. Am. Chem. Soc.*, **125**, 9602–9603.
14. Kessler, S.N., Neuburger, M., and Wegner, H.A. (2012) *J. Am. Chem. Soc.*, **134**, 17885–17888.
15. Duguet, N., Slawin, A.M.Z., and Smith, A.D. (2009) *Org. Lett.*, **11**, 3858–3861.
16. Çelebi-Ülçüm, N., Lam, Y., Richmond, E., Ling, K.B., Smith, A.D., and Houk, K.N. (2011) *Angew. Chem. Int. Ed.*, **50**, 11478–11482.
17. Richmond, E., Duguet, N., Slawin, A.M.Z., Lébl, T., and Smith, A.D. (2012) *Org. Lett.*, **14**, 2762–2765.
18. Hamura, T., Iwata, S., and Suzuki, K. (2011) *Chem. Commun.*, **47**, 6891–6893.
19. Wang, X., Abrahams, Q.M., Zavalij, P.Y., and Doyle, M.P. (2012) *Angew. Chem. Int. Ed.*, **51**, 5907–5910.
20. Hoye, T.R., Baire, B., Niu, D., Willoughby, P.H., and Woods, B.P. (2012) *Nature*, **490**, 209–212.
21. Karmakar, R., Mamidipalli, P., Yun, S.Y., and Lee, D. (2013) *Org. Lett.*, **15**, 1938–1941.
22. Suzuki, T., Miyajima, Y., Suzuki, K., Iwakiri, K., Koshimizu, M., Hirai, G., Sodeoka, M., and Kobayashi, S. (2013) *Org. Lett.*, **15**, 1748–1751.
23. Ramadhar, T.R., Kawakami, J.I., Lough, A.J., and Batey, R.A. (2010) *Org. Lett.*, **12**, 4446–4449.
24. Arns, S. and Barriault, L. (2006) *J. Org. Chem.*, **71**, 1809–1816.
25. Barriault, L. and Sauer, E.L.O. (2004) *J. Am. Chem. Soc.*, **126**, 8569–8575.
26. Parr, B.T., Li, Z., and Davies, H.M.L. (2011) *Chem. Sci.*, **2**, 2378–2382.
27. (a) Martin, C.L., Overman, L.E., and Rohde, J.M. (2008) *J. Am. Chem. Soc.*, **130**, 7568–7569; (b) Dunn, T.B., Ellis, J.M., Kofink, C.C., Manning, J.R., and Overman, L.E. (2009) *Org. Lett.*, **11**, 5658–5661; (c) Knight, S.D., Overman, L.E., and Pairaudeau, G. (1995) *J. Am. Chem. Soc.*, **117**, 5776–5788.
28. (a) Bennett, D.J. and Hamilton, N.M. (2000) *Tetrahedron Lett.*, **41**, 7961–7964; (b) Agami, C., Couty, F., Lin, J., Mikeaeloff, A., and Poursoulis, M. (1993) *Tetrahedron*, **49**, 7239–7250.
29. McCormack, M.P., Shalumova, T., Tanski, J.M., and Waters, S.P. (2010) *Org. Lett.*, **12**, 3906–3909.
30. Pal, R., Clark, R.J., Manoharan, M., and Alabugin, I.V. (2010) *J. Org. Chem.*, **75**, 8689–8692.
31. Cheung, L.L.W. and Yudin, A.K. (2009) *Org. Lett.*, **1**, 1281–1284.
32. Cheung, L.L.W. and Yudin, A.K. (2010) *Chem. Eur. J.*, **16**, 4100–4109.
33. Weston, M.H., Parvez, M., and Back, T.G. (2010) *J. Org. Chem.*, **75**, 5402–5405.
34. Dhayalan, V., Clement, J.A., Jagan, R., and Mohanakrishnan, A.K. (2009) *Eur. J. Org. Chem.*, 531–546.
35. Barriault, L. and Denissova, I. (2002) *Org. Lett.*, **4**, 1371–1374.
36. Yin, B., Zeng, G., Cai, C., Ji, F., Huang, L., Li, Z., and Jiang, H. (2012) *Org. Lett.*, **14**, 616–619.
37. Hulot, C., Amiri, S., Blond, G., Schreiner, P.R., and Suffert, J. (2009) *J. Am. Chem. Soc.*, **131**, 13387–13398.
38. Charpenay, M., Boudhar, A., Blond, G., and Suffert, J. (2012) *Angew. Chem. Int. Ed.*, **51**, 4379–4382.
39. Miller, A.K. and Trauner, D. (2005) *Angew. Chem. Int. Ed.*, **44**, 4602–4606.
40. Drew, S.L., Lawrence, A.L., and Sherburn, M.S. (2013) *Angew. Chem. Int. Ed.*, **52**, 4221–4224.
41. von Essen, R., Frank, D., Sünnemann, H.W., Vidović, D., Magull, J., and de Meijere, A. (2005) *Chem. Eur. J.*, **11**, 6583–6592.
42. Webster, R., Gaspar, B., Mayer, P., and Trauner, D. (2013) *Org. Lett.*, **15**, 1866–1869.
43. Mak, X.Y., Crombie, A.L., and Danheiser, R.L. (2011) *J. Org. Chem.*, **76**, 1852–1873.
44. Creech, G.S. and Kwon, O. (2010) *J. Am. Chem. Soc.*, **132**, 8876–8877.
45. (a) Zincke, T. (1903) *Liebigs Ann. Chem.*, **330**, 361–374; (b) Zincke, T.

and Wurker, W. (1904) *Liebigs Ann. Chem.*, **333**, 296–345.
46. Vanderwal, C.D. (2011) *J. Org. Chem.*, **76**, 9555–9567.
47. Steinhardt, S.E., Silverston, J.S., and Vanderwal, C.D. (2008) *J. Am. Chem. Soc.*, **130**, 7560–7561.
48. Steinhardt, S.E. and Vanderwal, C.D. (2009) *J. Am. Chem. Soc.*, **131**, 7546–7547.
49. Paton, R.S., Steinhardt, S.E., Vanderwal, C.D., and Houk, K.N. (2011) *J. Am. Chem. Soc.*, **133**, 3895–3905.
50. Wu, Y.K., McDonald, R., and West, F.G. (2011) *Org. Lett.*, **13**, 3584–3587.
51. Robinson, J.M., Sakai, T., Okano, K., Kitawaki, T., and Danheiser, R.L. (2010) *J. Am. Chem. Soc.*, **132**, 11039–11041.
52. Sakai, T. and Danheiser, R.L. (2010) *J. Am. Chem. Soc.*, **132**, 13203–13205.
53. Saaby, S., Baxendale, I.R., and Ley, S.V. (2005) *Org. Biomol. Chem.*, **3**, 3365–3368.
54. Presset, M., Coquerel, Y., and Rodriguez, J. (2010) *Org. Lett.*, **12**, 4212–4215.
55. Blond, G., Bour, C., Salem, B., and Suffert, J. (2008) *Org. Lett.*, **10**, 1075–1078.
56. Leibeling, M., Koester, D.C., Pawliczek, M., Schild, S.C., and Werz, D.B. (2010) *Nat. Chem. Biol.*, **6**, 199–201.
57. Leibeling, M., Koester, D.C., Pawliczek, M., Kratzert, D., Dittrich, B., and Werz, D.B. (2010) *Bioorg. Med. Chem.*, **18**, 3656–3667.
58. Leibeling, M., Milde, B., Kratzert, D., Stalke, D., and Werz, D.B. (2011) *Chem. Eur. J.*, **17**, 9888–9892.
59. Leibeling, M. and Werz, D.B. (2012) *Chem. Eur. J.*, **18**, 6138–6141.
60. (a) Tietze, L.F., Böhnke, N., and Dietz, S. (2009) *Org. Lett.*, **11**, 2948–2950; (b) Tietze, L.F., Dietz, S., Böhnke, N., Düfert, M.A., Objartel, I., and Stalke, D. (2011) *Eur. J. Org. Chem.*, 6574–6580.

7
Modern Domino Reactions Containing a Michael Addition Reaction

Scott G. Stewart

7.1
Introduction

The Michael reaction [1, 2], or in general a nucleophilic 1,4-addition to α,β-unsaturated carbonyl compounds, is one of the most important and well-described transformations in organic chemistry. Combining this thoroughly researched reaction, which provides the choice of many reaction conditions, with other synthetic reactions is a logical combination for domino processes. The variety of new C–C, C–N, C–O, and C–S bond-forming processes is vast, many of which will be shown in this chapter. Enantioselective Michael reactions are also well understood now, with a range of catalysts such as chiral allylic sulfoxides, chiral phase-transfer catalysts, and hetero-bimetallic catalysts to carry out these specific reactions. Several reactions involving transition metals (strictly 1,4-conjugate addition) or conditions in which many cross-coupling reactions occur (i.e., with base) have been used in Michael reaction processes. The use of such transition-metal reactions such as cross-coupling reactions is rapidly growing, and as such applications combining a Michael reaction have also been exploited.

One of the most recent and influential factors in the application of Michael processes in domino reactions has been the advent and use of organocatalysts. The application of these catalysts results in asymmetric organocatalytic domino reactions, which in many cases evoke remarkable enantiomeric excess and diastereoselectivity. Invariably, such reactions will have a growing influence on natural product synthesis and processes in the pharmaceutical and fine chemical industry, which in many cases require enantiopure products. Several organocatalysts have been applied in domino reactions, which include Michael processes, with the most popular being proline [3, 4] and those catalysts derived from proline such as the diphenylprolinol silyl ether **1** (the Jørgensen–Hayashi catalyst). The groups of Jørgensen and List reported early domino reactions with imidazolidine and imidazolidinone catalysts, respectively [5, 6]. Other frequently used compounds as Michael domino catalysts include the cinchonidine-based compounds such as **2**, thiourea-derived compounds such as Takemoto's catalyst **3**, or cinchonidine derivatives bearing a thiourea moiety which are proposed to initiate hydrogen bonding

Domino Reactions: Concepts for Efficient Organic Synthesis, First Edition. Edited by Lutz F. Tietze.

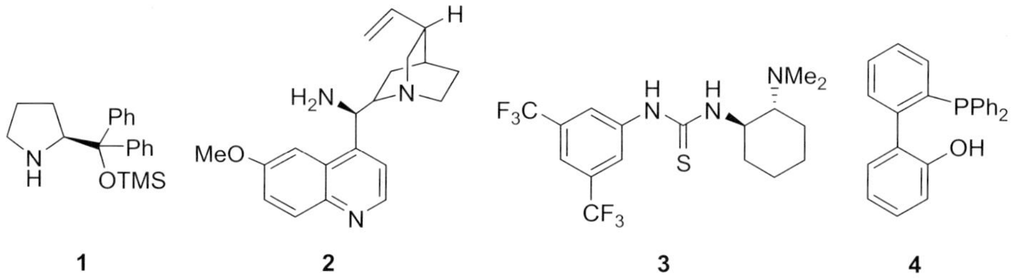

Figure 7.1 Examples of commonly used organocatalysts.

to Michael precursor carbonyl species. Moreover, catalysts containing both a Lewis base and a Brønsted acid moiety as in **4** will also be dealt with, along with some N-heterocyclic carbenes (NHCs). However, it should be stressed that this book contains another chapter dealing with organocatalysis (Chapter 10), and therefore only selected examples of this topic using a Michael reaction will be presented here (Figure 7.1).

As mentioned, the applications of chiral secondary amines in domino processes have been numerous, mainly because of the dual mechanism of activation allowing easy incorporation of other reactions [7, 8]. The first of these processes involves an iminium–enamine (Scheme 7.1a), and the second an enamine–iminium activation (Scheme 7.1b), both of which can be in operation in domino reactions involving Michael reactions. These modes of action rely on the lower energy lowest unoccupied molecular orbital (LUMO) for the iminium ion and a rise in the highest occupied molecular orbital (HOMO) energy for enamine activation, which are discussed in further detail in Chapter 10 of this book [9, 10]. In the

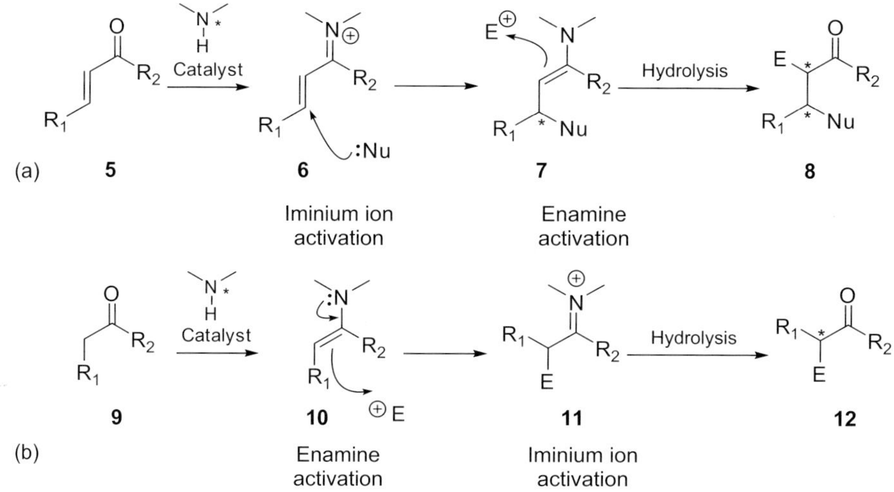

Scheme 7.1 (a) Iminium–enamine and (b) enamine–iminium activation pathways in organocatalysis Ref. [11].

iminium–enamine activation, the α-β-unsaturated aldehyde or ketone used in many Michael reactions is activated through the formation of an iminium species **6**, which leads to the intermediate **7** by addition of a nucleophile; it then follows an attack of an electrophile at **7** to afford compound **8** containing two new stereogenic centers. Substrates that have an enolizable site, such as ketone **9**, will react with secondary amines and the resulting product **10** in turn can be trapped by an electrophile, eventually generating a stereogenic center within compound **12** [7, 11].

The first organocatalyzed domino Michael/aldol reaction was used in the total synthesis of the natural product (−)-huperzine A [12]. This process was catalyzed by (−)-cinchonidine to convert a β-keto ester and methacrolein to a 5,9-methanocycloocta[*b*]pyridine tricyclic core. Two excellent reviews dealing with these processes have been recently published [8, 11], and in the field of natural products several total syntheses utilize a domino process involving a Michael reaction step [12–24]. This chapter will mainly cover the recent work in the development of novel domino reactions containing a Michael reaction published since the beginning of 2006, as the many such domino reactions were collated prior to this date by Tietze *et al.* in a comprehensive book [25].

7.2
Formation of Acyclic Products

Early efficient domino Michael processes for the formation of acyclic products concentrated on domino Michael/aldol type reactions using a Lewis acid-mediated reaction of an enone species with halide sources as a nucleophile, followed by a reaction of an aldehydes for a bis-α,β-functionalization process [26, 27]. Later, a seminal publication in the field of organocatalyzed domino reactions containing a Michael addition was reported by MacMillan *et al.* [28], introducing imidazolidinone catalysts in this context and describing the combined iminium–enamine process plus proposing a mechanistic pathway. In particular, when the α,β-unsaturated aldehyde **13** is exposed to the imidazolidinone catalyst **15**, an iminium species is generated (compound **6** as described in Scheme 7.1), which in turn reacts in a Michael-type manner with a range of nucleophiles including those indicated in Scheme 7.2 (furans, thiophenes, indoles, and butenolides). The resulting addition adduct (generic **7**, Scheme 7.1) is proposed to enter a second catalytic cycle, wherein enamine activation allows reaction with the electrophilic species, chlorinated quinone **14**. Described in detail, these reactions occur in a syn-selective manner to give compounds such as **16** in high yields and excellent enantiomeric excess. Another significant publication in the field of domino reactions, which included organocatalyzed Michael reactions, was reported for the preparation of enantiopure amino-thiols in moderate to good yields and high enantioselectivities from α,β-unsaturated aldehydes [29]. This work, described by Jørgensen, explores the use of the diphenylprolinol silyl ether catalyst **1** in such Michael domino processes; a significant number of publications highlighting the use of this catalyst by this group and others are discussed in great detail later in this chapter.

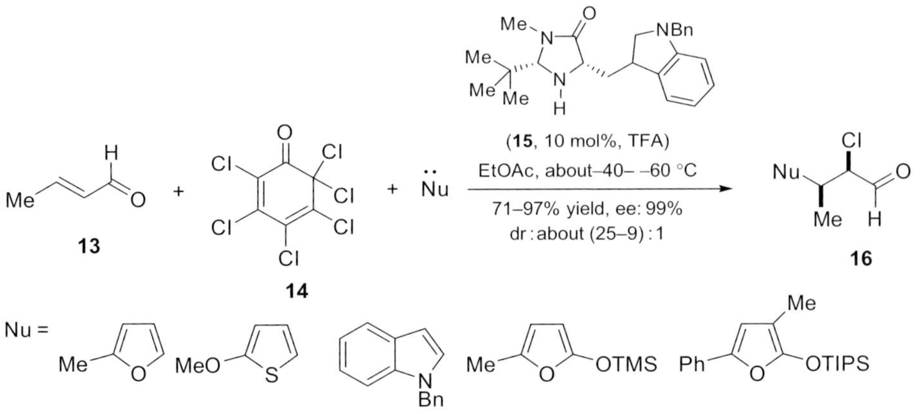

Scheme 7.2 MacMillan *et al.* introducing imidazolidinone catalysts and the merged iminium–enamine process.

Amphoteric molecules, containing both nucleophilic and electrophilic nodes, were used in a domino aza-Michael/aldol reaction by the group of Yudin [30] (Scheme 7.3). In this process, an isolable aziridine aldehyde dimer **17** was used as a precursor, and the authors exploited the difference in dissociation kinetics

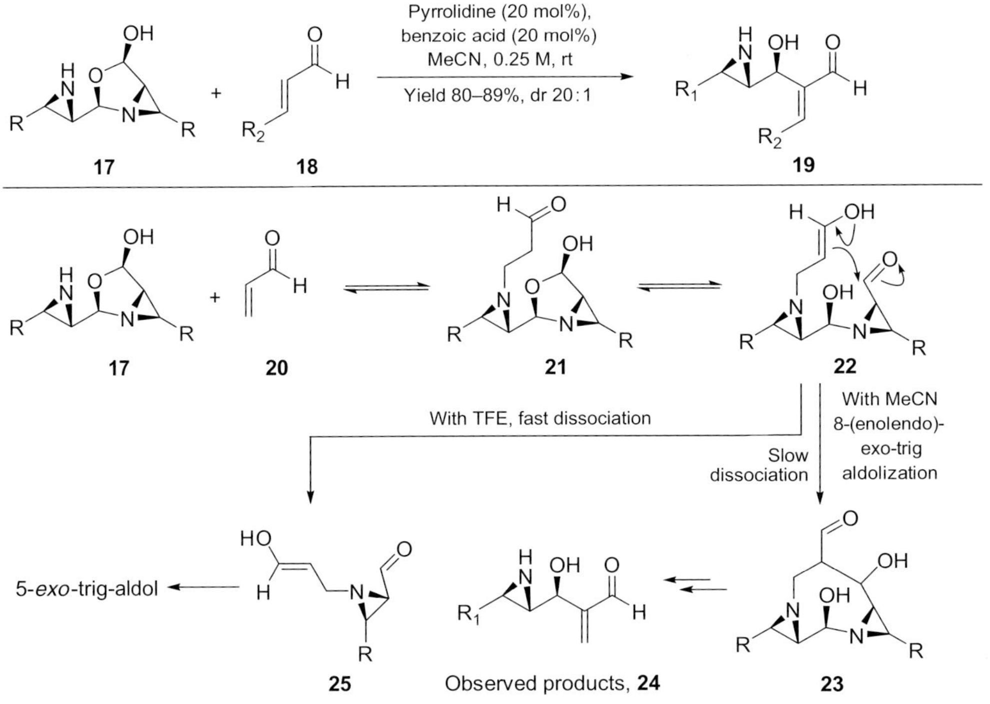

Scheme 7.3 Aza-Michael/aldol reactions reported by the group of Yudin.

to provide access to a domino aza-Michael/aldol reaction pathway. When using polar aprotic solvents, access to an 8-(enol-endo)-exo-trig reaction pathway with unsaturated aldehydes **18** to enone **24** is more likely than the known 5-(enol-endo)-exo-trig cyclization pathway through aziridine **25**. Importantly, benzoic acid was also essential to encourage the rate of reaction, while an iminium activation between pyrrolidine and the aldehyde **18** provides an electrophile for the aza(aziridine nitrogen)-Michael addition to afford the aza-Michael adduct **21** and eventually the product amino-hydroxy-α,β-unsaturated aldehyde **24**. The high diastereoselectivity of this process is proposed to be a result of the rigid stereochemical environment predetermined by the dimeric intermediate.

In expanding the number Michael/aldol domino processes, chiral *N*-sulfinimines **27** could be subjected to a magnesium thiolate in the presence of an acrylate **26** to undergo an asymmetric thio-Michael/nucleophilic addition reaction (Scheme 7.4) [31]. Both aromatic sulfinimines as well as aliphatic sulfinimines were successful in this domino reaction to give (*R*)-phenylthiomethyl-β-(*N*-sulfinylamino) esters **28** occurring in high diastereomeric excess. When chiral lithium amide (*R*)-**31** and compound **30** were used (using sequential addition of *n*-butyllithium in tetrahydrofuran at −78 °C to the amine), a stereoselective Ireland–Claisen rearrangement and asymmetric Michael addition occurred to afford the γ-substituted δ-amino acids **32** with excellent diastereo- and enantioselectivities (Scheme 7.5) [32]. Also described in this paper is a further synthetic process involving a lactamization and reduction sequence of **32** into 2,3-disubstituted piperidines.

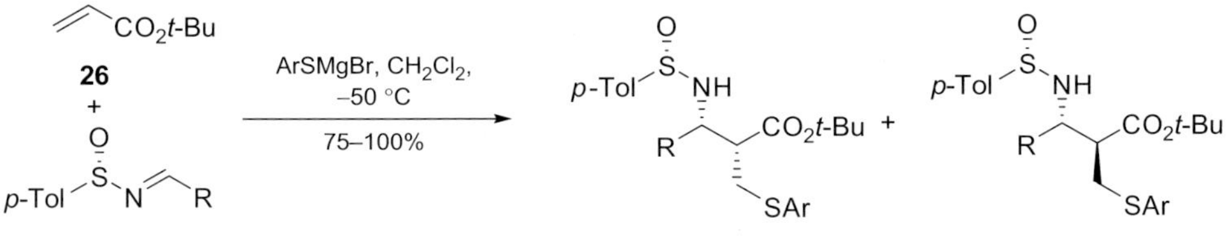

Scheme 7.4 Asymmetric thio-Michael/nucleophilic addition domino reaction.

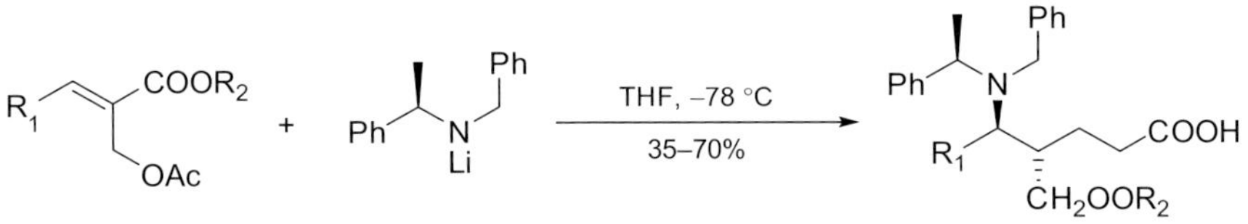

Scheme 7.5 Ireland–Claisen rearrangement/Michael addition domino reaction.

Also in the field of alkyl chain production, a Michael-type addition/Dieckmann reaction sequence has been used to synthesize α-(hydroxymethyl)glutamic acid **37** (HMG) (Scheme 7.6). Treatment of the bicyclic serine equivalent (3*S*,7*R*,7a*S*)-**33**

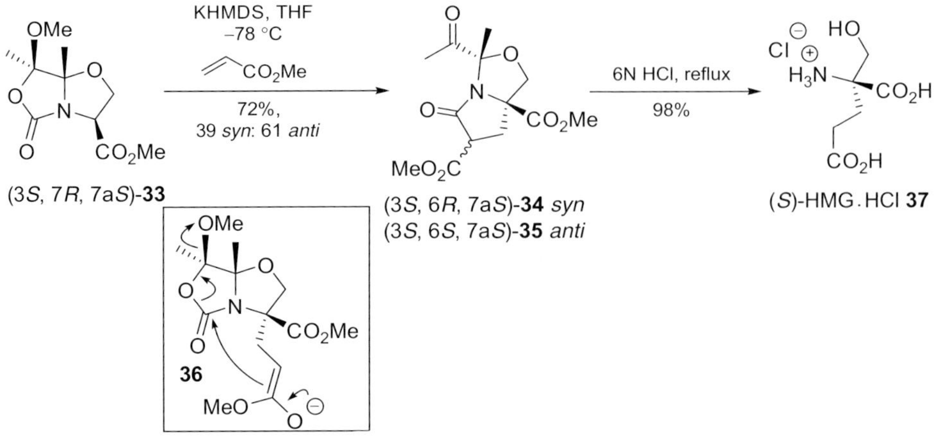

Scheme 7.6 Synthesis of α-(hydroxymethyl)glutamic acid (HMG).

with potassium hexamethyldisilazide (KHMDS) and methyl acrylate afforded the domino products **34** and **35** in both (syn and anti forms) [33]. Compound **36** is proposed to be the intermediate of the Michael addition, leading to the Dieckmann reaction sequence. The product of this process could be hydrolyzed to the HMG (*S*)-**37**.HCl, as in this sequence, could also be repeated for the corresponding (*R*)-enantiomer.

The prostaglandin family of compounds display a wide range of biological activities including the regulation of inflammatory response, lowering of blood pressure, and contraction of smooth muscle. While the synthesis of prostaglandins represented a challenge in the 1970s, Shibasaki and coworkers [20] applied a three-component domino reaction in an enantioselective synthesis of the five-membered ring contained within 11-deoxy-prostaglandin $F_{1\alpha}$ ($PGF_{1\alpha}$) (Scheme 7.7). Similarly, an approach to $PGF_{1\alpha}$ **44** was also proposed through this methodology. In one reaction vessel, the racemic silyl ether **38** initially undergoes Michael addition with dibenzyl methylmalonate. The use of AlLi-bis[(*S*)-binaphthoxide **41** allows for the activation of the enone and formation a new stereogenic center at the point of nucleophilic attack (illustrated in **43**). Following this initial process to an aldol addition, with the third component aldehyde **39**, provides compound **42** containing the desired substitution pattern in a 12 : 1 mixture of diastereomers in 75% yield (based on malonate **40**). This compound was considered a key intermediate for further synthesis of prostaglandins, especially $PGF_{1\alpha}$. A Michael/aldol process has also been used in a similar process for the substitution of cyclopentene-3,5-dione monoacetals *en route* to the (−)-prostaglandin E1 methyl ester [34]. This same approach involving the reaction of the conjugate addition of an acyl anion equivalent to a butenolide, followed by the *in situ* trapping of the resulting enolate, has been employed in the synthesis of other natural products [35, 36].

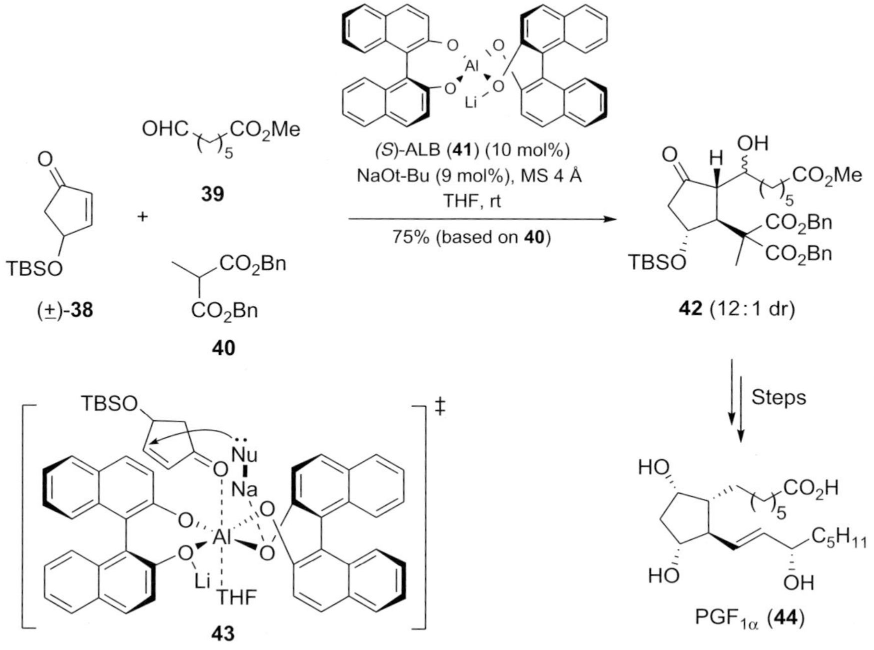

Scheme 7.7 Shibasaki's three-component domino reaction to $PGF_{1\alpha}$.

7.3 Formation of Carbocycles

A domino Michael/α-alkylation reaction between bromoacetoacetates and α,β-unsaturated aldehydes provides the 2-formylcyclopropane derivatives in high yields with excellent diastereoselectivity and up to 99% ee (for further details see Chapter 10) [37]. Similarly, an approach using enones **45** and bromonitromethane **46** has been used in organocatalytic nitro-cyclopropanation reaction to give **47** (Scheme 7.8)

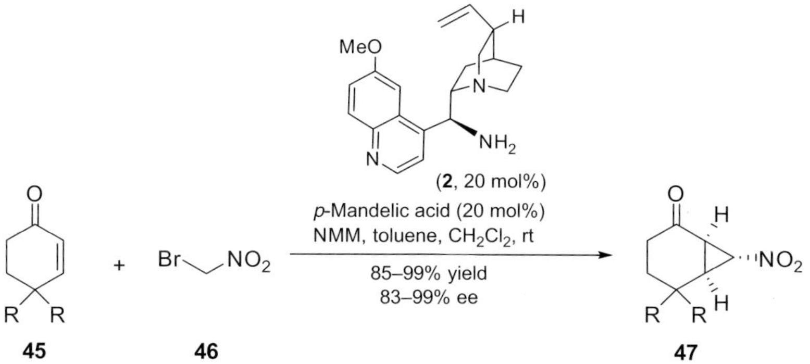

Scheme 7.8 Wang's domino Michael-cyclization reaction.

[38]. The reaction, catalyzed by the 9-amino-9-deoxyepiquinine **2**, was proposed to go through an iminium–enamine activation process.

Early work by Tomioka and coworkers [39] described a two-component Michael/aldol process to cyclopentenes. Furthermore, rhodium-assisted Michael/aldol processes to cyclopentanes and cyclohexanes have also been reported [40]. Later, a Michael addition reaction in combination with an adehyde α-alkylation reaction was reported for the highly stereoselective formation of γ-nitroaldehydes **50** [41]. In this publication, a series of aliphatic aldehydes **49** (at R_1) and (*E*)-5-iodo-1-nitropent-1-ene **48** were reacted in the presence of the organocatalyst **1** and benzoic acid in dimethyl sulfoxide (DMSO) to afford the resulting cyclopentene ring system **50** (Scheme 7.9). The diastereo- and enantioselective process follows the proposed mechanism beginning with enamine activation of the aldehyde to **51** by the catalyst **1** (blocking the *re* face), and Michael addition of **48** occurs at its more accessible *si* face. The full enamine–enamine mechanism, illustrated in Scheme 7.9, provided

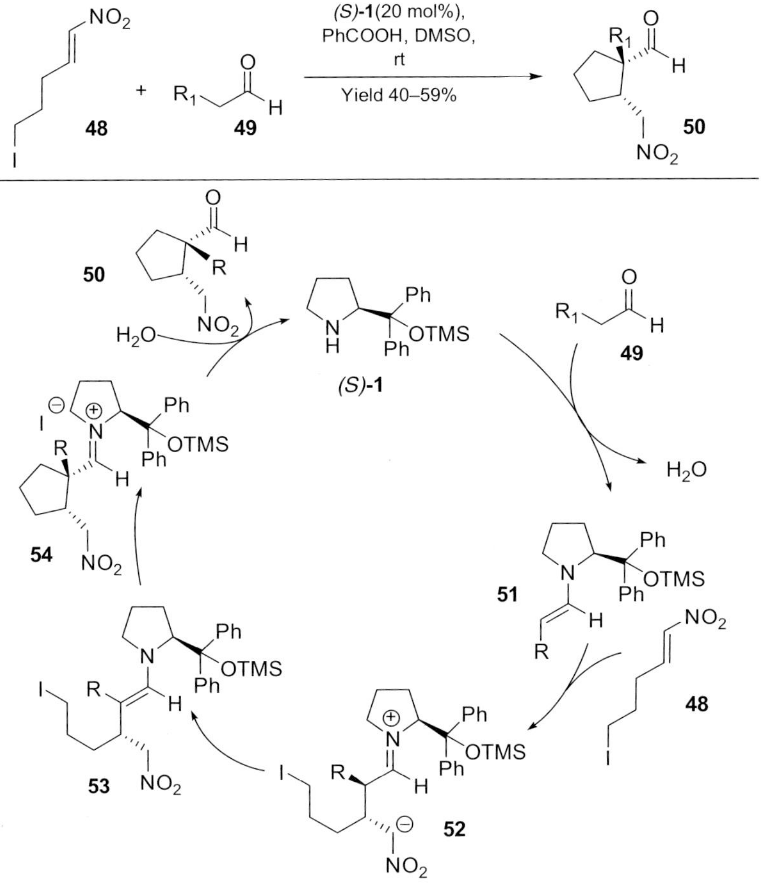

Scheme 7.9 Synthesis of cyclopentanes: enamine–enamine catalytic cycle.

the aldehydes **50** (following hydrolysis of **54**) in moderate to good yields, with varying diastereoselectivity but in most cases high enantiomeric excess.

A domino Michael/aldol reaction was used in the synthesis of fully substituted cyclopentanes [42] with five contiguous stereogenic centers (for further details see Chapter 10). A range of substrates for the cyclopentane ring construction have been reported across two papers by the group of Hong. Iminium activation with nitroalkane nucleophile attack from the *re* face is assumed to govern the configuration of the product [43]. A similar cyclopentane ring construction was observed in the domino double Michael reaction of ethyl-2-acetyl-5-oxohexanoate ($E:Z$, 6 : 1) and nitroolefins [44]. In this case, the cinchona-derived 9-amino-9-deoxyepiquinine **2** was the chosen catalyst, and the mode of action was through multiple H-bonds to the two nitrogen centers. This organocatalyst was used to afford cyclopentane in high enantioselectivity and diastereoselectivity (for further details see Chapter 10). Also reported by the same group was the reaction of ethyl 2-acetyl-4-oxo-4-phenylbutanoate, which allowed the incorporation of an additional quaternary carbon center into the cyclopentane ring system [45]. Another study by the group of Ma provided polysubstituted cyclopentenones from an iminium–enamine domino process or Michael/Michael domino reaction [46]. The starting material, a triester unit **56**, is the proposed nucleophilic unit reacting with the iminium ion species, which is formed between **1** and the unsaturated aldehyde **55** (Scheme 7.10). Following the first Michael addition, **58** is a proposed intermediate, and the mechanism illustrated is followed to the desired product **57**. The low catalyst loading coupled with the excellent enantiomeric excess is a significant feature of this work.

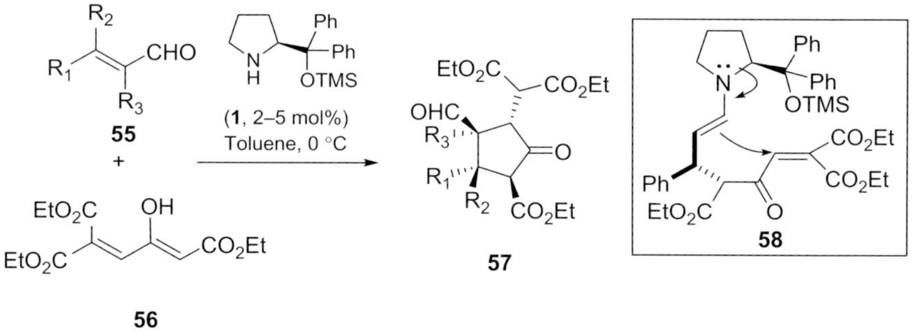

Scheme 7.10 Double Michael addition to cyclopentanones.

Domino Michael/Michael/aldol transformations were earlier used to efficiently prepare highly substituted compounds, with a cyclohexane unit as part of a decalin core [47]. Later, in 2007, the Jørgensen group [48] reported an asymmetric organocatalytic nitro-Michael/Henry reaction, leading to pentasubstituted cyclohexanes **62** (Scheme 7.11). In this example, utilizing a reaction between the (*S*)-2-bis(3,5-bistrifluoromethylphenyl-pyrrolidine catalyst **61** and aldehyde **59**, the resulting iminium ion precursor is attacked in a Michael manner by the incoming

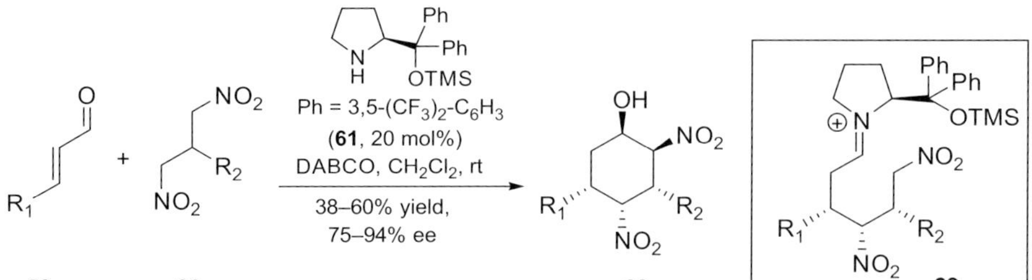

59 60 62 63

Scheme 7.11 Jørgensen's organocatalyzed asymmetric nitro-Michael/Henry reaction.

dinitroalkane **60**. Under basic conditions (1,4-diazabicyclo[2.2.2]octane, DABCO), the 1,3-dinucleophilic properties of **60** form the intermediate **63**, which then undergoes an intramolecular Henry reaction. The combination of a range of R groups (commonly R_1 alkyl and R_2 aryl) on both substrates has been extensively studied, resulting in good diastereomeric and good to excellent enantiomeric excess. During this period, Hayashi [49] also described an enantioselective Michael/Henry reaction using diphenylprolinol silyl ether **1** (Jørgensen–Hayashi catalyst). In this example, the reaction between pentane-1,5-dial **65** and nitroalkene **64** provides the cyclohexane derivatives **66** with four contiguous stereocenters in high enantiomeric excess (Scheme 7.12). Hayashi also proposed an initial enamine formation (pentane-1,5-dial plus catalyst) followed by a reaction with nitro-aldehyde to provide a zwitterion **67**, which undergoes a Henry reaction and hydrolysis to the product **66**. Continuing with this work, the group of Headley and Ni [50] reported a similar process but carried out in water. Following the theme of a domino Michael/Henry reaction, this process was used to prepare hexasubstituted cyclohexanes, with all six carbons containing a stereogenic center (Scheme 7.13). This study completed following the establishment of a diastereo- and enantioselective organocatalytic Michael addition of 1,2-ketoamides **68** to nitroolefins **69**, providing anti selectivity using the thiourea **3** as a hydrogen-bonding catalyst [51]. This work was preceded by a study into a catalyzed conjugate addition of *R*-ketoanilides to nitroolefins through a multiple H-bonded transition state for the observed stereochemistry. Interestingly, the domino reaction completion time was 4 days.

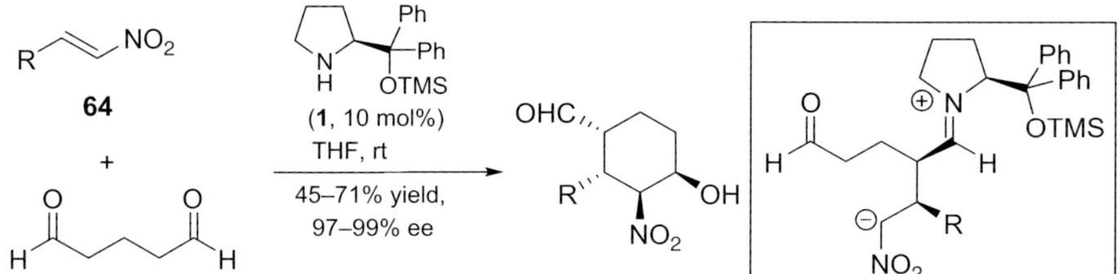

65 66 67

Scheme 7.12 Hayashi's Michael/Henry reaction to chiral nitrocyclohexanecarbaldehydes.

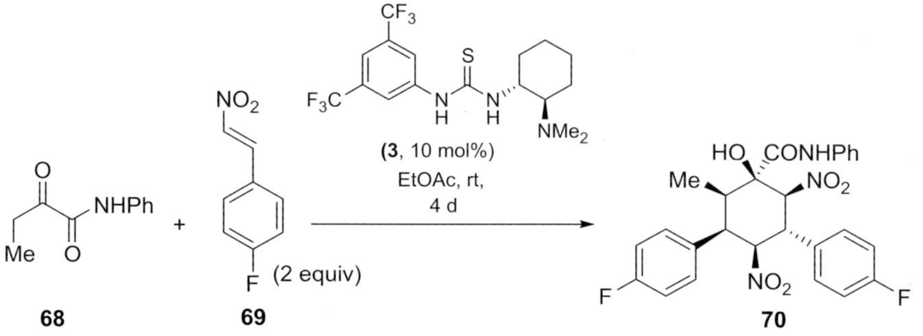

Scheme 7.13 Domino Michael/Michael/Henry reaction.

The synthesis of tetrasubstituted cyclohexene carbaldehydes was reported by the group of Enders [52] in an early seminal publication dealing with organocatalytic domino processes. In this example, aldehyde **71**, a nitroalkene **72**, and an α,β-unsaturated aldehyde **73** were used in this multicomponent reaction, while the Jørgensen–Hayashi catalyst **1** was proposed to catalyze each reaction step (Scheme 7.14). Initially, enamine formation between (*S*)-**1** and the aldehyde **71** occurs, which is quenched through the addition of the nitroalkene **72**, resulting in the formation of aldehyde **74**. As hydrolysis re-liberates **1**, it can form a second imine species **73** for subsequent reaction with nitroalkene **74**, in a second Michael process, resulting in **75**. Finally, an intramolecular aldol condensation occurs, resulting in

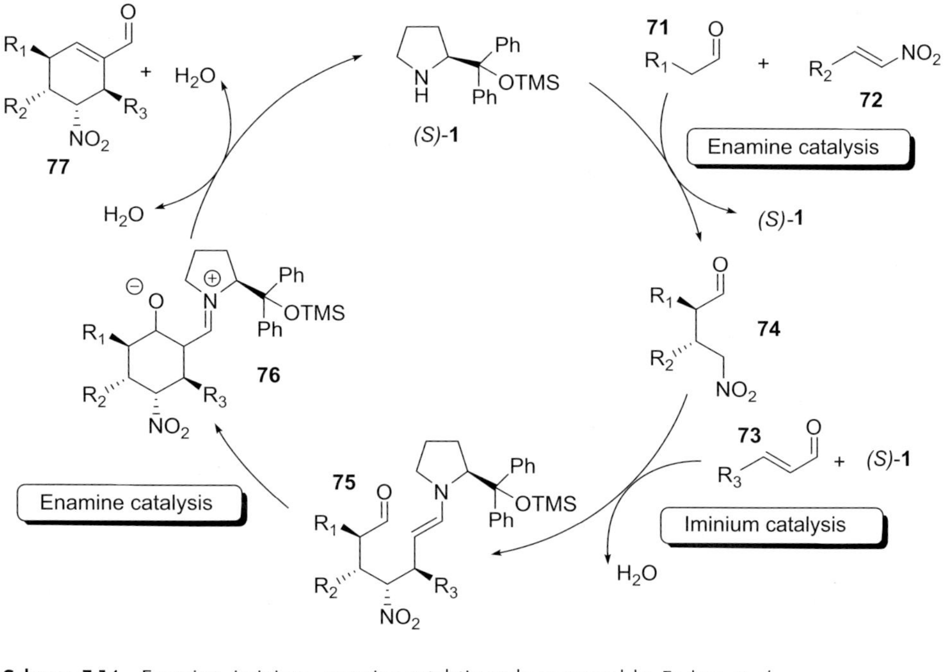

Scheme 7.14 Enamine–iminium–enamine catalytic cycle proposed by Enders *et al.*

the generation of the products **77** upon hydrolysis. This reaction, typically carried out in toluene at 0 °C, generates the product in reasonable yields considering the complexity of the overall process. Remarkably, the chemo-, diastereo-, and enantioselectivity of this three-component domino reaction and enantiomeric excess are consistently 99%, while the diastereomeric ratio is also good and a range of R_1 and R_3 groups are tolerated. Furthermore, as a proof of principle employing the (*R*)-proline-derived catalyst of **1** afforded the corresponding enantiomeric product. In a subsequent study, unsaturated aldehydes were first reacted with Hantzsch dihydropyridine (HEH) for an initial hydrogenation reaction prior to forming similar tetrasubstituted cyclohexene carbaldehydes [53]. The pathway of such reactions could follow a double domino route (hydrogenation/Michael) to aliphatic aldehydes or quadruple domino reaction (hydrogenation/Michael/Michael/aldol condensation) to a tetrasubstituted cyclohexene carbaldehyde. An impressive asymmetric organocatalytic four-component quadruple domino reaction has also been reported by Gong [54]. In this oxa-Michael/Michael/Michael/aldol condensation, trisubstituted cyclohexene carbaldehydes are formed from an alcohol, an acrolein, and a nitroalkene. Through a range of alcohols, the pyrrole-based catalyst (with catalytic benzoic acid as an additive) delivers yields ranging from 40 to 60% with excellent diastereomeric excess. Recently, also Pericàs [55] has reported a domino Michael/Knoevenagel process to afford highly enantioenriched cyclohexane derivatives using a polystyrene-immobilized diarylprolinol silyl ether catalyst.

In pursuit of a rapid and cost-effective synthesis of the anti-influenza drug (−)-oseltamivir (**83**) (Tamiflu™, as phosphate), Hayashi and coworkers have developed an innovative synthesis involving three one-pot processes (Scheme 7.15) [56]. A large-scale synthesis of Tamiflu is considered so profoundly important today with

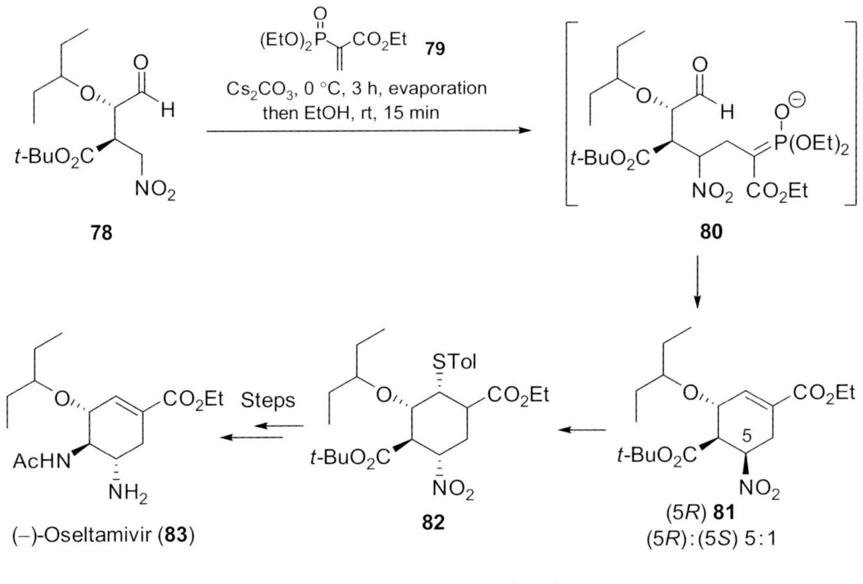

Scheme 7.15 Hayashi's synthesis of (−)-oseltamivir phosphate.

the current outbreak of the bird flu (H5N1) in several countries as well as the probable mutation of the H5N1 virus into a strain capable of human-to-human transmission. Initially, the enantiopure precursor **78**, containing two stereogenic centers, was prepared from inexpensive starting materials. Treating aldehyde **78** with vinylphosphonate **79** initiates a domino Michael/Wittig–Horner reaction to afford a 5 : 1 mixture of (5*R*)-**81** and (5*S*)-**81**, with compound **80** presumably as an intermediate. Unfortunately, in this domino reaction the undesired (5*R*)-diastereomer predominates. However, the Hayashi group overcame this imbalance in the next step, which involves an equilibration of this C5-stereo center via a Michael reaction with *p*-toluene thiol to give **82**. Following two more single-pot processes, the desired (−)-oseltamivir (**83**) was obtained in a remarkable overall yield of 57%. A later report on three and two one-pot sequences was published by the same group; in the early stages of this process, a domino Michael/Horner–Wadsworth–Emmons reaction occurs similar to the process described above [56, 57]. Another influenza neuraminidase inhibitor (ABT-341) has also been prepared via a domino reaction, in this case a domino nitro-Michael/Horner–Wadsworth–Emmons (HWE) reaction [58].

The combination of an imine derived from the reaction of acrolein organocatalyst **1** with simple indoles **84** and nitroalkenes **85** affords the 3-(cyclohexenylmethyl)-indoles **86** (Scheme 7.16) [59]. In this reaction, the indole **84** initiates the Friedel–Crafts-type reaction followed by a Michael reaction with nitroalkenes **85** to the intermediate **87**. From this process, a hydrolysis takes place and the resulting compound enters another catalytic cycle involving a Michael/aldol condensation reaction similar to those reported previously.

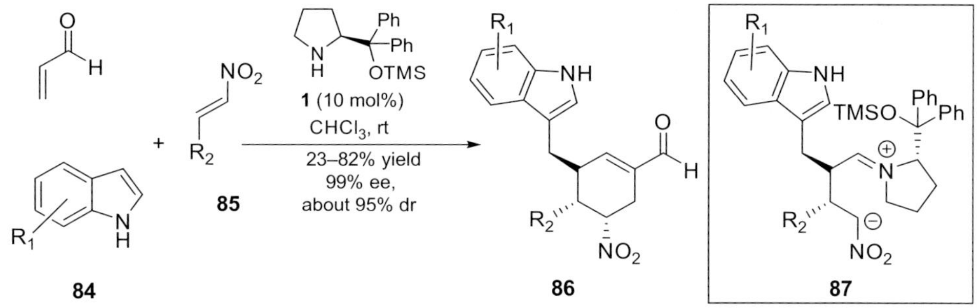

Scheme 7.16 Friedel–Crafts-type/Michael/Michael/aldol condensation.

A diastereo- and enantioselective Michael addition combined with a Darzens condensation reaction can be used to form two products of interest in the field of medicinal and natural products chemistry [60]. Additionally, depending on the workup conditions, an optically active epoxycyclohexanone, **92**, can be prepared through an S_N2 reaction, or the E1cB reaction pathway to **91** can be accessed (Scheme 7.17). In the early stages of the proposed mechanism, a planar iminium ion is suggested between the 2-[bis(3,5-bistrifluoromethylphenyl) trimethylsilanyloxymethyl]pyrrolidine **61** and the aldehyde moiety of compound **88**, which is subsequently attacked by the β-ketoester **89**. For the synthesis of the epoxide, a

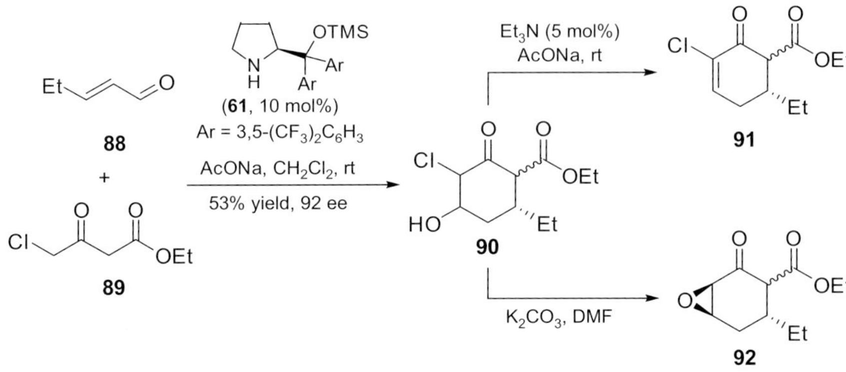

Scheme 7.17 Domino Michael/aldol and intramolecular S_N2 reactions.

range of derivatives were prepared in high diastereomeric and enantiomeric excess; for example, a simple modification of **89** to the allyl ester resulted in a 46% yield of the product in >99% de and 92% ee. In another study, a self-assembly of two precatalysts, one a pyrrolidine **93** and the other a primary amino acid **94**, was proposed to catalyze a domino reaction between salicylic aldehyde and cyclohexenone (Scheme 7.18) [61]. In the formation of the active species (**95** and **96**), it is proposed that the cyclohexenone is activated by the pyrrolidine **93** and the salicylic aldehyde by the primary amino acid **94**. This mechanistic proposal was verified by mass spectrometry experiments and resulted in the identification of key intermediates. The resulting species undergoes the domino oxa-Michael addition and intramolecular Mannich reaction, eventually leading to tetrahydroxanthenones **97**.

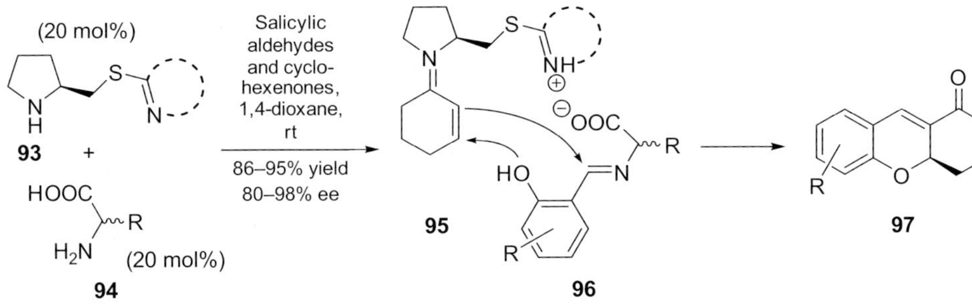

Scheme 7.18 Xu and Xu's asymmetric oxa-Michael/Mannich reaction.

Early work by Jørgensen *et al.* involved an impressive organocatalytic asymmetric two-component domino Michael/aldol reaction, paving the way for a range of other organocatalyzed domino reactions, as highlighted in the earlier book [25]. In that work, the construction of cyclohexanones of type **101** was described with up to four stereogenic centers with excellent enantio- and diastereoselectivity from unsaturated ketones **99** and acyclic β-ketoesters **98** in the presence of 10 mol% of the phenylalanine-derived imidazolidine catalyst **100** (Scheme 7.19) [5]. The

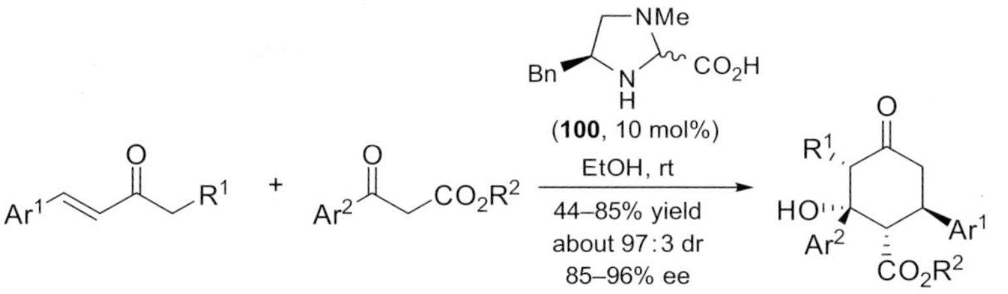

98 99 101

Scheme 7.19 Jørgensen's domino Michael/aldol reaction.

synthetic utility of these domino reactions was demonstrated by further reaction to a series of optically active building blocks. The same group prepared 6-substituted 3-diethoxyphosphoryl-2-oxocyclohex-3-enecarboxylates organocatalytically through a domino Michael/Knoevenagel condensation of ethyl 4-diethoxyphosphoryl-3-oxobutanoate and aliphatic aldehydes or aromatic aldehydes [62]. More Recently, the cyclohexane ring system was prepared through a formal carbo[3 + 3]cycloaddition or domino Michael reaction/Knoevenagel condensation reaction (for further details see Chapter 10) [63].

The bisanthraquinones are a growing class of natural product compounds that have been isolated from a range of sources. (+)-Rugulosin (**111**) is present in *Penicillium rugulosum*, the *Penicillia* contaminating rice, maize, and cereal crops [64]. In the key bond-forming process, MOM (methoxymethyl) protected and enantiopure **102** is initially oxidized to the respective anthraquinone **103**, which immediately undergoes dimerization to the intermediate **106** (Scheme 7.20) [65]. Interestingly, in some instances the authors have reported only the isolation of compound **106** when using a crude sample of **102**. In the domino reaction sequence, the alternating oxidations and double Michael reactions through compound **104** to **105** and **106** provide three new bonds linking the two anthraquinones. The final stage of the synthesis is achieved through the treatment of compound **110** with concentrated HCl. In this well-designed example of domino reaction in natural product total synthesis, many of the intermediate compounds have been identified in slight modifications of this final procedure.

The Gravel group has reported the preparation of a series of indanes though a domino Stetter/Michael reaction. In this example, NHCs (thiazolium salts) were found to catalyze the domino process from dienones and aldehydes, forming two new carbon–carbon bonds and three new contiguous stereogenic centers [66]. This work was followed up by the same group, this time to produce bis-indanes in a domino Stetter/aldol/Michael (SAM) reaction [67]. This process involves taking aldehyde enone **111** and treating this with an NHC **112** to generate a Breslow intermediate, which participates in a Stetter reaction to an enolate intermediate **114**. Following the proposed mechanism, aldol condensation, and in turn an eventual Michael reaction, results in the formation of the spiro-derivatives **113** in good yields and excellent diastereoselectivity (Scheme 7.21). This

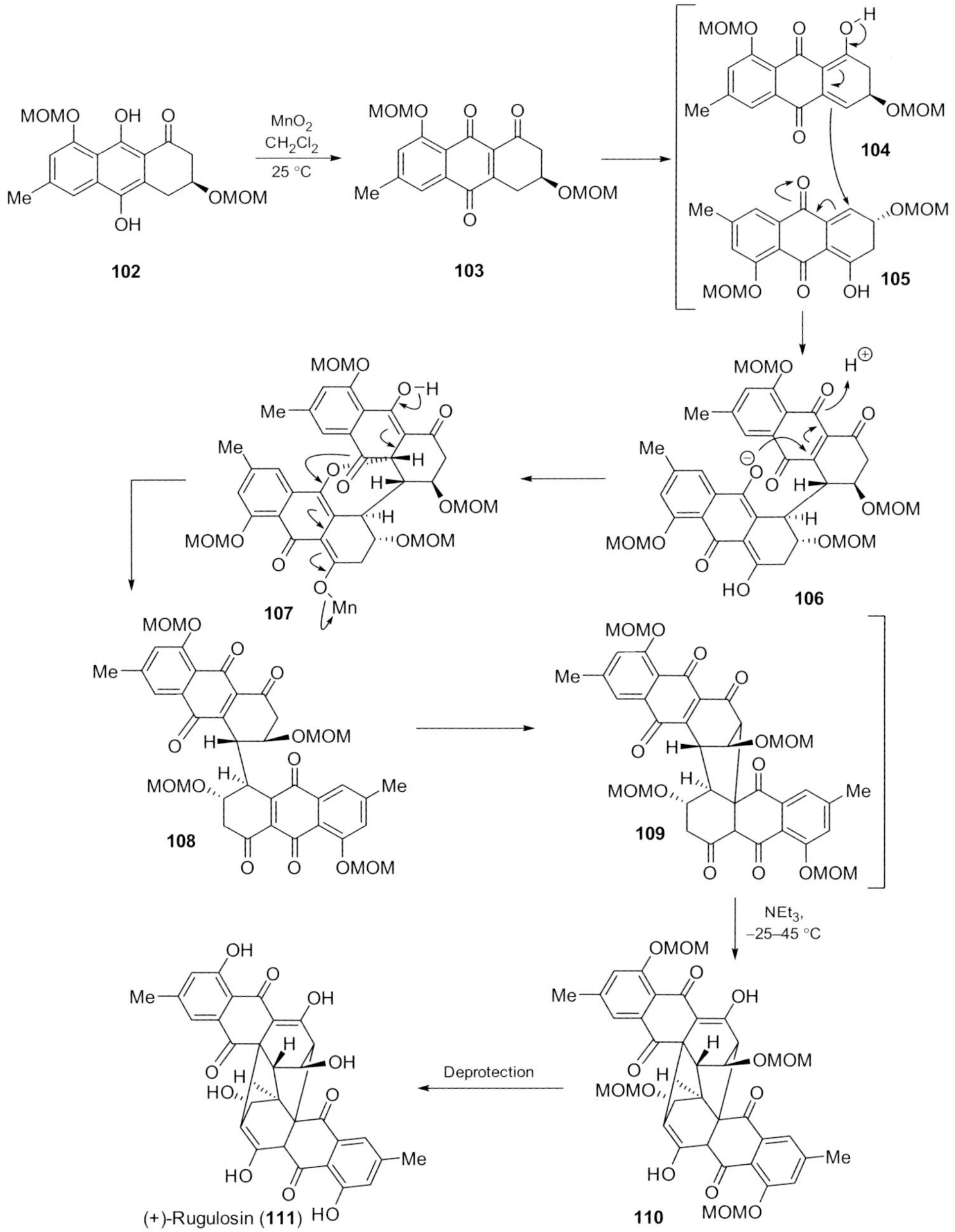

Scheme 7.20 Nicolaou's synthesis of (+)-rugulosin.

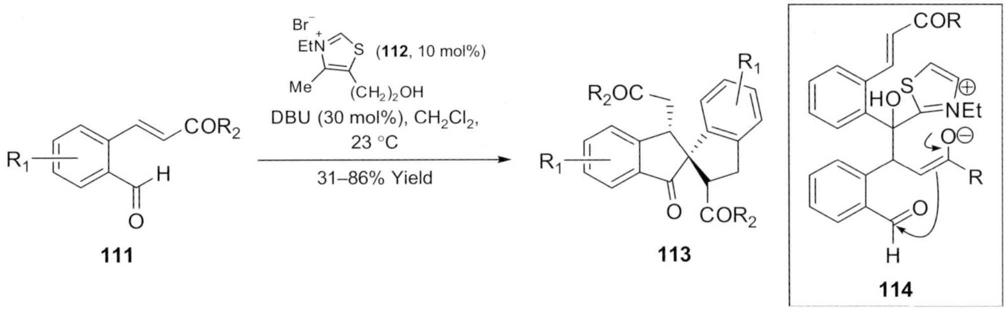

Scheme 7.21 Domino Stetter/Aldol/Michael reaction.

work was also extended to an NHC-catalyzed cross-SAM reaction and a domino Stetter/aldol/aldol (SAA) process.

An organocatalytic (Jørgensen–Hayashi catalyst 1) domino Michael/Michael/aldol condensation was used to prepare a hexahydronaphthalenone (+)-**121** *en route* to the natural product (+)-galbulin (Scheme 7.22). In this case, it was proposed that an iminium salt **118** (activated from the *re* face) and an enamine species **117** were preformed. Following a kinetic asymmetric transformation (KAT) of the racemic precursor **115**, the intermediate **119** is first formed through an intermolecular Michael reaction. A second intramolecular Michael reaction occurs and the intermediate **120** forms, and finally an acid-initiated aldol condensation

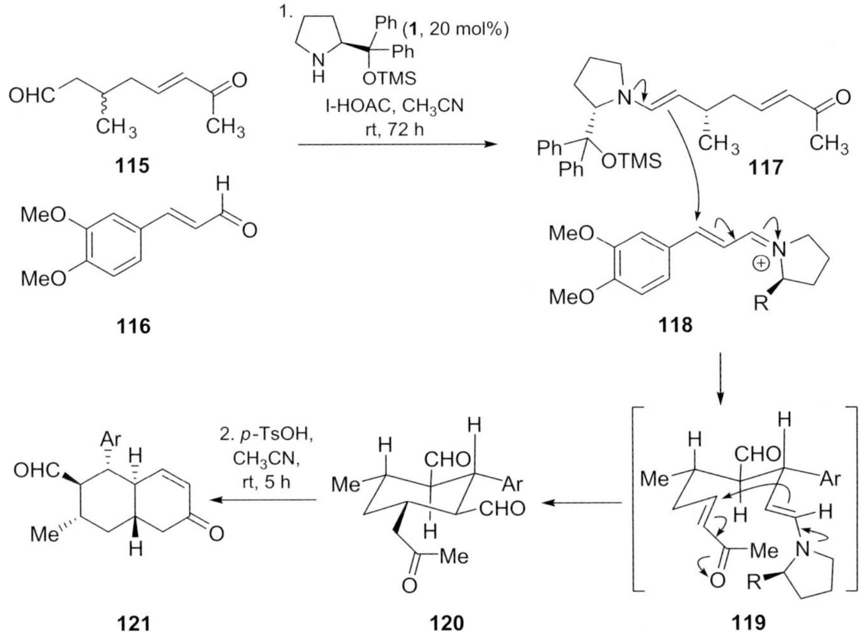

Scheme 7.22 Domino Michael/Michael/aldol condensation reaction.

provides the hexahydronaphthalenone (+)-**121**. This compound is later converted to the decalin-based natural product [68]. The combination of a Simmons–Smith reaction and a Michael process was also used by Alexakis *et al.* [69] in the synthesis of 3/5 and 3/6 fused ring systems. The group of Brenner [70] also prepared substituted bicyclic systems through a Michael/Michael domino process of conjugated β-keto esters and unsaturated aldehydes.

Work by the groups of Padwa [71], Hagiwara [72], and Spitzner [73] have all dealt with domino Michael addition reactions to bi- and tricyclic octenes(ane), which were discussed in the first domino book by Tietze [25]. Michael processes forming tricyclo[6.2.2.0]dodecane and tricyclo[5.3.1.0]undecane systems have been also described by the group of Fukumoto [74]. Later, a domino Michael/Henry reaction was used to synthesize the medicinally important bicyclo[3.2.1]octane ring system in an enantioselective manner [75]. The cinchona alkaloid-derived catalyst was applied in this example, where a new model based on density functional theory calculations involving four H-bond interactions between the chiral organocatalyst and the substrates was proposed (for further details see Chapter 10). A similar bicyclic core was arrived at through an enantioselective domino Michael/aldol reaction by the Rueping group (Scheme 7.23) [76]. In this case, α,β-unsaturated aldehydes **123** containing β-electron-donating or electron-withdrawing groups were reacted with 1,2-cyclohexadione **122** to give a bicyclic compound **124** containing four new stereocenters in a range of yields and high enantiomeric excess (90–98%). The product **124** was also converted to tetrahydrochromenones (via a base-induced retro-aldol reaction) and a polysubstituted cycloheptanone (via reduction and oxidative cleavage). For further details on the substrate scope, see Chapter 8. Later studies by the same group have allowed the synthesis of more substituted bicyclo[3.2.1]octan-8-ones through higher substitution of the β-nitrostyrene using a cinchonidine-based thiourea catalyst [77]. Similar ring systems have been prepared earlier by a domino Michael/epoxide ring-opening process [78].

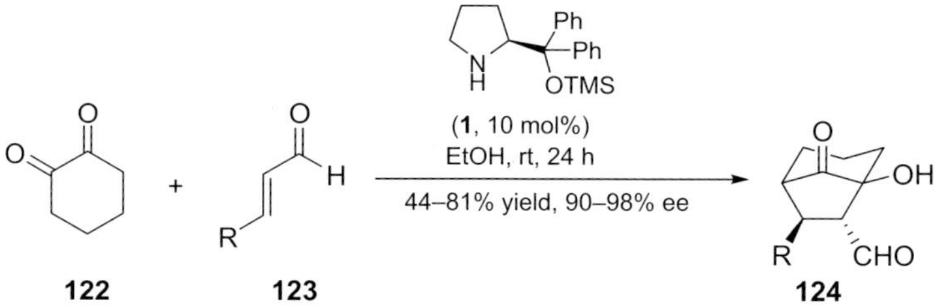

Scheme 7.23 Enantioselective domino Michael/aldol reaction by the Rueping group.

7.4
Formation of *O*-Heterocycles

Recently, Zhang prepared a series of four *trans*-perhydroindolic acid isomers as organocatalysts to oversee a reaction to 3,4-dihydro-2*H*-pyrans **128**. In this process, a chiral asymmetric domino Michael addition was initiated by an enamine intermediate **129**, which was initially formed between the catalyst **127** and the aldehyde **126** (Scheme 7.24) [79]. The resulting imine was then utilized as an electrophile for the second step cyclization and attack by oxygen, leading to the pyran ring system. The yield as well as diastereomeric and enantiomeric excesses for this reaction sequence were excellent. The related tetrahydropyrano[2,3-*c*]pyrazoles have also been prepared through an initial domino Michael/acetalization reaction between 3-trifluoro-methylpyrazolone **130** and crotonaldehyde to first provide the acetal **131** in excellent yields (Scheme 7.25) [80]. In this sequence, developed by the Enders group, rapid equilibrium between the two diastereomers and the open-chain form prevented an accurate determination of the enantiomeric excess, thus an immediate one-pot Wittig reaction was performed to afford **132**. It should also be noted that a variety of aldehydes, 3-trifluoromethyl pyrazolones, and Wittig reagents were used, and the iminium and enamine activation and catalytic cycle were assumed to be in operation. The same research group has reported domino oxa-Michael/Michael reaction of a series of nitrovinylphenols **133**, and acrolien gave trans-3,4-disubstituted chromans **135** in moderate to excellent yields (62–91%) and excellent stereoselectivities (Scheme 7.26). Further preparation of benzopyrano[3,4-c]pyrrolidine was also achieved [81]. Similarly, an organocatalytic Michael/hemiacetalization reaction sequence using saturated aldehydes and (*E*)-2-(2-nitrovinyl)phenols afforded the cis-3,4-disubstituted chromanols which could

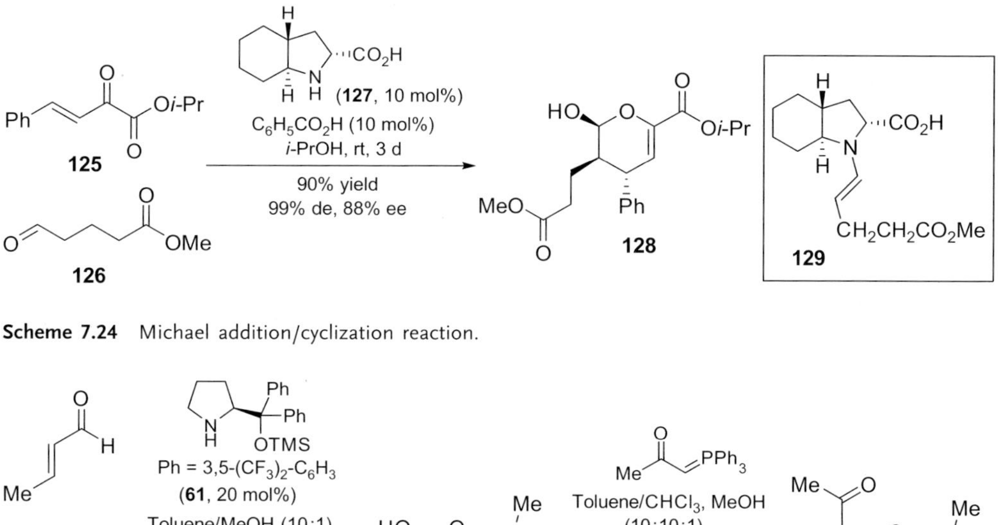

Scheme 7.24 Michael addition/cyclization reaction.

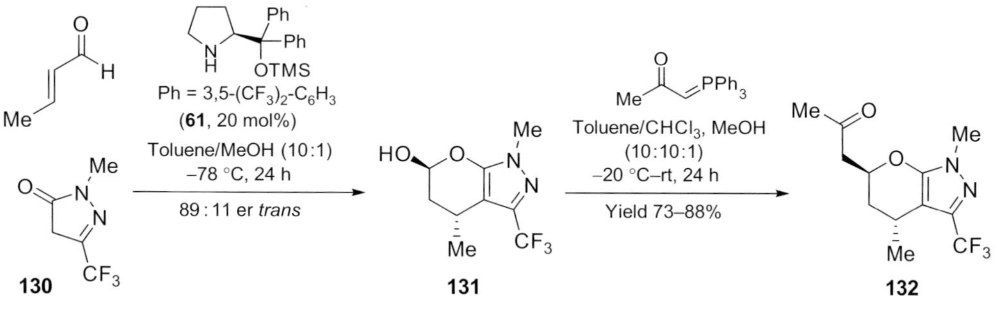

Scheme 7.25 Domino Michael/acetalization reaction.

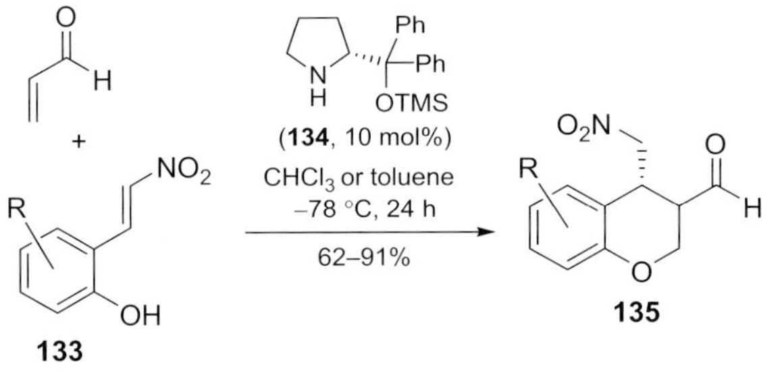

Scheme 7.26 Enders's domino oxa-Michael/Michael reaction.

also be further transformed (through dehydroxylation) to chromans [82]. A high-yielding oxa-Michael/Michael reaction to **137** followed by a domino Michael/aldol condensation with crotonaldehyde to **138** was achieved by the group of Hong (Scheme 7.27). This dual domino reaction was the first enantioselective total synthesis of the natural product marine meroterpene (+)-conicol [83].

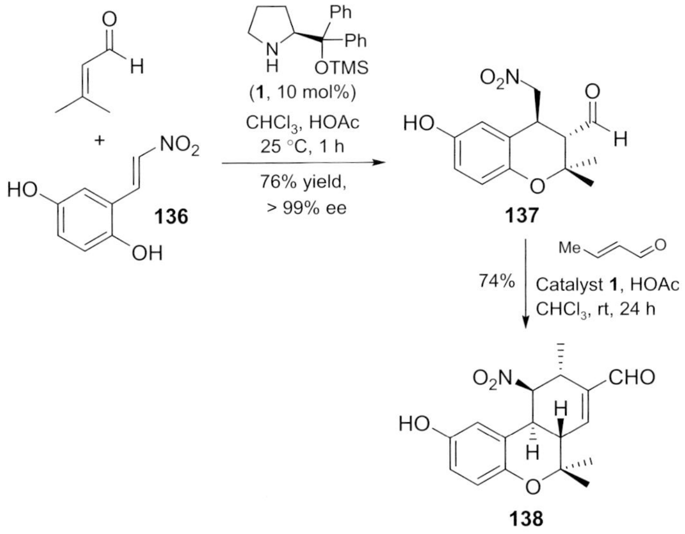

Scheme 7.27 Michael/Michael/Michael–aldol condensation reaction.

The structure of tetronasin (**142**), an ionophore of an acyltetronic acid containing twelve stereogenic centers, many of which are contiguous, presents a demanding synthetic challenge. The natural product exhibits a range of biological activities, including antibiotic as well as antiparasitic, and acts as a growth-promoting agent in ruminants. The preparation of this compound, produced from *Streptomyces longisporoflavus*, was accomplished through a convergent synthesis devised by the Ley group (Scheme 7.28) [15]. In the later stages of this synthesis, the tetraene **139** was treated with potassium hexamethyldisilazide to first initiate a conjugate

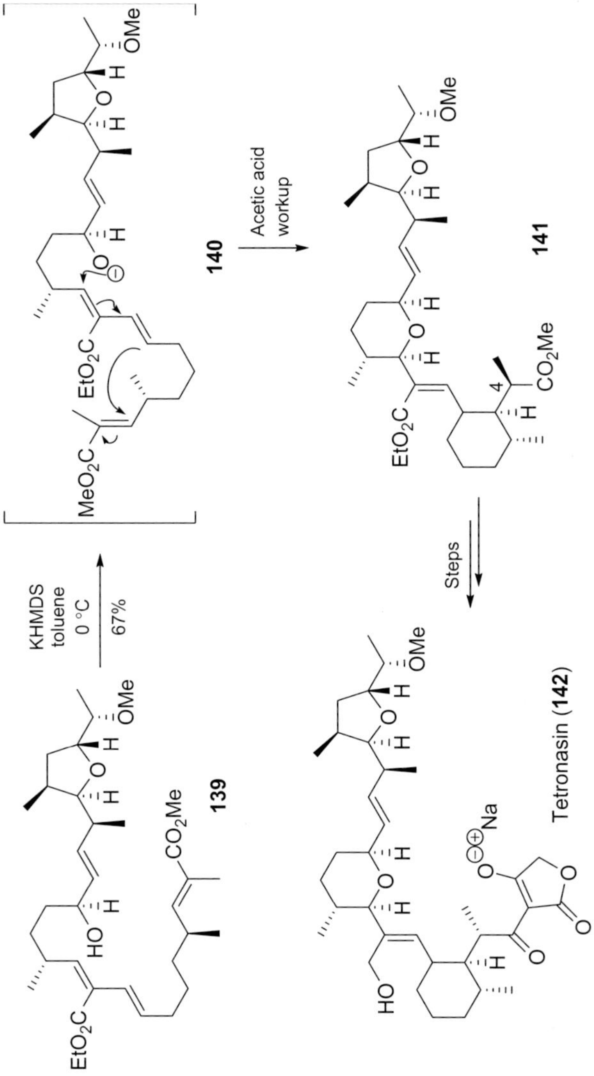

Scheme 7.28 Ley's synthesis of tetronasin.

addition of the secondary alcohol to the dienic ester. Following a subsequent re-equilibration of the addition product (enolate), a second cyclization occurs in a Michael-type manner (abbreviated in Scheme 7.28) to give compound **141** in 67% yield. In this domino double Michael addition reaction, a tetrahydropyran and a cyclohexane ring are formed in a single process with stereocontrol at each of the new stereogenic centers. The configuration of the methyl substituent at C-4, however, was later epimerized to generate the correct stereochemistry required for the natural product **142**.

Another natural product synthesis features that of a key domino ene–yne coupling/Michael reaction to a tetrahydropyran ring (Scheme 7.29) [22]. An in-depth Trost domino ene–yne/oxa-Michael reaction methodology process was first carried out prior to using the key substrates **143** and **144**, with $[CpRu(MeCN)_3]PF_6$ **145** as the catalyst. The product **146** contained almost all the carbons of the natural product target (−)-exiguolide as well as most of its asymmetric centers.

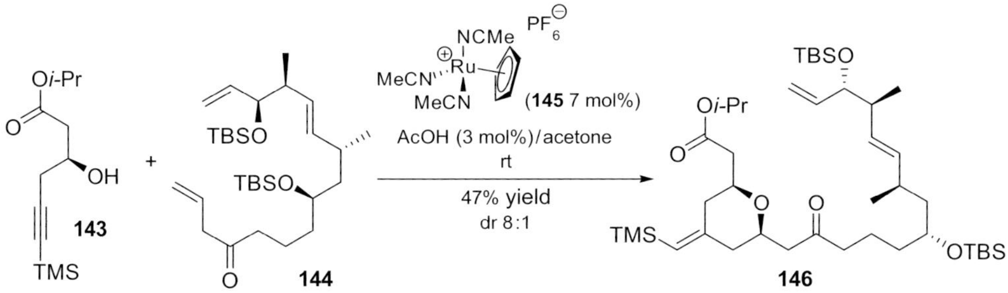

Scheme 7.29 Roulland's synthesis of (−)-exiguolide.

In 2004, the group of Sorensen examined domino reactions initiated by intramolecular Michael addition reactions in the preparation of a *Trichoderma harziznum* natural product harziphilone (**153**) [17]. In this procedure, which could also be labeled an isomerization, the enantiopure diol **147** is treated with catalytic amounts of DABCO (**148**) at room temperature to produce the proposed "Balyis–Hillman like zwitterion **149**" in a reversible reaction (Scheme 7.30). Subsequent intramolecular 1,4-addition and proton transfer results in a second zwitterion **151**, also containing the DABCO tether. This intermediate is then proposed to either undergo β-elimination to give **152** followed by a 6π-electrocyclic reaction to the natural product **153** or to the same product through immediate intramolecular substitution of **151**. Surprisingly, this isomerization (or domino reaction), given its complexity, was carried out over 24 h, whereas initial trials of the model compounds were carried out in 10 days.

A domino nitro-Michael/Henry reaction was conceived by Seeberger *et al.* to afford a 2-nitropyranoside system **156**, regarded as an excellent carbohydrate building block (Scheme 7.31) [84]. In this case, the reaction between hydroxyaldehyde **154** and nitroalkene **155** was investigated, with the base LiHMDS performing well

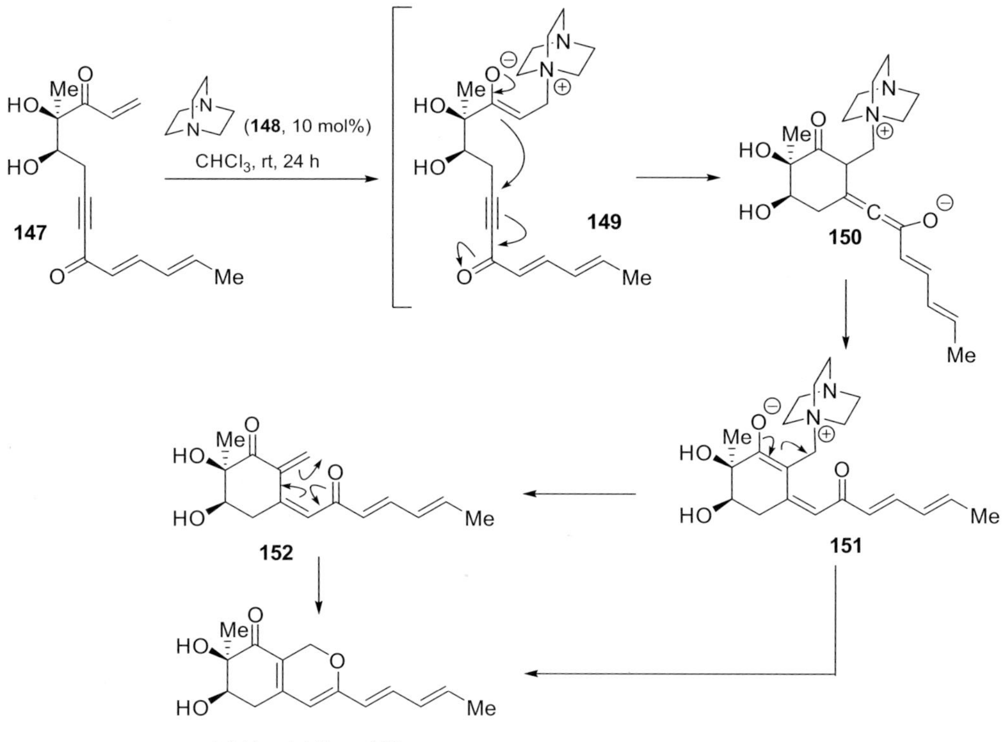

Scheme 7.30 Sorensen's synthesis of (+)-harziphilone.

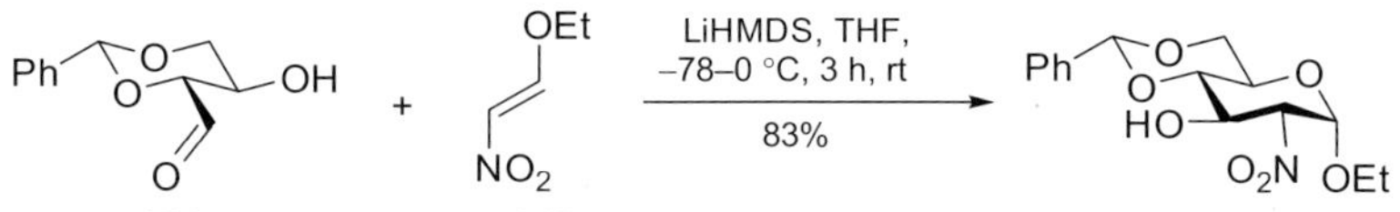

Scheme 7.31 Seeberger's domino nitro-Michael/Henry reaction.

to give the desired product in 83% yield. Remarkably, only one of eight possible stereoisomers was produced (α-anomer).

In a paper that described a detailed preparation of air-stable cinchona alkaloid-derived chiral quaternary ammonium phenoxides, the Mukaiyama group also used these to enantioselectively prepare 3,4-dihydropyran-2-ones **160** [85]. A low loading of organocatalyst **159** at low temperatures, in a series of solvents, resulted in the formation of the optically active lactone products in high yields with excellent control of enantio- and diastereoselectivity. This process was proposed to go through a phenoxide-ion-catalyzed domino Michael addition and lactonization catalytic cycle as illustrated below. Many variations of the ketene silyl acetals and α,β-unsaturated ketones were combined in this domino process (Scheme 7.32). Earlier,

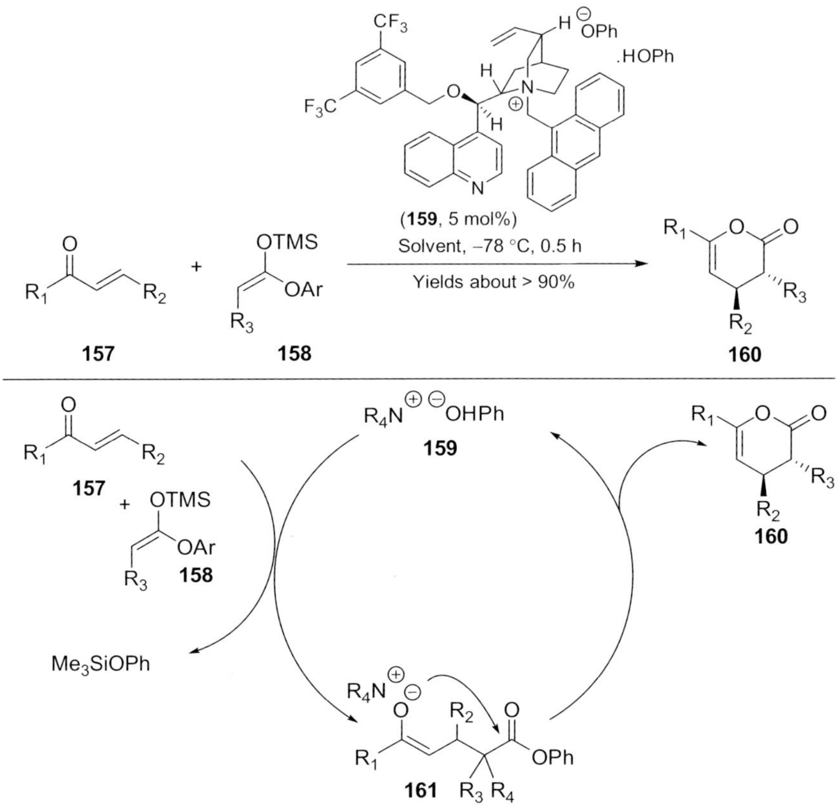

Scheme 7.32 Phenoxide-ion-catalyzed domino Michael addition and lactonization.

a work involving a Michael-addition-initiated threefold domino process was also developed by the group of Carreño to afford complex cage compounds containing lactones from quinamines with OTMS furans [86]. In a separate study, NHCs **164** were used as organocatalysts in a domino Michael/Michael process with both base (1,8-diazabicyclo[5.4.0]undec-7-ene, DBU) and $MgSO_4$ present (Scheme 7.33) [87]. The initial process is the formation of the Breslow intermediate **166**, which then reacts in a Michael-type manner with one enone moiety of **163**.

An earlier domino reaction to tetrahydro-2*H*-chromenes involving the treatment of pyran-4-one **167** with a silyl triflate followed by the addition of an α,β-unsaturated ketone **169** and 2,6-lutidine in CH_2Cl_2 furnished substituted tetrahydro-2*H*-chromenes of type **174** in up to 98% yield (Scheme 7.34) [88]. The sequence, relying of positively charged heterocycles, is assumed to start with an aldol type reaction of the oxenium ion **168** and enol ether **170**, being formed from **169**, to give **171**, which is followed by two 1,4-additions with **172** as the proposed intermediate. In this domino process, three new C–C bonds and four stereogenic centers are built up in a highly efficient and selective manner, with the exception of the acyclic double bond. A Michael addition/acetalization sequence

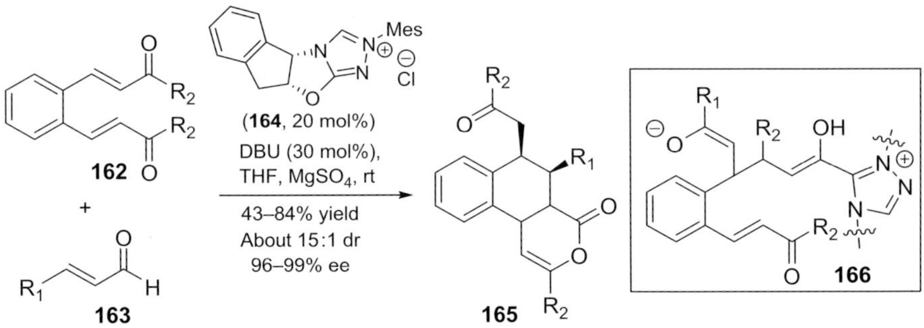

Scheme 7.33 Chi's annulation of di(enone) and enal substrates catalyzed by NHCs.

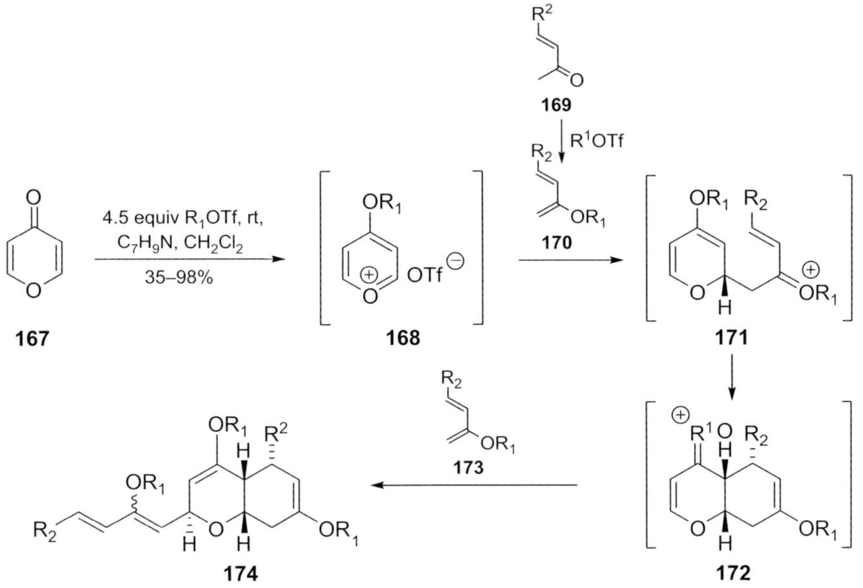

Scheme 7.34 Beifuss's multiple functionalization of positively charged heteroarenes.

was used to create chromenones by Rueping (Scheme 7.35) [89]. In this domino process, the initial asymmetric oxidative iminium activation step occurs between an oxidized allylic alcohol **175** (tetrapropylammonium perruthenate, TPAP/*N*-methylmorpholine-*N*-oxide, NMO) and the amine catalyst **61**. Secondary Michael addition and cyclization reaction proceeds with high selectivity for both aliphatic and aromatic allylic alcohols. Likewise, the diethyl bromomalonate **177**, through a similar initial Michael addition and cyclization process, affords the corresponding cyclopropane **180**. An interesting domino Michael/enolization/cyclization reaction was used between two dicarbonyl containing compounds, diketone, and pyruvates. This enantioselective process, catalyzed by indane thiourea catalysts, is assumed to go through a process where the two oxygens within behave in a nucleophile

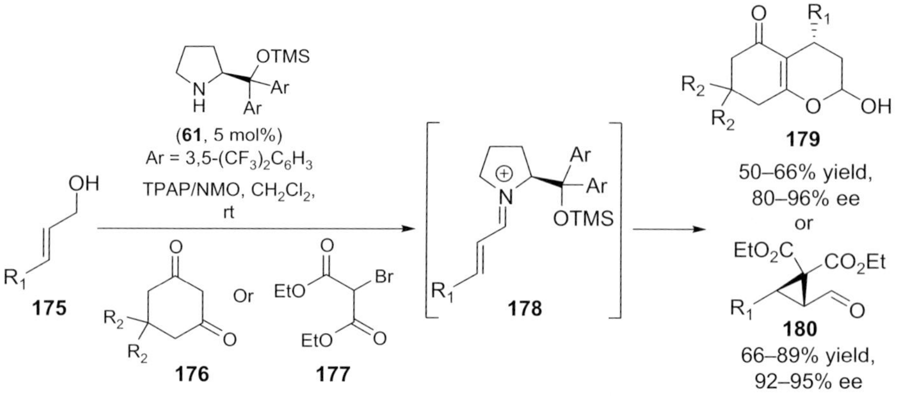

Scheme 7.35 Rueping's catalytic asymmetric oxidative iminium domino reactions.

manner (see Chapter 10 for further details) [90]. Wang also used a similar catalyst for an adjustment of the relative positions of amine and thiourea to (suggesting a strongly hydrogen-bonded intermediate system) to prepare pyranochromenes using a Michael/oxa-Michael/tautomerization reaction of malononitrile and enones [91]. A domino retro-Michael/aldol/lactonization reaction was also used in the formation of a variety of 7-hydroxy-6*H*-benzo[*c*]chromen-6-ones [92].

An asymmetric domino oxa-Michael/aldol condensation reaction was reported independently by Córdova [93] and Wang [94] occurring between 2-hydroxybenzaldehyde **181** and β-substituted α,β-unsaturated aldehydes **182** (Scheme 7.36). Both papers explored a range of derivatives through R_1 and R_2 using two simple organocatalysts (**183** and **1**, both utilizing iminium ion activation) to provide good yields of compound 184 in good to high enantiomeric excesses.

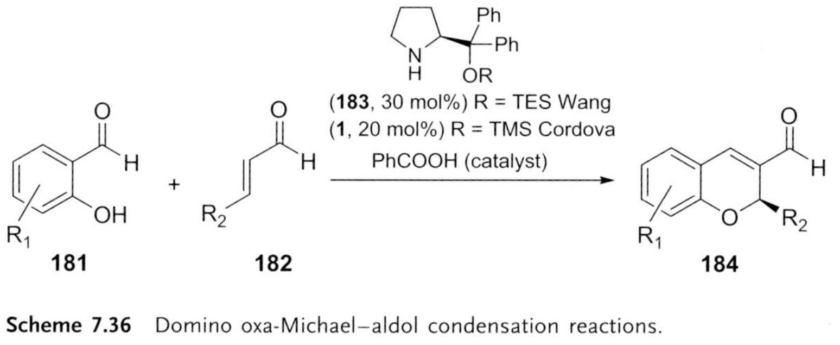

Scheme 7.36 Domino oxa-Michael–aldol condensation reactions.

The Rueping group also applied the previously described TPAP/NMO oxidation process to reactions involving propargyl alcohols **186** and (*E*)-2-(2-nitrovinyl)-phenols **185**. Again, the iminium ion is preformed, and a proposed allenamine intermediate **188** is formed through the initial oxa-Michael process (Scheme 7.37) [95]. A second Michael reaction (intramolecular) then forms the reported bicyclic 4*H*-chromenes ring system **187**. This combination brings about a combined

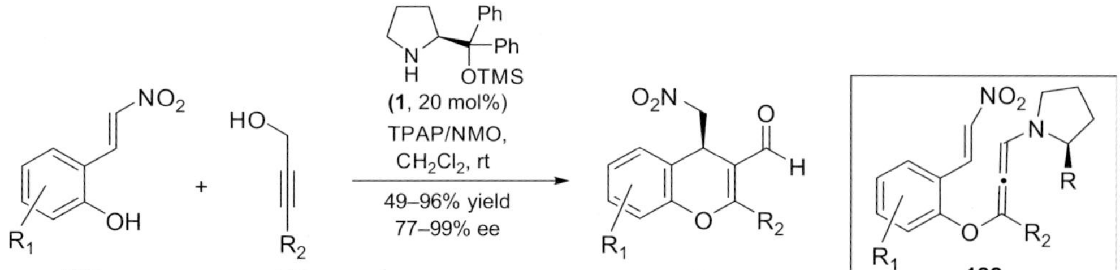

185 186 187 188

Scheme 7.37 Oxidative iminium–allenamine reaction to 4*H*-chromenes.

catalytic reaction with a metal- and an organocatalytic cycle. Completing this process again, this time with **187** as the substrate, the tricyclic 4*H*-chromenes could be achieved in a highly enantioselective manner. Such ring systems were also prepared through a domino Knoevenagel/Michael/hetero-Thorpe–Ziegler three-component reaction to access a library of compounds that are structural analogs of the antimitotic lignin podophyllotoxin and its reported semisynthetic derivatives (Scheme 7.38) [96]. The Knoevenagel process between polyalkoxybenzaldehydes **189** and malononitrile **190** initiated the domino process which was systematically completed with a large amount of substrates, resulting in chromene 192 a range of yields depending on the substitution pattern. NHCs have been used to prepare 3,4-dihydrocoumarins through a domino Michael addition/acylation reaction [97].

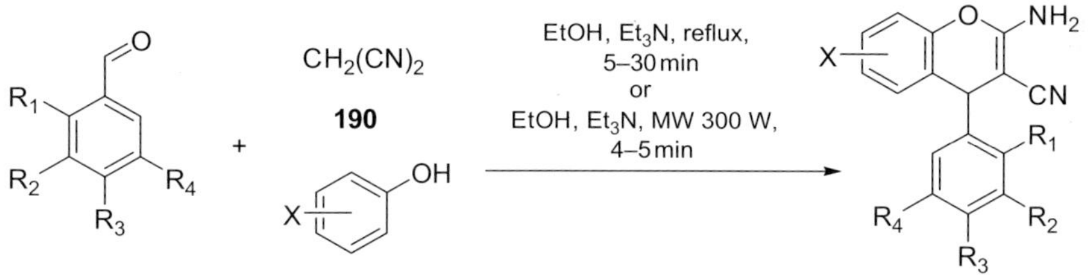

189 190 191 192

Scheme 7.38 Preparation of polyalkoxy-substituted 4-Aryl-4*H*-chromenes.

In 2005, as highlighted in the first book of Tietze [25], the group of Bräse devised a vinylogous aldol/oxa-Michael domino reaction [98, 99]. This study was followed by the first total synthesis of the secondary metabolite diversonol (195) and blennolide C, the former synthetic pathway beginning with a domino oxa-Michael/aldol condensation process between 4-hydroxycyclohexenone and a functionalized salicylic aldehyde **193** to a tetrahydroxanthenone **194** taking 7 days (Scheme 7.39) [100, 101]. A further methodical investigation was conducted by the same group on the reaction of salicylaldehydes **196** and **197** with prenal under the control of one of the Jørgensen catalysts **134** [102]. Each of the tricyclic lactols **196** and **197** was prepared by the two enantiomeric catalysts **134** and **61** (Scheme 7.40). Additionally, **198** and **199** were prepared through the synthesis of lachnone C and

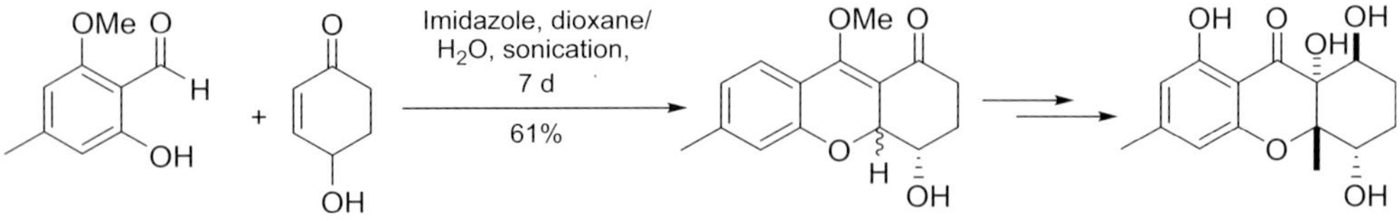

Scheme 7.39 Brase's domino oxa-Michael–aldol condensation.

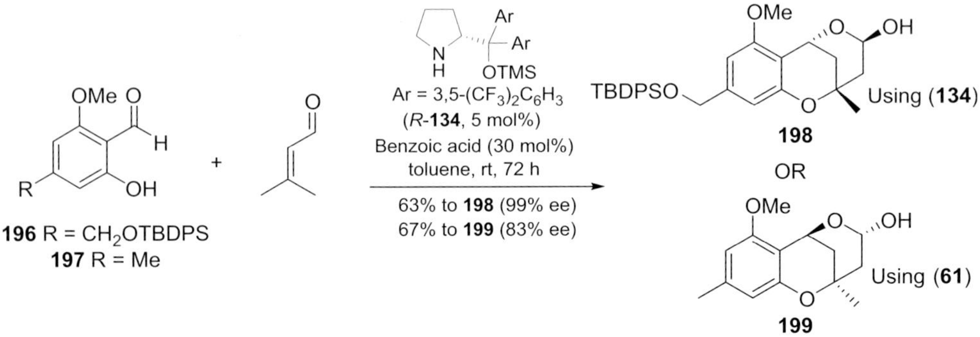

Scheme 7.40 Domino reactions for an asymmetric syntheses of diversonol and lachnone C.

diversonol, respectively. A different approach to this type of ring system, employing a domino aldol/oxa-Michael reaction, was used as the main step in the synthesis of α-tocopherol (see Chapter 8 for further details) [103].

Spirocyclohexene compounds bearing 1,3-thiazolidinediones were prepared through a double Michael domino reaction of 1,3-thiazolidinedione, malononitrile, and aromatic aldehydes and organic amines [104]. Later, in the field of preparing the spirocyclohexene scaffold, a single carbon quaternary stereocenter of a 1′,3-spiro-2′-oxocyclohexan-3,4-dihydrocoumarin **202** was constructed through a domino Michael/acetalization sequence. Following treatment of 2-oxocyclohexanecarbaldehyde **200** with the Takemoto catalyst **3** in the presence of the phenol **201**, the reaction was completed in under 30 min (Scheme 7.41) [105].

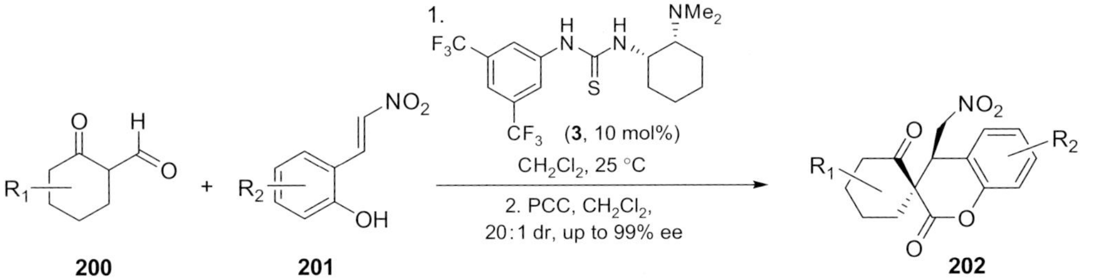

Scheme 7.41 Hong's domino sequence to 1′,3-spiro-2′-oxocyclohexanes.

A Brønsted base and a H-bonding mode binding from the $N(Me)_2$ group and the two NHs of the catalyst were proposed to bring together the two substrates. Following oxidation (pyridinium chlorochromate, PCC), excellent diastereo (>20 : 1), and enantioselectivities (up to >99% ee) were obtained. Alternatively, the cyclohexene core within a spirocyclohexano-1,3-thiazole was formed through a double Michael addition/spirocyclization reaction of 1,3-thiazolidinedione, malononitrile, aromatic aldehydes, and various bases [104]. Spiro-lactones also were previously prepared via a Michael addition process of methylenecyclopentenone and ester enolates followed by an aldol and a lactonization step [106].

In the domino Michael/alkylation reaction applied to the synthesis of 3-(2*H*)-furanones, the ethyl 4-bromoacetoacetate **203** and nitrostyrene **204** were first trialed with a range of catalysts. In this instance, the so-called modified Feist Bénary reaction was completed with an *L*-threonine bifunctional tertiary amine/thiourea catalyst **205** to produce the furanone **206** in excellent yield and high enantioselectivity (Scheme 7.42) [107]. In another report, the furan ring as part of other bicyclic or tricyclic systems was also prepared through an enantioselective Michael addition/nucleophilic substitution reaction (Scheme 7.43) [108]. When diketones and (*E*)-β,β-bromonitrostyrenes **207** were stirred, again with a bifunctional thiourea

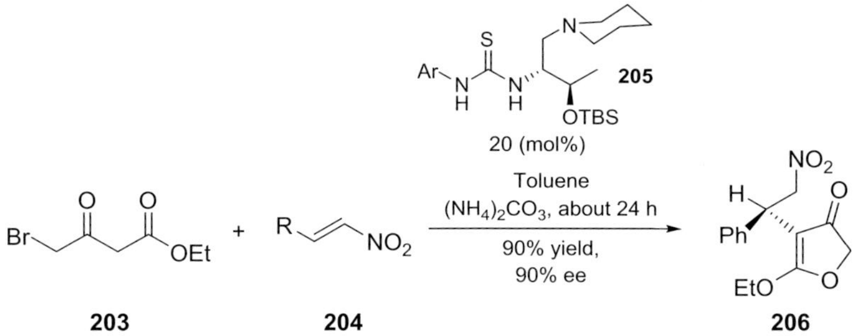

Scheme 7.42 Domino Michael/alkylation to furanones.

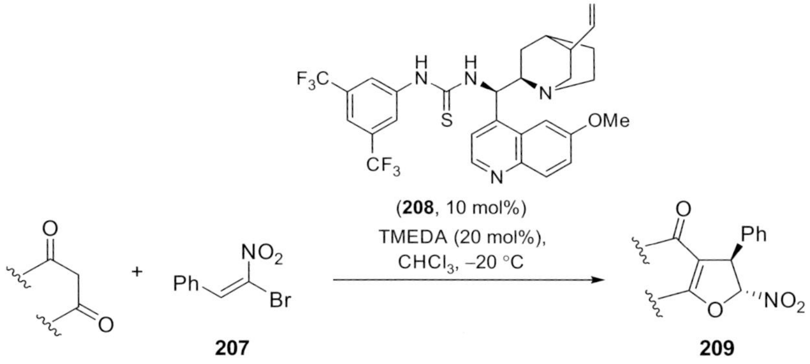

Scheme 7.43 Domino Michael addition/nucleophilic substitution reaction.

organocatalyst **208**, the formation of *trans*-tetrasubstituted dihydrofuranes **209** occurred.

A more substituted tetrahydropyran ring system was targeted through a domino process by the Menche group. The domino oxa-Michael/Tsuji–Trost reaction begins with treating a nitro olefin with a homoallylic carbonate substrate **211**, where R = Ph or *i*-Pr (Scheme 7.44) [109]. Through the trail of a range of bases and palladium catalysts, several reactions conditions were used to produce the product through the oxa-Michael addition intermediate **212**. In an example of this process, the two diastereoisomers (with a small amount of a third) could be produced in 63% yield in the ratio 1.6 (**213**) : 1 (**214**). A domino reaction was also reported by Rychnovsky *et al.* to form the tetrahydropyran ring system as single diastereoisomer (Scheme 7.45) [110]. The process, a domino Michael addition to intermediate **217** followed by a Cope/aldol type reaction, provided the tetrahydropyrans **216**. A range of R groups were explored with methyl vinyl ketone and various vinyl ethers **215**. The reaction, also described as a Mukaiyama/Michael domino process, was mediated by 2,6-di-*tert*-butyl-4-methylpyridine (DTBMP) and titanium tetrabromide.

The tetrahydropyranol ring system was also formed in a domino Michael-ketalization reaction. The reaction between cyclohexanone and nitro alkenols in the presence of proline-based organocatalysts was used to successfully prepare fused cyclohexyl pyran derivatives [111]. Rodrigues and Coquerel reported an NHC-catalyzed Michael-based spirocyclization from 1,3-dicarbonyl compounds bearing

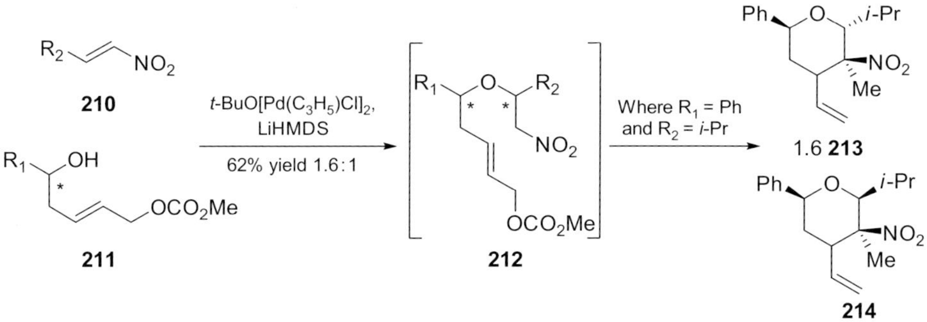

Scheme 7.44 Menche's oxa-Michael/Tsuji–Trost reaction.

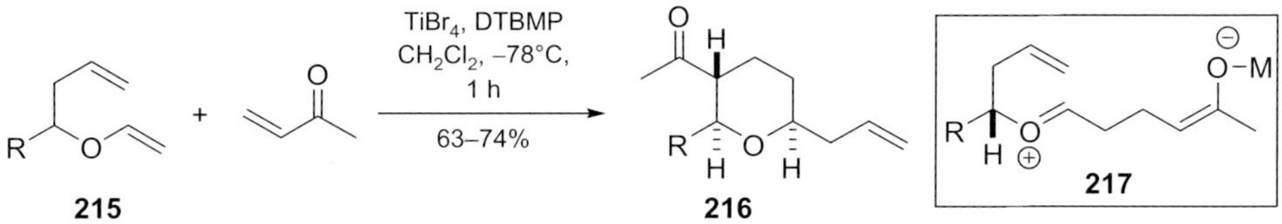

Scheme 7.45 Domino Michael/Cope/aldol reaction.

a vinyl nitrile tether. Linked to this work, the same *N,N*-diaryl-1,3-imidazol(in)-2-ylidene **219** (NHC) acts as an excellent organocatalyst in high yields for a three-component domino Michael/aldol/retro-Dieckmann (MARDi) reaction. Thus, **218** was treated with crotonaldehyde in methanol in the presence of the 20 mol% of NHC **219** to provide bicyclo[3.2.1]octanols **220** (41%, dr $OH_{ax}/OH_{equiv} = 2.9:1$) and the expected cycloheptanol **221** (39%) as a single diastereomer (Scheme 7.46) [112, 113]. The MARDi domino process was also successfully applied to the synthesis of azepanes, azepines, and thiepines.

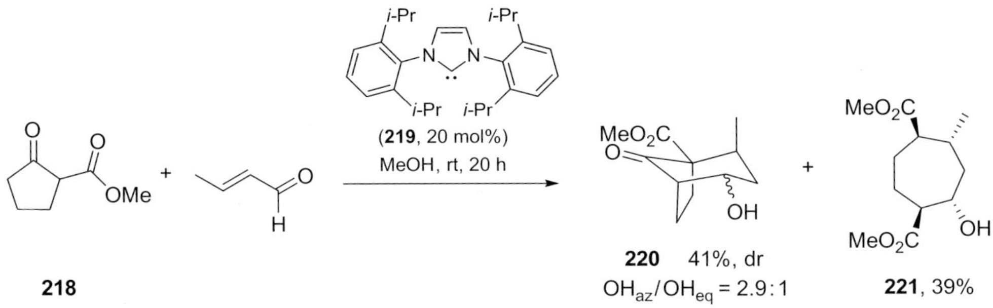

Scheme 7.46 Domino Michael/aldol/retro-Dieckmann (MARDi) reaction.

The precursor dihydroxyacetone dimer **223** and aldehyde **222** underwent a domino sequence to afford the interesting hexahydrofuro[3,4-*c*]furane in excellent yields [114]. In this example by Vicario, in the oxa-Michael/aldol/hemiacetalization process, an iminium ion species formed between organocatalyst **1** and enal **222** reacts with the structurally interesting dihydroxyacetone dimer **223**, providing the intermediate enamine which undergoes an intramolecular aldol reaction (Scheme 7.47). The high stereocontrol of the reaction (about 90–99% ee and 10 : 1 dr) was proposed to involve the reversibility of oxa-Michael addition and a predicted fast aldol condensation and/or dynamic kinetic resolution process where the chiral catalyst **1** accelerates the aldol reaction for one diastereoisomer over the other. For a mechanistic rationale of this reaction please, see Chapter 8.

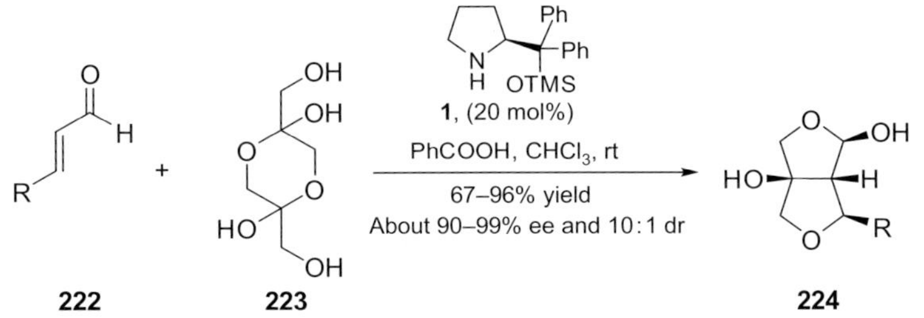

Scheme 7.47 Domino reaction to polysubstituted furofuranes.

7.5
Formation of *N*-Heterocycles

The synthesis of *N*-heterocycles is, in many cases, governed by a biologically active alkaloid target. A representative example is the double Michael addition catalyzed by a thiourea-based catalyst, later named Takemoto's catalyst (**3**), in the synthesis of (−)-epibatidine **229** (Scheme 7.48). The Takemoto group first investigated a simple double Michael reaction and later moved to a reaction between the pyridine derivative **225** and the γ,δ-unsaturated-β-ketoester **226** to construct three contiguous stereogenic centers found within **227** [115]. This compound was taken through to the natural product in another seven steps.

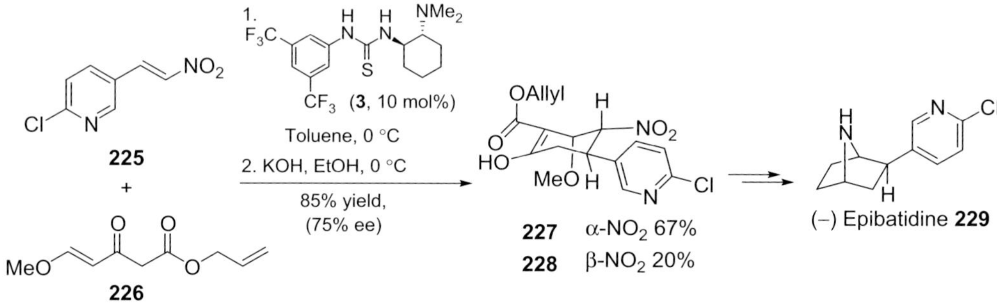

Scheme 7.48 Enantioselective total synthesis of (−)-epibatidine.

The inexpensive and readily available diisopropylamine was used in the iminium-ion-forming process for domino Michael addition/aldol condensation reaction to *N*-hydroxypyrrole **232** (Scheme 7.49). When a variety of different functional groups at R_3 of the α,β-unsaturated aldehydes **231** were tried (phenylethyl, benzoxypropyl, and 2-(*tert*-butoxycarbonylamino)ethyl), the process was efficient in forming the highly substituted hydroxypyrroles **232**. However, aromatic groups in this position were unsuccessful. The regioselectivity of the domino reaction in water and methods for the later synthesis of 1*H*-pyrrole were also reported [116].

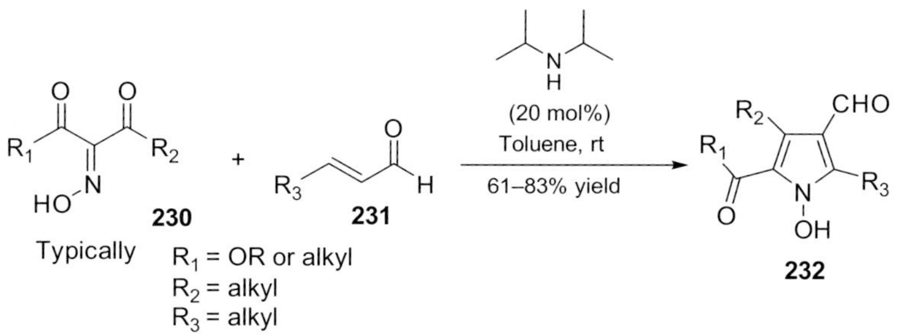

Scheme 7.49 Michael addition/aldol condensation to pyrroles.

A domino Mannich/aza-Michael reaction was applied to the synthesis of 2,5-cis-configured polysubstituted pyrrolidines from γ-malonate-substituted α,β-unsaturated esters with N-protected arylaldimines [117]. In this report, bifunctional thioureas were trialed with the Takemoto catalyst, being the most efficient with respect to yield as well as enantiomeric and diastereomeric excess. In a separate approach, the García–Tellado group approached the pyrrole ring system **234**, beginning with a tertiary skipped diyne **233** and a primary amine (Scheme 7.50). A 5-endo-digonal ring-cyclization step to **235** followed by a [3,3]-sigmatropic rearrangement forming **236** allowed the introduction of tetrasubstitution on these *N*-heterocycles in good yields through heating or microwave irradiation.

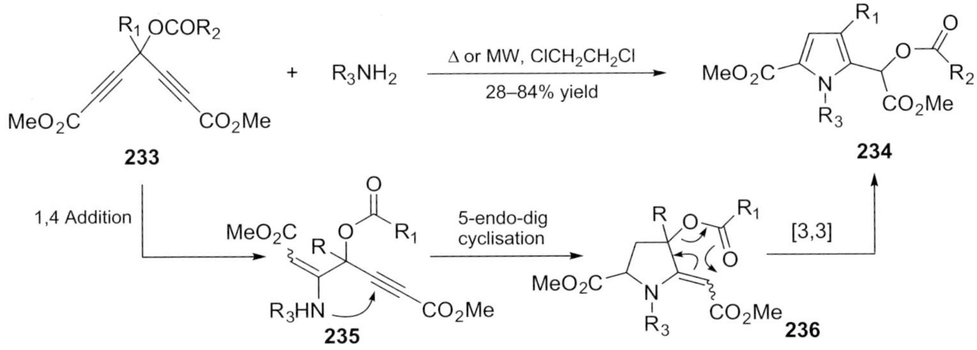

Scheme 7.50 Domino process to tetrasubstituted pyrroles.

Pyrrolidine-3-carboxylic acid (β-proline) were prepared in a diasteroselective manner through a domino Michael addition/carbocyclization reaction of ester **237** (Scheme 7.51) [118]. A strong salt effect controlled the relative stereoselectivity between the two newly formed carbon centers where the optimum conditions (organocopper/zinc mixed reagents prepared from of zinc salts (3 equiv) and saltfree organolithium compounds) led to a dr > 95 : 5 and 57% yield of **239** [119]. A piperidinone ring system was rapidly produced by a diastereoselective domino Mannich/Michael reaction using a galactosylamine as a chiral auxiliary (Scheme 7.52) [120]. With an enantiomeric synthesis of pure nupharamine natural

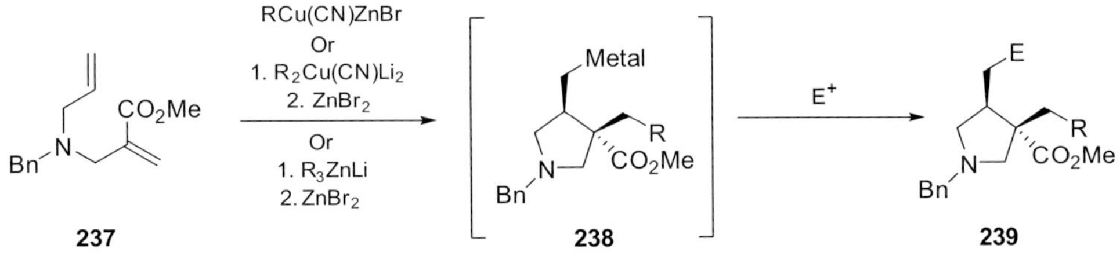

Scheme 7.51 Michael addition/carbocyclization reaction.

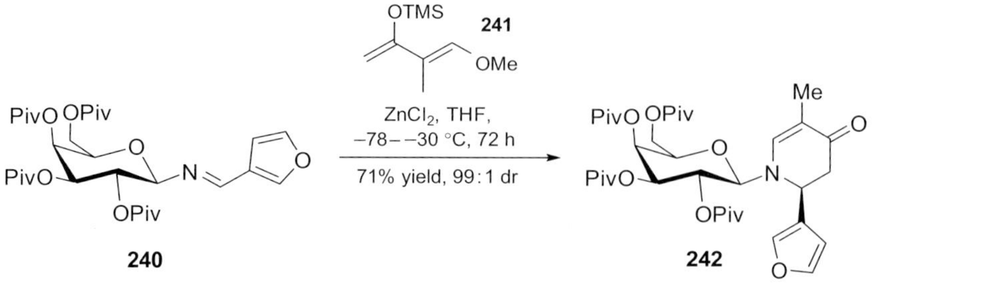

Scheme 7.52 Domino Mannich/Michael reaction.

product as the goal, this work in the early stages reports the reaction of *N*-galactosylimine **240** and the methyl-substituted Danishefsky's diene **241**. This reaction, sterically controlled by the 2-pivaloyloxy group (shielding the *re* face), forms the stereogenic center within heterocycle **242** in high selectivity. Later, a third ring is further constructed, forming the indolizidine natural product target.

In 2010, two independent studies involving Michael-incorporated domino reaction for the synthesis of multifunctionalized piperidines were reported. First, 2,6-disubstituted piperidines **245** were prepared through a domino imino-aldol/aza-Michael reaction, which involved preparing the enolate of α-arylmethylidene-β-keto ester **243** and reacting this with 2-aryl-*N*-sulfonylaldimines **244** (Scheme 7.53) [121]. Following the aldol reaction, the intermediate **246** was proposed to react from the *si* face as shown to eventually produce the piperidine diastereoselectively. This publication also details how enantiopure piperidines can be synthesized from chiral sulfinyl imines. The second and recent domino synthesis of piperidines **248** utilizes organocatalytic techniques to ensure enantiomeric excesses in the range 93–99% [122]. The one-pot diphenylprolinol silyl ether **1** catalyzed Michael reaction of crotonaldehyde and nitroalkene arrives at the intermediate **247**, which is followed by the an aza-Henry reaction/hemiaminalization process to a proposed alcohol intermediate, followed by Lewis acid-mediated allylation or cyanation (Scheme 7.54). A later report of polysubstituted piperidine **251** preparation was carried out through a Michael addition of **249** and **250** and an aminalization reaction process, remarkably in water [123]. This process was also carried out to form tetrahydropyrans

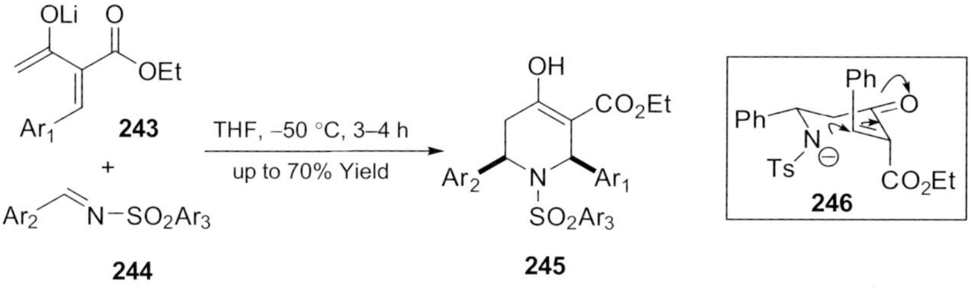

Scheme 7.53 Enantioselective domino synthesis of piperidines.

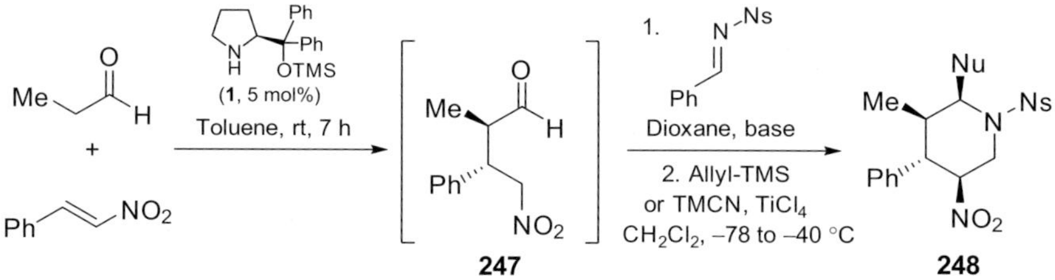

Scheme 7.54 Enantio- and diastereoselective synthesis of piperidines.

from hydroxymethyl-substituted nitroolefins and aldehydes (Scheme 7.55). Zhai and colleagues [124] have also contributed a Michael-type addition/iminium ion formation/cyclization to this field of domino reactions. By treating the enaminone **52** with acrolein, in the presence of the Lewis acid $BF_3{\cdot}OEt_2$ as reaction promoter, they observed a smooth formation of the pentacycle **254**, probably via the iminium ion **253** (Scheme 7.56).

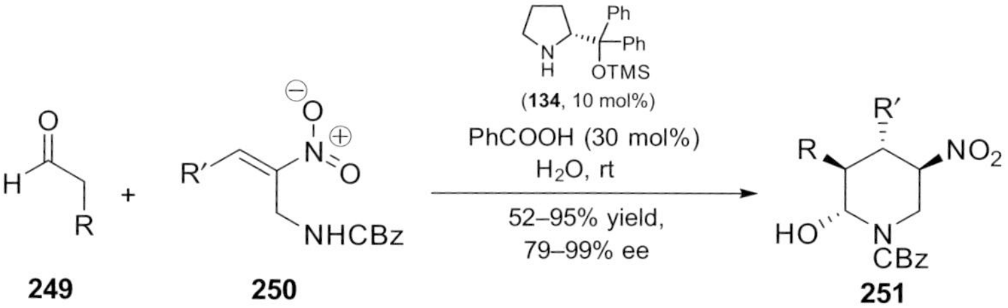

Scheme 7.55 Ma's synthesis of polysubstituted piperidines.

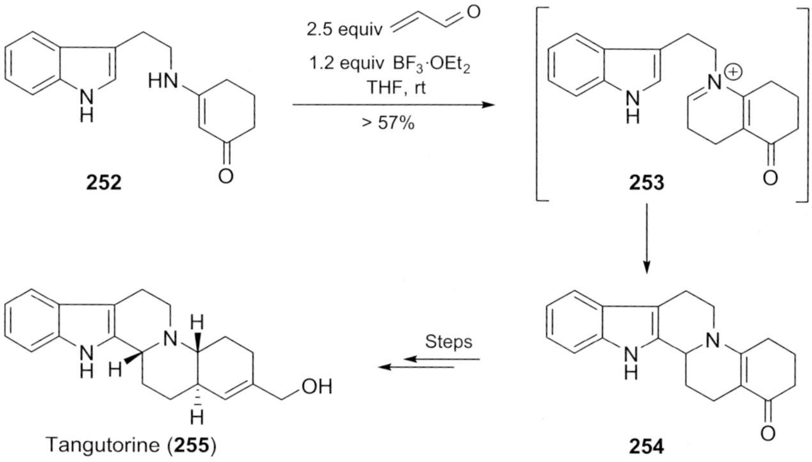

Scheme 7.56 Domino reaction applied to the synthesis of (±)-tangutorine.

Lewis base–Brønsted acid (LBBA) organocatalysts have found many uses in organic synthesis. One such example is the application in an aza-Morita–Baylis–Hillman (aza-MBH) reaction [125]. The group of Huang applied the organocatalyst 2'-(hydroxy-biphenyl-2-yl)-diphenylphosphane **4** to a domino reaction aza-MBH/Michael/aldol/dehydration domino sequence from imines **256** to tetrahydropyridines **257** (Scheme 7.57) [126]. The initial stages of the reaction process were proposed to involve phosphonium enolate formed between the catalyst **4** and acrolein. Likewise, enals **259** and enaminones **258** react to give tetrahydropyridin-2-ols **260** through an organocatalytic process (Scheme 7.58) [127]. In this example, the iminium activation mode is assumed trough reaction with secondary amines for the initial Michael addition. The domino reaction is concluded with a hemiaminalization process to the tetrahydropyridin-2-ol **260**.

A novel domino Heck/aza-Michael reaction has been developed and applied to the synthesis of a diverse set of 1-substituted tetrahydro-β-carbolines **262** (Scheme 7.59) [128]. This domino process involves a Heck reaction at the indole-2-position of a

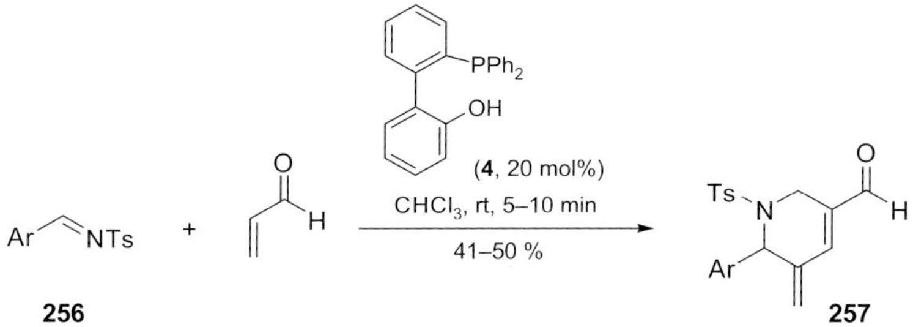

256 **257**

Scheme 7.57 Domino aza-Morita–Baylis–Hillman reaction.

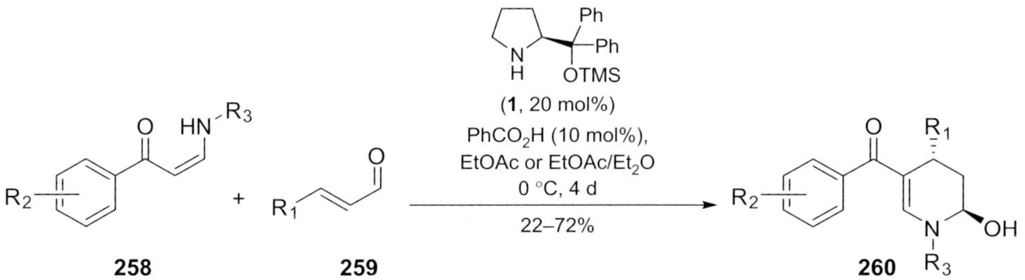

258 **259** **260**

Scheme 7.58 Domino reaction to tetrahydropyridin-2-ols.

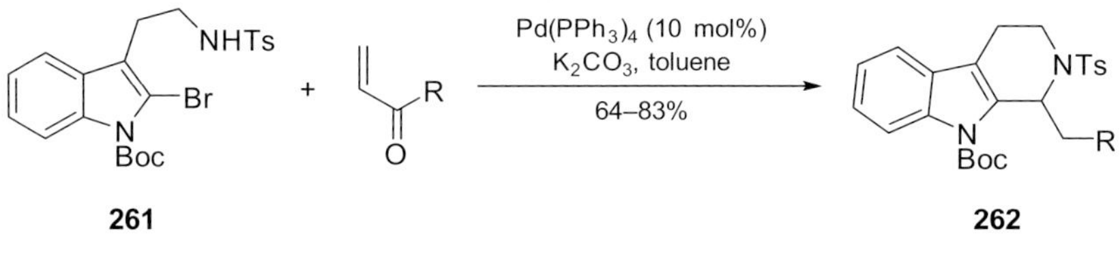

261 **262**

Scheme 7.59 Domino Heck/aza-Michael addition reaction.

C2 halogenated tryptamine precursor **261**, followed by intramolecular aza-Michael addition. In this example, by Stewart and Pfeffer, the order of these reaction was controlled by the type of protecting group on the N^{10} nitrogen, while a range of acrylate systems were reported in good to excellent yields. Others have also reported similar Heck/Michael processes [129–131].

The Barbas III group has synthesized a series of compounds bearing a bispirooxindole ring system using a single multifunctional organocatalyst **265** (Scheme 7.60). In this excellent example of a domino process leading to complex structures, the 3-substituted oxindole **263** reacts first with the Michael acceptor methyleneindolinones **264** followed by an intramolecular aldol reaction which is under the control of the *S*-binaphthyl diamine catalyst **265** containing a binaphthyl primary amine, a thiourea, and a tertiary amine moiety. The stereocontrol observed in a range of derivatives with four stereocenters, three of which are quaternary carbon stereocenters, is excellent [132].

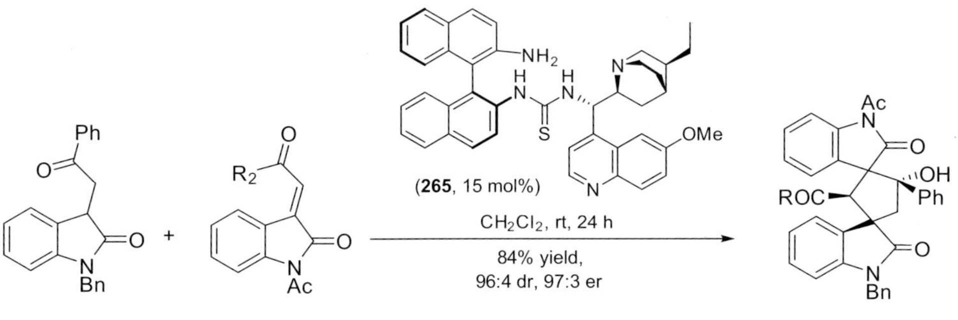

263 **264** **266**

Scheme 7.60 Domino Michael/aldol reaction to bispirooxindoles.

The spiro[4*H*-pyran-3,3′-oxindole] core was also prepared in excellent yields with high enantiomeric excesses through a domino Knoevenagel/Michael/cyclization process. This example, by the Yuan group, utilized the organocatalyst cupreine [133]. Moreover, in the same year, an asymmetric Michael/ketone aldol/dehydration domino reaction was also reported [134]. This process also concentrated on using cinchona-based primary catalysts such as **269** (Scheme 7.61). The organocatalyst in this example was proposed to initiate the enamine–iminium activation mode and produce the spiro oxindole **270** in excellent yields and enantiomeric excess. Spiro-chromanones bearing an additional *S*-heterocycle were prepared by Wang using a bifunctional urea complex through a domino thia-Michael/aldol domino reaction [135].

The group of Rueping has developed an enantioselective synthesis of chromenones through a Michael reaction and cyclization process between diketones and unsaturated aldehydes [136]. Furthermore, this reaction can be also followed with the addition of tryptamine to afford indoloquinolizidines [137]. An aza-MBH reaction was used in the enantioselective synthesis of isoindolines. In this

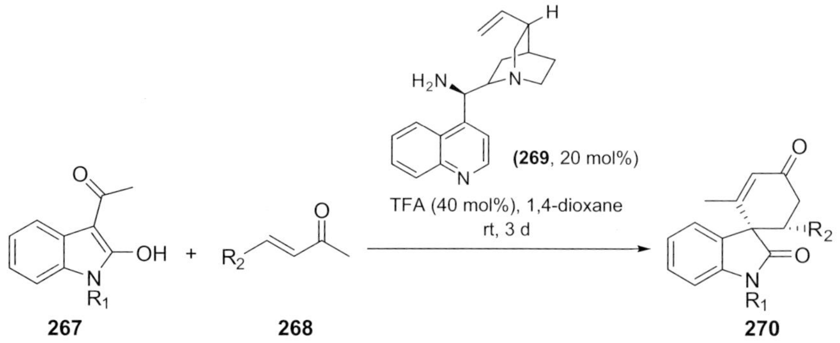

Scheme 7.61 Domino Michael/ketone aldol/dehydration process.

case, Sasai *et al.* [138] prepared (*S*)-2-diphenylphosphanyl-[1,1']binaphthalenyl-2-ol **273** as the organocatalyst, containing both a Lewis base and Brønsted acid moiety. In the proposed pathway, an initial reaction of chiral enolate **275** (stabilized by the Brønsted acid) and *N*-tosylimine **272** is proposed for the intermediate **276** (Scheme 7.62). Following this, the proposed mechanism implies an aza-MBH reaction step to give **277**, followed by a proton transfer from the α position of the carbonyl group and a retro-Michael reaction. This process affords the desired products **274** and regenerates the organocatalyst **273**. Typically for this process, the yields range from 66 to 98%, with 68–93% ee.

An aza-Michael/aldol reaction analogous to the oxa-Michael process discussed earlier was used in the synthesis of 1,2-dihydroquinolidines **279** [139]. In this example for dihydroquinolines, phenyl or alkyl α,β-unsaturated aldehydes **278**

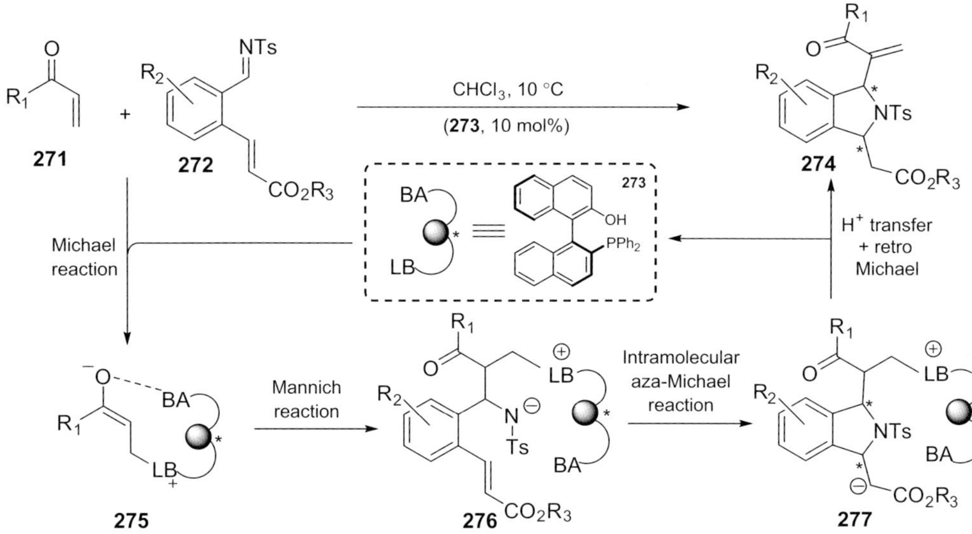

Scheme 7.62 Enantioselective domino aza-MBH/intramolecular aza-Michael reaction.

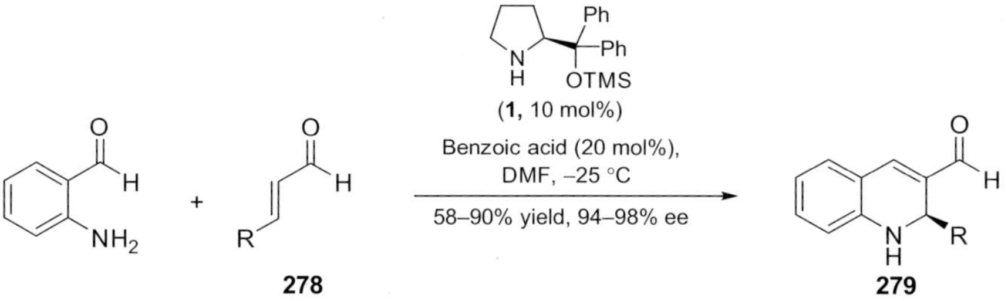

Scheme 7.63 Domino aza-Michael–aldol reaction.

were initially successful (Scheme 7.63). This work was confirmed by a short study on Ar-substituted 2-aminobenzaldehyde with ester substituted aldehydes. In another study, 3-nitro-1,2-dihydroquinolines were prepared by Xu [140] through a domino aza-Michael/Henry reaction of 2-aminobenzaldehyde with nitrostyrene in the presence of thiourea and benzoic acid as catalyst. Furthermore, in an earlier work, Yadav and coworkers observed diastereoselective domino Michael addition/Friedel–Crafts/S_N-type cyclizations, which allowed the synthesis of bridged tetrahydroquinolines such as **282** (Scheme 7.64). In this example, a mixture of anilines **280** and δ-hydroxy-α,β-unsaturated aldehydes **281** were treated with catalytic amounts of the mild $Bi(OTf)_3$ or $InCl_3$ Lewis acid to provide the structurally complex **282** [141]. An asymmetric domino aza-Michael/Mannich reaction was also used in the synthesis of tetrahydroimidazopyrimidine [142]. In this case, the *N*-arylidene-1*H*-imidazol-2-amines were stirred with unsaturated aldehydes in the presence of an organocatalyst and benzoic acid (see Chapter 10 for further details).

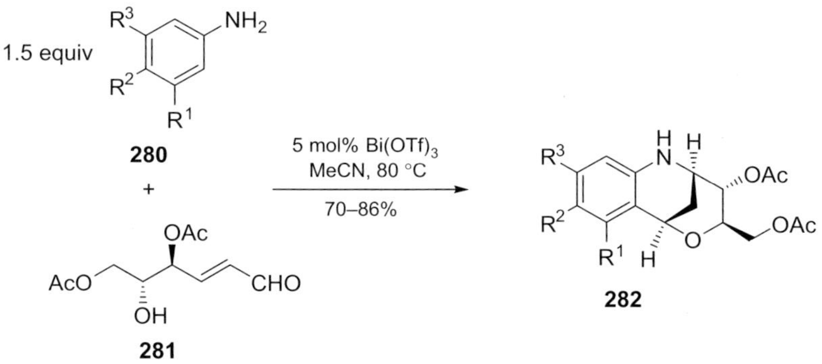

Scheme 7.64 Domino Michael addition/Friedel–Crafts/S_N-type cyclizations.

7.6 Formation of *S*-Heterocycles

In the field of constructing *S*-heterocycles, a reaction involving a low loading of Takemoto's catalyst **3** was used in the synthesis of succinimide-containing thiochromanes **285** containing three chiral centers in good to high enantiomeric excesses (Scheme 7.65) [143]. The Michael/aldol reactions between 2-mercaptobenzaldehydes **283** and maleimides **285** also proceeded in good yields and good diastereomeric ratio. In another study, Córdova *et al.* reported a domino thia-Michael/aldol reaction between 2-mercaptoacetophenone and various enals **286** (Scheme 7.66) [144]. In this case, the Jørgensen-designed catalyst **61** performed well, and, when 2-nitro-benzoic acid was added, the reaction performed at low temperature, giving high diastereomeric ratios, enantiomeric excesses, and yields. This reaction could be applied through a range of unsaturated aldehydes (e.g., R = Ph, CO_2Et, Alkyl). Similarly, the group of Wang approached a series of thiochromane ring systems through a domino organocatalytic and a symmetric sulfa-Michael/aldol reaction [145]. Following extensive methodology using a series of bifunctional amine-thiourea organocatalysts, a cinchona alkaloid-derived candidate **289** was used (Scheme 7.67). Treatment of the 2-mercaptobenzaldehyde with β-alkyl substituted α,β-unsaturated *N*-pyrazole **288** in the presence of **289** resulted in the desired thiochromane **289** in good yields and excellent diastereomeric ratio and enantiomeric excess. The work could be extended to β-alkyl-substituted α,β-unsaturated *N*-pyrazoles. Interestingly, several biologically active thiochromane ring systems exist, in an obvious impetus for this work.

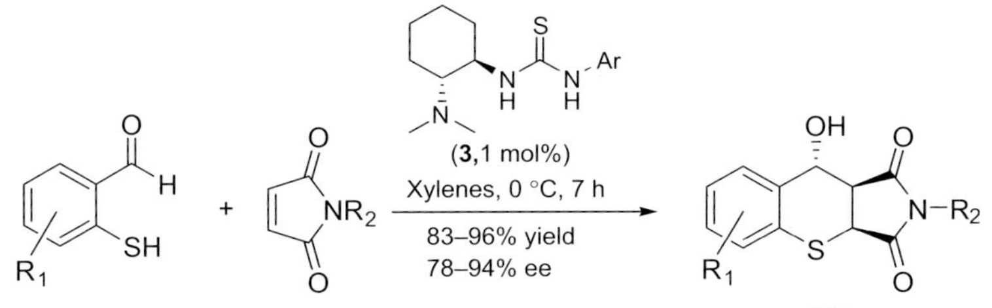

283 **284** **285**

Scheme 7.65 Domino Michael/aldol reaction to succinimide-benzothiopyrans.

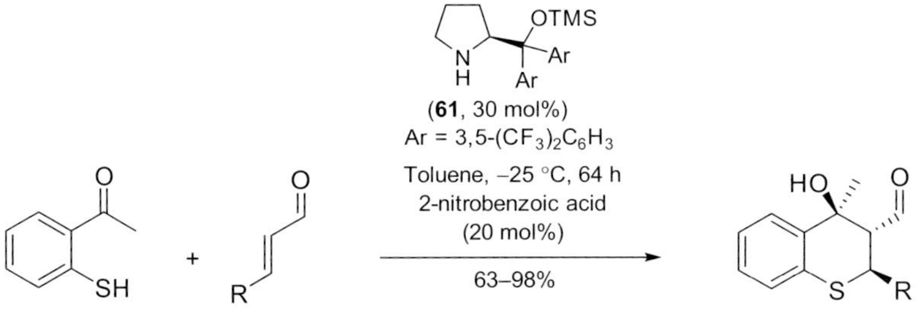

286 **287**

Scheme 7.66 Domino thia-Michael/aldol reaction.

The process of enamine formation/aldol condensation/Michael addition/6-exo-trig cyclization/elimination cyclization was used to prepare thienothiopyrans. As

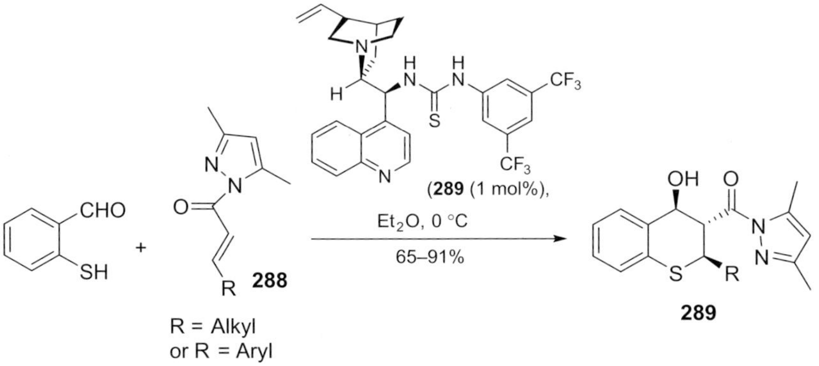

Scheme 7.67 Asymmetric domino sulfa-Michael/aldol reaction.

an example for the formation of this ring system, four stereocenters were produced in one process beginning with ethyl 2-[(2-oxo-2-arylethyl)sulfonyl]acetate or ethyl 2-[(2-ethoxy-2-oxoethyl)sulfonyl]acetate, aromatic aldehydes, and 5-aryltetrahydro-3-thiophenone [146]. The dihydrothiophene ring system was also prepared in high yields and excellent enantioselectivities with an organocatalytic domino thia-Michael/aldol condensation from a 1,4-dithiane-2,5-diol precursor [147]. Organocatalytic domino reactions were also used to prepare chiral tetrahydrothiophenes **291** through a Michael/aldol domino reaction. In this reaction, the (*E*)-methylpentenals **290** and mercaptophenylethanone were stirred in the presence of an organocatalyst **61** and benzoic acid in toluene (Scheme 7.68) [148]. As in other similar domino reactions, an iminium ion species between **290** and **61** is proposed prior to the thia-Michael process. Using a range of R groups, the yields ranged from 44 to 74%, while the enantiomeric excesses were consistently high at >90%. Following this work, a domino double Michael addition was used to prepare similar ring systems **293** starting from α,β-unsaturated aldehydes **292** with ethyl 4-mercapto-2-butenoate (Scheme 7.69, also see Chapter 8 for the scope of the substrate (R)). The excellent enantiomeric excess and yields, for a range of aldehydes, for this transformation is a salient feature of this work [149]. Thieno[2,3-*b*]thiopyran-4-ones have also been prepared in a one-pot HWE olefination and a base-induced intramolecular domino

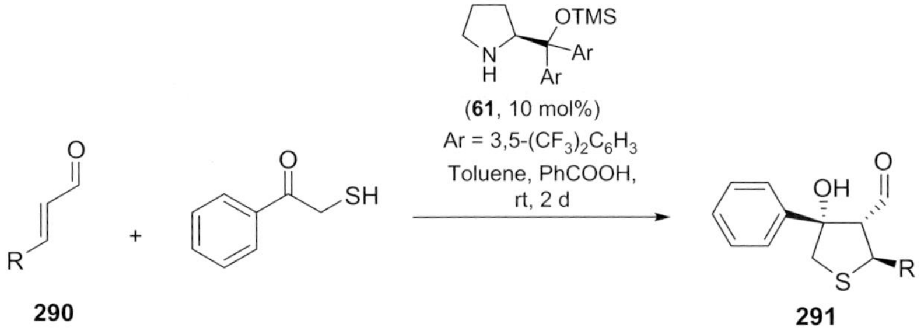

Scheme 7.68 Domino approach to prepare chiral tetrahydrothiophenes.

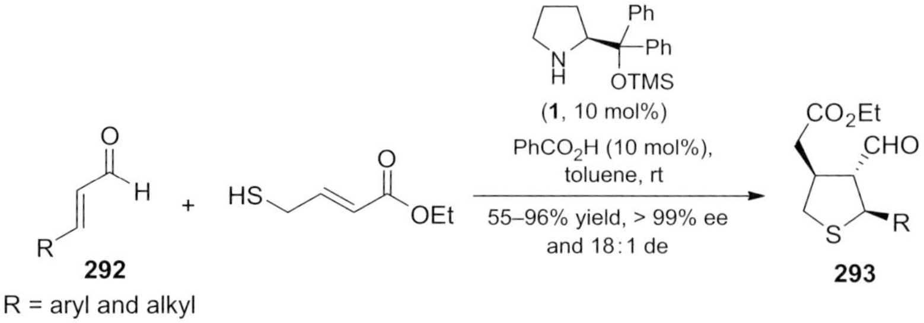

Scheme 7.69 Wang's domino double Michael addition to tetrahydrothiophenes.

cyclization/thio-Michael addition [150]. A domino thia-Michael/Michael process also has been applied in the preparation of tetrahydrothiophenes [151].

7.7
Formation of Heterocycles Containing Nitrogen and Oxygen

A chiral bicyclophosphinate ring system **297** was prepared through a hydrophosphinylation/Michael/Michael reaction combination, beginning with the electrophilic and nucleophilic allenyl *H*-phosphinates **294** (Scheme 7.70) [152]. In this manner, Virieux *et al.* were able to first generate the key intermediate **295** for a double Michael process with a nitrogen- and oxygen-based nucleophile. The yields for this domino process were moderate to good; however, low double diastereo-differentiation was also observed.

Brière and coworkers gained rapid access to the pyrazolines ring system **300** through a Michael-based domino reaction sequence. This enantioselective process

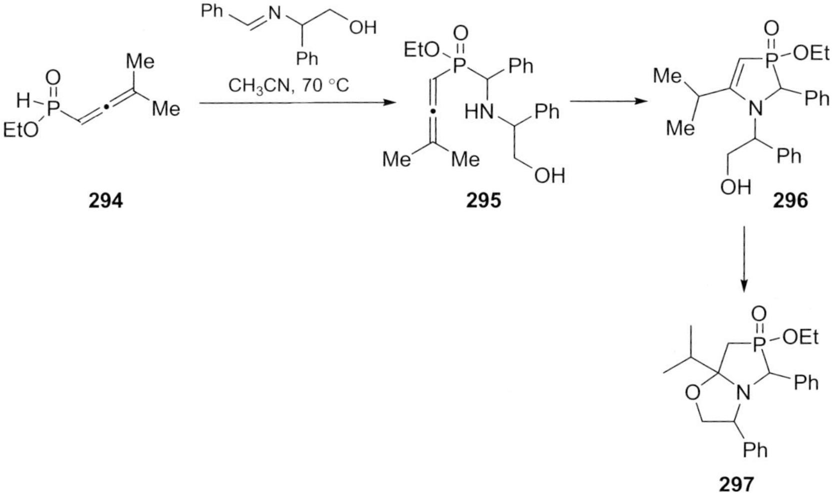

Scheme 7.70 Domino hydrophosphinylation/Michael/Michael reaction.

between chalcone **298** and *N-tert*-butyloxycarbonyl hydrazine was overseen by the presence of base and chiral ammonium salts derived from cinchona alkaloids (Scheme 7.71). These catalysts were termed *enantioselective phase-transfer catalysts*. Following catalyst optimization, the derived **299** was found to be the most enantioselective in this domino aza-Michael addition/cyclo-condensation reaction [153].

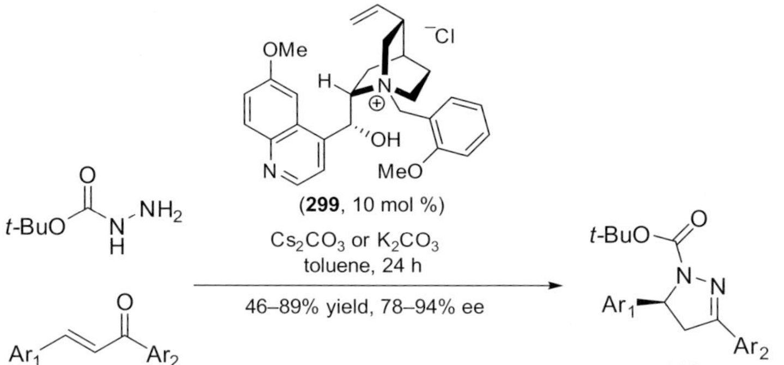

298 **300**

Scheme 7.71 Brière's synthesis of pyrazolines.

A solid-supported carbonate reagent was used to initiate the formation of 2*H*-1,4-benzoxazin-2-ones **303** from β-nitroacrylate **301** and 2-aminophenol **302** (Scheme 7.72) [154]. This hetero-Michael addition/intramolecular transesterification/elimination of HNO_3/[1,3]-proton shift domino reaction proceeded in moderate to excellent yields. In another study, a series of various functionalized pyrrolo[1,2-*r*]pyrazines **305** were made synthetically available through the reaction of vinyl azides **304** and 1*H*-2-pyrrolecarbaldehyde (Scheme 7.73). The deprotonated pyrrole initially reacts in a Michael manner with the vinyl azides, with nitrogen as

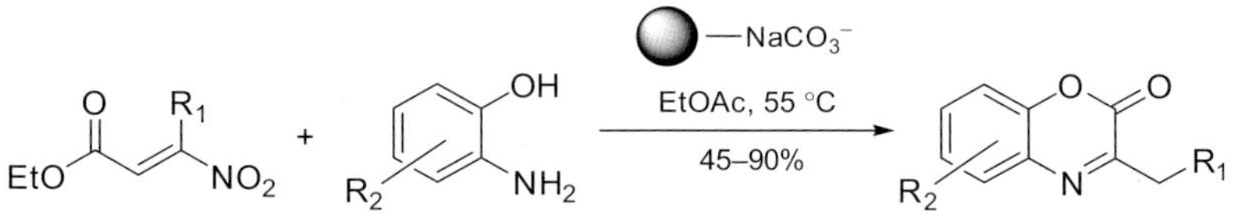

301 **302** **303**

Scheme 7.72 Solid support applied in domino reactions to benzoxazin-2-ones.

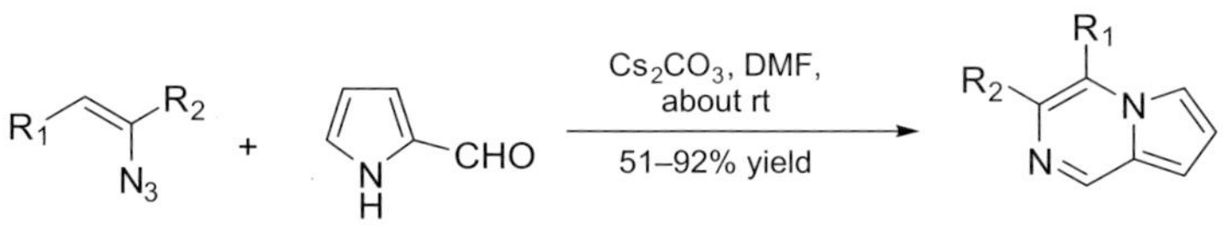

304 **305**

Scheme 7.73 Michael addition intramolecular condensation reaction.

the leaving group. The sequence is then followed by an intramolecular condensation to the bicyclic ring system **305**.

A domino Michael addition/S_N sequence has been elaborated by the group of de Meijere [155]. In this example, basic treatment of 2-chloro-2-cyclopropylidenacetates **306** with carboxamides **307** forms 4-spirocyclopropane annulated oxazoline-5-carboxylates **310** (Scheme 7.74). As intermediates, the carbanion **308** and anion **309** have been proposed.

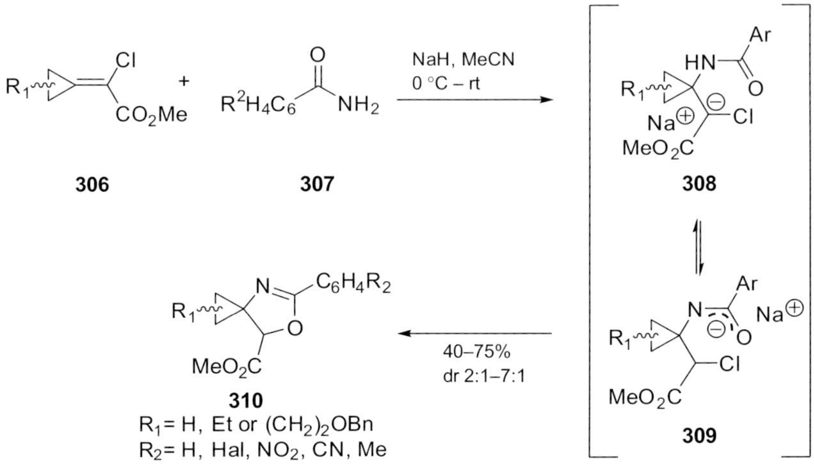

Scheme 7.74 de Meijere's domino synthesis of oxazoline-5-carboxylates.

References

1. Michael, A. (1887) *J. Prak. Chem.*, **35**, 349–356.
2. Michael, A. (1894) *J. Prak. Chem.*, **49**, 20–25.
3. Bui, T. and Barbas, C.F. III, (2000) *Tetrahedron Lett.*, **41**, 6951–6954.
4. Gryko, D. (2005) *Tetrahedron: Asymmetry*, **16**, 1377–1383.
5. Halland, N., Aburel, P.S., and Jørgensen, K.A. (2004) *Angew. Chem. Int. Ed.*, **43**, 1272–1277.
6. Yang, J.W., Hechavarria Fonseca, M.T., and List, B. (2005) *J. Am. Chem. Soc.*, **127**, 15036–15037.
7. Enders, D., Grondal, C., and Hüttl, M.R.M. (2007) *Angew. Chem. Int. Ed.*, **46**, 1570–1581.
8. Grondal, C., Jeanty, M., and Enders, D. (2010) *Nat. Chem.*, **2**, 167–178.
9. Ahrendt, K.A., Borths, C.J., and MacMillan, D.W.C. (2000) *J. Am. Chem. Soc.*, **122**, 4243–4244.
10. Northrup, A.B. and MacMillan, D.W.C. (2002) *J. Am. Chem. Soc.*, **124**, 6798–6799.
11. Pellissier, H. (2012) *Adv. Synth. Catal.*, **354**, 237–294.
12. Kaneko, S., Yoshino, T., Katoh, T., and Terashima, S. (1998) *Tetrahedron*, **54**, 5471–5484.
13. Hagiwara, H., Kobayashi, K., Miya, S., Hoshi, T., Suzuki, T., Ando, M., Okamoto, T., Kobayashi, M., Yamamoto, I., Ohtsubo, S., Kato, M., and Uda, H. (2002) *J. Org. Chem.*, **67**, 5969–5976.
14. Kaneko, S., Nakajima, N., Shikano, M., Katoh, T., and Terashima, S. (1998) *Tetrahedron*, **54**, 5485–5506.
15. Steven, V.L., Brown, D.S., Andrew Clase, J., Fairbanks, A.J., Lennon, I.C., Osborn, H.M.I., StokeséOwen, E.S.E., and Wadsworth,

D.J. (1998) *J. Chem. Soc., Perkin Trans. 1*, 2259–2276.
16. Nicolai, C., Sabine, L., Angelika, B., Harald, S., and Christian, R. (2005) *Angew. Chem. Int. Ed.*, **44**, 820–822.
17. Stark, L.M., Pekari, K., and Sorensen, E.J. (2004) *Proc. Natl. Acad. Sci. U.S.A.*, **101**, 12064–12066.
18. Austin, J.F., Kim, S.-G., Sinz, C.J., Xiao, W.-J., and MacMillan, D.W.C. (2004) *Proc. Natl. Acad. Sci. U.S.A.*, **101**, 5482–5487.
19. Watanabe, H., Onoda, T., and Kitahara, T. (1999) *Tetrahedron Lett.*, **40**, 2545–2548.
20. Yamada, K.-I., Arai, T., Sasai, H., and Shibasaki, M. (1998) *J. Org. Chem.*, **63**, 3666–3672.
21. Kant, J., Roth, J.A., Fuller, C.E., Walker, D.G., Benigni, D.A., and Farina, V. (1994) *J. Org. Chem.*, **59**, 4956–4966.
22. Cook, C., Liron, F., Guinchard, X., and Roulland, E. (2012) *J. Org. Chem.*, **77**, 6728–6742.
23. Brown, P.D., Willis, A.C., Sherburn, M.S., and Lawrence, A.L. (2012) *Org. Lett.*, **14**, 4537–4539.
24. Tatsuta, K., Tanaka, Y., Kojima, M., and Ikegami, H. (2002) *Chem. Lett.*, **31**, 14–15.
25. Tietze, L.F., Brasche, G., and Gericke, K. (2006) *Domino Reactions in Organic Synthesis*, 1st edn, Wiley-VCH Verlag GmbH, Weinheim.
26. Han, Z., Uehira, S., Shinokubo, H., and Oshima, K. (2001) *J. Org. Chem.*, **66**, 7854–7857.
27. Yagi, K., Turitani, T., Shinokubo, H., and Oshima, K. (2002) *Org. Lett.*, **4**, 3111–3114.
28. Huang, Y., Walji, A.M., Larsen, C.H., and MacMillan, D.W.C. (2005) *J. Am. Chem. Soc.*, **127**, 15051–15053.
29. Marigo, M., Schulte, T., Franzén, J., and Jørgensen, K.A. (2005) *J. Am. Chem. Soc.*, **127**, 15710–15711.
30. Hili, R. and Yudin, A.K. (2009) *J. Am. Chem. Soc.*, **131**, 16404–16406.
31. Kamimura, A., Okawa, H., Morisaki, Y., Ishikawa, S., and Uno, H. (2007) *J. Org. Chem.*, **72**, 3569–3572.
32. Garrido, N.M., García, M., Díez, D., Sánchez, M.R., Sanz, F., and Urones, J.G. (2008) *Org. Lett.*, **10**, 1687–1690.
33. Aydillo, C., Jiménez-Osés, G., Avenoza, A., Busto, J.S.H., Peregrina, J.S.M., and Zurbano, M.A.M. (2011) *J. Org. Chem.*, **76**, 6990–6996.
34. Arnold, L.A., Naasz, R., Minnaard, A.J., and Feringa, B.L. (2002) *J. Org. Chem.*, **67**, 7244–7254.
35. Ogiku, T., Yoshida, S.-I., Kuroda, T., Takahashi, M., Ohmizu, H., and Iwasaki, T. (1992) *Bull. Chem. Soc. Jpn.*, **65**, 3495–3497.
36. Pelter, A., Ward, R.S., and Storer, N.P. (1994) *Tetrahedron*, **50**, 10829–10838.
37. Ibrahem, I., Zhao, G.-L., Rios, R., Vesely, J., Sundén, H., Dziedzic, P., and Córdova, A. (2008) *Chem. Eur. J.*, **14**, 7867–7879.
38. Lv, J., Zhang, J., Lin, Z., and Wang, Y. (2009) *Chem. Eur. J.*, **15**, 972–979.
39. Inoue, H., Nagaoka, Y., and Tomioka, K. (2002) *J. Org. Chem.*, **67**, 5864–5867.
40. Cauble, D.F., Gipson, J.D., and Krische, M.J. (2003) *J. Am. Chem. Soc.*, **125**, 1110–1111.
41. Enders, D., Wang, C., and Bats, J.W. (2008) *Angew. Chem. Int. Ed.*, **47**, 7539–7542.
42. Hong, B.-C., Dange, N.S., Hsu, C.-S., Liao, J.-H., and Lee, G.-H. (2011) *Org. Lett.*, **13**, 1338–1341.
43. Hong, B.-C., Dange, N.S., Hsu, C.-S., and Liao, J.-H. (2010) *Org. Lett.*, **12**, 4812–4815.
44. Tan, B., Shi, Z., Chua, P.J., and Zhong, G. (2008) *Org. Lett.*, **10**, 3425–3428.
45. Tan, B., Chua, P.J., Zeng, X., Lu, M., and Zhong, G. (2008) *Org. Lett.*, **10**, 3489–3492.
46. Ma, A. and Ma, D. (2010) *Org. Lett.*, **12**, 3634–3637.
47. Paulsen, H., Antons, S., Brandes, A., Lögers, M., Müller, S.N., Naab, P., Schmeck, C., Schneider, S., and Stoltefuß, J. (1999) *Angew. Chem. Int. Ed.*, **38**, 3373–3375.
48. Reyes, E., Jiang, H., Milelli, A., Elsner, P., Hazell, R.G., and Jørgensen, K.A. (2007) *Angew. Chem. Int. Ed.*, **46**, 9202–9205.

49. Hayashi, Y., Okano, T., Aratake, S., and Hazelard, D. (2007) *Angew. Chem. Int. Ed.*, **46**, 4922–4925.
50. Chintala, P., Ghosh, S.K., Long, E., Headley, A.D., and Ni, B. (2011) *Adv. Synth. Catal.*, **353**, 2905–2909.
51. Baslé, O., Raimondi, W., Duque, M.S., Bonne, D., Constantieux, T., and Rodriguez, J. (2010) *Org. Lett.*, **12**, 5246–5249.
52. Enders, D., Hüttl, M.R.M., Grondal, C., and Raabe, G. (2006) *Nature*, **441**, 861–863.
53. Rueping, M., Haack, K., Ieawsuwan, W., Sunden, H., Blanco, M., and Schoepke, F.R. (2011) *Chem. Commun.*, **47**, 3828–3830.
54. Zhang, F.-L., Xu, A.-W., Gong, Y.-F., Wei, M.-H., and Yang, X.-L. (2009) *Chem. Eur. J.*, **15**, 6815–6818.
55. Alza, E., Sayalero, S., Cambeiro, X.C., Martín-Rapún, R., Miranda, P.O., and Pericàs, M.A. (2011) *Synlett*, **2011**, 464–468.
56. Ishikawa, H., Suzuki, T., and Hayashi, Y. (2009) *Angew. Chem. Int. Ed.*, **48**, 1304–1307.
57. Ishikawa, H., Suzuki, T., Orita, H., Uchimaru, T., and Hayashi, Y. (2010) *Chem. Eur. J.*, **16**, 12616–12626.
58. Weng, J., Li, J.-M., Li, F.-Q., Xie, Z.-S., and Lu, G. (2012) *Adv. Synth. Catal.*, **354**, 1961–1970.
59. Enders, D., Wang, C., Mukanova, M., and Greb, A. (2010) *Chem. Commun.*, **46**, 2447–2449.
60. Marigo, M., Bertelsen, S., Landa, A., and Jørgensen, K.A. (2006) *J. Am. Chem. Soc.*, **128**, 5475–5479.
61. Xia, A.-B., Xu, D.-Q., Luo, S.-P., Jiang, J.-R., Tang, J., Wang, Y.-F., and Xu, Z.-Y. (2010) *Chem. Eur. J.*, **16**, 801–804.
62. Albrecht, T., Richter, B., Vila, C., Krawczyk, H., and Jørgensen, K.A. (2009) *Chem. Eur. J.*, **15**, 3093–3102.
63. Hayashi, Y., Toyoshima, M., Gotoh, H., and Ishikawa, H. (2008) *Org. Lett.*, **11**, 45–48.
64. Bouhet, J.C., Pham Van Chuong, P., Toma, F., Kirszenbaum, M., and Fromageot, P. (1976) *J. Agric. Food Chem.*, **24**, 964–972.
65. Nicolaou, K.C., Yee Hwee, L., Charles, D.P., and Jared, L.P. (2005) *Angew. Chem. Int. Ed.*, **44**, 7917–7921.
66. Sánchez-Larios, E. and Gravel, M. (2009) *J. Org. Chem.*, **74**, 7536–7539.
67. Sánchez-Larios, E., Holmes, J.M., Daschner, C.L., and Gravel, M. (2010) *Org. Lett.*, **12**, 5772–5775.
68. Hong, B.-C., Hsu, C.-S., and Lee, G.-H. (2012) *Chem. Commun.*, **48**, 2385–2387.
69. Alexakis, A. and March, S. (2002) *J. Org. Chem.*, **67**, 8753–8757.
70. McGarraugh, P.G. and Brenner, S.E. (2009) *Org. Lett.*, **11**, 5654–5657.
71. Padwa, A., Murphree, S.S., Ni, Z., and Watterson, S.H. (1996) *J. Org. Chem.*, **61**, 3829–3838.
72. Hagiwara, H., Sakai, H., Kirita, M., Hoshi, T., Suzuki, T., and Ando, M. (2000) *Tetrahedron*, **56**, 1445–1449.
73. Gutke, H.-J., Oesterreich, K., Spitzner, D., and Braun, N.A. (2001) *Tetrahedron*, **57**, 997–1004.
74. Ihara, M., Makita, K., Tokunaga, Y., and Fukumoto, K. (1994) *J. Org. Chem.*, **59**, 6008–6013.
75. Tan, B., Lu, Y., Zeng, X., Chua, P.J., and Zhong, G. (2010) *Org. Lett.*, **12**, 2682–2685.
76. Rueping, M., Kuenkel, A., Tato, F., and Bats, J.W. (2009) *Angew. Chem. Int. Ed.*, **48**, 3699–3702.
77. Rueping, M., Kuenkel, A., and Fröhlich, R. (2010) *Chem. Eur. J.*, **16**, 4173–4176.
78. Pettus, T.R.R., Inoue, M., Chen, X.-T., and Danishefsky, S.J. (2000) *J. Am. Chem. Soc.*, **122**, 6160–6168.
79. Shen, J., Liu, D., An, Q., Liu, Y., and Zhang, W. (2012) *Adv. Synth. Catal.*, **354**, 3311–3325.
80. Enders, D., Grossmann, A., Gieraths, B., Düzdemir, M., and Merkens, C. (2012) *Org. Lett.*, **14**, 4254–4257.
81. Wang, C., Yang, X., Raabe, G., and Enders, D. (2012) *Adv. Synth. Catal.*, **354**, 2629–2634.
82. Enders, D., Wang, C., Yang, X., and Raabe, G. (2010) *Adv. Synth. Catal.*, **352**, 2869–2874.
83. Hong, B.-C., Kotame, P., Tsai, C.-W., and Liao, J.-H. (2010) *Org. Lett.*, **12**, 776–779.

84. Adibekian, A., Timmer, M.S.M., Stallforth, P., van Rijn, J., Werz, D.B., and Seeberger, P.H. (2008) *Chem. Commun.*, 3549–3551.
85. Tozawa, T., Nagao, H., Yamane, Y., and Mukaiyama, T. (2007) *Chem. Asian J.*, **2**, 123–134.
86. Carreño, M.C., Luzón, C.G., and Ribagorda, M. (2002) *Chem. Eur. J.*, **8**, 208–216.
87. Fang, X., Jiang, K., Xing, C., Hao, L., and Chi, Y.R. (2011) *Angew. Chem. Int. Ed.*, **50**, 1910–1913.
88. Beifuss, U., Goldenstein, K., Döring, F., Lehmann, C., and Noltemeyer, M. (2001) *Angew. Chem. Int. Ed.*, **40**, 568–570.
89. Rueping, M., Sundén, H., and Sugiono, E. (2012) *Chem. Eur. J.*, **18**, 3649–3653.
90. Ren, Q., Gao, Y., and Wang, J. (2011) *Org. Biomol. Chem.*, **9**, 5297–5302.
91. Ren, Q., Gao, Y., and Wang, J. (2010) *Chem. Eur. J.*, **16**, 13594–13598.
92. Appel, B., Saleh, N.N.R., and Langer, P. (2006) *Chem. Eur. J.*, **12**, 1221–1236.
93. Sundén, H., Ibrahem, I., Zhao, G.-L., Eriksson, L., and Córdova, A. (2007) *Chem. Eur. J.*, **13**, 574–581.
94. Li, H., Wang, J., E-Nunu, T., Zu, L., Jiang, W., Wei, S., and Wang, W. (2007) *Chem. Commun.*, 507–509.
95. Rueping, M., Dufour, J., and Maji, M.S. (2012) *Chem. Commun.*, **48**, 3329–3331.
96. Shestopalov, A.M., Litvinov, Y.M., Rodinovskaya, L.A., Malyshev, O.R., Semenova, M.N., and Semenov, V.V. (2012) *ACS Comb. Sci.*, **14**, 484–490.
97. Phillips, E.M., Wadamoto, M., Roth, H.S., Ott, A.W., and Scheidt, K.A. (2008) *Org. Lett.*, **11**, 105–108.
98. Lesch, B., Toräng, J., Vanderheiden, S., and Bräse, S. (2005) *Adv. Synth. Catal.*, **347**, 555–562.
99. Lesch, B., Toräng, J., Nieger, M., and Bräse, S. (2005) *Synthesis*, **2005**, 1888–1900.
100. Nising, C.F., Ohnemüller, U.K., and Bräse, S. (2006) *Angew. Chem. Int. Ed.*, **45**, 307–309.
101. Gérard, E.M.C. and Bräse, S. (2008) *Chem. Eur. J.*, **14**, 8086–8089.
102. Bröhmer, M.C., Bourcet, E., Nieger, M., and Bräse, S. (2011) *Chem. Eur. J.*, **17**, 13706–13711.
103. Liu, K., Chougnet, A., and Woggon, W.-D. (2008) *Angew. Chem. Int. Ed.*, **47**, 5827–5829.
104. Sun, J., Zhang, L.-L., Xia, E.-Y., and Yan, C.-G. (2009) *J. Org. Chem.*, **74**, 3398–3401.
105. Hong, B.-C., Kotame, P., and Lee, G.-H. (2011) *Org. Lett.*, **13**, 5758–5761.
106. Kittivarapong, S., Rajviroonkit, S., Siwapinyoyos, T., Thebtaranonth, C., and Thebtaranonth, Y. (1995) *Synthesis*, **1995**, 136–138.
107. Dou, X., Han, X., and Lu, Y. (2012) *Chem. Eur. J.*, **18**, 85–89.
108. Rueping, M., Parra, A., Uria, U., Besselièvre, F.O., and Merino, E.B. (2010) *Org. Lett.*, **12**, 5680–5683.
109. Wang, L. and Menche, D. (2012) *J. Org. Chem.*, **77**, 10811–10823.
110. Bolla, M.L., Patterson, B., and Rychnovsky, S.D. (2005) *J. Am. Chem. Soc.*, **127**, 16044–16045.
111. Chandrasekhar, S., Mallikarjun, K., Pavankumarreddy, G., Rao, K.V., and Jagadeesh, B. (2009) *Chem. Commun.*, 4985–4987.
112. Boddaert, T., Coquerel, Y., and Rodriguez, J. (2011) *Chem. Eur. J.*, **17**, 2266–2271.
113. Coquerel, Y., Filippini, M.-H., Bensa, D., and Rodriguez, J. (2008) *Chem. Eur. J.*, **14**, 3078–3092.
114. Reyes, E., Talavera, G., Vicario, J.L., Badía, D., and Carrillo, L. (2009) *Angew. Chem. Int. Ed.*, **48**, 5701–5704.
115. Hoashi, Y., Yabuta, T., and Takemoto, Y. (2004) *Tetrahedron Lett.*, **45**, 9185–9188.
116. Tan, B., Shi, Z., Chua, P.J., Li, Y., and Zhong, G. (2009) *Angew. Chem. Int. Ed.*, **48**, 758–761.
117. Enders, D., Göddertz, D.P., Beceño, C., and Raabe, G. (2010) *Adv. Synth. Catal.*, **352**, 2863–2868.
118. Denes, F., Perez-Luna, A., and Chemla, F. (2006) *J. Org. Chem.*, **72**, 398–406.
119. Denes, F., Chemla, F., and Normant, J.F. (2002) *Eur. J. Org. Chem.*, **2002**, 3536–3542.
120. Stoye, A., Quandt, G., Brunnhöfer, B., Kapatsina, E., Baron, J., Fischer, A.,

Weymann, M., and Kunz, H. (2009) *Angew. Chem. Int. Ed.*, **48**, 2228–2230.
121. Ghorai, M.K., Halder, S., and Das, R.K. (2010) *J. Org. Chem.*, **75**, 7061–7072.
122. Urushima, T., Sakamoto, D., Ishikawa, H., and Hayashi, Y. (2010) *Org. Lett.*, **12**, 4588–4591.
123. Wang, Y., Zhu, S., and Ma, D. (2011) *Org. Lett.*, **13**, 1602–1605.
124. Luo, S., Zhao, J., and Zhai, H. (2004) *J. Org. Chem.*, **69**, 4548–4550.
125. Shi, M., Chen, L.-H., and Li, C.-Q. (2005) *J. Am. Chem. Soc.*, **127**, 3790–3800.
126. Meng, X., Huang, Y., and Chen, R. (2008) *Chem. Eur. J.*, **14**, 6852–6856.
127. Wan, J.-P., Loh, C.C.J., Pan, F., and Enders, D. (2012) *Chem. Commun.*, **48**, 10049–10051.
128. Priebbenow, D.L., Henderson, L.C., Pfeffer, F.M., and Stewart, S.G. (2010) *J. Org. Chem.*, **75**, 1787–1790.
129. Dyker, G. and Grundt, P. (1996) *Tetrahedron Lett.*, **37**, 619–622.
130. Khan, M.W. and Reza, A.F.G.M. (2005) *Tetrahedron*, **61**, 11204–11210.
131. Rolfe, A., Young, K., and Hanson, P.R. (2008) *Eur. J. Org. Chem.*, **2008**, 5254–5262.
132. Tan, B., Candeias, N.R., and Barbas, C.F. (2011) *Nat. Chem.*, **3**, 473–477.
133. Chen, W.-B., Wu, Z.-J., Pei, Q.-L., Cun, L.-F., Zhang, X.-M., and Yuan, W.-C. (2010) *Org. Lett.*, **12**, 3132–3135.
134. Wang, L.-L., Peng, L., Bai, J.-F., Huang, Q.-C., Xu, X.-Y., and Wang, L.-X. (2010) *Chem. Commun.*, **46**, 8064–8066.
135. Gao, Y., Ren, Q., Wu, H., Li, M., and Wang, J. (2010) *Chem. Commun.*, **46**, 9232–9234.
136. Rueping, M., Sugiono, E., and Merino, E. (2008) *Chem. Eur. J.*, **14**, 6329–6332.
137. Rueping, M., Volla, C.M.R., Bolte, M., and Raabe, G. (2011) *Adv. Synth. Catal.*, **353**, 2853–2859.
138. Takizawa, S., Inoue, N., Hirata, S., and Sasai, H. (2010) *Angew. Chem. Int. Ed.*, **49**, 9725–9729.
139. Sundén, H., Rios, R., Ibrahem, I., Zhao, G.-L., Eriksson, L., and Córdova, A. (2007) *Adv. Synth. Catal.*, **349**, 827–832.
140. Wang, Y.-F., Zhang, W., Luo, S.-P., Li, B.-L., Xia, A.-B., Zhong, A.-G., and Xu, D.-Q. (2009) *Chem. Asian J.*, **4**, 1834–1838.
141. Yadav, J.S., Reddy, B.V.S., Parimala, G., and Krishnam Raju, A. (2004) *Tetrahedron Lett.*, **45**, 1543–1546.
142. Li, H., Zhao, J., Zeng, L., and Hu, W. (2011) *J. Org. Chem.*, **76**, 8064–8069.
143. Zu, L., Xie, H., Li, H., Wang, J., Jiang, W., and Wang, W. (2007) *Adv. Synth. Catal.*, **349**, 1882–1886.
144. Zhao, G.-L., Vesely, J., Rios, R., Ibrahem, I., Sundén, H., and Córdova, A. (2008) *Adv. Synth. Catal.*, **350**, 237–242.
145. Dong, X.-Q., Fang, X., Tao, H.-Y., Zhou, X., and Wang, C.-J. (2012) *Chem. Commun.*, **48**, 7238–7240.
146. Indumathi, S., Perumal, S., and Menéndez, J.C. (2009) *J. Org. Chem.*, **75**, 472–475.
147. Tang, J., Xu, D.Q., Xia, A.B., Wang, Y.F., Jiang, J.R., Luo, S.P., and Xu, Z.Y. (2010) *Adv. Synth. Catal.*, **352**, 2121–2126.
148. Brandau, S., Maerten, E., and Jørgensen, K.A. (2006) *J. Am. Chem. Soc.*, **128**, 14986–14991.
149. Li, H., Zu, L., Xie, H., Wang, J., Jiang, W., and Wang, W. (2007) *Org. Lett.*, **9**, 1833–1835.
150. Corbet, M. and Zard, S.Z. (2008) *Org. Lett.*, **10**, 2861–2864.
151. Yu, C., Zhang, Y., Song, A., Ji, Y., and Wang, W. (2011) *Chem. Eur. J.*, **17**, 770–774.
152. Fourgeaud, P., Daydé, B., Volle, J.-N., Vors, J.-P., Van der Lee, A., Pirat, J.-L., and Virieux, D. (2011) *Org. Lett.*, **13**, 5076–5079.
153. Mahé, O., Dez, I., Levacher, V., and Brière, J.-F. (2010) *Angew. Chem. Int. Ed.*, **49**, 7072–7075.
154. Ballini, R., Palmieri, A., Talaq, M.A., and Gabrielli, S. (2009) *Adv. Synth. Catal.*, **351**, 2611–2614.
155. Nötzel, M.W., Tamm, M., Labahn, T., Noltemeyer, M., Es-Sayed, M., and de Meijere, A. (2000) *J. Org. Chem.*, **65**, 3850–3852.

8
Aldol Reactions in Domino Processes

Christoph Schneider and Michael Boomhoff

8.1
Introduction

The aldol reaction is among the most powerful carbon–carbon bond-forming processes available to synthetic chemists, and, accordingly, it comes as no surprise that it has been included in various domino transformations. Actually, it is ideally suited to participate in such reactions based upon the nature of both the starting material (an enol or enolate) and the product (a β-hydroxy carbonyl compound). The latter compound may easily react further at the just-formed aldolate, such as in lactonization, acetalization, as well as Prins and Tishchenko reactions, making the aldol reaction the initial step. On the other hand, the starting enolate may readily be derived from a conjugate addition onto an α,β-unsaturated carbonyl compound in a variety of different processes, moreover through transition-metal-catalyzed isomerization of allylic alcohols, rearrangement reactions, and cycloaddition processes. This chapter will attempt to cover the most significant achievements in this area. However, we are aware that only a fraction of the large number of reports can be presented here because of the page limit. For the same reason, we have focused primarily on the chemistry reported since 2000 and had to leave out other important types of aldol reactions, such as the Henry reaction and aldol condensation processes.

8.2
Domino Processes with the Aldol Reaction as First Step

8.2.1
Aldol-Lactonization Reactions

When the aldol product contains a highly reactive carboxylic acid derivative, a spontaneous lactonization to furnish β-lactones (2-oxetanones) may follow immediately.

Romo and coworkers [1] have extensively studied the domino Mukaiyama-aldol-lactonization reaction of aldehydes and thiopyridyl silyl *S,O*-ketene acetals,

Domino Reactions: Concepts for Efficient Organic Synthesis, First Edition. Edited by Lutz F. Tietze.

Table 8.1 Diastereoselective synthesis of *trans*-β-lactones [1].

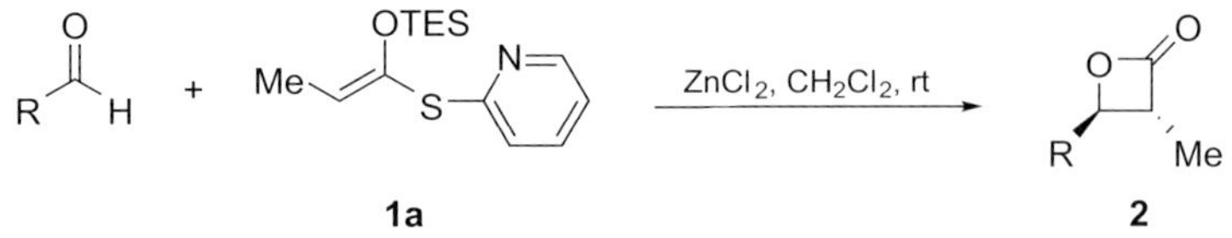

Entry	R	Time (h)	*trans/cis*	Yield (%)
1	$PhCH_2CH_2$-	22	>19 : 1	70
2	$TBSOCH_2CH_2$-	24	>19 : 1	51
3	$BnOCH_2$-	24	>19 : 1	74
4	$CH_3(CH_2)_6$-	23	>19 : 1	60

providing a highly diastereoselective approach to a variety of either trans-1,2- or cis-1,2-disubstituted β-lactones depending upon the Lewis acid employed. Thus, the triethylsilyl thiopyridyl ketene acetal **1a**, which proved optimal in this respect, delivered *trans*-β-lactones **2** in moderate to good yields and as single stereoisomers with $ZnCl_2$ as Lewis acid (Table 8.1).

The reaction was shown to proceed in a stereoconvergent manner, as both the *Z*- and the *E*-ketene acetal stereoisomers furnished the *trans*-β-lactone exclusively. On the other hand, when the same reaction was conducted with $SnCl_4$ as Lewis acid, the *cis*-β-lactone **3** was obtained in moderate yield and high stereoselectivity [2]. Here, only α-unbranched aldehydes gave acceptable yields (Scheme 8.1).

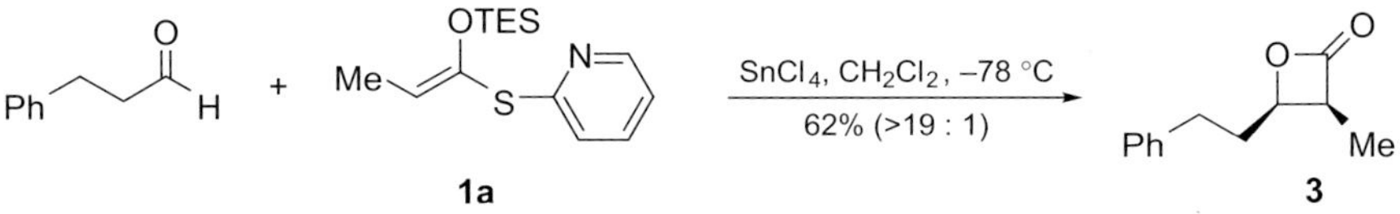

Scheme 8.1 Diastereoselective synthesis of *cis*-β-lactones [2].

The $ZnCl_2$-mediated reaction was studied in more detail, revealing that it most likely proceeded through a highly asynchronous, yet concerted, boat-like transition state and that the bidentate Lewis acid facilitated not only the carbon–carbon bond-forming event but also helped to take off the thiopyridyl group through coordination to the sulfur atom. Various natural products and drugs such as (−)-panclicin D and orlistat have been synthesized in this manner [3].

This reaction scheme can be further extended when γ-keto aldehydes **4** are employed as reaction partners to thiopyridyl ketene acetal **1b** (Scheme 8.2) [4]. In that case, the formed β-lactone **5** with the TIPS (triisopropylsilyl) group still attached to the lactone oxygen is attacked intramolecularly by the ketone group to furnish, upon lactone opening, a five-membered oxocarbenium ion **6**. The *in situ*

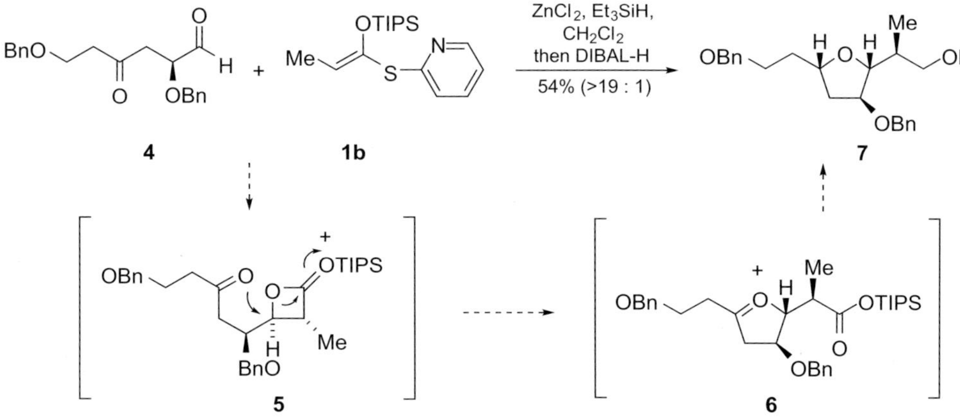

Scheme 8.2 Domino aldol/lactonization/oxoniumion formation/reduction reaction [4].

reduction of **6** with Et_3SiH was proposed to take place via inside attack onto an envelope-shaped oxocarbenium ion to provide the 2,5-disubstituted tetrahydrofuran **7** in moderate yield and excellent cis diastereoselectivity after DIBAL-H (diisobutylaluminum hydride) reduction of the silyl ester (Scheme 8.2). As by-product, the corresponding furan was isolated in varying amounts. The same process was utilized to set up a quaternary stereogenic center with excellent stereocontrol when the intermediate oxocarbenium ion **6** was treated with a carbon nucleophile such as allyl trimethylsilane. With δ-keto aldehydes, this process was readily applied to the synthesis of 2,6-*cis*-tetrahydropyrans.

Isotetronic acids were readily assembled through an organocatalyzed domino aldol-lactonization reaction starting from α-keto carboxylic acids and aldehydes [5]. As chiral organocatalyst, the proline-derived benzimidazol pyrrolidine **8** was employed, which proved superior to proline itself. Electron-poor aldehydes were both more reactive and selective. Thus, as a typical example, pyruvate was converted into isotetronic acid **10** in 77% yield and 83% ee upon treatment with *para*-nitrobenzyldehyde (**9**) and 10 mol% **8** (Scheme 8.3).

A conceptually related process was developed for the synthesis of sterically highly congested spirooxindoles **13** from isothiocyanato imides **11** and *N*-methylisatins **12**

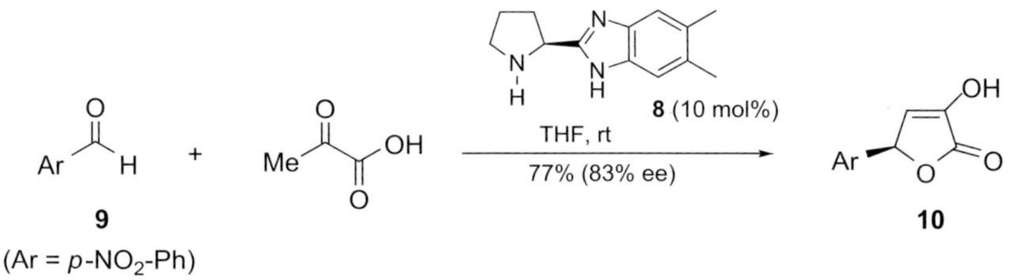

Scheme 8.3 Domino aldol-lactonization reaction for the assembly of isotetronic acids [5].

in an organocatalyzed domino aldol-thiolactamization reaction with the appended isothiocyano moiety (Scheme 8.4) [6]. As chiral organocatalyst, a rosin-derived amine-thiourea catalyst **14** was employed, which gave **13** with typically excellent yields and very high enantio- and diastereoselectivity.

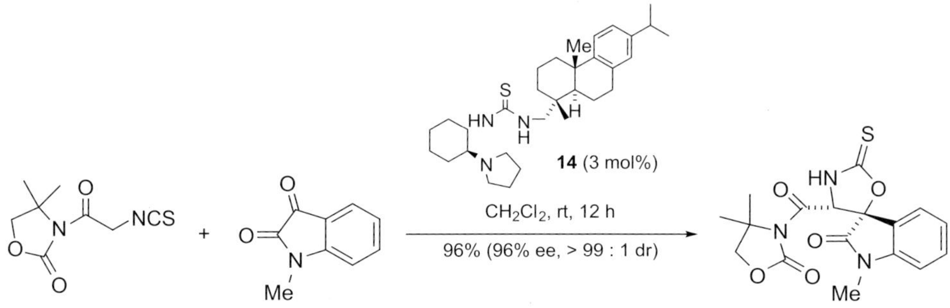

11 **12** **13**

Scheme 8.4 Domino aldol/thiolactamization reaction [6].

8.2.2
Aldol/Prins Reactions

When the aldol reaction furnishes an intermediate oxonium ion, a Prins cyclization may ensue in the case where a suitable internal nucleophile is present that intercepts this oxonium ion. Rychnovsky *et al.* have developed this strategy into a powerful tool for the straightforward synthesis of tetrahydropyrans [7]. Thus, enol ether **15** attached to an allyl silane reacted with various aldehydes under BF_3 activation to produce 2,6-*cis*-substituted-4-methylene tetrahydropyrans **16** in good to very good yields (Table 8.2).

Whereas the initial aldol stereoselectivity was very low, eventually resulting in the final epimeric mixture of products, the stereoselectivity of the Prins cyclization was remarkably high, furnishing the 2,6-cis configuration exclusively. This methodology was successfully employed in a stereoselective synthesis of the cytotoxic marine natural product leucascandrolide A by the Rychnovsky group.

A modified version of this process employs simple alkenes as internal nucleophiles, and the Prins cyclization is facilitated by the use of $TiBr_4$ as Lewis acid, which not only activates the aldehyde toward enol ether attack but also delivers a nucleophilic bromine onto the developing carbocation [8]. Thus, various 2,6-dialkyl-4-bromo-tetrahydropyrans **17** were obtained as single diastereomers within the tetrahydropyran ring and in good yields (Scheme 8.5).

Building on this precedence, Loh and coworkers developed a Mukaiyama–aldol–Prins cyclization of aldehyde-based silyl enol ethers **18** and unsaturated acetals **19** to access highly functionalized five-membered rings **20** with excellent stereocontrol [9]. Here again, the nucleophilic bromine liberated from $TiBr_4$ helps in triggering the Prins cyclization event of the alkene (Scheme 8.6). This process can easily be

Table 8.2 Domino aldol/Prins reaction toward tetrahydropyran synthesis [7].

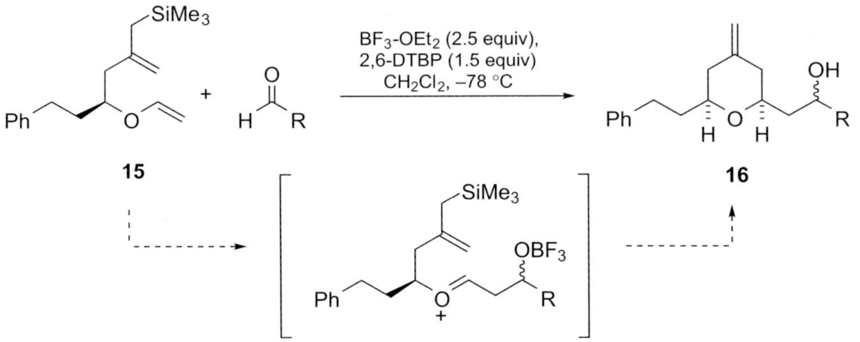

Entry	R	Epimer ratio	Yield (%)
1	*i*-Pr-	1 : 1	98
2	Ph-	1.2 : 1	84
3	PhCH=CH–	1.4 : 1	87
4	c-C_6H_{11}–	1.7 : 1	72
5	$TBSOCH_2CH_2$–	1.8 : 1	87

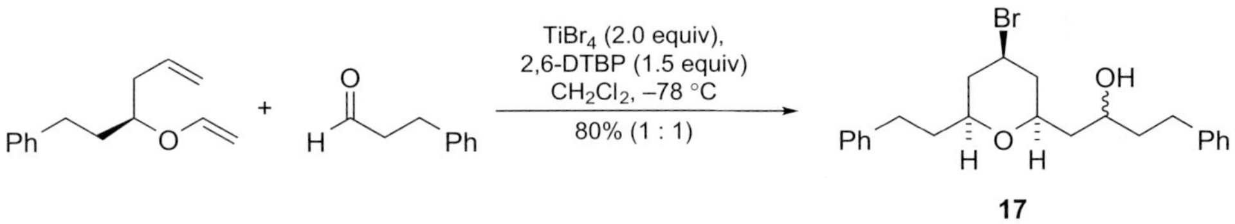

Scheme 8.5 Domino aldol/Prins reaction of unsaturated enol ethers [8].

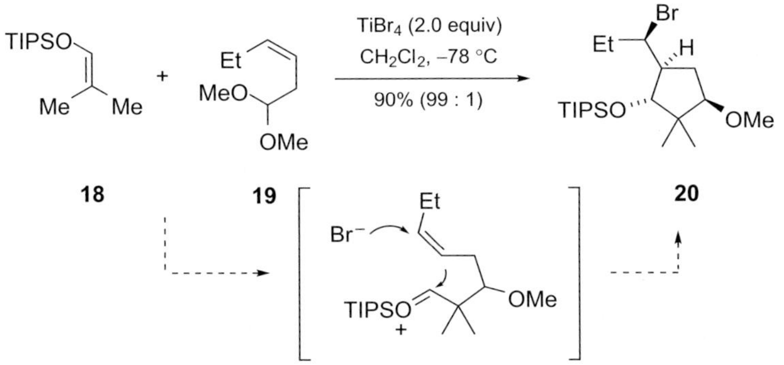

Scheme 8.6 Domino Mukaiyama–aldol–Prins reaction of silyl enol ethers and unsaturated acetals [9].

modified to the synthesis of six-membered tetrahydropyrans [9b]. For this purpose, terminal alkenes are employed as reaction partners in the Prins cyclization, which preferentially react at the terminal carbon atom because of the greater stability of a secondary over a primary carbocation.

8.2.3 Aldol/Acetalization Reactions

With dialdehydes as enol components, the aldol reaction may be followed by an acetalization event. Thus, Hayashi and coworkers [10] established a highly enantioselective proline-catalyzed domino aldol-acetalization of aqueous tetrahydro-2*H*-pyran-2,6-diol (**21**) and aldehydes (Scheme 8.7). In this sequence, the intermediate aldolate product **22** is trapped by the additional aldehyde moiety to generate a lactol, which through addition of methanol and acid is converted into the double acetal **23**.

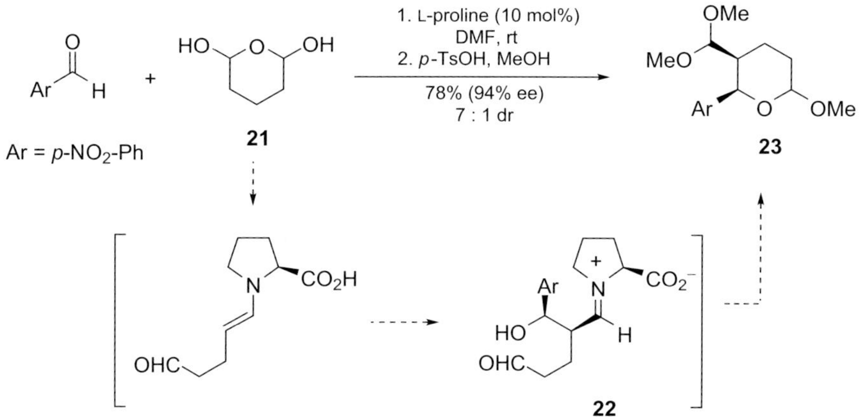

Scheme 8.7 Domino aldol/acetalization reaction [10].

It turns out that this sequence works best for electron-poor aromatic aldehydes, whereas the parent benzaldehyde delivered the product in only moderate yield; results with aliphatic aldehydes were not reported. The dimethylacetal moiety was selectively hydrolyzed into the aldehyde group, and the neighboring stereogenic center could be partially epimerized into the thermodynamically more stable 5,6-trans stereoisomer.

Hydroxyproline-catalyzed aldol reactions of acetone with salicyl aldehydes were recently shown to furnish an equilibrating mixture of the aldol product **24** and the cyclized lactol **25** in good yield and 77% ee [11]. By subsequent treatment of this 1 : 1 mixture with *p*-TsOH and methanol, 2-methoxy-2-methylchromanol (**26**) was obtained in 55% yield with 77% ee as a single diastereomer (Scheme 8.8).

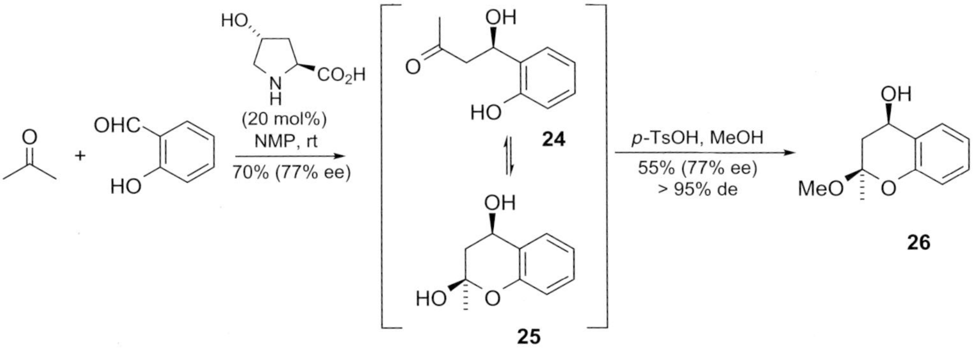

Scheme 8.8 Synthesis of methoxy chromanols via domino aldol/acetalization reaction [11].

8.2.4 Aldol–Tishchenko Reactions

The large number of polyketide-derived natural products typically contain 1,3-diol motifs. A very powerful strategy to access these building blocks is based upon the domino aldol–Tishchenko reaction, which employs a second aldehyde as hydride donor to reduce the ketone moiety directly into a second hydroxy group. Mahrwald and Costisella [12], Woerpel [13], and their coworkers were among the first to take advantage of this strategy and discover the highly diastereoselective aldol–Tishchenko reactions of dialkylketones catalyzed by titanium and lithium alkoxides, respectively. Morken and coworkers [14] established the first asymmetric aldol–Tishchenko process with a chiral yttrium–salen complex. Isobutyraldehyde was reacted with various aromatic aldehydes to provide 1,3-diol monoesters such as **27** in moderate to good yields and moderate enantioselectivity (Scheme 8.9).

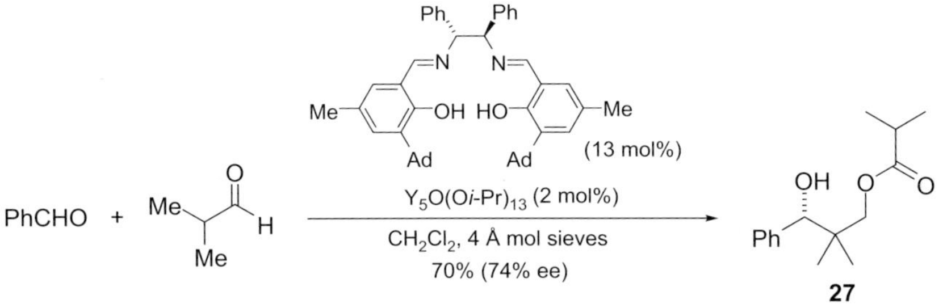

Scheme 8.9 Domino aldol–Tishchenko reaction developed by Morken *et al.* [14].

The first highly enantioselective cross-aldol–Tishchenko reaction of alkyl aryl ketones **28** and aldehydes was developed by the group of Shibasaki *et al.* [15]. With a chiral heterobimetallic lanthanide-based catalyst, this reaction was shown to proceed in typically very good enantioselectivity to furnish, after saponification of

Table 8.3 Domino aldol–Tishchenko reaction developed by Shibasaki *et al.* [15].

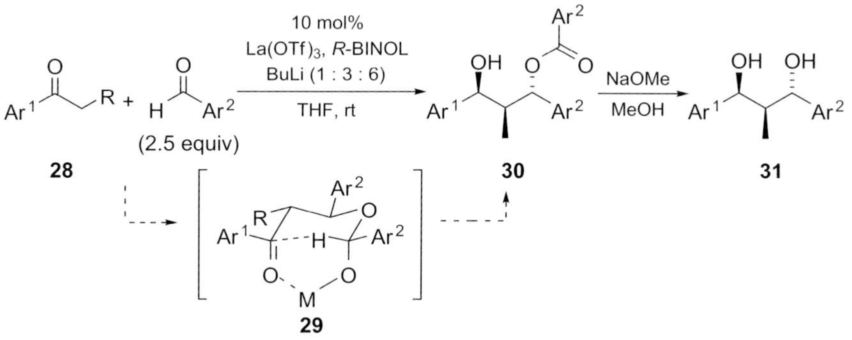

Entry	Ar^1	R	Ar^2	Yield of **31** (%)	ee (%)
1	4-CF_3–C_6H_4	Me	4-Cl–C_6H_4	95	93
2	4-CF_3–C_6H_4	Me	4-Br–C_6H_4	96	95
3	4-CF_3–C_6H_4	Me	Ph	95	91
4	4-Br–C_6H_4	Me	4-Br–C_6H_4	70	85
5	4-CF_3–C_6H_4	Et	4-Br–C_6H_4	90	88
6	4-CF_3–C_6H_4	Pr	4-Br–C_6H_4	88	87

the Tishchenko products **30**, the free 1,3-diols **31** in good to excellent overall yield (Table 8.3). Because of the mechanism of the Tishchenko reduction, which most likely proceeds through the depicted chair-like metal-bridged transition state **29** with intramolecular hydride delivery, the product 1,3-diolmonoesters **30** were obtained exclusively as 1,3-anti diastereomers. Interestingly, the absolute configuration of the products does not arise from the initial aldol addition step, which was proven to furnish all four stereoisomers with no selectivity. Instead, a rapid retro-aldol cleavage equilibrates the aldol stereoisomers, and then a slow and highly enantio- and diastereoselective Tishchenko reduction through the most favorable transition state **29** with all substituents located in equatorial positions establishes the correct relative and absolute configuration of the products.

Mlynarski *et al.* [16] developed ytterbium-catalyzed enantio- and diastereoselective aldol–Tishchenko reactions of symmetrical dialkyl ketones as enol components for the first time. As chiral ytterbium ligand, they employed the amino alcohol **32**, which gave rise to aldol–Tishchenko products such as **33** with up to 86% ee (Scheme 8.10). As documented by control experiments and very similar to the above discussed processes, the rate- and stereo-determining step in this reaction was proven to be the Tishchenko reduction with a rapid pre-retro-aldol equilibrium of the initially formed aldol products. This process may be utilized for reactions of alkyl aryl ketones as well, broadening its scope significantly.

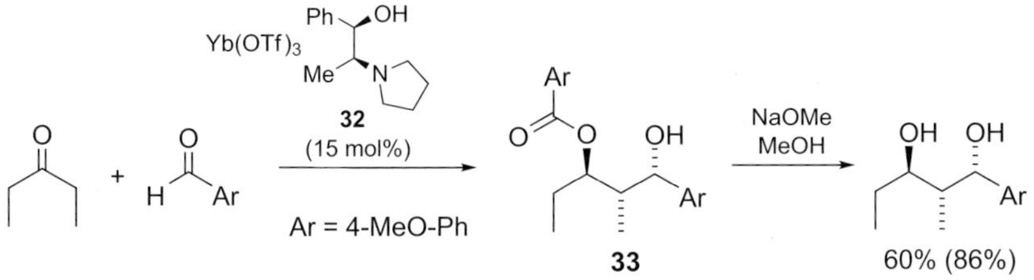

Scheme 8.10 Domino aldol–Tishchenko reaction developed by Mlynarski *et al.* [16].

Schneider and coworkers [17] established a different entry into aldol–Tishchenko reactions. Taking advantage of the well-known tendency of ketone aldol products to undergo an exothermic retro-aldol cleavage with *in situ* formation of metal enolates, they treated diacetone alcohol with zirconium *tert*-butoxide as catalyst (10 mol%) and various aliphatic aldehydes (2 equiv) and obtained a mixture of regioisomeric 1,3-*anti*-diol monoesters in good yields and complete diastereoselectivity (Scheme 8.11). The Brønsted basic and Lewis acidic zirconium alkoxide catalyzed both the retro-aldol–cross-aldol reaction and the subsequent Tishchenko reaction with high efficiency, but unfortunately gave rise to some acyl migration as well. Independently, Nevalainen and Simpura [18] showed that the same process can be catalyzed with aluminum alkoxides as catalysts, obtaining similar results.

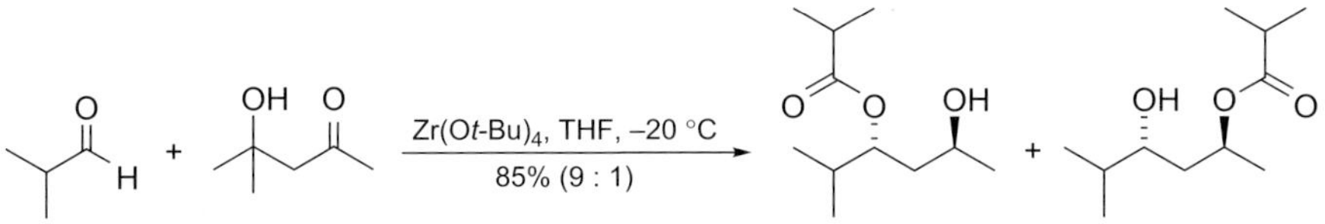

Scheme 8.11 Zirconium-catalyzed aldol–Tishchenko reaction of ketone aldol products [17].

It turned out that the undesired acyl group migration was less pronounced for *α*-branched acyl groups and also with bulkier zirconium alkoxide catalysts. Thus, with the modified zirconium catalyst **34** carrying a sterically very hindered diol as bidentate ligand, a general process by the reaction of an aldehyde and **35** was established by the Schneider group, furnishing a broad range of 1,3-*anti*-diol monoesters **36a/36b** in typically good to excellent yields, with complete diastereoselectivity and only very minor amounts of the regioisomer **36b** being formed even for straight-chain aldehydes (Table 8.4) [17b]. Interestingly, the pinacolone-derived ketone aldol product **35c** ($R^2 = t$-Bu) was the only substrate that furnished an aldol–Tishchenko product with benzaldehyde (entry 6, Table 8.4).

Both (−)-(4*R*,5*R*)- or (+)(4*S*,5*S*)-2,2-dimethyl-α,α,α′,α′-tetraphenyl-1,3-dioxolane-4,5-dimethanol (TADDOL) and 1,1′-bi-2-naphthol (BINOL) ligands were investigated as chiral ligands in this process to establish an enantioselective aldol–Tishchenko reaction based upon this concept. However, only moderate enantioselectivities of up to 60% ee were obtained [19].

Table 8.4 Zirconium-catalyzed aldol–Tishchenko reactions as developed by Schneider *et al.* [17b].

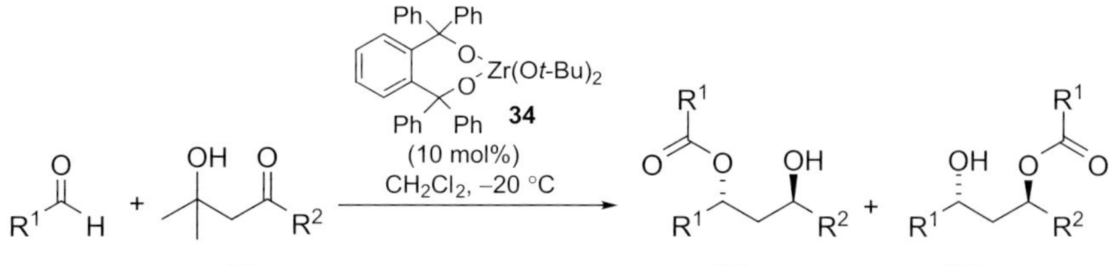

35 36a 36b

Entry	R^1	R^2 (**35**)	Yield (%)	Ratio **36a/36b**
1	*n*-C_6H_{13}	Me (**35a**)	91	5 : 1
2	*i*-Pr	Me (**35a**)	79	20 : 1
3	c-C_6H_{11}	Me (**35a**)	86	20 : 1
4	*i*-Pr	*i*-Pr (**35b**)	94	—
5	*n*-C_6H_{13}	*t*-Bu (**35c**)	88	15 : 1
6	Ph	*t*-Bu (**35c**)	86	20 : 1
7	*n*-C_6H_{13}	Ph (**35d**)	79	10 : 1

8.2.5
Vinylogous Aldol/Michael Reactions

When a vinylogous aldol reaction occurs, an α,β-unsaturated carbonyl compound is generated, which may react in a subsequent intramolecular Michael addition. In 2005, the group of Bräse [20] introduced a domino reaction of several salicylaldehydes and senecioaldehydes to produce tricyclic hemiacetals **39** (Scheme 8.12). Only one stereoisomer was formed in the presence of sodium carbonate with various salicylaldehydes. γ-Deprotonation of senecioaldehyde followed by a vinylogous aldol

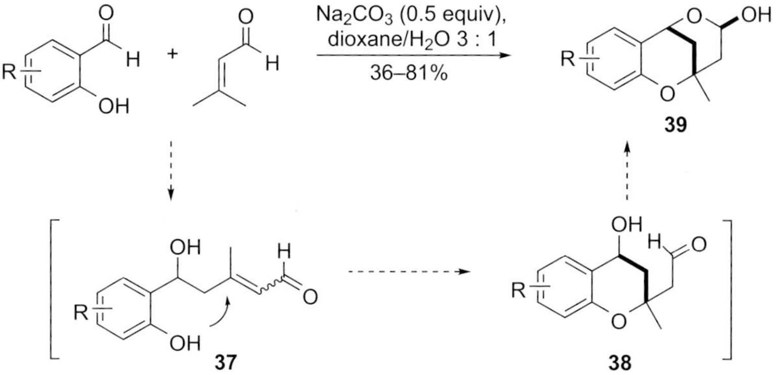

Scheme 8.12 Domino vinylogous aldol/Michael reaction by Brase *et al.* [20].

reaction furnished diol **37**, which reacted to form chroman **38** in an intramolecular oxa-Michael reaction. Eventually, cyclization of the initially formed alcohol onto the aldehyde moiety gave rise to lactol **39**. This reaction can be rendered enantioselective using chiral organocatalysts, and was shown nicely in the total synthesis of various natural products [21].

In this context, Woggon and coworkers [21e] applied an asymmetric vinylogous aldol/Michael reaction in the total synthesis of α-tocopherol (**44**). They converted enantiomerically pure phytenal and substituted salicylaldehyde **40** into lactol **42** (Scheme 8.13). Using prolinolether *R*-**41** as the chiral organocatalyst, they obtained the lactol in 60% yield with a diastereomeric ratio of 97 : 3, which was measured after oxidation to the lactone **43**.

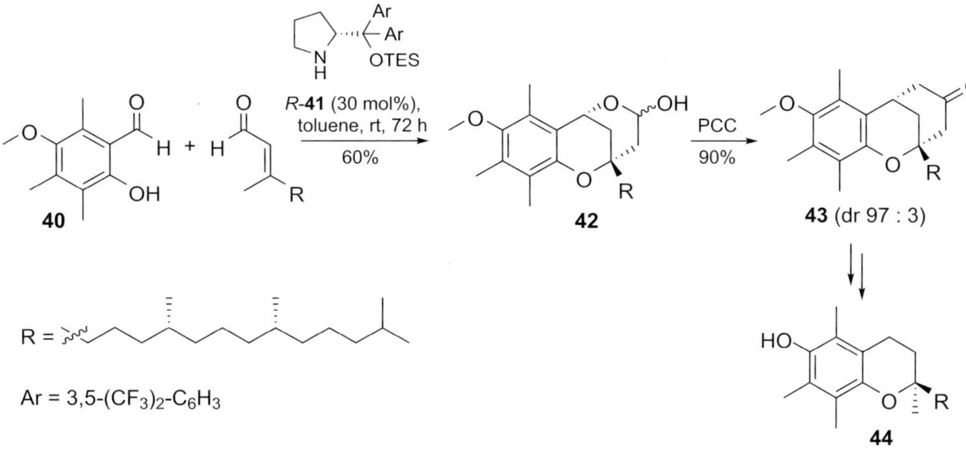

Scheme 8.13 Asymmetric organocatalytic domino reaction toward α-tocopherol [21e].

8.3 Domino Processes with the Aldol Reaction as Subsequent Step

8.3.1 Conjugate Addition/Aldol Reactions

A common strategy to generate an enolate for a subsequent aldol reaction is based upon the conjugate addition of nucleophiles onto α,β-unsaturated carbonyl compounds. A variety of different processes, including the addition of carbon, oxygen, sulfur, and nitrogen nucleophiles, are known. Moreover, reductive aldol reactions have been employed in which α,β-unsaturated carbonyl compounds were reduced in the initial step. Representative examples are given in the following section.

8.3.1.1 Addition of Carbon Nucleophiles

1,2-Cyclohexadione was recently shown to react with α,β-unsaturated aldehydes to form bicyclo[3.2.1]octanes **46** under Lewis base catalysis [22]. The best results were

Table 8.5 Domino Michael/aldol reaction toward bicyclo[3.2.1]octanes [22].

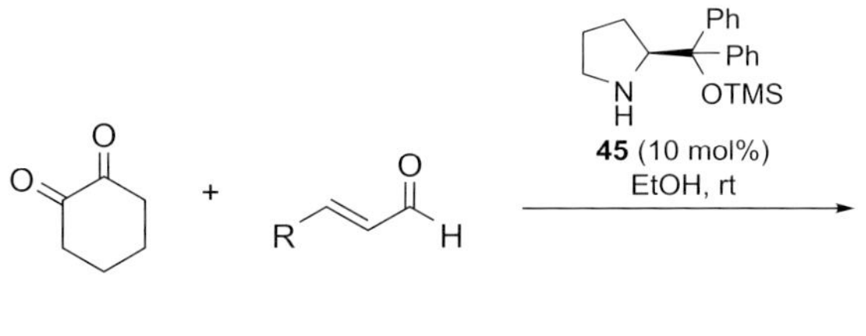

Entry	R	Yield (%)	ee (%)
1	Ph	77	96
2	4-MeO–C_6H_4	79	95
3	2-NO_2–C_6H_4	66	98
4	4-N$(Me)_2$–C_6H_4	60	93
5	2-Furanyl	81	95
6	*n*-Bu	44	98

obtained when diphenylprolinol silyl ether **45** was used as the catalyst in ethanol, and the products were obtained throughout as single diastereomers in good yields and excellent enantioselectivities (Table 8.5). Upon 1,4-addition of the dione (enol-tautomer) onto the activated iminium ion species, a reactive enamine was formed, which subsequently underwent an intramolecular aldol reaction. Rueping and coworkers were able to convert these tricyclic compounds into synthetically useful cycloheptanones and tetrahydrochromenones.

A very interesting stereoselective approach toward highly functionalized cyclohexanones was independently established by several working groups [23]. Here, an acyclic 1,3-dicarbonyl system and an α,β-unsaturated compound reacted in an organocatalytic asymmetric domino Michael/aldol reaction. In this context, Jørgensen and coworkers applied the phenylalanine-derived imidazolidine **47** as the catalyst, furnishing cyclohexanones **49** with high enantio- and diastereoselectivities and moderate to good yields (Table 8.6). The high diastereoselectivities observed in this process arise from the rapid *in situ* epimerization of the intermediate **48** and the preference of the large substituents for equatorial positions within **49**. The organocatalyst is assumed to activate the Michael acceptor by the formation of an iminium ion, deprotonate the β-ketoester, and act as a base in the intramolecular aldol reaction of the intermediate **48** at the same time [23b]. However, in a slightly different approach, Gryko [23a] revealed that L-proline may be employed as catalyst in the intramolecular aldol reaction through an enamine mechanism.

Organometallic reagents are well known for 1,4-additions and can hence be utilized to catalyze domino Michael/aldol reactions. Based on a methodology published by Noyori *et al.* [24], the group of Feringa established a catalytic enantioselective protocol for the synthesis of (−)-prostaglandin E_1 methyl ester [25]. There,

Table 8.6 Organocatalytic domino Michael/aldol reaction by Jørgensen *et al.* [23b].

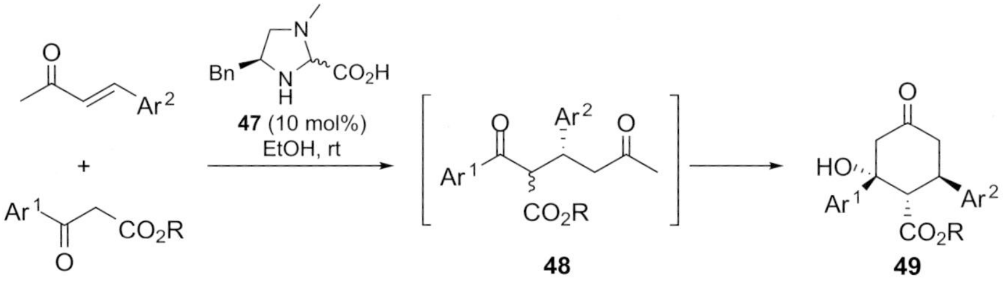

Entry	Ar^1	R	Ar^2	Yield (%)	ee (%)
1	Ph	Bn	Ph	80	95
2	Ph	Bn	4-Cl-Ph	80	93
3	Ph	Bn	2-NO_2-Ph	56	96
4	Ph	Bn	2-Furyl	40	85
5	Ph	Bn	2-Pyrimidyl	84	89
6	4-F–C_6H_4	Me	Ph	44	92
7	4-MeO–C_6H_4	Et	Ph	22	90

cyclopentenones were treated with organozinc reagents under copper catalysis in the presence of various chiral phosphoramidite ligands. After intensive investigations, with respect to the domino reaction process, they revealed that the acetal group in the 4-position of the enone had a great impact on the enantioselectivity of the aldol reaction as well as on the stability of the product. The use of 5,5-diphenyl-substituted 1,3-dioxane **50** was hence shown to be advantageous for the outcome of the reaction. Finally, aldol product **51** could be isolated as a *trans–anti*:*trans–syn* mixture of 87 : 13 in good yield and was further converted into the corresponding *syn*-diol **52** by means of a stereoselective reduction of the ketone. At this stage, the enantiomeric excess of the main diastereomer was determined as 94% (Scheme 8.14).

Krische and coworkers employed a rhodium-catalyzed conjugate addition in combination with an intramolecular enolate trapping by a ketone (Scheme 8.15). Starting from simple keto enones **53**, they obtained five- and six-membered carbocycles **54** with high levels of relative and absolute stereochemical control and created three new stereogenic centers in the domino process accordingly [26b].

On the basis of this strategy, a conceptually related desymmetrization reaction of enone-diones was introduced by the same group [26a]. An initial enone carbometalation furnished a rhodium enolate, which efficiently differentiated between the two carbonyl groups of the neighboring dione. Compound **55** as a representative example could be converted into the aldol product **57** in 87% yield and excellent selectivities. Only a single diastereomer out of 16 potential stereoisomers was formed selectively (Scheme 8.16).

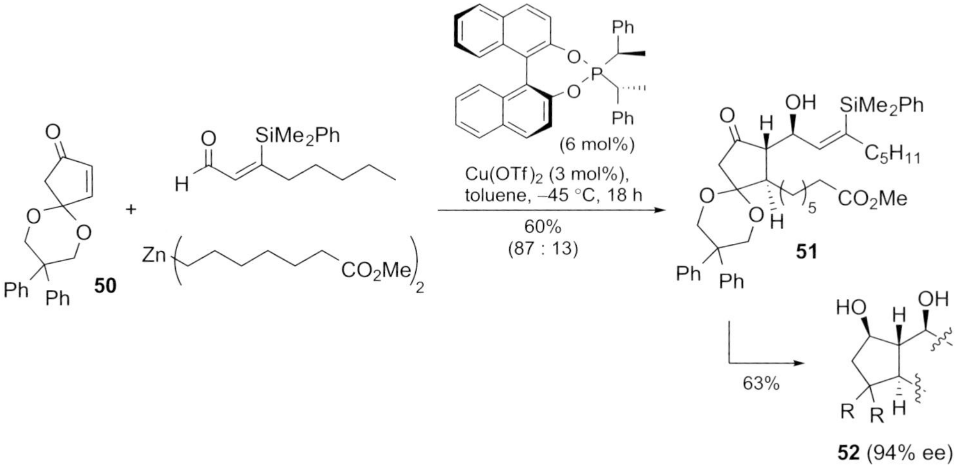

Scheme 8.14 Catalytic enantioselective domino Michael/aldol reaction as developed by Feringa *et al.* [25].

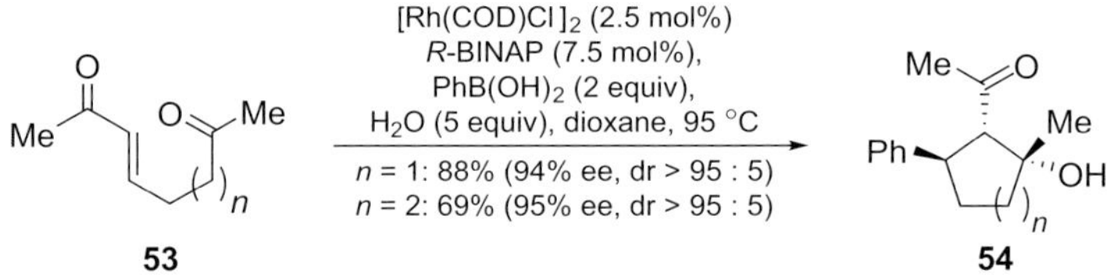

Scheme 8.15 Carbometallative aldol cycloreduction by Krische [26b].

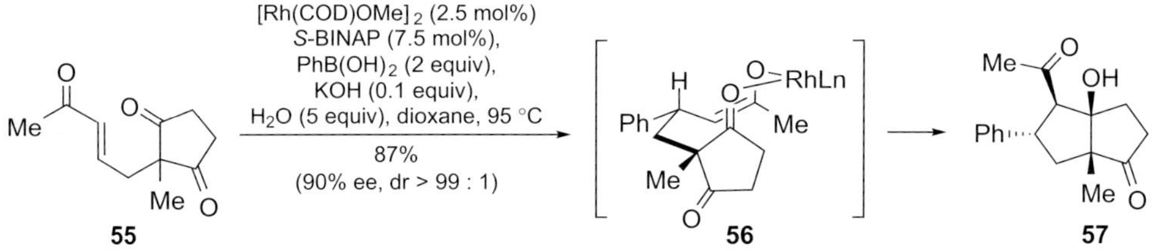

Scheme 8.16 Desymmetrization of enone-diones via a Michael/aldol cyclization [26a].

The authors proposed a Zimmerman–Traxler type transition state **56** bearing a *Z*-enolate, which was based on the fact that mainly η^3-haptomeric rhodium enolates exist after conjugate additions. Additionally, this concept could be extended to parallel kinetic resolution reactions with racemic unsymmetrically substituted diones.

An intramolecular domino Michael/aldol reaction starting from the chiral Cope products **58** was developed by the group of Schneider [27]. After 1,4-addition of

Table 8.7 Stereoselective synthesis of highly functionalized cyclohexanes by Schneider *et al.* [27].

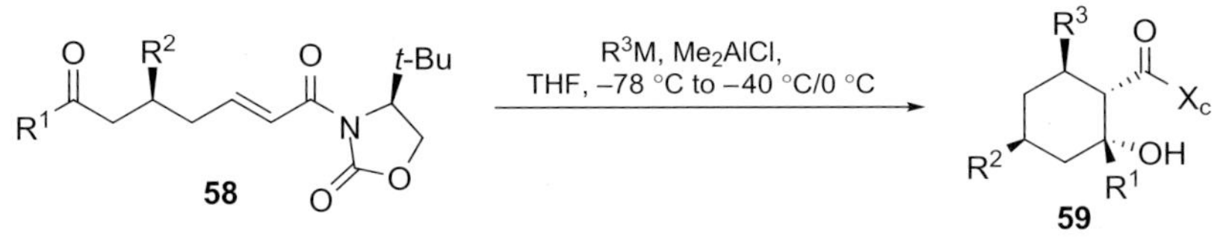

Entry	R^1	R^2	R^3	M	Yield (%)	dr
1	H	Me	*n*-Bu	Cu-LiI	41	>95 : 5
2	H	Ph	*n*-Bu	Cu-LiI	42	>95 : 5
3	Me	Me	*n*-Bu	Cu-LiI	83	>95 : 5
4	Me	Me	Allyl	Cu-$MgBr_2$	81	>95 : 5
5	Me	Me	Ph	Cu-LiI	51	>95 : 5

various organocopper and organoaluminum reagents, they could isolate highly functionalized cyclohexanes **59** in moderate to good yields with excellent stereocontrol (Table 8.7). They were able to convert both aldehydes and ketones to the corresponding secondary and tertiary alcohols.

Hayashi and coworkers [28] established an intermolecular domino 1,4-addition/aldol reaction. The reaction of vinyl ketones with aldehydes and several arylboranes in the presence of a rhodium catalyst was carried out smoothly, which eventually furnished aldol products **60** in generally very good yields and syn selectivities (Scheme 8.17).

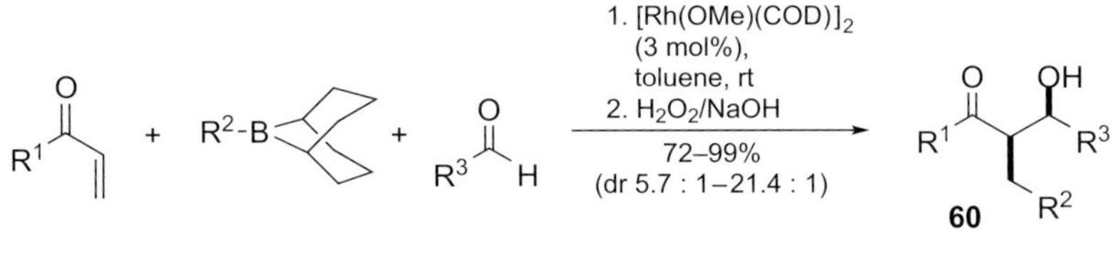

Scheme 8.17 Intermolecular rhodium-catalyzed domino reaction by Hayashi *et al.* [28].

8.3.1.2 Addition of Sulfur Nucleophiles

The access toward tetrahydrothiophenes via domino thia-Michael/aldol reactions has been well studied recently [29]. Jørgensen *et al.* [29d] introduced an organocatalytic approach toward optically active and highly substituted tetrahydrothiophenes by the treatment of α,β-unsaturated aldehydes with 2-mercapto ketones. The corresponding products were obtained as single diastereomers in moderate yields and very good enantioselectivities (Table 8.8). Depending on the additive, they

Table 8.8 Organocatalytic domino thia-Michael/aldol reaction by Jørgensen *et al.* [29d].

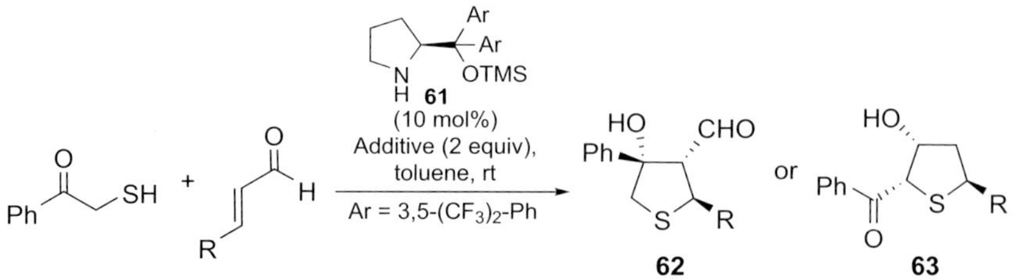

62 Additive: $PhCO_2H$ 63 Additive: $NaHCO_3$

Entry	R	Yield of **62** (%)	ee (%)	Yield of **63** (%)	ee (%)
1	Me	59	90	59	74
2	Et	73	95	44	72
3	*i*-Pr	56	94	61	80
4	*n*-Bu	62	90	66	64
5	CH_2CH_2OTBS	44	96	61	70

could isolate either tetrahydrothiophene carbaldehydes **62** or tetrahydrothiophenyl ketones **63**. Presumably, the outcome of the reaction is related to the hydrolysis of the enamine intermediate that is formed after the initial 1,4-addition.

A common methodology to access enantiomerically pure thiochromanes is the domino thia-Michael/aldol reaction of 2-mercaptobenzaldehydes and α,β-unsaturated carbonyl compounds [30]. An elegant asymmetric version with a bifunctional organocatalyst was developed by the group of Wang *et al.* [30e]. Several β-aryl α,β-unsaturated imides were reacted with 2-mercapto benzaldehydes to form the corresponding thiochromanes **65** in generally very good yields with high enantio- and diastereoselectivities (Scheme 8.18). Most notably, excellent results in this domino Michael/aldol process were achieved using only 1 mol% of the cinchona alkaloid-derived thiourea catalyst **64**.

The authors proposed a multiple hydrogen-bonding activation of both substrates, with hydrogen-bonding activation and LUMO (lowest unoccupied molecular orbital) lowering of the imide and deprotonation of the thiol by the amine to enhance its nucleophilicity. To overcome the problem of spontaneous dehydration and thus the loss of two additional stereogenic centers, the use of imides instead of aldehydes was essential [31]. Another approach to avoid dehydration was developed by Cordova *et al.* [30d], who employed 2-mercapto acetophenone to provide benzothiopyrans with a tertiary aldol structural motif. The group of Wang [30f] could further extend this strategy to synthesize enantio-enriched thiochromeno[2,3-*c*]maleimides by using *N*-aryl maleimides as Michael acceptors.

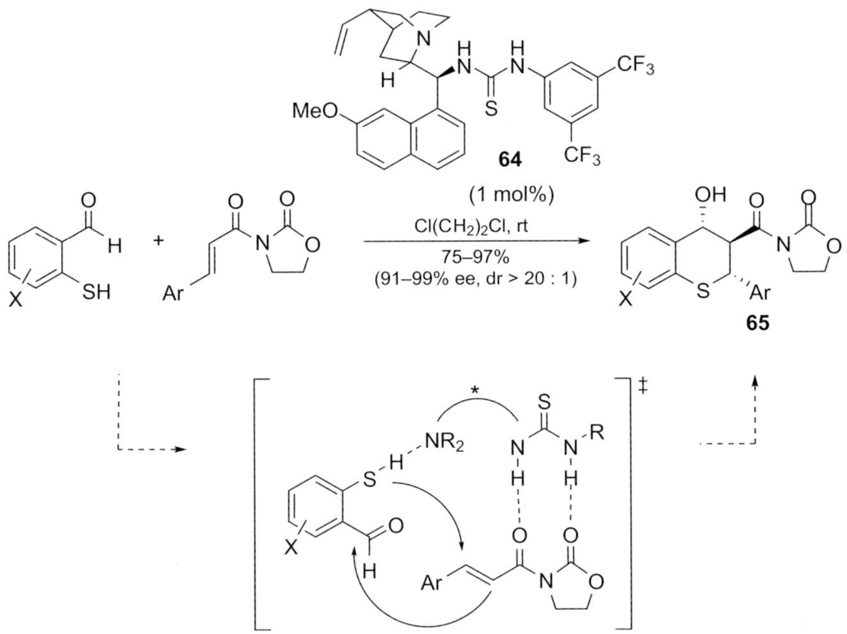

Scheme 8.18 Domino thia-Michael/aldol process accessing thiochromanes by Wang *et al.* [30e].

8.3.1.3 **Addition of Oxygen and Nitrogen Nucleophiles**

An organocatalytic oxa-Michael/aldol-hemiacetalization sequence was reported by Carrillo *et al.* [32]. They converted dimeric dihydroxyacetone into hexahydrofuro[3,4-*c*]furans **66** upon reaction with several α,β-unsaturated aldehydes (Table 8.9). Using diphenylprolinol silyl ether **45** as chiral organocatalyst, they consistently obtained the corresponding bicyclic products with four stereogenic centers with excellent yields and enantioselectivities as well as high diastereoselectivities. An initial conjugate addition of dihydroxyacetone onto the activated iminium ion generated an enamine, which intramolecularly attacked the ketone to form a quaternary stereogenic center. This was subsequently followed by an intramolecular hemiacetalization step.

Propargyl aldehydes undergo domino oxa-Michael/aldol reactions with hydroxyphenylketones to furnish 4*H*-chromenes in the presence of prolinol silyl ether **67**. In this context, the group of Wang [33b] developed a catalytic asymmetric methodology that generated 4*H*-chromenes **68** in excellent yields and enantioselectivities (Scheme 8.19a). Predominantly, electron-rich phenol derivatives and aryl-substituted propargyl aldehydes were applied and converted into the products with a quaternary center in the 4-position. During their investigations, they found the parent salicylaldehyde as an inappropriate reaction partner, which was less reactive under those conditions.

Table 8.9 Asymmetric organocatalytic domino oxa-Michael/aldol/hemiacetalization by Carrillo *et al.* [32].

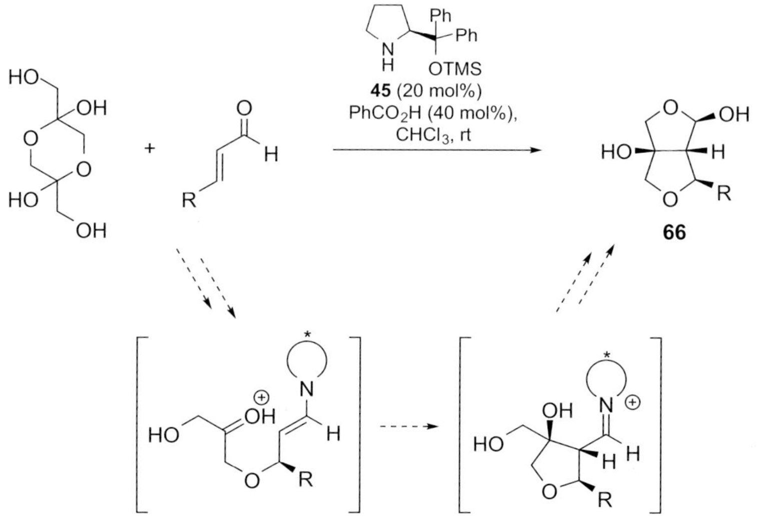

Entry	R	Yield (%)	ee (%)	dr
1	*n*-Pr	96	99	>10 : 1
2	Me	89	92	7 : 1
3	Et	86	95	7 : 1
4	*n*-C_8H_{17}	76	98	>10 : 1
5	*Z*-EtCH=CHCH$_2$CH$_2$	83	94	>10 : 1
6	Ph	76	98	>10 : 1

On the basis of this concept, the same group recently introduced an asymmetric catalytic aza-Michael version of this reaction type which provided the corresponding 1,4-dihydroquinolines **70** (Scheme 8.19b) [33a]. Electron-withdrawing protecting groups on the nitrogen atom were essential for a successful reaction, as electron-releasing groups furnished the undesired aromatized quinolines. In many cases, only 1 mol% of the C_2-symmetric pyrrolidine catalyst **69** was required to convert mostly amino chalcone derivatives into highly enantio-enriched 1,4-dihydroquinolines in yields of up to 99%.

Various bioactive synthetic targets contain aminoindanol backbones and, therefore, many different strategies to access them have been developed. Among them, the Friedel–Crafts reaction is a common strategy, although harsh reaction conditions and the requirement for electron-rich aromatic rings limit the scope of this reaction type.

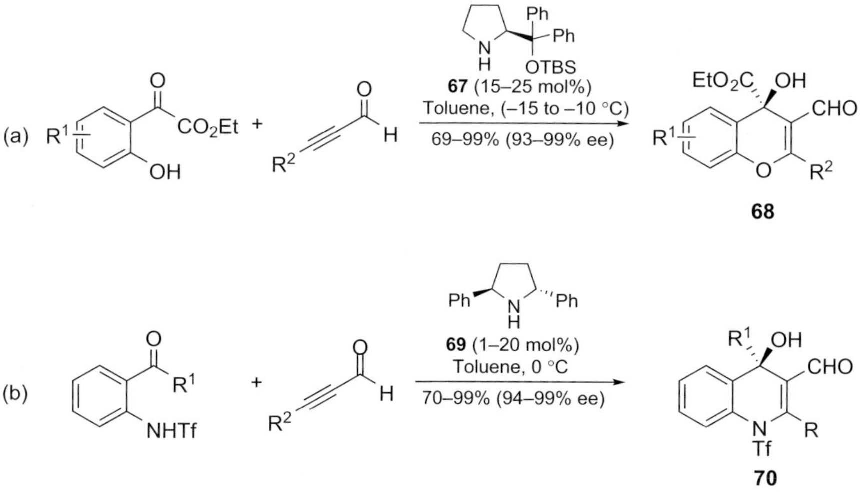

Scheme 8.19 (a,b) Organocatalytic domino oxa- and aza-Michael/aldol reactions by Wang *et al.* [33].

Sun and coworkers established an alternative diastereoselective approach to 3-amino-1-indanols that proceeds under mild conditions. They employed chalcone-derived aldehydes **71** and treated them with weak nitrogen nucleophiles and catalytic amounts of base (Scheme 8.20) [34]. Electron-rich and -deficient substrates as well as several amines were tolerated and gave rise to **72** in very good yields and diastereoselectivities up to >30 : 1. Sulfonamides were shown to be the best choice with respect to their ability to add to the α,β-unsaturated ketone. The generated enolate subsequently attacked the aldehyde in an intramolecular aldol reaction. The alternative reaction pathway involving an initial aldehyde attack followed by an intramolecular oxa-Michael addition could be suppressed by using DBU (1,8-diazabicyclo[5.4.0]undec-7-ene) as the appropriate base.

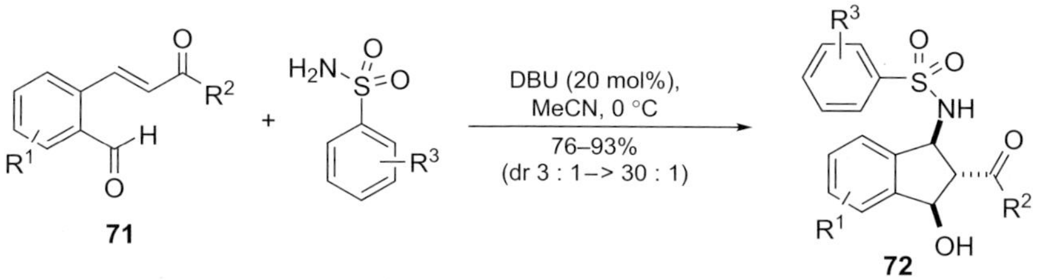

Scheme 8.20 Domino aza-Michael/aldol reaction providing aminoindanols by Sun *et al.* [34].

8.3.1.4 Iodo-Aldol Reactions

α-Methylene-β-hydroxy carbonyl derivatives are known as useful and important building blocks for the synthesis of natural products and biologically active molecules. One approach toward this class of compounds was reported by Ryu *et al.* [35b] and complements the access using the Morita–Baylis–Hillman (MBH) reaction. For the synthesis of optically active β-iodo MBH esters **74**, a catalytic asymmetric iodo-aldol reaction was investigated starting from aldehydes, propiolates, and TMSI (trimethylsilyl iodide) as iodine source in the presence of chiral oxazaborolidinium salts **73** (Table 8.10). The desired products were isolated in high yields and with excellent Z/E ratio and high enantiomeric excess. Even aliphatic aldehydes were converted under those conditions, albeit with slightly lower yields but still very good selectivities. Taking advantage of the vinyl iodide moiety, they could furthermore reveal the synthetic potential of the β-iodo MBH esters **74** in different cross-coupling reactions.

Recently, the group extended this methodology to α-substituted acroleins and applied their reaction nicely to the short synthesis of various enantio-enriched γ-butyrolactone natural products [35a].

During their studies toward the total synthesis of kibdelone C, Porco *et al.* developed an intramolecular iodo-Michael/aldol sequence [36]. Treatment of ynoate **75** with 2 equiv of MgI_2 gave rise to **76a** in 78% yield along with 9% of the minor diastereomer **76b** (Scheme 8.21). The authors proposed a reversible formation of two diastereomeric β-iodoallenoates, which arise from the conjugate iodo addition. These are believed to subsequently undergo the aldol reaction through two different transition states, leading to the diastereomers **76a** and **76b**. However,

Table 8.10 Asymmetric catalytic iodo–aldol reaction by Ryu *et al.* [35b].

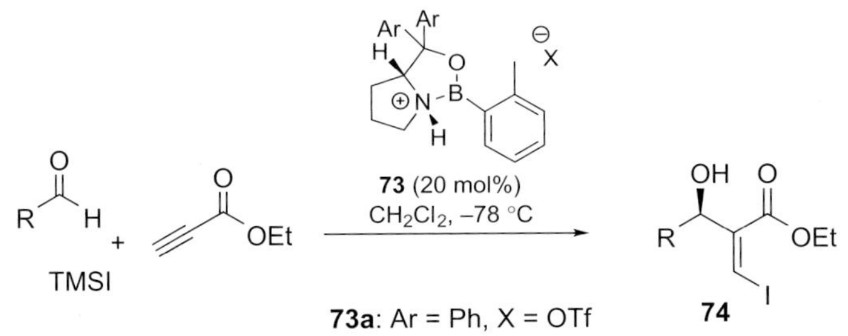

74

73a: Ar = Ph, X = OTf
73b: Ar = Ph, X = NTf_2
73c: Ar = 3,5-dimethylphenyl, X = OTf

Entry	R	Yield (%)	ee (%)	Z/E	Catalyst
1	Ph	95	94	>99 : 1	**73c**
2	4-F–C_6H_4	92	92	99 : 1	**73c**
3	2-Br–C_6H_4	95	90	>99 : 1	**73a**
4	2-Naphthyl	65	91	98 : 2	**73a**
5	*n*-Pr	72	93	96 : 4	**73b**
6	*i*-Pr	50	90	95 : 5	**73b**

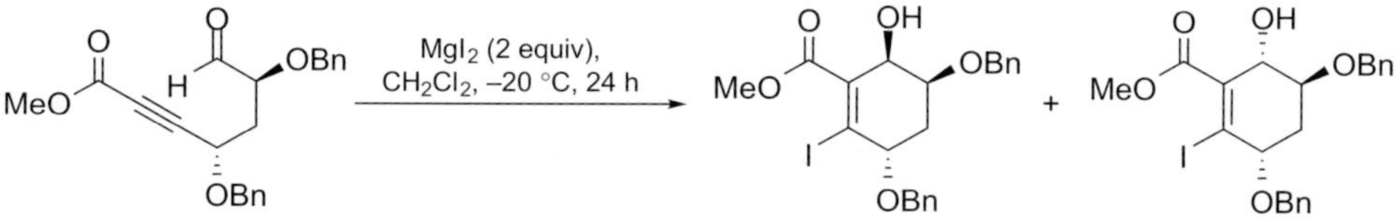

Scheme 8.21 Diastereoselective iodo-Michael/aldol reaction by Porco *et al.* [36].

under the reaction conditions used, they could not exclude the possibility of a thermodynamically controlled retro-iodo-aldol reaction.

8.3.1.5 Reductive Aldol Reactions

The formal conjugate addition of a hydride to α,β-unsaturated carbonyl compounds with a subsequent aldol reaction of the *in situ* formed enolate has been frequently employed in organic synthesis. A broad range of procedures have been developed using various metals (e.g., rhodium, cobalt, iridium, ruthenium, copper) and different reductants (typically silanes, boranes, or elemental hydrogen) [37].

The hydrogenative reductive aldol reaction, in which elemental hydrogen is employed as the reductant, emerged in the beginning of this century with the pioneering work of Krische [38e]. Mild basic conditions and the absence of any by-products are advantages of this type of enolate formation. Furthermore, a regioselective enolization of unsymmetric ketones can be achieved, which is not possible using the deprotonation strategy with saturated analogs.

After the introduction of the first hydrogenative reductive aldol reaction of vinyl ketones with catalytic amounts of a cationic rhodium species [38e], the group of Krische [38b-d] improved the syn diastereoselectivity using tri-2-furylphosphine ligands and, in 2008, introduced the first catalytic enantioselective process. After careful ligand screening and optimization, they identified the monodentate TADDOL-based phosphonite **77** as the ligand of choice [38a]. The 2-benzothienyl, diethyl ketal, and dimethyl carbinol units worked best with respect to reactivity and enantioselectivity. Several aldehydes with different labile functionalities were tolerated and reacted with methyl and ethyl vinyl ketone to give the respective aldol products **78** in very good yields and selectivities (Scheme 8.22). The authors proposed a cyclic Zimmerman–Traxler type transition state of the *Z*-enolate to explain the formation of *syn*-aldol products.

Morken and coworkers [39b] developed the first asymmetric reductive aldol reaction with silanes as reductants in combination with a chiral rhodium catalyst. α,β-Unsaturated esters were reacted with several aldehydes to provide the corresponding aldol products **79** in good yields and enantio- and diastereoselectivities (Scheme 8.23). Both aliphatic and aromatic aldehydes could be converted into aldol products **79** under these conditions. Furthermore, the group reported an iridium-catalyzed asymmetric version that tolerated various protected hydroxyaldehydes [39a]. On the basis of this precedence, a highly enantio- and diastereoselective

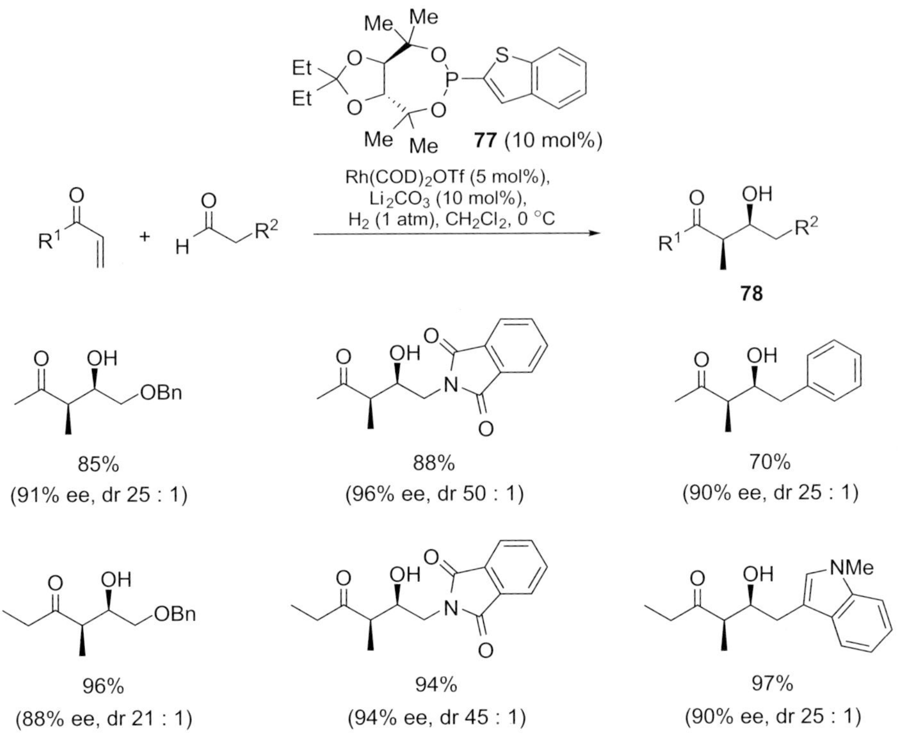

Scheme 8.22 Intermolecular enantioselective hydrogenative reductive aldol reaction by Krische *et al.* [38a].

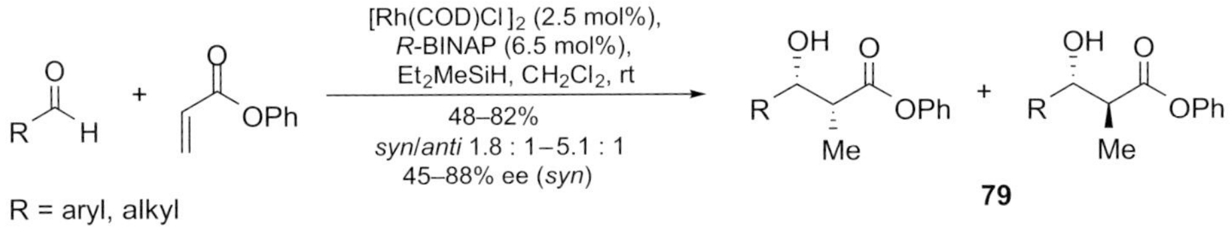

Scheme 8.23 First asymmetric catalytic reductive aldol reaction by Morken *et al.* [39b].

copper-catalyzed reaction of aromatic ketones and unsaturated esters was reported by Riant and coworkers [40].

In contrast to the hydrogenative version, which is limited to unsaturated ketones and aldehydes, the reductive aldol reaction using silanes so far has enabled only coupling of α,β-unsaturated esters. However, Sugiura and Nakajima [41b] introduced the first organocatalytic reductive aldol reaction of α,β-unsaturated ketones with trichlorosilanes as reductants. They were able to render this reaction enantioselective by using chiral bisphosphine oxide **80** as Lewis base (Scheme 8.24) [41a]. The *syn*-aldol products **81** were isolated in yields of up to 92% with very good stereoselectivities, which arise from a selective Lewis-base-catalyzed *Z*-silyl enolate formation after conjugate reduction.

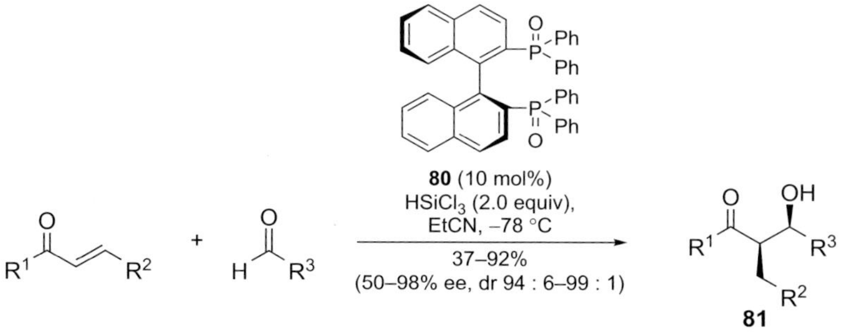

Scheme 8.24 Organocatalytic reductive aldol reaction by Sugiura and Nakajima [41a].

8.3.2
Isomerization/Aldol Reactions

Quite a number of transition-metal complexes are capable of isomerizing allylic alcohols into transition-metal–enol complexes through an internal redox process. Those can then be trapped *in situ* with aldehydes in an aldol reaction. Motherwell and coworkers developed a $RhCl(PPh_3)_3$-catalyzed domino isomerization–aldol reaction of secondary allyl alcohols, which gave rise to a mixture of *syn*- and *anti*-aldol products, with only small amounts of the regioisomeric aldol product occasionally being formed (Scheme 8.25) [42].

Besides various iron and ruthenium complexes [43, 44], nickel-based catalysts have recently been shown to be highly reactive in this respect as well [45]. Thus, a nickel catalyst prepared *in situ* from equimolar amounts of $NiCl_2$(dppe) (dppe, 1,2-bis(diphenylphosphino)ethane) and $LiBHEt_3$ (5 mol% each), which presumably resulted in the formation of NiHCl(dppe) as the active catalyst, efficiently catalyzed the isomerization/aldol event of allyl alcohols **82** and aldehydes **83** in combination with the Lewis acid $MgBr_2$ (5 mol%) to furnish aldol products **84** in typically excellent yields and variable isomeric ratios (Table 8.11). Allyl alcohols with a terminal alkene reacted much faster than those with an internal olefin, and the aldol reaction occurred exclusively on the side of the former allyl alcohol.

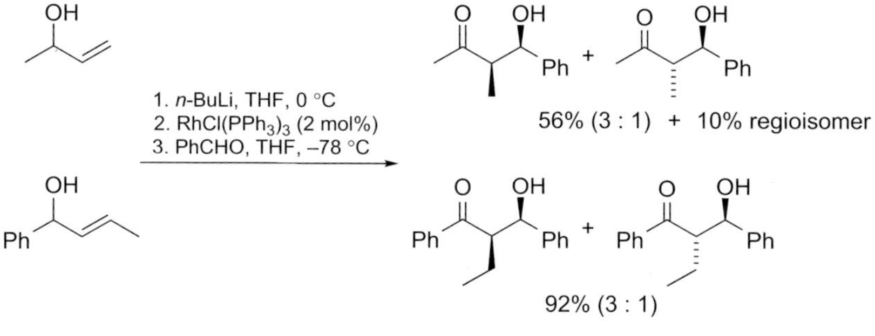

Scheme 8.25 Rhodium-catalyzed domino isomerization–aldol reaction.

Table 8.11 Domino isomerization/aldol reaction of allyl alcohols [45].

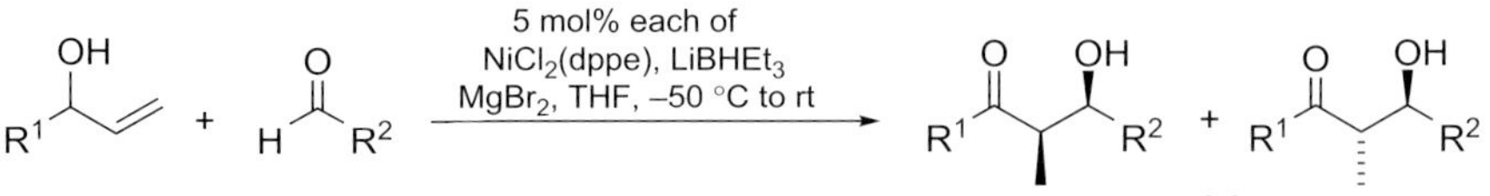

Entry	R^1	R^2	Yield (%)	*syn/anti*
1	Ph	Ph	97	60 : 40
2	Ph	*i*-Pr	86	50 : 50
3	*i*-Pr	Ph	68	75 : 25
4	Et_2CH	Ph	90	70 : 30
5	CMe_2CO_2Et	Ph	93	>95 : 5
6	*t*-Bu	Ph	89	90 : 10

8.3.3
Wittig Rearrangement/Aldol Reactions

Chiral α-benzyloxy and α-allyloxy acetates **85** upon deprotonation with Bu_2BOTf and NEt_3 undergo a smooth 1,2-Wittig rearrangement to furnish α-benzyl- or α-allyl glycolates **86** *in situ,* which upon further deprotonation with a second equivalent of base generate boron enolates that are easily trapped with aldehydes in an aldol reaction. Wolfe and coworkers [46] have established this strategy for a highly stereoselective synthesis of glycolate aldol products **87**, which are obtained in good yields and with a quaternary and a tertiary chiral center (Table 8.12). As chiral auxiliary, they found 2-phenyl-1-cyclohexanol most suitable, which delivered highly functionalized products with excellent diastereoselectivity and was easily cleaved off through $LiAlH_4$ reduction to furnish 1,2,3-triols with up to 95% ee.

8.3.4
Cycloaddition/Aldol Reactions

In the course of a total synthesis of the diterpenoic tetraol (+)-aphidicolin, Deslongchamps and coworkers developed a domino transannular Diels–Alder/aldol reaction of macrocyclic trienyl oxo aldehyde **88**, which closed four rings and assembled the entire skeleton of the natural product in one transformation. Heating **88** in a sealed tube in toluene at 230 °C for 24 h in the presence of triethylamine as proton scavenger first initiated the transannular Diels–Alder reaction, which was followed by the intramolecular aldol reaction to deliver tetracycle **90** in excellent yield and as a single diastereomer (Scheme 8.26) [47]. Three new carbon–carbon σ-bonds and four new stereogenic centers were established with complete stereocontrol.

Table 8.12 Wittig rearrangement/aldol reaction as developed by Wolfe *et al.* [46].

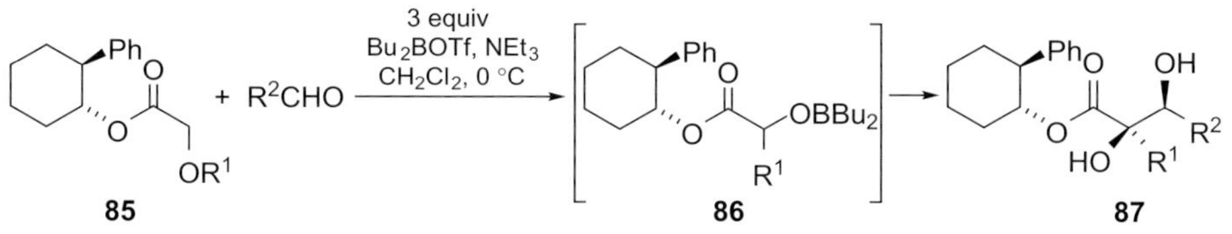

Entry	R^1	R^2	Yield (%)	*syn/anti*	ee after LAH-reduction (%)
1	Bn	Ph	83	20 : 1	89
2	Bn	C_9H_{19}	88	>20 : 1	95
3	Bn	*c*-C_6H_{11}	71	20 : 1	91
4	Allyl	Ph	83	20 : 1	90
5	Allyl	C_9H_{19}	79	8 : 1	75
6	Allyl	*i*-Pr	68	20 : 1	89

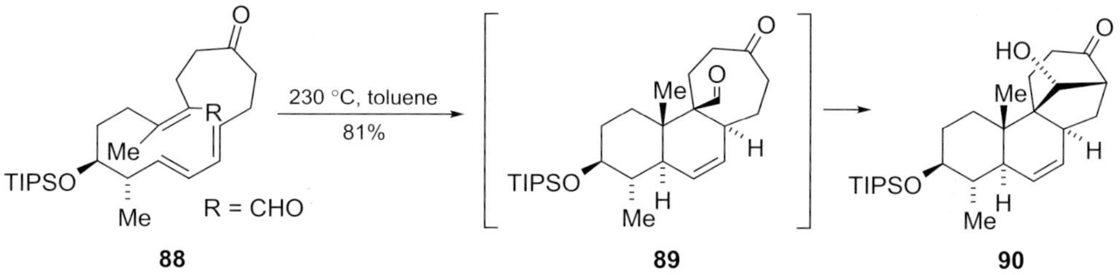

Scheme 8.26 Transannular Diels–Alder/aldol reaction as developed by Deslongchamps *et al.* [47].

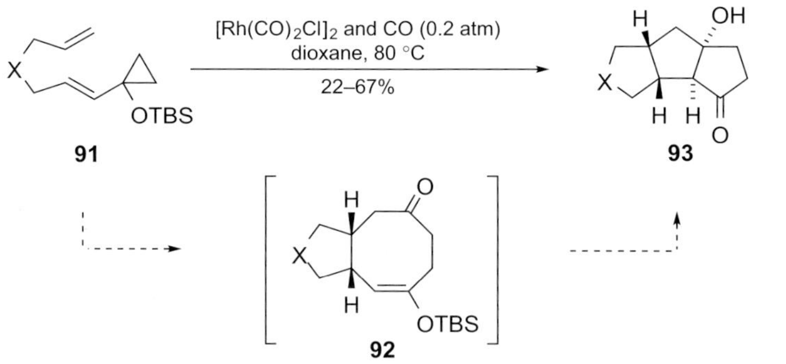

Scheme 8.27 Rh(I)-catalyzed [5+2+1]-cycloaddition–aldol reaction of dienylcyclopropanes and CO [49].

Rh(I)-catalyzed [5+2+1]-cycloadditions of vinylcyclopropanes, alkynes, and CO followed by a transannular aldol reaction were reported by Wender and coworkers [48], which produced diquinane skeletons in good yields. In an elegant extension of this strategy, Yu and coworkers [49] established a two-component coupling reaction of dienyl cyclopropanes **91** and CO, which furnished linear triquinanes **93** in moderate to good yields via the intermediate cyclooctanone **92** (Scheme 8.27). Heteroatoms in the tether were readily tolerated, giving rise to heterocyclic products. The triquinane natural product hirsutene was synthesized along this route in 11% overall yield, requiring only eight steps from commercially available starting materials.

8.4 Conclusion and Outlook

The aldol reaction has been employed numerously in domino-type transformations to furnish complex and highly functionalized molecules in just one synthetic operation. In combination with subsequent lactonization, acetalization, as well as Prins and Tishchenko processes, the aldol step is used as a prelude for the synthesis of the requisite starting material, which is then converted into the final product *in situ* through one of the indicated reactions. Alternatively, conjugate addition, isomerization, rearrangement, and cycloaddition reactions may be used to generate the enolate component, which is then converted into the final product in the subsequent aldol step. Both strategies have been very successfully implemented into modern reaction design and the assembly of complex natural products. It is almost certain that new combinations and more valuable applications of such domino-type aldol reactions are yet to come in the future.

References

1. (a) Yang, H.W. and Romo, D. (1997) *J. Org. Chem.*, **62**, 4–5; (b) Zhao, C., Mitchell, T.A., Vallakati, R., Perez, L.M., and Romo, D. (2012) *J. Am. Chem. Soc.*, **134**, 3084–3094.
2. Wang, Y., Zhao, C., and Romo, D. (1999) *Org. Lett.*, **1**, 1197–1199.
3. (a) Yang, H.W., Zhao, C., and Romo, D. (1997) *Tetrahedron*, **53**, 16471–16488; (b) Ma, G., Zancanella, M., Oyola, Y., Richardson, R.D., Smith, J.W., and Romo, D. (2006) *Org. Lett.*, **8**, 4497–4500.
4. (a) Mitchell, T.A., Zhao, C., and Romo, D. (2008) *J. Org. Chem.*, **73**, 9544–9551; (b) Mitchell, T.A., Zhao, C., and Romo, D. (2008) *Angew. Chem. Int. Ed.*, **47**, 5026–5029.
5. (a) Vincent, J.M., Margottin, C., Berlande, M., Cavagnat, D., Buffeteau, T., and Landais, Y. (2007) *Chem. Commun.*, 4782–4784; (b) Dambruoso, P., Massi, A., and Dondoni, A. (2005) *Org. Lett.*, **7**, 4657–4660.
6. Jiang, X., Cao, Y., Wang, Y., Liu, L., Shen, F., and Wang, R. (2010) *J. Am. Chem. Soc.*, **132**, 15328–15333.
7. Kopecky, D.J. and Rychnovsky, S.D. (2001) *J. Am. Chem. Soc.*, **123**, 8420–8421.
8. Patterson, B., Marumoto, S., and Rychnovsky, S.D. (2003) *Org. Lett.*, **5**, 3163–3166.
9. (a) Li, H. and Loh, T.P. (2008) *J. Am. Chem. Soc.*, **130**, 7194–7195; (b) Li, H. and Loh, T.P. (2010) *Org. Lett.*, **12**, 2679–2681.

10. Hazelard, D., Ishikawa, H., Hashizume, D., Koshino, H., and Hayashi, Y. (2008) *Org. Lett.*, **10**, 1445–1448.
11. Ramachary, D.B. and Sakthidevi, R. (2009) *Chem. Eur. J.*, **15**, 4516–4522.
12. Mahrwald, R. and Costisella, B. (1996) *Synthesis*, 1087–1089.
13. Bodnar, P.M., Shaw, J.T., and Woerpel, K.A. (1997) *J. Org. Chem.*, **62**, 5674–5675.
14. Mascarenhas, C.M., Miller, S.P., White, P.S., and Morken, J.P. (2001) *Angew. Chem. Int. Ed.*, **113**, 621–623; *Angew. Chem. Int. Ed.*, 2001, **40**, 601–603.
15. Gnanadesikan, V., Horiuchi, Y., Ohshima, T., and Shibasaki, M. (2004) *J. Am. Chem. Soc.*, **126**, 7782–7783.
16. (a) Mlynarski, J., Jankowska, J., and Ralie, B. (2005) *Chem. Commun.*, 4854–4856; (b) Mlynarski, J., Rakiel, B., Stodulski, M., Suszczynska, A., and Frelek, J. (2006) *Chem. Eur. J.*, **12**, 8158–8167.
17. (a) Schneider, C. and Hansch, M. (2001) *Chem. Commun.*, 1218–1219; (b) Schneider, C., Hansch, M., and Weide, T. (2005) *Chem. Eur. J.*, **11**, 3010–3021.
18. (a) Simpura, I. and Nevalainen, V. (2001) *Tetrahedron Lett.*, **42**, 3905–3908; (b) Simpura, I. and Nevalainen, V. (2003) *Tetrahedron*, **59**, 7535–7546.
19. (a) Schneider, C. and Hansch, M. (2003) *Synlett*, 837–840; (b) Schneider, C., Hansch, M., and Sreekumar, P. (2006) *Tetrahedron: Asymmetry*, **17**, 2738–2742.
20. Lesch, B., Toraeng, J., Vanderheiden, S., and Bräse, S. (2005) *Adv. Synth. Catal.*, **347**, 555–562.
21. (a) Bröhmer, M.C., Bourcet, E., Nieger, M., and Bräse, S. (2011) *Chem. Eur. J.*, **17**, 13706–13711; (b) Liu, K. and Woggon, W.D. (2010) *Eur. J. Org. Chem.*, 1033–1036; (c) Volz, N., Bröhmer, M.C., Toräng, J., Bräse, S., and Nieger, M. (2009) *Indian J. Chem., Sect. B*, **48**, 1699–1703; (d) Volz, N., Bröhmer, M.C., Bräse, S., and Nieger, M. (2009) *Synlett*, 550–553; (e) Liu, K., Chougnet, A., and Woggon, W.D. (2008) *Angew. Chem. Int. Ed.*, **47**, 5827–5829.
22. Rueping, M., Kuenkel, A., Tato, F., and Bats, J.W. (2009) *Angew. Chem. Int. Ed.*, **48**, 3699–3702.
23. (a) Gryko, D. (2005) *Tetrahedron: Asymmetry*, **16**, 1377–1383; (b) Halland, N., Aburel, P.S., and Jørgensen, K.A. (2004) *Angew. Chem. Int. Ed.*, **43**, 1272–1277; (c) Pulkkinen, J., Aburel, P.S., Halland, N., and Jørgensen, K.A. (2004) *Adv. Synth. Catal.*, **346**, 1077–1080.
24. Suzuki, M., Yanagisawa, A., and Noyori, R. (1988) *J. Am. Chem. Soc.*, **110**, 4718–4726.
25. (a) Arnold, L.A., Naasz, R., Minnaard, A.J., and Feringa, B.L. (2002) *J. Org. Chem.*, **67**, 7244–7254; (b) Arnold, L.A., Naasz, R., Minnaard, A.J., and Feringa, B.L. (2001) *J. Am. Chem. Soc.*, **123**, 5841–5842.
26. (a) Bocknack, B.M., Wang, L.C., and Krische, M.J. (2004) *Proc. Natl. Acad. Sci. U.S.A.*, **101**, 5421–5424; (b) Cauble, D.F., Gipson, J.D., and Krische, M.J. (2003) *J. Am. Chem. Soc.*, **125**, 1110–1111.
27. (a) Schneider, C. and Reese, O. (2002) *Chem. Eur. J.*, **8**, 2585–2594; (b) Schneider, C. and Reese, O. (2000) *Angew. Chem. Int. Ed.*, **39**, 2948–2950.
28. Yoshida, K., Ogasawara, M., and Hayashi, T. (2002) *J. Am. Chem. Soc.*, **124**, 10984–10985.
29. (a) Duan, S.W., Li, Y., Liu, Y.Y., Zou, Y.Q., Shi, D.Q., and Xiao, W.J. (2012) *Chem. Commun.*, **48**, 5160–5162; (b) Ling, J.B., Su, Y., Zhu, H.L., Wang, G.Y., and Xu, P.F. (2012) *Org. Lett.*, **14**, 1090–1093; (c) Luo, G., Zhang, S., Duan, W., and Wang, W. (2009) *Tetrahedron Lett.*, **50**, 2946–2948; (d) Brandau, S., Maerten, E., and Jørgensen, K.A. (2006) *J. Am. Chem. Soc.*, **128**, 14986–14991.
30. (a) Dong, X.Q., Fang, X., Tao, H.Y., Zhou, X., and Wang, C.J. (2012) *Chem. Commun.*, **48**, 7238–7240; (b) Pavan, M.P., Reddy, M.N., Kumar, N.N.B., and Swamy, K.C.K. (2012) *Org. Biomol. Chem.*, **10**, 8113–8118; (c) Gao, Y., Ren, Q., Wu, H., Li, M., and Wang, J. (2010) *Chem. Commun.*, **46**, 9232–9234; (d) Zhao, G.L., Vesely, J., Rios, R., Ibrahem, I., Sunden, H., and Cordova, A. (2008) *Adv. Synth. Catal.*, **350**, 237–242; (e) Zu, L., Wang, J., Li, H., Xie, H., Jiang, W., and Wang, W. (2007) *J. Am. Chem. Soc.*, **129**,

1036–1037; (f) Zu, L., Xie, H., Li, H., Wang, J., Jiang, W., and Wang, W. (2007) *Adv. Synth. Catal.*, **349**, 1882–1886.

31. (a) Wang, W., Li, H., Wang, J., and Zu, L. (2006) *J. Am. Chem. Soc.*, **128**, 10354–10355; (b) Rios, R., Sunden, H., Ibrahem, I., Zhao, G.L., Eriksson, L., and Cordova, A. (2006) *Tetrahedron Lett.*, **47**, 8547–8551.
32. Reyes, E., Talavera, G., Vicario, J.L., Badia, D., and Carrillo, L. (2009) *Angew. Chem. Int. Ed.*, **48**, 5701–5704.
33. (a) Zhang, X., Song, X., Li, H., Zhang, S., Chen, X., Yu, X., and Wang, W. (2012) *Angew. Chem. Int. Ed.*, **51**, 7282–7286; (b) Liu, C.L., Zhang, X.S., Wang, R., and Wang, W. (2010) *Org. Lett.*, **12**, 4948–4951.
34. Qian, H., Zhao, W., Sung, H.H.Y., Williams, I.D., and Sun, J. (2013) *Chem. Commun.*, **49**, 4361–4363.
35. (a) Lee, S.I., Jang, J.H., Hwang, G.S., and Ryu, D.H. (2013) *J. Org. Chem.*, **78**, 770–775; (b) Senapati, B.K., Hwang, G.S., Lee, S., and Ryu, D.H. (2009) *Angew. Chem. Int. Ed.*, **48**, 4398–4401.
36. Sloman, D.L., Bacon, J.W., and Porco, J.A. (2011) *J. Am. Chem. Soc.*, **133**, 9952–9955.
37. (a) Han, S.B., Hassan, A., and Krische, M.J. (2008) *Synthesis*, 2669–2679; (b) Ngai, M.Y., Kong, J.R., and Krische, M.J. (2007) *J. Org. Chem.*, **72**, 1063–1072.
38. (a) Bee, C., Han, S.B., Hassan, A., Iida, H., and Krische, M.J. (2008) *J. Am. Chem. Soc.*, **130**, 2746–2747; (b) Han, S.B. and Krische, M.J. (2006) *Org. Lett.*, **8**, 5657–5660; (c) Jung, C.K., Garner, S.A., and Krische, M.J. (2006) *Org. Lett.*, **8**, 519–522; (d) Jung, C.K. and Krische, M.J. (2006) *J. Am. Chem. Soc.*, **128**, 17051–17056; (e) Jang, H.Y., Huddleston, R.R., and Krische, M.J. (2002) *J. Am. Chem. Soc.*, **124**, 15156–15157.
39. (a) Zhao, C.X., Duffey, M.O., Taylor, S.J., and Morken, J.P. (2001) *Org. Lett.*, **3**, 1829–1831; (b) Taylor, S.J., Duffey, M.O., and Morken, J.P. (2000) *J. Am. Chem. Soc.*, **122**, 4528–4529.
40. Deschamp, J., Chuzel, O., Hannedouche, J., and Riant, O. (2006) *Angew. Chem. Int. Ed.*, **45**, 1292–1297.
41. (a) Sugiura, M., Sato, N., Sonada, Y., Kotani, S., and Nakajima, M. (2010) *Chem. Asian J.*, **5**, 478–481; (b) Sugiura, M., Sato, N., Kotani, S., and Nakajima, M. (2008) *Chem. Commun.*, 4309–4311.
42. Gazzard, L.J., Motherwell, W.B., and Sandham, D.A. (1999) *J. Chem. Soc., Perkin Trans. 1*, 979–993.
43. (a) Crevisy, C., Wietrich, M., LeBoulaire, V., Uma, R., and Gree, R. (2001) *Tetrahedron Lett.*, **42**, 395–398; (b) Uma, R., Gouault, N., Crevisy, C., and Gree, R. (2003) *Tetrahedron Lett.*, **44**, 6187–6190.
44. Uma, R., Davies, M., Crevisy, C., and Gree, R. (2001) *Tetrahedron Lett.*, **42**, 3069–3072.
45. Cuperly, D., Petrignet, J., Crevisy, C., and Gree, R. (2006) *Chem. Eur. J.*, **12**, 3261–3274.
46. (a) Bertrand, M.B. and Wolfe, J.P. (2006) *Org. Lett.*, **8**, 4661–4663; (b) Giampietro, N.C., Kampf, J.W., and Wolfe, J.P. (2009) *J. Am. Chem. Soc.*, **131**, 12556–12557.
47. (a) Belanger, G. and Deslongchamps, P. (2000) *Org. Lett.*, **2**, 285–287; (b) Belanger, G. and Deslongchamps, P. (2000) *J. Org. Chem.*, **65**, 7070–7074.
48. Wender, P., Gamber, G.G., Hubbard, R.D., and Zhang, L. (2002) *J. Am. Chem. Soc.*, **124**, 2876–2877.
49. Jiao, L., Yuan, C., and Yu, Z.X. (2008) *J. Am. Chem. Soc.*, **130**, 4421–4430.

9
Oxidations and Reductions in Domino Processes

Govindasamy Sekar, Iyyanar Karthikeyan, and Dhandapani Ganapathy

9.1
Introduction

Oxidation and reduction reactions play a vital role in the field of synthetic organic chemistry. Mild, selective, and economical catalytic oxidations as well as reduction reactions are recent developments in modern synthetic organic chemistry [1]. These transformations in combination with other reactions in a domino fashion give synthetically challenging organic products or intermediates in a very short and economical way [2]. After the first book by Tietze *et al.* [2a] in 2006 about domino reactions, where domino reactions initiated by oxidation or reduction reactions are reviewed as a separate chapter, enormous developments have taken place in the synthesis of a multitude of important organic compounds using oxidative or reductive domino reactions.

Since oxidation and reduction reactions can provide many organic compounds with reactive functional groups such as aldehydes, ketones, enones, amines, alcohol, allylic alcohols, and so on, further transformations can easily be added to give a domino process. Depending upon the position of the oxidation or reduction reaction in the domino process, this chapter is divided into three classes: first, the domino reaction is initiated by an oxidation or reduction reaction; second, the domino reaction has the oxidation or reduction step in the middle; and third, the domino reaction is terminated by an oxidation or reduction reaction. Most of the oxidation and reduction reactions come under the category of anionic domino process, as they provide nucleophilic or electrophilic functionalities and only very few oxidation and reduction reactions proceed with cationic domino process.

Domino Reactions: Concepts for Efficient Organic Synthesis, First Edition. Edited by Lutz F. Tietze.

9.2
Domino Reactions Initiated by Oxidation or Reduction Reaction

9.2.1
Domino Reactions Initiated by an Oxidation Reaction

This section deals with oxidation reactions as an initiative step of the domino process in combination with many other organic transformations. In 2004, Snaith and coworkers [3] demonstrated a simple and efficient method to synthesize 3-substituted 4-piperidinones **4** using a domino oxidation/carbonyl-ene/oxidation reaction (Scheme 9.1). This domino reaction comprises oxidation of unsaturated alcohol **1** using PCC (pyridinium chlorochromate) to give the corresponding aldehyde **2**, followed by a carbonyl-ene type reaction to yield the secondary cyclic alcohol **3**, which was oxidized under the reaction conditions to give 3-substituted 4-piperidinones **4** in good yields.

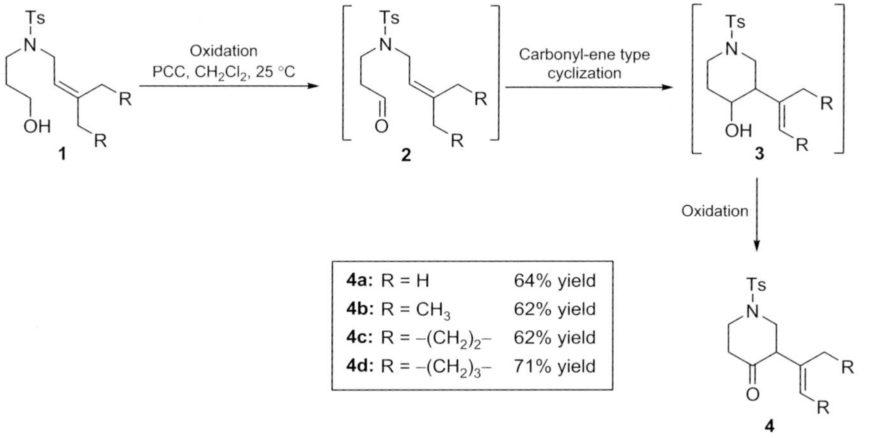

4a: R = H		64% yield
4b: R = CH_3		62% yield
4c: R = $-(CH_2)_2-$		62% yield
4d: R = $-(CH_2)_3-$		71% yield

Scheme 9.1 Domino oxidation/cyclization/oxidation process for the synthesis of **4**.

Porco *et al.* [4] reported a synthesis of the interleukin-1β-converting-enzyme inhibitor EI-1941-2 (**9**) using Bobbitt's reagent **10** (Scheme 9.2). In this synthesis, the key step is the selective oxidation of primary alcohol **5** to give the corresponding aldehyde **6**, which undergoes 6π-electrocyclization to provide **7** followed by oxidation to yield α-pyrone **8** in 58% yield. The compound **8** was further converted into the target molecule **9** in a few steps.

A domino reaction involving two consecutive oxidations followed by a hetero Diels–Alder reaction led to the formation of bicyclic dihydropyran **14** in 74% yield (Scheme 9.3). In this process, developed by Koo *et al.* [5] using $Mn(OAc)_3$ and $Cu(OAc)_2$, the β-ketoester **11** having a geranyl group at the α-position is converted into the α-ketoester **12**, which undergoes further oxidation to form the tertiary alcohol **13** and is followed by an intramolecular hetero Diels–Alder reaction to provide the desired bicyclic trans-fused dihydro cyclopenta[*c*]pyran **14**. In 2012,

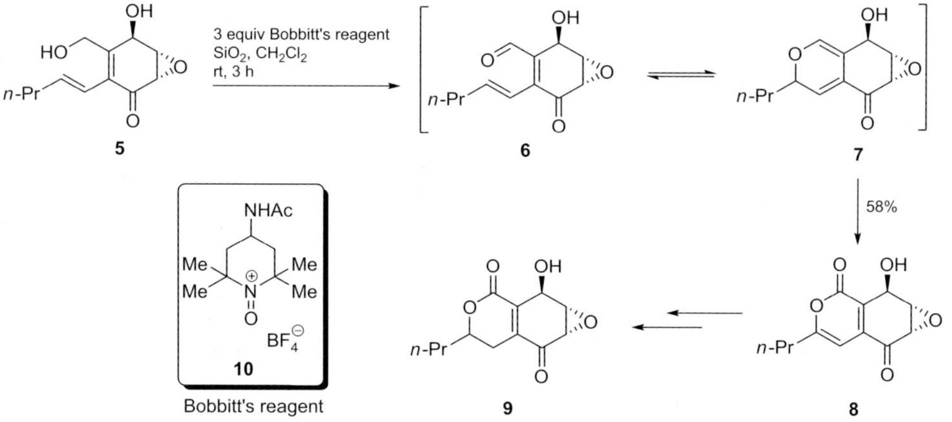

Scheme 9.2 Total synthesis of **9** by domino oxidation/oxa-electrocyclization/oxidation process.

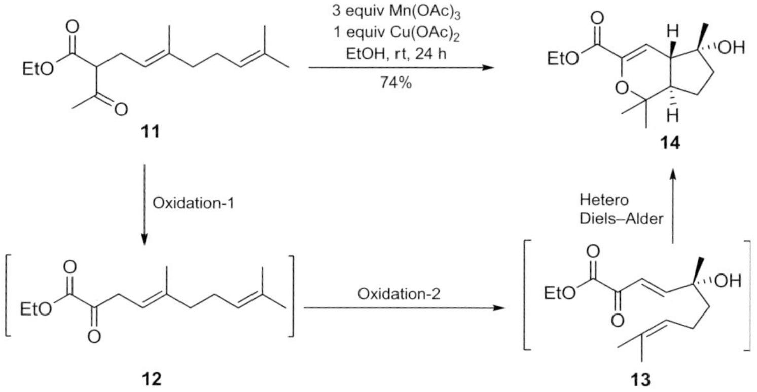

Scheme 9.3 Mn(III) initiated oxidation/oxidation/hetero Diels–Alder reaction of **11**.

Koo *et al.* [6] published the mechanistic evidence for this domino reaction. In addition, there are several reports available in the literature dealing with similar oxidative cyclizations to synthesize spiro-lactones [7], azoles [8], and isotwistanes [9].

In 2009, Donohoe *et al.* reported a total synthesis of the natural products (+)-*cis*-sylvaticin **19** and (+)-sylvaticin **20** using a Sharpless asymmetric dihydroxylation of diene **15** with an osmium catalyst as key step to give the diol **16**. This was followed by an oxidative cyclization to yield the functionalized tetrahydrofuran (THF) **18** stereoselectively through the cyclic intermediate **17** (Scheme 9.4) [10].

Very recently, Donohoe *et al.* [11] used a similar domino process in the efficient synthesis of the ABC spiroacetal **23** ring system of pectenotoxin-4 **24**. In this reaction, triol **21** was converted into the *bis*-THF compound **22** with a yield of 69% using potassium osmate in an aqueous acetonitrile solution with pyridine-*N*-oxide (PNO) as a re-oxidant and a Lewis acid as promoter. The resulting compound **22**

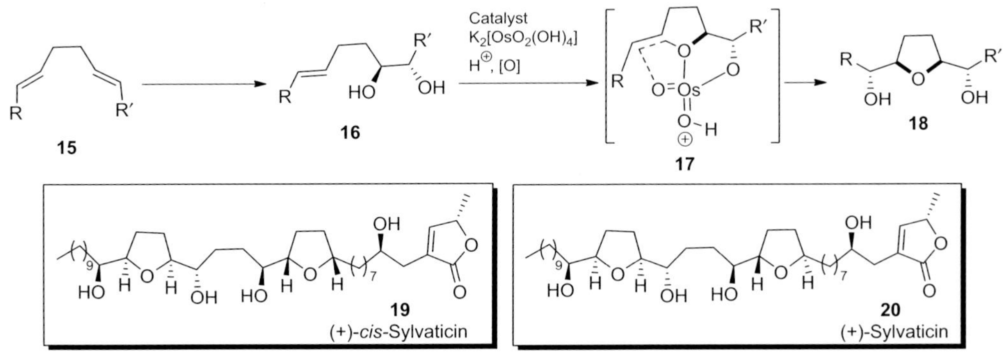

Scheme 9.4 Osmium-catalyzed oxidation/cyclization for the formation of THF derivative.

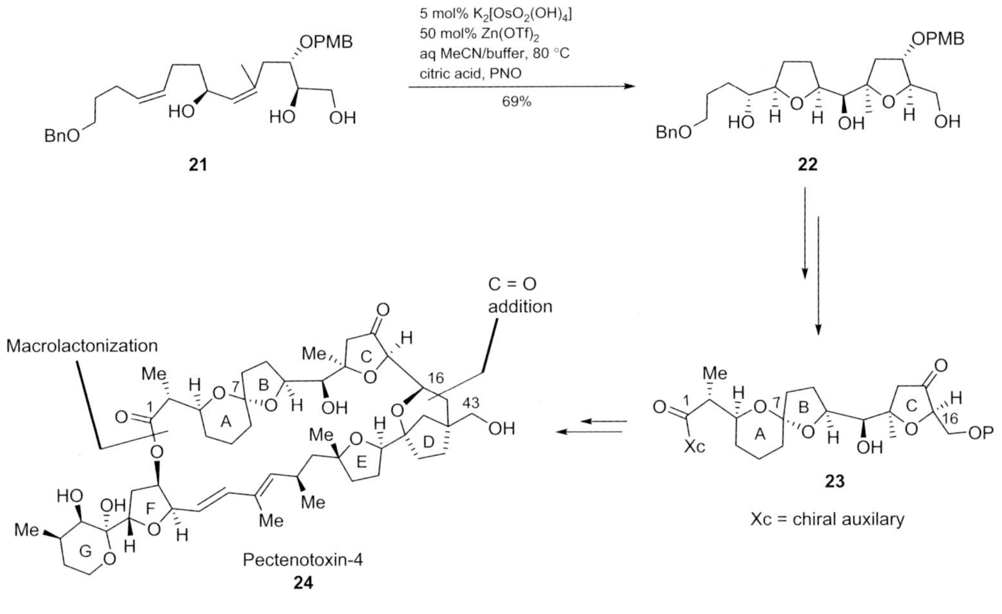

Scheme 9.5 Domino oxidative cyclization in the total synthesis of **24**.

was easily converted into the ABC spiroacetal **23** ring system of pectenotoxin-4 **24** in a few steps (Scheme 9.5).

In 2009, Beifuss *et al.* [12] developed a laccase (isolated from *Agaricus bisporus*) catalyzed domino reaction of the heterocyclic 1,3-dicarbonyl compounds **27** and catechol **25**. In this domino process, first, catechol **25** was oxidized to give *o*-benzoquinone **26**, which was then followed by a 1,4-addition of the heterocyclic 1,3-dicarbonyl compounds **27** to provide various types of annulated benzofuran derivatives **28** such as benzofuro pyridinones ($R_1 = R_2 = H$), benzofuro quinolinones ($R_1 = R_2 = C_6H_4$), and thio coumestans ($X = S$) in high yield (55–98%) (Scheme 9.6).

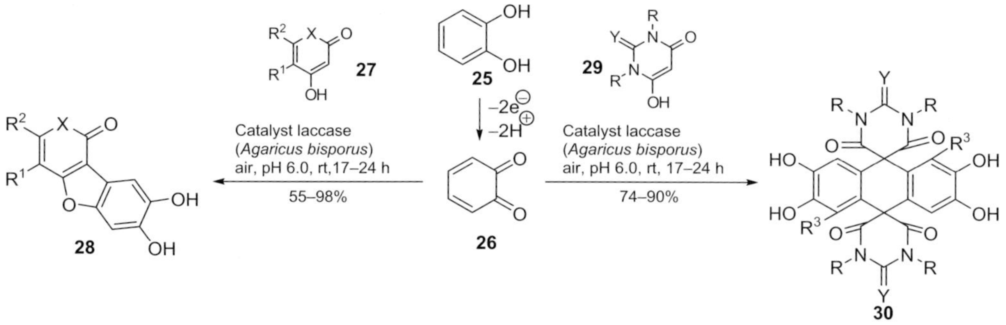

Scheme 9.6 Synthesis of annulated benzofurans and polycyclic dispiropyrimidinones.

When the same domino reaction was performed with *N*,*N*-dialkyl barbituric acid **29** as a Michael donor and catechol **25**, the reaction did not provide the expected benzofuran derivatives. Interestingly, a new type of domino reaction took place, which gave polycyclic dispiropyrimidinone **30** in excellent yield.

The first step of the domino process again was the laccase-catalyzed oxidation of catechol **25** to *o*-quinone **26**, which reacted through Michael addition with *N*,*N*-dimethyl barbituric acid **31** to yield the intermediate **32**. This compound underwent a further oxidation followed by cyclization through two Michael additions. Finally, laccase-catalyzed oxidation provided polycyclic dispiropyrimidinone **35** in excellent yield of 90% (Scheme 9.7). Recently, the same authors reported the synthesis of

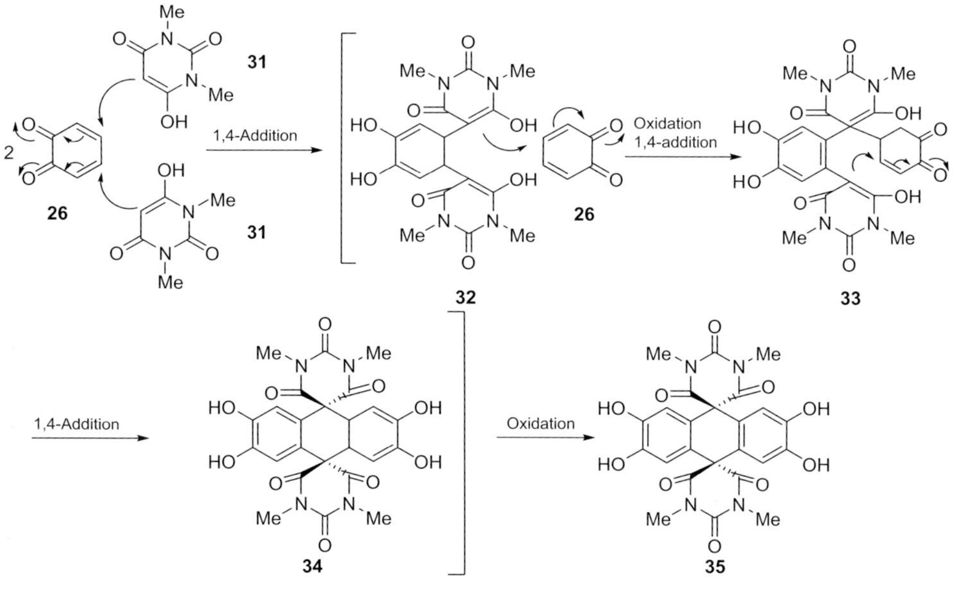

Scheme 9.7 Proposed mechanism for the formation of polycyclic dispiropyrimidinone **35**.

substituted *p*-benzoquinones in a regioselective manner by using a laccase-catalyzed oxidation followed by a 1,4-addition [13]. A similar kind of domino process was reported by Rueping *et al.* [14] using a MnO_2-promoted oxidation followed by an enantioselective Michael addition.

Metal-free iodine-promoted oxidative cyclization domino process was successfully utilized by Jiang *et al.* [15] to synthesize poly substituted oxazoles. Similarly, very efficient iodine-promoted oxidative cyclization domino process was reported by Wu *et al.* [16] for the synthesis of 2-acylbenzothiazoles **38** via a sp^3 C–H functionalization (Scheme 9.8).

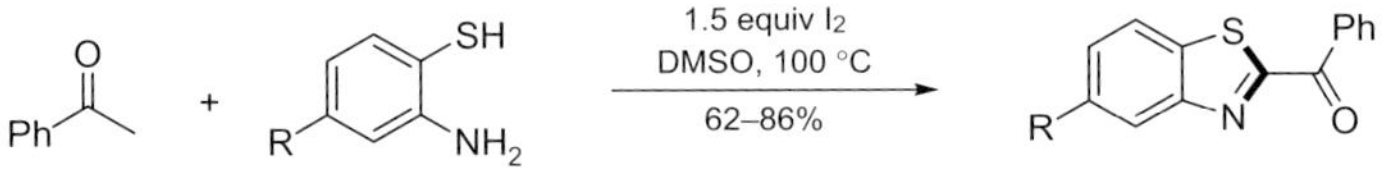

Scheme 9.8 Synthesis of 2-acyl benzothiazole through metal-free sp^3 C–H functionalization.

In this reaction, initially, acetophenone **36** was converted to α-iodo acetophenone **39** by iodine, which was further transformed into phenyl glyoxal **40** by DMSO (dimethyl sulfoxide). The formed aldehyde then reacted with 2-aminothiophenol **37** to provide intermediate **41**, which was further converted to **42** through an intramolecular 1,2-addition. Then, **42** underwent iodine-catalyzed oxidative dehydrogenation to afford the desired product 2-acyl benzothiazole **38** (Scheme 9.9). Very recently, the same authors reported the synthesis of 2,5-disubstituted oxazoles using a similar domino process [17].

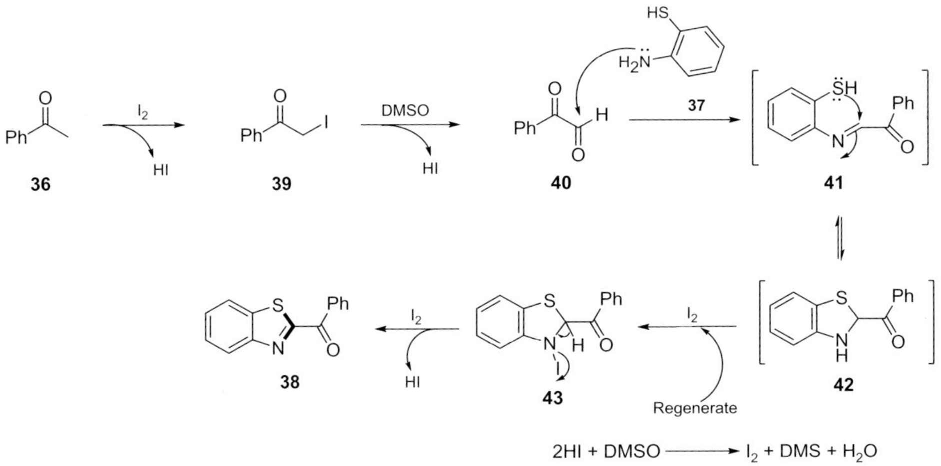

Scheme 9.9 Proposed mechanism for the synthesis of 2-acyl benzothiazole.

9.2.2
Domino Reactions Initiated by Reduction Reaction

Domino reduction/aldol cyclization was applied for the synthesis of (±)-fredericamycin A **44**, a potent antitumor and antibiotic agent, by Kelly *et al.* [18] in 1988 (Scheme 9.10). DIBAL-H (diisobutylaluminum hydride) reduction of lactone **45** afforded an intermediate **46** having both enolate and aldehyde. Subsequent cyclization via aldol reaction gave a diastereomeric mixture of spiro hydroxy ketone **47**, which was further converted into (±)-fredericamycin A **44** by sequential organic transformations. A similar strategy was employed by Mehta *et al.* [19] for the synthesis of bicyclo[3.3.1]nonan-9-one core of hyperforin, an antidepressant.

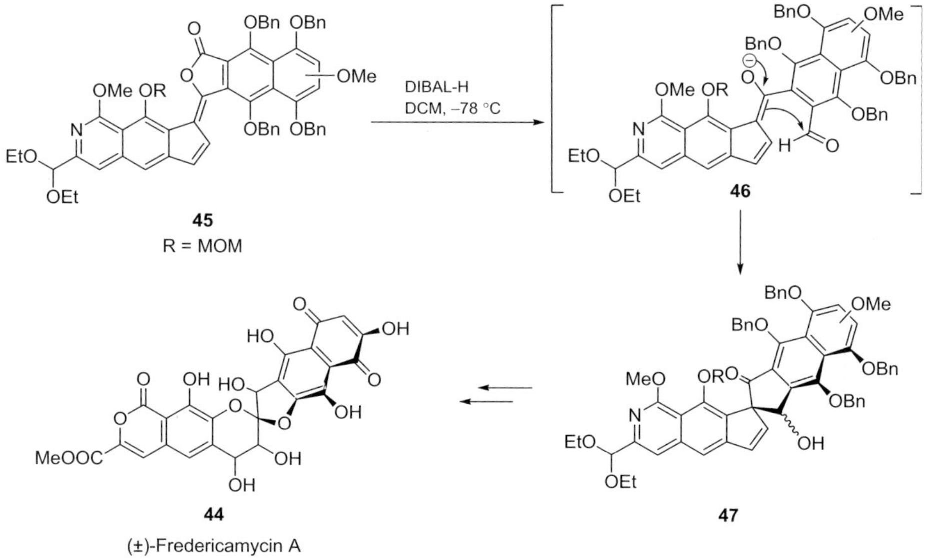

Scheme 9.10 Synthesis of the spiro fragment **47** of (±)-fredericamycin A via domino reduction/aldol reaction.

Conjugate reduction of α,β-unsaturated carbonyl compounds and subsequent carbon–carbon bond formation provides a versatile domino process of carbocycle and heterocyle synthesis [20]. These reactions proceed by the formation of enolate **49** through conjugate reduction of α,β-unsaturated carbonyl compound **48** with a reducing agent (M–H) followed by trapping of the enolate with an electrophile **50** to give **51** (Scheme 9.11).

A first example of a domino conjugate reduction/C–C bond formation was reported by Revis and Hilty in 1987. In this reaction, methyl methacrylate **52** was treated with acetone and trimethyl silane in the presence of $RhCl_3 \cdot 3H_2O$ as catalyst to afford 3-trimethylsiloxy-2,2,3-trimethylbutyrate **53** in excellent yield (Scheme 9.12) [21].

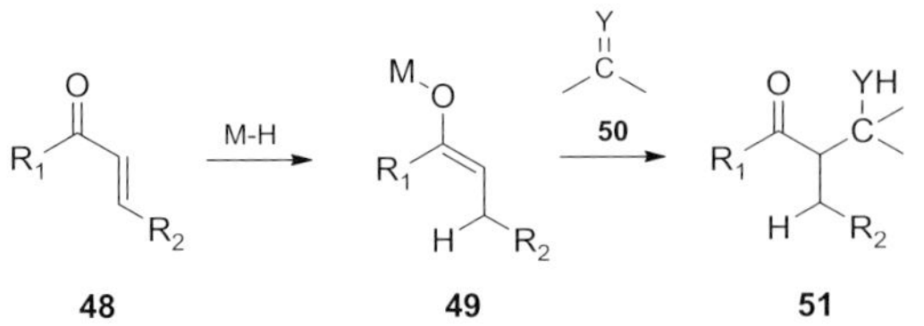

Scheme 9.11 Domino conjugate reduction/C–C bond formation.

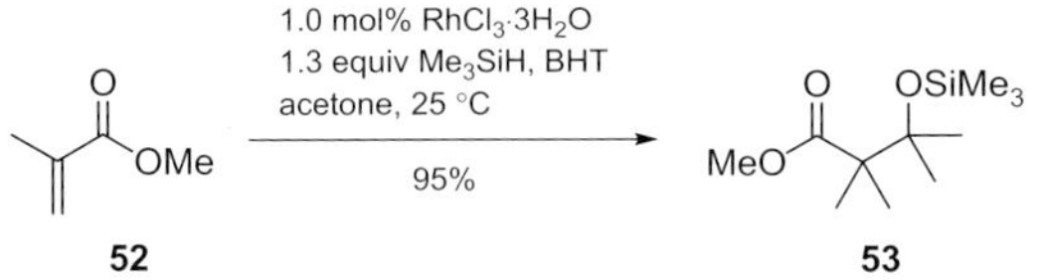

Scheme 9.12 Synthesis of β-siloxy esters by domino conjugate reduction/C–C bond formation.

Recently, extensive efforts have been made in the development of domino conjugate reduction/inter and intramolecular aldol reaction catalyzed by various metal complexes based on rhodium, iridium, cobalt, indium, nickel, and copper [22]. In 1999, Morken and Taylor [23] reported a rhodium-catalyzed conjugative reductive/aldol reaction between aldehyde and acrylate using Cl_2MeSiH as reducing agent with a maximum diastereoselectivity of 23 : 1.

Following this result, Chiu *et al.* [24] developed the conjugate reduction/intramolecular aldol reaction using stoichiometric amounts of Stryker's reagent. An intermolecular copper-catalyzed enantio- and diastereoselective reductive aldol reaction of methyl acrylate **55** with aldehyde **54** to afford diastereomeric β-hydroxy ester **56** and **57** was reported by Riant *et al.* [25, 26] (Scheme 9.13). In the same year, Shibasaki *et al.* [27] reported a similar result using ketones, which gave good to moderate enantioselectivity. Krische and coworkers developed a cobalt-catalyzed domino conjugate reduction/aldol and Michael reaction for the synthesis of cyclohexanols **61** and cyclohexanes **62** from enone aldehyde **59** and dienone **60**, respectively, with excellent diastereoselectivity (Scheme 9.14). Interestingly, the products obtained from reductive aldol cyclization

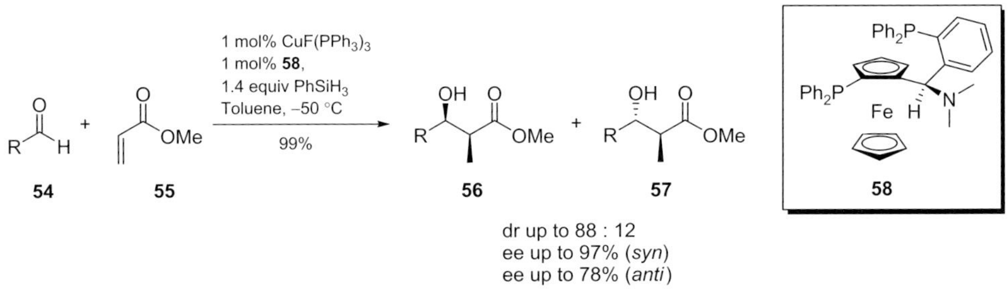

Scheme 9.13 Copper-catalyzed enantioselective domino reductive aldol reaction.

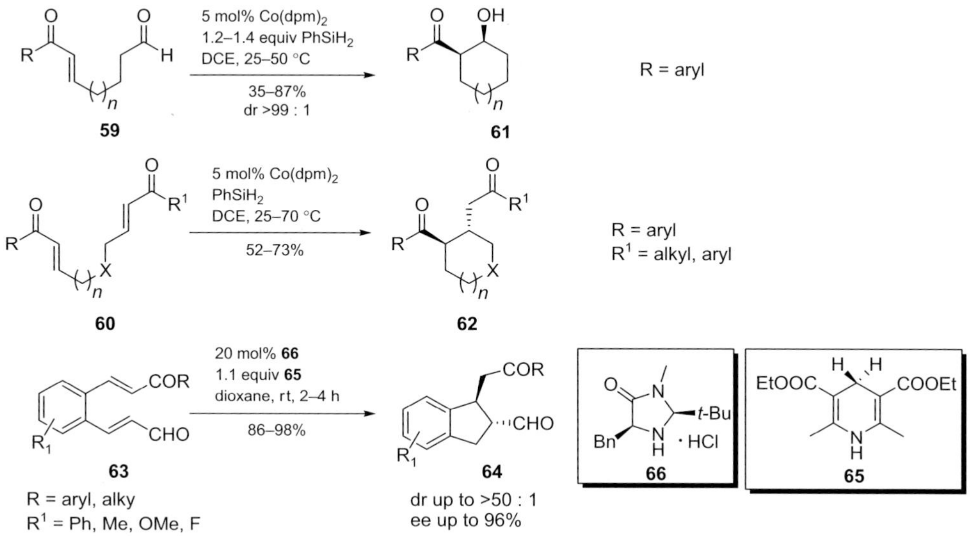

Scheme 9.14 Domino reductive aldol or Michael cyclization.

exhibited syn stereochemistry, whereas exclusively anti stereochemistry was observed for the reductive Michael cyclization [28].

List *et al.* reported an organocatalytic asymmetric conjugate reduction/Michael cyclization using an organic reducing agent. This process involved the reaction of enal–enone **63** with Hantzsch ester **65** in the presence of imidazolidinone salt **66** to give *anti*-ketoaldehyde **64** in a highly diastereo- and enantioselective manner and excellent yield [29].

Reductive generation of enolates from enone using elemental hydrogen as reducing agent followed by C–C bond formation using a rhodium catalyst was reported by Krische *et al.* [30] using domino reduction/inter or intramolecular aldol cyclization. Chiu *et al.* applied this reductive aldol cyclization methodology using Stryker's reagent $[(Ph_3P)CuH]_6$ as reducing agent for the asymmetric synthesis of lucinone **69**, an antispasmodic drug. They used ketoenone **67** as starting material, which was transformed into **69** via **68** (Scheme 9.15) [31].

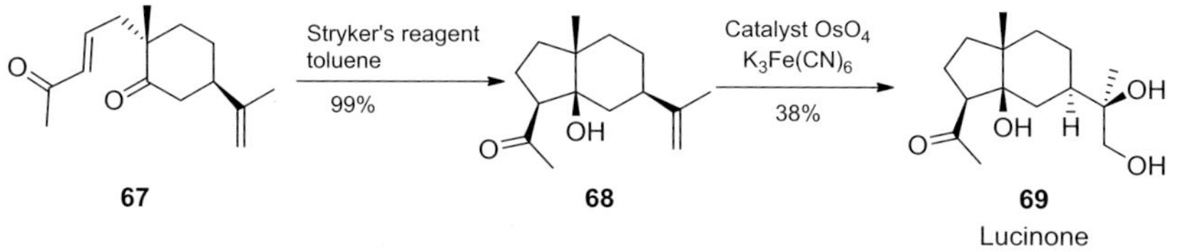

Scheme 9.15 Synthesis of lucinone.

A general mechanism of domino conjugate reduction/aldol reactions applying the *in situ* formation of metal hydrides **71** from metal complex **70** by reduction with

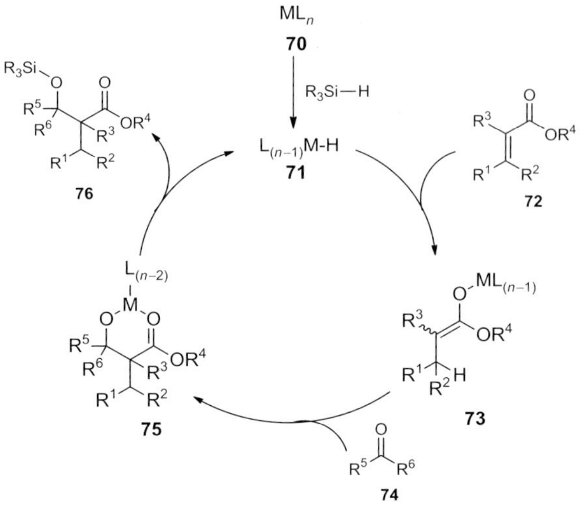

Figure 9.1 General mechanism for domino conjugate reduction/aldol reaction.

a reducing agent such as R_3SiH is shown in Figure 9.1. The conjugate reduction of enone **72** with the metal hydride leads to an enolate intermediate **73**, which reacts with an electrophile such as ketone **74** to generate metal aldolate **75**. Metathesis with silane at this stage completes the catalytic cycle, with the formation of silylated β-hydroxy esters **76** and regeneration of the catalyst.

Domino reduction/Ireland–Claisen rearrangement was used for the synthesis of γ,δ-unsaturated carboxylic acids by Miller and Morken [32], who applied this methodology for the synthesis of the C_{10}–C_{24} ketone fragment **80** of the natural product inostamycin (Scheme 9.16). In this strategy, reduction of allylic acrylate **77**

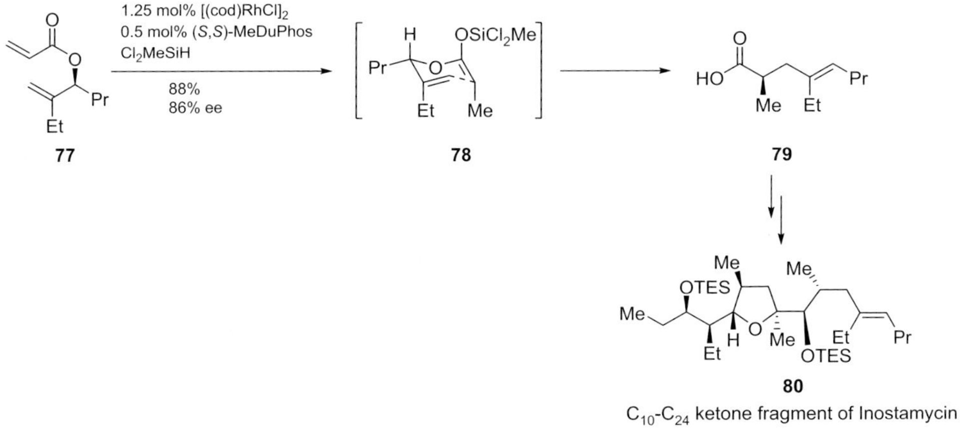

Scheme 9.16 Domino reductive conjugate reduction/Ireland–Claisen rearrangement.

with the Rh/Duphos complex and dichloromethyl silane afforded an intermediate (*E*)-silyl ketene acetal. Subsequent rearrangement via the six-membered chair-like transition state **78** gave the unsaturated carboxylic acid **79** [33].

β-Lactams have been of great synthetic interest because of their use as antibiotic agents, protease inhibitors, and chlolesterol absorption inhibitors [34]. Morken *et al.* [35] disclosed an efficient iridium-catalyzed diastereoselective synthesis of *trans*-β-lactams **83** from acrylates **82** and imines **81** through a reductive Mannich addition/cyclization (Table 9.1). The yields ranged from 58 to 80% and the diastereoselectivity exceeded 20 : 1.

Pyrrolizidinones have been found to be potent anti-inflammatory and antidepressant drugs [36]. Recently, Sukhorukov *et al.* developed a novel domino reductive ring-opening/double cyclization method for the synthesis of pyrrolizidinone **89**, which is considered as a highly potent second-generation phospho diesterase IVb inhibitor.

Table 9.1 Domino conjugate reduction/Mannich addition/cyclization for the synthesis of *trans*-β-lactams.

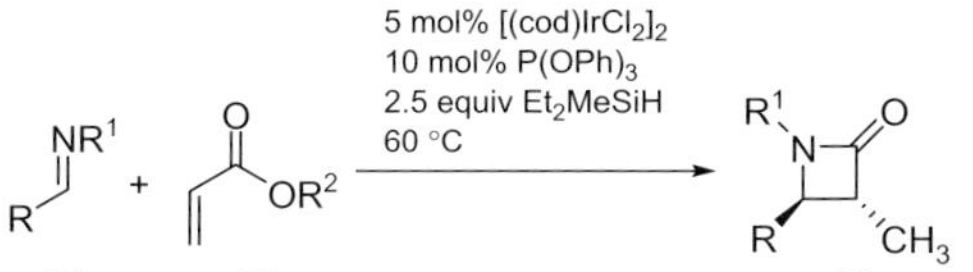

81 **82** **83**

Entry	Imine	Product	Yield (%)	*trans:cis*
1	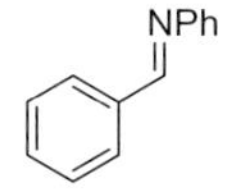	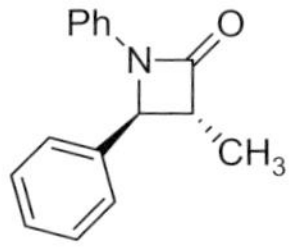	68	>20 : 1
2	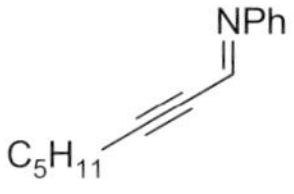	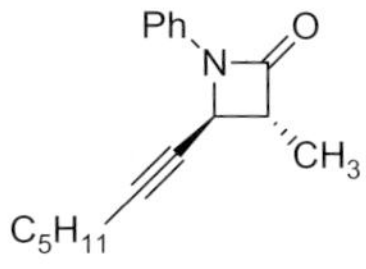	58	>20 : 1
3	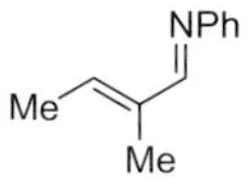	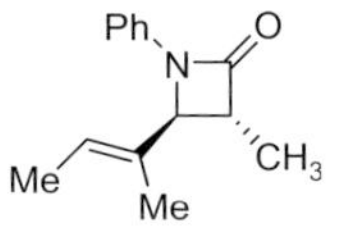	60	>20 : 1
4	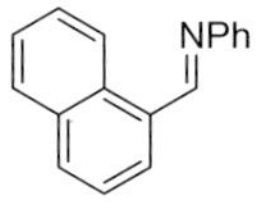	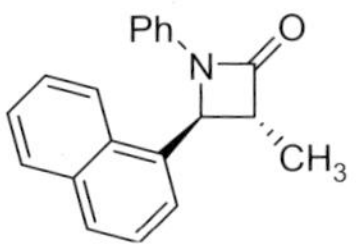	80	>20 : 1

The domino process is triggered by reductive cleavage of an N–O bond in nitronate **84** using Adams catalyst to give **86** containing a reactive aldehyde and amine functionality via **85**. Compound **86** then leads to cyclic imine **87**, which under the reaction conditions is hydrogenated to give cyclic amine **88**. The final step of the domino process is a cyclization to afford **89** (Scheme 9.17) [37].

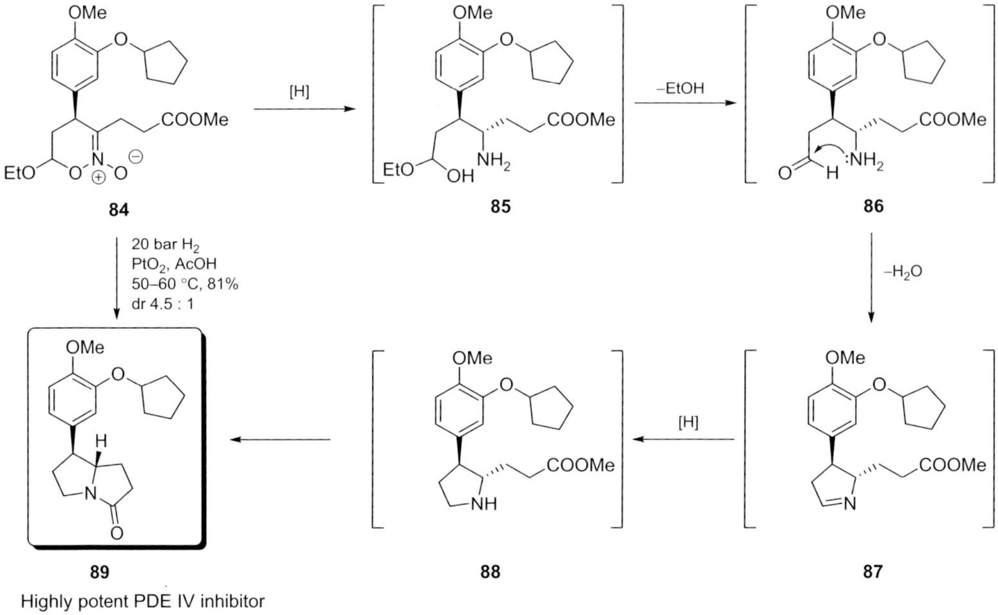

Scheme 9.17 Domino reductive ring-opening/double cyclization synthesis of phosphodiesterase (PDE) IV inhibitor.

Scholarisine A **92**, a monoterpenoid indole alkaloid with a great potency for selective inhibition of SGLT2, was synthesized by using a domino nitrile reduction/epoxide opening cyclization as key step by Smith *et al.* Compound **90** containing an epoxy and a nitrile functionality was reduced with hydrogen in the presence of Rh/Al_2O_3 to give hydroxy amine **91**, which was further manipulated to afford scholarisine A **92** (Scheme 9.18) [38].

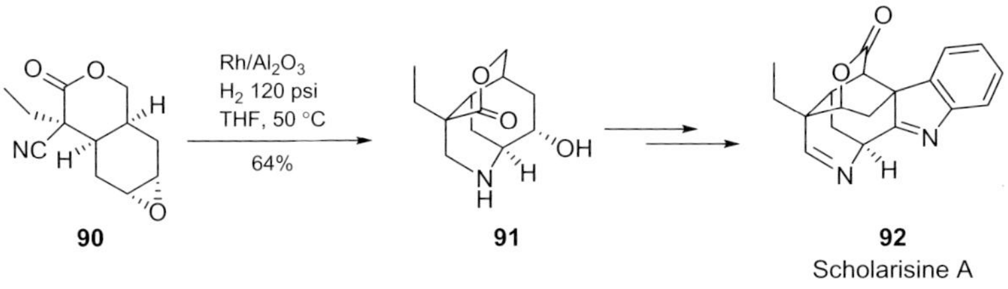

Scheme 9.18 Synthesis of scholarisine A.

A domino nitrile reduction/cyclization process was utilized by Kawasaki *et al.* [39] in the total synthesis of (−)-fructigenine A **95** and (−)-5-acetylardemin **96** from nitrile **93** by reduction with $LiAlH_4$ to give bicyclic compound **94** (Scheme 9.19). This reductive cyclization was also applied for the total synthesis of pseudophrynaminol [40], hexahydropyrrolo[2,3-b]indoles [41], and flustramines A, B, and C [42, 43].

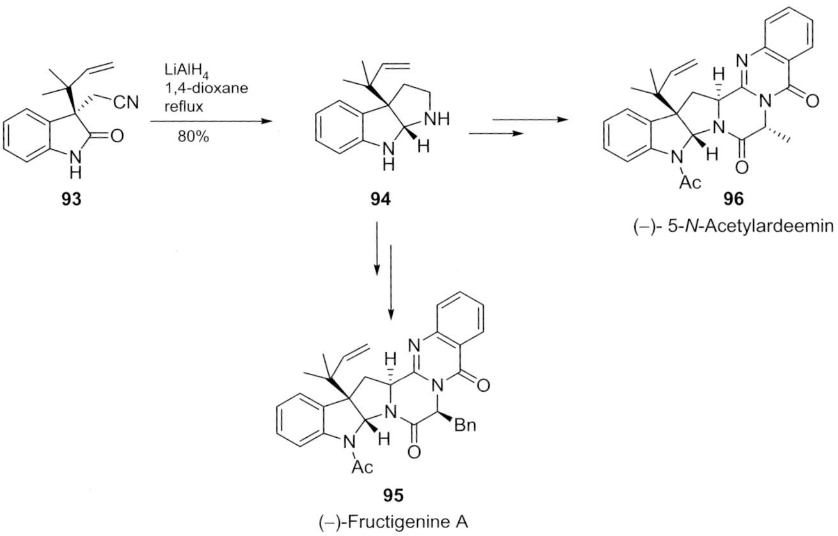

Scheme 9.19 Synthesis of (−)-fructigenine and (−)-5-acetylardeemin by nitrile reduction/cyclization.

The heterocyclic spiro-oxindole moiety is found in a large number of alkaloids, which are important pharmaceutical compounds [44]. Several synthetic protocols are reported in the literature for the synthesis of these compounds [45]. Recently, Tao *et al.* [46] developed a direct asymmetric domino Michael/reduction/cyclization sequence for the construction of spiro[2*H*-pyran-3,4′-indoline] **99** from isatylidene malononitrile **97** and acetone **98** in the presence of the chiral compound **100** (Scheme 9.20).

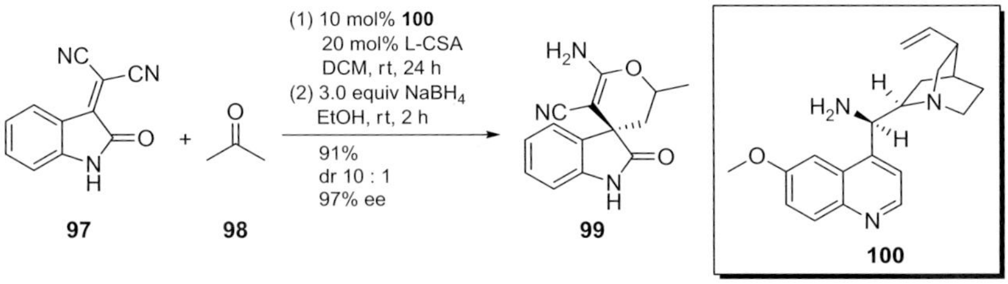

Scheme 9.20 One-pot domino Michael/reduction/cyclization synthesis of spiro-oxindole.

In 1980, Kagan *et al.* [47] reported SmI_2 as an electron-reducing agent in organic synthesis, which it is known to promote a wide range of reduction and coupling reactions in the synthesis [48]. Recently, Reisman *et al.* [49] utilized SmI_2-mediated domino reductive dialdehyde double cyclization as the key step in the total synthesis of (−)-maoecrystal Z **103**, which exhibits *in vitro* cytotoxicity against A2780 ovarian cancer cell lines (Scheme 9.21). In this domino process, dialdehyde **101** was treated with SmI_2 and LiCl in the presence of *t*-BuOH. The reaction was initiated by one-electron reduction of aldehyde followed by radical cyclization to afford diol **102** with high diastereoselectivity. Further, SmI_2 has been widely used for other natural product syntheses employing similar domino processes [50].

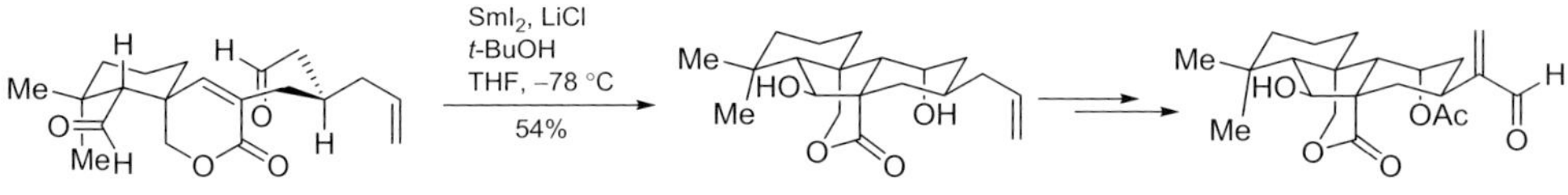

Scheme 9.21 Domino SmI_2-mediated reductive dialdehyde double cyclization: synthesis of (−)-maoecrystal Z.

Reaction of azides with triphenylphoshine generates imino phosphoranes, which on further treatment under various conditions lead to amines (Staudinger reaction) [51], imines (Staudinger/aza-Wittig) [52], and pyrroles [53] including several natural products [54]. Thus, the domino Staudinger/aza-Wittig reaction is an efficient procedure for the synthesis of N-containing heterocycles [55]. Recently, Kiessling *et al.* [56] reported an asymmetric synthesis of L-pyrolysine. A novel domino Staudinger/aza-Wittig/1,5-phosphonium rearrangement/fragmentation was reported by Langer *et al.* [57] in 2003 (Scheme 9.22). For

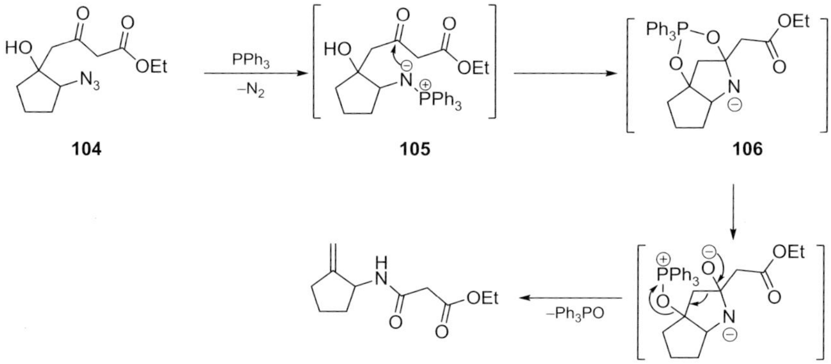

Scheme 9.22 Domino Staudinger/aza-Wittig-1,5/phosphonium rearrangement/fragmentation.

this domino process, 1-azido-2-hydroxycyclopentane **104** was treated with triphenylphosphine to afford imino phosphoranes **105**. Then a betaine intermediate **107** was formed by intramolecular nucleophilic attack of the nitrogen atom on the carbonyl carbon and an intramolecular 1,5-rearrangement via **106**. Finally, the fragmentation of **107** led to 1-acetamido-2-methylidenecyclopentane **108**.

Reductions of aromatic nitro compounds provide a simple and general access to various heterocyclic compounds through the domino process (Scheme 9.23). Quinolines are important skeletal moieties present in various natural products and biologically active compounds [58]. Most common methods of their preparation involve condensation of *o*-amino benzaldehydes with an enolizable carbonyl compound (Friedländer synthesis) [59]. Miller *et al.* [60] reported an efficient synthesis of quinolines **109**, in which a reduction of *o*-nitroaryl carbaldehyde by $SnCl_2$ followed by condensation with an enolizable carbonyl compound in the presence of $ZnCl_2$ yielded **109** through a domino process. In 2001, Bunce *et al.* [61] reported a domino nitroarene reduction/reductive amination sequence for the preparation of tetrahydroquinoline-4-carboxylic ester **110** with excellent yields.

A domino Heck/reduction/cyclization sequence has been developed for the synthesis of oxindoles **111** [62]. This process involves a Heck cross-coupling of 2-(2-nitroaryl)acrylate with aryldiazonium tetrafluroborate followed by the reduction of the nitro group under hydrogenative conditions and finally cyclization of the formed amino group with an ester. Aliamo *et al.* [63] developed an indium-catalyzed domino nitroarene reduction/aldimine formation/aza-Diels–Alder reaction for the preparation of dihydropyridin-4-one **112** and a domino nitroarene reduction/aldimine formation/Mannich-type reaction for the synthesis of β-amino esters **113**. Recently, Nguyen *et al.* [64] introduced FeS as a redox catalyst in the domino synthesis of benzimidazoles and benzoxazoles **114**, which are important building blocks for the construction of various pharmaceuticals, natural products, and agrochemicals [65]. $SnCl_2$-mediated reduction/hetrocyclization of 2-nitroacyl benzenes have been used for the preparation of anthranils **115** [66]. A similar domino nitro reduction/cyclization process was utilized for the preparation of 2-arylbenoxazoles **116** from *o*-nitrophenol and benzyl alcohols involving a hydrogen transfer mechanism [67].

The indole nucleus is a ubiquitous moiety present in various bioactive natural products and pharmaceuticals [68]. Bunce *et al.* synthesized 2-alkylindole-3-carboxylate ester **117** by treating a nitroarene and an enol with excess of Fe/AcOH. In this domino process, first, the nitro group gets reduced to give an amine moiety, which then undergoes a Michael addition to afford an unsaturated ester; the final step is the elimination of water to yield **117** [69]. Recently, Banwell *et al.* applied a domino nitroarene reduction protocol for the total synthesis of 1,5-methanoazocino[4,3-*b*]indole **118** framework of the uleine and strychnos alkaloids (Scheme 9.24). Pd/C reduction of nitroarene **119** in methanolic solution afforded annulated indole **120**, which presumably arose from the reduction of both nitro and the carbon–carbon double bond. Raney-nickel reduction produced the amine analog **121**. However, Raney-cobalt reduction resulted in the tetracyclic compound

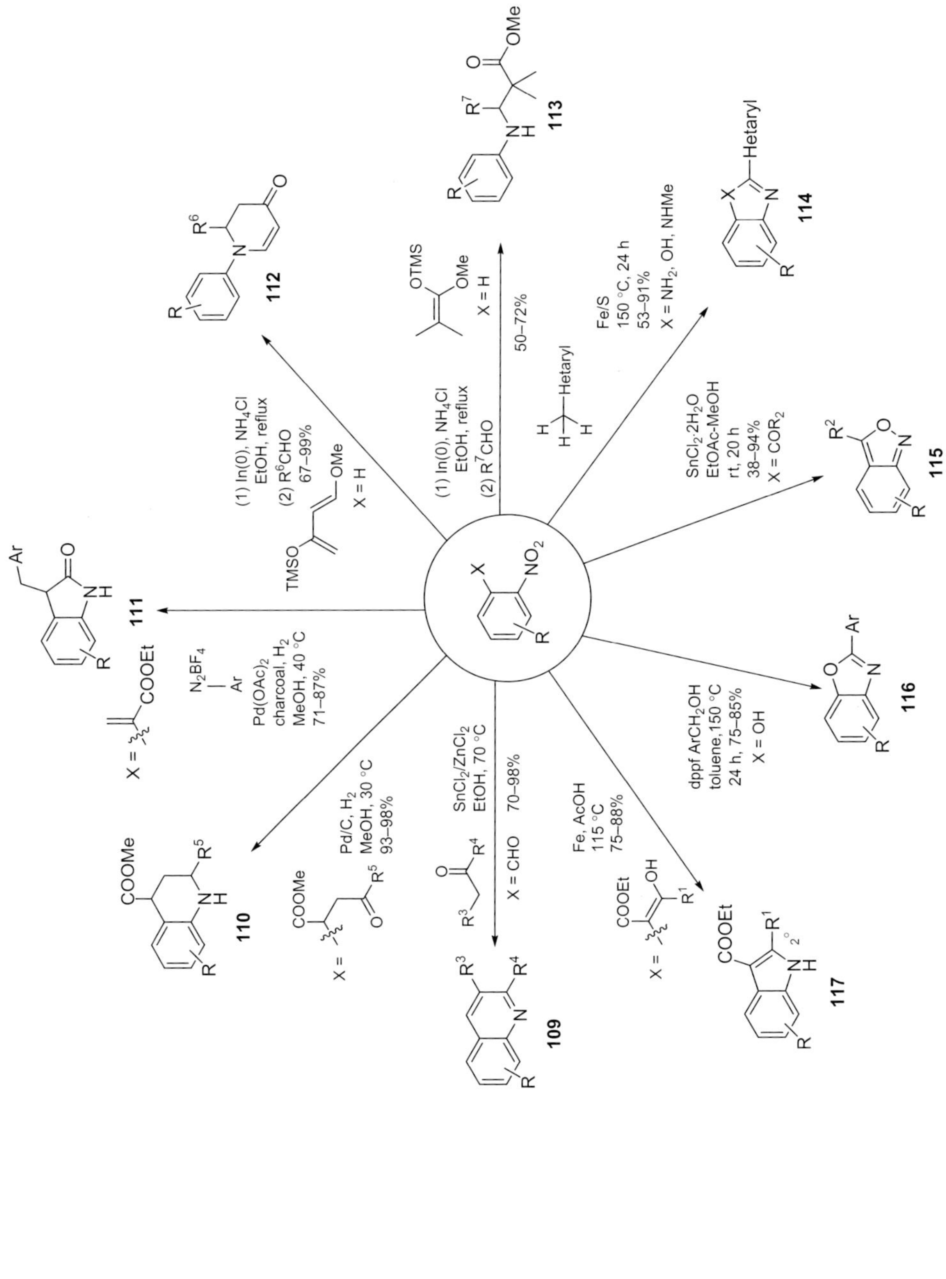

Scheme 9.23 Nitroarene reduction initiated domino process in various heterocycles synthesis.

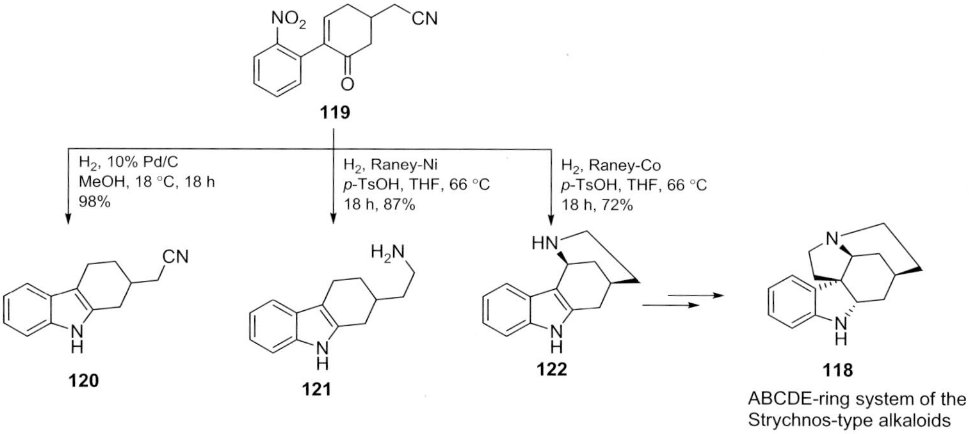

Scheme 9.24 Synthesis of 1,5-methanoazocino[4,3-*b*]indole **118** framework of the uleine and strychnos alkaloids.

122, which was further transformed to the ABCDE-ring system of the strychnos-type alkaloids **118** [70]. They also reported a similar strategy for the synthesis of (±)-limaspermidine and (±)-1-acetylaspidoalbidine [71].

Reduction-triggered pericyclic reaction has been used for the biomimetic total synthesis of endiandric acids A–G [72]. Endiandric acids are polycyclic natural products isolated from the Australian plant *Endiandra introrsa* and exist naturally as a racemic mixture which is found to have very high antibacterial effect [73]. In 1980, Black and coworkers [74] proposed a hypothesis that these polycyclic systems are formed in nature by a series of electrocyclization reactions. On the basis of this hypothesis, Nicolaou *et al.* [72c] in 1982 reported an excellent domino reduction/electrocyclization process for the synthesis of endiandric acid A **127** (Scheme 9.25). Selective hydrogenation using a Lindlar catalyst of diacetylenic diol **123** afforded the bicyclic diol **126**. The reaction presumably proceeds via polyene **124**, which then undergoes a spontaneous 8π-electrocyclization to give **125**. It follows a second 6π-electrocyclization to afford bicyclic **126**, which is further converted to endiandric acid A **127**.

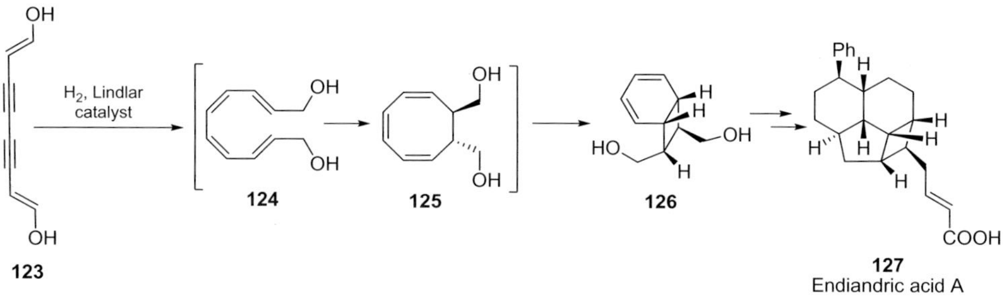

Scheme 9.25 Synthesis of endiandric acid A by domino reduction/double electrocyclization.

9.3
Domino Reactions Having Oxidation in Middle of the Sequence

An efficient domino approach to synthesize esters from aldehydes was reported by Scheidt *et al.* (Table 9.2) [75]. The domino process was carried on by reaction of aldehydes **128** with N-heterocyclic carbene (NHC) **132** used in catalytic amounts to generate alcohol **129** as the first intermediate, which was oxidized in the presence of MnO_2 to give the carbonyl compound **130** as the second intermediate. The final conversion to the desired ester **131** proceeded by a formal nucleophilic substitution of **130** with various primary or secondary alcohols present in the reaction mixture with regeneration of the carbene. Indeed, in this domino reaction, the oxidation reaction takes place in the middle of the domino sequence. Interestingly, the domino process also allows the conversion of optically pure aldehydes to the corresponding methyl esters in excellent yield without losing their optical purity. Scheidt *et al.* [76] also reported on the synthesis of esters from allylic and benzylic alcohols using catalytic amounts of NHCs. Another interesting domino reaction

Table 9.2 Synthesis of esters using NHC with MnO_2.

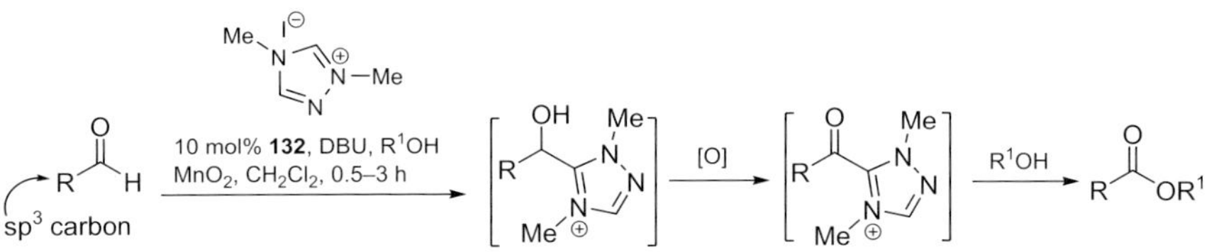

128 129 130 131

Entry	Substrate	Product	Yield (%)
1	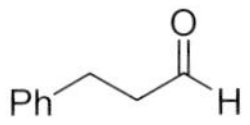	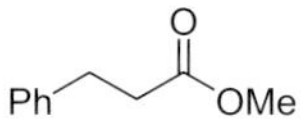	98
2	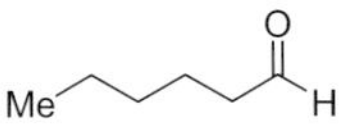	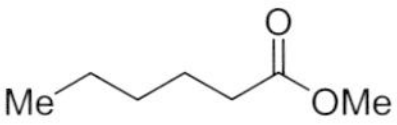	91
3	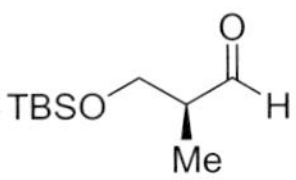	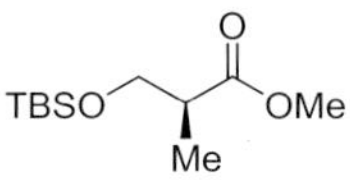	91
4	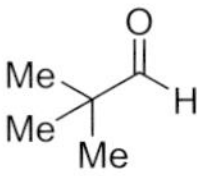	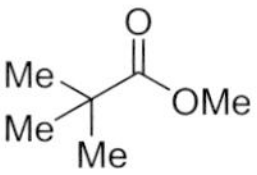	56
5	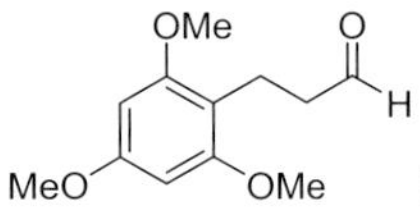	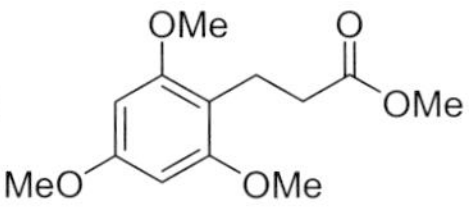	99

is the domino Knoevenagel/oxidation/hetero Diels–Alder sequence developed by Kikelj *et al.* [77] to synthesize various oxygen- and nitrogen-containing heterocycles.

9.4 Domino Reactions Terminated by Oxidation or Reduction Reaction

9.4.1 Domino Reactions Terminated by Oxidation Reaction

The first oxidative palladium-catalyzed cyclization of 1,6-enyne was reported in 2007 by Tse *et al.* [78]. In this domino process, cyclization of 1,6-enynes **133a–f** was followed by oxidation using (bisacetoxyiodo) benzene (BAIB) to provide analogs of bicyclo[3.1.0]hexane **134a–f** derivatives as the final products in good yields. The reaction does not proceed without the presence of BAIB in the reaction mixture (Table 9.3).

The proposed mechanism involves an acetoxy palladation of the triple bond in **133** to produce a vinyl-Pd(II) intermediate **135**, which further undergoes an intramolecular insertion to give **136** (Figure 9.2). This is followed by insertion through carbo palladation with the formation of bicyclic intermediate **138**, which is immediately oxidized by $PhI(OAc)_2$ to give the Pd(IV) intermediate **139** (path a). Another pathway (path b) with the formation of intermediate **139** via **137** has also

Table 9.3 Synthesis of bicyclo[3.1.0]hexane derivatives.

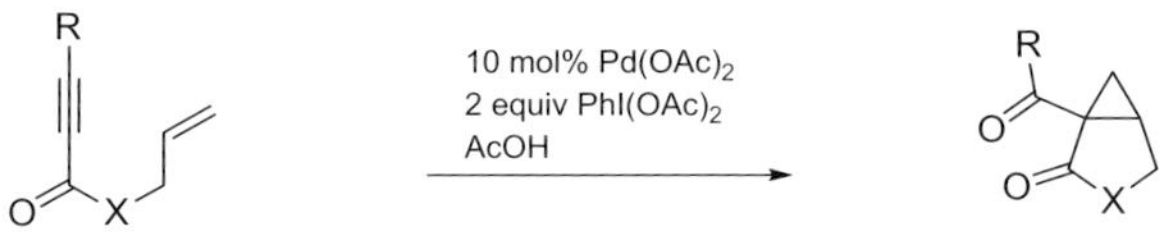

133 a–f

134 a–f

Entry	R	X	T (°C)	Time (h)	Yield (%)[a]
1	Ph	O	RT	60	79
2	Me	O	RT	54	83
3	Ph	O	80	2	0[b]
4	Ph	O	80	2	63
5	Me	O	80	2	51
6	*n*-Bu	O	80	3	75
7	2-$MeOC_6H_4$	O	80	2	80
8	*n*-Bu	NTs	80	2	83
9	Ph	NTs	80	3	77

[a] Isolated yield.
[b] Without $PhI(OAc)_2$.

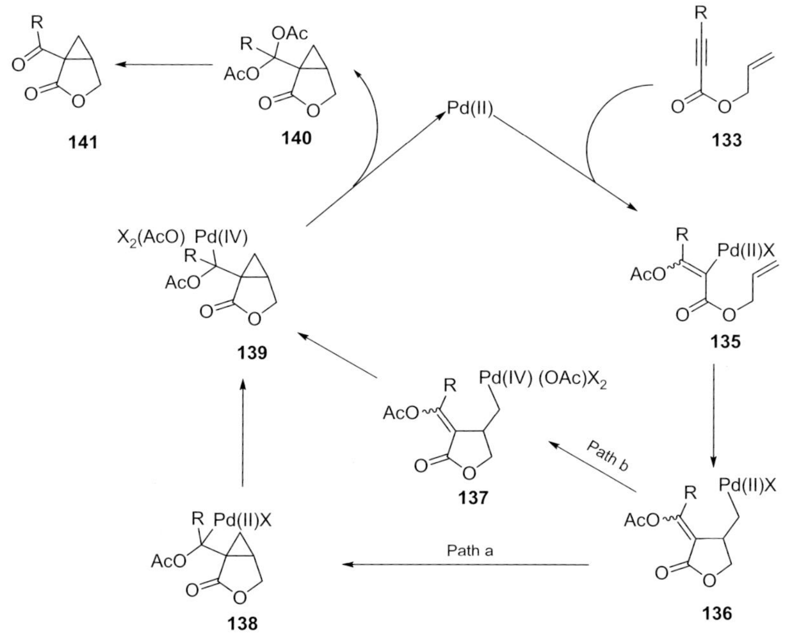

Figure 9.2 The proposed mechanism for 1,6-enyne cyclization/oxidation domino process.

been considered. C–O bond formation in **139** produces **140** with regeneration of the Pd(II) catalyst. Finally, hydrolysis of **140** yields the desired bicyclic compound **141**.

For the synthesis of quinazolinones **147**, a domino Ullmann-type coupling/aerobic oxidative C–H amidation/oxidation process was successfully utilized by Fu *et al.* [79].

In this domino process, benzylamine was reacted with 2-halobenzamides through Ullman-type coupling to yield the N-arylated intermediate **144** (Scheme 9.26). Then, a copper-catalyzed aerobic oxidation of **144** afforded the intermediate **145**, which underwent an intramolecular nucleophilic addition of the amide to provide **146**. Under the used reaction conditions, the intermediate **146** was oxidized to provide the desired products **147** in excellent yield. This protocol was also utilized by Fu *et al.* [80] by employing easily available α-amino acids instead of benzylamines to synthesize similar analogs. Very recently, the same method was applied for the synthesis of pyrimido[4,5-*b*]carbazoles by Nagarajan *et al.* [81].

9.4.2
Domino Reactions Terminated by Reduction Reaction

The Wittig reaction is an ideal procedure for the formation of carbon–carbon double bonds in organic synthesis, which includes the synthesis of α,β-unsaturated carbonyl compounds [82]. Recently, Williams and coworkers described a domino

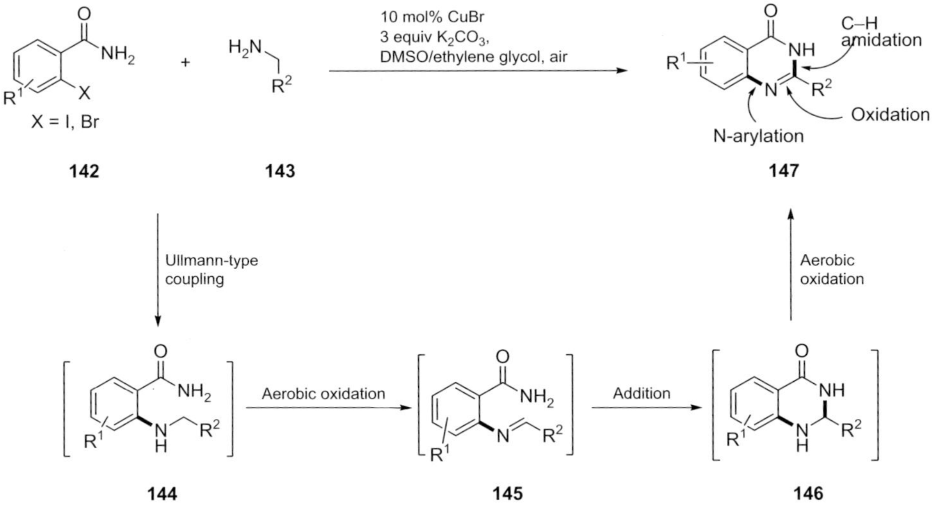

Scheme 9.26 Synthesis of quinazolinone derivatives from 2-halobenzamides.

process initiated by oxidation and terminated by conjugate reduction in combination with a Wittig reaction [83]. Here, the domino reaction proceeds by the "borrowing hydrogen approach" [84]. In this process, hydrogen is temporarily removed from a secondary alcohol such as **148** by an iridium complex to give an aldehyde **149** as intermediate. This then undergoes a Wittig reaction with a phosphonium ylide to afford α,β-unsaturated compound **150**, which on further reduction by the "borrowed hydrogen" affords alkane **151** (Scheme 9.27). Krische *et al.* [85] utilized this "borrowing hydrogen" approach for several other C–C bond formations. A similar domino Wittig reaction/conjugate reduction sequence was reported by Zhou *et al.* [86] in 2013.

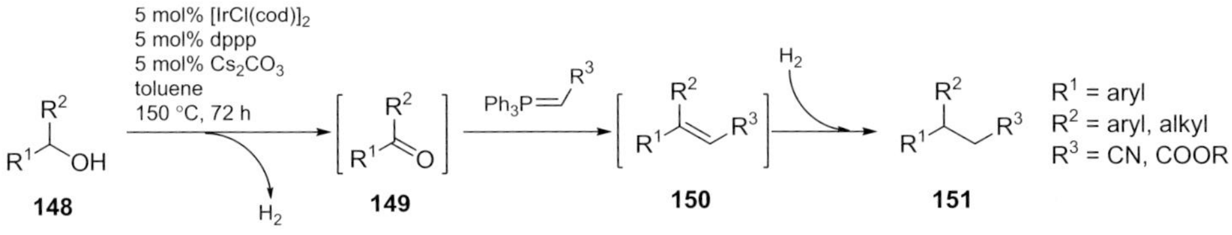

Scheme 9.27 Domino oxidation/Wittig/conjugate reduction.

A domino lactone ring-opening/reduction protocol was applied for the synthesis of THFs **153** from lactones **152** by Romo's group [87] using a Lewis acid such as TESOTf (triethylsilyl trifluoromethanesulfonate) and Et_3SiH as reducing agent (Table 9.4). In this process, two new stereogenic centers are formed with high diastereoselectivity. A possible mechanism is shown in Figure 9.3. First, a Lewis acid activation of the lactone carbonyl group in **154** leads to **155**, which is followed by an attack of the nonbonding electrons at the keto group with opening of the lactone moiety to give a five-membered oxenium ion **156**. Because of the 1,3-diaxial interaction as exemplified in conformation **157**, inside attack of hydride on the oxenium ion leads to THF **158**. As side product in the process, a small amount of the furan **161** is formed, most likely by an acid-catalyzed elimination of HO–Bn in **157** via **159** and **160**.

Another report deals with this class of Lewis acid-mediated domino ring-opening/reduction. Tong *et al.* [88] described a diastereoselective formation of cis-fused pyrano pyranone **163** by reaction of spiroacetal **162** with $TiCl_4$ and Et_3SiH in 32–65% yield (Scheme 9.28). Two possible pathways have been proposed for the bicyclic ether formation with different stereochemistry. In path a, the Lewis acid coordinates with the enone carbonyl in **164**, which is followed by ring expansion and reduction to give the anti-bicyclic ether **167** through oxenium ions **165** and **166**. The explanation of the stereochemical outcome leading to the observed cis product **170** (path b) was proposed, where ring-opening of **164** by the Lewis acid affords oxenium ion **168**, which by reduction followed by an oxa-Michael cyclization yields **169**.

Table 9.4 Domino lactone ring-opening/reduction.

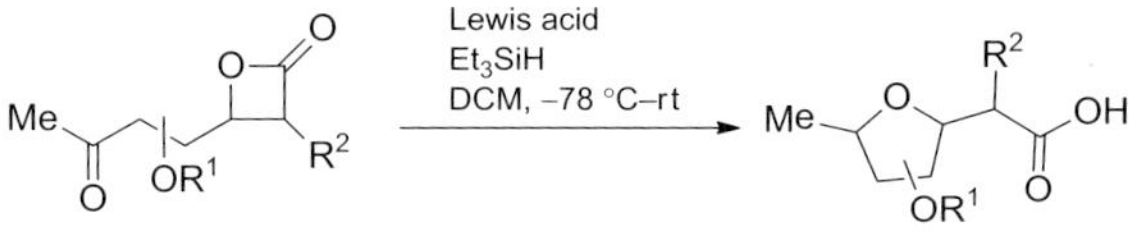

152 **153**

Entry	Lactone	Lewis acid	Tetrahydrofuran	Yield (%)	dr
1	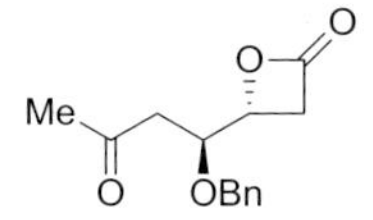 *anti* (>19 : 1)	TESOTf	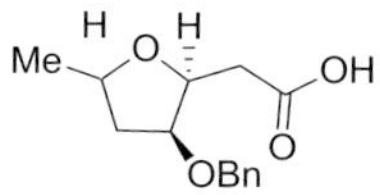	82	14 : 1
2	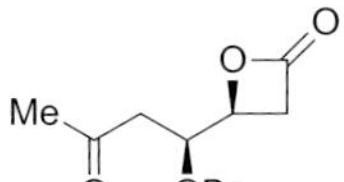 *syn* (>19 : 1)	TESOTf	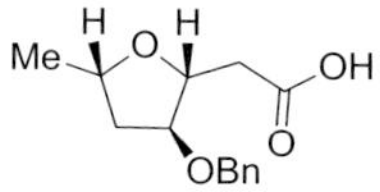	78	>19 : 1
3	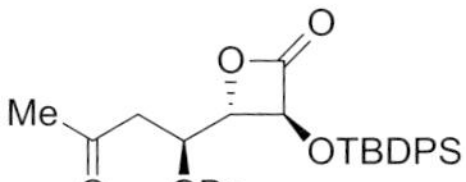 *anti* (18 : 1)	$BF_3 \cdot OEt_2$	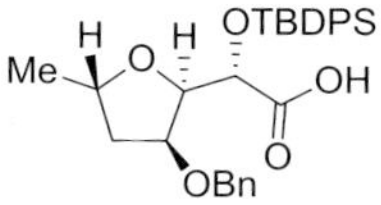	51	18 : 1
4	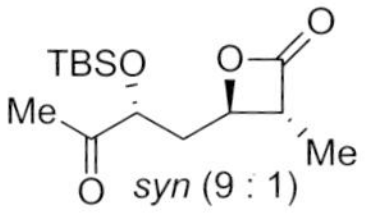 *syn* (9 : 1)	$TiCl_4$	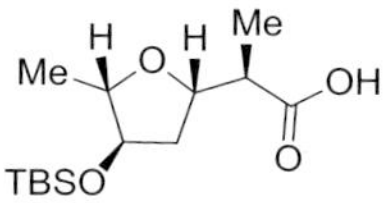	84	9 : 1

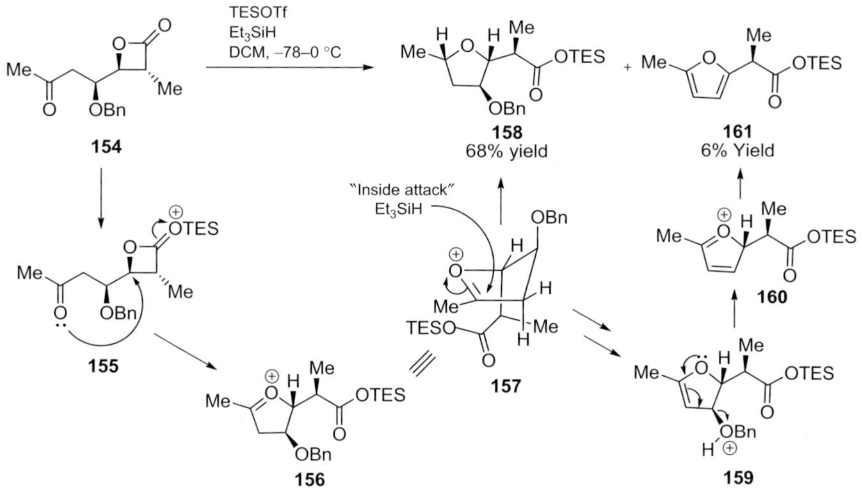

Figure 9.3 Proposed mechanism for the formation of **158** and **161**.

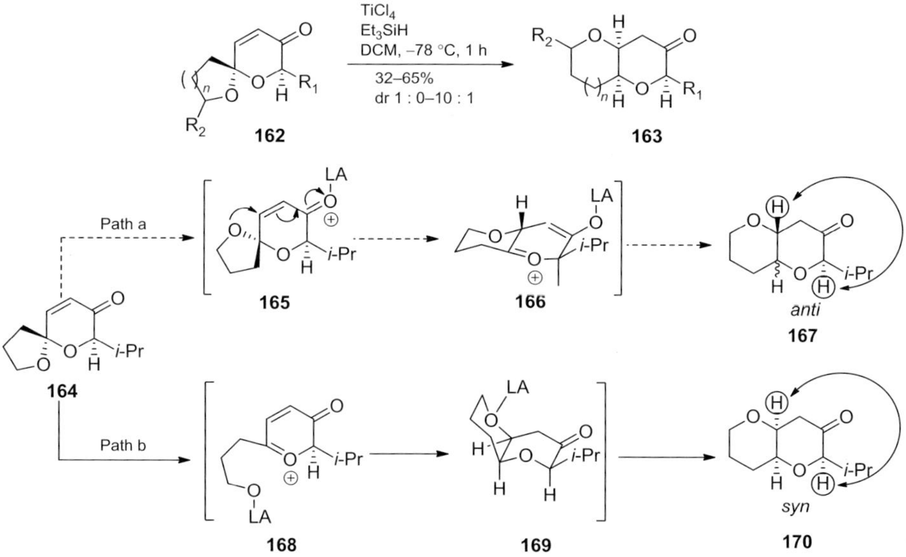

Scheme 9.28 Domino ring-opening/reduction synthesis of cyclic ethers.

A domino hydrometallation/cyclization/reduction process was developed for the synthesis of carbocyclic amino alcohols by Gais *et al.* Treatment of exocyclic alkenyl sulfoximes **171** with DIBAL-H leads to the hydroaluminated intermediate **172**, which subsequently cyclizes to give keto sulfoxyimines **173**.

Further reduction of the carbonyl moiety affords the carbocyclic amino alcohols **174**. Using this protocol, three consecutive stereogenic centers are formed with good diastereoselectivity (Scheme 9.29) [89].

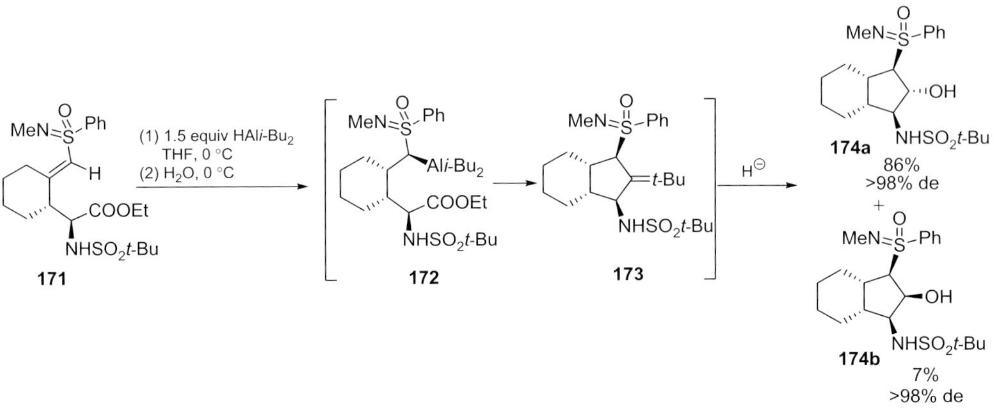

Scheme 9.29 Domino hydrometallation/cyclization/reduction for the synthesis of bicyclic β-amino alcohol.

Yao *et al.* reported a chemo- and stereoselective asymmetric total synthesis of (−)-lannotinidine B **179** through two reductive amination in a domino fashion. Thus, the tricarbonyl compound **175** was treated with benzylamine in the presence of $NaBH(OAc)_3$ to give **178** via **176** and **177** (Scheme 9.30) [90]. Similar reductive amination strategies have been applied for several other natural products [22a].

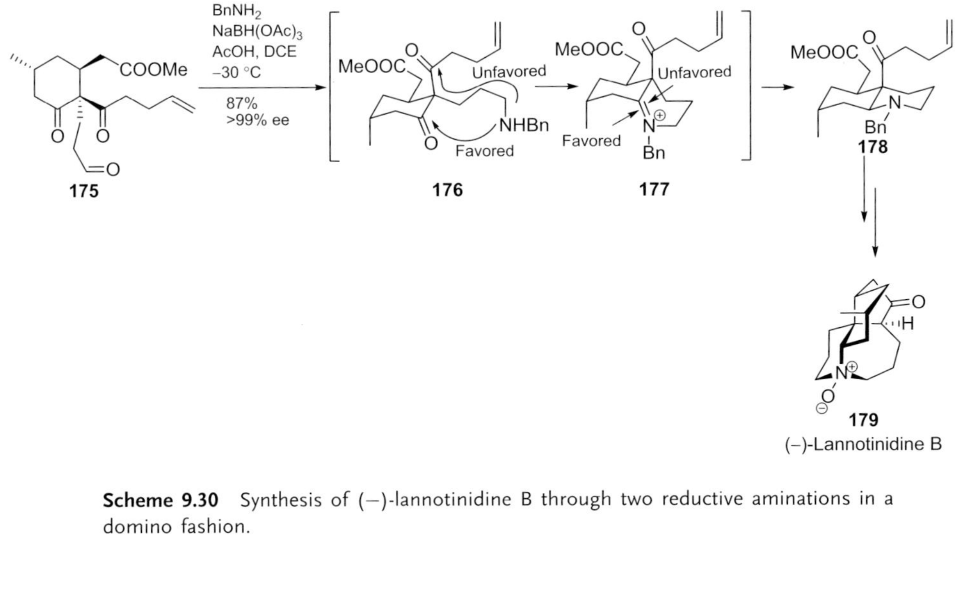

Scheme 9.30 Synthesis of (−)-lannotinidine B through two reductive aminations in a domino fashion.

9.5 Conclusion

In this chapter, we presented the importance of oxidation and reduction reactions as part of a domino process in organic synthesis. Particularly, in the synthesis of complex organic molecules, this allows for short, simple, economical, as well as highly selective approaches in which several bonds are formed in one process. Domino reactions initiated by oxidation or reduction reactions have seen enormous growth in the last years. However, the protocols dealing with domino reactions having an oxidation or a reduction in the middle of a reaction sequence or as the terminating step are yet to be explored to their full potential.

Acknowledgments

We thank IIT Madras, Chennai, India, for infrastructure facilities. The DST and CSIR (India) are thanked for financial support. I. Karthikeyan and D. Ganapathy thank the CSIR for Senior Research Fellowships.

References

1. (a) Backvall, J.E. (2010) *Mordern Oxidation Methods*, Wiley-VCH Verlag GmbH, Weinheim; (b) Donohue, T.J. (2000) *Oxidation and Reduction in Organic Synthesis*, Oxford Science, New York; (c) Andersson, P.G. and Munslow, I.J. (2008) *Modern Reduction Method*, Wiley-VCH Verlag GmbH, Weinheim; (d) Corma, A. and Garcia, H. (2002) *Chem. Rev.*, **102**, 3837–3892; (e) Punniyamurthy, T., Velusamy, S., and Iqbal, J. (2005) *Chem. Rev.*, **105**, 2329–2363.
2. (a) Tietze, L.F., Brasche, G., and Gericke, K.M. (2006) *Domino Reactions in Organic Synthesis*, Wiley-VCH Verlag GmbH, Weinheim; (b) Tietze, L.F. and Beifuss, U. (1993) *Angew. Chem., Int. Ed. Engl.*, **105**, 137–170; See, also 1993) *Angew. Chem., Int. Ed. Engl.*, **32**, 131–163; (c) Presset, M., Coquerel, Y., and Rodriguez, J. (2013) *Chem. Rev.*, **113**, 525–595; (d) Tietze, L.F. (1996) *Chem. Rev.*, **96**, 115–136; (e) Nicolaou, K.C., Edmonds, D.J., and Bulger, P.G. (2006) *Angew. Chem. Int. Ed.*, **45**, 7134–7186; (f) Wille, U. (2013) *Chem. Rev.*, **113**, 813–853; (g) Pellissier, H. (2013) *Chem. Rev.*, **113**, 442–524.
3. Bahia, P.S. and Snaith, J.S. (2004) *J. Org. Chem.*, **69**, 3226–3229.
4. Kleinke, S., Li, C., Rabasso, N., and Porco, J.A. Jr., (2006) *Org. Lett.*, **8**, 2847–2850.
5. Li, Z., Jung, H., Park, M., Lah, M.S., and Koo, S. (2011) *Adv. Synth. Catal.*, **353**, 1913–1917.
6. Wang, C., Li, Z., Ju, Y., and Koo, S. (2012) *Eur. J. Org. Chem.*, 6976–6985.
7. Fujioka, H., Komatsu, H., Nakamura, T., Miyoshi, A., Hata, K., Ganesh, J., Murai, K., and Kita, Y. (2010) *Chem. Commun.*, **46**, 4133–4135.
8. Chaudhari, P.S., Pathare, S.P., and Akamanch, K.G. (2012) *J. Org. Chem.*, **77**, 3716–3723.
9. Yates, P. and Macas, T.S. (1988) *Can. J. Chem.*, **66**, 1–10.
10. Donohoe, T.J., Harris, R.M., Williams, O., Hargaden, G.C., Burrows, J., and Parker, J. (2009) *J. Am. Chem. Soc.*, **131**, 12854–12861.
11. Donohoe, T.J. and Lipinski, R.M. (2013) *Angew. Chem. Int. Ed.*, **52**, 2491–2494.

12. Hajdok, S., Conrad, J., Leutbecher, H., Strobel, S., Schleid, T., and Beifuss, U. (2009) *J. Org. Chem.*, **74**, 7230–7237.
13. Hajdok, S., Conrad, J., and Beifuss, U. (2012) *J. Org. Chem.*, **77**, 445–459.
14. Rueping, M., Sunden, H., Hubener, L., and Sugiono, E. (2012) *Chem. Commun.*, **48**, 2201–2203.
15. Jiang, H.-F., Huang, H.-W., Cao, H., and Qi, C.-R. (2010) *Org. Lett.*, **12**, 5561–5563.
16. Zhu, Y.-P., Lian, M., Jia, F.-C., Liu, M.-C., Yuan, J.-J., Gao, Q.-H., and Wu, A.-X. (2012) *Chem. Commun.*, **48**, 9086–9088.
17. Gao, Q.-H., Fei, Z., Zhu, Y.-P., Lian, M., Jia, F.-C., Liu, M.-C., She, N.-F., and Wu, A.-X. (2013) *Tetrahedron*, **69**, 22–28.
18. Kelly, T.R., Bell, S.H., Ohashi, N., and Chong, R.J.A. (1988) *J. Am. Chem. Soc.*, **110**, 6471–6480.
19. Mehta, G. and Bera, M.K. (2008) *Tetrahedron Lett.*, **49**, 1417–1420.
20. (a) Deschamp, J. and Riant, O. (2009) *Org. Lett.*, **11**, 1217–1220; (b) Deschamp, J., Hermant, T., and Riant, O. (2012) *Tetrahedron*, **68**, 3457–3467; (c) Donohoe, T.J., House, D., and Ace, K.W. (2003) *Org. Biomol. Chem.*, **1**, 3749–3757.
21. Revis, A. and Hilty, T.K. (1987) *Tetrahedron Lett.*, **28**, 4809–4812.
22. (a) Randl, S. and Blechert, S. (2003) *J. Org. Chem.*, **68**, 8879–8882; (b) Shibata, I., Kato, H., Ishida, T., Yasuda, M., and Baba, A. (2004) *Angew. Chem. Int. Ed.*, **43**, 711–714; (c) Nishiyama, H., Shiomi, T., Tsuchiya, Y., and Matsuda, I. (2005) *J. Am. Chem. Soc.*, **127**, 6972–6973; (d) Fuller, N.O. and Morken, J.P. (2005) *Synlett*, 1459–1461.
23. Taylor, S.J. and Morken, J.P. (1999) *J. Am. Chem. Soc.*, **121**, 12202–12203.
24. (a) Ito, J.-I., Shiomi, T., and Nishiyama, H. (2006) *Adv. Synth. Catal.*, **348**, 1235–1240; (b) Zhao, C.-X., Duffey, M.O., Taylor, S.J., and Morken, J.P. (2001) *Org. Lett.*, **3**, 1829–1831; (c) Joensuu, P.M., Murray, G.J., Fordyce, E.A.F., Luebbers, T., and Lam, H.W. (2008) *J. Am. Chem. Soc.*, **130**, 7328–7338; (d) Villanueva, M.I., Rupnicki, L., and Lam, H.W. (2008) *Tetrahedron*, **64**, 7896–7901; (e) Lumby, R.J.R., Joensuu, P.M., and Lam, H.W. (2007) *Org. Lett.*, **9**, 4367–4370; (f) Lumby, R.J.R., Joensuu, P.M., and Lam, H.W. (2008) *Tetrahedron*, **64**, 7729–7740; (g) Chiu, P., Szeto, C.-P., Geng, Z., and Cheng, K.-F. (2001) *Org. Lett.*, **3**, 1901–1903; (h) Chiu, P. and Leung, S.K. (2004) *Chem. Commun.*, 2308–2309; (i) Lam, H.W., Murray, G.J., and Firth, J.D. (2005) *Org. Lett.*, **7**, 5743–5746.
25. Chuzel, O., Deschamp, J., Chausteur, C., and Riant, O. (2006) *Org. Lett.*, **8**, 5943–5946.
26. Deschamp, J., Chuzel, O., Hannedouche, J., and Riant, O. (2006) *Angew. Chem. Int. Ed.*, **45**, 1292–1297.
27. Zhao, D., Oisaki, K., Kanai, M., and Shibasaki, M. (2006) *Tetrahedron Lett.*, **47**, 1403–1407.
28. Baik, T.-G., Luis, A.L., Wang, L.-C., and Krische, M.J. (2001) *J. Am. Chem. Soc.*, **123**, 5112–5113.
29. Yang, J.W., Fonseca, M.T.H., and List, B. (2005) *J. Am. Chem. Soc.*, **127**, 15036–15037.
30. (a) Jang, H.-Y., Huddleston, R.R., and Krische, M.J. (2002) *J. Am. Chem. Soc.*, **124**, 15156–15157; (b) Huddleston, R.R. and Krische, M.J. (2003) *Org. Lett.*, **5**, 1143–1146.
31. (a) Chiu, P., Szeto, C.P., Geng, Z., and Cheng, K.F. (2001) *Tetrahedron Lett.*, **42**, 4091–4093; (b) Villaescusa, C.L., Diaz, L.A.M., Faure, R., Debrauwer, L., Elias, R., and Balansard, G. (1995) *Phytochemistry*, **40**, 1193–1195.
32. Miller, S.P. and Morken, J.P. (2002) *Org. Lett.*, **4**, 2743–2745.
33. Fuller, N.O. and Morken, J.P. (2005) *Org. Lett.*, **7**, 4867–4869.
34. (a) Liu, A. (1995) Dandong City Biological Products and Immune Technology, Application Research Centre, People's Republic of China; (b) Lu, C. (2005) *Hebei Yike Daxue Xuebao*, **26**, 707–708; (c) Vatmurge, N.S., Hazra, B.G., Pore, V.S., Shirazi, F., Chavan, P.S., and Deshpande, M.V. (2008) *Bioorg. Med. Chem. Lett.*, **18**, 2043–2047.
35. Townes, J.A., Evans, M.A., Queffelec, J., Taylor, S.J., and Morken, J.P. (2002) *Org. Lett.*, **4**, 2537–2540.
36. Brackeen, M.F., Stafford, J.A., Cowan, D.J., Brown, P.J., Domanico, P.L.,

Feldman, P.L., Rose, D., Strickland, A.B., Veal, J.M., and Verghese, M. (1995) *J. Med. Chem.*, **38**, 4848–4854.

37. Sukhorukov, A.Y., Boyko, Y.D., Nelyubina, Y.V., Gerard, S., Ioffe, S.L., Tartakovsky, V.A., Malleret, L., and Belaaouaj, A. (2012) *J. Org. Chem.*, **77**, 5465–5469.

38. Adams, G.L., Carroll, P.J., and Smith, A.B. (2012) *J. Am. Chem. Soc.*, **134**, 4037–4040.

39. Takiguchi, S., Iizuka, T., Kumakura, Y.S., Murasaki, K., Ban, N., Higuchi, K., and Kawasaki, T. (2010) *J. Org. Chem.*, **75**, 1126–1131.

40. Kawasaki, T., Ogawa, A., Takashima, Y., and Sakamoto, M. (2003) *Tetrahedron Lett.*, **44**, 1591–1593.

41. Kawasaki, T., Ogawa, A., Terashima, R., Saheki, T., Ban, N., Sekiguchi, H., Sakaguchi, K., and Sakamoto, M. (2005) *J. Org. Chem.*, **70**, 2957–2966.

42. (a) Kawasaki, T., Shinada, M., Ohzono, M., Ogawa, A., Terashima, R., and Sakamoto, M. (2008) *J. Org. Chem.*, **73**, 5959–5964; (b) Kawasaki, T., Shinada, M., Kamimura, D., Ohzono, M., and Ogawa, A. (2006) *Chem. Commun.*, 420–422.

43. Kawasaki, T., Terashima, R., Sakaguchi, K.E., Sekiguchi, H., and Sakamoto, M. (1996) *Tetrahedron Lett.*, **37**, 7525–7528.

44. (a) Ding, K., Lu, Y., Coleska, Z.N., Qiu, S., Ding, Y., Gao, W., Stuckey, J., Krajewski, K., Roller, P.P., Tomita, Y., Parrish, D.A., Deschamps, J.R., and Wang, S. (2005) *J. Am. Chem. Soc.*, **127**, 10130–10131; (b) Suchy, M., Kutschy, P., Monde, K., Goto, H., Harada, N., Takasugi, M., Dzurilla, M., and Balentova, E. (2001) *J. Org. Chem.*, **66**, 3940–3947; (c) Miller, K.A., Tsukamoto, S., and Williams, R.M. (2009) *Nat. Chem.*, **1**, 63–68, S63/61–S63/66; (d) Fensome, A., Adams, W.R., Adams, A.L., Berrodin, T.J., Cohen, J., Huselton, C., Illenberger, A., Kern, J.C., Hudak, V.A., Marella, M.A., Melenski, E.G., McComas, C.C., Mugford, C.A., Slayden, O.D., Yudt, M., Zhang, Z., Zhang, P., Zhu, Y., Winneker, R.C., and Wrobel, J.E. (2008) *J. Med. Chem.*, **51**, 1861–1873; (e) Bignan, G.C., Battista, K., Connolly, P.J., Orsini, M.J., Liu, J., Middleton, S.A., and Reitz, A.B. (2005) *Bioorg. Med. Chem. Lett.*, **15**, 5022–5026.

45. (a) Wei, H.-L., Piou, T., Dufour, J., Neuville, L., and Zhu, J. (2011) *Org. Lett.*, **13**, 2244–2247; (b) Li, J., Liu, Y., Li, C., and Jia, X. (2011) *Chem. Eur. J.*, **17**, 7409–7413.

46. Zhao, H., Lan, Y.B., Liu, Z.-M., Wang, Y., Wang, X.-W., and Tao, J.-C. (2012) *Eur. J. Org. Chem.*, 1935–1944.

47. Girard, P., Namy, J.L., and Kagan, H.B. (1980) *J. Am. Chem. Soc.*, **102**, 2693–2698.

48. (a) Molander, G.A. and Harris, C.R. (1996) *Chem. Rev.*, **96**, 307–338; (b) Molander, G.A. and Sono, M. (1998) *Tetrahedron*, **54**, 9289–9302; (c) Machrouhi, F. and Namy, J.-L. (1998) *Tetrahedron*, **54**, 11111–11122.

49. Cha, Y., Yeoman, J.T.S., and Reisman, S.E. (2011) *J. Am. Chem. Soc.*, **133**, 14964–14967.

50. (a) Baker, T.M., Sloan, L.A., Choudhury, L.H., Murai, M., and Procter, D.J. (2010) *Tetrahedron: Asymmetry*, **21**, 1246–1261; (b) Huang, J., Yang, J.R., Zhang, J., and Yang, J. (2012) *J. Am. Chem. Soc.*, **134**, 8806–8809; (c) Li, Z., Nakashige, M., and Chain, W.J. (2011) *J. Am. Chem. Soc.*, **133**, 6553–6556; (d) Helm, M.D., Da, S.M., Sucunza, D., Findley, T.J.K., and Procter, D.J. (2009) *Angew. Chem. Int. Ed.*, **48**, 9315–9317.

51. (a) Gololobov, Y.G., Zhmurova, I.N., and Kasukhin, L.F. (1981) *Tetrahedron*, **37**, 437–472; (b) Staudinger, H. and Meyer, J. (1919) *Helv. Chim. Acta*, **2**, 635–646.

52. Menand, M., Blais, J.C., Valery, J.M., and Xie, J. (2006) *J. Org. Chem.*, **71**, 3295–3298.

53. Fesenko, A.A. and Shutalev, A.D. (2013) *J. Org. Chem.*, **78**, 1190–1207.

54. (a) Eguchi, S. (2005) *ARKIVOC*, **II**, 98–119; (b) Eguchi, S., Suzuki, T., Okawa, T., Matsushita, Y., Yashima, E., and Okamoto, Y. (1996) *J. Org. Chem.*, **61**, 7316–7319; (c) Williams, D.R., Shamim, K., Reddy, J.P., Amato, G.S., and Shaw, S.M. (2003) *Org. Lett.*, **5**, 3361–3364.

55. (a) De, M.F., Crosignani, S., Banfi, L., Riva, R., and Basso, A. (2010) *J. Comb. Chem.*, **12**, 613–616; (b) Wang, Y., Feng, J., and Ju, X. (2010) *Wuhan Gongcheng*

Daxue Xuebao, **32**, 37–39; (c) Xie, H., Yu, J.B., and Ding, M.-W. (2011) *Eur. J. Org. Chem.*, **2011**, 6933–6938.
56. Wong, M.L., Guzei, I.A., and Kiessling, L.L. (2012) *Org. Lett.*, **14**, 1378–1381.
57. (a) Langer, P., Freifeld, I., and Shojaei, H. (2003) *Chem. Commun.*, 3044–3045; (b) Freifeld, I., Shojaei, H., Dede, R., and Langer, P. (2006) *J. Org. Chem.*, **71**, 6165–6170.
58. (a) Renteria, I.B., Barranco, P.G., Garcia, A., Banik, B.K., and Rivera, G. (2012) *Curr. Med. Chem.*, **19**, 4377–4398. (b) Mahajan, J.P., Suryawanshi, Y.R., and Mhaske, S.B. (2012) *Org. Lett.*, **14**, 5804–5807; (c) Montalban, A.G. (2011) *Quinolines and isoquinolines in natural product synthesis*, Wiley-VCH Verlag GmbH & Co. KGaA, pp. 299–339; (d) Yang, X., Feng, Y., Duffy, S., Avery, V.M., Camp, D., Quinn, R.J., and Davis, R.A. (2011) *Planta Med.*, **77**, 1644–1647.
59. (a) Luo, W., Mu, Q., Qiu, W., Liu, T., Yang, F., Liu, X., and Tang, J. (2011) *Tetrahedron*, **67**, 7090–7095; (b) Shen, Q., Wang, L., Yu, J., Liu, M., Qiu, J., Fang, L., Guo, F., and Tang, J. (2012) *Synthesis*, **44**, 389–392; (c) Marco-Contelles, J., Perez-Mayoral, E., Samadi, A., Carreiras, M.D.C., and Soriano, E. (2009) *Chem. Rev.*, **109**, 2652–2671.
60. McNaughton, B.R. and Miller, B.L. (2003) *Org. Lett.*, **5**, 4257–4259.
61. Bunce, R.A., Herron, D.M., Johnson, L.B., and Kotturi, S.V. (2001) *J. Org. Chem.*, **66**, 2822–2827.
62. Felpin, F.-X., Ibarguren, O., Nassar-Hardy, L., and Fouquet, E. (2009) *J. Org. Chem.*, **74**, 1349–1352.
63. Alaimo, P.J., Brien, R.O. III, Johnson, A.W., Slauson, S.R., O'Brien, J.M., Tyson, E.L., Marshall, A.L., Ottinger, C.E., Chacon, J.G., Wallace, L., Paulino, C.Y., and Connell, S. (2008) *Org. Lett.*, **10**, 5111–5114.
64. Nguyen, T.B., Ermolenko, L., and Al-Mourabit, A. (2013) *J. Am. Chem. Soc.*, **135**, 118–121.
65. (a) LaBarbera, D.V. and Skibo, E.B. (2004) *Bioorg. Med. Chem.*, **13**, 387–395; (b) Labarbera, D.V. and Skibo, E.B. (2005) *Bioorg. Med. Chem.*, **13**, 387–395; (c) McKee, M.L. and Kerwin, S.M. (2008) *Bioorg. Med. Chem.*, **16**, 1775–1783.
66. Chauhan, J. and Fletcher, S. (2012) *Tetrahedron Lett.*, **53**, 4951–4954.
67. Wu, M., Hu, X., Liu, J., Liao, Y., and Deng, G.J. (2012) *Org. Lett.*, **14**, 2722–2725.
68. (a) Buszek, K.R., Brown, N., and Luo, D. (2009) *Org. Lett.*, **11**, 201–204; (b) Fu, L. (2010) *Top. Heterocycl. Chem.*, **26**, 433–480; (c) Hanessian, S. and Stoffman, E.J.L. (2013) *Can. J. Chem.*, **91**, 13–20; (d) Liu, W., Lim, H.J., and RajanBabu, T.V. (2012) *J. Am. Chem. Soc.*, **134**, 5496–5499.
69. Bunce, R.A., Randall, M.H., and Applegate, K.G. (2002) *Org. Prep. Proced. Int.*, **34**, 483–489.
70. Reekie, T.A., Banwell, M.G., and Willis, A.C. (2012) *J. Org. Chem.*, **77**, 10773–10781.
71. Tan, S.H., Banwell, M.G., Willis, A.C., and Reekie, T.A. (2012) *Org. Lett.*, **14**, 5621–5623.
72. (a) Nicolaou, K.C., Petasis, N.A., Uenishi, J., and Zipkin, R.E. (1982) *J. Am. Chem. Soc.*, **104**, 5557–5558; (b) Nicolaou, K.C., Petasis, N.A., and Zipkin, R.E. (1982) *J. Am. Chem. Soc.*, **104**, 5560–5562; (c) Nicolaou, K.C., Petasis, N.A., Zipkin, R.E., and Uenishi, J. (1982) *J. Am. Chem. Soc.*, **104**, 5555–5557.
73. Bandaranayake, W.M., Banfield, J.E., Black, D.S.C., Fallon, G.D., and Gatehouse, B.M. (1980) *J. Chem. Soc., Chem. Commun.*, 162–163.
74. Bandaranayake, W.M., Banfield, J.E., and Black, D.S.C. (1980) *J. Chem. Soc., Chem. Commun.*, 902–903.
75. Maki, B.E. and Scheidt, K.A. (2008) *Org. Lett.*, **10**, 4331–4334.
76. Maki, B.E., Chan, A., Phillips, E.M., and Scheidt, K.A. (2007) *Org. Lett.*, **9**, 371–374.
77. Zidar, N. and Kikelj, D. (2011) *Helv. Chim. Acta*, **94**, 859–867.
78. Tong, X., Beller, M., and Tse, M.K. (2007) *J. Am. Chem. Soc.*, **129**, 4906–4907.
79. Xu, W., Jin, Y., Liu, H., Jiang, Y., and Fu, H. (2011) *Org. Lett.*, **13**, 1274–1277.
80. Xu, W. and Fu, H. (2011) *J. Org. Chem.*, **76**, 3846–3852.

81. Sreenivas, D.K., Ramkumar, N., and Nagarajan, R. (2012) *Org. Biomol. Chem.*, **10**, 3417–3423.

82. Maryanoff, B.E. and Reitz, A.B. (1989) *Chem. Rev.*, **89**, 863–927.

83. (a) Burling, S., Paine, B.M., Nama, D., Brown, V.S., Mahon, M.F., Prior, T.J., Pregosin, P.S., Whittlesey, M.K., and Williams, J.M.J. (2007) *J. Am. Chem. Soc.*, **129**, 1987–1995; (b) Edwards, M.G., Jazzar, R.F.R., Paine, B.M., Shermer, D.J., Whittlesey, M.K., Williams, J.M.J., and Edney, D.D. (2004) *Chem. Commun.*, 90–91; (c) Edwards, M.G. and Williams, J.M.J. (2002) *Angew. Chem. Int. Ed.*, **41**, 4740–4743.

84. Watson, A.J.A. and Williams, J.M.J. (2010) *Science*, **329**, 635–636.

85. (a) Bower, J.F., Skucas, E., Patman, R.L., and Krische, M.J. (2007) *J. Am. Chem. Soc.*, **129**, 15134–15135; (b) Han, H. and Krische, M.J. (2010) *Org. Lett.*, **12**, 2844–2846; (c) Patman, R.L., Chaulagain, M.R., Williams, V.M., and Krische, M.J. (2009) *J. Am. Chem. Soc.*, **131**, 2066–2067.

86. (a) Cao, J.J., Zhou, F., and Zhou, J. (2010) *Angew. Chem. Int. Ed.*, **49**, 4976–4980; (b) Chen, L., Shi, T.D., and Zhou, J. (2013) *Chem. Asian J.*, **8**, 556–559.

87. Mitchell, T.A. and Romo, D. (2007) *J. Org. Chem.*, **72**, 9053–9059.

88. Zhu, L., Song, L., and Tong, R. (2012) *Org. Lett.*, **14**, 5892–5895.

89. Acikalin, S., Raabe, G., Runsink, J., and Gais, H.J. (2011) *Eur. J. Org. Chem.*, **2011**, 5991–6008.

90. Ge, H.M., Zhang, L.-D., Tan, R.X., and Yao, Z.-J. (2012) *J. Am. Chem. Soc.*, **134**, 12323–12325.

10
Organocatalysis in Domino Processes

Hélène Pellissier

10.1
Introduction

The economic interest in combinations of asymmetric catalytic processes with domino reactions is obvious. While the end of the last century has been dominated by the use of metal [1] and biocatalysis [2], a change in perception occurred during the last decade when several reports confirmed that relatively simple organic molecules, such as proline, could be highly effective and remarkably enantioselective catalysts of a variety of fundamentally important transformations [3]. This rediscovery has initiated an explosive growth of research activities in organocatalysis, in both industry and academia. Organocatalysts have several important advantages: they are usually robust, inexpensive, readily available, and nontoxic. Enantioselective organocatalytic processes have reached maturity in recent years with an impressive and steadily increasing number of publications on the applications of this type of reactions, which paint a comprehensive picture of their real possibilities in organic synthesis [4]. Hence, the application of chiral organocatalysts has permitted the preparation of a number of highly valuable chiral products with the exclusion of any trace of hazardous metals and with several advantages from an economic and environmental point of view [5]. The ability of organocatalysts to promote a wide range of reactions by different activation modes makes organocatalysis ideal for its application in domino reactions, which proceed in a one-pot procedure to build complex frameworks from simple starting compounds [4k,l,u, 6]. These organocatalyzed domino reactions are often highly efficient and follow, in some way, different biomimetic pathways, with the same principles that are found in biosynthesis in Nature. In this context, chemists have devoted much effort in the development of new and powerful strategies in domino reactions that avoid the use of costly and time-consuming protection/deprotection processes as well as purification procedures of the intermediates. The efficiency of enantioselective domino processes can be evaluated by the bonds formed, the stereoselectivity achieved, and the complexity of the newly formed molecules. The goal of this chapter is to cover the last efforts of the chemical community in the development of novel enantioselective organocatalytic domino reactions published since

Domino Reactions: Concepts for Efficient Organic Synthesis, First Edition. Edited by Lutz F. Tietze.

the beginning of 2006, since the general field of asymmetric domino reactions has been previously collected by Tietze [7] in a book and by this author in two reviews [6m,n], all published in 2006. The domino reactions are cataloged on the basis of the reaction types involved in the first synthetic (two) step(s). In order to facilitate presentation, the chapter is divided into two parts. The first part deals with organocatalytic, enantioselective domino reactions involving one or two components, whereas the second part includes multicomponent (domino) reactions involving at least three substrates [7, 8]. For the sake of this chapter, Tietze's [9] definition of domino reactions is suitable, qualifying domino reactions as reactions involving two or more bond-forming transformations that take place under the same reaction conditions, without adding additional reagents and catalysts, and in which the subsequent reactions result as a consequence of the functionality formed by bond formation or fragmentation in the previous step.

10.2
One- and Two-Component Domino Reactions

Asymmetric organocatalytic domino reactions are often catalyzed by chiral amines, especially chiral secondary amines, since they are capable of both enamine and iminium catalysis [10]. Their capability to tolerate numerous functional groups makes them ideal for the design of new asymmetric organocatalytic domino reactions, although their scope is mainly limited to carbonyl systems. The combination of iminium and enamine activations in a single organocatalyzed operation can serve as an attractive platform in the design of new catalytic domino processes, based on the involvement of intermediate enamine and iminium species, in which several bond-forming steps take place in a single operation. Since the first organocatalytic domino reaction reported by Bui and Barbas [11] in 2000, the use of these two powerful methodologies based on lowest unoccupied molecular orbital (LUMO)-lowering iminium ion activation and highest occupied molecular orbital (HOMO)-raising enamine activation has been studied intensively, significantly contributing to the success of the rapidly developing area of asymmetric organocatalysis. Generally, chiral secondary amines, such as diarylprolinol ethers or MacMillan's [12] chiral imidazolidinones, are most often used to activate α,β-unsaturated aldehydes (or ketones) by forming electron-deficient iminium ions, which render the β-carbons more electrophilic than their carbonyl precursors for nucleophilic attack (Scheme 10.1). All the reactions based on this strategy have in common the use of an α,β-unsaturated aldehyde or ketone and are catalyzed by a chiral secondary amine. The general mechanism of the iminium–enamine activation mode is depicted in Scheme 10.1. An α,β-unsaturated aldehyde (or ketone) is activated by a secondary chiral amine catalyst, forming an iminium ion intermediate, which then undergoes a conjugate addition with a nucleophile. The resulting enamine can then undergo a second reaction with an electrophile to afford the double-substituted product, which generally contains two novel stereocenters.

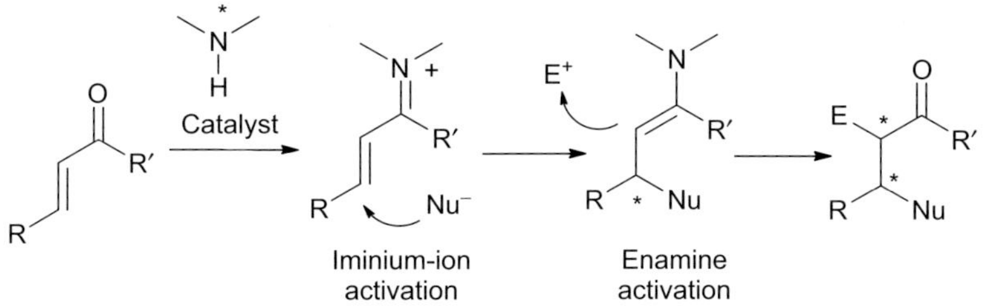

Scheme 10.1 Mechanism of iminium–enamine activation mode.

On the other hand, chiral secondary amines are also used for promoting the formation of electron-rich enamines from enolizable aldehydes or ketones, which then react with various electrophiles to afford the products [13]. For example, proline is generally less effective in iminium catalysis particularly with enals or enones. Enamine (–iminium) activation, depicted in Scheme 10.2, has gained significant attention for controlling the absolute configuration in the α-functionalization of aldehydes and ketones by a variety of electrophilic reagents.

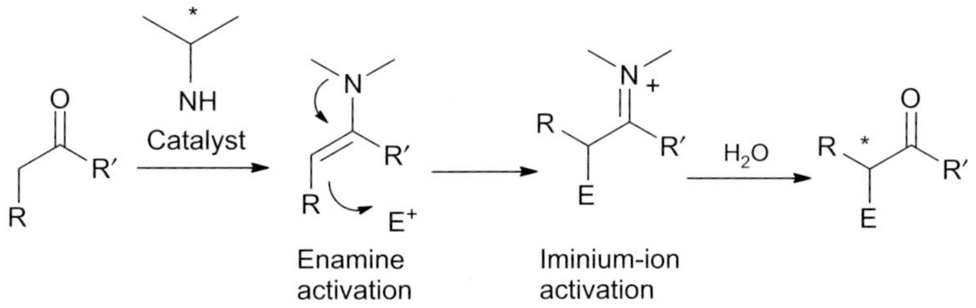

Scheme 10.2 Mechanism of enamine–iminium activation mode.

The reactions will be organized by the reaction types involved in the first synthetic (two) step(s) and also by the type of activation mode of their components.

10.2.1
Domino Reactions Initiated by the Michael Reaction

10.2.1.1 Domino Michael/Michael Reactions

The nucleophilic 1,4-addition of stabilized carbon nucleophiles to electron-poor olefins, generally α,β-unsaturated carbonyl compounds, is known as *Michael addition*, although it was first reported by Komnenos [14] in 1883. Michael-type reactions can be considered as one of the most powerful and reliable tools for the stereocontrolled formation of carbon–carbon and carbon–heteroatom bonds [15], as has been demonstrated by the huge number of examples in which it has been applied as a key strategic transformation in total synthesis. As a consequence, in recent years, many different organocatalytic versions of this important transformation have been reported, using a wide variety of nucleophiles and conjugate acceptors [16]. The iminium–enamine approach is currently the strategy most employed in developing new asymmetric organocatalytic domino reactions. All examples based on this concept have in common the use of enal or enone systems. Early examples based on the iminium–enamine concept were successively developed by Barbas and Bui [11] with the L-proline-catalyzed synthesis of the Wieland–Miescher ketone through domino Michael/aldol reaction of methyl vinyl ketone and 2-methyl 1,3-cyclohexanedione; by Jørgensen *et al.* [17] with Michael/aldol reactions of β-ketoesters and enones catalyzed by imidazolidine catalyst; and then, at nearly the same time, by Gryko [18] with L-proline-catalyzed domino Michael/aldol reactions of methyl vinyl ketone and 1,3-diketones; by List *et al.* [19] with intramolecular reductive Michael cyclizations of enal enones catalyzed by imidazolidinone catalysts; and by MacMillan *et al.* [20] with domino Michael/intermolecular halogenation reactions catalyzed by imidazolidinone catalysts. A more recent work reported by Wang *et al.* [21] describes the synthesis of highly functionalized tetrasubstituted chiral five-membered carbocycles based on a domino Michael/Michael reaction of γ-malonate-α,β-unsaturated esters with α,β-unsaturated aldehydes. This process was catalyzed by chiral diphenylprolinol trimethylsilyl ether (10 mol%) in ethanol media, affording through the iminium–enamine activation mode the corresponding cyclopentanes bearing three stereogenic centers in high yields (87–95%), diastereoselectivities of 80–90% de, and enantioselectivities of 90–99% ee. Later, Cordova *et al.* [22] developed a related process based on the use of the same organocatalyst, which allowed the construction of cyclopentanes through a domino nitro-Michael/Michael sequence. Instead of malonate derivatives, these authors employed a γ-nitro-α,β-unsaturated ester as the nucleophile for the initial Michael addition, providing nitrogen-, formyl-, and ester-functionalized cyclopentanes with four stereogenic centers with good yields (70–88%) and excellent enantioselectivities (97–99% ee), but with moderate diastereoselectivities (40–60% de). In 2009, Brenner and McGarraugh [23] reported the generation of highly substituted fused chiral carbocycles based on a new enantioselective Michael/Michael domino reaction of conjugated β-ketoesters in which the olefin was part of a carbocycle with α,β-unsaturated aldehydes. This process, catalyzed by chiral diphenylprolinol trimethylsilyl ether, allowed the corresponding domino Michael/Michael products to be achieved through iminium–enamine approach as mixtures of two diastereomers in good yields (61–87%) and a high diastereomeric ratio of ≥91 : 9. The major products were obtained in excellent enantioselectivities of ≥96% ee as a mixture of two epimers (α and β). In 2010, the same catalyst (2 mol%) was employed by Ma [24] to promote an enantioselective synthesis of polysubstituted cyclopentanones through domino Michael/Michael reactions of a range of α,β-unsaturated aldehydes with a keto ester bearing a highly electron-deficient olefin unit. This process allowed the formation of four contiguous stereocenters in the cyclopentanone ring with high enantioselectivities of 90% to >99% ee. An iminium–enamine activation mode was proposed to explain the results. Thus, the authors assumed that, after the Michael addition of an anion generated from the keto ester onto iminium, the newly formed enamine moiety in the enamine intermediate attacked the highly electron-deficient olefin part to afford the intramolecular Michael adduct. The scope of the process was extended to various β-, α,β-, and β,β-substituted α,β-unsaturated aldehydes, including cyclic α,β-unsaturated aldehydes, which led to the formation of the corresponding major stereomers in good yields (70–90%), excellent enantioselectivities, and moderate to high diastereoselectivities (22–92% de). In marked contrast with the Michael addition of carbon nucleophiles, the addition of non-carbon nucleophiles, such as amines [16b], thiols [25], phosphorus [26], and alcohols [27] (hetero-Michael addition), has gained considerably less interest in the past decades. This holds especially true for the conjugate addition of thiols, that is, the thia-Michael addition. As an example, Wang *et al.* [28] reported the synthesis of chiral tetrahydrothiophenes based on a domino thia-Michael/Michael reaction of α,β-unsaturated aldehydes with *trans*-ethyl 4-mercapto-2-butenoate catalyzed by chiral diphenylprolinol trimethylsilyl ether, which evolved through the iminium–enamine activation mode. After the first thio-Michael addition, the enamine intermediate promoted a conjugated attack to the α,β-unsaturated ester, furnishing the thiophene ring. A variety of aromatic, heteroaromatic, and aliphatic enals could be employed, providing the corresponding highly functionalized chiral tetrahydrothiophenes with three new stereogenic centers in good yields (55–96%) and diastereoselectivities of up to 88% de, combined with excellent enantioselectivities (94–99% ee). The iminium–enamine activation mode can also be envisaged to explain the domino oxa-Michael/Michael reaction occurring between 3-methylbut-2-enal and (*E*)-2-(2-nitrovinyl)-benzene-1,4-diol upon catalysis with chiral diphenylprolinol trimethylsilyl ether, which afforded the corresponding almost enantiopure oxa-Michael/Michael cycloadduct in 76% yield [29]. This domino reaction constituted the key step of the first total synthesis of the natural, biologically active product (+)-conicol. As an extension of the iminium–enamine activation mode, Wang *et al.* have recently developed a novel asymmetric domino oxa-Michael/Michael reaction catalyzed by a closely related chiral catalyst, such as diphenylprolinol *tert*-butyldimethylsilyl ether, involving an unprecedented chiral iminium–allenamine cascade. This process was a viable one-pot approach to the synthetically and biologically significant chiral 4*H*-chromenes in high yields (93–98%) and excellent enantioselectivities of at least 99% ee. A broad substrate scope was successfully employed in this methodology, including aromatic and aliphatic alkynals as Michael acceptors and 2-(*E*)-(2-nitrovinyl)-phenols as Michael donors/acceptors with significant structural variation. The iminium–enamine activation mode could also be involved in a highly diastereo- and enantioselective domino Michael/Michael/aldol reaction of α,β-unsaturated aldehydes with (*E*)-7-oxooct-5-enal or (*E*)-7-oxonon-5-enal developed by Hong *et al.*, in 2009 [31]. Upon catalysis with the chiral diphenylprolinol trimethylsilyl ether, this process provided a quick access to various highly functionalized and almost enantiopure octahydro-6-oxo-1-phenylnaphthalene-2-carbaldehydes bearing four stereogenic centers in high yields (72–86%) and enantioselectivities of at least 99% ee. In 2012, these authors applied this methodology to (*E*)-3-methyl-7-oxooct-5-enal as the ketoaldehyde component, reacting under the same reaction conditions with an aromatic enal to give the corresponding domino product, which constituted a key intermediate in the first total synthesis of (+)-galbulin [32]. In 2011, the same authors reported a nice synthesis of chiral cyclohexanedicarbaldehydes based on the enantioselective domino Michael/Michael reaction of α,β-unsaturated aldehydes with ethyl (*E*)-7-oxohept-2-enoate catalyzed by the same catalyst [33]. Remarkably, almost diastereo- and enantiopure domino products were achieved in all cases of aromatic as well as α,β-unsaturated aldehydes studied, as shown in Scheme 10.3.

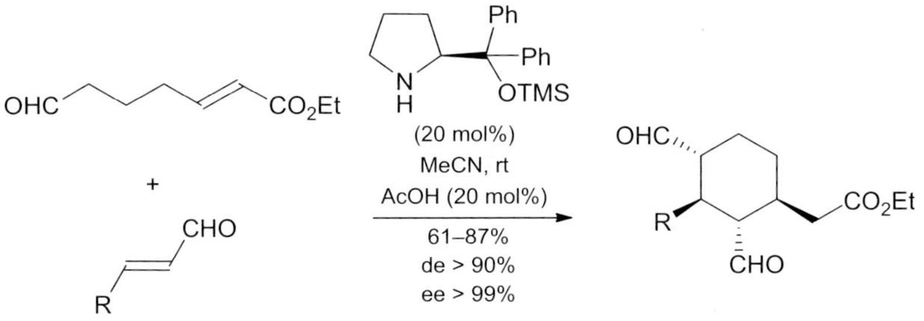

Scheme 10.3 Domino Michael/Michael reaction through iminium–enamine activation mode.

Along with α,β-unsaturated aldehydes, enones have been also widely employed as Michael acceptor components in recent enantioselective organocatalyzed domino Michael/Michael processes. In this context, Deng *et al.* [34] developed the first example of an asymmetric domino reaction catalyzed by chiral primary amines, in 2007. In this work, a chiral primary aminocatalyst derived from quinine (20 mol%) promoted a Michael/Michael/retro-Michael cascade, occurring between α,α-dicyanoalkenes and α,β-unsaturated ketones such as benzylideneacetone, affording the corresponding almost enantiomerically pure polysubstituted 2-cyclohexen-1-ones. This highly efficient domino process evolved through iminium–enamine activation mode, and involved two reagents that acted alternatively and selectively as the Michael donor and acceptor under readily controllable conditions. The reaction scope was quite substantial, and almost complete diastereo- and enantioselectivities of >99% de and >98% ee, respectively, were observed along with moderate yields (46–54%). Asymmetric organocatalytic domino reactions are not limited to amine catalysis. Indeed, significant contributions have also been made in the field of hydrogen-bonding and Bronsted acid catalysis [4d,h, 35]. These catalysts activate the substrates by forming a hydrogen bond (LUMO-lowering) and are able to promote several C–C and C–heteroatom bond-forming reactions. The interaction between the catalyst and the substrate is noncovalent, and the chiral ion pair is the intrinsic activated species. In particular, chiral thiourea-based derivatives and phosphoric acid derivatives are well known for their application as effective Bronsted acid organocatalysts. A recent example of this type of activation was reported by Gong and Wei [36], who investigated a series of Takemoto-type catalysts to promote the domino Michael/Michael reaction of Nazarov reagents with methyleneindolines to provide the corresponding spiro[4-cyclohexanone-1,3'-oxindoline] derivatives. The use of a bifunctional (thio)urea catalyst (10 mol%)

allowed the formal [4+2] cycloadducts to be obtained in high yields (76–97%) and diastereo- and enantioselectivities of 80–96% de and 91–97% ee, respectively. In the same area, Wang *et al.* [37] have disclosed a new enantioselective domino Michael/oxa-Michael/tautomerization reaction induced by a novel indane amine thiourea catalyst via hydrogen-bonding catalysis. This sequence afforded highly functionalized chiral pyranochromenes in high enantioselectivities of 80–99% ee combined with good to high yields (72–95%) by reaction of malononitrile, which was used both as a nucleophile and an electrophile, with enones such as (*E*)-3-benzylidenechroman-4-ones. In another context, N-heterocyclic carbenes [38], prepared *in situ* from diarylimidazolium salts, were shown to be highly effective catalysts for the generation of reactive homoenolates from α,β-unsaturated aldehydes [39]. A recent application of this type of catalysts was reported by Chi *et al.* [40] with a new annulation of benzodi(enone)s and α,β-unsaturated aldehydes mediated by a chiral aminoindanol-derived triazolium catalyst. In the presence of 1,8-diazabicyclo[5.4.0]undec-7-ene (DBU) combined with $MgSO_4$ as an additive, the reaction afforded the corresponding tricyclic domino Michael/Michael products with good yields (61–82%) and excellent diastereo- and enantioselectivities of up to 90% de and 99% ee, respectively. The scope of the reaction of α,β-unsaturated aldehydes could be extended to unsymmetric benzo(dienone)s bearing two different substitutives. In these cases of substrates, the corresponding domino products were obtained with an exceptionally high regioselectivity of 100% in almost all cases of substrates, combined with excellent diastereo- and enantioselectivities of up to 90% de and 99% ee, respectively. In addition to α,β-unsaturated aldehydes and α,β-unsaturated ketones or related compounds, nitroalkenes have been also employed as Michael acceptor components in a number of enantioselective organocatalytic domino Michael/Michael reactions. In this context, one of the milestones in organocatalysis was the first enantioselective addition of malonates to nitrostyrenes [41] catalyzed by chiral thioureas, reported by Takemoto, in 2003 [42]. In this work, the authors disclosed the domino Michael/Michael reaction of γ,δ-unsaturated-β-ketoesters with nitroalkenes catalyzed by a chiral bifunctional thiourea and 1,1,3,3-tetramethylguanidine. Interestingly, the conjugate addition of the ketoesters to nitroalkenes took place first, and then an intramolecular Michael addition catalyzed by the base provided the corresponding final cyclohexanes. The process afforded highly functionalized chiral cyclohexanones in high yields (63–87%) combined with diastereo- and enantioselectivities of up to >99% de and 92% ee, respectively. Furthermore, the reaction was applied to the total synthesis of the naturally occurring and biologically active (−)-epibatidine. In 2008, Zhong *et al.* [43] developed powerful domino Michael/Michael reactions of diethyl 5-acetylhex-2-enedionate with nitrostyrenes catalyzed by a chiral cinchona alkaloid. The process consisted in a Michael addition of the ketoester to nitrostyrene followed by a subsequent intramolecular cyclization via a second Michael addition of the nitro compound to the α,β-unsaturated ester. This reaction was possible because of the low reactivity of unsaturated esters as Michael acceptors in comparison with nitrostyrenes. As shown in Scheme 10.4, the process furnished, through a hydrogen-bonding-activation mode, the corresponding tetrasubstituted cyclopentanes with high yields and in almost diastereo- and enantiopure form.

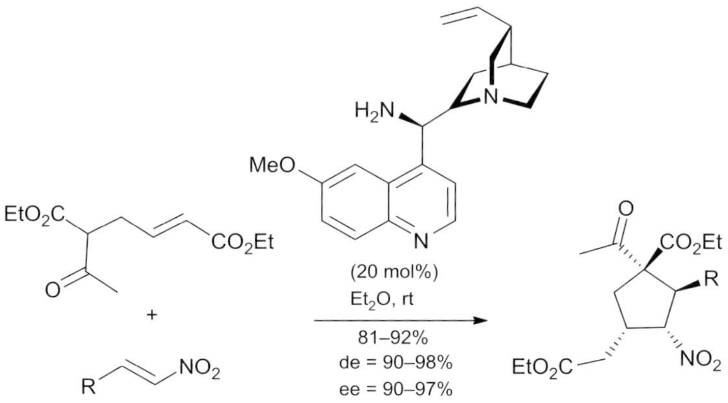

Scheme 10.4 Domino Michael/Michael reaction through hydrogen-bonding-activation mode.

In the same year, Wang *et al.* reported the synthesis of chiral thiochromenes based on an enantioselective domino Michael/Michael reaction catalyzed by another chiral cinchona alkaloid thiourea (2 mol%) [44]. The process occurred through dynamic kinetic resolution [45] between nitroalkenes and *trans*-3-(2-mercaptop)henyl)-2-propenoic acid ethyl ester derivatives, providing the corresponding chiral thiochromanes in good yields (32–99%), excellent diastereoselectivity of >94% de, and high enantioselectivities of 92–99% ee. In 2010, Xiao *et al.* [46] reported the domino thia-Michael/Michael reaction of thiols with nitroolefin enoates, providing polyfunctionalized chroman derivatives in a highly stereoselective manner. This reaction, catalyzed by a bifunctional chiral thiourea (3 mol%), allowed three consecutive stereogenic centers including one quaternary to be generated with high enantioselectivities of 88–96% ee combined with moderate to high diastereoselectivities (34–90% de). The authors proposed that the catalyst activated nitroolefin enoates through hydrogen-bonding activation. Later, these authors described a remarkable enantioselective domino aza-Michael/Michael reaction of nitroolefin enonates with benzotriazoles catalyzed with the same thiourea, which afforded densely functionalized chiral chromans containing a quaternary stereogenic center in good yields (73–91%) and diastereo- and enantioselectivities of 70–92% de and 90–96% ee, respectively [47]. In addition, these authors developed domino aza-Michael/Michael reactions of anilines with nitroolefin enoates upon catalysis with another chiral bifunctional thiourea (10 mol%), which afforded chiral 4-aminobenzopyrans bearing three consecutive stereogenic centers including one quaternary [48]. As shown in Scheme 10.5, these products were isolated with high yields (71–96%) and excellent diastereo- and enantioselectivities of >90% de and 93% to >99% ee, respectively. Also in 2011, another domino thia-Michael/Michael reaction was described by Wang *et al.* [49] upon catalysis by a chiral bifunctional

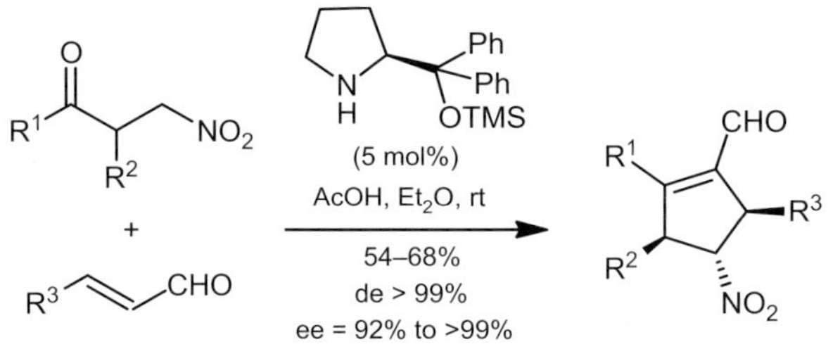

Scheme 10.5 Domino Michael/aldol reaction through iminium–enamine activation mode.

thiourea (20 mol%). This novel hydrogen-bonding-mediated highly stereoselective domino reaction of *trans*-ethyl-4-mercapto-2-butenoate with (*E*)-β-nitrostyrenes led to biologically significant trisubstituted tetrahydrothiophenes bearing three stereogenic centers in good yields (50–82%) and high enantio- and diastereoselectivities of 93–97% ee and 84–96% de, respectively. Aromatic nitroolefins bearing electron-neutral, electron-withdrawing, and electron-donating groups were well tolerated. The authors demonstrated that a dynamic kinetic resolution occurred in the domino reaction in addition to the direct stereocontrol of the substrates by the bifunctional catalyst through hydrogen bonds. In 2012, Cobb *et al.* [50] developed a novel domino Michael/Michael reaction between nitrohex-4-enoates and nitroolefins catalyzed by a chiral bifunctional thiourea. The process generated highly functionalized chiral cyclohexanes containing four contiguous stereocenters including one quaternary one. These products were achieved in moderate to good yields (28–78%) and diastereoselectivities of 34–90% de, and were almost enantiopure (87–99% ee). On the other hand, Melchiorre *et al.* [51] used chiral primary amines to activate α,β-unsaturated enones toward a well-defined enamine–iminium activation mode. Thus, these authors have developed domino Michael/Michael reactions between enones and nitroalkenes which, upon catalysis by a chiral primary amine in combination with 2-fluorobenzoic acid as an additive, afforded the corresponding formal Diels–Alder adducts bearing three or four stereogenic centers with very high enantioselectivities of up to 99% ee. The authors assumed the initial formation of an enamine generated *in situ* by the condensation of the catalyst onto the enone. The nitroalkene acted as a Michael acceptor through intercepting this enamine, which provided the resulting carbon nucleophile. The latter then selectively engaged in an intramolecular, iminium-catalyzed conjugate addition to afford the final cyclohexanones. In all cases of substrates studied, the preferential formation of the formal endo trans products was observed with high diastereoselectivities of 86–90% de. The extension of this methodology to other Michael acceptors, such as *trans*-α-cyanocinnamate and *N*-benzyl maleimide, led to the corresponding cyclohexanones and bicyclic compounds, respectively, which were obtained in good yields (40–92%) and enantioselectivities of 90% to >99% ee. A closely related methodology had been previously developed by Cordova *et al.* [52], in 2007. In this study, the authors employed chiral diamine (*S*)-(+)-1-(2-pyrrolidinylmethyl)pyrrolidine as catalyst to induce the domino Michael/Michael reaction of α,β-unsaturated cyclic ketones with nitroolefins, which afforded the corresponding bicyclic Diels–Alder products containing four stereogenic centers in good to excellent yields (69–95%), excellent diastereoselectivities (>92% de), and good enantioselectivities (65–86% ee). On the other hand, Wang *et al.* have recently proposed a novel activation mode to explain the highly enantioselective domino oxa-Michael/Michael reaction of 2-hydroxy cinnamaldehydes with nitroalkenes catalyzed with chiral diphenylprolinol trimethylsilyl ether [53]. This reaction provided a series of highly functionalized useful chiral chromanes through an unprecedented activation mode, involving an aminal, which was formed very quickly from the reaction of 2-hydroxy cinnamaldehydes with the catalyst, in an almost quantitative yield. The authors assumed that this aminal was produced through an intramolecular O-addition to an intermediate iminium. This aminal served as a nucleophile for the first Michael addition to the nitroalkene to produce an iminium intermediate, which released the catalyst for the next cycle reaction and generated an intermediate enal. Finally, an intramolecular Michael reaction in this enal gave rise to the final products which were obtained in good yields (60–85%) and diastereoselectivities of 34–82% de, combined with high enantioselectivities of up to 98% ee.

10.2.1.2 **Domino Michael/Aldol Reactions**

In recent years, a number of enantioselective organocatalyzed domino Michael/aldol reactions have been developed, employing α,β-unsaturated aldehydes or α,β-unsaturated ketones or derivatives as Michael acceptor components. Among the reactions of α,β-unsaturated aldehydes are those using 2-mercapto-, 2-hydroxy-, or 2-aminobenzaldehydes as Michael donors. As an example, the groups of Wang [54] and Cordova [55] developed, almost at the same time, in 2006, an organocatalytic synthesis of chiral thiochromenes, evolving through a domino thia-Michael/aldol reaction. This process was based on the reaction occurring between α,β-unsaturated aldehydes and 2-mercaptobenzaldehydes upon catalysis with a chiral diarylprolinol trimethylsilyl ether and benzoic acid as an additive. Carrying out the reaction in toluene or in $CHCl_3$ at $-15\,^{\circ}C$, chiral thiochromenes derived from aromatic and alkyl enals were achieved in good yields (55–97%) combined with high enantioselectivities of 89–98% ee. Soon after, the same groups of Cordova [56] and Wang [57] described related processes that gave access to a range of chiral chromanes. For example, the oxygenated analogous compounds were synthesized using the same approach based on a domino oxa-Michael/aldol reaction of enals with 2-hydroxybenzaldehydes. Upon catalysis with 20 mol% of chiral diphenylprolinol trimethylsilyl ether, Cordova *et al.* obtained moderate to high yields (20–92%) combined with high enantioselectivities (92–98% ee) by performing the reaction in toluene at room temperature in the presence of 2-nitrobenzoic acid as an additive. On the other hand, Wang *et al.* employed the same catalyst in the presence of benzoic acid as an additive in 1,2-dichloroethane at room temperature, which also allowed an excellent outcome of the reaction, since the chiral chromanes were achieved in good yields (64–98%) combined with good to high enantioselectivities (75–99% ee). In 2009, Wang *et al.* [58] extended the scope of this domino process to other enals and 2-hydroxybenzaldehydes, demonstrating that the reaction could also be performed in diethyl ether at room temperature, providing high yields and enantioselectivities of upto 99% ee. In 2011, Xu *et al.* [59] investigated the same reactions by using a recyclable, sterically hindered chiral diarylprolinol trimethylsilyl ether catalyst containing two tertiary amine groups. The chiral chromenes were achieved in moderate to high yields (40–95%) and enantioselectivities (53–94% ee). In addition, Cordova *et al.* [60] reported the application of their domino thia-Michael/aldol reaction to mercaptobenzofenone. Indeed, the reaction of this substrate with α,β-unsaturated aldehydes in the presence of a chiral diarylprolinol trimethylsilyl ether as catalyst and 2-nitrobenzoic acid as an additive in toluene at −25 °C afforded, by avoiding the dehydration step, the corresponding chiral thiochromenes bearing three contiguous stereocenters with good yields (63–98%), excellent enantioselectivities (96–99% ee), and good diastereoselectivities (82–88% de). This methodology constituted a simple, catalytic, and highly stereoselective entry to pharmaceutically valuable benzothiopyran derivatives. The same authors have also adapted this process for the synthesis of chiral 1,2-dihydroquinolidines [61]. The development of conjugate addition of an amine to an electron-deficient α,β-unsaturated aldehyde represented an unprecedented organocatalytic process since, generally, an amine is a much weaker nucleophile than a thiol or an alcohol. In fact, this methodology represented the first asymmetric organocatalytic domino aza-Michael/aldol reaction of primary amines with α,β-unsaturated aldehydes. Indeed, the domino aza-Michael/aldol reaction of 2-aminobenzaldehydes and α,β-unsaturated aldehydes provided the corresponding chiral 1,2-dihydroquinolidines in moderate to high yields (31–90%) and excellent enantioselectivities of 96–99% ee, when employing chiral diphenylprolinol trimethylsilyl ether as catalyst (20 mol%) in dimethylformamide (DMF) at −25 °C. Some months later, Wang *et al.* [62] reported the same sequence employing 2-N-protected-aminobenzaldehydes in a basic medium (NaOAc), rendering also good results. In this case, the process was catalyzed by 20 mol% of (*S*)-diphenylprolinol triethylsilyl ether in 1,2-dichloroethane, providing a range of biologically useful chiral 1,2-dihydroquinolines in high yields (83–98%) and enantioselectivities (88–95% ee). In 2008, Hamada *et al.* [63] developed a related domino reaction induced by the same catalyst. In this case, the reaction occurred between 4-methoxycarbonylanthranilaldehyde and 6-*N*-Boc-2-hexenal in acetonitrile at −20 °C in the presence of acetic acid as an additive to give the corresponding 1,2-dihydroquinolidine in 88% yield and excellent enantioselectivity of up to 98% ee. In 2010, Wang *et al.* [64] reported the synthesis of other chiral chromenes based on a highly enantioselective domino oxa-Michael/aldol reaction, evolving through iminium–allenamine activation mode induced by chiral diphenylprolinol *tert*-butyldimethylsilyl ether (15 mol%). High yields (75–99%) and remarkable enantioselectivities (98–99% ee) were observed for the products arising from the reaction of a range of alkynals with ethyl 2-(2-hydroxyphenyl)-2-oxoacetate. The concept of hetero-Michael/aldol reaction was also put into practice by Jørgensen *et al.* [65] in the formation of highly functionalized chiral tetrahydrothiophenes, which constituted a family of compounds very useful in biochemistry, pharmaceutical science, and nanoscience.

Interestingly, Jørgensen *et al.* demonstrated that an appropriate choice of the additive allowed controlling the regioselectivity of the reaction (aldol step). When aliphatic α,β-unsaturated aldehydes and a thiol were put under the effect of 10 mol% of a chiral diarylprolinol trimethylsilyl ether as catalyst and benzoic acid as additive, the corresponding tetrahydrothiophene carbaldehydes were obtained in moderate to good yields (44–74%), high enantioselectivities (90–96% ee), and total diastereocontrol. This outcome involved the normal pathway in this type of domino reactions (thia-Michael addition over the iminium ion, and a subsequent aldol reaction between the just formed enamine and ketone moiety). On the other hand, when the reaction was carried out under basic conditions ($NaHCO_3$) instead of acidic (benzoic acid), the aldol step was thermodynamically controlled by the substrate, without catalyst induction, affording the corresponding (tetrahydrothiophen-2-yl)phenyl methanones as single diastereomers in good yields (43–66%) and enantioselectivities (64–82% ee). Another contribution of Wang *et al.* [66] was focused on the synthesis of chiral cyclopentenes. On the basis of a domino Michael addition/aldol reaction, catalyzed by chiral diphenylprolinol triethylsilyl ether (10 mol%) followed by dehydration, occurring between aromatic enals and dimethyl 2-oxoethylmalonate, a range of densely functionalized chiral cyclopentenes were achieved in good yields (63–89%) and high enantioselectivities (91–97% ee). In 2008, a highly diastereo- and enantioselective domino Michael/aldol/dehydration reaction between α,β-unsaturated aldehydes and 5-oxoalkanals was successfully developed by Hong *et al.* [67]. Evolving through the iminium–enamine activation mode by using chiral diphenylprolinol trimethylsilyl ether (50 mol%) in the presence of acetic acid as an additive, this process afforded densely functionalized cyclohexenes in good yields (53–66%), trans diastereoselectivity of >99% de, and good to excellent enantioselectivities (74% to >99% ee). The iminium–enamine activation mode was also applied by Rueping *et al.* [68] to the domino Michael/aldol reaction of α,β-unsaturated aldehydes with 1,2-cyclohexadione, which provided, using the same catalyst (10 mol%), the corresponding chiral bicyclo[3.2.1]octane-6-carbaldehydes in good yields (44–80%) and excellent enantioselectivities (94–98% ee). These bicyclic products were obtained as single diastereomers. Asymmetric domino oxa-Michael/aldol reactions have also been recently developed by several groups. For example, Vicario *et al.* [69] developed domino oxa-Michael/aldol/hemiacetalization reactions of α,β-unsaturated aldehydes with the dihydroxyacetone dimer, which led to the corresponding hexahydrofuro[3,4-*c*]furans. This process consisted of an initial oxa-Michael reaction, a subsequent intramolecular aldol reaction, and, finally, a hemiacetalization step, and proceeded with the generation of four new stereocenters upon catalysis with chiral diphenylprolinol trimethylsilyl ether. Remarkably, the intramolecular aldol reaction step involved the participation of a ketone as the internal electrophile, thereby generating a quaternary stereocenter. High enantioselectivities (90–99% ee) were obtained in combination with good diastereoselectivity of >82% de and with good yields (71–98%). The iminium–enamine activation mode was also applied by Enders *et al.* [70] to asymmetric domino aza-Michael/aldol reactions catalyzed with the same catalyst.

These novel processes opened efficient and enantioselective entry to the tricyclic pyrrolo indole core, which is a characteristic structural unit of many bioactive natural products. Thus, a series of chiral 3-substituted 3*H*-pyrrolo[1,2-*a*]indoles were achieved in moderate to good yields (40–71%) and high enantioselectivities (85% to >99% ee) by reaction of 1*H*-indole-2-carbaldehyde and α,β-unsaturated aldehydes. Later, these reactions were reinvestigated by Wang *et al.* by using toluene as the solvent and in the presence of molecular sieves at room temperature. Under these conditions, these authors obtained the same products albeit in better yields (61–84%) combined with comparable enantioselectivities (81% to >99% ee) [71]. Furthermore, several groups have successfully developed asymmetric domino thia-Michael/aldol reactions based on the iminium–enamine activation mode. For example, Wang *et al.* [72] used a chiral diarylprolinol trimethylsilyl ether as catalyst (5 mol%) to promote the synthesis of chiral highly functionalized tetrahydrothiophenes by reaction of α,β-unsaturated aldehydes with ethyl-mercapto-2-oxopropanoate. Notably, three consecutive stereogenic centers including one chiral quaternary carbon center were efficiently created in this process. It must be noted that a subsequent reduction of the aldehyde function of the domino process was necessary for convenient chromatography of the products. The corresponding diols were obtained with moderate to good yields (31–59%) with high enantioselectivities (91–97% ee) and good diastereoselectivities (78% to >90% de). In 2010, Xu *et al.* [73] condensed 2-mercaptoacetaldehyde, generated from 1,4-dithiane-2,5-diol under equilibrium conditions, to α,β-unsaturated aldehydes in the presence of chiral diphenylprolinol trimethylsilyl ether as catalyst (20 mol%) associated to *p*-nitrobenzoic acid as an additive, which provided through the iminium–enamine activation mode the corresponding chiral domino thia-Michael/aldol products in good yields (74–86%) and high enantioselectivities (94% to >99% ee). In 2010, Cho *et al.* [74] investigated the enantioselective domino Michael/aldol reaction of α,β-unsaturated aldehydes with 2-trihaloacetylpyrroles by using a chiral diarylprolinol trimethylsilyl ether as catalyst (20 mol%). The process employed benzoic acid as an additive and provided chiral functionalized bicyclic pyrrolizines in good yields (60–81%), high diastereoselectivity of >90% de, and high enantioselectivities (90–98% ee). Highly diastereo- and enantioselective domino Michael/aldol reactions of α,β-unsaturated aldehydes with imidazole derivatives evolving through the iminium–enamine activation mode were reported by Ye *et al.* in 2011 [75]. The chiral domino bicyclic products were achieved in good yields (65–95%) and good diastereoselectivities (72–94% de), and enantioselectivities of 87% to >98% ee by using chiral diphenylprolinol trimethylsilyl ether as catalyst (20 mol%). On the other hand, a number of organocatalytic enantioselective domino Michael/aldol reactions have involved the condensation of nitroalkanes onto α,β-unsaturated aldehydes. For example, Enders *et al.* [76] showed that chiral diphenylprolinol trimethylsilyl ether could highly efficiently induce the domino nitroalkane-Michael addition/aldol reaction. This process based on the iminium–enamine activation mode provided the efficient synthesis of chiral trisubstituted 3,4-dihydronaphthalenes bearing two stereogenic centers in good yields (40–75%) and excellent diastereo- and enantioselectivities of >98% de and 91% to >99% ee, respectively, starting from 2-(nitromethyl)benzaldehyde and various α,β-unsaturated aldehydes. Moreover, the condensation of β-nitroketones onto α,β-unsaturated aldehydes in the presence of the same catalyst combined with acetic acid as an additive was shown by Hong *et al.* [77] to afford the corresponding domino Michael/aldol products through the iminium–enamine activation mode. As shown in Scheme 10.5, these fully substituted cyclopentenes were obtained in moderate to good yields albeit as single diastereomers and with excellent enantioselectivities of up to 99% ee.

In addition to α,β-unsaturated aldehydes, α,β-unsaturated ketones and derivatives have also been widely employed as Michael acceptors in recent enantioselective organocatalyzed domino Michael/aldol reactions. As an example, Cordova *et al.* [78] have developed a simple catalytic synthesis of tetrahydrothioxanthenones and tetrahydroxanthenones by applying the iminium–enamine activation mode to a domino Michael/aldol of reaction of enones. Indeed, the domino Michael/aldol reaction of 2-mercaptobenzaldehydes or salicylic aldehyde derivatives with cyclic enones proceeded in a highly chemoselective manner, furnishing the corresponding tetrahydrothioxanthenones or tetrahydroxanthenones, respectively, in good yields (51–78%) and moderate to good enantioselectivities (48–91% ee). The intermediate chiral alcohols were isolated as single diastereomers when a rapid column chromatography eluent system was employed. The mechanism proposed involved the iminium activation of the cyclic enone by a chiral pyrrolidine catalyst. Subsequent stereoselective nucleophilic conjugate attack on the β-carbon by the thiol or alcohol resulted in a chiral enamine intermediate, which performed an intramolecular 6-exo-trig aldol addition from the same face as the incoming thiol or alcohol. Hydrolysis of the resulting iminium intermediate gave the intermediate chiral alcohol. Elimination of water afforded tetrahydrothioxanthenones or tetrahydroxanthenones. On the other hand, a simple chiral primary/secondary diamine catalyst was successfully employed by Zhao *et al.* [79] to induce the highly enantioselective domino Michael/aldol/dehydration reaction between benzoylacetate and α,β-unsaturated ketones to furnish the corresponding functionalized chiral cyclohexenones. Enantioselectivities of at least 99% ee for both major and minor diastereomers were obtained combined with good to high yields (55–98%) and moderate diastereoselectivities (42–62% de). The enantioselective catalytic construction of chiral fluorinated quaternary carbon centers, which are a class of versatile and important monofluorinated synthons used in organic synthesis, is still a very challenging subject in organic chemistry. In this context, the same authors have used another chiral primary/secondary amine as catalyst (10 mol%) in combination with *p*-nitrobenzoic acid as cocatalyst to induce a closely related domino Michael/aldol/dehydration reaction occurring between α-fluoro-β-keto esters and α,β-unsaturated ketones, providing functionalized fluorinated cyclohexenones with moderate to excellent diastereoselectivities (72–98% de) and excellent enantioselectivities of 98% to >99% ee combined with good yields (60–80%) [80]. In order to access the chiral spiro[cyclohex-2-enone-oxindole] derivatives, Wang *et al.* developed a novel domino Michael/aldol reaction of oxindole derivatives with α,β-unsaturated ketones based on the iminium–enamine activation mode induced by a chiral cinchona-derived primary amine (20 mol%) [81]. The best results were observed when the reaction was performed in the presence of an additive such as trifluoroacetic acid (TFA), providing the domino products in good yields (63–99%) and with excellent levels of diastereoselectivities of >90% de in all cases of substrates studied, along with high enantioselectivities of 92–96% ee. On the other hand, hydrogen-bonding-mediated catalysis was used by Wang *et al.* [82] in 2007 to perform highly enantio- and diastereoselective domino Michael/aldol reactions. They were efficiently catalyzed by a cinchona alkaloid thiourea, using as low as 1 mol% of catalyst loading, via synergistic, noncovalent hydrogen-bonding activation of both the Michael donor and acceptor. The domino reaction of 2-mercaptobenzaldehydes with α,β-unsaturated oxazolidinones afforded the corresponding chiral thiochromanes bearing three stereogenic centers in good to excellent yields and enantioselectivities of upto 99% ee, as shown in Scheme 10.6.

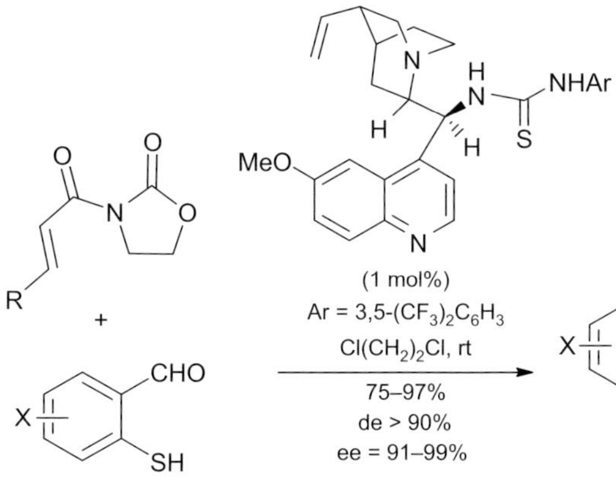

Scheme 10.6 Domino Michael/aldol reaction through hydrogen-bonding activation mode.

Soon after, the same group reported another enantioselective domino Michael/aldol reaction, occurring this time between 2-mercaptobenzaldehydes and maleimides, these last being much less explored substrates [83]. They managed to incorporate succinimides into complex benzothiopyrans, generating three stereogenic centers. The domino process was catalyzed by a bifunctional chiral amine thiourea used at only 1 mol% of catalyst loading through a hydrogen-bonding-mediated activation mechanism. The domino products were achieved in good yields (86–96%) and enantioselectivities (75–94% ee), combined with moderate to high diastereoselectivities (50–90% de). In 2010, another domino thia-Michael/aldol reaction was described by Wang *et al.* [84] upon catalysis by a chiral bifunctional thiourea (5 mol%). This novel hydrogen-bonding-mediated highly stereoselective domino reaction of various benzylidenechroman-4-ones with 2-mercaptobenzaldehyde allowed an efficient construction of chiral spiro chromanone-thiochromans. These potential biologically interesting products were obtained in excellent yields (95–98%) and general excellent enantioselectivities

(95–99% ee) combined with good diastereoselectivities (84–96% de). The hydrogen-bonding-activation mode was also recently utilized by Barbas *et al.* [85] to explain the formation of the chiral complexes (1*S*,2*S*,3*R*,4*S*)-bispirooxindoles on the basis of a novel domino Michael/aldol reaction occurring between 3-substituted oxindoles and a range of methyleneindolinones. This process was catalyzed by a novel multifunctional organocatalyst (15 mol%) containing tertiary and primary amines and thiourea moieties to activate the substrates simultaneously, providing extraordinary levels of stereocontrol over four stereocenters, three of which were quaternary carbon centers. Indeed, this novel methodology provided a facile access to a number of potent, biologically active, multisubstituted bispirocyclooxindole derivatives with high yields (56–92%), low to high diastereoselectivities of 26–98% de, and high enantioselectivities of 90% to >98% ee. Even more interestingly, these authors proposed the possibility of achieving the opposite enantiomers of these multifunctionalized products. Indeed, they demonstrated that performing the reaction in the presence of a reconfigured catalyst in which the *S*-diamine component was kept and the tertiary amine and the thiourea configurations were changed when compared with the firstly employed catalyst gave rise to the expected enantiomer (1*R*,2*R*,3*S*,4*R*)-bispirooxindole in good yield (76%) combined with high diastereoselectivity of 88% de and high enantioselectivity of 90% ee. The possibility of achieving both enantiomers of these novel complex products will be highly useful to investigate their biological activity. In 2012, a novel domino thia-Michael/aldol reaction of 2-mercaptobenzaldehyde with α,β-unsaturated *N*-acylpyrazoles was reported by Wang *et al.* [86]. This highly efficient protocol, catalyzed by a chiral cinchona alkaloid thiourea used at as low as 1 mol% of catalyst loading, allowed a range of highly substituted and biologically active thiochromanes bearing three stereogenic centers to be achieved in good yields (65–94%) and high diastereo- and enantioselectivities of 86–98% de and 93–99% ee, respectively. In 2007, Tang *et al.* [87] reported the first example of an asymmetric formal [3+3] annulation of cyclic ketones with enones. In this work, the enantioselective domino Michael/aldol reaction evolved through the enamine–iminium activation mode upon catalysis with chiral *N*-(pyrrolidin-2-ylmethyl)trifluoromethanesulfonamide (20 mol%), producing a range of chiral bicyclic [3.3.1] adducts bearing four stereogenic centers in good yields (66–92%) and good to high enantioselectivities of 80–94% ee. On the other hand, asymmetric domino Michael/aldol reactions have also been developed on the basis of enamine activation. As an example, Cordova *et al.* [88] reported a highly enantioselective domino Michael/aldol reaction catalyzed by chiral diphenylprolinol trimethylsilyl ether (10 mol%), giving access to highly functionalized cyclohexanes bearing four stereogenic centers. These chiral products were generated in good yields (72–95%) and enantioselectivities (85–98% ee), albeit with moderate to good diastereoselectivities (66–82% de) by the reaction of glutaraldehyde with various alkylidene malonates.

10.2.1.3 Domino Michael/Intramolecular Heterocyclization Reactions

A large number of enantioselective domino Michael/intramolecular heterocyclization reactions have been successfully developed in the last years, using three types of Michael acceptors, namely α,β-unsaturated aldehydes, α,β-unsaturated ketones or their derivatives, and nitroolefins. The reactions of these three possible types of reactants will be successively developed in this section. In the context of reactions involving α,β-unsaturated aldehydes as Michael acceptors, Jørgensen *et al.* [89] reported, in 2005, the first asymmetric organocatalytic epoxidation of α,β-unsaturated aldehydes, employing simple peroxides such as H_2O_2 as the oxygen source. This reaction proceeded via the iminium–enamine activation mode, using a chiral diarylprolinol trimethylsilyl ether as catalyst. First of all, the chiral iminium ion formed was attacked by nucleophilic oxygen at the electrophilic β-carbon, generating the first carbon–oxygen bond and leading to an enamine intermediate. Afterward, this nucleophilic enamine attacked the electrophilic peroxygen atom, forming the α,β-epoxy aldehyde and regenerating the catalyst. It is noteworthy that the reaction worked well in a wide range of solvents at room temperature, obtaining the best results when dichloromethane was used with 10 mol% of catalyst. The domino reaction tolerated a wide range of β-substituents in the enal moiety, such as variously substituted aromatic rings, alkylic substituents, and functionalized carbons such as esters or protected alcohols. The epoxy aldehydes were achieved in good yields (75–98%) and enantioselectivities of up to 99% ee. Soon after, Cordova *et al.* [90] performed a similar reaction using chiral diphenyl prolinol trimethylsilyl ether as catalyst, which provided excellent results in terms of conversion (59–82% yields) and enantioselectivities (91–97% ee). In 2007, the same authors reported a one-pot combination of amine and heterocyclic carbene catalysis, which enabled the synthesis of β-hydroxy and β-aminoesters from α,β-unsaturated aldehydes with high enantioselectivities [91]. Indeed, iminium activation of enals followed by enantioselective Michael addition of nucleophiles, such as H_2O_2 or CbzNHOAc, followed by cyclization furnished the corresponding epoxy or aziridine aldehydes. Next, the base-generated heterocyclic carbene catalyst induced the C–O or C–N bond-cleavage ring-opening followed by concomitant oxidation of aldehyde and subsequent esterification. The final β-hydroxy and β-aminoesters were achieved in good yields (59–82%) and high enantioselectivities (91–97% ee). In 2008, Wang [92] and List published a nice epoxidation of α,β-unsaturated aldehydes through asymmetric counterion-directed catalysis. This asymmetric induction mode involved an achiral secondary amine which formed a captionic achiral iminium ion with the enal. The interaction of this cation with an anionic chiral phosphoric acid (the chiral counterion) created a chiral environment. Then, *tert*-butyl hydroperoxide performed an asymmetric epoxidation through an iminium–enamine mechanism, as previously described. Aromatic enals were epoxidated with excellent diastereo- and enantioselectivities of up to >98% de and 96% ee, respectively. However, the reaction of aliphatic enals provided the corresponding epoxides with high diastereoselectivities of up to 88% de, albeit with moderate enantioselectivities (≤70% ee). In 2007, Cordova *et al.* [93] developed the first asymmetric organocatalytic synthesis of aziridines between aliphatic enals and acylated hydroxycarbamates, which was catalyzed by chiral diphenylprolinol trimethylsilyl ether in chloroform. As in the case with epoxidation, the choice of the nitrogen source was crucial for the success of the domino reaction. It was demonstrated that acylated hydroxycarbamates were the best substrates, providing chiral 2-formylaziridines in moderate to good yields (54–78%) with high diastereoselectivities (60–82% de) and enantioselectivities (84–99% ee). More recently, Hamada *et al.* [94] reported an interesting variation on the enantioselective aziridination of α,β-unsaturated aldehydes, which was based on the use of *N*-arenesulfonylcarbamates as the nitrogen source and chiral diphenylprolinol triethylsilyl ether as catalyst in the presence of 3 equiv of a base such as NaOAc or Na_2CO_3. This new protocol improved the yields (51–99%) and the diastereoselectivity (80–98% de), maintaining excellent enantioselectivity (91–99% ee) in comparison with the previous methodology reported by Cordova. Moreover, it was also noteworthy that this methodology expanded the aldehyde scope, allowing the aziridination of aromatic enals. On the other hand, Melchiorre *et al.* [95] developed, in 2008, the aziridination of α,β-unsaturated ketones. This domino process was catalyzed by a chiral primary amine derived from cinchona alkaloids as a salt with D-*N*-Boc-phenylglycine (20 mol%), and occurred efficiently with both linear and cyclic substrates, providing the corresponding chiral aziridines in high yields (82–96%) combined with excellent diastereoselectivity of >90% de and good to excellent enantioselectivities (73–99% ee). On the other hand, several groups have demonstrated that asymmetric domino Michael/hemiacetal (or hemiaminal) formation reactions constituted powerful methodologies to afford a number of other chiral heterocycles. As an early example, Cordova *et al.* [96] developed a very elegant synthesis of 5-hydroxyisoxazolidines based on the addition of N-protected hydroxyamines to α,β-unsaturated aldehydes. The authors disclosed that, in a first step, the amine attacked the β-position of the iminium ion, this reaction being an equilibrium one. This equilibrium was pushed to the final products because of hemiacetal formation between the hydroxyl moiety at the nitrogen atom and the aldehyde. The reaction worked well with any unsaturated aldehyde (aromatic and aliphatic), achieving the final compounds in good yields (75–94%) and high enantioselectivities of 91–99% ee. Moreover, the greatness of this reaction was clearly shown by the synthesis of chiral β-aminoacids from α,β-unsaturated aldehydes in only two steps. In the same year, Rios and Cordova [97] developed a synthesis of chiral highly functionalized pyrrolidines, following a similar approach. Thus, 2-amidomalonates reacted with α,β-unsaturated aldehydes as a result of a malonate addition followed by hemiaminal formation between the corresponding amide and aldehyde. This domino process furnished chiral pyrrolidines from aromatic α,β-unsaturated aldehydes in good yields (67–77%), moderate to good diastereoselectivities (≤ 82% de), and excellent enantioselectivities (95–99% ee). In 2008, Jørgensen [98] and Rueping *et al.* [99] independently reported the enantioselective addition of 1,3-cycloalkanediones to α,β-unsaturated aldehydes catalyzed by a chiral diarylprolinol trimethylsilyl ether, which furnished the corresponding 3,4-dihydropyrans through a domino Michael/hemiketalization reaction. These domino products were achieved in good yields (48–95%), diastereoselectivities of 44% to >90% de, and enantioselectivities of 82–97% ee. The same year, Rueping *et al.* developed the enantioselective synthesis of 1,4-pyranonaphthoquinones based on a domino Michael/hemiacetalization reaction of 2-hydroxy-1,4-naphthoquinone with α,β-unsaturated aliphatic as well as aromatic aldehydes induced by the same catalyst [100]. Biologically interesting quinines were achieved in good yields (43–90%) and high enantioselectivities (90–99% ee). Furthermore, these products could be easily converted into the corresponding valuable 1,4-naphthoquinones without loss of enantiopurity. In 2009, Franzen and Fisher [101] reported an easy entry to the synthesis of chiral quinolizidines bearing three stereogenic centers by the reaction of activated indol-substituted amide and α,β-unsaturated aldehydes. The authors disclosed that, in a first step, the α,β-unsaturated aldehydes were activated by chiral diphenylprolinol trimethylsilyl ether to form the corresponding iminium intermediates, which added according to a Michael process the activated amide to form the corresponding intermediates. Next, these intermediates cyclized spontaneously under the reaction conditions to give hemiaminals. A further epimerization of their stereochemically labile stereocenter at C3 established the more thermodynamically stable trans configuration. These domino products were further treated with added catalytic amounts of HCl, resulting in the formation of the corresponding acyliminium ions which cyclized to provide the final indolo[2,3*a*]quinolizidines with good yields (53–69%) and diastereoselectivities of 48–80% de, combined with good enantioselectivities (87–95% ee). In the same context, Rios *et al.* [102] developed the Michael addition of more simple nucleophiles, such as amidomalonates, to a range of α,β-unsaturated aldehydes by using the same catalyst, followed by the formation of the corresponding hemiaminals in high yields (86–94%) and high enantioselectivities of 90–99% ee. In all cases of the substrates studied, these piperidines were produced as a mixture of two diastereomers with 3 : 1–5 : 1 ratio. One of the formed chiral piperidine was applied as intermediate in the synthesis of (−)-paroxetine, a blockbuster antidepressive drug. In 2010, Chen *et al.* [103] reported a facile method for the asymmetric synthesis of chiral pyrimidinones via a domino aza-Michael/hemiaminal formation reaction of α,β-unsaturated aldehydes with *N,N'*-dibenzyloxyurea as dinitrogen source. The domino reaction, induced by the same catalyst in the presence of acetic acid, afforded the corresponding hemiaminals, which were subsequently submitted to dehydroxylation by treatment with Et_3SiH/BF_3 which was directly added to the reaction, thus providing the corresponding pyrimidinones in both high yields (83–95%) and enantioselectivities of 83–97% ee. In 2011, highly functionalized chiral quinolizidines were achieved by Zhao *et al.* [104] on the basis of an enantioselective domino reaction including a domino Michael/intramolecular cyclization of β-ketoamides and α,β-unsaturated aldehydes. The domino products were not isolated but subsequently submitted to a Pictet–Spengler cyclization by treatment with TFA to give chiral indolo[2,3-α]quinolizidines and benzo[α]quinolizidines as single diastereomers (de >99%) in moderate to good yields (56–90%) and good to excellent enantioselectivities of 77–97% ee. The authors had previously reported the synthesis of the related chiral quinolizidines through a two-step sequence, including an enantioselective domino Michael/intramolecular hemiaminalization/dehydration reaction of β-ketoesters with α,β-unsaturated aldehydes catalyzed by a chiral diarylprolinol trimethylsilyl ether as a first step followed by, after addition of tryptamine and benzoic acid, a Pictet–Spengler cyclization as a second step which provided chiral indoloquinolizidines as single diastereomers in good yields (51–95%) and high enantioselectivities (67–96% ee) [105]. A total synthesis of the natural product diazonamide A was accomplished by MacMillan *et al.*, [106] in 2011. This work featured a highly stereoselective synthesis of the C-10 quaternary center and the central furanoindoline core enabled by an iminium-catalyzed Michael/intramolecular cyclization reaction. The process, which involved propynal as Michael acceptor, was catalyzed by a chiral imidazolidinone salt, affording the corresponding chiral, tetracyclic, densely functionalized domino product in 86% yield and diastereoselectivity of >90% de. On the other hand, a novel general methodology based on iminium activation mode was recently developed by Kanger *et al.* to access highly enantiomerically enriched 1,4-dihydropyridines, which are a well-known class of biologically active heterocycles as well as analogs of nicotinamide adenine dinucleotide (NADH) coenzymes [107]. These products were generated by reaction of aliphatic and aromatic α,β-unsaturated aldehydes with enaminones (or β-enaminoesters) employed as nucleophiles. Instead of being Michael acceptors, the amine-activated double bond of enaminones (or β-enaminoester) enabled a conjugate addition to α,β-unsaturated aldehydes via iminium intermediates, followed by proton transfer and spontaneous iminium–enamine transformation. The following cascade of reactions involved the hydrolysis of the iminium intermediates, leading to an intramolecular cyclization, which gave rise to six-membered heterocycles. After a proton transfer has taken place, these intermediates dehydrated to give the energetically more favored final 1,4-dihydropyridines. This cascade of reactions was induced by the bulky 3,5-ditrifluoromethylphenylprolinol trimethylsilyl ether (20 mol%) combined with benzoic acid as an additive, and provided good enantioselectivities of 76–93% ee and good yields of 70–86%. The process had a wide scope, affording a range of 1,4-dihydropyridines with various substituents at four different positions. In 2010, Zlotin *et al.* [108] demonstrated that a novel chiral ionic liquid bearing an α,α-diphenylprolinol unit was able to act as an efficient catalyst in domino reactions between α,β-unsaturated aldehydes and N-protected hydroxylamines, involving an aza-Michael addition followed by an intramolecular hemiacetalization reaction. Indeed, the corresponding 5-hydroxy-3-arylisoxazolidines were achieved in good yields of 78–94% and with moderate to high enantioselectivities (67% to >99% ee). The ionic-liquid-supported catalyst could be easily recycled and reused for at least four times without significant loss of chemical yield or enantioselectivity. In 2011, Kim *et al.* developed the enantioselective domino Michael/hemiacetalization reaction of 2-hydroxybenzalacetones with malonates catalyzed by chiral diphenylprolinol trimethylsilyl ether [109]. Evolving through an iminium activation mode, this powerful process afforded a range of almost enantiopure 4-substituted chroman-2-ols (94–99% ee) in good yields of 62–98%. A novel approach to chiral pyrazolidines has been very recently reported by Vicario *et al.*, employing α,β-unsaturated aldehydes and *N,N'*-disubstituted hydrazines as components

[110]. It consisted in an enantioselective domino aza-Michael/hemiaminalization reaction catalyzed by a chiral diarylprolinol trimethylsilyl ether, providing through iminium activation the corresponding pyrazolidin-3-ols bearing two stereogenic centers in 50–99% yields, high enantioselectivities of 89–97% ee, and good diastereoselectivities of 82–90% de. These chiral products could be further converted into the corresponding pyrazolines through sequential deprotection/dehydration, or oxidized to the corresponding pyrazolidin-3-ones. On the other hand, the dienamine activation mode consists in the *in situ* formation of dienamines from α,β-unsaturated aldehydes or ketones in the presence of a chiral amine organocatalyst [111]. This activation mode has been recently applied by Enders *et al.* [112] to the asymmetric synthesis of functionalized dihydrocoumarins and chromans based on a domino reaction of 2-hydroxy-nitrostyrenes with α,β-unsaturated aldehydes. Indeed, this domino Michael/intramolecular hemiacetalization reaction evolved through the *in situ* formation of the dienamine of the α,β-unsaturated aldehyde in the presence of chiral diphenylprolinol trimethylsilyl ether. This dienamine underwent a Michael addition into 2-hydroxy-nitrostyrene to give the corresponding nitromethane intermediate, which was subsequently submitted to an intramolecular hemiacetalization to give the domino product. This product was further oxidized by treatment with pyridinium chlorochromate (PCC) to provide the corresponding dihydrocoumarin in good yield (81%) and diastereoselectivity (80% de) combined with excellent enantioselectivity (98% ee). A number of recent works have also involved enones or their derivatives as Michael acceptors in asymmetric domino reactions involving Michael and intramolecular heterocyclization reactions. As an early example, Mukaiyama *et al.* [113] have described chiral quaternary ammonium phenoxides derived from cinchona alkaloids and proved them to be useful novel organocatalysts. For example, a chiral cinchonidine-derived catalyst, bearing both a sterically hindered *N*1-9-anthracenylmethyl group and a strongly electron-withdrawing 9-*O*-3,5-bis(trifluoromethyl)benzyl group, was found to be highly effective for the Michael addition of ketene silyl acetals derived from phenyl carboxylates and α,β-unsaturated enones followed by lactonization. The corresponding optically active 3,4-dihydropyran-2-one derivatives were achieved in high yields (84–99%) and high control of enantio- and diastereoselectivity of 84–98% de and 85–96% ee, respectively. In 2009, Xie *et al.* [114] developed an enantioselective synthesis of highly functionalized chiral 2-amino-2-chromene derivatives based on a domino reaction occurring between α,β-unsaturated enones, such as 2-hydroxybenzalacetones, and malononitrile. This novel process was catalyzed by a cinchona-alkaloid-derived primary amine, such as 9-amino-9-deoxyepiquinine, in combination with (*R*)-1,1′-binaphth-2,2′-diyl hydrogen phosphate. Moderate to high enantioselectivities of 75–96% ee were obtained in combination with good yields of 53–84% for a range of β-substituted 2-hydroxybenzalacetones. Once again, the iminium activation mode was proposed by the authors to explain the results. In the same year, Zhao *et al.* [115] reported the first enantioselective domino Michael/hemiketalization reaction of β,γ-unsaturated α-keto esters with α-substituted cyano ketones. This process employed a novel tyrosine-derived

chiral bifunctional thiourea–tertiary-amine catalyst (2 mol%), and provided the corresponding densely functionalized chiral dihydropyrans in high yields (70–95%) and high enantioselectivities of 87–96% ee. These authors also studied the enantioselective domino Michael/hemiketalization/dehydration reaction of β,γ-unsaturated α-keto esters with malononitrile, which provided, when catalyzed by 5 mol% of another chiral bifunctional thiourea bearing a tertiary-amine, the corresponding highly functionalized 4*H*-pyrans in good yields (50–68%) and enantioselectivities (72–88% ee) [116]. On the other hand, Calter and Wang [117] have reported the synthesis of novel cinchona-alkaloid-derived pyrimidines which were shown to be effective chiral catalysts for the domino Michael/hemiketalization reaction of cyclic diketones with β,γ-unsaturated α-ketoesters. This methodology allowed bicyclic lactols to be achieved as an equilibrating mixture of anomers in good yields (64–97%) and excellent enantioselectivities (94–99% ee). Another asymmetric domino Michael/hemiacetalization reaction was reported by Ma *et al.*, in 2010 [118]. In this case, it involved the reaction of cyclic β-oxo aldehydes with β,γ-unsaturated α-keto esters, resulting in the formation of spiro-dihydropyrans as mixtures of two anomers in good yields (62–99%) upon catalysis with hydroquinidine-2,5-diphenyl-4,6-pyrimidinediyl diether ($(DHQD)_2PYR$). The obtained mixtures of anomers were subsequently oxidized by treatment with PCC to give the corresponding α-spirolactones as single diastereomers in high enantioselectivities of 84–97% ee. In 2010, Brière *et al.* [119] described the first organocatalytic enantioselective synthesis of biologically important pyrazolines based on a domino aza-Michael/cyclocondensation reaction of chalcones with *N*-Boc hydrazines catalyzed by a chiral cinchona alkaloid (10 mol%). The reaction was performed under phase-transfer catalysis, providing the pyrazolines in good to high yields (60–89%) and enantioselectivities of 87–92% ee. As heterocycles containing a trifluoromethyl group are representatives of a major structure type in agricultural and medicinal chemistry, Shibata *et al.* developed a novel enantioselective synthesis of trifluoromethyl-substituted 2-isoxazolines based on a domino Michael/cyclization/dehydration reaction of hydroxylamine with a range of (*E*)-trifluoromethylated enone derivatives [120]. This process, which employed *N*-3,5-bis(trifluoromethylbenzyl)quinidinium bromide as a chiral phase-transfer catalyst (10 mol%) combined with CsOH as a base, provided a series of trifluoromethyl-substituted 2-isoxazolines in high yields (83–99%) and enantioselectivities of 82–94% ee. In 2011, Wang *et al.* [121] reported the use of a simple, bifunctional chiral indane amine–thiourea catalyst (10 mol%) to induce an enantioselective domino Michael/hemiketalization/retro-Henry sequence of β,γ-unsaturated ketoesters with α-nitroketones. This process provided a new route to chiral 5-nitro-pent-2-enoates, which were obtained in high yields of 91–99% and moderate to excellent enantioselectivities of 75–97% ee. The catalyst was supposed by the authors to activate the β,γ-unsaturated ketoester using the amine and thiourea groups through the enolization and hydrogen-bonding formation. In addition, another chiral piperidine-based thiourea–tertiary amine catalyst was applied by Yan *et al.* to induce the conjugate addition of malononitriles to conformationally restricted dienones, which was followed by an intramolecular cyclization, providing the corresponding chiral bicyclic pyran derivatives in good to excellent yields (48–99%) and enantioselectivities of 83–99% ee [122]. It must be noted that these products constituted potential inhibitors against *Mycobacterium tuberculosis*. Furthermore, the results obtained with conformationally restricted dienones were significantly different from those obtained from the reaction of conformationally flexible dienones, which generated upon catalysis with 9-amino-9-deoxyepiquinine the corresponding cyclohexanones in moderate to high yields (35–87%) and excellent enantioselectivities (95–99% ee), as previously reported by the same authors [123]. A nice asymmetric synthesis of trifluoromethyl-substituted dihydropyrans was reported by Yan *et al.*, in 2011 [124]. It was based on an enantioselective domino Michael/hemiketalization reaction of ethyl 4,4,4-trifluoroacetoacetate and other trifluoromethyl-substituted nucleophiles with β,γ-unsaturated α-ketoesters catalyzed by a chiral quinine-derived thiourea. By using this methodology, a number of chiral trifluoromethyl-substituted dihydropyrans bearing three contiguous stereocenters could be produced in remarkable levels of yields as well as diastereo- and enantioselectivities, as shown in Scheme 10.7.

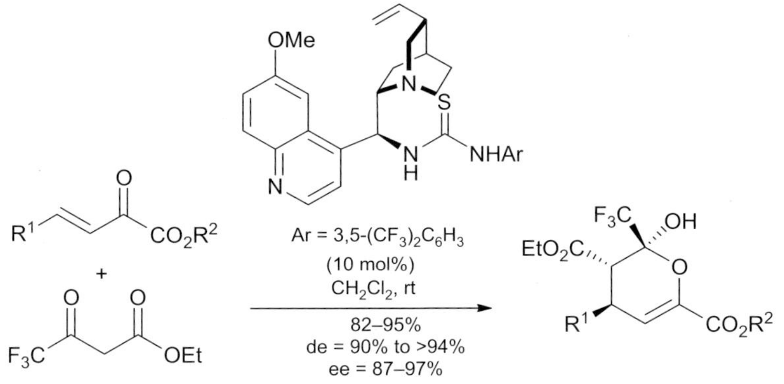

Scheme 10.7 Synthesis of trifluoromethyl-substituted dihydropyrans through domino Michael/hemiketalization reaction.

On the other hand, a highly enantioselective synthesis of chiral multisubstituted polyfunctionalized 2,3-dihydropyrroles was described by Wang *et al.*, in 2011 [125]. This unique approach was catalyzed by a novel tertiary amine–thiourea based on rosin and evolved through enantioselective domino Michael/cyclization reaction of 2-aminomalonates with α,α-dicyanoolefins. These densely functionalized heterocyclic compounds were achieved in good yields (62–97%) and enantioselectivities of 28–97% ee. Asymmetric domino Michael/hemiketalization reactions have also been developed starting from nitroalkenes as Michael acceptors. As an example, Chandrasekhar *et al.* reported an enantioselective synthesis of chiral cycloalkane fused tetrahydropyrans based on the reaction of cyclohexanone with various nitroalkenes [126]. This process was induced through the enamine activation mode by chiral pyrrolidine-triazole (20 mol%), affording the corresponding domino

Michael/hemiketalization products bearing four stereogenic centers in moderate to good yields (41–62%) and high enantioselectivities of 91–99% ee. In the same area, Gong [127] and Hong *et al.* [128] independently reported asymmetric domino Michael/hemiacetalization reactions of 2-hydroxynitrostyrenes with aliphatic aldehydes catalyzed by chiral diphenylprolinol trimethylsilyl ether (20 mol%) combined with benzoic acid in 95% EtOH or water only as solvent. These reactions afforded, after a subsequent oxidation with PCC, the corresponding 3,4-disubstituted dihydrocoumarins as a mixture of cis and trans diastereomers. In all cases of substrates studied, the cis diastereomers were obtained as major products with low to good diastereoselectivities of 2–76% de combined with high yields (50–89%) and high enantioselectivities of 87 to >99% ee. Similar reactions were also investigated by Ramachary *et al.* by using the same catalyst, which afforded enantiomeric 3,4-disubstituted dihydrocoumarins in good to high yields (65% to >95%) and both excellent cis diastereo- and enantioselectivities of up to >99% de and >99% ee, respectively. In this case, the reactions were performed in dichloromethane as solvent in the presence of 20 mol% of catalyst combined with benzoic acid as cocatalyst [129]. Furthermore, Enders *et al.* [130] developed in the same year enantioselective domino Michael/hemiketalization reaction of 2-hydroxynitrostyrenes with acyclic β-keto esters to afford functionalized chiral 4*H*-chromenes after subsequent dehydration by treatment of the domino products with *p*-toluenesulfonic acid (PTSA). This remarkable process was catalyzed by a chiral thiourea derived from norpseudoephedrine (10 mol%), providing the oxidized domino products in high yields (76–95%) and moderate to excellent enantioselectivities (30–98% ee). In 2012, these authors investigated the related domino Michael/hemiketalization reaction of 2-hydroxynitrostyrenes with cyclic β-keto esters as well as cyclic 1,3-diketones [131]. A chiral quinine thiourea (10 mol%) was demonstrated to be the most effective catalyst to afford the corresponding tricyclic polyfunctionalized chromans, after a subsequent dehydration. A range of tricyclic complex products were achieved in good yields (55–76%) and high enantioselectivities of 89–99% ee. In the same area, remarkably, chiral 1′,3-spiro-2′-oxocyclohexan-3,4-dihydrocoumarins were obtained by Hong *et al.* [132] on the basis of a domino Michael/hemiacetalization reaction of 2-hydroxynitrostyrenes with 2-oxocyclohexanecarbaldehyde catalyzed by a chiral bifunctional thiourea bearing a tertiary-amine (2–10 mol%), providing, after a subsequent oxidation by treatment with PCC, the corresponding lactones in good yields (51–72%) and high diastereo- and enantioselectivities of up to >90% de and 80% to >99% ee, respectively. In 2011, Ma *et al.* [133] reported an organocatalytic approach to chiral polysubstituted piperidines through Michael addition of various aldehydes onto trisubstituted nitroalkenes, followed by intramolecular hemiaminalization catalyzed by chiral diphenylprolinol trimethylsilyl ether (10 mol%) associated to benzoic acid as an additive. This domino reaction performed in water allowed four contiguous stereocenters to be formed in excellent general enantioselectivity of up to >99% ee, combined with high yields (80–91%) and diastereoselectivities of 80 to >90% de. Finally, Enders *et al.* [134] developed, in 2011, a highly efficient enantioselective domino Michael/hemiaminalization reaction of aldehydes and (*E*)-2-(2-nitrovinyl)-1*H*-indoles, which provided the corresponding almost

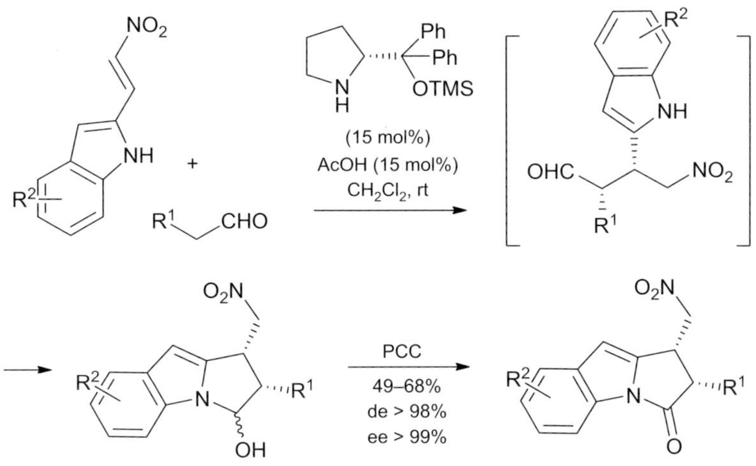

Scheme 10.8 Synthesis of 1*H*-pyrrolo[1,2*a*]indol-3(2*H*)-ones through domino Michael/hemiaminalization reaction.

diastereo- and enantiopure 1*H*-pyrrolo[1,2*a*]indol-3(2*H*)-ols in moderate to good yields (Scheme 10.8). The domino products obtained through catalysis by chiral diphenylprolinol trimethylsilyl ether in the presence of acetic acid as an additive were subsequently oxidized by treatment with PCC to give the corresponding 1*H*-pyrrolo[1,2*a*]indol-3(2*H*)-ones.

10.2.1.4 **Domino Michael/Intramolecular Alkylation Reactions**

The groups of Cordova and Wang [135] developed, almost simultaneously, asymmetric organocatalytic domino reactions involving a Michael addition followed by an intramolecular alkylation for the synthesis of chiral functionalized cyclopropanes. Indeed, highly diastereo- and enantioselective cyclopropanation of α,β-unsaturated aldehydes and 2-bromomalonates based on Michael addition followed by α-alkylation of the enamine intermediate was achieved, furnishing the final cyclopropanes. The domino process, evolving through the iminium–enamine activation mode, was catalyzed by chiral diphenylprolinol trimethylsilyl ether, which provided the best results with aromatic enals, leading to the corresponding cyclopropanated aldehydes in a total trans diastereoselectivity, good to high yields (50–87%), and excellent enantioselectivities (93–99% ee), as shown in Scheme 10.9. On the other hand, when aliphatic α,β-unsaturated aldehydes were used, the *trans*/*cis* ratio diminished from 90 : 10 to 94 : 6, while maintaining the high enantiocontrol.

In 2009, Rios *et al.* reported a nice variation of this domino process by employing 2-bromoketoesters instead of 2-bromomalonates. This change allowed the synthesis of chiral cyclopropanes containing a quaternary stereocenter with high diastereocontrol of up to >92% de, high yields, and excellent enantioselectivities of up to 99% ee [136]. In 2008, Cordova *et al.* reported a novel nitrocyclopropanation of α,β-unsaturated aldehydes employing bromonitromethane [137]. The

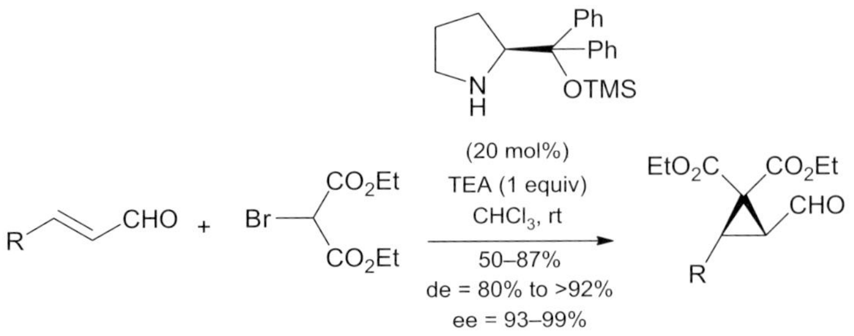

Scheme 10.9 Synthesis of cyclopropanes through iminium–enamine activation mode.

domino process was catalyzed by 20 mol% of chiral diphenylprolinol trimethylsilyl ether in the presence of 1 equiv of a base, such as triethanolamine (TEA), and afforded the corresponding chiral 1-nitro-2-formylcyclopropanes in moderate to good yields (29–63%), none to moderate diastereoselectivities of upto 50% de, and high enantioselectivities of up to 99% ee. Employing the same domino Michael/α-alkylation reaction, these authors developed an enantioselective synthesis of cyclopentanones bearing three stereogenic centers [138]. Indeed, replacing 2-bromomalonate by 4-bromoacetoacetate, under the effect of 20 mol% of the same catalyst combined with 1 equiv of potassium carbonate, allowed cyclopentanones containing three novel stereocenters to be achieved in good yields (58–88%), good diastereoselectivities (72–84% de), and high enantioselectivities (93–99% ee). In 2010, de Figueiredo *et al.* [139] revisited the previously reported enantioselective cyclopropanation of α-substituted α,β-unsaturated aldehydes with bromomalonate through domino a Michael/alkylation reaction catalyzed by the simple chiral diphenylprolinol trimethylsilyl ether. This method allowed the efficient formation of chiral cyclopropanes bearing a quaternary stereogenic center at the α-position of the aldehydes, giving a nice extension to this reaction with good yields of up to 81% and excellent enantioselectivities of up to 97% ee. Finally, Kim *et al.* [140] have reported an enantioselective domino intramolecular 1,5-hydride transferring closure reaction catalyzed by a chiral diarylprolinol triethylsilyl ether (30 mol%) associated with an equimolecular quantity of (−)-camphorsulfonic acid (CSA). Indeed, the intramolecular reaction of a series of cinnamaldehydes provided the corresponding ring-fused tetrahydroquinolines in high enantioselectivities of 90–99% ee, combined with moderate to good yields (53–75%) and diastereoselectivities of 14–99% de. In addition to enals, enones or their derivatives have also been successfully employed in asymmetric cyclopropanations through domino Michael/intramolecular alkylation reactions. As an early example, Ley *et al.* [141] reported the organocatalytic intermolecular cyclopropanation reaction of 2-cyclohexen-1-one with bromonitromethane catalyzed by chiral 5-(pyrrolidin-2-yl)-1*H*-tetrazole (15 mol%). The corresponding chiral cyclopropane was achieved in 80% yield and good enantioselective control (77% ee). In 2008, the scope of this reaction was expanded by using a variety of cyclic and acyclic enones [142]. In 2009, Wang *et al.* [143] reported the enantioselective synthesis of functionalized nitrocyclopropanes through the domino Michael/alkylation reaction of bromonitromethane with various cyclic α,β-unsaturated ketones. This process, induced by chiral amino-9-deoxyepiquinine (20 mol%) combined with *p*-methyl mandelic acid as cocatalyst in the presence of *N*-methyl morpholine (NMM) as base, allowed complete diastereoselectivity, good yields (85–99%), and high levels of enantioselectivities of 83 to >99% ee to be reached through the iminium–enamine activation mode. In 2011, Lattanzi *et al.* [144] developed the first enantioselective cyclopropanation to spirocyclopropanes from the domino Michael/alkylation reaction of 2-arylidene-1,3-indanediones and dimethyl bromomalonate, using chiral diphenylprolinol trimethylsilyl ether as catalyst and potassium carbonate as an additive. The chiral spirocyclopropanes were achieved in high yields (73–96%) and moderate to good enantioselectivities (70–85%). In the same year, Bartoli *et al.* reported the enantioselective nitrocyclopropanation of oxindoles with bromonitromethane, which was induced through the hydrogen-bonding-activation mode by chiral 9-*epi*-9-thiourea-9-deoxydihydroquinidine in the presence of $NaHCO_3$ [145]. The corresponding spiro nitrocyclopropyl oxindoles were achieved in good to high yields (74–82%), good diastereoselectivities (75–90% de), and high enantioselectivities of 91–98% ee. An intramolecular version of enantioselective domino Michael/intramolecular alkylation reaction was developed by Gaunt *et al.* [146] as a rare example with an enantioselective intramolecular cyclopropanation using chiral modified cinchona alkaloid organocatalysts. This domino Michael/intramolecular alkylation reaction afforded chiral functionalized [4.1.0]-bicycloheptanes in high yields (>70%) and general excellent enantioselectivity of >95% ee.

In addition to α,β-unsaturated aldehydes, α,β-unsaturated ketones, and their derivatives, nitroalkenes have also been used as Michael acceptors in enantioselective domino Michael/intramolecular alkylation reactions. For example, Yan *et al.* [147] developed, in 2009, a novel methodology for the cyclopropanation of nitroalkenes, based on the addition of 2-bromomalonates to nitroalkenes catalyzed by chiral cinchona alkaloids. The domino process proceeded with good yields (47–78%) and almost complete diastereo- and enantioselectivities in all cases of substrates studied. In another context, Gong *et al.* [148] developed an enantioselective formal [3+2] cycloaddition reaction of isocyanoesters to nitroolefins catalyzed by the same type of catalysts. In this approach, isocyanoesters underwent a Michael addition to nitroalkenes, and a subsequent intramolecular alkylation afforded the dihydropyrroles after protonation. As shown in Scheme 10.10, the chiral highly functionalized domino products were obtained in good yields, moderate to high diastereoselectivities, and high enantioselectivities.

Another asymmetric domino Michael/alkylation methodology was independently developed by Xie *et al.* [149] and Rueping *et al.* [150] for the synthesis of a range of chiral 2,3-dihydrofurans from reaction of α-bromonitroalkenes with 1,3-dicarbonyl compounds. In both cases, the catalysts were chiral bifunctional thioureas (30 mol%) associated to a base, such as diisopropylethylamine (DIPEA) or tetramethylethylenediamine (TMEDA). For example, the reaction of bicyclic 1,3-dicarbonyl compounds, such as 4-hydroxyl(thio)coumarins, led to the corresponding tricyclic 2,3-dihydrofurans in yields of 89–98%, general

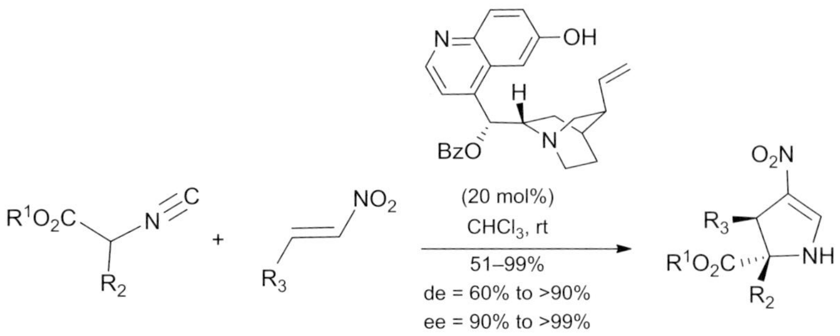

Scheme 10.10 Synthesis of dihydropyrroles through hydrogen-bonding-activation mode.

excellent diastereoselectivity of >99% de, and good enantioselectivities (82–92% ee). The scope of this process was extended to the synthesis of chiral bicyclic 2,3-dihydrofurans by the reaction of α-bromonitroalkenes with cyclohexane-1,3-diones, providing comparable diastereoselectivities (>99% de) combined with general excellent yields (98–99%) and good enantioselectivities of 84–88% ee. The scope of the domino reaction was also extended to acyclic 1,3-dicarbonyl compounds, which afforded the corresponding chiral monocyclic 2,3-dihydrofurans in comparable diastereoselectivities (> 99% de), high yields (86–88%), and good enantioselectivities (76–87% ee). In 2012, Lu *et al.* [151] described an enantioselective synthesis of 3(2*H*)-furanones based on a domino Michael/alkylation reaction of ethyl 4-bromoacetoacetate and nitroolefins induced by a novel chiral tertiary-amine thiourea catalyst derived from L-threonine (20 mol%). A range of biologically important 3(2*H*)-furanones were achieved in good to high yields (72–90%) and enantioselectivities of 87–94% ee.

10.2.1.5 **Domino Michael/(aza)–Henry Reactions**

The Henry reaction has been combined with the Michael reaction in successful enantioselective domino sequences [152]. For example, Jørgensen *et al.* [153] reported, in 2007, an organocatalytic synthesis of pentasubstituted cyclohexanes based on a Michael reaction of nitroalkanes to α,β-unsaturated aldehydes followed by an intramolecular Henry reaction. This domino process, evolving through iminium activation, was catalyzed by a chiral diarylprolinol trimethylsilyl ether in the presence of 1,4-diazabicyclo[2.2.2]octane (DABCO) and led to pentasubstituted cyclohexanols with remarkable control over five contiguous stereocenters. The domino products were achieved in moderate to good yields (40–65%) and diastereoselectivities of 20–72% de, combined with good to high enantioselectivities of 75–94% ee. Later, Ruano *et al.*[154] described a sequence constituted of a Michael addition of 1,3-diones to α,β-unsaturated aldehydes as a first step, which was followed, after the addition of nitromethane and tetrabutylammonium fluoride (TBAF), by a domino intermolecular Henry/intramolecular Henry reaction constituting the second step of the sequence . This sequential Michael domino intermolecular Henry/intramolecular Henry reaction induced by the same catalyst produced the corresponding cyclohexanes bearing five stereogenic centers with high enantio- and diastereoselectivities of >96% de and 92–99% ee, respectively, for a wide range of substrates. In 2010, Lu and Liu [155] developed domino Michael/aza-Henry reactions by employing a quinidine-derived tertiary-amine–thiourea catalyst. The authors showed that installing an electron-withdrawing sulfone group on the amino group of 2-aminobenzaldehydes allowed their activation, since the corresponding sulfonated 3-nitro-1,2-dihydroquinolines were generally obtained in better yields (75–92%) and enantioselectivities (70–90% ee). More recently, Xu *et al.* [156] developed the access to chiral 2,3,4-trisubstituted tetrahydroquinolines based on domino Michael/aza-Henry reactions involving nitromethane and chalcones. When these reactions were catalyzed with a chiral bifunctional thiourea, they furnished the potential biologically active polysubstituted tetrahydroquinolines in excellent yields (94–98%) and enantioselectivities (98% to > 99% ee), combined with moderate to high diastereoselectivities of 50–90% de. On the other hand, a number of domino Michael/(aza)-Henry reactions have employed nitroolefins as Michael acceptors. For example, Takemoto *et al.* [157] reported, in 2008, the use of a chiral thiourea derived from *trans*-1,2-diaminocyclohexane (10 mol%) to induce the domino Michael/aza-Henry reaction of α-amino malonate imines with nitroolefins to give the corresponding highly functionalized pyrrolidines bearing four stereogenic centers with good yields (52–86%) and diastereoselectivities of 82–96% de, combined with moderate to high enantioselectivities of 54–92% ee. Also in 2008, Xu *et al.* [158] developed a novel domino oxa-Michael/Henry reaction of 2-hydroxybenzaldehydes with nitroalkenes, providing the corresponding 3-nitro-2*H*-chromenes in 35–87% yields and enantioselectivities of 48–91% ee. The process was efficiently catalyzed by a chiral pyrrolidine in the presence of salicylic acid. It must be noted that Zhao *et al.* [159] had previously developed, in 2007, a similar reaction for the synthesis of chiral thiochromanes through domino oxa-Michael/Henry reaction of 2-mercaptobenzaldehydes with nitrostyrenes catalyzed by a simple chiral cupreine. The corresponding chiral 2-aryl-3-nitrothiochroman-4-ols were achieved in good yields (84–97%) with enantioselectivities of up to 86% ee and diastereoselectivities of up to 56% de. In addition, several groups have developed domino Michael/aza-Henry reactions catalyzed by chiral thioureas. For example, Xu *et al.* [160] reported the reaction of 2-aminobenzaldehydes with various nitroolefins upon catalysis with a chiral primary amine–thiourea catalyst to provide a variety of 3-nitro-1,2-dihydroquinolines in moderate yields (37–70%) and moderate to high enantioselectivities (52–90% ee). The authors assumed synergistic activation of both reactants through stereoselective covalent activation and hydrogen-bonding interactions. In another context, a highly diastereo- and enantioselective synthesis of multisubstituted and densely functionalized cyclopentanes bearing four contiguous stereogenic centers was described by Zhong *et al.*, in 2008 [161]. It was based on the enantioselective domino Michael/Henry reaction of ethyl 2-acetyl-4-oxo-4-phenylbutanoate and nitroolefins induced by a chiral cinchona-alkaloid-derived diamine. As shown in Scheme 10.11, remarkable results were obtained since the yields in domino products ranged from 90 to 95%, the enantioselectivities were up to 96% ee, and the general diastereoselectivity was >99% de. The scope of this methodology was applied to the synthesis of chiral functionalized

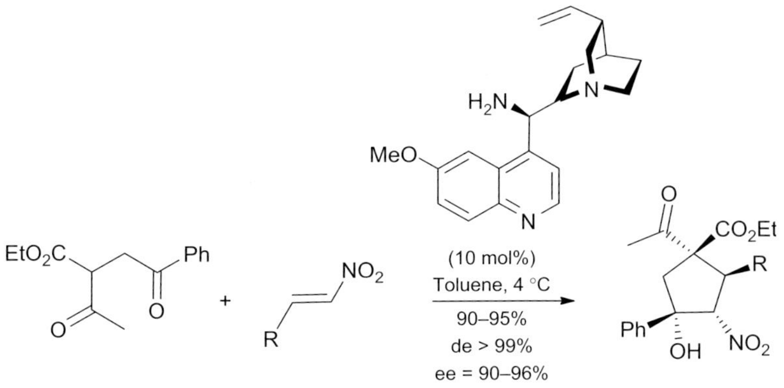

Scheme 10.11 Synthesis of cyclopentanes through hydrogen-bonding-activation mode.

cyclohexanes from the reaction of ethyl 2-acetyl-5-oxohexanoate and nitroolefins [162]. This domino Michael/Henry reaction provided excellent enantioselectivities (97–99% ee) combined with high diastereoselectivities (86–98% de) and high yields (88–94%).

Later, the groups of Zhao [163] and Rueping *et al.* [164] independently developed a novel domino Michael/Henry reaction induced by chiral bifunctional cinchona thioureas, which provided access to a range of chiral synthetically unique and medicinally important chiral bicyclo[3.2.1]octan-8-ones. The reaction occurred between cyclohexane-1,2-diones and nitroalkenes. The best results were obtained by using a chiral quinine-derived thiourea as catalyst. The complex polyfunctionalized products were obtained as mixtures of two diastereomers bearing four stereogenic centers with good diastereoselectivities of 76–90% de, combined with good yields (67–90%) and enantioselectvities of 93–99% ee. Nevertheless, it must be noted that this process was limited to cyclohexane-1,2-diones, since the attempted reactions of both butane-2,3-dione and 3-methylcyclopentane-1,2-dione failed under the same conditions. In the same area, Zhong *et al.* [165] developed highly enantio- and diastereoselective domino Michael/Henry reactions of nitroalkenes with 1,3-dicarbonyl compounds, such as 2-carboxymethyl-1,4-cyclohexanedione, which provided a range of other chiral bicyclo[3.2.1]octan-8-ones bearing four stereogenic centers including two quaternary stereocenters. These products were produced in high yields and excellent enantioselectivities of up to 96% ee, with a remarkable general diastereoselectivity of >98% de, as shown in Scheme 10.12. The process was induced by a chiral cinchona alkaloid thiourea catalyst, which was employed at a low catalyst loading of 5 mol%. Theoretical density functional theory (DFT) calculations on the transition states have been carried out to reveal origins of the stereoselectivities. They have shown that the induction of chirality was achieved through a novel type of activation mode in which the thiourea group and an acidic proton in the phenyl ring of the catalyst activated the 1,3-dicarbonyl substrate together, and at the same time a tertiary-amine activated the nitro group, which promoted the domino reaction smoothly with excellent stereoselectivity (Scheme 10.12).

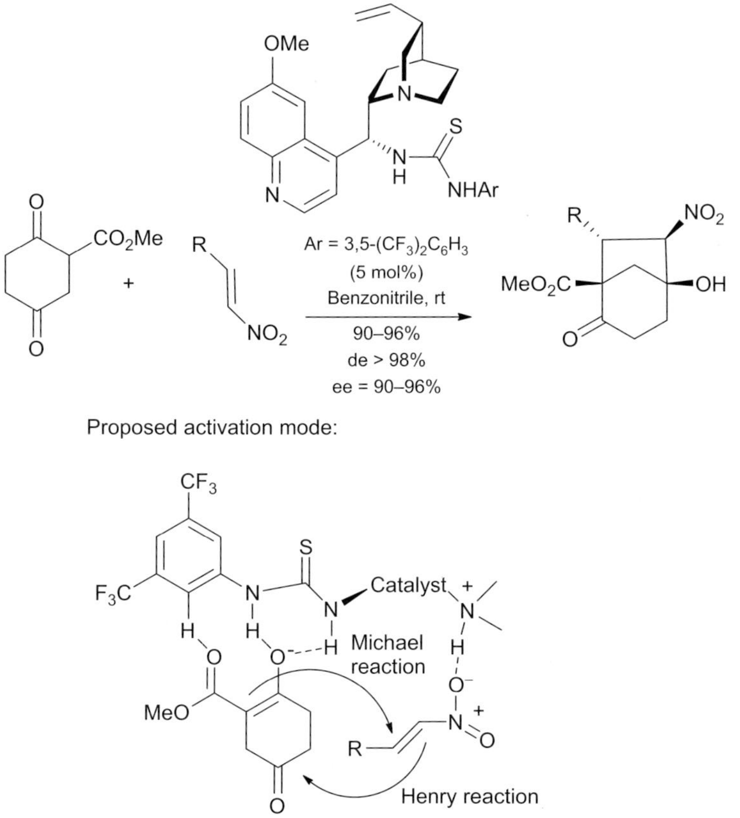

Scheme 10.12 Synthesis of tetrasubstituted bicyclo[3.2.1]octan-8-ones through hydrogen-bonding-activation mode.

In 2011, a new type of pyrrolidine-based chiral organocatalyst was developed by Headley *et al.* [166] and found to be very effective for inducing the domino Michael/Henry reaction of nitroolefins with pentane-1,5-dial. Performed in aqueous media and in the presence of benzoic acid as an additive, the domino process provided the corresponding tetrasubstituted and densely functionalized cyclohexanes in good yields (64–85%) and general excellent enantioselectivities of >98% ee, albeit with moderate diastereoselectivities of ≤50% de. Furthermore, the catalyst could be recycled up to four times resulting in good yields and up to seven times with good enantioselectivity.

10.2.1.6 **Domino Michael/Knoevenagel Reactions**

An early example of an asymmetric domino Michael/Knoevenagel reaction was reported by Jørgensen *et al.* [167] in 2006, providing the synthesis of chiral cyclohexenones from the reaction of *tert*-butyl 3-oxobutyrate and α,β-unsaturated aldehydes catalyzed by chiral diphenylprolinol silyl ether derivative. Soon after, the same authors developed a similar domino reaction starting from 4-chloro-ketoesters [168]. This process furnished highly functionalized epoxycyclohexanones in good yields and excellent diastereoselectivity of >99% de in all cases of substrates studied, combined with high enantioselectivities (85–97% ee). In this case, the domino Michael/aldol products were directly submitted to an intramolecular S_N2 reaction by treatment with a base, such as K_2CO_3, to give the final epoxycyclohexanone derivatives. In 2009, Hayashi *et al.* [169] developed a highly enantioselective formal [3+3] cycloaddition reaction of α,β-unsaturated aldehydes and dimethyl 3-oxopentanedioate catalyzed by chiral diphenylprolinol *t*-butyldimethylsilyl ether in combination with benzoic acid as an additive via a domino Michael/Knoevenagel reaction. This reaction proceeded with good yields and constituted a clean process to afford chiral substituted cyclohexenones after a subsequent reduction performed with sodium borohydride. As shown in Scheme 10.13, excellent enantioselectivities (94–99% ee) were obtained for a range of aromatic and heteroaromatic aldehydes. In 2011, Pericas *et al.* [170] reported the suitability of polymer-supported diarylprolinol silylethers for these reactions. Indeed, comparable results were achieved with these catalysts immobilized onto polystyrenes both under batch and continuous flow conditions. Excellent enantioselectivities of up to >99% ee were obtained for aromatic aldehydes; moreover, this methodology was extended to aliphatic aldehydes, which provided the corresponding chiral substituted cyclohexanols with enantioselectivities of up to 87% ee (with R = *n*-Bu).

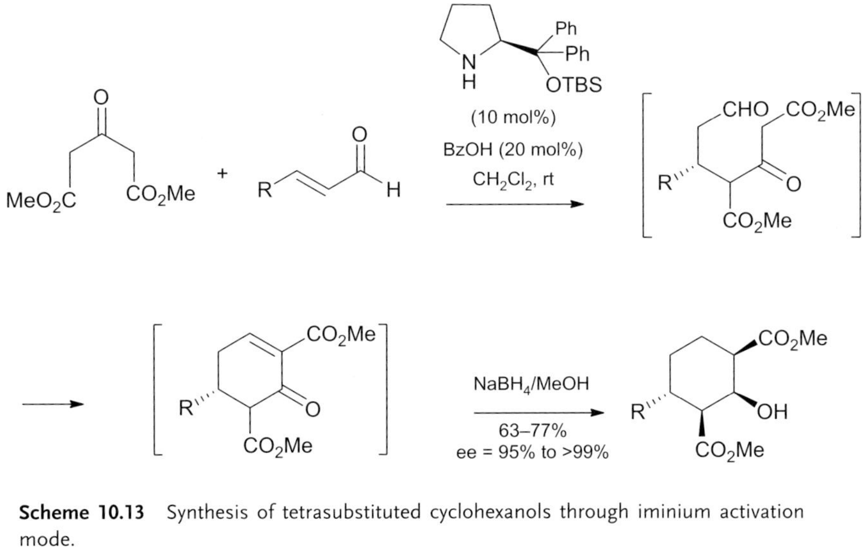

Scheme 10.13 Synthesis of tetrasubstituted cyclohexanols through iminium activation mode.

Independent of the work published by Hayashi *et al.*, Jørgensen *et al.* [171] have reported the synthesis of chiral 6-aryl 3-diethoxyphosphoryl-2-oxocyclohex-3-enecarboxylates based on domino Michael/Knoevenagel reactions of ethyl 4-diethoxyphosphoryl-3-oxobutanoate with aromatic α,β-unsaturated aldehydes. These reactions were catalyzed by a chiral diarylprolinol trimethylsilyl ether in combination with benzoic acid as cocatalyst. Both excellent diastereo- and enantioselectivities of up to >90% de and up to 98% ee, respectively, were achieved for the corresponding products in association with high yields in the case of aromatic aldehydes. Surprisingly, aliphatic aldehydes turned out to be unreactive under these conditions. However, the use of an additive, such as hydroquinine (DHQ), allowed the corresponding 6-alkylated 3-diethoxyphosphoryl-2-oxocyclohex-3-enecarboxylates to be obtained in excellent enantioselectivities (94–96% ee), albeit with moderate diastereoselectivities (50–60% de).

10.2.1.7 Domino Michael/aza-Morita–Baylis–Hillman Reactions

Since the early example of proline-catalyzed enantioselective domino Michael/aldol reaction reported by Bui and Barbas in 2000 11, a number of organocatalyzed domino Michael/Morita–Baylis–Hillman reactions have been successfully developed by several groups. In 2006, Hong *et al.* [172] reported an interesting example of an iminium–enamine sequence catalyzed by unmodified L-proline. This process involved 2 equiv of crotonaldehyde, which reacted through a domino Michael/aldol reaction to afford the corresponding chiral cyclohexenecarbaldehydes as a 53 : 47 mixture of two diastereomers obtained in 69% yield. While the diastereoselectivity of the reaction was low (6% de), the two epimers were obtained in high optical purity of 80% and 97% ee, respectively. In 2008, Jørgensen *et al.* [173] reported the domino Michael/Morita–Baylis–Hillman reaction of Nazarov reagents with α,β-unsaturated aldehydes catalyzed with chiral diphenylprolinol trimethylsilyl ether (10 mol%) in the presence of benzoic acid as an additive, which provided the corresponding highly substituted cyclohexanones in good yields (45–69%), low to good diastereoselectivities (20–84% de), and high enantioselectivities of 86–98% ee. Since 4-aminochromanes are a class of structures integrated in hundreds of natural and bioactive compounds, the same authors later proposed a novel access to this type of products based on a domino Michael/aza-Baylis–Hillman reaction occurring between alkynals and salicyl *N*-tosylimine [174]. This highly enantioselective reaction was catalyzed by a chiral diarylprolinol trimethylsilyl ether, which was supposed to activate the alkynals, forming iminium intermediates that underwent an oxa-Michael addition with the salicyl *N*-tosylimine. The resulting allenamine intermediate reacted with the imine intramolecularly, leading to the final 4-amino-4*H*-chromenes and regenerating the catalyst. It must be noted that this reaction took place in less than 2 h with high yields and general excellent enantioselectivities (94–99% ee), as shown in Scheme 10.14.

10.2.1.8 Domino Michael/Mannich Reactions

In 2007, Vicario *et al.* [175] reported an enantioselective [3+2] cycloaddition reaction of azomethine ylides with α,β-unsaturated aldehydes, which can also be described as a domino Michael/Mannich reaction of *in situ* generated azomethine ylides from imines with α,β-unsaturated aldehydes catalyzed by chiral diphenylprolinol (20 mol%) in the presence of water in tetrahydrofuran (THF). This remarkable process, evolving through the iminium–enamine activation mode, afforded the corresponding highly functionalized polysubstituted pyrrolidines in good to high

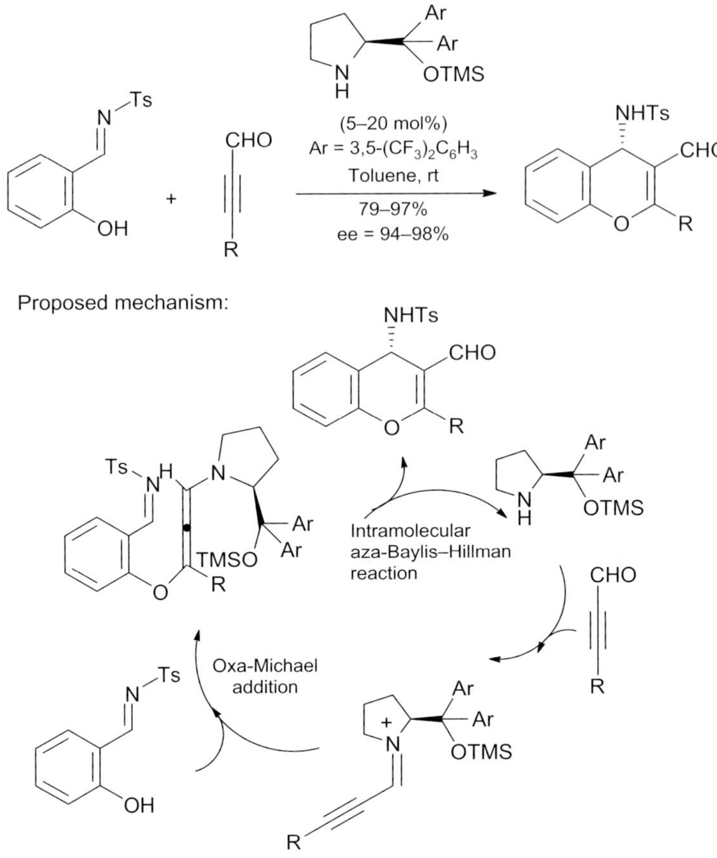

Scheme 10.14 Synthesis of 4-amino-4*H*-chromenes through iminium–allenamine activation mode.

yields (57–91%) and diastereo- and enantioselectivities of 82–90% de and 85–99% ee, respectively. Linear and branched aliphatic aldehydes as well as different imine substrates were tolerated. Later, Xu *et al.* [176] reported a novel type of bifunctional catalyst generated by the self-assembly of chiral pyrrolidine and the readily available primary amino acid D-*t*-leucine. The protonation of the aromatic nitrogen atom of the chiral pyrrolidine by this amino acid spontaneously led to ion-pair assemblies possessing dual activation centers which, similar to catalysis by enzymes, enabled the catalysis of the electrophilic and nucleophilic substrates simultaneously. This system was applied to induce novel domino oxa-Michael/Mannich reactions of salicylic aldehydes with cyclohexenones to afford the corresponding chiral tetrahydroxanthenones in high yields (87–95%) and enantioselectivities of 80–98% ee. A synthesis of highly substituted chiral tetrahydroimidazopyrimidine derivatives containing three stereogenic centers was described by Hu *et al.*, in 2011 [177]. These biologically important products were obtained for the first time through organocatalytic enantioselective domino aza-Michael/Mannich reaction of α,β-unsaturated aldehydes and *N*-arylidene-1*H*-imidazol-2-amines. The reaction was efficiently catalyzed by chiral diphenylprolinol trimethylsilyl ether in the presence of benzoic acid as additive, providing the domino products in good yields and general excellent diastereo- and enantioselectivities of up to >90% de and > 99% ee, respectively, as shown in Scheme 10.15.

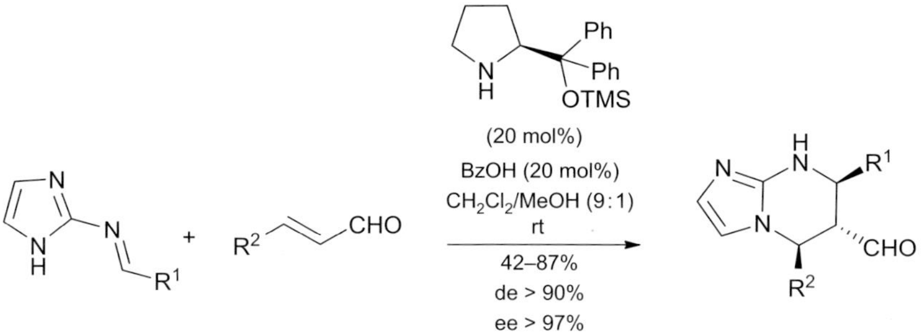

Scheme 10.15 Synthesis of tetrahydroimidazopyrimidines through iminium–enamine activation mode.

The same year, You *et al.* [178] reported an intramolecular domino Michael/Mannich reaction of indolyl methyl enones catalyzed by a chiral quinine-derived primary amine (20 mol%), affording a series of highly enantioenriched tetracyclic compounds in moderate to high yields (22–84%), low to good diastereoselectivities (12–78% de), and low to excellent enantioselectivities (5–98% ee). This novel polycyclization, evolving through the iminium–enamine activation mode, has been applied to the total synthesis of an analog of the natural product (+)-kreysiginine. In addition, Barbas *et al.* [179] have recently reported the enantioselective synthesis of carbazole spirooxindoles in good yields (75–99%) and enantioselectivities (88–97% ee) based on a domino Michael/Mannich-type reaction occurring between two indole derivatives upon catalysis with a chiral bifunctional thiourea (15 mol%).

10.2.1.9 **Other Domino Reactions Initiated by the Michael Reaction**

In 2008, Cordova *et al.* [180] reported an enantioselective domino aminosulfenylation reaction of α,β-unsaturated aldehydes catalyzed by chiral diphenylprolinol trimethylsilyl ether (20 mol%). Although the diastereocontrol was fairly low (10–48% de) for aromatic enals, they were easily separated by simple flash chromatography, and both the syn and anti adducts were obtained in excellent enantioselectivities (93–99% ee). In 2009, Chen *et al.* reported the synthesis of a newly designed bulky chiral secondary amine derived from proline, which was applied as catalyst (20–30 mol%) to the domino Michael/Wittig reaction of various α,β-unsaturated aldehydes with (3-carboxy-2-oxopropylidene)triphenylphosphorane [181]. The process, performed in the presence of $LiClO_4$ and DABCO, afforded the corresponding formal [3+3] cycloadducts in good to high yields (56–85%)

and moderate to excellent diastereoselectivities (50–96% de), combined with high enantioselectivities (86–99% ee). In another context, Zhao and Gogoi [182] developed the first enantioselective synthesis of biologically active 6-amino-5-cyanodihydropyrano[2,3-*c*]pyrazoles through a cinchona-alkaloid-catalyzed domino Michael/Thorpe–Ziegler type reaction between 2-pyrazolin-5-ones and benzylidenemalononitriles. The authors investigated a series of cinchona alkaloid catalysts, such as quinine, cupreine, 9-*epi*-cupreine, quinidine, and cupreidine, to induce this process. The best results were obtained by using cupreine (5 mol%), which provided high yields (89–96%) combined with low to excellent enantioselectivities (20–96% ee). A novel asymmetric domino reaction of cyclic ketones with (*E*)-2-nitroallylic acetates catalyzed by a chiral pyrrolidine–thiourea (20 mol%) combined with an equal amount of 4-methoxybenzoic acid as cocatalyst was developed by Li *et al.* [183]. The process afforded chiral bicyclic cyclohexanones bearing four stereogenic centers with enantioselectivities of 77–98% ee and moderate to high yields (27–78%). In 2011, Wang *et al.* [184] demonstrated that cyclohexan-1,2-dione could be used as a dual C–O nucleophile toward α,β-unsaturated pyruvates acting as dual electrophiles. As shown in Scheme 10.16, they developed an enantioselective domino Michael/enolization/cyclization reaction catalyzed by a chiral thiourea, providing chiral dihydro-2*H*-pyranes in high yields and excellent enantioselectivities.

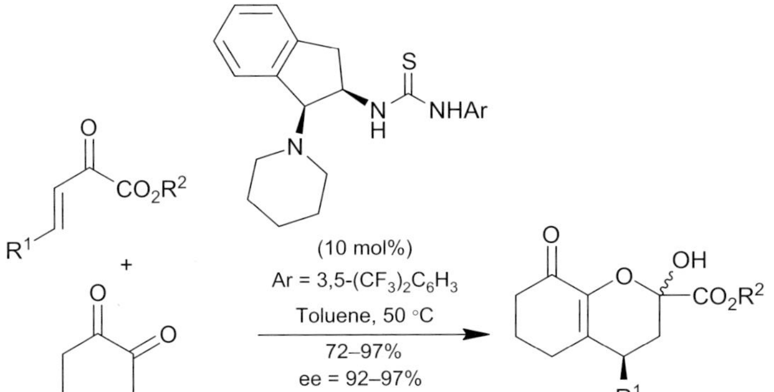

Scheme 10.16 Synthesis of dihydro-2*H*-pyranes through domino Michael/enolization/cyclization reaction.

Although a number of examples of asymmetric domino reactions have been developed, most rely on a single catalyst to perform generally two sequential operations. More recently, multiple catalyst systems for this type of reactions have been proposed [185]. Although these reactions showcase the potential power in this field, relatively few asymmetric versions exist [186]. In this context, Rovis and Lathrop [187] have developed a multicatalytic asymmetric domino reaction for the preparation of chiral α-hydroxycyclopentanones containing three contiguous stereocenters. This process was based on the reaction of 1,3-diketones with α,β-unsaturated aldehydes, which generated through Michael addition the corresponding aldehydes bearing a tethered ketone, induced by a chiral diarylprolinol trimethylsilyl ether (10 mol%) followed by hydrolysis. These intermediates then underwent an intramolecular crossed-benzoin reaction in the presence of a second carbene achiral catalyst to afford final cyclopentanones via a formal [3+2] process. In spite of moderate diastereoselectivities (28–70% de), these highly functionalized products were obtained in good enantioselectivities of 81–95% ee. These authors have developed another multicatalytic domino reaction occurring between salicylaldehydes and electron-deficient alkynes, such as dimethyl acetylenedicarboxylate (DMAD) [188]. In this case, the reaction proceeded via amine-catalyzed Michael addition, followed by an intramolecular Stetter reaction promoted by an N-heterocyclic carbene. A tertiary-amine, such as quinuclidine or DABCO, activated DMAD via the corresponding ammonium intermediate toward nucleophilic attack to give an intermediate aldehyde. Subsequent chiral carbene-promoted Stetter reaction set a quaternary stereocenter and yielded the corresponding final functionalized benzofuranones in good yields and good to high enantioselectivities of up to 94% ee. In addition, an N-heterocyclic carbene catalyst was very recently combined by the same authors with chiral diphenylprolinol trimethylsilyl ether to promote the diastereo- and enantioselective domino Michael/benzoin cyclization reaction between aliphatic aldehydes and activated enones, providing the corresponding complex cyclopentanones in moderate to excellent yields (35–99%), moderate to high diastereoselectivities (56–94% de), and moderate to excellent enantioselectivities (51–98% ee) [189]. An enantioselective formal aza-Nazarov cyclization reaction was described by Tius *et al.* in 2010 [190]. It involved the domino aza-Michael/water condensation reaction of an azirine catalyzed by a chiral 1,2-diaminocyclohexane derivative (20 mol%) in the presence of water, which led to the corresponding functionalized piperidone in low yield (28%), albeit with complete diastereo- and enantioselectivities. In 2011, Enders *et al.* [191] reported the synthesis of chiral polyfunctionalized cyclopentanones with three contiguous stereogenic centers based on an enantioselective domino Michael/cross-benzoin reaction of β-oxo sulfones with α,β-unsaturated aldehydes. This process was efficiently catalyzed by a dual chiral secondary amine/N-heterocyclic carbene catalytic system, which provided cyclopentanones in good yields (33–99%), modest to excellent diastereoselectivities (16–98% de), and very good enantiomeric excesses (86–96% ee).

10.2.2 Domino Reactions Initiated by Other Reactions

10.2.2.1 Domino Reactions Initiated by the Indirect Mannich Reaction

The classic direct Mannich reaction discovered in 1912 [192a] is an aminoalkylation of carbonylic compounds involving ammonia (or a primary or secondary amine derivative), a non-enolizable aldehyde (usually formaldehyde) or a ketone, and an enolizable carbonyl compound, leading to β-aminocarbonyl derivatives [192b–e]. The indirect version of the Mannich reaction is the corresponding two-component reaction of a preformed iminium salt [193] and an enolizable carbonyl compound.

An early example of asymmetric domino reactions initiated by the Mannich reaction was successfully developed by Ohsawa *et al.*, in 2003 [194]. In 2006, the same authors reported an enantioselective domino Mannich/Michael reaction catalyzed by L-proline, which occurred between 3,4-dihydro-β-carboline and 3-ethyl-3-buten-2-one [195]. Remarkably, the corresponding *ent*-dihydrocorynantheol was formed in 85% yield as a single stereomer (ee = de >99%). In 2008, Xu *et al.* [196] described a proline-catalyzed enantioselective construction of 2,3-disubstituted tetrahydropyridines based on a domino Mannich-type/intramolecular hemiaminalization/dehydration reaction. This process occurred between preformed *N*-*p*-methoxyphenyl aldimines and the inexpensive aqueous tetrahydro-2*H*-pyran-2,6-diol in dimethyl sulfoxide (DMSO) at room temperature. It afforded chiral tetrahydropyridines as almost single diastereomers (>90% de) in moderate to good yields (21–74%) and good to excellent enantioselectivities of 69% to >99% ee. A more recent example was reported by Carter *et al.*, allowing a new route to azabicyclo[2.2.2]octanes, which have generated considerable attention because of their presence in numerous alkaloid natural products. Indeed, this work dealt with a synthesis of these chiral products based on an asymmetric domino Mannich/aza-Michael reaction of cyclohexenone and a series of aromatic imines promoted by a *p*-dodecylphenylsulfonamide-modified L-proline catalyst (30 mol%) [197]. The process performed in the absence of a solvent strongly favored the exo products, which were achieved in high levels of enantioselectivities of 91–99% ee combined with moderate to good yields (51–63%). The use of other ketones in place of cyclohexenone, such as cyclopentenone or acyclic enones, led to complex mixtures of products. Furthermore, attempts to extend the scope of the reaction to aliphatic imines did not result in the formation of the corresponding azabicyclo[2.2.2]octanes. In 2010, Enders *et al.* [198] developed another domino Mannich/aza-Michael reaction, involving γ-malonate-substituted α,β-unsaturated ester and *N*-Boc-protected arylaldimines as the substrates. As shown in Scheme 10.17, the reaction catalyzed by a chiral bifunctional thiourea afforded the corresponding 2,5-cis-configured polysubstituted pyrrolidines in good to excellent yields (76–99%) and enantioselectivities (75–94% ee), combined with an excellent diastereoselectivity of >95% de in all cases of substrates studied.

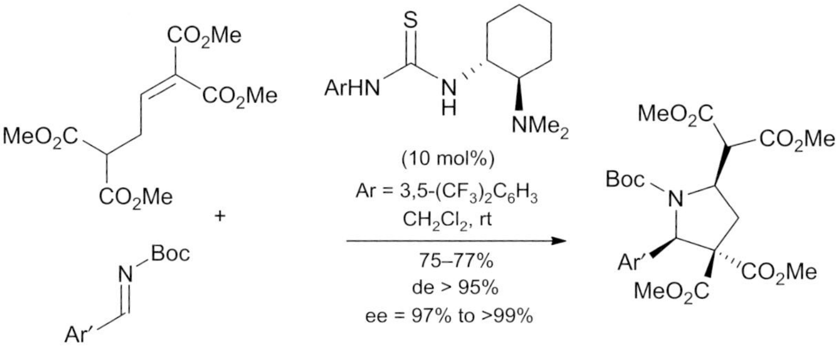

Scheme 10.17 Synthesis of pyrrolidines through domino Mannich/aza-Michael reaction.

Another recent example of asymmetric domino reaction initiated by the Mannich reaction was reported by Rueping and Lin, in 2010 [199]. This work constituted the first catalytic asymmetric synthesis of 4-aminobenzopyrans and was based on a domino Mannich/ketalization reaction of electron-rich alkenes with *o*-hydroxy benzaldimines induced by a chiral *N*-triflylphosphoramide derived from (*S*)-BINOL (1,1′-Bi-2-naphthol) (5 mol%). For example, the reaction of 2,3-dihydro-2*H*-furan with *o*-hydroxy benzaldimines provided the corresponding furanobenzopyrans in good yields (46–89%) and enantioselectivities of 74–96% ee combined with low to high diastereoselectivities (12–99% de). The scope of this methodology was extended to 3,4-dihydro-2*H*-pyran, which gave rise to the corresponding pyranobenzopyrans with comparable results. In 2011, Wang *et al.* [200] developed an expeditious assembly of 2-amino-4*H*-chromene skeleton based on a domino elimination/Mannich/intramolecular ring cyclization/tautomerization reaction catalyzed by a chiral indane–amine–thiourea (10 mol%). A base, such as Li_2CO_3, was employed to generate the Mannich substrate from *tert*-butyl (2-hydroxyphenyl)(phenylsulfonyl) methylcarbamate through elimination. Then, the Mannich reaction of the *in situ* generated imine with malononitrile occurred to give the intermediate Mannich product, which was subsequently submitted to an intramolecular oxa-nucleophilic addition of the nitrile group. The formed intermediate imine finally tautomerized to provide final 2-amino-4*H*-chromene in 94% yield and 88% ee. In 2011, Feng *et al.* [201] developed a highly enantioselective synthesis of 3,4-diaminochroman-2-ones through the domino Mannich-type reaction of *o*-hydroxy aromatic aldimines and azlactones. The process was efficiently catalyzed by an (*S*)-pipecolic acid-derived guanidine (10 mol%), leading to a range of *cis*-3,4-diaminochroman-2-ones in high yields (85–99%) with diastereo- and enantioselectivities of 70–98% de and 90–95% ee, respectively. Furthermore, the authors demonstrated that changing the catalyst for a chiral bisguanidium salt $3{\cdot}HBAr^F{}_4$ allowed the corresponding trans domino products to be achieved in comparable excellent results.

10.2.2.2 Domino Reactions Initiated by the (Aza)-Morita–Baylis–Hillman Reaction

The aza-Morita–Baylis–Hillman reaction is known to be a useful and atom-economical C–C bond-forming reaction of electron-deficient alkenes with imines usually catalyzed by Lewis bases [202]. It formally involves a sequence of reactions including a Michael addition, a Mannich reaction, a proton transfer, and a retro-Michael reaction (β-elimination). Although there are many reports in the field of the enantioselective aza-Morita–Baylis–Hillman reaction, only rare examples of asymmetric domino reactions initiated by this reaction have been reported. In 2010, Sasai *et al.* [203] developed the first organocatalyzed asymmetric domino aza-Morita–Baylis–Hillman/aza-Michael reaction of α,β-unsaturated carbonyl compounds with *N*-tosylimines, allowing an easy access to chiral cis-1,3-disubstituted isoindolines as single diastereomers. The process was induced by a H_8-BINOL-derived catalyst and provided these products in high yields and enantioselectivities, as shown in Scheme 10.18.

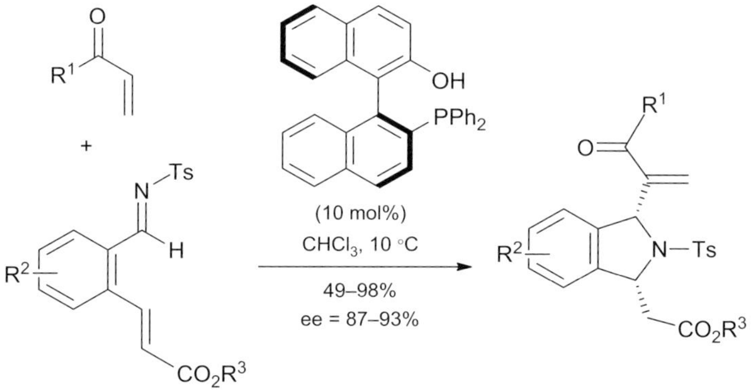

Scheme 10.18 Synthesis of isoindolines through domino aza-Morita–Baylis–Hillman/aza-Michael reaction.

Later, Sasai *et al.* [204] developed another example of domino reaction initiated by the aza-Morita–Baylis–Hillman reaction followed successively by an aza-Michael reaction, an aldol reaction, and a dehydration. This novel process, induced by a chiral acid–base organocatalyst (20 mol%), involved acrolein and various *N*-tosylarylimines as the substrates, leading to the corresponding highly functionalized tetrahydropyridines in moderate yields (40–60%) and good enantioselectivities of 80–88% ee. Attempts to extend the scope of the reaction to other activated alkenes such as methylvinylketone failed.

10.2.2.3 Domino Reactions Initiated by the Friedel–Crafts Reaction

The asymmetric Friedel–Crafts reaction is one of the most powerful methods to synthesize optically active aromatic compounds, and chiral Bronsted acids have recently emerged as efficient catalysts [205]. In 2009, You *et al.* reported the synthesis of chiral fluorenes through a domino double Friedel–Crafts reaction catalyzed with a chiral phosphoric acid (5 mol%). This reaction involved indoles and 3′,5′-dimethoxybiphenyl-2-carbaldehyde as substrates, and afforded the corresponding fluorenes in moderate to high yields (66–96%) and enantioselectivities of 73–96% ee. Another domino reaction initiated by a Friedel–Crafts reaction promoted by chiral diphenylprolinol trimethylsilylether (10 mol%) was developed by Wang *et al.* [206], affording chiral chromanes in good yields (79–93%), albeit with moderate diastereo- and enantioselectivities (50–56% de and 84–86% ee) by reaction of α,β-unsaturated aldehydes with 1-naphthol, according to a domino Friedel–Crafts/cyclization reaction. In addition, a novel domino Friedel–Crafts/Michael/aromatization reaction, reported by Xiao *et al.* [207], provided a range of chiral highly functionalized tetrahydrocarbazoles upon catalysis with a chiral diarylprolinol trimethylsilyl ether (20 mol%). These products were obtained in good to high yields (70–97%) with high enantio- and diastereoselectivities of 92% to >99% ee and 92–99% de, respectively, through the reaction of α,β-unsaturated aldehydes with 2-vinylindoles, evolving through the iminium–enamine activation mode. Furthermore, these authors had previously developed a route to this type of products through another domino reaction initiated by the Friedel–Crafts reaction, occurring between nitroolefins and 2-propenylindoles. In this case, the reaction was induced by a chiral bis-sulfonamide catalyst, which, through hydrogen-bonding activation, afforded the corresponding highly substituted tetrahydrocarbazoles in moderate to good yields (42–86%) combined with good to high diastereo- and enantioselectivities (66–98% de, 82–98% ee) [208]. Enantioselective domino reactions exploiting the electrophilic properties of indole intermediates have been developed by several groups [209]. For example, Gouverneur *et al.* demonstrated an approach to chiral tricyclic indoles by using fluorine electrophiles to quaternarize indole C-3 and trapping the C-2 intermediate iminium species by a C-3-tethered nucleophile [210]. As shown in Scheme 10.19, the domino fluorination/intramolecular cyclization reaction was catalyzed by hydroquinidine 1,4-phthalazinediyl diether ((DHQD)$_2$PHAL) in the presence of K_2CO_3, providing the corresponding tricyclic indoles in moderate to excellent yields and excellent diastereo- and enantioselectivities.

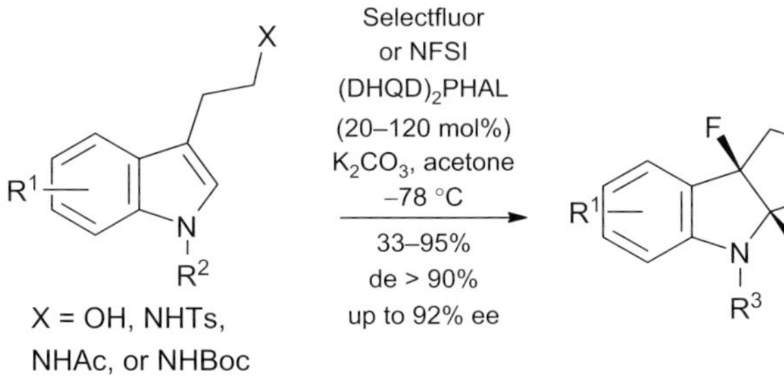

Scheme 10.19 Synthesis of tricyclic indoles through domino Friedel–Crafts-type/intramolecular cyclization reaction.

10.2.2.4 **Miscellaneous Domino Reactions**

In 2007, Landais *et al.* reported an enantioselective domino aldolization/lactonization reaction of α-oxocarboxylic acids and various aldehydes, which provided, when catalyzed by a chiral benzimidazole proline-derived catalyst, the corresponding chiral isotetronic derivatives in moderate to high yields (44–90%) and good to high enantioselectivities of 76–90% ee. The same year, Yamamoto *et al.* [211] reported a domino *N*-nitroso aldol/Michael reaction of a dienamine with aromatic nitroso compounds catalyzed by a chiral BINOL derivative (30 mol%). The process afforded the corresponding chiral 3-aza-2-oxa-bicycloketones in 52–90% yields, excellent regioselectivity, and good enantioselectivities (80–90% ee). In addition, Ramachary and Sakthidevi [212] later successfully developed the domino aldol/hemiacetalization reaction of simple ketones with a range of 2-hydroxybenzaldehydes catalyzed by *trans*-4-OH-L-proline. The reaction provided the corresponding chiral lactols in moderate yields and enantioselectivities (26–90% ee). In 2011, a domino aldolization/cyclization reaction was developed by Zhao and Guang [213], involving α-isothiocyanato imides and activated carbonyl compounds, such as isatin derivatives, as components. The reaction was efficiently catalyzed by a chiral amine thiourea derived from cinchona alkaloids (10 mol%), providing chiral spirobicyclic thiocarbamates in high yields (90–99%) and enantioselectivities of 94–98% ee, combined with moderate diastereoselectivities of 40–72% de. An enantioselective vinylogous aldol/oxa-Michael reaction constituted the key step of total syntheses of the natural products diversonol and lachnone C13706, which were reported by Bräse *et al.*, in 2011 [214]. This domino process, occurring between salicylaldehyde and prenal, was induced by a chiral diarylprolinol trimethylsilyl ether (30 mol%) and led to the corresponding enantiopure tricyclic lactol in 63% yield. In 2008, Jørgensen *et al.* [215] developed a novel domino vinylic substitution/cyclization reaction for the synthesis of chiral halo-substituted pyrrolidin-2-ones, which constitute flexible starting materials for the preparation of chiral γ-lactams. The reaction occurred between α,β-dihalogenated acrylate esters as the electrophiles and 1,2-nucleophiles in the presence of a chiral dihydrocinchonine-derived phase-transfer catalyst (3 mol%) to provide the corresponding chiral C-5 quaternary 3-halo-3-pyrrolin-2-ones in good yields (62–90%) and enantioselectivities of 85 to >99% ee. The domino process evolved through the stereospecific substitution of the chlorine atom with retention of configuration by the 1,2-nucleophile, followed by immediate ring closure, resulting in the selective formation of the corresponding halo-substituted pyrrolidin-2-ones. In 2008, Zhong *et al.* [216] developed a novel, practical, and highly enantio- and diastereoselective domino α-aminoxylation/aza-Michael reaction for the synthesis of chiral functionalized tetrahydro-1,2-oxazines by using simple L-proline as catalyst (5 mol%) in the presence of 1 equiv of triethylammonium bicarbonate. The reaction occurred between aryl nitroso compounds and bifunctional enemalonate aldehydes. In the first step, O-alkylation took place in α-position of the aldehyde; then an intramolecular aza-Michael reaction took place, closing the ring and affording chiral tetrahydro-1,2-oxazines with good yields (44–90%) and good to high diastereoselectivities of 80 to >98% de, combined with complete enantioselectivity. In this process, the bifunctional enemalonate aldehydes acted successively as a nucleophile through an enamine intermediate and as an electrophile through the enemalonate moiety. The same year, Chen *et al.* [217] reported the first organocatalytic formal inverse-electron-demand aza-Diels–Alder reaction of *N*-tosyl-1-aza-1,3-butadienes and aldehydes. Actually, this remarkable reaction evolved through the formation of an enamine from the aldehyde component and chiral diphenylprolinol trimethylsilyl ether employed as catalyst. Then, this enamine underwent an aza-Diels–Alder cycloaddition to give the corresponding chiral piperidine derivatives as single diastereomers in good to excellent yields and general excellent enantioselectivities, as shown in Scheme 10.20. Later, the same authors developed the related enantioselective domino inverse-electron-demand aza-Diels–Alder/hemiacetalization reaction of *N*-tosyl-1-aza-1,3-butadienes with aliphatic dialdehydes, such as glutaraldehyde, induced by the same catalyst [218]. This powerful process allowed the corresponding almost diastereo- and enantiopure δ-lactone[2,3-*b*]piperidine derivatives to be achieved in good yields, as shown in Scheme 10.20.

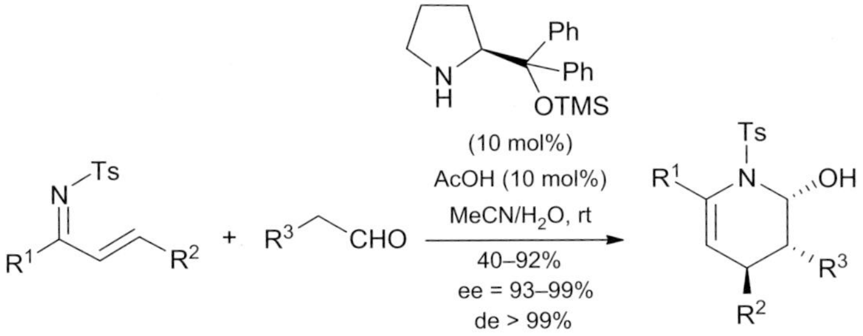

Scheme 10.20 Synthesis of piperidines through domino enamine formation/inverse-electron-demand aza-Diels–Alder reaction.

The first organocatalytic enantio- and diastereoselective 1,3-dipolar cycloaddition [219] of azomethine ylides with nitroalkenes was reported by Zhang *et al.*, in 2008 [220]. This domino azomethine ylide formation/1,3-dipolar cycloaddition process was efficiently catalyzed by thiourea alkaloid derivatives. Indeed, benzophenone imines reacted with various nitroalkenes, providing the corresponding *trans*-pyrrolidines in good to high yields (49–77%) and diastereoselectivities (70% to >98% de), combined with moderate enantioselectivities (≤65% ee). A number of domino reactions initiated by pericyclic reactions have been reported [221]. A recent example was developed by MacMillan *et al.* [222] in the context of a total synthesis of (+)-minfiensine, the key step of which was a novel domino Diels–Alder amine cyclization reaction. The reaction involved propynal and 2-vinylindole as the substrates and a chiral secondary amine as the catalyst (5–15 mol%) associated to tribromoacetic acid, leading to the corresponding domino tetracyclic pyrroloindoline, which was subsequently converted by reduction with sodium borohydride into the final product in 87% yield combined with 96% ee. In another context, Dixon *et al.* [223] reported that the reaction of enol lactones with tryptamines, performed in the presence of a chiral phosphoric acid (10 mol%), led to the formation of intermediate ketoamides. The latter were subsequently submitted to a dehydrative condensation which gave the corresponding *N*-acyliminium ions, which finally cyclized to provide the corresponding chiral tetracyclic products in moderate to excellent yields (78–99%) and enantioselectivities (84–99% ee). In 2009, Bode and Kaeobamrung [224] reported that a chiral triazolium-derived N-heterocyclic carbene catalyst (10 mol%) promoted the annulation of α,β-unsaturated aldehydes and an α-hydroxy enone to afford the corresponding cyclopentane-fused lactones with general excellent enantioselectivity of 99% ee, good yields (35–65%), and moderate diastereoselectivities (≤ 40% de). Surprisingly, the authors found that using an imidazolium-derived N-heterocyclic carbene catalyst to induce the reaction of α,β-unsaturated aldehydes with the same α-hydroxy enone achieved structurally different cyclopentane-fused lactones with the same enantioselectivity of 99% ee, better yields (68–85%), and moderate diastereoselectivities (50–66% de). An organocatalytic, enantioselective synthesis of highly biologically active dihydroquinazolinones was described by Rueping *et al.* [225] based on a domino condensation/amine addition reaction involving the simplest and readily available starting materials, such as 2-aminobenzamide and various benzaldehydes. This extremely simple and practical process was catalyzed by a chiral phosphoric acid derived from BINOL (10 mol%), providing the required products in good yields (85–93%) and enantioselectivities of 80–92% ee. On the other hand, a range of chiral 4,5-disubstituted isoxazoline-*N*-oxide products were achieved by Jørgensen *et al.* [226] through two novel, interesting domino Henry reaction/cyclization reactions of *in situ* generated α-brominated aldehydes with ethyl nitroacetate and of *in situ* generated epoxides with *t*-butyl nitroacetate, respectively. Remarkable enantioselectivities of 93–99% ee were obtained combined with good yields (65–90%) and moderate to good diastereoselectivities of 46–90% de in both cases of domino reactions, which involved a chiral diarylprolinol trimethylsilyl ether as catalyst (10–20 mol%). In 2009, Jacobsen and Klausen [227] described a domino imine formation/Pictet–Spengler reaction of aliphatic as well as aromatic aldehydes with tryptamines. This process was catalyzed by a chiral thiourea (20 mol%) combined with benzoic acid as an additive, providing chiral tetrahydro-β-carbolines which constitute ubiquitous structural motifs in biologically active natural products. These products were achieved in good yields (39–94%) and high enantioselectivities of 85–99% ee. Even though this reaction involved the use of a metal oxidant such as $Cu(OTf)_2$ and, consequently, is somewhat out of the subject of this chapter which does not include dual catalysis, it is important to remind us of the first and remarkable enantioselective catalytic cyclization strategy for accessing steroidal and terpenoidal frameworks using organocatalysis, which was reported by MacMillan and Rendler in 2010 [228]. This singly occupied molecular orbital (SOMO)-activated polyene cyclization of unsaturated aldehydes provided, in the presence of a chiral imidazolidinone as cocatalyst (20–30 mol%), the corresponding chiral polycycles in high yields and enantioselectivities of up to 91% ee. The complex chiral formed products arose from radical bi- and tricyclization. The organocatalyst was supposed to activate the radical aldehyde through the formation of the corresponding α-imino radical intermediate upon oxidation with $Cu(OTf)_2$. The authors have extended the scope of this powerful procedure to extended ring systems, such as tetra-, penta-, and hexacycles, bearing up to 11 contiguous stereocenters, of which 5 were all-carbon quaternary centers. While iminium catalysis, involving the transformation of iminium ions to enamines, has been extensively studied in the context of a number of organocatalytic processes, the reverse of this process, involving the conversion of enamines to iminium species, has been achieved only very recently by Wang *et al.* [229]. In this study, these authors have developed an unprecedented method for enantioselective β-functionalization of simple aldehydes, which involved the oxidative conversion of enamines derived from aldehydes to iminium species, which then underwent Michael addition to a nucleophile. This process employed chiral diphenylprolinol trimethylsilyl ether as catalyst (20 mol%) combined with the nontoxic *o*-iodoxybenzoic acid (IBX) as oxidant. This methodology has been employed to develop a novel domino oxidation/Michael/cyclization reaction, occurring between aliphatic aldehydes and 4-hydroxycoumarin as carbon nucleophile, which afforded biologically interesting coumarin-containing chiral pyrans in good yields (81–84%) and useful levels of enantioselectivity of 77–91% ee. In addition to carbon nucleophiles, O- and N-centered nucleophilic reactants participated by reaction with aldehydes, such as 3-phenylpropionaldehyde, in domino oxidation/Michael/aldol/dehydration processes, leading to the corresponding chiral bicyclic products in moderate yields (26–55%) and enantioselectivities of 85–92% ee. In the same area, Rueping *et al.* [230] developed, in 2012, enantioselective domino oxidation/Michael/intramolecular alkylation reaction of allylic alcohols with diethyl bromomalonate, providing the corresponding cyclopropanes in high yields (66–88%) and enantioselectivities of 93–98% ee. The reaction was performed in the presence of MnO_2 as the oxidant and chiral diphenylprolinol trimethylsilyl ether (20 mol%) as catalyst. Moreover, a domino oxidation/oxa-Michael/Michael reaction could be employed by the same authors to synthesize chiral bi- and tricyclic 4*H*-chromenes [231]. The domino process occurred between propargylic alcohols and (*E*)-2-(2-nitrovinyl)-phenols in the presence of the same catalyst and tetrapropylammonium peruthenate (TPAP)/*N*-methylmorpholine-*N*-oxide (NMO) as the catalytic oxidative system. Good yields (53–96%) and enantioselectivities (77–99% ee) were reached in the case of bicyclic chromenes. Also, the same authors reported, in 2011, a domino Nazarov cyclization/bromination reaction of dienones with 2,4,4,6-tetrabromocyclohexa-2,5-dienone as a bromine source [232]. The process catalyzed by a chiral BINOL-derived *N*-triflylphosphoramide (5 mol%) yielded chiral highly functionalized trans-4,5-substituted 5-bromocyclopentenones in good yields (43–82%) and enantioselectivities (89–97% ee), combined with moderate diastereoselectivities (34–90% de). In 2006, Akiyama *et al.* [233] employed the pyridinium salt of a chiral BINOL-derived phosphoric acid as catalyst (3 mol%) to induce a formal aza-Diels–Alder reaction of aldimines with Brassard's diene, providing the corresponding α,β-unsaturated δ-lactams in good yields (63–91%) and high enantioselectivities of 92–99% ee. In addition, the same authors reported an enantioselective inverse-electron-demand aza-Diels–Alder reaction of aldimines with enol ethers, employing a BINOL-derived chiral phosphoric acid as catalyst (10 mol%) [234]. The domino process gave access to chiral disubstituted tetrahydroquinolines as single cis diastereomers in high yields (59–89%) and enantioselectivities (88–97% ee). In 2006, Bode *et al.* [235] performed the first N-heterocyclic carbene-catalyzed formal aza-Diels–Alder reaction. This reaction was performed in the presence of a novel chiral triazolium salt based on the *cis*-1,2-aminoindanol platform, which served as an efficient precatalyst for the N-heterocyclic carbene-catalyzed redox generation of enolate dienophiles from enals, which were exceptionally reactive. These species underwent $LUMO_{diene}$-controlled Diels–Alder reactions with *N*-sulfonyl-α,β-unsaturated imines in good yields and with exceptional diastereo- and enantioselectivities, providing chiral *cis*-3,4-disubstituted dihydropyridinone products as almost single stereomers, as shown in Scheme 10.21.

Later, these authors reported the synthesis of chiral cis-1,3,4-trisubstituted cyclopentenes based on an enantioselective N-heterocyclic carbene-catalyzed domino annulation of α,β-unsaturated aldehydes and 4-oxoenoates [236]. The domino products were achieved in low to high yields (25–93%), in moderate to

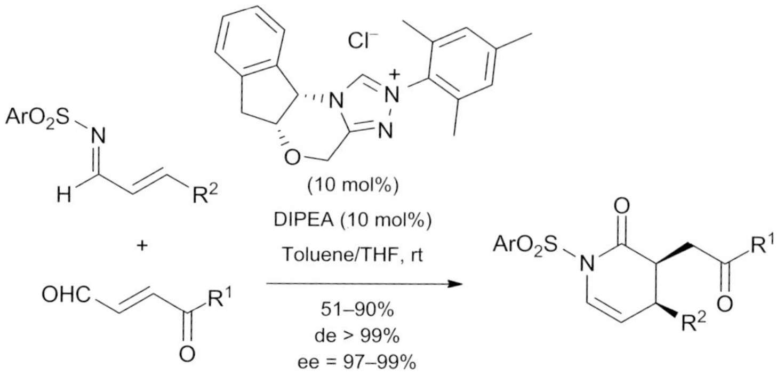

Scheme 10.21 Synthesis of tetrahydroquinolines through N-heterocyclic carbene-catalyzed formal aza-Diels–Alder reaction.

high diastereoselectivities (80–90% de) and with excellent enantioselectivities (96–99% ee). The authors proposed that the reaction evolved through an intermolecular crossed-benzoin reaction, followed by tautomerization and intramolecular aldol reaction, and, finally, acyl addition and decarboxylation. In 2007, Scheidt and Chan [237] developed a highly stereoselective formal [3+3] cycloaddition between of α,β-unsaturated aldehydes and azomethine imines catalyzed by an N-heterocyclic carbene derived from *N*-mesityl benzimidatzolium salt (20 mol%). The domino process provided the corresponding substituted pyridazinones in good yields (51–93%) and high relative diastereoselectivity (>90% de). Again, the N-heterocyclic carbene was effective in generating the homoenolate species from the α,β-unsaturated aldehyde, which further underwent an addition/acylation sequence with an azomethine imine to afford new bicyclic heterocycles. An intramolecular asymmetric synthesis of α,α-disubstituted cyclopentenes was achieved by the same authors in good yields and good to high enantioselectivities of up to 94% ee, based on N-heterocyclic carbene-catalyzed desymmetrization of 1,3-diketones, evolving through domino aldol/intramolecular acylation/decarboxylation reaction [238]. In 2011, Wang *et al.* [239] reported an efficient synthesis of medicinally important chiral 2-amino-4*H*-chromene-3-carbonitriles, based on a domino reaction occurring between *N*-Boc-α-amido sulfones and malononitrile. The process was catalyzed by a chiral indane–amine–thiourea (10 mol%) in the presence of Li_2CO_3 as a base, providing chiral 2-amino-4*H*-chromene-3-carbonitriles in good yields (50–94%) and enantioselectivities (52–89% ee). Good results were reported by Hayashi *et al.* [240] for a domino desulfonylative *N*-tosyl imine formation/Mannich reaction between an α-amidosulfone and aliphatic aldehydes. The reaction employed a chiral diarylprolinol trimethylsilyl ether as catalyst (10 mol%) combined with $NaHCO_3$ as base, and produced the chiral domino products in good yields (50–94%) with excellent levels of diastereoselectivities of >90% de and enantioselectivities of 80–99% ee. A further treatment of these domino products by $NaBH_4$ generated the corresponding enantiopure aziridines.

10.3
Multicomponent Reactions

One of the challenges in organocatalysis is to implement various reaction strategies in a multicomponent reaction, which is a domino reaction involving at least three substrates, to achieve multiple bond formation in one operation. This strategy is atom-economical and avoids the necessity of protecting groups and isolation of intermediates. Its goal is to mimic Nature in its highly selective sequential transformations. Even though the history of multicomponent reactions dates back to the second half of the nineteenth century with the reactions of Strecker [241], Hantzsch [242], and Biginelli [243], it was only in the last decades with the work of Ugi *et al.* [244] that the concept of multicomponent reaction has emerged as a powerful tool in synthetic chemistry. The first organocatalytic enantioselective domino multicomponent reaction was developed by Barbas *et al.* [245] in 2001, involving an enantioselective Knoevenagel/Michael reaction between benzaldehyde, acetone, and diethylmalonate, catalyzed by L-proline.

10.3.1
Multicomponent Reactions Initiated by the Michael Reaction

10.3.1.1 Michael Reactions of α,β-Unsaturated Aldehydes

The first multicomponent reaction initiated by a Michael addition combining enamine and iminium ion activations was reported by Jørgensen *et al.*, in 2005 [246]. This domino thia-Michael/amination reaction proceeded through conjugate addition of nucleophilic thiols to the iminium ion intermediate of a chiral diarylprolinol trimethylsilyl ether and an α,β-unsaturated aldehyde, followed by amination of the resulting enamine intermediate with azodicarboxylates. The domino products were subsequently reduced by treatment with BH_4Na, and then underwent base-induced cyclization into the final corresponding oxazolidinones. Good results in terms of yields (38–72%), diastereoselectivities (76–90% de), and enantioselectivities (97–99% ee) were achieved in the presence of an additive, such as benzoic acid, for the formation of the corresponding polyfunctionalized oxazolidinones bearing two adjacent stereogenic centers. In 2006, Enders *et al.* [247] developed a triple domino reaction for the construction of tetrasubstituted cyclohexene carbaldehydes. In this work, the authors paved the way for the sequential creation of three bonds by using a highly enantioselective combination of enamine–iminium–enamine catalysis in a triple domino reaction. This three-component domino process involved a linear aldehyde, a nitroalkene, and an α,β-unsaturated aldehyde as components and chiral diphenylprolinol trimethylsilyl ether as catalyst. It proceeded through a domino Michael/Michael/aldol reaction, affording four stereogenic centers generated in three consecutive carbon–carbon bond formations with high diastereoselectivities and complete enantioselectivities. Indeed, from 16 possible stereomers, only 2 epimers located in the α position of the nitro group were formed in a ratio ranging from 2 : 1 to 99 : 1. A range of substrates were tolerated, allowing diverse chiral polyfunctional cyclohexenes to be achieved in 30–58% yield. In the first step, the catalyst activated the linear aldehyde through enamine formation to give the first Michael addition to the nitroolefin. Then, the catalyst was liberated by hydrolysis, being able to form an iminium ion with the enal to catalyze the second Michael addition of the nitroalkane. During this addition, a novel enamine intermediate was formed, which cyclized through an intramolecular aldol condensation to afford the final cyclohexenes with moderate to excellent diastereoselectivities of 34–98% de and complete enantioselectivity. Later, the methodology of this three-component reaction was extended by these authors through combination with a highly stereoselective intramolecular Diels–Alder reaction [248]. The domino Michael/Michael/aldol products were submitted to a subsequent intramolecular Diels–Alder cycloaddition through the addition of Me_2AlCl to afford the corresponding almost enantiopure tricyclic carbaldehydes in moderate to good yields (35–56%) and diastereoselectivities (42–88% de). In this remarkable domino reaction, five C–C bonds were formed with creation of up to eight new stereogenic centers. A number of other enantioselective domino Michael/Michael/aldol reactions have been developed by several groups. In 2007, Jørgensen *et al.* [249] disclosed a new approach for a three-component reaction evolving through iminium–iminium–enamine catalysis. This process was catalyzed by a chiral diarylprolinol trimethylsilyl ether, and involved two α,β-unsaturated aldehydes and an activated methylene compound, such as malononitrile or a cyanoacetate. The reaction began with the Michael addition of the activated methylene compound onto the first α,β-unsaturated aldehyde through iminium activation, which produced the corresponding enamine, which was further hydrolyzed into the corresponding aldehyde with the regenerating catalyst. This novel intermediate was subsequently added to the second α,β-unsaturated aldehyde activated through iminium catalysis to give the corresponding enamine. This enamine was then intramolecularly condensed to the aldehyde function to finally afford the corresponding cyclohexen-1-ene-carbaldehyde. A range of chiral domino functionalized products could be generated in good yields (47–52%), diastereoselectivities of 20–96% de, and general excellent enantioselectivities of >97% ee through this novel domino Michael/Michael/aldol reaction. A highly efficient enamine–iminium–enamine sequence was described by Melchiorre *et al.*, in 2008 [250]. In this work, a three-component domino Michael/Michael/aldol reaction occurred between an aldehyde, an α,β-unsaturated aldehyde, and a *trans*-α-cyanocinnamate, which provided through a triple cascade catalyzed by chiral diphenylprolinol trimethylsilyl ether the corresponding tetrasubstituted cyclohexene carbaldehydes in moderate to good yields (32–52%), moderate to high diastereoselectivities of 34–90% de, and general excellent enantioselectivities of >98% ee. The application of polymer-supported pyrrolidine catalysis in directing the asymmetric domino Michael/Michael/aldol three-component reaction of cinnamaldehyde, *trans*-β-nitrostyrene, and propionaldehyde was reported by Schore *et al.*, in 2008 [251]. The advantages of catalyst immobilization included the ease of both product separation and purification as well as reusability of the catalyst. While Enders *et al.* [247] achieved 40% yield and enantioselectivity of >99% ee by using chiral diphenylprolinol trimethylsilyl ether, the solid-phase variant proceeded with significantly decreased yield (15%) and selectivity (89% ee) in the case of the nonfunctionalized, polymer-supported catalyst. Superior results in comparison with those of Enders' work were, however, reached by employing a more sterically constrained and functionalized polymer-supported catalyst. Indeed, in this case, 45% yield combined with enantioselectivity of 99% ee was achieved. A remarkable three-component domino reaction of nitromethane and 2 equiv of α,β-unsaturated aldehydes was reported by Enders *et al.*, in 2009 [252]. Upon catalysis with chiral diphenylprolinol trimethylsilyl ether, this novel triple domino reaction based on two subsequent Michael additions followed by an intramolecular aldol condensation provided the corresponding almost diastereo- and enantiopure 5-nitrocyclohexene carbaldehydes in moderate to good yields with general diastereo- and enantioselectivities of >98% de and >99% ee, respectively, as shown in Scheme 10.22. The mechanism proposed by the authors is depicted in this scheme. A first equivalent of α,β-unsaturated aldehyde was activated as its corresponding iminium ion by chiral diphenylprolinol trimethylsilyl ether and then underwent the first nitromethane Michael addition to afford a nitroalkane enamine intermediate. Subsequent hydrolysis generated a nitroaldehyde intermediate and the catalyst, which could promote a second Michael addition with a second equivalent of iminium, furnishing an enamine intermediate. This enamine reacted via an intramolecular aldol cyclization to give a novel iminium intermediate. After hydrolysis, the catalyst was regenerated, and the intermediate alcohol could be dehydrated to afford the final products.

Another highly enantioselective three-component reaction, also based on a domino Michael/Michael/aldol sequence induced by the same catalyst associated to benzoic acid as cocatalyst, was reported by Rios *et al.*, in 2010 [253]. In this work, 2 equiv of α,β-unsaturated aldehydes reacted with oxindoles to afford the corresponding almost enantiopure spirooxindoles in good yields (71–90%) and general high diastereoselectivities of >92% de in combination with high enantioselectivities of 93 to >99% ee. Remarkably, the scope of this reaction could be extended to a range of other heterocycles, such as benzofuranones, pyrazolones, or azlactones, rendering the corresponding spiro compounds in comparable yields and stereoselectivities. In 2011, Melchiorre *et al.* reported the synthesis of a range of other almost enantiopure spirocyclic benzofuranones in moderate to good yields (52–70%) and high diastereoselectivity of >90% de [254]. These products were generated from a three-component domino Michael/Michael/aldol/dehydration sequence involving aldehydes, α,β-unsaturated aldehydes, and benzofuranone-based compounds upon catalysis with chiral diphenylprolinol trimethylsilyl ether in combination with *ortho*-fluorobenzoic acid. In this case, the sequence started with the Michael addition of the aldehydes onto the Michael acceptors, evolving through enamine activation to give intermediate aldehydes. Then, a Michael/aldol sequence occurred between these aldehydes and α,β-unsaturated aldehydes through iminium–enamine activation, which led to aldehyde intermediates, which subsequently dehydrated to afford the desired spirocyclic benzofuranones, as shown in Scheme 10.23. These authors had previously developed the corresponding three-component domino Michael/Michael/aldol/dehydration sequence from aldehydes,

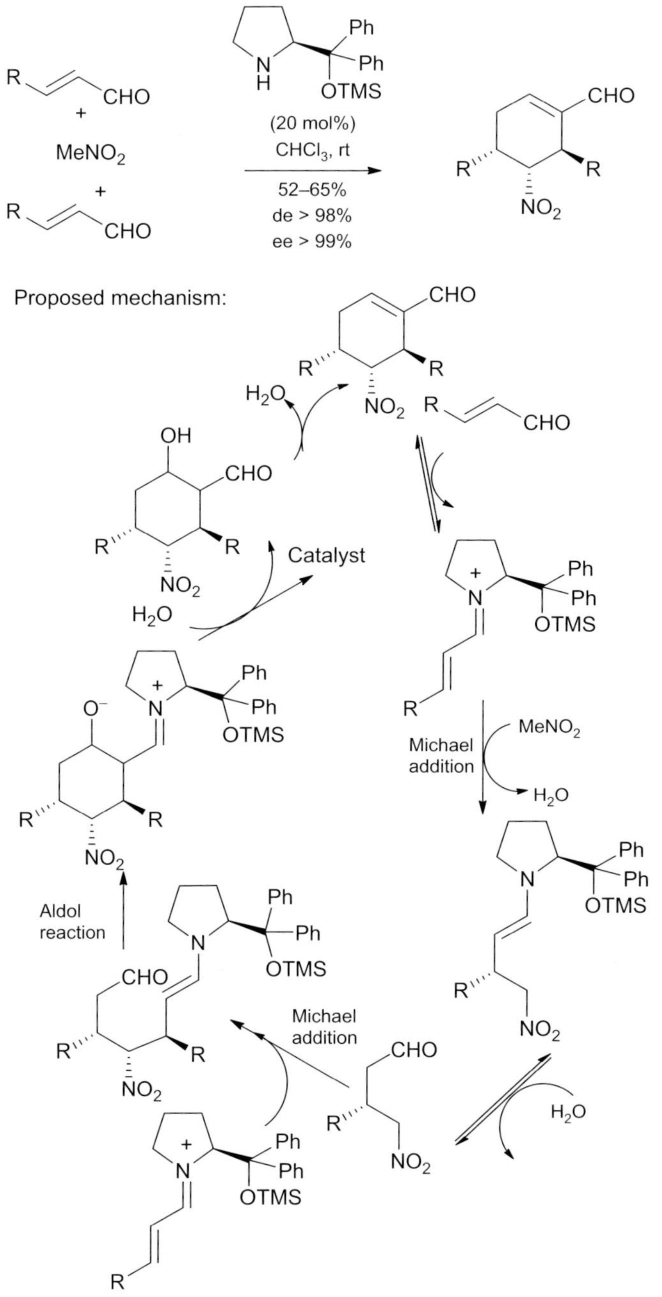

Scheme 10.22 Three-component domino Michael/Michael/aldol reaction.

α,β-unsaturated aldehydes, and oxindole-based compounds under comparable conditions [255]. This process provided the corresponding spirocyclic oxindoles in good yields, high diastereoselectivity of >90% de, and excellent enantioselectivity of > 99% ee, as shown in Scheme 10.23. Comparable results were also obtained by Chen *et al.* by using the same catalyst in combination with benzoic acid in acetonitrile as solvent [256].

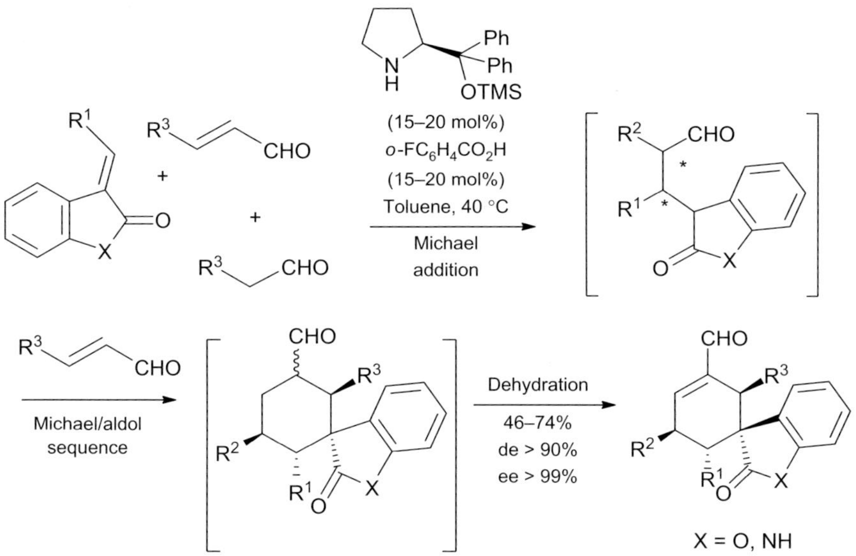

Scheme 10.23 Three-component domino Michael/Michael/aldol/dehydration reaction.

Furthermore, remarkable enantioselective domino reactions involving three successive Michael additions followed by an aldol reaction were recently reported by several groups. As an example, Hong *et al.* [257] employed chiral diphenylprolinol trimethylsilyl ether to induce a highly diastereo- and enantioselective quadruple domino oxa-Michael/Michael/Michael/aldol reaction, building four new bonds and five stereocenters in one step, which provided an expedited access to highly functionalized and almost enantiopure tetrahydro-6*H*-benzo[*c*]chromene. This process was based on the reaction of 2-(*E*)-(2-nitrovinyl)-phenol with cinnamaldehyde performed in the presence of an additive, such as acetic acid, and occurring through the iminium–enamine mode activation. Both the diastereo- and enantioselectivities were excellent, since they were found to be up to >94% de and >99% ee, respectively. Also in 2009, a novel four-component domino oxa-Michael/Michael/Michael/aldol reaction was described by Gong *et al.* based on an iminium–enamine–iminium–enamine sequence [258]. This process, catalyzed by a combination of chiral diphenylprolinol trimethylsilyl ether used at a low catalyst loading of 5 mol% and benzoic acid as an additive, afforded a range of almost enantiopure cyclohexene carbaldehydes in moderate to good yields (49–57%), high diastereoselectivity of >90% de, and general excellent enantioselectivity of >99% ee. The first intermolecular oxa-Michael addition of the alcohol to an iminium ion, derived from activation of 1 equiv of acrolein by chiral catalyst, led to the formation of the corresponding enamine intermediate, which underwent a second intermolecular Michael addition onto nitroalkene, providing a nitromethane intermediate. Then, a third intermolecular Michael addition of this nitromethane to a second equivalent of iminium ion produced an enamine intermediate, which was unstable and easily reacted through an intramolecular aldol condensation to give the final trisubstituted cyclohexene carbaldehyde and regenerate the catalyst. In 2010, the efficient assembly of hydroindane derivatives incorporating a spirooxindole motif was achieved by Chen *et al.* [259] through another enantioselective three-component domino reaction of two different molecules of α,β-unsaturated aldehydes with (*E*)-4-(1-methyl-2-oxoindolin-3-ylidene)-3-oxobutanoates catalyzed by the same catalyst combined with benzoic acid as cocatalyst. According to a Michael/Michael/Michael/aldol sequence, the process provided the corresponding almost enantio- and diastereopure complex fused tetracyclic products bearing six contiguous stereogenic centers in moderate to excellent yields (30–97%), as shown in Scheme 10.24. The authors have proposed a quadruple iminium–enamine–iminium–enamine catalysis to explain the results. As depicted in Scheme 10.24, after the first domino Michael/Michael reaction of butanoate and 1 equiv of α,β-unsaturated aldehyde was complete, from which an aldehyde intermediate was generated, the second domino Michael/aldol reaction occurred between this intermediate and the second equivalent of α,β-unsaturated aldehyde, leading to the final product.

In addition, highly efficient and simpler enantioselective domino Michael/aldol reactions have been developed. For example, Jørgensen *et al.* [260] reported the synthesis of chiral bicyclic cyclohexanones based on a three-component reaction of α,β-unsaturated aldehydes with 2 equiv of a tricarbonylic compound catalyzed by chiral diphenylprolinol trimethylsilyl ether, in 2008. The domino process provided high levels of diastereo- and enantioselectivities of 84–99% de and 89–96% ee, respectively. It remarkably allowed the formation of 1 out of 64 stereomers by creating six stereocenters. The process seems to occur through a domino Michael/aldol/dehydration/Michael reaction, producing the domino products in 38–93% yields. Later, another example was reported by Rios *et al.* [261] dealing with an interesting asymmetric synthesis of spiropyrazolones induced by the same catalyst. In this case, the domino Michael/aldol reaction occurred between an enolizable aldehyde, an α,β-unsaturated aldehyde, and an unsaturated pyrazolone to afford the corresponding chiral spiropyrazolones bearing four stereocenters in moderate to good yields (21–75%) and high diastereo- and enantioselectivities of 82–92% de and >98% ee, respectively. In another context, Cao *et al.* [262] have developed highly enantioselective three-component domino Michael/alkylation/Wittig reaction of α,β-unsaturated aldehydes with 2 equiv of arsonium ylides catalyzed by a chiral recyclable dendritic catalyst derived from diphenylprolinol trimethylsilyl ether. The use of this 2-trimethylsilanyloxy-methyl-pyrrolidine-based dendritic catalyst allowed the corresponding (*E*)-cyclopropanes to be achieved in good yields (51–86%), and

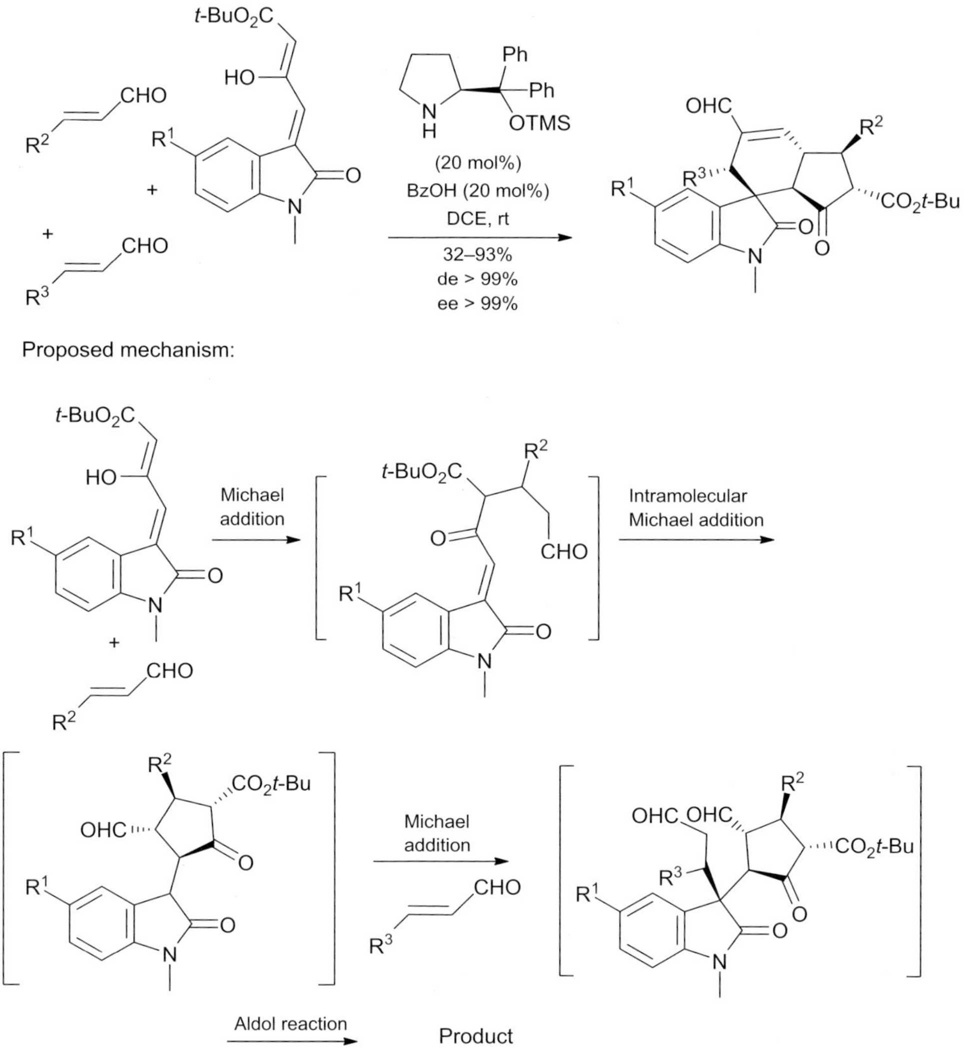

Scheme 10.24 Three-component domino Michael/Michael/Michael/aldol reaction.

diastereo- and enantioselectivities of 60–98% de and 71–99% ee, respectively. It was demonstrated that the catalyst was recoverable and reusable without any loss in activity. In the same year, Cordova *et al.* [263] reported the highly enantioselective synthesis of cycloheptane derivatives. This remarkable four-component process furnished tricyclic products with excellent yields and enantioselectivities, and moreover, in a diastereopure form, as shown in Scheme 10.25. On the other hand, an unprecedented example of a regiospecific and highly chemoselective one-pot organocatalytic domino synthesis of bis-oxazolidines with a functionalized seven-membered carbocycle core was reported. The reaction was efficiently catalyzed by chiral diphenylprolinol trimethylsilyl ether and provided a direct entry to nearly diastereo- and enantiomerically pure cycloheptanes, in which the formation of five new bonds and five stereocenters were controlled.

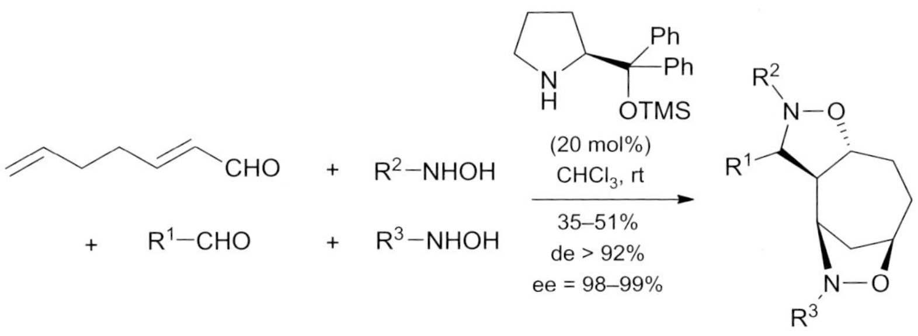

Scheme 10.25 Four-component aza-Michael-initiated domino reaction.

In addition, Melchiorre *et al.* [264] developed two asymmetric three-component domino Michael/amination reactions catalyzed by a chiral cinchona-derived primary amine and evolving through the iminium–enamine activation mode. The first sequence involved α,β-disubstituted aldehydes, indoles, and azodicarboxylates, which provided chiral highly functionalized indole products in moderate to good yields (31–80%), moderate diastereoselectivities (50–78% de), and excellent enantioselectivities of up to 99% ee. The second three-component domino thia-Michael/amination reaction occurred between α,β-disubstituted enals, azodicarboxylates, and thiols, affording the corresponding domino products in moderate yields (27–54%) and moderate to high diastereoselectivities (60–90% de), combined with good to excellent enantioselectivities (72 to >99% ee).

10.3.1.2 **Michael Reactions of Other α,β-Unsaturated Carbonyl Compounds**

In 2008, Rueping and Antonchick [265] described a highly enantioselective three-component reaction of an enamine, a vinyl ketone, and a Hantzsch ester catalyzed by a chiral BINOL-derived phosphoric acid. This powerful process provided an efficient access to chiral tetrahydropyridines and azadecalinones from simple starting materials, involving five successive type of reactions, starting with Michael addition of the enamine to the enone, intramolecular aza-cyclization to give an intermediate six-membered hemiaminal, dehydration, isomerization/protonation, and, finally, transfer hydrogenation, all catalyzed by the same organocatalyst. This three-component domino Michael/hemiaminalization/dehydration/isomerization/transfer hydrogenation reaction allowed a range of chiral tetrahydropyridines and azadecalinones to be achieved in good yields (42–89%) and high enantioselectivities of 89–99% ee. Later, Carter and Yang [266] developed the synthesis of chiral highly functionalized bicyclo[2.2.2]-octanes containing up to four contiguous stereocenters starting from cyclic enones, such as cyclohexenone.

Indeed, the three-component reaction between cyclohexenone, benzylamine, and aldehydes provided, upon catalysis with a proline-derived chiral sulfonamide, the corresponding highly functionalized bicyclo[2.2.2]-octanes in moderate yields (31–69%), high diastereoselectivity of >90% de in all cases of substrates studied, and high enantioselectivities of 84–92% ee. These products arose from a domino Michael/Mannich reaction. First, the enamine, which was *in situ* generated from benzyl amine and aldehyde, was added onto the iminium ion arisen from the activation of cyclohexenone by the catalyst to give the corresponding enamine intermediate. After interconversion of this enamine, an intramolecular Mannich reaction cyclization provided a bicyclic intermediate, which after hydrolysis regenerated the catalyst and furnished the final product. These authors have extended the scope of this methodology to various acyclic enones, which afforded, by reaction with benzyl amine and aldehydes, the corresponding densely functionalized cyclohexenones bearing two contiguous stereogenic centers, arising from the corresponding three-component domino Michael/Mannich reaction [267]. In this case, the reaction was performed in 1,2-dichloroethane (DCE) as solvent in the presence of molecular sieves, providing moderate to good yields (47–84%) and diastereoselectivities of 78% to >90% de, combined with enantioselectivities of 85–95% ee. In addition, Sugiura *et al.* [268] have developed a diastereo- and enantioselective domino reductive aldol reaction of enones and aldehydes with trichlorosilane upon catalysis by (*S*)-BINAPO (diphenyl-[1,1'-binaphthalene]-2-2'-diyl ester). This methodology afforded the corresponding chiral β-hydroxy ketones with good to high syn diastereo- and enantioselectivities of up to 98% de and 98% ee, respectively. As shown in Scheme 10.26, the authors have proposed that the catalyst induced the conjugate reduction with trichlorosilane through a six-membered transition state with the enone in the s-cis conformation to give the (*Z*)-trichlorosilyl enolate exclusively. The high observed syn selectivity arose from the subsequent aldol process, which proceeded through a chair-like cyclic transition state.

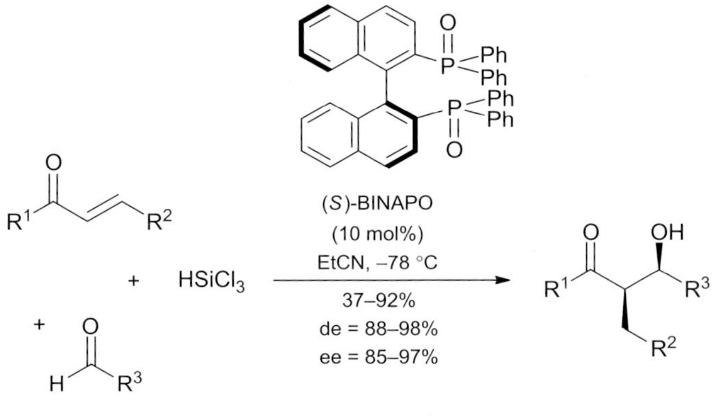

Scheme 10.26 Domino reductive Michael/aldol reaction.

10.3.1.3 Michael Reactions of Nitroolefins

One of the milestones in organocatalysis was the first enantioselective addition of malonates to nitrostyrenes catalyzed by chiral thioureas reported by Takemoto *et al.* [42, 269] in 2003, which was the starting point for multiple reactions with nitrostyrenes. In 2008, the groups of Chen [270] and Rios [271] reported, almost at the same time, the first asymmetric three-component formal 1,3-dipolar cycloaddition of aldehydes, α-aminomalonates, and nitroalkenes catalyzed by chiral thioureas. The domino reaction began with the formation of an imine from the aldehyde and the α-aminomalonate. This imine then reacted with the nitroalkene through a Michael addition and subsequently an aza-Henry reaction occurred, affording the corresponding highly substituted pyrrolidines as single diastereomers, in good to high yields (56–93%) and moderate to high enantioselectivities of 60–91% ee. A three-component domino Michael/Michael/aldol reaction was employed as key step by Enders *et al.* [272] for the stereoselective synthesis of polysubstituted mono- and bicyclic cyclohexene carbaldehydes, in 2008. This domino process involved aldehydes, α,β-unsaturated aldehydes, and nitroalkenes, and was induced by chiral diphenylprolinol trimethylsilyl ether. The process, evolving through the enamine–iminium–enamine activation mode, began with the formation of the corresponding enamine of the aldehyde. This enamine underwent a Michael addition to the nitroalkene to give an intermediate nitroaldehyde. The latter then was condensed onto the α,β-unsaturated aldehyde, providing an intermediate which finally cyclized through aldol condensation to give the domino products in none to good yields of up to 60%, low to good diastereoselectivities (30–78% de), and general excellent enantioselectivities of >97% ee. When nitrostyrenes and dimethylmalonate were reacted with α,β-unsaturated aldehydes in the presence of a combination of chiral diphenylprolinol triethylsilyl ether and a chiral cinchona alkaloid catalyst through a three-component reaction, Xu *et al.* [273] showed that they afforded, according to a Michael/Michael/aldol domino reaction, chiral functionalized cyclohexanes in moderate to good yields (52–74%) and high enantioselectivities of 96 to >99% ee, albeit with moderate diastereoselectivities (≤ 54% de). The first Michael addition occurred between dimethylmalonate and nitroalkenes, affording the corresponding nitroalkanes, which were subsequently added onto α,β-unsaturated aldehydes to give the corresponding aldehydes. These aldehyde intermediates then cyclized into the final cyclohexanes through an intramolecular aldol condensation. Importantly, the authors demonstrated that it was possible to achieve different stereoisomers by changing the combination of catalysts used in the domino reaction. In 2010, another three-component domino Michael/Michael/aldol reaction was reported by Enders *et al.* [274], upon catalysis with chiral diphenylprolinol trimethylsilyl ether. In this case, the domino reaction started with a first Michael addition of aldehydes bearing a thioester onto nitroalkenes to give the corresponding nitroalkanes, which then added according to a second Michael addition onto α,β-unsaturated aldehydes to afford dialdehydes, which subsequently cyclized through aldol condensation followed by dehydration to give the domino products. The latter were not isolated but directly submitted to an intramolecular thia-Michael addition which was initiated by addition of K_2CO_3

to finally provide the corresponding chiral functionalized thiadecalins. As shown in Scheme 10.27, this two-step triple cascade followed by a thia-Michael reaction provided products in moderate to good yields (32–60%) and excellent diastereo- and enantioselectivities of >94% de and >99% ee, respectively.

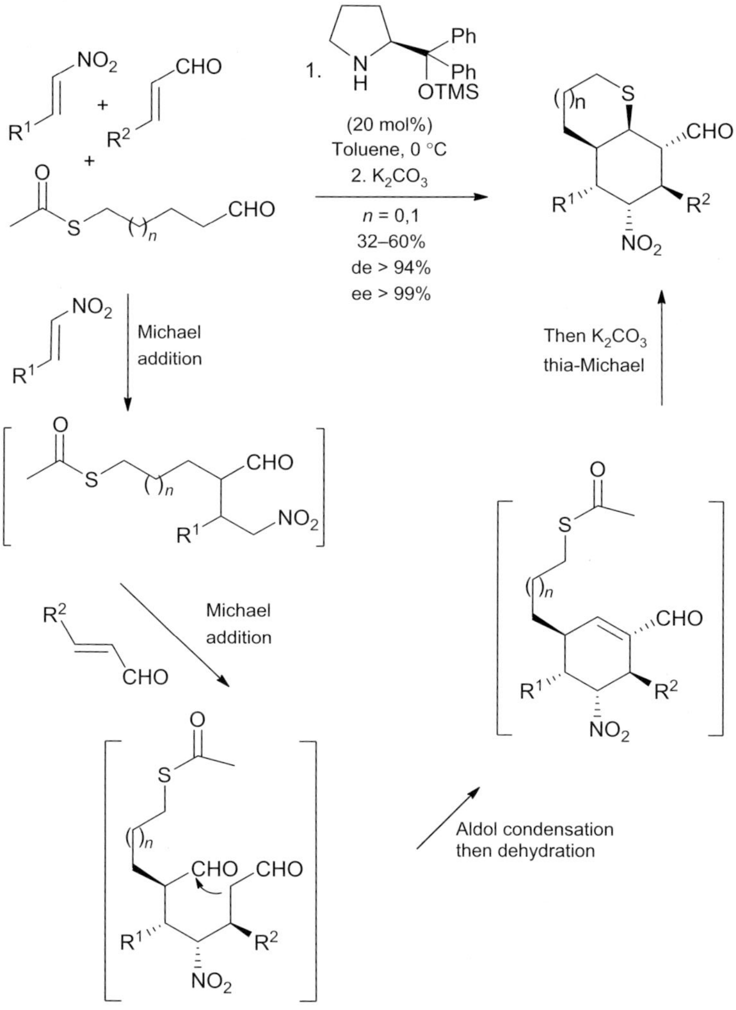

Scheme 10.27 Three-component domino Michael/Michael/aldol reaction followed by thia-Michael reaction.

Polymer-supported organocatalysts are valuable materials for conducting catalytic asymmetric transformations under simple and environmentally begnin conditions [275]. A novel polymer-supported diphenylprolinol trimethylsilyl ether was synthesized by Hansen *et al.* [276], and further applied as catalyst to induce a diastereo- and enantioselective three-component domino Michael/Michael/aldol reaction between cinnamaldehyde, *n*-propyl alcohol, and a nitroalkene, affording the corresponding diastereo- and enantiopure tetrasubstituted cyclohexene carbaldehyde in 32% yield. In another context, an unprecedented enantioselective three-component domino Michael/Michael/Wittig reaction was recently reported by Hong *et al.* [277], providing a direct protocol for the construction of chiral trisubstituted cyclohexenecarboxylates bearing three contiguous stereocenters. These products with all cis stereochemistry were generated from the reaction of nitroalkenes, α,β-unsaturated aldehydes, and a stabilized Wittig reagent, upon catalysis with chiral diphenylprolinol trimethylsilyl ether in the presence of acetic acid. Excellent enantioselectivities of up to 99% ee were achieved in combination with moderate to good yields (34–59%) and none to moderate diastereoselectivities (0–34% de). As an extension of this methodology, an enantioselective three-component domino Michael/Michael/Horner–Wadsworth–Emmons reaction was developed by Hayashi *et al.*, in 2010 [278]. Employing the same catalyst, this process was applied to the total synthesis of the important influenza drug oseltamivir, as well as the drug candidate TRIP-triisopropylphenyl [279]. In 2010, Rodriguez *et al.* [280] reported an enantioselective synthesis of a polyfunctionalized hexasubstituted cyclohexane through a three-component domino Michael/Michael/Henry-type reaction catalyzed by a chiral bifunctional amine thiourea. In this process, 2 equiv of a nitroalkene reacted with an α-ketoanilide to give the product bearing six contiguous stereogenic centers in 61% yield and high diastereo- and enantioselectivity of >90% de and 98% ee, respectively. A remarkable microwave-assisted four-component enantioselective domino Michael/Henry/Michael/aldol reaction was developed by Enders *et al.*, in 2010 [281]. Employing chiral diphenylprolinol trimethylsilyl ether as catalyst, 3 equiv of acetaldehyde reacted with nitroalkenes according to a quadruple cascade to afford the corresponding trisubstituted cyclohexene carbaldehydes in moderate yields (25–45%) and good to high enantioselectivities (89 to >99% ee) combined with diastereoselectivity of >96% de. On the other hand, chiral highly functionalized piperidines were synthesized by Hayashi *et al.* [282] on the basis of a highly efficient four-component two-step enantioselective reaction catalyzed by the same catalyst. In the first part of this sequence, a Michael addition of aldehydes onto nitroalkenes occurred, providing the corresponding nitroalkanes. Then, the latter reacted with imines through an aza-Henry reaction to give aldehyde intermediates, which were spontaneously submitted to hemiaminalization to afford the domino Michael/aza-Henry/hemiaminalization products. These domino products were not isolated but were directly engaged in a subsequent Lewis acid-mediated allylation or cyanation to provide the corresponding 2-allyl and 2-cyano piperidines, respectively. This last allylation or cyanation step was initiated by a further addition of $TiCl_4$ and allyltrimethylsilane or trimethylsilyl cyanide, respectively. The densely functionalized products bearing five contiguous stereocenters were isolated as single diastereomers in good yields (66–81%) and excellent enantioselectivities (93–99% ee), as shown in Scheme 10.28.

It must be noted that Xu *et al.* [283] had previously reported the enantioselective synthesis of fully substituted chiral *N*-tosylpiperidines arisen from the domino Michael/aza-Henry/hemiaminalization reaction of aldehydes, tosylimines, and

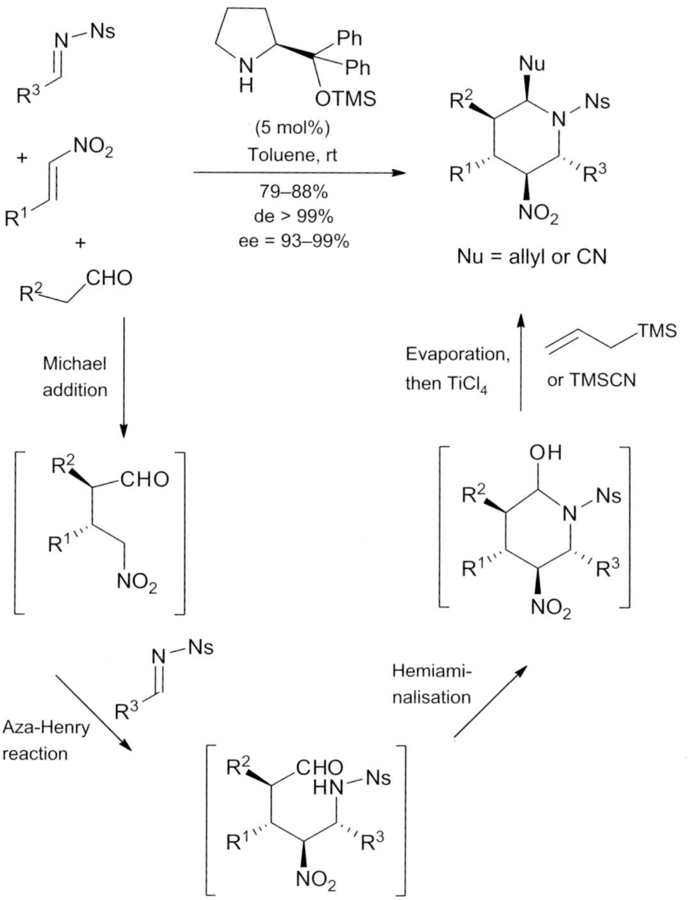

Scheme 10.28 Three-component domino Michael/aza-Henry/hemiaminalization reaction followed by $TiCl_4$-mediated allylation/cyanation.

nitroalkenes based on a multiple-organocatalyst-promoted three-component reaction. This process was induced by the combination of chiral diphenylprolinol triethylsilyl ether with a chiral thiourea, and afforded mixtures of the corresponding α- and β-diastereomeric hemiaminals, which were obtained in moderate to good yields (47–71%) and enantioselectivity of >99% ee. In 2011, an approach to chiral polyfunctionalized octahydroquinolines incorporating three contiguous stereocenters was reported by Yadav *et al.* [284]. These products were generated as single diastereomers in high yields and excellent enantioselectivities through enantioselective three-component domino Michael/aza-Henry/hemiaminalization/dehydration reaction of 1,3-cyclohexanedione, nitroalkenes, and *N*-tosyl aldimines upon catalysis with chiral diphenylprolinol trimethylsilyl ether (Scheme 10.29).

The same year, Hayashi *et al.* [285] reported an asymmetric synthesis of substituted tetrahydropyranols through a sequence of reactions beginning with

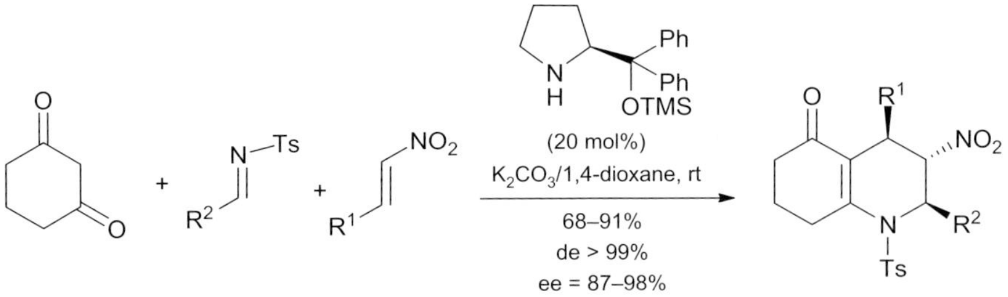

Scheme 10.29 Three-component domino Michael/aza-Henry reaction.

a Michael addition of an aldehyde to a nitroalkene, followed by a domino Henry/acetalization reaction of the formed Michael adduct with ethyl glyoxylate to give the corresponding chiral tetrahydropyranols. When catalyzed by chiral diphenylprolinol trimethylsilyl ether, the process provided the domino products as α/β diastereomeric mixtures with enantioselectivities of up to 99% ee and in good to high yields. The scope of this methodology was extended using aldehydes instead of ethyl glyoxylate, which provided, by reaction with the previously formed Michael adducts through domino Henry/acetalization/isomerization reaction, the corresponding tetrahydropyranols in even better yields (50–91%) and enantioselectivities (95–99% ee), albeit with comparable diastereoselectivities (0–86% de). A chiral tertiary-amine-modified diarylprolinol trimethylsilyl ether was successfully employed by Xu *et al.* [58] to induce a four-component domino Michael/Michael/Horner–Wadsworth–Emmons/thia-Michael reaction between ethyl β-nitroacrylate, aldehydes, toluenethiol, and ethyl 2-(diethoxyphosphoryl)-acrylate. Remarkably, the process furnished the corresponding almost stereomerically pure, highly functionalized, polysubstituted cyclohexanes in good yields, as shown in Scheme 10.30.

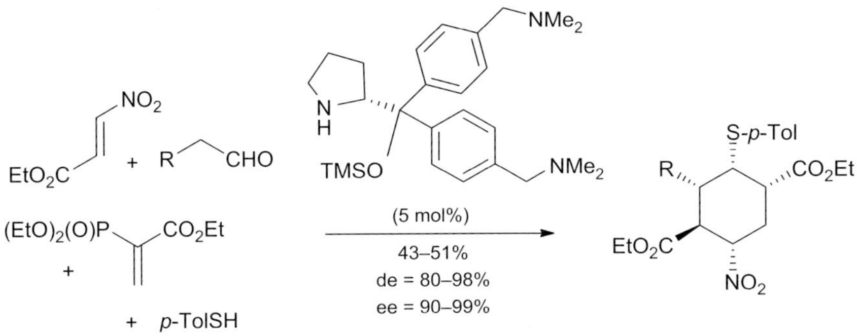

Scheme 10.30 Four-component domino Michael/Michael/Horner–Wadsworth–Emmons/thia-Michael reaction.

In addition, a four-component domino Michael/nitro-Mannich/lactamization reaction was recently developed by Dixon *et al.* [286]. It occurred between a nitroalkene, an activated lactam, formaldehyde, and allylamine or a cyclic imine to give the corresponding chiral bicyclic spirolactam or tetracyclic spirolactam, respectively, in good yields (62–67%) and high enantioselectivities (90–92% ee) by using a cinchona-derived chiral bifunctional thiourea as catalyst.

10.3.2
Multicomponent Reactions Initiated by the Knoevenagel Reaction

The Knoevenagel condensation was described in 1894 by Emil Knoevenagel [287]. It consists in the condensation of aldehydes or ketones with active methylene compounds in the presence of a base [288]. The first organocatalytic enantioselective domino multicomponent reaction, which was developed by Barbas *et al.* [245] in 2001, was initiated by a Knoevenagel reaction. It involved an enantioselective domino Knoevenagel/Michael reaction between benzaldehyde, acetone, and diethylmalonate catalyzed by L-proline. Since then, the Knoevenagel reaction has been used to initiate various other types of enantioselective organocatalytic domino multicomponent reactions. In particular, domino Knoevenagel/hetero-Diels–Alder reactions have been extensively studied by Tietze *et al.* [289] and, consequently this reaction is often referred to as the *Tietze multicomponent reaction*. The first organocatalytic diastereospecific and enantioselective direct asymmetric domino Knoevenagel/Diels–Alder reaction was described by Barbas *et al.*, in 2003 [290]. This three-component process produced highly substituted spiro [5,5]undecane-1,5,9-triones from 4-substituted-3-buten-ones, aldehydes, and 2,2-dimethyl-1,3-dioxane (Meldrum's acid). A chiral amino acid catalyzed the Knoevenagel condensation of the aldehyde to Meldrum's acid to give an intermediate alkylidene derivative of Meldrum's acid, which further underwent a concerted [4+2] cycloaddition with the 2-amino-1,3-butadiene generated *in situ* from the enone and amino acid. The corresponding enantiopure spirocyclic ketones were remarkably obtained in high yields, as shown in Scheme 10.31. These products constitute attractive starting materials for the synthesis of natural and biologically active products. Later, these authors extended the scope of this reaction by performing hetero-domino Knoevenagel/Diels–Alder/epimerization reactions which provided access to highly

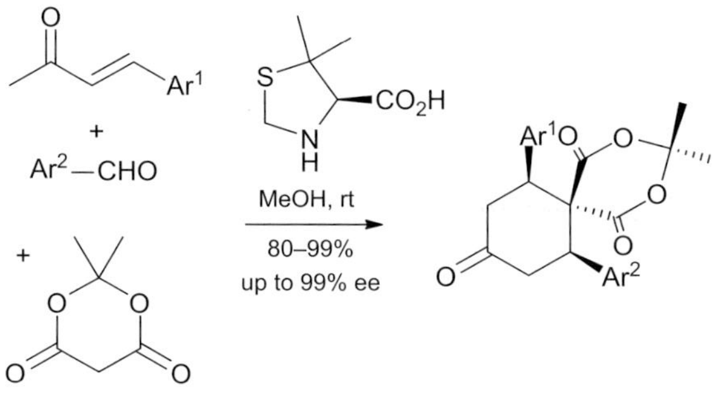

Scheme 10.31 Three-component domino Knoevenagel/Diels–Alder reaction.

substituted prochiral spiro [cyclohexane-1,2′-indan]-1′,3′,4-triones, employing 1,3-indanedione instead of Meldrum's acid [291]. These products were remarkably achieved in excellent yields (90–98%) and as single stereomers. In 2011, Feng *et al.* [292] reinvestigated these reactions by using chiral 9-amino-9-deoxy-*epi*-quinine as catalyst in the presence of 4-bromophenol as an additive. A range of pharmacological multisubstituted spiro[5,5]undecane-1,5,9-triones were achieved in good yields (53–81%) and excellent diastereoselectivities of > 98% de, and with good enantioselectivities (80–97% ee). A four-component version of this domino three-component Knoevenagel/Diels–Alder reaction, providing identical spirotrione products via a domino Wittig/Knoevenagel/Diels–Alder reaction, was also previously reported by Barbas and Ramachary [293]. In this case, the required α,β-unsaturated aldehyde was *in situ* generated via a Wittig reaction of a phosphorane and an aromatic aldehyde. Although the versatility of the Knoevenagel/Diels–Alder reaction was improved by this four-component methodology, lower enantioselectivities were observed (42–70% ee).

In 2008, Ramachary *et al.* [294] reported the synthesis of chiral densely functionalized cyclohexanes through the dienamine activation mode. This three-component domino Knoevenagel/Diels–Alder reaction occurred between benzaldehyde, methyl cyanoacetate, and an α,β-unsaturated enone in the presence of L-proline as catalyst. The process consisted in the *in situ* formation of the dienamine of the starting α,β-unsaturated enone in the presence of L-proline. This dienamine subsequently underwent a Diels–Alder cycloaddition with the Knoevenagel product *in situ* generated from the starting aldehyde and methyl cyanoacetate. The corresponding highly substituted domino product was not isolated, but was directly submitted to reaction with Hantzsch ester and a second equivalent of methyl cyanoacetate to finally afford the final cyclohexane bearing four stereogenic centers in 73% yield, and almost complete diastereo- and enantioselectivities. A more recent example was reported by Yuan *et al.* [295], who developed the first enantioselective organocatalytic three-component domino Knoevenagel/Michael/cyclization reaction, giving access to a range of chiral spiro[4*H*-pyran-3,3′-oxindoles]. This novel and highly efficient sequence was catalyzed by cupreine, and involved malononitrile, a 1,3-diketone, and isatylidene malononitrile derivatives as the components. As shown in Scheme 10.32, the spiro heterocyclic products were generally obtained in high yields and enantioselectivities of up to 96% ee. The authors proposed the mechanism depicted in Scheme 10.32, in which the sequence started with the Knoevenagel condensation of malononitrile onto isatylidene malononitrile derivatives to give the corresponding alkylidene malononitrile intermediates. Then, Michael addition of the 1,3-diketone to these intermediates led to dicyano intermediates, in equilibrium with enol intermediates, which finally cyclized to afford the final products. More recently, Macaev *et al.* [296] investigated the reaction of isatine with malononitrile and acetylacetone catalyzed by brevicolline (10 mol%). The corresponding spiro[4*H*-pyran-3,3′-oxindole] was achieved in 62% yield and enantioselectivity of 94% ee.

Another four-component reaction initiated by a Knoevenagel condensation was reported by Xu *et al.* [297]. This process was catalyzed by simple L-proline and

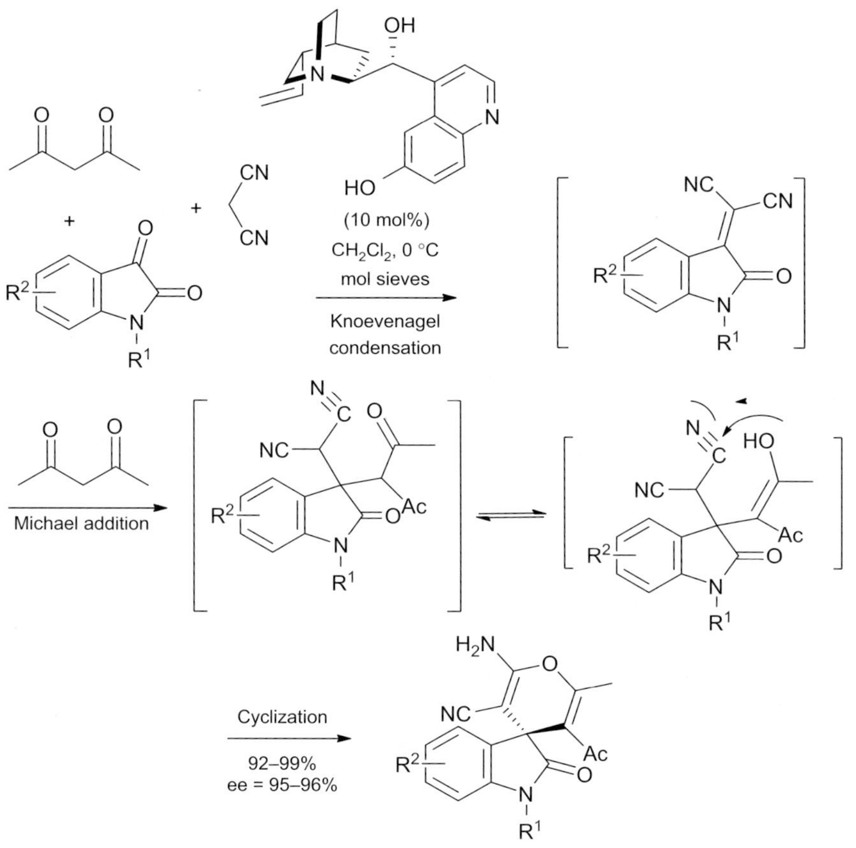

Scheme 10.32 Three-component domino Knoevenagel/Michael/cyclization reaction.

involved formaldehyde, 2 equiv of a β-keto ester, and p-substituted aniline as the substrates, providing the corresponding piperidines in both moderate yields of 24–65% and enantioselectivities of 55–74% ee. In 2012, a chiral cinchona-alkaloid-derived primary amine was associated by Wang *et al.* [298] to an (*R*)-BINOL-derived phosphoric acid to induce a three-component domino Knoevenagel/Michael reaction between isatins, malononitrile, and acetone, providing the corresponding domino products in high yields of 85–97% and enantioselectivities of 88–99% ee. In addition, chiral thioureas derived from quinine were identified by Yan *et al.* [299] as the most efficient catalysts to induce a three-component reaction of isatins, malononitrile, and α-aryl-isocyanoacetates. This process allowed a number of chiral 3,3′-dihydropyrrylspirooxindoles to be achieved in high yields (82–95%) and none to excellent enantioselectivities of up to 99% ee, combined with moderate diastereoselectivities (≤72% de). These tricyclic products arose from a domino Knoevenagel/Michael/cyclization reaction.

10.3.3 Multicomponent Reactions Based on the Mannich Reaction

The Mannich reaction, discovered by Carl Mannich [192] in 1912, is a condensation of amine derivatives, enolizable carbonyl compounds, and non-enolizable aldehydes. The first example of a direct organocatalytic three-component Mannich reaction was reported in 2000 by List [300], using L-proline as catalyst, and was followed by a number of excellent works from several groups [301]. Among the works reported since 2006, Bolm and Rodriguez [302] developed, in 2006, an L-proline-catalyzed Mannich reaction between cyclohexanone, formaldehyde, and various anilines. Employing only 0.5 mol% of this catalyst in DMSO under microwave irradiation, the corresponding Mannich products were achieved with excellent enantioselectivities of 96–98% ee, in combination with good yields (40–93%), as shown in Scheme 10.33.

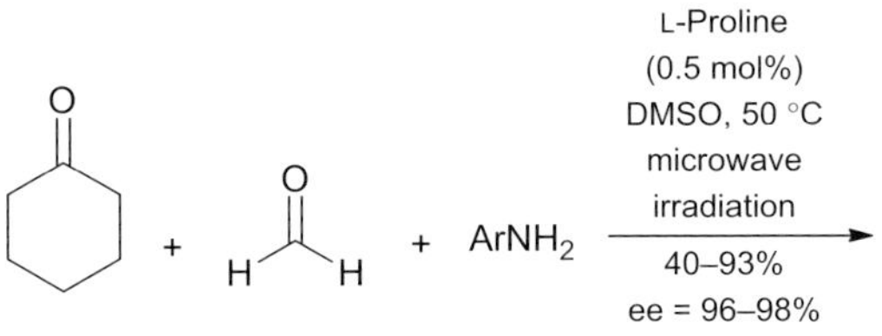

Scheme 10.33 Three-component Mannich reaction catalyzed by L-proline under microwave irradiation.

In the same year, Cordova *et al.* [303] studied the three-component Mannich reaction of dimethyl-1,3-dioxan-5-one with *p*-anisidine and various aldehydes. Among the various amino acids investigated as catalysts, the best results were obtained by using L- or D-proline in DMSO, which provided the protected *syn*-amino sugar products in good yields (40–84%), moderate to high diastereoselectivities (50–90% de), and good to excellent enantioselectivities (48–99% ee). These authors also developed an interesting sequential Mannich/Horner–Wittig–Emmons reaction catalyzed by L-proline [304]. The first step of the sequence consisted in an enantioselective three-component Mannich reaction between two different aldehydes and *p*-anisidine, which was followed, after subsequent addition of methyl diethylphosphonoacetate, by olefination to give the corresponding chiral vicinal amino alcohols in good yields (45–88%) and syn diastereoselectivities (60–90% de), combined with high enantioselectivities (95–96% ee). Also in 2006, Choudary *et al.* [305] introduced the use of ultrasonic conditions to increase the yields of L-proline-catalyzed Mannich reaction of hydroxyacetone, aromatic aldehydes, and *p*-anisidine. Both electron-rich and electron-deficient aromatic aldehydes led, under these conditions, to the corresponding *syn*-aminoalcohols in high yields (85–98%), moderate to high cis diastereoselectivities (50–92% de), and enantioselectivities of 66–99% ee. Besides proline, a number of other chiral catalysts were successfully investigated to induce this type of reactions. For example, L-proline-derived tetrazole was applied by Barbas *et al.* [306] to the catalysis of the Mannich reaction of protected amino ketones with *p*-anisidine and aldehydes to give the corresponding 1,2-diamines from azido ketones, and 1,4-diamines from phthalimido ketones. A range of chiral 1,2-azidoamines were achieved in good yields (60–96%), moderate to good syn diastereoselectivities (40–82% de), and moderate to excellent enantioselectivities (29–99% ee). On the other hand, a number of chiral 1,4-diamines, arising from reactions of phthalimido acetone with various aldehydes, and *p*-anisidine were prepared in good yields (53–95%) and good to excellent enantioselectivities (57–97% ee). Later, these authors also investigated the use of chiral linear α-amino acids, such as L-tryptophane or *O*-*t*-Bu-L-threonine, to induce the Mannich reaction between aldehydes, α-hydroxyaldehydes, and *p*-anisidine [307]. The reaction allowed the corresponding α-hydroxy-β-aminoketones to be formed in good yields (67–95%), low to high anti diastereoselectivities of 13–90% de, and enantioselectivities of 53–96% ee. In the same year, Gong *et al.* reported the use of BINOL- and H_8-BINOL-derived chiral phosphoric acids in the three-component Mannich reaction between aliphatic ketones, aldehydes, and aromatic amines [308]. Only a catalytic amount as low as 0.5 mol% was sufficient to promote the process, which provided the corresponding Mannich products in moderate to excellent yields (42–99%), anti diastereoselectivities of 54–96% de, and enantioselectivities of 70–98% ee. The first catalytic enantioselective Mannich reaction in purely aqueous media was reported by Lu *et al.*, in 2007 [309]. A threonine-derived catalyst proved to be more effective than those derived from serine in the Mannich reaction of aldehydes, 2-benzyloxyacetone, and *p*-anisidine. The scope of the process was explored with various aliphatic as well as aromatic aldehydes to afford the corresponding β-amino- α-hydroxy ketones in good to high yields (50–98%) and moderate to high anti diastereo- and enantioselectivities of 50–90% de and 62–97% ee, respectively. The selectivities of electron-rich aromatic aldehydes were much poorer ($\leq$ 62% ee), and rather disappointing yields were found using aliphatic aldehydes (50–70%). In 2008, Wu *et al.* [310] reported a comparable Mannich reaction in aqueous media catalyzed by a hydrophobic *O*-silyl serine catalyst. The reaction between aromatic aldehydes, electron-rich aromatic amines, and a variety of carbonyl donors resulted in good enantioselectivities of up to 92% ee for the corresponding Mannich adducts. Interestingly, the application of cyclohexanone as the donor in this Mannich reaction resulted in the formation of the corresponding syn product, while the use of benzyloxyacetone provided the corresponding anti adduct selectively. Protected dihydroxyacetone derivatives were employed for the synthesis of amino sugars and polyhydroxylated amino acids via a Mannich reaction by several groups, whereas the use of an unmodified dihydroxyacetone donor was not reported until 2007 [311]. The main advantage of such an approach was the immensely reduced cost (>230-fold) by using dihydroxyacetone as a direct biomimetic donor. As reported by Cordova *et al.*, Mannich reactions with dihydroxyacetone afforded the corresponding products with yields of up to 90% and excellent enantioselectivities of >99% ee. Interestingly, reactions with amino acid or peptide catalysts bearing a primary amine moiety were anti-selective, whereas those with proline were syn-selective. In order to improve the solubility of proline-derived catalysts in nonpolar solvents, Carter and Yang [312] developed, in 2009, several novel chiral proline derivatives, which were further applied to induce three-component Mannich reactions. A β-amino alcohol was obtained with moderate syn diastereoselectivity of 67% de when *p*-nitrobenzaldehyde was used as the aldehyde component, whereas the use of α,α-dimethoxyacetaldehyde favored the formation of the corresponding anti diastereomer in 82% de and modest enantioselectivity of 20% ee. The stereochemical outcome of the Mannich reaction could be altered by changing the steric features at the 5-position of proline. In this context, Wu *et al.* [313] developed proline mimetic catalysts for the asymmetric Mannich three-component reaction of hydroxyacetone. For example, when the reaction of this ketone with electron-deficient aromatic aldehydes and unsubstituted anilines was induced by L-proline bearing a *trans*-carboxylic acid function at the 5-position, it provided the corresponding Mannich products in superior diastereo- and enantioselectivities of up to 88% de and 98% ee, respectively. In 2009, Zhu *et al.* [314] developed an enantioselective four-component Mannich reaction catalyzed by a chiral H_8-BINOL-derived phosphoric acid. The process was performed between aldehydes, anilines, enecarbamates, and ethanol, providing the corresponding aminoethers. These Mannich products were *in situ* reduced by treatment with $NaBH_3CN$ into the corresponding chiral anti-1,2-disubstituted 1,3-diamines, which were achieved through the two-step sequence in good to excellent yields (55–97%) and enantioselectivities of 76–99% ee combined with anti diastereoselectivities of >90% de. Chiral ionic liquids have the advantages of easy reusability, nonvolatility, and thermal stability, allowing the products to be easily isolated. In this context, chiral ionic liquid catalysts were investigated in three-component Mannich reactions [315]. As an example, a chiral ionic-liquid-catalyzed Mannich reaction was reported by Wang *et al.*, in 2010 [316]. This L-proline-derived ionic liquid $[EMIm]^+[Pro]^-$ was shown to promote the Mannich reaction of aliphatic ketones, para-substituted aromatic aldehydes, and amines to afford the corresponding (*S*)-configured β-amino carbonyls in predominantly good to excellent yields (48–99%) and diastereoselectivities (48–99%), combined with enantioselectivities of 90–99% ee. In addition, the synthesis of (*R*)-configured β-amino carbonyls from aromatic amines, aliphatic aldehydes, and aliphatic ketones was reported by Xu *et al.* [317]. By using the chiral ionic liquid $[DEMImBF_4]^+[Pro]^-$, predominantly high yields (72–96%) and moderate to excellent enantioselectivities (28–99% ee) were obtained for the corresponding (*R*)-configured β-amino carbonyls. In 2011, Fu *et al.* introduced other L-threonine derivatives as novel catalysts for enantioselective direct anti Mannich reactions of hydroxyacetone with aldehydes and anilines [318]. When the reaction was promoted by an L-threonine-derived catalyst in 1-methyl-2-pyrrolidinone (NMP) as the solvent, it afforded the corresponding *anti*-1,2-amino alcohols in good yields (75–93%) and enantioselectivities of 94–99% ee, albeit with variable diastereoselectivities of 2–92% de. In addition, Yang *et al.* [319] developed a highly enantioselective three-component Mannich reaction of *in situ* generated imines from 2-aminobenzoxazol and aldehydes with malonates in the presence of a chiral cinchona alkaloid thiourea catalyst. The corresponding chiral β-amino ester derivatives were produced in good to high yields (65–73%) and high enantioselectivities of 87–99% ee. A recyclable siloxy-L-serine catalyst was developed by Teo and

Yong [320] to induce Mannich reactions performed in ionic liquids. Indeed, the reaction of cyclic as well as acyclic ketones with aldehydes and *p*-anisidine afforded, under these conditions, the corresponding β-amino ketones as mixtures of syn and anti diastereomers in moderate to high yields (50–91%) and enantioselectivities (55–91% ee), albeit with moderate diastereoselectivities (20–88% de). On the other hand, Tao *et al.* [321] have developed chiral isosteviol-derived amino acids to be used in Mannich reaction of hydroxyacetone, aromatic aldehydes, and anilines. When the reaction was performed with an L-proline isosteviol-derived amino acid as catalyst, it afforded the corresponding *syn*-1,2-amino alcohols with excellent stereoselectivities of up to 98% de and >99% ee, whereas the use of an L-serine isosteviol-derived amino acid as catalyst allowed the corresponding *anti*-1,2-amino alcohols to be achieved in comparable excellent stereoselectivities of up to 94% de and 99% ee, as shown in Scheme 10.34. In 2012, these authors extended the scope of this methodology to cyclohexanone, which provided the corresponding chiral syn Mannich products with excellent diastereo- and enantioselectivities of up

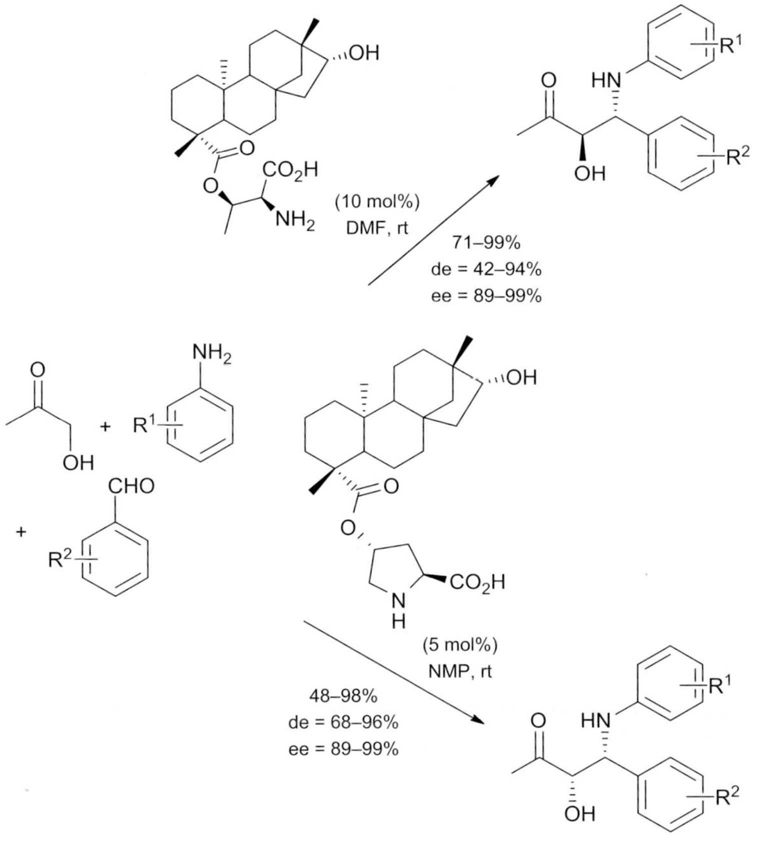

Scheme 10.34 Three-component Mannich reaction catalyzed by isosteviol-derived amino acids.

to 96% de and >99% ee, respectively, when the same L-proline isosteviol-derived amino acid was employed as catalyst [322].

10.3.4 Multicomponent Reactions Based on the Biginelli Reaction

The Biginelli reaction is a three-component reaction allowing the synthesis of 3,4-dihydropyrimidin-2-(1*H*)-ones or -thiones by reacting urea or thiourea, a 1,3-dicarbonyl derivative, and an aldehyde. It was discovered by Pietro Biginelli in 1893 [243]. The heterocyclic pyrimidinone products are known to exhibit a wide range of important pharmacological properties and make up a large family of medicinally relevant compounds. The asymmetric version of the Biginelli reaction is of significant contemporary interest. It must be noted that there are only a few reports in the literature on the synthesis of chiral dihydropyrimidinones by employing organocatalysts. The first organocatalytic enantioselective Biginelli reaction was described by Gong *et al.*, in 2006 [323]. Indeed, these authors reported excellent results for the Biginelli reaction of aldehydes, (thio)urea, and β-ketoesters catalyzed by a chiral phosphoric acid [324]. The corresponding Biginelli products were achieved in good to high yields (51–86%) and generally high enantioselectivities of 88–97% ee. In the course of developing an enantioselective synthesis of the potent melanin-concentrating hormone receptor antagonist SNAP-7941, Schauss and Goss used the same chiral phosphoric acid to induce the key Biginelli reaction of urea, methylacetoacetate, and 3,4-difluorobenzaldehyde [325]. The required key product was produced in 96% yield and 79% ee and further converted into the expected SNAP-7941. In 2008, Feng *et al.* [326]. investigated these reactions by employing a combination of a chiral *trans*-4-hydroxyproline-derived secondary amine and a Bronsted acid, such as 2-chloro-4-nitrobenzoic acid, as the catalyst system The dual-catalyzed process was performed at 25 °C with 5 mol% of catalyst loading of each of the two catalysts in 1,4-dioxane and in the presence of an additive such as *t*-$BuNH_2$·TFA. Under these conditions, the Biginelli products were obtained in moderate to good yields (34–73%) and good to excellent enantioselectivities (70–98% ee). In the same year, Yadav *et al.* reported an interesting, chiral ionic-liquid-catalyzed, efficient, and unprecedented version of the Biginelli reaction using new variants of its active methylene component, such as 2-phenyl-1,3-oxazol-5one and 2-methyl-2-phenyl-1,3-oxathiolan-5-one. The reaction of these carbonyl compounds with aromatic aldehydes, and (thio)ureas afforded, when catalyzed by chiral ionic liquid, such as L-prolinium sulfate, the corresponding 5-amino or 5-mercaptoperhydropyrimidines in good yields (42–93%) and good to high enantioselectivities of 81–95% ee [327]. Later, a diastereoselective Biginelli-type reaction catalyzed by L-proline was developed by Tripathi *et al.* [328]. This process occurred between 3,4-dihydro-(2*H*)-pyran, aromatic aldehydes, and (thio)urea, providing the corresponding hexahydropyrano pyrimidinones in relative diastereoselectivity of >90% de and in good yields (83–90%). These products constituted potent precursors for many biologically active molecules and fused oxazines, which are the lead molecules for tuberculosis chemotherapy. In 2009, Lee *et al.*

[329] studied the efficiency of L-proline ester salts to promote the Biginelli reaction between *m*-nitrobenzaldehyde, urea, and ethyl acetoacetate. The corresponding 3,4-dihydropyrimidin-2-(1*H*)-one was obtained in 87% yield and enantioselectivity of 40% ee by employing the salt of L-proline-derived *tert*-butyl ester as catalyst. Higher enantioselectivities of up to 81% ee combined with good yields (63–88%) were reported by Wang *et al.* [330] by using chiral substituted 5-(pyrrolidin-2-yl)tetrazoles as catalysts for the Biginelli reaction of various substituted benzaldehydes, urea (or thiourea), and a range of β-ketoesters. The best results (68–81% ee) were obtained with 4-*N*-tosyl-5-(pyrrolidin-2-yl)tetrazole bearing an *n*-butyl group on the nitrogen atom. Comparable results (14–81% yields, 51–78% ee) were also reported by Zhao and Ding [331] for similar reactions but catalyzed by a chiral primary amine derived from quinine. In addition, these authors have included aliphatic aldehydes as substrates of the Biginelli reaction, providing comparable results. For example, 43% yield combined with enantioselectivity of 72% ee was achieved for the reaction of *n*-heptanal with urea and ethyl acetoacetate. On the other hand, chiral bifunctional primary amine thiourea catalysts derived from D-glucose were demonstrated by Chen *et al.* to be highly efficient to promote highly enantioselective Biginelli reaction of a range of benzaldehydes, (thio)urea, and ethyl acetoacetate [332]. The best results were obtained by using a gluco-2-aminocyclohexylthiourea as catalyst in combination with 2,4,6-trichlorobenzoic acid, and *tert*-butylammonium trifluoroacetate as additives. Under these conditions, the corresponding dihydropyrimidines were achieved in moderate to high yields (51–93%) and enantioselectivities of up to 99% ee. In order to explain the stereoselectivity of the reaction, the authors proposed that both hydrogen-bonding interactions and enamine activation of the β-ketoester occurred in the transition states. It must be noted that a low catalyst loading of only 5 mol% was sufficient to reach these results, but poor enantioselectivities were obtained for aliphatic aldehydes, such as *n*-butyl alcohol, which provided the corresponding product in only 15% ee. More recently, these authors investigated the same reaction under phase-transfer conditions [333]. Indeed, by using the same catalyst at 15 mol% of catalyst loading combined with 15 mol% of TfOH and 10 mol% of t-BuNH$_2$·TFA as an additive, the reaction could be performed in aqueous media, providing the Biginelli products in good to high yields (62–93%) and high enantioselectivities (87–99% ee). On the other hand, a series of chiral phosphoric acids derived from 3,3′-disubstituted binaphthols were investigated as catalysts for the Biginelli reaction with thioureas by Gong *et al.* [334]. The authors showed that the size of the 3,3′-substituents of the catalysts had a considerable impact on the control of the stereochemistry of the reaction. Indeed, by tuning these substituents, the stereochemistry of the reaction could be reversed, allowing the synthesis of both enantiomers of the same dihydropyrimidine. The best results were achieved by using a chiral phosphoric acid, which allowed a range of dihydropyrimidines to be produced in moderate to good yields (26–98%) and moderate to excellent enantioselectivities (26–99% ee) from the reactions of various β-ketoesters and aldehydes with thiourea. It must be noted that some aliphatic aldehydes also gave good results, such as cyclohexylcarbaldehyde, which allowed an enantioselectivity of 90% ee to be obtained in combination with 54% yield. The

scope of this methodology was extended to the reaction of a range of aldehydes and thioureas with various cyclic and acyclic ketones, providing the corresponding dihydropyrimidines in good yields (33–86%) and enantioselectivities of 91–99% ee through a Biginelli-like reaction. In 2011, Saha and Moorthy [335] reported the synthesis of a series of novel sterically hindered organocatalysts derived from L-proline. From a systematic investigation of these catalysts and screening studies, the chiral catalyst depicted in Scheme 10.35, containing a strong hydrogen-bonding site, was selected to be the most efficient to promote the Biginelli condensation of aromatic as well as aliphatic aldehydes with ethyl acetoacetate and urea in a remarkably high enantioselectivity (94–99% ee), as shown in Scheme 10.35. This sterically hindered catalyst was used in combination with pentafluorobenzoic acid (PFBA) and tritylammonium trifluoroacetate as additives, allowing a range of chiral dihydropyrimidines to be produced in moderate to good yields (44–68%).

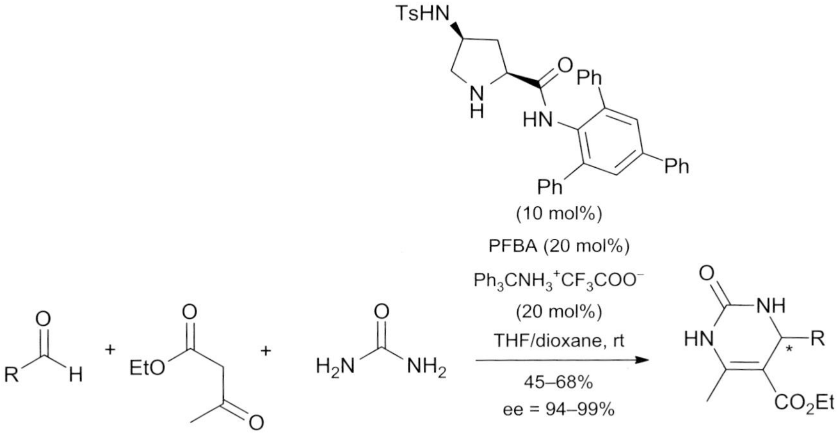

Scheme 10.35 Biginelli reaction catalyzed by sterically hindered prolinamide.

10.3.5 Multicomponent Reactions Based on the Hantzsch Reaction

Another venerable and old multicomponent reaction is the so-called Hantzsch reaction, which was first reported in 1882 [242, 336], allowing the synthesis of 1,4-dihydropyridines through the reaction of enamines, aldehydes, and 1,3-dicarbonyl compounds. An asymmetric, three-component, Hantzsch-type reaction for the construction of dihydropyridines was developed by Jørgensen *et al.* [337] in 2008, representing the first organocatalytic enantioselective one-pot synthesis of chiral dihydropyridines. The reaction of α,β-unsaturated aldehydes, β-diketones or β-ketoesters, and primary amines afforded, by using a chiral diarylprolinol trimethylsilyl ether as catalyst, the corresponding chiral 1,4-dihydropyridines in moderate to good yields (31–60%) and moderate to high enantioselectivities of 38–95% ee. The scope of the reaction was explored with α,β-unsaturated aldehydes bearing aliphatic and aromatic groups as well as heteroatoms and double bonds. Moreover, both acetoacetone and methyl acetoacetate consistently gave high enantioselectivities in combination with various substrates. The primary amine tolerated aliphatic and aromatic substituents, although the highest reactivity and yield were observed for aniline. In 2008, Gong *et al.* [338] simultaneously reported an almost identical Hantzsch-type three-component reaction for the generation of oppositely configured dihydropyridines under chiral Bronsted acid catalysis by using a chiral phosphoric acid as catalyst. Advantageously, a broader range of substrates was tolerated and the products were mostly obtained in considerably high yields (31–93%) and enantioselectivities (66–98% ee). Finally, the first highly enantioselective four-component version of the Hantzsch reaction was developed by Gestwicki and Evans [339] in 2009, providing access to a range of pharmaceutically important dihydropyridines. The reaction was catalyzed by a BINOL-derived phosphoric acid, and occurred between ethylacetoacetate, ammonium acetate, dimedone, and an aromatic aldehyde. As shown in Scheme 10.36, a series of 1,4-dihydropyridines were produced in both excellent yields and enantioselectivities in almost all cases of substrates studied. Conversely, reactions involving aliphatic aldehydes proceeded in good yields (66–94%) albeit with no enantio-enrichment.

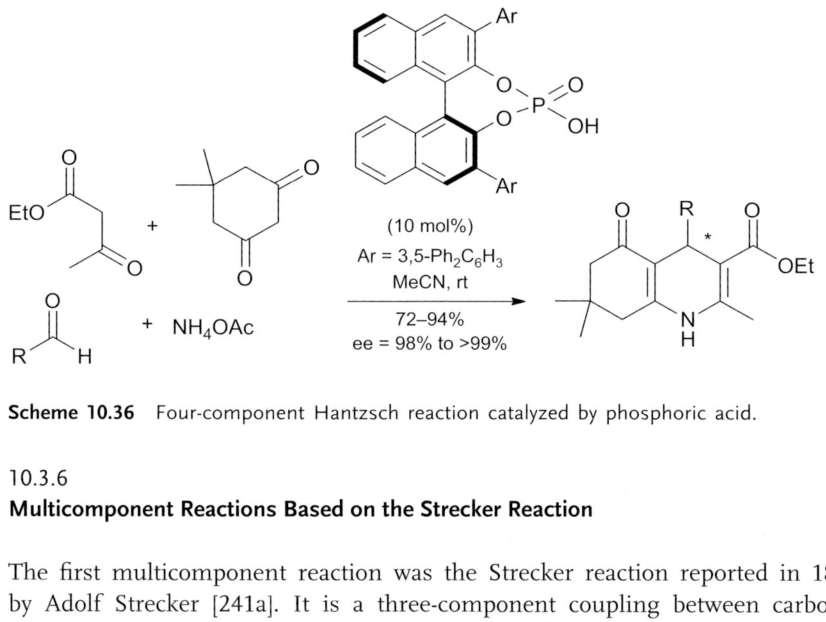

Scheme 10.36 Four-component Hantzsch reaction catalyzed by phosphoric acid.

10.3.6
Multicomponent Reactions Based on the Strecker Reaction

The first multicomponent reaction was the Strecker reaction reported in 1850 by Adolf Strecker [241a]. It is a three-component coupling between carbonyl derivatives, amines, and cyanide source, such as hydrogen cyanide, to provide α-aminonitriles which constitute potent starting materials to achieve important α-amino acids by simple hydrolysis. The mechanism of the Strecker reaction involves the initial formation of an imine from condensation of the amine component to the carbonyl component, after which addition of the cyanide component to this imine intermediate follows. Although the first enantioselective, metal-catalyzed Strecker reaction was reported by Kobayashi *et al.* in 2000 [340], the first organocatalytic version was not described until 2007 by Pan and List [341]. An advantage of this novel procedure was that the highly toxic and volatile hydrogen cyanide was replaced by acyl cyanide. This three-component acyl-Strecker reaction between acetyl cyanide, amines, and aldehydes was catalyzed by a chiral thiourea, and provided the corresponding α-amino nitriles in good to excellent yields and enantioselectivities of up to 94% ee, as shown in Scheme 10.37. Both aromatic and aliphatic aldehydes were tolerated in addition to amine inputs. Furthermore, the formed α-amino nitriles were subsequently converted into the corresponding α-amino acids and their salts through acid-mediated hydrolysis and hydrogenolysis without loss of optical purity.

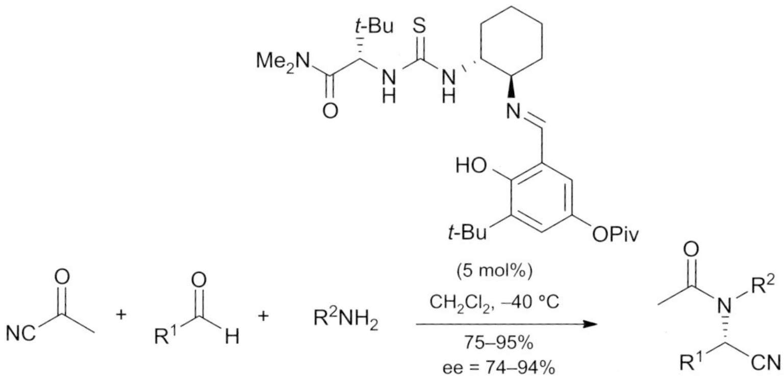

Scheme 10.37 Three-component acyl-Strecker reaction catalyzed by thiourea.

In the same year, C_2-symmetrical chiral bisformamides were demonstrated by Feng *et al.* [342] to induce the Strecker three-component reaction. The catalytic efficiency of the reaction was highly dependent on the match between the chiral centers of the diamine and proline parts of the catalyst. A disappointing enantioselectivity of 12% ee was observed when the L-proline formamide part of the catalyst was combined with (*R,R*)-diamine bisformamide instead of (*S,S*)-diamine bisformamide (61% ee). Optimization of the reaction conditions showed that a formamide-based catalyst was the most efficient in polar aprotic solvents such as 1,2-dichloroethane. Indeed, the use of protic solvents, such as methanol, resulted in the formation of racemic α-aminonitriles. The scope of the process was investigated with various aromatic electron-rich and electron-deficient aldehydes, which allowed the corresponding chiral α-aminonitriles to be achieved in good to high yields of 74–99% and moderate to good enantioselectivities (43–86% ee). Later, these authors developed the same Strecker reaction induced by chiral amino acid-derived *N,N'*-dioxides as effective catalysts [343]. The combination of the Lewis basic *N*-oxide and amide moieties provided a highly efficient bifunctional organocatalyst. Once again, the catalytic efficiency was shown to be highly dependent on the match between the chiral centers of the amide and proline parts of catalyst, giving better results for L-proline in combination with an (*S*,*S*)-diamide. Reduced enantioselectivity was obtained after expanding or decreasing the size of the cyclohexyl substituent of the catalyst, while the proline-hydroxyl proved to complement the stereoselectivity. The process was compatible with a broad range of aromatic and aliphatic aldehydes, affording the corresponding α-aminonitriles in good yields (60–99%) and enantioselectivities of 72–95% ee.

10.3.7
Multicomponent Reactions Based on the Petasis Reaction

As a multicomponent reaction, the Petasis reaction enjoys a history of nearly two decades, since it was discovered by N. A. Petasis, in 1993 [344]. It involves the condensation of amines, carbonyl derivatives, and aryl- or vinylboronic acids for the preparation of amine derivatives. This reaction evolved to be a powerful synthetic tool in the last decade, owing to the high accessibility of the reagents and mild reaction conditions. Its most important application is in the preparation of chiral α-amino acids by using glyoxylic acid as the aldehyde component. Although remarkable success have been achieved in the asymmetric version based on the use of chiral auxiliaries, to date there are only few examples concerning the catalytic enantioselective Petasis reaction. Most of the time, these rare enantioselective versions of the Petasis reaction are limited to substrates including quinolines and ethyl glyoxylates. The first organocatalytic enantioselective Petasis reaction was reported by Schaus and Lou, in 2008 [345]. The reaction between styryl boronates, secondary amines, and ethyl glyoxylate was catalyzed by chiral diphenol (*S*)-2,2′-diphenyl-(4-biphenanthrol) (VAPOL). Vaulted biaryl phenols, such as (*S*)-VAPOL, gave superior results for the asymmetric Petasis reaction in comparison with chiral BINOL derivatives. Advantageously, the catalyst could be recovered and recycled without loss of activity and optical purity. The scope of the process was explored with various electron-rich and electron-deficient styryl boronates, providing the corresponding α-amino acids in high yields and enantioselectivities, as shown in Scheme 10.38. Moreover, high functional group tolerance was observed for the secondary amine.

On the other hand, Takemoto *et al.* [346] developed the Petasis-type reaction of substituted quinolines, styrylboronic acids, and phenyl chloroformate catalyzed by a newly designed chiral thiourea catalyst. Various quinolines and styrylboronic acids were tolerated to afford the corresponding dihydroquinolines in moderate yields (28–75%) and good enantioselectivities of 82–97% ee. In 2012, Yuan *et al.* [347] developed a three-component Petasis reaction between salicylaldehydes, amines, and organoboric acids catalyzed by a newly designed chiral BINOL-derived thiourea. A broad range of chiral alkylaminophenols could be achieved in good yields (48–92%) and good to high enantioselectivities of 62–95% ee.

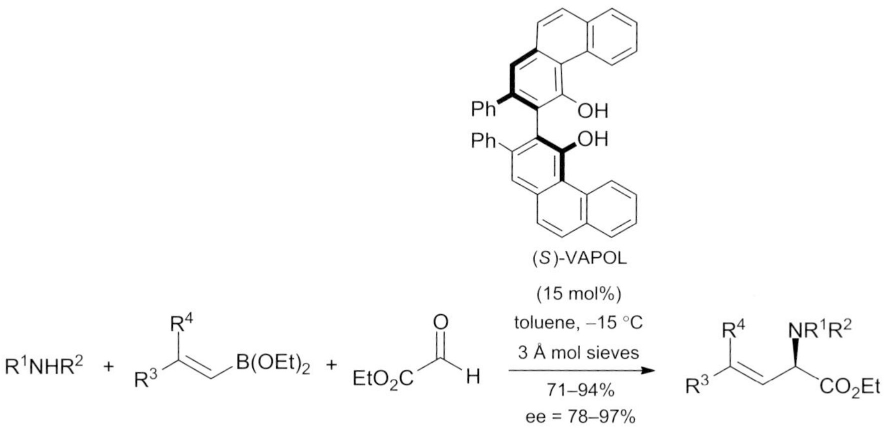

Scheme 10.38 Three-component Petasis reaction catalyzed by (*S*)-VAPOL.

10.3.8
1,3-Dipolar Cycloaddition-Based Multicomponent Reactions

A 1,3-dipolar cycloaddition is a reaction between a 1,3-dipole and a dipolarophile, such as a substituted alkene or alkyne, to construct five-membered heterocycles, which can bear up to four stereocenters often with a high degree of stereocontrol [348]. The resulting heterocycles, such as isoxazolidines and pyrrolidines, constitute valuable intermediates for the synthesis of various natural products, including amino acids and alkaloids. 1,3-Dipolar-cycloaddition-based multicomponent reactions usually proceed via an *in situ* formation of a 1,3-dipole which then reacts with the dipolarophile component. In 2007, Cordova *et al.* [349] described a nice highly chemo-, diastereo-, and enantioselective organocatalytic one-pot 1,3-dipolar cycloaddition-based three-component reaction between *N*-arylhydroxylamines, aldehydes, and α,β-unsaturated aldehydes. This process, catalyzed by chiral diphenylprolinol trimethylsilyl ether, constituted a versatile asymmetric entry to a variety of valuable isoxazolidines in high yields, excellent diastereoselectivity of >92% de, and associated remarkable enantioselectivities ranging from 94 to 99% ee, as shown in Scheme 10.39. Since the domino aldehyde products were susceptible to be unstable, they were *in situ* reduced into the corresponding alcohols by treatment with $NaBH_4$.

As an extension of this methodology, these authors employed the same catalyst to induce a three-component synthesis of chiral pyrrolidines [350]. The first step of the domino process consisted in an imine formation between 2-aminomalonate and an aldehyde. Its corresponding azomethine ylide, prepared *in situ*, reacted with α,β-unsaturated aldehydes, furnishing the final products in good yields (51–63%) and diastereo- and enantioselectivities of 50–82% de and 92–98% ee, respectively. Various aromatic and aliphatic enals were compatible acceptors in this process, as were a broad range of benzaldehydes. In 2008, a three-component reaction of aldehydes, dialkyl maleates, and α-amino esters was described by Gong *et al.* [351]. The

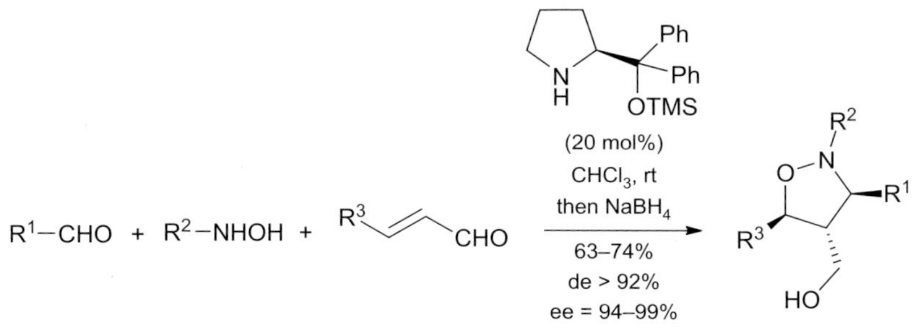

Scheme 10.39 1,3-Dipolar cycloaddition-based three-component reaction catalyzed by proline derivative.

process, catalyzed by a chiral biphosphoric acid, afforded the corresponding 1,3-dipolar cycloadducts in good to excellent yields (67–97%) and enantioselectivities of 76–99% ee. Several aldehydes including aromatic, aliphatic, and α,β-unsaturated were compatible with this reaction. It seemed highly sensitive to the nature of the dipolarophile and particularly suffered from the presence of bulky substituents on the maleate component. By the incorporation of an unsymmetrical α-amino ester, four contiguous stereogenic centers were created with high enantioselectivities (81–97% ee). The spiro[pyrrolidin-3,3′-oxindole] ring system constitutes the core structural element found in a large family of natural alkaloids and unnatural compounds exhibiting important biological activities. Although these are elegant and creative strategies toward the construction of spirooxindole architectures, the first direct catalytic approach to access chiral spiro[pyrrolidin-3,3′-oxindoles] was reported by Gong *et al.* only in 2009 [352]. It was based on a three-component 1,3-dipolar cycloaddition of a broad range of methyleneindolinones with aldehydes, and amino diester catalyzed by a chiral BINOL-derived phosphoric acid. A series of chiral spirooxindoles were achieved in good to high yields (59–97%) with an unusual regiochemistry and high enantioselectivities of 81–98% ee. In 2011, Cordova *et al.* developed the first highly diastereoselective and enantioselective organo-cocatalytic dynamic three-component reaction between aldehydes, protected α-cyanoglycine esters, and α,β-unsaturated aldehydes [353]. This powerful process afforded cyano-, formyl-, and ester-functionalized α-quaternary proline derivatives bearing four contiguous stereocenters in high enantioselectivities of >95% ee, combined with good yields (56–88%) and diastereoselectivities of 82–90% de. Finally, Jørgensen *et al.* described, in 2012, a synthesis of chiral hexahydropyrrolo-isoquinolines based on a 1,3-dipolar cycloaddition of *in situ* generated dihydroisoquinolinium ylides from the corresponding imines and α-bromoesters or ketones to α,β-disubstituted aldehydes [354]. This domino imine formation–1,3-dipolar cycloaddition reaction was catalyzed by chiral diphenylprolinol trimethylsilyl ether in the presence of a base such as Na_2CO_3. The corresponding domino aldehyde products were further converted into the corresponding alkenes by treatment with a Wittig–Horner reagent. The final pyrrolo-isoquinolines were achieved in good yields (62–89%) and high enantioselectivities of 90–99% ee, along with moderate to good diastereoselectivities (26–80% de).

10.3.9
Miscellaneous Multicomponent Reactions

The modern concept of multicomponent reaction is intimately related to the reactions developed with isocyanide reagents [355]. The first isocyanide-based multicomponent reaction was discovered by Passerini, in 1921 [356]. It was not until 2003 that Denmark and Fan [357] developed an enantioselective organocatalytic version of this reaction. In the same area, the Ugi four-component reaction is the reaction of a carbonyl compound (usually an aldehyde), an amine, an isocyanide, and a carboxylic acid (or an alcohol) to yield α-amino acid derivatives. This reaction, first described in 1959, has been more widely studied and used than any other multicomponent reaction [244a]. The Ugi three-component reaction is a variant of the general reaction in which either two of the usual groups are included in the same reagents, or the condensation of the carbonyl compound with the amine takes place before the addition of the isocyanide and acid derivatives. Even though the imine condensation could be performed in a one-pot process and its isolation was not necessary, it must be included in this category. Although a number of diastereoselective versions of this reaction have been successfully developed, enantioselective versions remain a significant challenge. In fact, the first example of enantioselective α-addition of α-isocyanides to imines was developed by Zhu *et al.*, in 2009 [358]. Indeed, these authors demonstrated that a chiral phosphoric acid was able to catalyze the three-component reaction of aldehydes, anilines, and an α-isocyanoacetamide, leading to the corresponding chiral 2-(1-aminoalkyl)-5-aminoxazoles in good yields (82–95%) and moderate to good enantioselectivities (66–90% ee). In 2008, Ma *et al.* [359] reported a highly enantioselective three-component reaction for the construction of trifluoromethyl-containing compounds. This domino Friedel–Crafts/aminoalkylation reaction occurred between indoles and imines *in situ* generated from trifluoroacetaldehyde methyl hemiacetal and aniline in the presence of a catalytic amount of a chiral BINOL-derived phosphoric acid bearing bulky 2,4,6-triisopropylphenyl groups at the 3,3′-positions. It afforded the corresponding chiral trifluoromethyl-containing indoles in high yields (80–99%) and excellent enantioselectivities (79–98% ee) when using 3,4,5-trimethoxyaniline as the amine component. Furthermore, the scope of the methodology was extended to difluoroacetaldehyde methyl hemiacetal, providing difluoromethyl-containing indoles in quantitative yields and enantioselectivities of >93% ee. In 2010, Enders *et al.* [360] reported a synthesis of chiral polyfunctionalized 3-(cyclohexenylmethyl)-indoles bearing three stereogenic centers based on a novel four-membered domino reaction, occurring between 3-(cyclohexenylmethyl)-indoles, nitroalkenes, and 2 equiv of acrolein. This novel process was based on a quadruple domino Friedel–Crafts/Michael/Michael/aldol condensation reaction catalyzed by chiral diphenylprolinol trimethylsilyl ether. As shown in Scheme 10.40, the products were produced in moderate to good yields (23–82%) and excellent diastereo- and enantioselectivities of up to >90% de and >99% ee, respectively. The authors proposed the catalytic cycle depicted in Scheme 10.40. In the first step, acrolein was activated by chiral catalyst through the

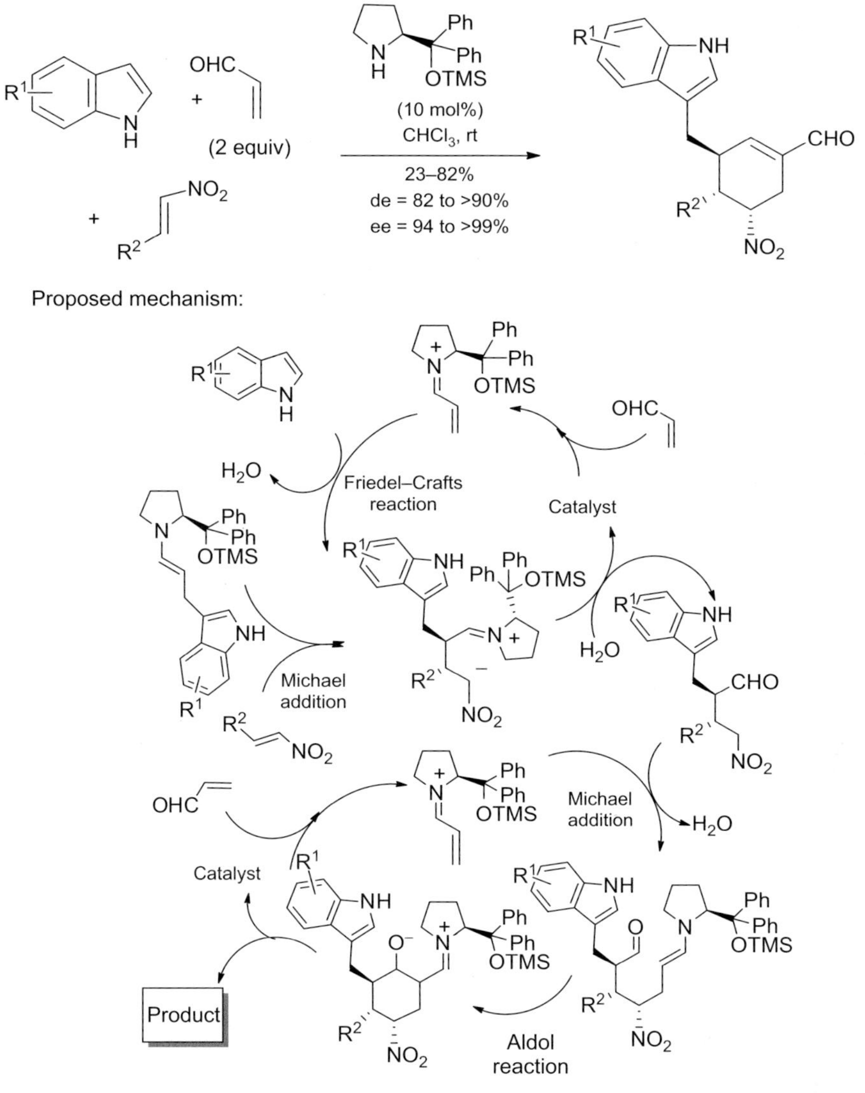

Scheme 10.40 Four-component domino Friedel–Crafts/Michael/Michael/aldol condensation reaction.

formation of the corresponding iminium ion, with which the indoles performed an intermolecular Friedel–Crafts-type reaction. The resulting enamines subsequently underwent an intermolecular Michael addition to nitroalkenes, affording novel iminium intermediates. Further hydrolysis of these intermediates led to the corresponding aldehydes, which reacted with a second equivalent of starting iminium ion to give the corresponding enamine intermediates. After an intramolecular enamine-mediated aldol reaction, alcoholate intermediates were formed, which underwent dehydration and hydrolysis to afford the final indoles, while the catalyst was regenerated.

In 2012, Chen and Roy [361] demonstrated the kinetic resolution of nitroallylic acetates via an enantioselective three-component coupling involving indoles, acrolein, and nitroolefin allylic acetates. This domino Friedel–Crafts/Michael/elimination reaction was catalyzed by the same catalyst and provided, through an iminium–enamine activation mode, the corresponding densely functionalized 3-substituted indoles in moderate yields (30–44%), albeit with remarkable general diastereo- and enantioselectivities of >98% de and 90–99% ee, respectively. It was found that the recovered nitroallylic acetates were obtained in enantiopurity ranging from 79 to 95% ee. A large number of organocatalysts have been shown to be highly efficient to induce Diels–Alder reactions, providing excellent yields and stereoselectivities [362]. Indeed, prolines, imidazolidinones, and other pyrrolidine-derived reagents have been demonstrated to be capable of promoting these reactions through enamine catalysis. Hydrogen-bonding activation is also possible with a variety of reagents, such as chiral Bronsted acids and diols. The first enantioselective aza-Diels–Alder three-component reaction was reported by Cordova *et al.*, in 2005 [363]. It was catalyzed by L-proline, and occurred between cyclohexenones, anilines, and formaldehyde. Diversely substituted aromatic amines were demonstrated to be excellent substrates, providing the corresponding bicyclic amines in good yields (70–82%) and high enantioselectivities (94–99% ee). To account for the stereochemical outcome of the reaction, a mechanism was proposed that involved an *in situ* enamine formation between the cyclohexenone derivative and L-proline. The intermediate imine formed from formaldehyde and aniline was attacked by this enamine intermediate from the less hindered *si* face. Subsequently, a 6-endo-trig-cyclization occurred to afford the final product. It must be noted that this approach suffered from a high catalyst loading (30 mol%) and the restriction to formaldehyde as the aldehyde component. A more widely applicable procedure was reported by Gong *et al.*, who developed the first Bronsted acid-catalyzed asymmetric direct aza-Diels–Alder reaction, in 2006 [364]. The process was catalyzed by a chiral H_8-BINOL-derived phosphoric acid and was shown to be compatible with various substituted aromatic aldehydes, providing by reaction with *p*-anisidine and 2-cyclohexenone the corresponding bicyclic piperidines in moderate to good yields (68–71%) and endo selectivities of 52–74% de, combined with good enantioselectivities of 83–85% ee. These N-heterocycles constitute key building blocks for a large family of biologically important compounds, such as alkaloids, amino acids, and aza-sugars. In 2011, Jørgensen *et al.* [365] disclosed a novel activation mode provided by organocatalysts. Thus, these authors demonstrated that the merger of chiral secondary amines and polyenals generated reactive trienamine intermediates, which readily participated in stereoselective Diels–Alder reactions with different types of dienophiles. The novelty of this activation strategy lies in the perfect chirality relay over a distance of up to eight bonds. An extension of this discovery was the development of the first domino multicomponent reaction, which combined trienamine catalysis with enamine activation. Chiral diphenylprolinol trimethylsilyl ether was able to first activate 2,4-hexadienal through its corresponding trienamine, which reacted with 3-olefinic oxindole employed as dienophile through a Diels–Alder cycloaddition to afford the corresponding enamine intermediate. This enamine then reacted with double-activated ethyl 2-(diethoxyphosphoryl)acrylate employed as electrophilic partner to give a 85 : 15 diastereomeric mixture of products in 89% yield. The noncontrollable stereocenter was easily eliminated through a further acetal protection/Horner–Wadsworth–Emmons reaction sequence, forming highly functionalized cyclohexene without epimerization of the remaining stereocenters. This chiral product was produced in 62% yield for the two steps, with enantioselectivity of 98% ee and diastereoselectivity of 70% de through combined trienamine and enamine activation modes. It is noteworthy that a double conjugate addition sequence could also account for the formed products. The Povarov reaction, which is an inverse-electron-demand aza-Diels–Alder reaction of 2-azadienes with electron-rich olefins, allows a rapid construction of polysubstituted tetrahydroquinolines [366]. Although three-component versions involving aniline, aldehydes, and olefins have been reported, enantioselective Povarov reactions remain rare. The first highly enantioselective example of this type of reactions was developed by Zhu *et al.*, in 2009 [367]. Indeed, the reaction of a wide variety of aromatic and aliphatic aldehydes as well as anilines with benzyl-*N*-vinylcarbamates in the presence of a BINOL-derived phosphoric acid afforded the corresponding chiral (2,4-*cis*)-4-amino-2-aryl(alkyl)-tetrahydroquinolines in good yields (64–90%) and remarkable enantioselectivities (92 to >99% ee). In 2011, the same authors extended the scope of this powerful methodology to β-substituted acyclic enecarbamates, which provided, by reaction with aromatic as well as aliphatic aldehydes and a wide range of anilines bearing electron-donating and electron-withdrawing groups, the corresponding 1,2,3,4-tetrahydroquinolines with three contiguous stereogenic centers in good to excellent yields (48–97%) and excellent diastereo- and enantioselectivities of >95% de and 87 to >99% ee, respectively [368]. It must be noted that this methodology has allowed, for the fisrt time, aliphatic aldehydes to be employed in the enantioselective Povarov reaction. In 2010, Gong *et al.* [369] reported a three-component synthesis of the tricyclic ring system julolidine by combining the enantioselective Povarov reaction with an intramolecular hydroamination reaction. In the same year, Ricci *et al.* [370] employed another BINOL-derived phosphoric acid, (*S*)-TRIP, to induce a domino Povarov/Friedel–Crafts reaction. The process began with an enantioselective two-component reaction of *N*-arylimines with 2- or 3-vinylindoles, providing the corresponding Povarov intermediates, which were subsequently trapped with *p*-methoxy-phenylamine as a nucleophile to give the corresponding indole derivatives as mixtures of diastereomers in good yields, none to moderate diastereoselectivities, and good to excellent enantioselectivities (60–99% ee). In another area, a convenient enantioselective three-component reductive amination of ketones was developed by Benaglia *et al.* [371], starting from a mixture of a ketone, an aryl amine, and trichlorosilane upon catalysis with chiral picolinamides. For example, the reaction of acetophenone, 4-methoxyaniline, and trichlorosilane in the presence of a chiral picolinamide catalyst led to the corresponding chiral amine in 80% yield and enantioselectivity of 87% ee. An enantioselective synthesis of α-aminophosphonates by means of direct organocatalytic three-component hydrophosphonylation was successfully developed by JunAn *et al.*, in 2010 [372]. Indeed, upon catalysis with a chiral phosphoric acid, the reaction between an aromatic aldehyde or a cinnamaldehyde, 4-methoxyaniline, and diisopropylphosphate afforded the corresponding α-aminophosphonates in good yields (61–91%) and moderate to good enantioselectivities (31–87% ee). In 2007, List and Zhou [373] reported a highly enantioselective organocatalytic three-component process toward chiral cis-3-substituted cyclohexylamines through enamine, iminium, and Bronsted acid catalysis. A domino aldol/dehydration/Michael/reductive amination reaction occurred between 2,6-diketones, *p*-alkoxy aniline, and Hantzsch ester when treated by a catalytic amount of a BINOL-derived chiral phosphoric acid, affording a range of densely functionalized chiral cyclohexylamines in moderate to high yields (35–89%) and diastereoselectivities of 34–98% de, combined with good to high enantioselectivities (82–96% ee). A pseudo-three-component domino aza-ene type reaction/cyclization reaction was described by Terada *et al.*, in 2007 [374]. It enabled the rapid construction of enantioenriched piperidines in 68–99% yields by using chiral phosphoric acids derived from BINOL as catalysts. The potential of this process was highlighted through their ability to achieve a rapid increase in molecular complexity from simple enecarbamates and a broad range of aldimines while also controlling three stereogenic centers in a highly diastereo- and enantioselective manner (76–90% de, 97–99% ee). In 2010, Fréchet *et al.* [375] reported an enantioselective three-component domino aldol dehydration/Michael reaction performed in an aqueous buffer. This novel methodology involved two different aldehydes and nitromethane as the three substrates and employed two catalysts, such as L-proline (40 mol%) and chiral diphenylprolinol trimethylsilyl ether (1 mol%). In a two-phase system, aldehydes with a small size difference could be differentiated and reacted in a controlled manner to selectively form a single cross-product. Each catalyst mediated an individual reaction step in either the aqueous or organic phase. For example, the reaction between *n*-butyl alcohol and *n*-decanal with nitromethane in an aqueous buffer catalyzed by L-proline and chiral diphenylprolinol trimethylsilyl ether in the presence of oily droplets as the organic phase led selectively to a single domino product. Indeed, this major domino product arose from a two-step sequence beginning with the reaction of the relatively polar *n*-butyl alcohol with nitromethane catalyzed by L-proline in the aqueous phase to give the corresponding nitroalkene intermediate through aldolization followed by dehydration. This nitroalkene intermediate was then converted into the final product by Michael addition with *n*-decanal catalyzed by chiral diphenylprolinol trimethylsilyl ether in the organic phase. This methodology could be applied to various systems of aliphatic aldehydes, providing by reaction with nitromethane the corresponding domino products in good yields and excellent enantioselectivities of >90% ee, combined with good diastereoselectivities of 82–90% de. This remarkable system highlighted an often-ignored approach to develop chemoselective reactions by using properties other than chemical reactivity, such as polarity, inherent to the substrates or catalysts.

10.4 Conclusions

This chapter illustrated the power of enantioselective organocatalytic domino reactions, which have quickly become a powerful, fascinating, and highly efficient tool in organic chemistry. Indeed, these beautiful one-pot reactions can be considered as one of the most influential reaction classes of the last century. The combination of asymmetric organocatalysis with the concept of domino sequences has allowed reaching easily high molecular complexity with very often excellent levels of stereocontrol in environmentally friendly conditions, with simple operational procedures and advantages of savings in solvent, time, energy, and costs. With the various recently developed organocatalytic activation modes at hand, including the concept of bifunctional organocatalysts and multicatalytic systems [376], numerous novel domino processes can be envisaged with a tendency from simple domino reactions to triple and even quadruple domino reactions in cascade [377]. Major progress has been achieved in the last years through the creation of highly enantioselective versions of existing and novel organocatalytic domino and multicomponent reactions. For example, in the area of multicomponent reactions, the first catalytic enantioselective Passerini, Biginelli, Petasis, and Hantzsch multicomponent reactions were developed in recent years. Other improvements include significantly broadened scopes, novel techniques, and more environmentally benign methods in combination with entirely novel reactions. The most progressive change is, however, the shift from transition-metal catalysis toward organocatalysis. Transition-metal contamination of the products is thus avoided, which is of great importance for future applications in the pharmaceutical industry. These significant improvements are expected to influence the development of novel organocatalytic enantioselective reactions to a great extent and will make them even more useful tools for the total synthesis of natural and biologically important products. Undoubtedly, the future direction in this emerging field is to continue expanding the scope of organocatalytic domino reactions through the identification of novel modes of reactivity and apply these powerful, simple, and easy-to-perform strategies to the synthesis of biologically interesting molecules including natural products.

References

1. (a) Noyori, R. (1994) *Asymmetric Catalysts in Organic Synthesis*, John Wiley & Sons, Inc., New York. (b) Jacobsen, E.N., Pfaltz, A., and Yamamoto, H. (eds) (1999) *Comprehensive Asymmetric Catalysis*, Springer, Berlin. (c) Ojima, I. (ed.) (2000) *Catalytic Asymmetric Synthesis*, 2nd edn, Wiley-VCH Verlag GmbH, New York. (d) Beller, M. and Bolm, C. (eds) (2004) *Transition Metals for Organic Synthesis*, 2nd edn, Wiley-VCH Verlag GmbH, Weinheim. (e) Ramon, D.J. and Yus, M. (2006) *Chem. Rev.*, **106**, 2126–2208.
2. (a) Roberts, S.M. (ed.) (1999) *Biocatalysts for Fine Chemicals Synthesis*, Wiley-VCH Verlag GmbH, New York. (b) Drauz, K. and Waldmann, H. (eds) (2002) *Enzyme Catalysis in Organic Synthesis*, 2nd edn, Wiley-VCH Verlag GmbH, Weinheim. (c) Bommarius,

A.S. and Riebel, B.R. (2004) *Biocatalysis*, Wiley-VCH Verlag GmbH, Weinheim.

3. (a) Knoevenagel, E. (1896) *Chem. Ber.*, **29**, 172–174. (b) Eder, U., Sauer, G., and Wiechert, R. (1971) *Angew. Chem., Int. Ed. Engl.*, **10**, 496–497. (c) Hajos, Z.G. and Parrish, D.R. (1974) *J. Org. Chem.*, **39**, 1615–1621. (d) Ahrendt, K.A., Borths, C.J., and MacMillan, D.W.C. (2000) *J. Am. Chem. Soc.*, **122**, 4243–4244. (e) List, B., Lerner, R.A., and Barbas, C.F. (2000) *J. Am. Chem. Soc.*, **122**, 2395–2396.
4. (a) Dalko, P.I. and Moisan, L. (2001) *Angew. Chem. Int. Ed.*, **40**, 3726–3748. (b) Berkessel, A. and Gröger, H. (2005) *Asymmetric Organocatalysis–From Biomimetic Concepts to Powerful Methods for Asymmetric Synthesis*, Wiley-VCH Verlag GmbH, Weinheim. (c) Seayad, J. and List, B. (2005) *Org. Biomol. Chem.*, **3**, 719–724. (d) Taylor, M.S. and Jacobsen, E.N. (2006) *Angew. Chem. Int. Ed.*, **45**, 1520–1543. (e) Dalko, P.I. (2007) *Enantioselective Organocatalysis*, Wiley-VCH Verlag GmbH, Weinheim. (f) Dalko, P.I. (2007) *Chimia*, **61**, 213–218. (g) Pellissier, H. (2007) *Tetrahedron*, **63**, 9267–9331. (h) Doyle, A.G. and Jacobsen, E.N. (2007) *Chem. Rev.*, **107**, 5713–5743. (i) Gaunt, M.G., Johansson, C.C.C., McNally, A., and Vo, N.C. (2007) *Drug Discovery Today*, **2**, 8–27. (j) List, B. (ed.) (2007) *Chem. Rev.*, **107** (12), 5413–5883, Special Issue on Organocatalysis. (k) Enders, D., Grondal, C., and Hüttl, M.R.M. (2007) *Angew. Chem. Int. Ed.*, **46**, 1570–1581. (l) Guillena, G., Ramon, D.J., and Yus, M. (2007) *Tetrahedron: Asymmetry*, **18**, 693–700. (m) MacMillan, D.W.C. (2008) *Nature*, **455**, 304–308. (n) Yu, X. and Wang, W. (2008) *Chem. Asian. J.*, **3**, 516–532. (o) Dondoni, A. and Massi, A. (2008) *Angew. Chem. Int. Ed.*, **47**, 4638–4660. (p) Melchiorre, P., Marigo, M., Carlone, A., and Bartoli, G. (2008) *Angew. Chem. Int. Ed*, **47**, 6138–6171. (q) Peng, F. and Shao, Z. (2008) *J. Mol. Catal. A*, **285**, 1–13. (r) Barbas, C.F. (2008) *Angew. Chem. Int. Ed.*, **47**, 42–47. (s) Palomo, C., Oiarbide, M., and Lopez, R. (2009) *Chem. Soc. Rev.*, **38**, 632–653. (t) Bertelsen, S. and Jørgensen, K.A. (2009) *Chem. Soc. Rev.*, **38**, 2178–2189. (u) Alba, A.-N., Companyo, X., Viciano, M., and Rios, R. (2009) *Curr. Org. Chem.*, **13**, 1432–1474. (v) Bella, M. and Gasperi, T. (2009) *Synthesis*, **10**, 1583–1614. (w) Gong, L. (2010) Special topic: asymmetric organocatalysis. *Chin. Sci. Bull.*, **55** (17), 1729. (x) List, B. (2010) Asymmetric organocatalysis. *Top. Curr. Chem.*, **291**, 395–456. (y) Pellissier, H. (2010) *Recent Developments in Asymmetric Organocatalysis*, Royal Society of Chemistry, Cambridge. (z) Mahrwald, R. (2011) *Enantioselective Organocatalysed Reactions*, Vol. 1 and 2, Springer, Berlin. (aa) Ruiz, M., Lopez-Alvarado, P., Giorgi, G., and Menéndez, J.C. (2011) *Chem. Soc. Rev.*, **40**, 3445–3454. (ab) Giacalone, F., Gruttadauria, M., Agrigento, P., and Noto, R. (2012) *Chem. Soc. Rev.*, **41**, 2406–2447.
5. (a) de Figueiredo, R.M. and Christmann, M. (2007) *Eur. J. Org. Chem.*, **16**, 2575–2600. (b) Marquès-Lopez, E. and Herrera, R.P. (2010) *Nat. Prod. Rep.*, **27**, 1138–1167.
6. (a) Posner, G.H. (1986) *Chem. Rev.*, **86**, 831–844. (b) Ho, T.-L. (1992) *Tandem Organic Reactions*, John Wiley & Sons, Inc., New York. (c) Waldmann, H. (1992) *Nachr. Chem. Tech. Lab.*, **40**, 1133–1140. (d) Fukumoto, K. (1994) *Synth. Org. Chem. Jpn.*, **52**, 2–18. (e) Bunce, R.A. (1995) *Tetrahedron*, **51**, 13103–13159. (f) Parsons, P.J., Penkett, C.S., and Shell, A.J. (1996) *Chem. Rev.*, **96**, 195–206. (g) Balaure, P.C.F. and Filip, P.I.A. (2002) *Rev. Roum. Chim.*, **46** (8), 809–833. (h) Capdevila, E., Rayo, J., Carrion, F., Jové, I., Borrell, J.I., and Teixido, J. (2003) *Afinidad*, **506**, 317–337. (i) Tietze, L.F. and Rackelmann, N. (2004) *Pure Appl. Chem.*, **76**, 1967–1983. (j) Dalko, P.I. and Moisan, L. (2004) *Ang. Chem., Int. Ed. Engl.*, **43**, 5138–5175. (k) Fogg, D.E. and dos Santos, E.N. (2004) *Coord. Chem. Rev.*, **248**, 2365–2379. (l) Ramon, D.J. and Yus, M. (2005) *Ang. Chem., Int. Ed. Engl.*, **44**, 1602–1634. (m) Pellissier, H. (2006) *Tetrahedron*,

62, 2143–2173. (n) Pellissier, H. (2006) *Tetrahedron*, **62**, 1619–1665. (o) Chapman, C.J. and Frost, C.G. (2007) *Synthesis*, **1**, 1–21.

7. Tietze, L.F., Brasche, G., and Gericke, K. (2006) *Domino Reactions in Organic Synthesis*, Wiley-VCH Verlag GmbH, Weinheim.
8. (a) Tietze, L.F. and Modi, A. (2000) *Med. Res. Rev.*, **20**, 304–322. (b) Zhu, J. and Bienaymé, H. (eds) (2005) *Multicomponent Reactions*, Wiley-VCH Verlag GmbH, Weinheim. (c) Orru, R.V.A. and Ruijter, E. (eds) (2010) *Synthesis of Heterocycles Via Multicomponent Reactions*, Topics in Heterocyclic Chemistry, Vol. 1 and 2, Springer, Berlin.
9. (a) Tietze, L.F. and Beifuss, U. (1993) *Ang. Chem., Int. Ed. Engl.*, **32**, 131–163. (b) Tietze, L.F. (1996) *Chem. Rev.*, **96**, 115–136.
10. (a) List, B. (2006) *Chem. Commun.*, 819–824. (b) Erkkilä, A., Majander, I., and Pihko, P.M. (2007) *Chem. Rev.*, **107**, 5416–5470. (c) Mukherjee, S., Yang, J.W., Hoffmann, S., and List, B. (2007) *Chem. Rev.*, **107**, 5471–5569. (d) Yu, X. and Wang, W. (2008) *Org. Biomol. Chem.*, **6**, 2037–2046. (e) Palomo, C. and Mielgo, A. (2006) *Angew. Chem. Int. Ed.*, **45**, 7876–7880.
11. Bui, T. and Barbas, C.F. (2000) *Tetrahedron Lett.*, **41**, 6951–6954.
12. Lelais, G. and MacMillan, D.W.C. (2006) *Aldrichim. Acta*, **39**, 79–87.
13. Wolgang, N., Tanaka, F., and Barbas, C.F. (2004) *Acc. Chem. Res.*, **37** (8), 580–591.
14. Komnenos, T. (1883) *Justus Liebigs An. Chem.*, **218**, 145–169.
15. Perlmutter, P. (1992) *Conjugate Addition Reactions in Organic Synthesis*, Pergamon Press, Oxford.
16. (a) Guo, H.-C. and Ma, J.-A. (2006) *Angew. Chem. Int. Ed.*, **45**, 354–366. (b) Almasi, D., Alonso, D.A., and Najera, C. (2007) *Tetrahedron: Asymmetry*, **18**, 299–365. (c) Vicario, J.L., Badia, D., and Carrillo, L. (2007) *Synthesis*, **14**, 2065–2092. (d) Tsogoeva, S.B. (2007) *Eur. J. Org. Chem.*, **2007**, 1701–1716. (e) Tietze, L.F. and Düfert, A. (2010) in: *Catalytic Asymmetric Conjugate Reactions* (ed. A. Cordova), Wiley-VCH Verlag GmbH, Weinheim, p. 321.
17. (a) Halland, N., Aburel, P.S., and Jørgensen, K.A. (2004) *Angew. Chem. Int. Ed.*, **43**, 1272–1277. (b) Pulkkinen, J., Aburel, P.S., and Jørgensen, K.A. (2004) *Adv. Synth. Catal.*, **346**, 1077–1080.
18. Gryko, D. (2005) *Tetrahedron: Asymmetry*, **16**, 1377–1383.
19. Yang, J.W., Hechavarria Fonsecca, M.T., and List, B. (2005) *J. Am. Chem. Soc.*, **127**, 15036–15037.
20. Huang, Y., Walji, A.M., Larsen, C.H., and MacMillan, D.W.C. (2005) *J. Am. Chem. Soc.*, **127**, 15051–15053.
21. Zu, L., Li, H., Xie, H., Wang, J., Jiang, W., Tang, Y., and Wang, W. (2007) *Angew. Chem. Int. Ed.*, **46**, 3732–3734.
22. Zhao, G.-L., Ibrahem, I., Dziedzic, P., Sun, J., Bonneau, C., and Cordova, A. (2008) *Chem. Eur. J.*, **14**, 10007–10011.
23. McGarraugh, P.G. and Brenner, S.E. (2009) *Org. Lett.*, **11**, 5654–5657.
24. Ma, A. and Ma, D. (2010) *Org. Lett.*, **12**, 3634–3637.
25. Enders, D., Lüttgen, K., and Narine, A.A. (2007) *Synthesis*, 959–980.
26. Enders, D., Saint-Dizier, A., Lannou, M.I., and Lenzen, A. (2006) *Eur. J. Org. Chem.*, **2006**, 29–49.
27. (a) Nising, C.F. and Bräse, S. (2008) *Chem. Soc. Rev.*, **37**, 1218–1228. (b) Nising, C.F. and Bräse, S. (2012) *Chem. Soc. Rev.*, **41**, 988–999.
28. Li, H., Zu, L., Xie, H., Wang, J., Jiang, W., and Wang, W. (2007) *Org. Lett.*, **9**, 1833–1835.
29. Hong, B.-C., Kotame, P., Tsai, C.-W., and Liao, J.-H. (2010) *Org. Lett.*, **12**, 776–779.
30. Zhang, X., Zhang, S., and Wang, W. (2010) *Angew. Chem. Int. Ed.*, **49**, 1481–1484.
31. Hong, B.-C., Nimje, R.Y., and Liao, J.-H. (2009) *Org. Biomol. Chem.*, **7**, 3095–3101.
32. Hong, B.-C., Hsu, C.-S., and Lee, G.-H. (2012) *Chem. Commun.*, **48**, 2385–2387.
33. Hong, B.-C., Sadani, A.A., Nimje, R.Y., Dange, N.S., and Lee, G.-H. (2011) *Synthesis*, **12**, 1887–1895.
34. Xie, J.-W., Chen, W., Li, R., Zeng, M., Du, W., Yue, L., Chen, Y.-C., Wu, Y.,

Zhu, J., and Deng, J.-G. (2007) *Angew. Chem. Int. Ed.*, **46**, 389–392.

35. (a) Schreiner, P.R. (2003) *Chem. Soc. Rev.*, **32**, 289–296. (b) Pihko, P.M. (2004) *Angew. Chem. Int. Ed.*, **43**, 2062–2064. (c) Connon, S.J. (2006) *Angew. Chem. Int. Ed.*, **45**, 3909–3912. (d) Akiyama, T., Itoh, J., and Fuchibe, K. (2006) *Adv. Synth. Catal.*, **348**, 999–1010. (e) Akiyama, T. (2007) *Chem. Rev.*, **107**, 5744–5748.

36. Wei, Q. and Gong, L.-Z. (2010) *Org. Lett.*, **12**, 1008–1011.

37. Ren, Q., Gao, Y., and Wang, J. (2010) *Chem. Eur. J.*, **16**, 13594–13598.

38. Grossmann, A. and Enders, D. (2011) *Angew. Chem. Int. Ed.*, **50**, 2–14.

39. Enders, D., Niemeier, O., and Henseler, A. (2007) *Chem. Rev.*, **107**, 5606–5655.

40. Fang, X., Jiang, K., Xing, C., Hao, L., and Chi, Y.R. (2011) *Angew. Chem. Int. Ed.*, **50**, 1910–1913.

41. (a) Roca-Lopez, D., Sadaba, D., Delso, I., Herrera, R.P., Tejero, T., and Merino, P. (2010) *Tetrahedron: Asymmetry*, **21**, 2561–2601. (b) Vicario, J.L., Badia, D., Carrillo, L., and Reyes, E. (eds) (2010) *Organocatalytic Enantioselective Conjugate Addition Reactions: A Powerful Tool for the Stereocontrolled Synthesis of Complex Molecules*, Royal Society of Chemistry, Cambridge.

42. Okino, T., Hoashi, Y., and Takemoto, Y. (2003) *J. Am. Chem. Soc.*, **125**, 12672–12673.

43. Tan, B., Shi, Z., Chua, P., and Zhong, G. (2008) *Org. Lett.*, **10**, 3425–3428.

44. Wang, J., Xie, H., Li, H., Zu, L., and Wang, W. (2008) *Angew. Chem. Int. Ed.*, **47**, 4177–4179.

45. (a) Pellissier, H. (2003) *Tetrahedron*, **59**, 8291–8327. (b) Pellissier, H. (2008) *Tetrahedron*, **64**, 1563–1601. (c) Pellissier, H. (2011) *Tetrahedron*, **67**, 3769–3802. (d) Pellissier, H. (2011) *Adv. Synth. Catal.*, **353**, 659–676. (e) Pellissier, H. (2011) *Chirality from Dynamic Kinetic Resolution*, Royal Society of Chemistry, Cambridge.

46. Wang, X.-F., Hua, Q.-L., Cheng, Y., An, X.-L., Yang, Q.-Q., Chen, J.-R., and Xiao, W.-J. (2010) *Angew. Chem. Int. Ed.*, **49**, 8379–8381.

47. Hua, Q.-L., Li, C., Wang, X.-F., Lu, L.-Q., Chen, J.-R., and Xiao, W.-J. (2011) *ACS Catal.*, **1**, 221–226.

48. Wang, X.-F., An, J., Zhang, X.-X., Tan, F., Chen, J.-R., and Xiao, W.-J. (2011) *Org. Lett.*, **13**, 808–811.

49. Yu, C., Zhang, Y., Song, A., Ji, Y., and Wang, W. (2011) *Chem. Eur. J.*, **17**, 770–774.

50. Rajkumar, S., Shankland, K., Brown, G.D., and Cobb, A.J.A. (2012) *Chem. Sci.*, **3**, 584–588.

51. Wu, L.-Y., Bencivenni, G., Mancinelli, M., Mazzanti, A., Bartoli, G., and Melchiorre, P. (2009) *Angew. Chem. Int. Ed.*, **48**, 7196–7199.

52. Sunden, H., Rios, R., Xu, Y., Eriksson, L., and Cordova, A. (2007) *Adv. Synth. Catal.*, **349**, 2549–2555.

53. Zu, L., Zhang, S., Xie, H., and Wang, W. (2009) *Org. Lett.*, **11**, 1627–1630.

54. Wang, W., Li, H., Wang, J., and Zu, L. (2006) *J. Am. Chem. Soc.*, **128**, 10354–10355.

55. Rios, R., Sunden, H., Ibrahem, I., Zhao, G.-L., Eriksson, L., and Cordova, A. (2006) *Tetrahedron Lett.*, **47**, 8547–8551.

56. Sunden, H., Ibrahem, I., Zhao, G.-L., Eriksson, L., and Cordova, A. (2007) *Chem. Eur. J.*, **13**, 574–581.

57. Li, H., Wang, J., Nunu, T.E., Zu, L., Jiang, W., Wie, S., and Wang, W. (2007) *Chem. Commun.*, 507–509.

58. Luo, S.-P., Li, Z.-B., Wang, L.-P., Guo, Y., Xia, A.-B., and Xu, D.-Q. (2009) *Org. Biomol. Chem.*, **7**, 4539–4546.

59. Shen, H., Yang, K.-F., Shi, Z.-H., Jiang, J.-X., Lai, G.-Q., and Xu, L.-W. (2011) *Eur. J. Org. Chem.*, **2011**, 5031–5038.

60. Zhao, G.-L., Vesely, J., Rios, R., Ibrahem, I., Sunden, H., and Cordova, A. (2008) *Adv. Synth. Catal.*, **350**, 237–242.

61. Sunden, H., Rios, R., Ibrahem, I., Zhao, G.-L., Eriksson, L., and Cordova, A. (2007) *Adv. Synth. Catal.*, **349**, 827–832.

62. Li, H., Wang, J., Xie, H., Zu, L., Jiang, W., Duesler, E.N., and Wang, W. (2007) *Org. Lett.*, **9**, 965–968.

63. Yoshitomi, Y., Arai, H., Makino, K., and Hamada, Y. (2008) *Tetrahedron*, **64**, 11568–11579.

64. Liu, C., Zhang, X., Wang, R., and Wang, W. (2010) *Org. Lett.*, **12**, 4948–4951.

65. Brandau, S., Maerten, E., and Jørgensen, K.A. (2006) *J. Am. Chem. Soc.*, **128**, 14986–14991.

66. Wang, J., Li, H., Xie, H., Zu, L., Shen, X., and Wang, W. (2007) *Angew. Chem. Int. Ed.*, **46**, 9050–9053.

67. Hong, B.-C., Nimje, R.Y., Sadani, A.A., and Liao, J.-H. (2008) *Org. Lett.*, **10**, 2345–2348.

68. Rueping, M., Kuenkel, A., Tato, F., and Bats, J.W. (2009) *Angew. Chem. Int. Ed.*, **48**, 3699–3702.

69. Reyes, E., Talavera, G., Vicario, J.L., Badia, D., and Carrillo, L. (2009) *Angew. Chem. Int. Ed.*, **48**, 5701–5704.

70. Enders, D., Wang, C., and Raabe, G. (2009) *Synthesis*, **24**, 4119–4124.

71. Hong, L., Sun, W., Liu, C., Wang, L., and Wang, R. (2010) *Chem. Eur. J.*, **16**, 440–444.

72. Luo, G., Zhang, S., Duan, W., and Wang, W. (2009) *Tetrahedron Lett.*, **50**, 2946–2948.

73. Tang, J., Xu, D.Q., Xia, A.B., Wang, Y.F., Jiang, J.R., Luo, S.P., and Xu, Z.Y. (2010) *Adv. Synth. Catal.*, **352**, 2121–2126.

74. Bae, J.-Y., Lee, H.-J., Youn, S.-H., Kwon, S.-H., and Cho, C.-W. (2010) *Org. Lett.*, **12**, 4352–4355.

75. Li, W., Li, X., Wu, W., Liang, X., and Ye, J. (2011) *Chem. Commun.*, **47**, 8325–8327.

76. (a) Enders, D., Wang, C., and Bats, J.W. (2009) *Synlett*, **11**, 1777–1780. (b) Enders, D., Narine, A.A., Benninghaus, T.R., and Raabe, G. (2007) *Synlett*, **11**, 1667–1670.

77. (a) Hong, B.-C., Dange, N.S., Hsu, C.-S., and Miao, J.-H. (2010) *Org. Lett.*, **12**, 4812–4815. (b) Hong, B.-C., Dange, N.S., Hsu, C.-S., Liao, J.-H., and Lee, G.-H. (2011) *Org. Lett.*, **13**, 1338–1341.

78. (a) Rios, R., Sunden, H., Ibrahem, I., Zhao, G.-L., and Cordova, A. (2006) *Tetrahedron Lett.*, **47**, 8679–8682. (b) Rios, R., Sunden, H., Ibrahem, I., and Cordova, A. (2007) *Tetrahedron Lett.*, **48**, 2181–2184.

79. Yuang, Y.-Q., Chai, Z., Wang, H.-F., Chen, X.-K., Cui, H.-F., Zheng, C.-W., Xiao, H., Li, P., and Zhao, G. (2009) *Chem. Eur. J.*, **15**, 13295–13298.

80. Cui, H.-F., Yang, Y.-Q., Chai, Z., Li, P., Zheng, C.-W., Zhu, S.-Z., and Zhao, G. (2010) *J. Org. Chem.*, **75**, 117–122.

81. Wang, L.-L., Peng, L., Bai, J.-F., Huang, Q.-C., Xu, X.-Y., and Wang, L.-X. (2010) *Chem. Commun.*, **46**, 8064–8066.

82. Zu, L., Wang, J., Li, H., Xie, H., Jiang, W., and Wang, W. (2007) *J. Am. Chem. Soc.*, **129**, 1036–1037.

83. Zu, L., Xie, H., Li, H., Wang, J., Jiang, W., and Wang, W. (2007) *Adv. Synth. Catal.*, **349**, 1882–1886.

84. Gao, Y., Ren, Q., Wu, H., Li, M., and Wang, J. (2010) *Chem. Commun.*, **46**, 9232–9234.

85. Tan, B., Candeias, N.R., and Barbas, C.F. (2011) *Nat. Chem.*, **3**, 473–477.

86. Dong, X.-Q., Fang, X., Tao, H.-Y., Zhou, X., and Wang, C.-J. (2012) *Chem. Commun.*, **48**, 7238–7240.

87. Cao, C.-L., Sun, X.-L., Kang, Y.-B., and Tang, Y. (2007) *Org. Lett.*, **9**, 4151–4154.

88. (a) Zhao, G.-L., Dziedzic, P., Ullah, F., Eriksson, L., and Cordova, A. (2009) *Tetrahedron Lett.*, **50**, 3458–3462. (b) Hayashi, Y., Okano, T., Aratake, S., and Hazelard, D. (2007) *Angew. Chem. Int. Ed.*, **46**, 4922–4925.

89. Mariago, M., Franzen, J., Toulsen, T.B., Zhuang, W., and Jørgensen, K.A. (2005) *J. Am. Chem. Soc.*, **127**, 6964–6965.

90. Sunden, H., Ibrahem, I., and Cordova, A. (2006) *Tetrahedron Lett.*, **47**, 99–103.

91. Zhao, G.-L. and Cordova, A. (2007) *Tetrahedron Lett.*, **48**, 5976–5980.

92. (a) Wang, X. and List, B. (2008) *Angew. Chem. Int. Ed.*, **47**, 1119–1122. (b) Wang, X., Reisinger, C.M., and List, B. (2008) *J. Am. Chem. Soc.*, **130**, 6070–6071.

93. Vesely, J., Ibrahem, I., Zhao, G.-L., Rios, R., and Cordova, A. (2007) *Angew. Chem. Int. Ed.*, **46**, 778–781.

94. Arai, H., Sugaya, N., Sasaki, N., Makino, K., Lectard, S., and Hamada, Y. (2009) *Tetrahedron Lett.*, **50**, 3329–3332.

95. Pesciaioli, F., De Vincentiis, F., Galzerano, P., Bencivenni, G., Bartoli, G., Mazzanti, A., and Melchiorre,

P. (2008) *Angew. Chem. Int. Ed.*, **47**, 8703–8706.
96. (a) Ibrahem, I., Rios, R., Vesely, J., Zhao, G.-L., and Cordova, A. (2007) *Angew. Chem. Int. Ed.*, **48**, 849–851. (b) Ibrahem, I., Rios, R., Vesely, J., Zhao, G.-L., and Cordova, A. (2007) *Synthesis*, **7**, 1153–1157.
97. Rios, R., Ibrahem, I., Vesely, J., Sunden, H., and Cordova, A. (2007) *Tetrahedron Lett.*, **48**, 8695–8699.
98. Franke, P.T., Richter, B., and Jørgensen, K.A. (2008) *Chem. Eur. J.*, **14**, 6317–6321.
99. Rueping, M., Sugiono, E., and Merino, E. (2008) *Chem. Eur. J.*, **14**, 6329–6332.
100. Rueping, M., Sugiono, E., and Merino, E. (2008) *Angew. Chem. Int. Ed.*, **47**, 3046–3049.
101. (a) Franzen, J. and Fisher, A. (2009) *Angew. Chem. Int. Ed.*, **48**, 787–791. (b) Zhang, W. and Franzen, J. (2010) *Adv. Synth. Catal.*, **352**, 499–518.
102. (a) Valero, G., Schimer, J., Cisarova, I., Vesely, J., Moyano, A., and Rios, R. (2009) *Tetrahedron Lett.*, **50**, 1943–1946. (b) Cihalova, S., Valero, G., Schimer, J., Humpl, M., Dracinsky, M., Moyano, A., Rios, R., and Vesely, J. (2011) *Tetrahedron*, **67**, 8942–8950.
103. He, Z.-Q., Zhou, Q., Wu, L., and Chen, Y.-C. (2010) *Adv. Synth. Catal.*, **352**, 1904–1908.
104. Dai, X., Wu, X., Fang, H., Nie, L., Chen, J., Deng, H., Cao, W., and Zhao, G. (2011) *Tetrahedron*, **67**, 3034–3040.
105. Wu, X., Dai, X., Nie, L., Fang, H., Chen, J., Ren, Z., Cao, W., and Zhao, G. (2010) *Chem. Commun.*, **46**, 2733–2735.
106. Knowles, R.R., Carpenter, J., Blakey, S.B., Kayano, A., Mangion, I.K., Sinz, C.J., and MacMillan, D.W.C. (2011) *Chem. Sci.*, **2**, 308–311.
107. Noole, A., Borissova, M., Lopp, M., and Kanger, T. (2011) *J. Org. Chem.*, **76**, 1538–1545.
108. Maltsev, O.V., Kucherenko, A.S., Chimishkyan, A.L., and Zlotin, S.G. (2010) *Tetrahedron: Asymmetry*, **21**, 2659–2670.
109. Lee, Y., Seo, S.W., and Kim, S.-G. (2011) *Adv. Synth. Catal.*, **353**, 2671–2675.
110. Fernandez, M., Reyes, E., Vicario, J.L., Badia, D., and Carrillo, L. (2012) *Adv. Synth. Catal.*, **354**, 371–376.
111. Ramachary, D.B. and Reddy, Y.V. (2012) *Eur. J. Org. Chem.*, **5**, 865–887.
112. Enders, D., Yang, X., Wang, C., Raabe, G., and Runsik, J. (2011) *Chem. Asian J.*, **6**, 2255–2259.
113. (a) Tozawa, T., Nagao, H., Yamane, Y., and Mukaiyama, T. (2007) *Chem. Asian J.*, **2**, 123–134. (b) Tozawa, T., Yamane, Y., and Mukaiyama, T. (2006) *Chem. Lett.*, **35**, 56–57.
114. Xie, J.-W., Huang, X., Fan, L.-P., Xu, D.-C., Li, X.-S., Su, H., and Wen, Y.-H. (2009) *Adv. Synth. Catal.*, **351**, 3077–3082.
115. Zhao, S.-L., Zheng, C.-W., Wang, H.-F., and Zhao, G. (2009) *Adv. Synth. Catal.*, **351**, 2811–2816.
116. Zhao, S.-L., Zheng, C.-W., and Zhao, G. (2009) *Tetrahedron: Asymmetry*, **20**, 1046–1051.
117. Calter, M.A. and Wang, J. (2009) *Org. Lett.*, **11**, 2205–2208.
118. Yao, W., Pan, L., Wu, Y., and Ma, C. (2010) *Org. Lett.*, **12**, 2422–2425.
119. Mahè, O., Dez, I., Levacher, V., and Brière, J.-F. (2010) *Angew. Chem. Int. Ed.*, **49**, 7072–7075.
120. Matoba, K., Kawai, H., Furukawa, T., Kusuda, A., Tokunaga, E., Nakamura, S., Shiro, M., and Shibata, N. (2010) *Angew. Chem. Int. Ed.*, **49**, 5762–5766.
121. Gao, Y., Ren, Q., Siau, W.-Y., and Wang, J. (2011) *Chem. Commun.*, **47**, 5819–5821.
122. Hu, Z.-P., Lou, C.-L., Wang, J.-J., Chen, C.-X., and Yan, M. (2011) *J. Org. Chem.*, **76**, 3797–3804.
123. Li, X.-M., Wang, B., Zhang, J.-M., and Yan, M. (2011) *Org. Lett.*, **13**, 374–377.
124. Wang, J.-J., Hu, Z.-P., Lou, C.-L., Liu, J.-L., Li, X.-M., and Yan, M. (2011) *Tetrahedron*, **67**, 4578–4583.
125. Zhang, G., Zhang, Y., Jiang, X., Yan, W., and Wang, R. (2011) *Org. Lett.*, **13**, 3806–3809.
126. Chandrasekhar, S., Mallikarjun, K., Pavankumarreddy, G., Rao, K.V., and Jagadeesh, B. (2009) *Chem. Commun.*, 4985–4987.
127. Lu, D., Li, Y., and Gong, Y. (2010) *J. Org. Chem.*, **75**, 6900–6907.

128. Hong, B.-C., Kotame, P., and Liao, J.-H. (2011) *Org. Biomol. Chem.*, **9**, 382–386.

129. Ramachary, D.B., Prasad, M.S., and Madhavachary, R. (2011) *Org. Biomol. Chem.*, **9**, 2715–2721.

130. Enders, D., Urbanietz, G., and Raabe, G. (2011) *Synthesis*, **2**, 1905–1911.

131. Enders, D., Urbanietz, G., Hahn, R., and Raabe, G. (2012) *Synthesis*, **44**, 773–782.

132. Hong, B.-C., Kotame, P., and Lee, G.-H. (2011) *Org. Lett.*, **13**, 5758–5761.

133. Wang, Y., Zhu, S., and Ma, D. (2011) *Org. Lett.*, **13**, 1602–1605.

134. Enders, D., Wang, C., Yang, X., and Raabe, G. (2011) *Synlett*, **4**, 469–472.

135. (a) Xie, H., Zu, L., Li, H., Wang, J., and Wang, W. (2007) *J. Am. Chem. Soc.*, **129**, 10886–10894. (b) Rios, R., Sunden, H., Vesely, J., Zhao, G.-L., Dziedzic, P., and Cordova, A. (2007) *Adv. Synth. Catal.*, **349**, 1028–1032.

136. Companyo, X., Alba, A.-N., Cardenas, F., Moyano, A., and Rios, R. (2009) *Eur. J. Org. Chem.*, **2009**, 3075–3080.

137. Vesely, J., Zhao, G.-L., Bartoszewicz, A., and Cordova, A. (2008) *Tetrahedron Lett.*, **49**, 4209–4212.

138. Rios, R., Vesely, J., Sunden, H., Ibrahem, I., and Cordova, G.-L. (2007) *Tetrahedron Lett.*, **48**, 5835–5839.

139. Terrasson, V., van der Lee, A., de Figueiredo, R.M., and Campagne, J.M. (2010) *Chem. Eur. J.*, **16**, 7875–7880.

140. Kang, Y.K., Kim, S.M., and Kim, D.Y. (2010) *J. Am. Chem. Soc.*, **132**, 11847–11849.

141. Hansen, H.M., Longbottom, D.A., and Ley, S.V. (2006) *Chem. Commun.*, 4838–4840.

142. Waschdowski, V., Hansen, H.M., Longbottom, D.A., and Ley, S.V. (2008) *Synthesis*, **8**, 1269–1275.

143. Lv, J., Zhang, J., Lin, Z., and Wang, Y. (2009) *Chem. Eur. J.*, **15**, 972–979.

144. Russo, A., Meninno, S., Tedesco, C., and Lattanzi, A. (2011) *Eur. J. Org. Chem.*, **2011**, 5096–5103.

145. Pesciaioli, F., Righi, P., Mazzanti, A., Bartoli, G., and Bencivenni, G. (2011) *Chem. Eur. J.*, **17**, 2842–2845.

146. Johansson, C.C.C., Bremeyer, N., Ley, S.V., Owen, D.R., Smith, S.C., and Gaunt, M.J. (2006) *Angew. Chem. Int. Ed.*, **45**, 6024–6028.

147. Xuan, Y.-N., Nie, S.-Z., Dong, L.-T., Zhang, J.-M., and Yan, M. (2009) *Org. Lett.*, **11**, 1583–1586.

148. Guo, C., Xue, M.-X., Zhu, M.-K., and Gong, L.-Z. (2008) *Angew. Chem. Int. Ed.*, **47**, 3414–3417.

149. Fan, L.-P., Li, P., Li, X.-S., Xu, D.-C., Ge, M.-M., Zhu, W.-D., and Xie, J.-W. (2010) *J. Org. Chem.*, **75**, 8716–8719.

150. Rueping, M., Parra, A., Uria, U., Besselièvre, F., and Merino, E. (2010) *Org. Lett.*, **12**, 5680–5683.

151. Dou, X., Han, X., and Lu, Y. (2012) *Chem. Eur. J.*, **18**, 85–89.

152. (a) Boruwa, J., Gogoi, N., Saikia, P.P., and Barua, N.C. (2006) *Tetrahedron: Asymmetry*, **17**, 3315–3326. (b) Palomo, C., Oiarbide, M., and Laso, A. (2007) *Eur. J. Org. Chem.*, **2007**, 2561–2574. (c) Marquès-Lopez, E., Merino, P., Tejero, T., and Herrera, R.P. (2009) *Eur. J. Org. Chem.*, **2009**, 2401–2420. (d) Alvarez-Casao, Y., Marques-Lopez, E., and Herrera, R.P. (2011) *Symmetry*, **3**, 220–245.

153. Reyes, E., Jiang, H., Milelli, A., Elsner, P., Hazell, R.G., and Jørgensen, K.A. (2007) *Angew. Chem. Int. Ed.*, **46**, 9202–9205.

154. Garcia Ruano, J.L., Marcos, V., Suanzes, J.A., Marzo, L., and Aleman, J. (2009) *Chem. Eur. J.*, **15**, 6576–6580.

155. Liu, X. and Lu, Y. (2010) *Org. Biomol. Chem.*, **8**, 4063–4065.

156. Jia, Z.-X., Luo, Y.-C., and Xu, P.-F. (2011) *Org. Lett.*, **13**, 832–835.

157. Xie, J., Yoshida, K., Takasu, K., and Takemoto, Y. (2008) *Tetrahedron Lett.*, **49**, 6910–6913.

158. Xu, D.-Q., Wang, Y.-F., Luo, S.-P., Zhang, S., Zhong, A.-G., Chen, H., and Xu, Z.-Y. (2008) *Adv. Synth. Catal.*, **350**, 2610–2616.

159. Dodda, R., Goldman, J.J., Mandal, T., Zhao, C.-G., Broker, G.A., and Tiekink, E.R.T. (2008) *Adv. Synth. Catal.*, **350**, 537–541.

160. Zhang, W., Luo, S.-P., Li, B.-L., Xia, A.-B., Zhong, A.-G., and Xu, D.-Q. (2009) *Chem. Asian J.*, **4**, 1834–1838.

161. Tan, B., Chua, P.J., Zeng, X., Lu, M., and Zhong, G. (2008) *Org. Lett.*, **10**, 3489–3492.

162. Tan, B., Chua, P.J., Li, Y., and Zhong, G. (2008) *Org. Lett.*, **10**, 2437–2440.

163. Ding, D., Zhao, C.-G., Guo, Q., and Arman, H. (2010) *Tetrahedron*, **66**, 4423–4427.

164. Rueping, M., Kuenkel, A., and Fröhlich, R. (2010) *Chem. Eur. J.*, **16**, 4173–4176.

165. Tan, B., Lu, Y., Zeng, X., Chua, P.J., and Zhong, G. (2010) *Org. Lett.*, **12**, 2682–2685.

166. Chintala, P., Gosh, S.K., Long, E., Headley, A.D., and Ni, B. (2011) *Adv. Synth. Catal.*, **353**, 2905–2909.

167. (a) Carlone, A., Marigo, M., North, C., Landa, A., and Jørgensen, K.A. (2006) *Chem. Commun.*, 4928–4930. (b) Brandau, S., Landa, A., Franzen, J., Marigo, M., and Jørgensen, K.A. (2006) *Angew. Chem. Int. Ed.*, **45**, 4305–4309.

168. Marigo, M., Bertelsen, S., Landa, A., and Jørgensen, K.A. (2006) *J. Am. Chem. Soc.*, **128**, 5475–5479.

169. Hayashi, Y., Toyoshima, M., Gotoh, H., and Ishikawa, H. (2009) *Org. Lett.*, **11**, 45–48.

170. Alza, E., Sayalero, S., Cambeiro, X.C., Martin-Rapun, R., Miranda, P.O., and Pericas, M.A. (2011) *Synlett*, **4**, 464–468.

171. Albrecht, L., Richter, B., Vila, C., Krawczyk, H., and Jørgensen, K.A. (2009) *Chem. Eur. J.*, **15**, 3093–3102.

172. Hong, B.-C., Wu, M.-F., Tseng, H.-C., and Liao, J.-H. (2006) *Org. Lett.*, **8**, 2217–2220.

173. Cabrera, S., Aleman, J., Bolze, P., Bertelsen, S., and Jørgensen, K.A. (2008) *Angew. Chem. Int. Ed.*, **47**, 121–125.

174. Aleman, J., Nunez, A., Marzo, L., Marcos, V., Alvarado, C., and Garcia Ruano, J.L. (2010) *Chem. Eur. J.*, **16**, 9453–9456.

175. Vicario, J.L., Reboredo, S., Badia, D., and Carrillo, L. (2007) *Angew. Chem. Int. Ed.*, **46**, 5168–5170.

176. Xia, A.-B., Xu, D.-Q., Luo, S.-P., Jiang, J.-R., Tang, J., Wang, Y.-F., and Xu, Z.-Y. (2010) *Chem. Eur. J.*, **16**, 801–804.

177. Li, H., Zhao, J., Zeng, L., and Hu, W. (2011) *J. Org. Chem.*, **76**, 8064–8069.

178. Cai, Q., Zheng, C., Zhang, J.-W., and You, S.-L. (2011) *Angew. Chem. Int. Ed.*, **50**, 8665–8669.

179. Tan, B., Hernandez-Torres, G., and Barbas, C.F. (2011) *J. Am. Chem. Soc.*, **133**, 12354–12357.

180. Zhao, G.-L., Rios, R., Vesely, J., Eriksson, L., and Cordova, A. (2008) *Angew. Chem. Int. Ed.*, **47**, 8468–8472.

181. Liu, Y.-K., Ma, C., Jiang, K., Liu, T.-Y., and Chen, Y.-C. (2009) *Org. Lett.*, **11**, 2848–2851.

182. Gogoi, S. and Zhao, C.-G. (2009) *Tetrahedron Lett.*, **50**, 2252–2255.

183. Cao, C.-L., Zhou, Y.-Y., Zhou, J., Sun, X.-L., Tang, Y., Li, Y.-X., Li, G.-Y., and Sun, J. (2009) *Chem. Eur. J.*, **15**, 11384–11389.

184. Ren, Q., Gao, Y., and Wang, J. (2011) *Org. Biomol. Chem.*, **9**, 5297–5302.

185. (a) Lee, J.M., Na, Y., Han, H., and Chang, S. (2004) *Chem. Soc. Rev.*, **33**, 302–312. (b) Wasilke, J.-C., Obrey, S.J., Baker, R.T., and Bazan, G.C. (2005) *Chem. Rev.*, **105**, 1001–1020.

186. Simmons, B., Walji, A.M., and MacMillan, D.W.C. (2009) *Angew. Chem. Int. Ed.*, **48**, 4349–4354.

187. Lathrop, S.P. and Rovis, T. (2009) *J. Am. Chem. Soc.*, **131**, 13628–13630.

188. Filloux, C.M., Lathrop, S.P., and Rovis, T. (2010) *Proc. Natl. Acad. Sci. U.S.A.*, **107**, 20666–20671.

189. Ozboya, K.E. and Rovis, T. (2011) *Chem. Sci.*, **2**, 1835–1838.

190. Shimada, N., Ashburn, B.O., Basak, A.K., Bow, W.F., Vicic, D.A., and Tius, M.A. (2010) *Chem. Commun.*, **46**, 3774–3775.

191. Enders, D., Grossmann, A., Huang, H., and Raabe, G. (2011) *Eur. J. Org. Chem.*, **2011**, 4298–4301.

192. (a) Mannich, C. and Krosche, W. (1912) *Arch. Pharm.*, **250**, 647. (b) Arend, M., Westermann, B., and Risch, N. (1998) *Angew. Chem., Int. Ed. Engl.*, **37**, 1044–1070. (c) Cordova, A. (2004) *Acc. Chem. Res.*, **37**, 102–112. (d) Verkade, J.M.M., Van Hemert, L.J.C., Quaedflieg, P.J.L.M., and Rutjes, F.P.J.T. (2008) *Chem. Soc. Rev.*, **37**, 29–41. (e) Arrayas, R.G. and

Carretero, J.C. (2009) *Chem. Soc. Rev.*, **38**, 1940–1948.

193. Kinast, G. and Tietze, L.F. (1976) *Angew. Chem. Int. Ed.*, **88**, 261–262 *Angew. Chem., Int. Ed. Engl.* **1976**, 15, 239–240.

194. Itoh, T., Yokoya, M., Miyauchi, K., Nagata, K., and Ohsawa, A. (2003) *Org. Lett.*, **5**, 4301–4304.

195. Itoh, T., Yokoya, M., Miyauchi, K., Nagata, K., and Ohsawa, A. (2006) *Org. Lett.*, **8**, 1533–1535.

196. Han, R.-G., Wang, Y., Li, Y.-Y., and Xu, P.-F. (2008) *Adv. Synth. Catal.*, **350**, 1474–1478.

197. Yang, H. and Carter, R.G. (2009) *J. Org. Chem.*, **74**, 5151–5156.

198. Enders, D., Göddertz, D.P., Beceno, C., and Raabe, G. (2010) *Adv. Synth. Catal.*, **352**, 2863–2868.

199. Rueping, M. and Lin, M.-Y. (2010) *Chem. Eur. J.*, **16**, 4169–4172.

200. Ren, Q., Siau, W.-Y., Du, Z., Zhang, K., and Wang, J. (2011) *Chem. Eur. J.*, **17**, 7781–7785.

201. Dong, S., Liu, X., Zhang, Y., Lin, L., and Feng, X. (2011) *Org. Lett.*, **13**, 5060–5063.

202. Masson, G., Housseman, C., and Zhu, J. (2007) *Angew. Chem. Int. Ed.*, **46**, 4614–4628.

203. Takizawa, S., Inoue, N., Hirata, S., and Sasai, H. (2010) *Angew. Chem. Int. Ed.*, **49**, 9725–9729.

204. Takizawa, S., Inoue, N., and Sasai, H. (2011) *Tetrahedron Lett.*, **52**, 377–380.

205. (a) Poulsen, T. and Jørgensen, K.A. (2008) *Chem. Rev.*, **108**, 2903–2915. (b) You, S.-L., Cai, Q., and Zeng, M. (2009) *Chem. Soc. Rev.*, **38**, 2190–2201.

206. Hong, L., Wang, L., Sun, W., Wong, K., and Wang, R. (2009) *J. Org. Chem.*, **74**, 6881–6884.

207. Cao, Y.-J., Cheng, H.-G., Lu, L.-Q., Zhang, J.-J., Cheng, Y., Chen, J.-R., and Xiao, W.-J. (2011) *Adv. Synth. Catal.*, **353**, 617–623.

208. Wang, X.-F., Chen, J.-R., Cao, Y.-J., Cheng, H.-G., and Xiao, W.-J. (2010) *Org. Lett.*, **12**, 1140–1143.

209. Loh, C.C.J. and Enders, D. (2011) *Angew. Chem. Int. Ed.*, **51**, 46–48.

210. Lozano, O., Bessley, G., Martinez Del Campo, T., Thompson, A.L., Giuffredi, G.T., Bettati, M., Walker, M., Borman, R., and Gouverneur, V. (2011) *Angew. Chem. Int. Ed.*, **50**, 8105–8109.

211. Momiyama, N., Yamamoto, Y., and Yamamoto, H. (2007) *J. Am. Chem. Soc.*, **129**, 1190–1195.

212. Ramachary, D.B. and Sakthidevi, R. (2009) *Chem. Eur. J.*, **15**, 4516–4522.

213. Guang, J. and Zhao, C.-G. (2011) *Tetrahedron: Asymmetry*, **22**, 1205–1211.

214. Bröhmer, M.C., Bourcet, E., Nieger, M., and Bräse, S. (2011) *Chem. Eur. J.*, **17**, 13706–13711.

215. Poulsen, T.B., Dickmeiss, G., Overgaard, J., and Jørgensen, K.A. (2008) *Angew. Chem. Int. Ed.*, **47**, 4687–4990.

216. (a) Liu, M., Zhu, D., Lu, Y., Hou, Y., Tan, B., and Zhong, G. (2008) *Angew. Chem. Int. Ed.*, **47**, 10187–10191. (b) Zhu, D., Lu, M., Chua, P.J., Tan, B., Wang, F., Yang, X., and Zhong, G. (2008) *Org. Lett.*, **10**, 4585–4588.

217. Han, B., Li, J.-L., Ma, C., Zhang, S.-J., and Chen, Y.-C. (2008) *Angew. Chem. Int. Ed.*, **47**, 9971–9974.

218. He, Z.-Q., Han, B., Li, R., Wu, L., and Chen, Y.-C. (2010) *Org. Biomol. Chem.*, **8**, 755–757.

219. Kanemasa, S. (2010) *Heterocycles*, **82**, 87–200.

220. Xue, M.-X., Zhang, X.-M., and Gong, L.-Z. (2008) *Synlett*, **5**, 691–694.

221. (a) Poulin, J., Grisé-Bard, C.M., and Barriault, L. (2009) *Chem. Soc. Rev.*, **38**, 3092–3101. (b) Moyano, A. and Rios, R. (2011) *Chem. Rev.*, **111**, 4703–4832.

222. Jones, S.B., Simmons, B., and MacMillan, D.W.C. (2009) *J. Am. Chem. Soc.*, **131**, 13606–13607.

223. Muratore, M.E., Holloway, C.A., Pilling, A.W., Storer, R.I., Trevitt, G., and Dixon, D.J. (2009) *J. Am. Chem. Soc.*, **131**, 10796–10797.

224. Kaeobamrung, J. and Bode, J.W. (2009) *Org. Lett.*, **11**, 677–680.

225. Rueping, M., Antonchick, A.P., Sugiono, E., and Grenader, K. (2009) *Angew. Chem. Int. Ed.*, **48**, 908–910.

226. Jiang, H., Elsner, P., Jensen, K.L., Falcicchio, A., Marcos, V., and Jørgensen, K.A. (2009) *Angew. Chem. Int. Ed.*, **48**, 6844–6848.

227. Klausen, R.S. and Jacobsen, E.N. (2009) *Org. Lett.*, **11**, 887–890.
228. Rendler, S. and MacMillan, D.W.C. (2010) *J. Am. Chem. Soc.*, **132**, 5027–5029.
229. Zhang, S.-L., Xie, H.-X., Zhu, J., Li, H., Zhang, X.-S., Li, J., and Wang, W. (2011) *Nat. Commun.*, **2**, 591. doi: 10:1038, Article No. 211.
230. Rueping, M., Sunden, H., Hubener, L., and Sugiono, E. (2012) *Chem. Commun.*, **48**, 2201–2203.
231. Rueping, M., Dufour, J., and Maji, M.S. (2012) *Chem. Commun.*, **48**, 3406–3408.
232. Rueping, M. and Ieawsuwan, W. (2011) *Chem. Commun.*, **47**, 11450–11452.
233. Itoh, J., Fuchibe, K., and Akiyama, T. (2006) *Angew. Chem. Int. Ed.*, **45**, 4796–4798.
234. Akiyama, T., Morita, H., and Fuchibe, K. (2006) *J. Am. Chem. Soc.*, **128**, 13070–13071.
235. He, M., Struble, J.R., and Bode, J.W. (2006) *J. Am. Chem. Soc.*, **128**, 8418–8420.
236. Chiang, P.-C., Kaeobamrung, J., and Bode, J.W. (2007) *J. Am. Chem. Soc.*, **129**, 3520–3521.
237. Chan, A. and Scheidt, K.A. (2007) *J. Am. Chem. Soc.*, **129**, 5334–5335.
238. Wadamoto, M., Phillips, E.M., Reynolds, T.E., and Scheidt, K.A. (2007) *J. Am. Chem. Soc.*, **129**, 10098–10099.
239. Du, Z., Siau, W.-Y., and Wang, J. (2011) *Tetrahedron Lett.*, **52**, 6137–6141.
240. Hayashi, Y., Urushima, T., Sakamoto, D., Torii, K., and Ishikawa, H. (2011) *Chem. Eur. J.*, **17**, 11715–11718.
241. (a) Strecker, A. (1850) *Ann. Chem. Pharm.*, **75**, 27–45. (b) Wang, J., Liu, X., and Feng, X. (2011) *Chem. Rev.*, **111**, 6947–6983.
242. Hantzsch, A. (1882) *Justus Liebigs Ann. Chem.*, **215**, 1–82.
243. (a) Biginelli, P. (1891) *Ber. Dtsch. Chem. Ges.*, **24**, 1317–1319. (b) Biginelli, P. (1893) *Gazz. Chim. Ital.*, **23**, 360–413.
244. (a) Ugi, I., Meyr, R., Fetzer, U., and Steinbrückner, C. (1959) *Angew. Chem. Int. Ed.*, **71**, 386. (b) Orru, R.V.A. and de Greef, M. (2003) *Synthesis*, 1471–1499.
245. Betancort, J.M., Sakthivel, K., Thayumanavan, R., and Barbas, C.F. (2001) *Tetrahedron Lett.*, **42**, 4441–4444.
246. Marigo, M., Schulte, T., Franzen, J., and Jørgensen, K.A. (2005) *J. Am. Chem. Soc.*, **127**, 15710–15711.
247. Enders, D., Hüttl, M.R.M., Grondal, C., and Raabe, G. (2006) *Nature*, **441**, 861–863.
248. Enders, D., Hüttl, M.R.M., Runsink, J., Raabe, G., and Wendt, B. (2007) *Angew. Chem. Int. Ed.*, **46**, 467–469.
249. Carlone, A., Cabrera, S., Marigo, M., and Jørgensen, K.A. (2007) *Angew. Chem. Int. Ed.*, **46**, 1101–1104.
250. Penon, O., Carlone, A., Mazzanti, A., Locatelli, M., Sambri, L., Bartoli, G., and Melchiorre, P. (2008) *Chem. Eur. J.*, **14**, 4788–4791.
251. Varela, M.C., Dixon, S.M., Lam, K.S., and Schore, N.E. (2008) *Tetrahedron*, **64**, 10087–10090.
252. Enders, D., Jeanty, M., and Bats, J.W. (2009) *Synlett*, **19**, 3175–3178.
253. Companyo, X., Zea, A., Alba, A.-N.R., Mazzanti, A., Moyano, A., and Rios, R. (2010) *Chem. Commun.*, **46**, 6953–6955.
254. Cassani, C., Tian, X., Escudero-Adan, E.C., and Melchiorre, P. (2011) *Chem. Commun.*, **47**, 233–235.
255. Bencivenni, G., Wu, L.-Y., Mazzanti, A., Giannichi, B., Pesciaoli, F., Song, M.-P., Bartoli, G., and Melchiorre, P. (2009) *Angew. Chem. Int. Ed.*, **48**, 7200–7203.
256. Jiang, K., Jia, Z.-J., Chen, S., Wu, L., and Chen, Y.-C. (2010) *Chem. Eur. J.*, **16**, 2852–2856.
257. Kotame, P., Hong, B.-C., and Liao, J.-H. (2009) *Tetrahedron Lett.*, **50**, 704–707.
258. Zhang, F.-L., Xu, A.-W., Gong, Y.-F., Wei, M.-H., and Yang, X.-L. (2009) *Chem. Eur. J.*, **15**, 6815–6818.
259. Jiang, K., Jia, Z.-J., Yin, X., Wu, L., and Chen, Y.-C. (2010) *Org. Lett.*, **12**, 2766–2769.
260. Bertelsen, S., Johansen, R.L., and Jørgensen, K.A. (2008) *Chem. Commun.*, 3016–3018.
261. Zea, A., Alba, A.-N.R., Mazzanti, A., Moyano, A., and Rios, R. (2011) *Org. Biomol. Chem.*, **9**, 6519–6523.

262. Zhao, Y.-H., Zheng, C.-W., Zhao, G., and Cao, W.-G. (2008) *Tetrahedron: Asymmetry*, **19**, 701–708.

263. Vesely, J., Rios, R., Ibrahem, I., Zhao, G.-L., Eriksson, L., and Cordova, A. (2008) *Chem. Eur. J.*, **14**, 2693–2698.

264. Galzerano, P., Pesciaioli, F., Mazzanti, A., Bartoli, G., and Melchiorre, P. (2009) *Angew. Chem. Int. Ed.*, **48**, 7892–7894.

265. Rueping, M. and Antonchick, A.P. (2008) *Angew. Chem. Int. Ed.*, **47**, 5836–5838.

266. Yang, H. and Carter, R.G. (2010) *Tetrahedron*, **66**, 4854–4859.

267. Yang, H. and Carter, R.G. (2010) *Org. Lett.*, **12**, 3108–3111.

268. Sugiura, M., Sato, N., Sonoda, Y., Kotani, S., and Nakajima, M. (2010) *Chem. Asian J.*, **5**, 478–481.

269. Hoashi, Y., Yabuta, T., Yuan, P., Miyabe, H., and Takemoto, Y. (2006) *Tetrahedron*, **62**, 365–374.

270. Liu, Y.-K., Liu, H., Du, W., Yue, L., and Chen, Y.-C. (2008) *Chem. Eur. J.*, **14**, 9873–9877.

271. Crovetto, L. and Rios, R. (2008) *Synlett*, **5**, 1840–1844.

272. Enders, D., Hüttl, M.R.M., Raabe, G., and Bats, J.W. (2008) *Adv. Synth. Catal.*, **350**, 267–279.

273. Wang, Y., Han, R.-G., Zhao, Y.-L., Yang, S., Xu, P.-F., and Dixon, D.J. (2009) *Angew. Chem. Int. Ed.*, **48**, 9834–9838.

274. Enders, D., Schmid, B., Erdmann, M., and Raabe, G. (2010) *Synthesis*, **13**, 2271–2277.

275. Bergbreiter, D.E., Tian, J., and Hongfa, C. (2009) *Chem. Rev.*, **109**, 530–582.

276. Kristensen, T.E., Vestli, K., Jakobsen, M.G., Hansen, F.K., and Hansen, T. (2010) *J. Org. Chem.*, **75**, 1620–1629.

277. Hong, B.-C., Jan, R.-H., Tsai, C.-W., Nimje, R.Y., Liao, J.-H., and Lee, G.-H. (2009) *Org. Lett.*, **11**, 5246–5249.

278. Ishikawa, H., Suzuki, T., Orita, H., Uchimaru, T., and Hayashi, Y. (2010) *Chem. Eur. J.*, **16**, 12616–12626.

279. Ishikawa, H., Honma, M., and Hayashi, Y. (2011) *Angew. Chem. Int. Ed.*, **50**, 2824–2827.

280. Baslé, O., Raimondi, W., Sanchez Duque, M., Bonne, D., Constantieux, T., and Rodriguez, J. (2010) *Org. Lett.*, **12**, 5246–5249.

281. Enders, D., Krüll, R., and Bettray, W. (2010) *Synthesis*, **4**, 567–572.

282. Urushima, T., Sakamoto, D., Ishikawa, H., and Hayashi, Y. (2010) *Org. Lett.*, **12**, 4588–4591.

283. Wang, Y., Yu, D.-F., Liu, Y.-Z., Wie, H., Luo, Y.-C., Dixon, D.J., and Xu, P.-F. (2010) *Chem. Eur. J.*, **16**, 3922–3925.

284. Rai, A., Singh, A.K., Singh, P., and Yadav, L.D.S. (2011) *Tetrahedron Lett.*, **52**, 1354–1358.

285. Ishikawa, H., Sawano, S., Yasui, Y., Shibata, Y., and Hayashi, Y. (2011) *Angew. Chem. Int. Ed.*, **50**, 3774–3779.

286. (a) Jakubec, P., Cockfield, D.M., Helliwell, M., Raftery, J., and Dixon, D.J. (2012) *Belstein J. Org. Chem.*, **8**, 567–578. (b) Jakubec, P., Helliwell, M., and Dixon, D.J. (2008) *Org. Lett.*, **10**, 4267–4270.

287. Knoevenagel, E. (1894) *Ber. Dtsch. Chem. Ges.*, **27**, 2345–2346.

288. List, B. (2010) *Angew. Chem. Int. Ed.*, **49**, 1730–1734.

289. (a) Tietze, L.F., Meier, H., and Nutt, H. (1989) *Chem. Ber.*, **122**, 643–643. (b) Tietze, L.F., Rackelmann, N., and Müller, I. (2004) *Chem. Eur. J.*, **10**, 2722–2731. (c) Tietze, L.F., Böhnke, N., and Dietz, S. (2009) *Org. Lett.*, **11**, 2948–2950.

290. Ramachary, D.B., Chowdari, N.S., and Barbas, C.F. (2003) *Angew. Chem. Int. Ed.*, **42**, 4233–4237.

291. (a) Ramachary, D.B., Chowdari, N.S., and Barbas, C.F. (2003) *Synlett*, **12**, 1910–1914. (b) Ramachary, D.B., Anebouselvy, K., Chowdari, N.S., and Barbas, C.F. (2004) *J. Org. Chem.*, **69**, 5838–5849.

292. Shi, J., Min, Y., Wang, M., Lin, L., Liu, X., and Feng, X. (2011) *Tetrahedron*, **67**, 1781–1787.

293. Ramachary, D.B. and Barbas, C.F. (2004) *Chem. Eur. J.*, **10**, 5323–5331.

294. Ramachary, D.B., Reddy, Y.V., and Prakash, B.V. (2008) *Org. Biomol. Chem.*, **6**, 719–726.

295. Chen, W.-B., Wu, Z.-J., Pei, Q.-L., Cun, L.-F., Zhang, X.-M., and Yuan, W.-C. (2010) *Org. Lett.*, **12**, 3132–3135.

296. Macaev, F., Sucman, N., Shepeli, F., Zveaghintseva, M., and Pogrebnoi, V. (2011) *Symmetry*, **3**, 165–170.
297. Yu, D.-F., Wang, Y., and Xu, P.-F. (2011) *Tetrahedron*, **67**, 3273–3277.
298. Liu, L., Wu, D., Li, X., Wang, S., Li, H., Li, J., and Wang, W. (2012) *Chem. Commun.*, **48**, 1692–1694.
299. Wei, W.-T., Chen, C.-X., Lu, R.-J., Wang, J.-J., Zhang, X.-J., and Yan, M. (2012) *Org. Biomol. Chem.*, **10**, 5245–5252.
300. List, B. (2000) *J. Am. Chem. Soc.*, **122**, 9336–9337.
301. (a) Notz, W., Sakthivel, K., Bui, T., Zhong, G., and Barbas, C.F. (2001) *Tetrahedron Lett.*, **42**, 199–201. (b) Cordova, A., Notz, W., Zhong, G.F., Betancort, J.M., and Barbas, C.F. (2002) *J. Am. Chem. Soc.*, **124**, 1842–1843. (c) Hayashi, Y., Tsuboi, W., Ashimine, I., Urushima, T., Shoji, M., and Sakai, K. (2003) *Angew. Chem. Int. Ed.*, **42**, 3677–3680. (d) Notz, W., Watanabe, S., Chowdari, N.S., Zhong, G.F., Betancort, J.M., Tanaka, F., and Barbas, C.F. (2004) *Adv. Synth. Catal.*, **346**, 1131–1140. (e) Cobb, A.J.A., Shaw, D.M., and Ley, S.V. (2004) *Synlett*, **3**, 558–560. (f) Ibrahem, I., Casas, J., and Cordova, A. (2004) *Angew. Chem. Int. Ed.*, **43**, 6528–6531. (g) Cordova, A. (2004) *Chem. Eur. J.*, **10**, 1987–1997. (h) Hayashi, Y., Yamaguchi, J., Hibino, K., Sumiya, T., Urushima, T., Shoji, M., Hashizume, D., and Koshino, H. (2004) *Adv. Synth. Catal.*, **346**, 1435–1439. (i) Ibrahem, I. and Cordova, A. (2005) *Tetrahedron Lett.*, **46**, 2839–2843. (j) Ibrahem, I. and Cordova, A. (2005) *Tetrahedron Lett.*, **46**, 3363–3367. (k) Enders, D., Grondal, C., Vrettou, M., and Raabe, G. (2005) *Angew. Chem. Int. Ed.*, **44**, 4079–4083. (l) Ibrahem, I., Zou, W., Engqvist, M., Xu, Y.M., and Cordova, A. (2005) *Chem. Eur. J.*, **11**, 7024–7029.
302. Rodriguez, B. and Bolm, C. (2006) *J. Org. Chem.*, **71**, 2888–2891.
303. Ibrahem, I., Zou, W., Xu, Y., and Cordova, A. (2006) *Adv. Synth. Catal.*, **348**, 211–222.
304. Liao, W.-W., Ibrahem, I., and Cordova, A. (2006) *Chem. Commun.*, 674–676.
305. Kantam, M.L., Rajasekhar, C.V., Gopikrishna, G., Reddy, K.R., and Choudary, B.M. (2006) *Tetrahedron Lett.*, **47**, 5965–5967.
306. Chowdari, N.S., Ahmad, M., Albertshofer, K., Tanaka, F., and Barbas, C.F. (2006) *Org. Lett.*, **8**, 2839–2842.
307. Ramasastry, S.S.V., Zhang, H., Tanake, F., and Barbas, C.F. (2007) *J. Am. Chem. Soc.*, **129**, 288–289.
308. Guo, Q.-X., Liu, H., Guo, C., Luo, S.-W., Gu, Y., and Gong, L.-Z. (2007) *J. Am. Chem. Soc.*, **129**, 3790–3791.
309. (a) Cheng, L., Wu, X., and Lu, Y. (2007) *Org. Biomol. Chem.*, **5**, 1018–1020. (b) Xu, L.-W., Luo, J., and Lu, Y. (2009) *Chem. Commun.*, 1807–1821.
310. Teo, Y.-C., Lau, J.-J., and Wu, M.-C. (2008) *Tetrahedron: Asymmetry*, **19**, 186–190.
311. Dziedzic, P., Ibrahem, I., and Cordova, A. (2008) *Tetrahedron Lett.*, **49**, 803–807.
312. Yang, H. and Carter, R.G. (2009) *J. Org. Chem.*, **74**, 2246–2249.
313. (a) Gu, Q., Gong, J.J., Feng, J., Wu, X.Y., and Zhou, Q.L. (2008) *Chin. J. Chem.*, **26**, 1902–1906. (b) Gu, Q., Jiang, L.X., Yuan, K., Zhang, L., and Wu, X.Y. (2008) *Synth. Commun.*, **38**, 4198–4206.
314. Dagousset, G., Drouet, F., Masson, G., and Zhu, J. (2009) *Org. Lett.*, **11**, 5546–5549.
315. (a) Gu, Y.L., Ogawa, C., Kobayashi, J., Mori, Y., and Kobayashi, S. (2006) *Angew. Chem. Int. Ed.*, **45**, 7217–7220. (b) Gong, K., Fang, D., Wang, H.L., and Liu, Z.L. (2007) *Monatsh. Chem.*, **138**, 1195–1198.
316. Zheng, X., Qian, Y.-B., and Wang, Y. (2010) *Eur. J. Org. Chem.*, **2010**, 515–522.
317. Liu, B.Y., Xu, D.Q., Dong, J.F., and Yang, H.L. (2007) *Synth. Commun.*, **37**, 3003–3010.
318. (a) Wu, C., Fu, X., Ma, X., Li, S., and Li, C. (2010) *Tetrahedron Lett.*, **51**, 5775–5777. (b) Wu, C., Fu, X., and Li, S. (2011) *Tetrahedron: Asymmetry*, **22**, 1063–1073.

319. Bai, S., Liang, X., Song, B., Bhadury, P.S., Hu, D., and Yang, S. (2011) *Tetrahedron: Asymmetry*, **22**, 518–523.
320. Yong, F.-F. and Teo, Y.-C. (2011) *Synth. Commun.*, **41**, 1293–1300.
321. An, Y.-J., Qin, Q., Wang, C., and Tao, J.-C. (2011) *Chin. J. Chem.*, **29**, 1511–1517.
322. An, Y.-J., Wang, C.-C., Liu, Z.-P., and Tao, J.-C. (2012) *Helv. Chim. Acta*, **95**, 43–51.
323. Chen, X.-H., Yu, X.-Y., Liu, H., Cun, L.-F., and Gong, L.-Z. (2006) *J. Am. Chem. Soc.*, **128**, 14802–14803.
324. Gong, L.-Z., Chen, X.H., and Xu, X.-Y. (2007) *Chem. Eur. J.*, **13**, 8920–8926.
325. Goss, J.M. and Schauss, S.E. (2008) *J. Org. Chem.*, **73**, 7651–7656.
326. Xin, J., Chan, L., Hou, Z., Shang, D., Liu, X., and Feng, X. (2008) *Chem. Eur. J.*, **14**, 3177–3181.
327. Yadav, L.D.S., Rai, A., Rai, V.K., and Awasthi, C. (2008) *Tetrahedron*, **64**, 1420–1429.
328. Pandey, J., Anand, N., and Tripathi, R.P. (2009) *Tetrahedron*, **65**, 9350–9356.
329. Sohn, J.-H., Choi, H.-M., Lee, S., Joung, S., and Lee, H.-Y. (2009) *Eur. J. Org. Chem.*, **2009**, 3858–3862.
330. Wu, Y.-Y., Chai, Z., Liu, X.-Y., Zhao, G., and Wang, S.-W. (2009) *Eur. J. Org. Chem.*, **2009**, 904–911.
331. Ding, D. and Zhao, C.-G. (2010) *Eur. J. Org. Chem.*, **2010**, 3802–3805.
332. Wang, Y., Yang, H., Yu, J., Miao, Z., and Chen, R. (2009) *Adv. Synth. Catal.*, **351**, 3057–3062.
333. Wang, Y., Yu, J., Miao, Z., and Chen, R. (2011) *Org. Biomol. Chem.*, **9**, 3050–3054.
334. Li, N., Chen, X.-H., Song, J., Luo, S.-W., Fan, W., and Gong, L.-Z. (2009) *J. Am. Chem. Soc.*, **131**, 15301–15310.
335. Saha, S. and Moorthy, J.N. (2011) *J. Org. Chem.*, **76**, 396–402.
336. Jones, G. (1996) in: *Comprehensive Heterocyclic Chemistry II*, Vol. 5 (eds A.R. Katritzky, C.W. Rees, and E.F.V. Scriven), Pergamon, Oxford, pp. 167–243.
337. Franke, P.T., Johansen, R.L., Bertelsen, S., and Jørgensen, K.A. (2008) *Chem. Asian J.*, **3**, 216–224.
338. Jiang, J., Yu, J., Sun, X.-X., Rao, Q.-Q., and Gong, L.-Z. (2008) *Angew. Chem. Int. Ed.*, **47**, 2458–2462.
339. Evans, C.G. and Gestwicki, J.E. (2009) *Org. Lett.*, **11**, 2957–2959.
340. Hishitani, H., Komiyama, S., Hasegawa, Y., and Kobayashi, S. (2000) *J. Am. Chem. Soc.*, **122**, 762–766.
341. Pan, S.C. and List, B. (2007) *Org. Lett.*, **6**, 1149–1151.
342. Wen, Y., Xiong, Y., Chang, L., Huang, J., Liu, X., and Feng, X. (2007) *J. Org. Chem.*, **72**, 7715–7719.
343. Wen, Y., Gao, B., Fu, Y., Dong, S., Liu, X., and Feng, X. (2008) *Chem. Eur. J.*, **14**, 6789–6795.
344. (a) Petasis, N.A. and Akritopoulou, I. (1993) *Tetrahedron Lett.*, **34**, 583–586. (b) Candeias, R.N., Montalbano, F., Cal, P.M.S.D., and Gois, P.M.P. (2010) *Chem. Rev.*, **110**, 6169–6193.
345. Lou, S. and Schaus, S.E. (2008) *J. Am. Chem. Soc.*, **130**, 6922–6923.
346. (a) Yamaoka, Y., Miyabe, H., and Takemoto, Y. (2007) *J. Am. Chem. Soc.*, **129**, 6686–6687. (b) Miyabe, H. and Takemoto, Y. (2008) *Bull. Chem. Soc. Jpn.*, **81**, 785–795. (c) Takemoto, Y. (2010) *Chem. Pharm. Bull.*, **58**, 593–601.
347. Han, W.-Y., Wu, Z.-J., Zhang, X.-M., and Yuan, W.-C. (2012) *Org. Lett.*, **14**, 976–979.
348. Pellissier, H. (2007) *Tetrahedron*, **63**, 3235–3285.
349. Rios, R., Ibrahem, I., Vesely, J., Zhao, G.-L., and Cordova, A. (2007) *Tetrahedron Lett.*, **48**, 5701–5705.
350. Ibrahem, I., Rios, R., Vesely, J., and Cordova, A. (2007) *Tetrahedron Lett.*, **48**, 6252–6257.
351. Chen, X.-H., Zhang, W.-Q., and Gong, L.-Z. (2008) *J. Am. Chem. Soc.*, **130**, 5652–5653.
352. Chen, X.-H., Wie, Q., Luo, S.-W., Xiao, H., and Gong, L.-Z. (2009) *J. Am. Chem. Soc.*, **131**, 13819–13825.
353. Lin, S., Deiana, L., Zhao, G.-L., Sun, J., and Cordova, A. (2011) *Angew. Chem. Int. Ed.*, **50**, 7624–7630.
354. Fraile, A., Scarpino Schietroma, D.M., Albrecht, A., Davis, R.L., and Jørgensen, K.A. (2012) *Chem. Eur. J.*, **18**, 2773–2776.

355. (a) Dömling, A. and Ugi, I. (2000) *Angew. Chem. Int. Ed.*, **39**, 3168–3210. (b) Zhu, J. (2003) *Eur. J. Org. Chem.*, **2003**, 1133–1144. (c) Van Berkel, S.S., Bögels, B.G.M., Wijdeven, M.A., Westermann, B., and Rutjes, F.P.J.T. (2012) *Eur. J. Org. Chem.*, **2012**, 3543–3559.
356. Passerini, M. and Simone, L. (1921) *Gazz. Chim. Ital.*, **51**, 126–129.
357. Denmark, S.E. and Fan, Y. (2003) *J. Am. Chem. Soc.*, **125**, 7825–7827.
358. Yue, T., Wang, M.-X., Wang, D.-X., Masson, G., and Zhu, J. (2009) *Angew. Chem. Int. Ed.*, **48**, 6717–6721.
359. Zhang, G.-W., Wang, L., Nie, J., and Ma, J.-A. (2008) *Adv. Synth. Catal.*, **350**, 1457–1463.
360. Enders, D., Wang, C., Mukanova, M., and Greb, A. (2010) *Chem. Commun.*, **46**, 2447–2449.
361. Roy, S. and Chen, K. (2012) *Org. Lett.*, **14**, 2496–2499.
362. Merino, P., Marques-Lopez, E., Tejero, T., and Herrera, R.P. (2010) *Synthesis*, **1**, 1–26.
363. (a) Sunden, H., Ibrahem, I., Eriksson, L., and Cordova, A. (2005) *Angew. Chem. Int. Ed.*, **44**, 4877–4880. (b) Ibrahem, I., Zou, W., Casas, J., Sunden, H., and Cordova, A. (2006) *Tetrahedron*, **62**, 357–364.
364. Liu, H., Cun, L.-F., Mi, A.-Q., Jiang, Y.-Z., and Gong, L.-Z. (2006) *Org. Lett.*, **26**, 6023–6026.
365. Jia, Z.-J., Jiang, H., Li, J.-L., Gschwend, B., Li, Q.-Z., Yin, X., Grouleff, J., Chen, Y.-C., and Jørgensen, K.A. (2011) *J. Am. Chem. Soc.*, **133**, 5053–5061.
366. Povarov, L.S. (1967) *Russ. Chem. Rev.*, **36**, 656–670.
367. Liu, H., Dagousset, G., Masson, G., Retailleau, P., and Zhu, J. (2009) *J. Am. Chem. Soc.*, **131**, 4598–4599.
368. Dagousset, G., Zhu, J., and Masson, G. (2011) *J. Am. Chem. Soc.*, **133**, 14804–14813.
369. Wang, C., Han, Z.-Y., Luo, H.-W., and Gong, L.-Z. (2010) *Org. Lett.*, **12**, 2266–2269.
370. Bergonzini, G., Gramigna, L., Mazzanti, A., Fochi, M., Bernardi, L., and Ricci, A. (2010) *Chem. Commun.*, **46**, 327–329.
371. Guizzetti, S., Benaglia, M., Cozzi, F., and Annunziata, R. (2009) *Tetrahedron*, **65**, 6354–6363.
372. Lian, W., ShuMin, C., Wie, M., GuangWu, Z., Jing, N., and JunAn, M. (2010) *Chin. Sci. Bull.*, **55**, 1729–1731.
373. (a) Zhou, J. and List, B. (2007) *J. Am. Chem. Soc.*, **129**, 7498–7499. (b) Zhou, J. and List, B. (2007) *Synlett*, **18**, 2037–2040.
374. Terada, M., Machioka, K., and Sorimachi, K. (2007) *J. Am. Chem. Soc.*, **129**, 10336–10337.
375. Scroggins, S.T., Chi, Y., and Fréchet, J.M.J. (2010) *Angew. Chem. Int. Ed.*, **49**, 2393–2396.
376. (a) Zhou, J. (2010) *Chem. Asian J.*, **5**, 422–434. (b) Zhong, C. and Shi, X. (2010) *Eur. J. Org. Chem.*, **2010**, 2999–3025. (c) Rueping, M., Koenigs, R.M., and Atodiresei, I. (2010) *Chem. Eur. J.*, **16**, 9350–9365. (d) Shao, Z. and Zhang, H. (2009) *Chem. Soc. Rev.*, **38**, 2745–2755.
377. Westermann, B., Ayaz, M., and van Berkel, S.S. (2010) *Angew. Chem. Int. Ed.*, **49**, 846–849.

11
Metal-Catalyzed Enantio- and Diastereoselective C–C Bond-Forming Reactions in Domino Processes

Shinobu Takizawa and Hiroaki Sasai

Since the first definition of the domino process by Tietze [1] in 1993, the development of stereoselective domino reactions with metal catalysts [2] or organocatalysts [3] has drawn much attention in current organic synthesis. In particular, the asymmetric catalytic C–C bond-forming domino process is a very attractive methodology because of its ability to construct complex chiral skeletons directly form readily available substrates in two or more steps in a single operation. In the enantio- and diastereoselective domino reactions, efficiency of the reactions can be judged by the number of bonds formed, the number of newly created stereocenters, and the increase in molecular complexity. Lately, several reviews on this fascinating process regarding the development of asymmetric reaction and its applications have been published [4]. In this chapter, we mainly cover the developments from 2006 [1c] to recent advances in the catalytic metal-mediated, enantioselective, C–C bond-forming domino transformations from readily available substrates, namely α,β-unsaturated carbonyl compounds, to produce multiple stereogenic centers.

11.1
Domino Reaction Initiated by C–C Bond Formation

11.1.1
Domino Reaction Initiated by Conjugate Addition

Metal-catalyzed reactions have been of major importance in synthetic organic chemistry. Over the past decade, enantio- and diastereoselective metal-mediated domino catalysis has emerged as an effective tool to construct really highly complex molecules in one-pot processes [2, 4b,d]. Among them, enantioseletcive metal-catalyzed conjugate additions (in particular, Cu-catalyzed 1,4-addition to α,β-unsaturated carbonyl compounds) have been useful components of domino reactions [4d, 5]. The generated metal enolates **2** after the additions of nucleophiles readily react with a variety of electrophiles (Scheme 11.1). Enantioselectivity of **3** depends on the first addition of nucleophiles to the β-position of the unsaturated carbonyl compounds **1**.

Domino Reactions: Concepts for Efficient Organic Synthesis, First Edition. Edited by Lutz F. Tietze.

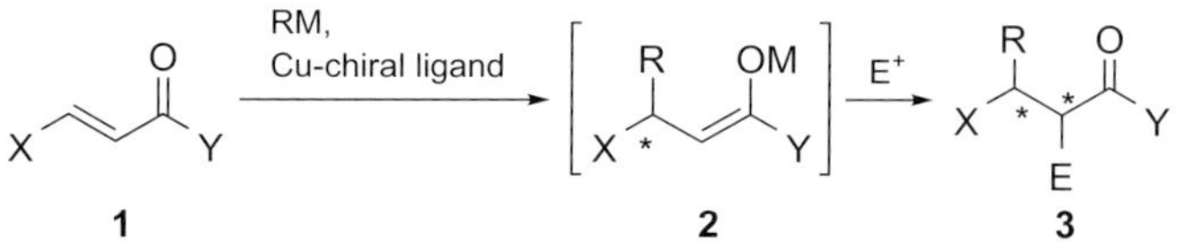

Scheme 11.1 Chiral Cu-catalyzed conjugate addition/electrophilic trapping sequence.

Since the first examples of electrophilic trapping of enolates produced by catalytic conjugate additions of organometallic reagents were described by Noyori [6] in 1988, there have been numerous examples of domino reactions using this methodology. In 1997, Feringa reported the first catalytic, enantioselective domino Michael/aldol reaction via zinc enolates as key intermediates using a chiral phosphoramidite ligand (such as **28** in Scheme 11.7) [7]. As a recent excellent example of zinc enolates in the domino process, Alexakis reported the Cu-catalyzed enantioselective conjugate addition of dialkylzinc **6** to bis-α,β-unsaturated carbonyl compounds **5**, followed by the intramolecular trapping of the intermediary zinc enolate through a second intramolecular conjugate addition in the presence of the chiral phosphoramidite ligand **4** [8]. This domino Michael/Michael process afforded the corresponding chiral cyclic products **7** having three stereocenters with excellent diastereoselectivities and high enantioselectivities (Scheme 11.2).

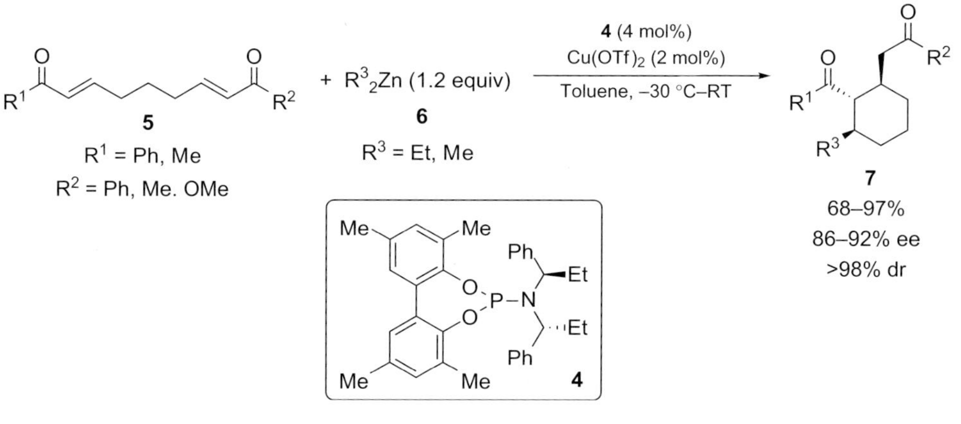

Scheme 11.2 Cu-catalyzed intramolecular Michael/Michael reaction [8].

Mikami [9] successfully utilized the Cu–phosphoramidite complex derived from chiral ligand **8** and $Cu(OTf)_2$ for the desymmetrization of cyclopentene-1,3-dione **9** (Scheme 11.3). The resulting enolates generated with dimethylzinc were efficiently trapped with butylaldehyde. Highly substituted cyclopentanedione **10** was formed with excellent diastereoselectivities and enantiomeric purities. This domino process was used for the formal synthesis of (+)-madindoline B (**12**) [10].

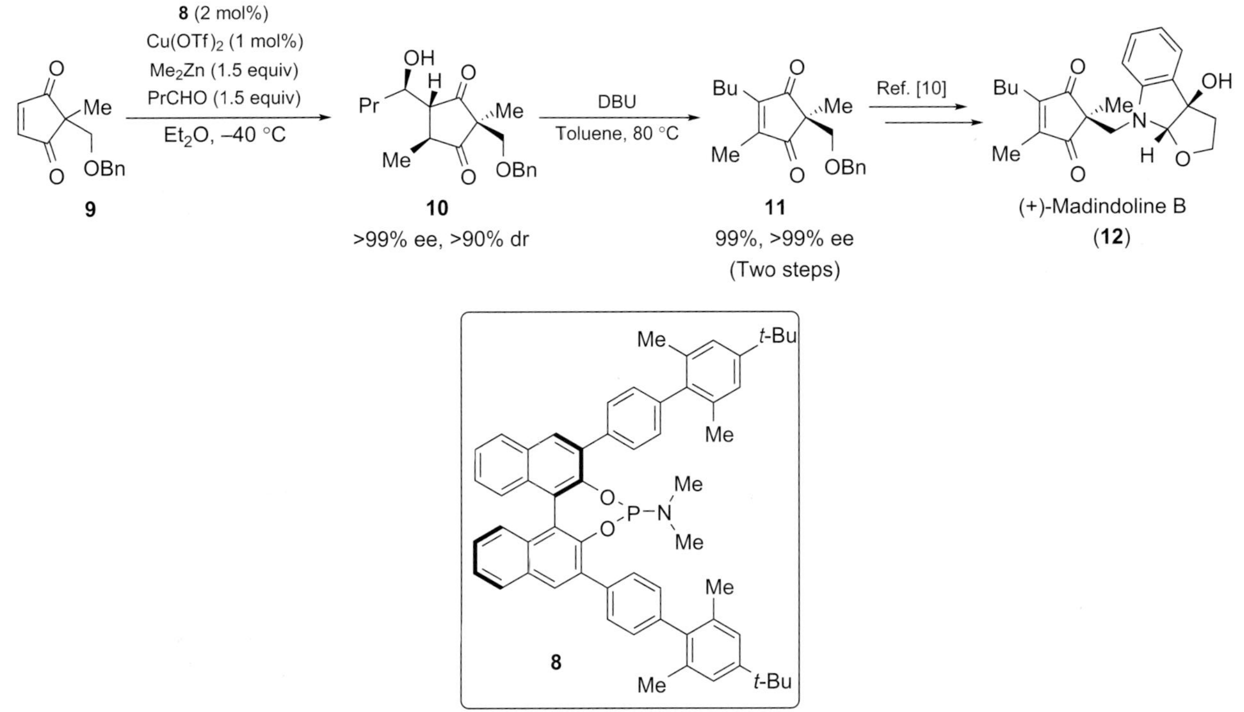

Scheme 11.3 Concise synthesis of precursor **11** of madindoline B (**12**) [9].

In 2006, Feringa developed the first catalytic asymmetric protocol for acyclic 1,4-addition/aldol reaction using Grignard reagents. The chiral Cu catalyst derived from Josiphos-type ligand **13** mediated addition of methyl magnesium bromide to linear α,β-unsaturated thioester **14** giving magnesium enolate **15**. The metal enolate **15** was consequently trapped with hexanal. Since the parent aldol product **16** was unstable, the thioester **14** was used and subsequently converted into the corresponding ester **17** in a one-pot process. This methodology was applied to the total synthesis of phaseolinic acid (**19**) (Scheme 11.4) [11].

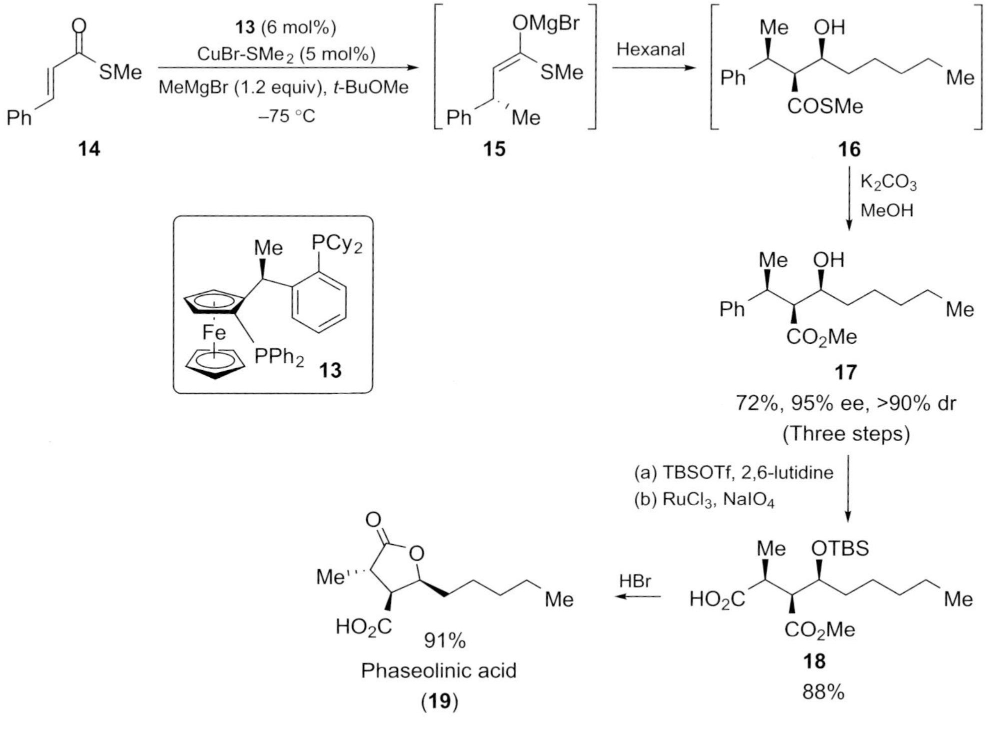

Scheme 11.4 Synthesis of phaseolinic acid (**19**) [11].

Feringa [12] also developed Cu-catalyzed conjugate addition of Grignard reagents **21** to γ-chloro-α,β-unsaturated carbonyl compounds **20**, which was followed by enolate trapping through intramolecular alkylation reaction to provide the corresponding *trans*-1-alkyl-2-substituted cyclopropanes **23** in high yields and enantioselectivities (Scheme 11.5). This novel and versatile methodology employed an extremely simple catalytic system based on commercially available (*R*)-Tol-BINAP, CuI, and Grignard reagents.

Feringa found that the Josiphos-type ligand **24** was the most effective one for addition of Grignard reagents to low-reactive coumarin **25**. The resulting ester enolate **26** also underwent a subsequent aldol reaction with benzaldehyde. The

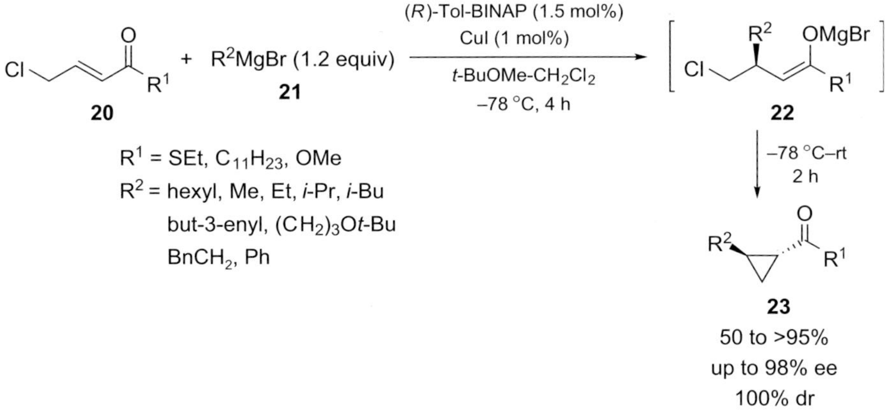

Scheme 11.5 Cu-catalyzed Michael/intramolecular alkylation reaction of **20** [12].

aldol product **27** was obtained in good yield and excellent enantioselectivity with 50% dr (Scheme 11.6) [13].

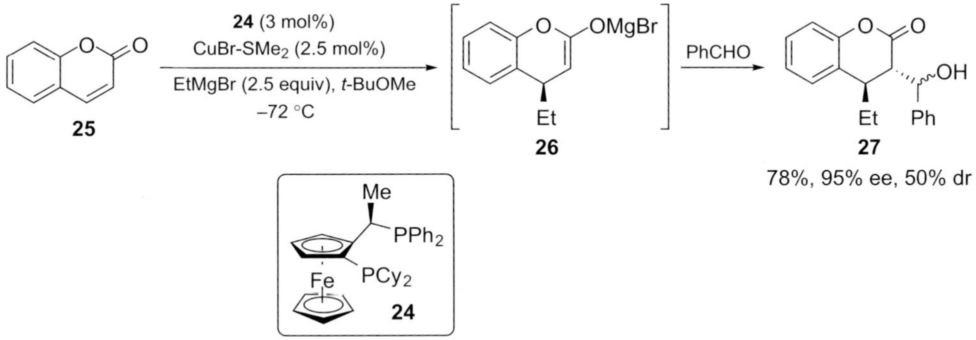

Scheme 11.6 Conjugate addition/aldol reaction of coumarin **25** [13].

Cramer [14] reported the enantioselective conjugate addition/electrophilic trapping sequence catalyzed by a Cu complex of phosphoramidite **28** using enone **30** and trimethylaluminum and alkyl halides (**32** or **33**). Aluminum enolates are known for their poor reactivity, impeding most direct electrophilic trapping reactions. The reactivity of aluminum enolates **31** is greatly enhanced by their conversion to the corresponding the ate complex with methyl lithium and further by the addition of hexamethylphosphoramide (HMPA) [15] (Scheme 11.7). This combination of an enantioselective conjugate addition/trapping sequence could provide a concise access to several anti-influenza active sesquiterpene natural product cores with a chiral quaternary carbon center [16].

The multimetal-mediated domino catalysis is one of the most powerful methodologies to give facile access to very complex chiral molecules. Feringa [17] developed domino 1,4-addition/allylation (Cu and Pd catalysis) followed by a ring-closing

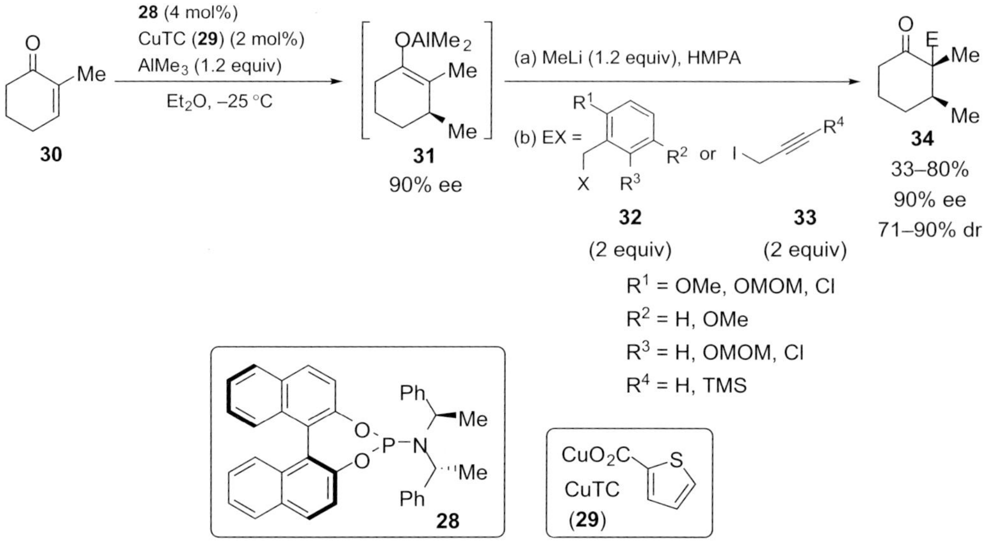

Scheme 11.7 One-pot addition/alkylation process of **30** [14].

metathesis (RCM, Ru catalysis) to get access to optically active cyclopentenone **38** with good diastereoselectivity and high enantioselectivity (Scheme 11.8). This catalytic asymmetric C–C bond formation provides alternative opportunities in a concise route to cyclopentenoid natural products.

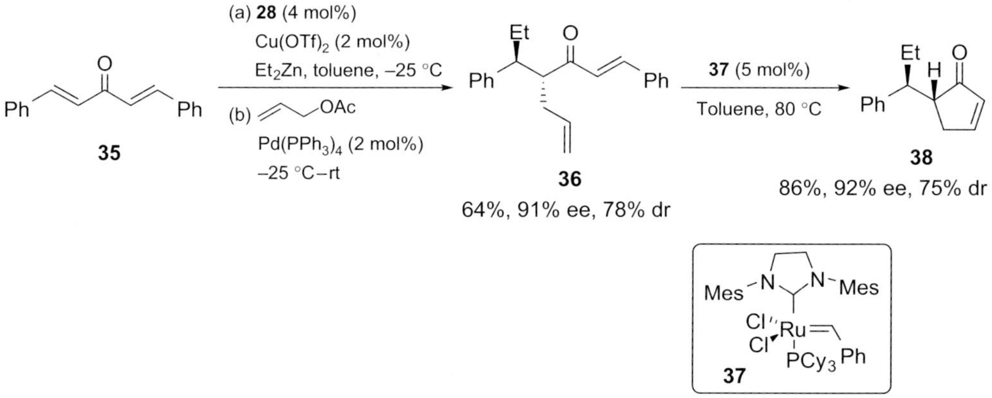

Scheme 11.8 Conjugate addition/allylation followed by RCM of **35** [17].

Alexakis also studied the enantioselective domino reactions catalyzed by a Cu complex of phosphoramidite **28**. The 1,4-addition products of this reaction can be employed in an enolate trapping with vinyl oxiran **40** with Pd(0) catalysis (Scheme 11.9) [18]. This domino process afforded the known precursor **43** in the synthesis of the anticancer agent clavularin B (**44**) [19].

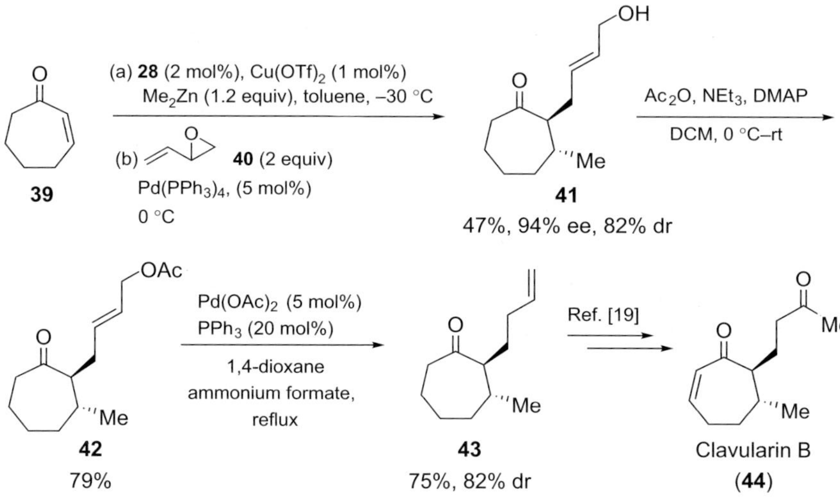

Scheme 11.9 Preparation of known precursor **43** in the synthesis of clavularin B (**44**) [18].

Alexakis developed the enantioselective domino reaction catalyzed by a Cu complex of phosphoramidite **45** using activated allylic compounds **47** with both leaving and electron-withdrawing groups. These designed electrophiles **47** could worked well even without Pd(0) catalysis (Scheme 11.10) [20].

In the enantioselective conjugate addition, the use of β,β-disubstituted enones to create quaternary chiral centers is more difficult because of the steric hindrance of the β-position [21]. Alexakis showed that Cu complexes with the chiral carbene ligand **49** are capable of catalyzing domino conjugate additions to the trisubstituted enone **50** in a one-pot process (Scheme 11.11). The magnesium enolates reacted poorly with alkyl halides, but their reactivity improved with use of an HMPA/THF

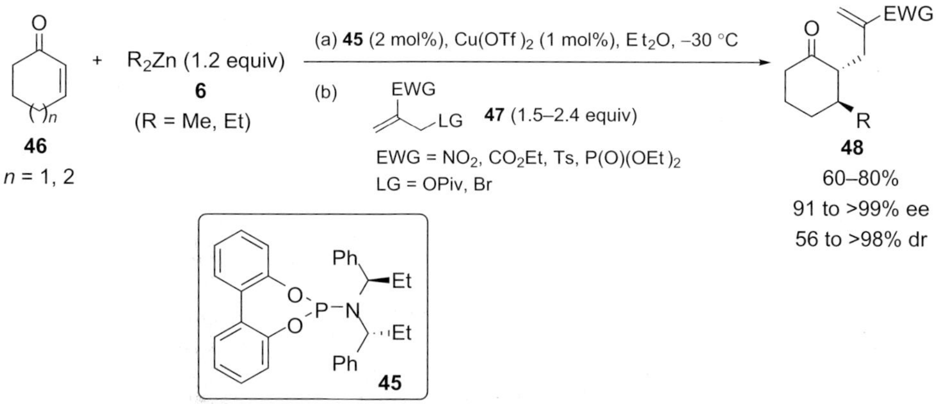

Scheme 11.10 Conjugate addition/trapping of electrophiles **46** without Pd catalysis [20].

(tetrahydrofuran) mixture as cosolvent [22]. The obtained dialkene substrate **51** was then cyclized by metathesis with the Grubbs II catalyst **37** to form bicyclic six/seven-membered ring **52** with high enantiomeric excess values but low cis/trans diastereoselectivity because of epimerization in the RCM step.

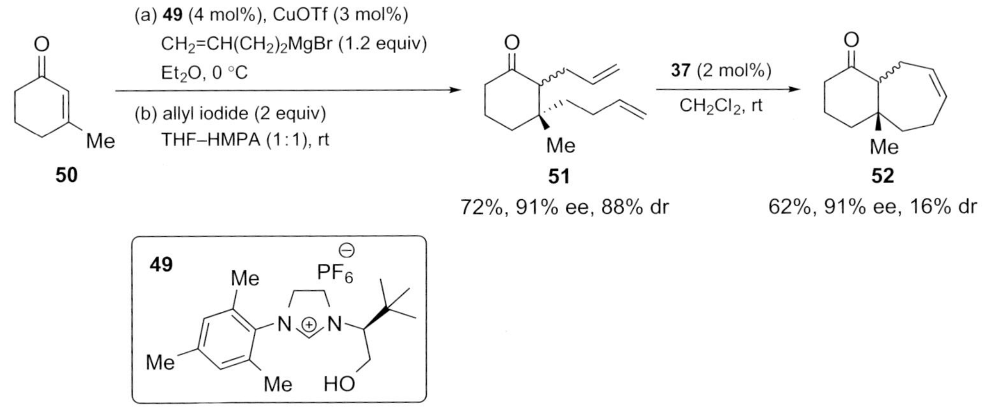

Scheme 11.11 Addition/trapping/ring-closing metathesis sequence of **50** [22].

A facile asymmetric introduction of a nitrogen atom at the α-position of a carbonyl compound is of great importance in the synthesis of natural products and drugs. In 2007, Liu [23] presented the first enantioselective domino conjugate addition/*N*-nitroso aldol reaction catalyzed by a Cu complex of phosphoramidite **53**. The reactions proceeded smoothly, furnishing the α-amino carbonyl compounds **55** in moderate yields with high enantioselectivities but low diastereoselectivities (Scheme 11.12).

Gonzalez-Gomez, Foubelo, and Yus showed that zinc enolates can also be useful nucleophiles in Mannich-type reactions (Scheme 11.13). A Cu complex of phosphoramidite **28** catalyzed conjugate additions of diethylzinc (**6a**) to cyclic

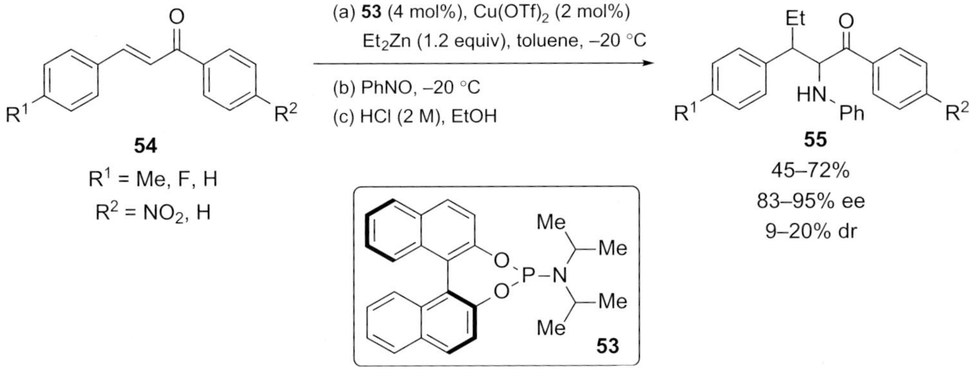

Scheme 11.12 Cu-catalyzed 1,4-addition/*N*-nitroso aldol reaction of **54** [23].

enones such as 2-cyclohexene-1-one (**45a**), which were followed by enolate trapping with chiral sulfinyl imine **56**. The configuration of the ethyl group was determined by the chiral ligand **28**. The approach of generating (3*S*)-enolate from the *si* face of the (R_s)-sulfinyl imine **56** (far from the bulky *t*-Bu group) in a six-membered chair-like transition state **57** leads to a completely stereochemically controlled product (R_s,*S*,*S*,*S*)-**58**. In this work, the stereoselectivity issue, controlled solely by the chiral catalyst, remained unexplored [24].

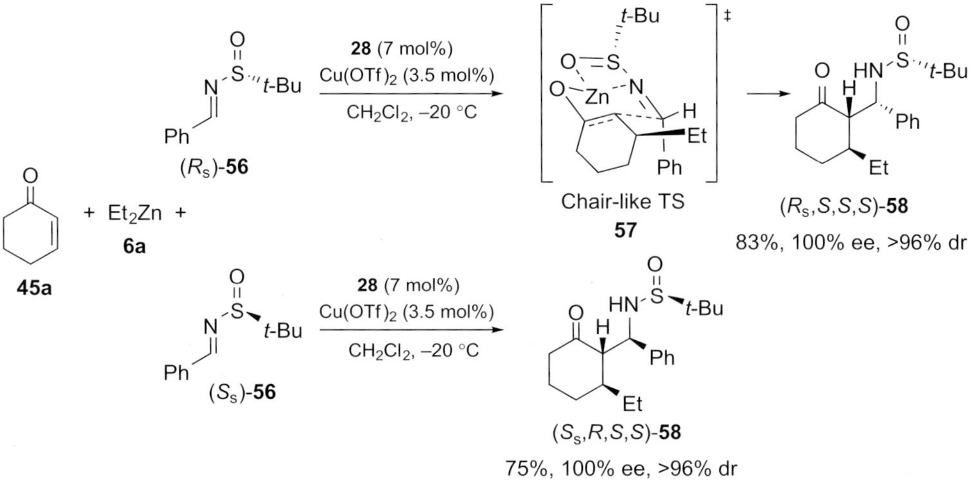

Scheme 11.13 Conjugate addition/Mannich reaction using the chiral sulfinyl imine **56** [24].

Sebesta reported that a Cu complex of Taniaphos-type ligand **59** catalyzed enantioselective addition of Grignard reagents to the cyclic enones/Mannich reaction sequence. The obtained chiral magnesium enolates were added to *N*-protected imines **60** directly (Scheme 11.14). The use of the diphenylphosphorane (DPP) protecting group for imine resulted in the formation of the desired domino products **61** in acceptable yields with excellent enantiomeric purities and diastereoselectivities [25a]. Computational studies of the possible transition states indicated that the eight-membered cyclic transition states **62a′**, in which Mg was bound to the oxygen atom of the DPP group, leads to approaching the (3*R*)-enolate from the *si* face of **60** to afford the highly stereochemically controlled β-aminocarbonyl compound (*R*,*R*,*R*)-**61**. Since then, Sebesta investigated the one-pot Mannich reaction of benzyl methylenecarbamate which was prepared from benzyl (isopropoxymethyl)carbamate (**65**) and $TiCl_4$ *in situ* (Scheme 11.15). In this way, the desired product **66** was obtained in moderate yields. However, diastereoselectivities were relatively low because of the Ti-mediated epimerization of the products **66** [25b].

Sebesta also studied the enantioselective conjugate additions of various Grignard reagents **21** to cyclic enones catalyzed by the Cu(I) complex of TADDOL-based

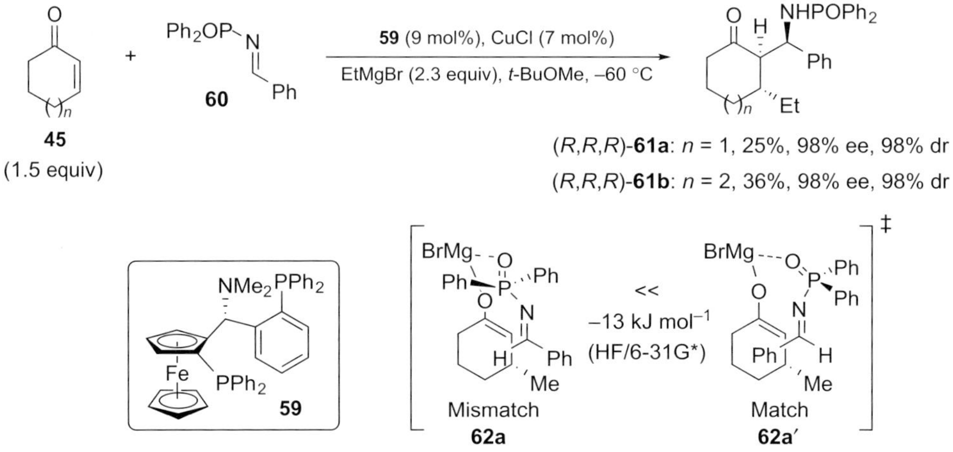

Scheme 11.14 Conjugate addition of Grignard reagent/Mannich reaction sequence [25a].

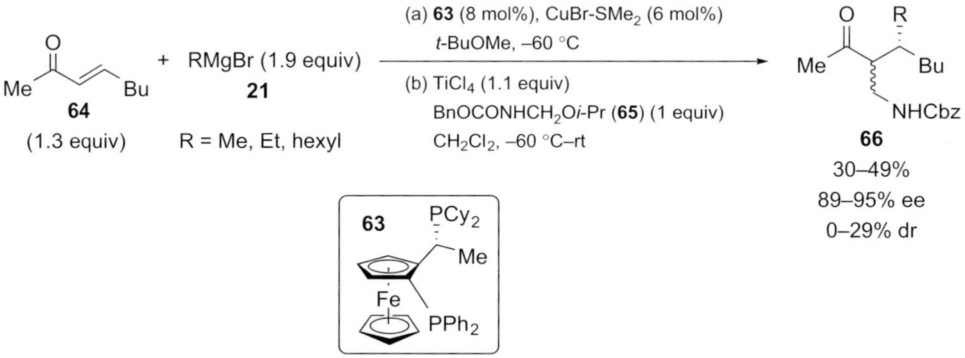

Scheme 11.15 One-pot conjugate addition/amidomethylation [25b].

((−)-(4*R*,5*R*)- or (+)(4*S*,5 *S*)-2,2-dimethyl-α,α,α′,α′-tetraphenyl-1,3-dioxolane-4,5-dimethanol) phosphane–phosphite ligand **67** (Scheme 11.16). Through trapping of the resulting chiral magnesium enolates with imines in one-pot procedures, the corresponding β-aminocarbonyl compounds **68** were obtained in moderate yields and high enantiomeric purities although with moderate diastereoselectivities [26].

An effective asymmetric domino cyclization to produce chiral *N*-heterocycles is also of great importance in the synthesis of chiral natural products and drugs. Huang succeeded in performing Cu-catalyzed asymmetric conjugate additions of diethylzinc reagent to acyclic enones **54** (Scheme 11.17), followed by enolate trapping with imine **70**. The key component for the successful execution of these transformations was the phosphane amide ligand **69**. The amino carbonyl compounds such as functionalized isoindolinones **71** were obtained in good yields, moderate diastereomeric ratios, and excellent enantiomeric purities [27].

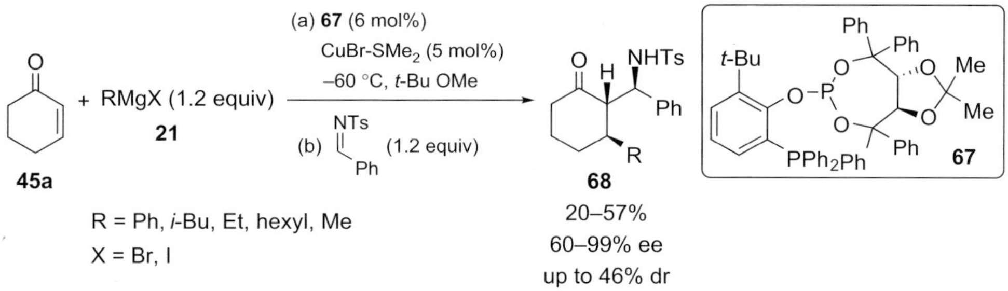

Scheme 11.16 Conjugate addition/Mannich reaction using TADDOL-based ligand **67** [26].

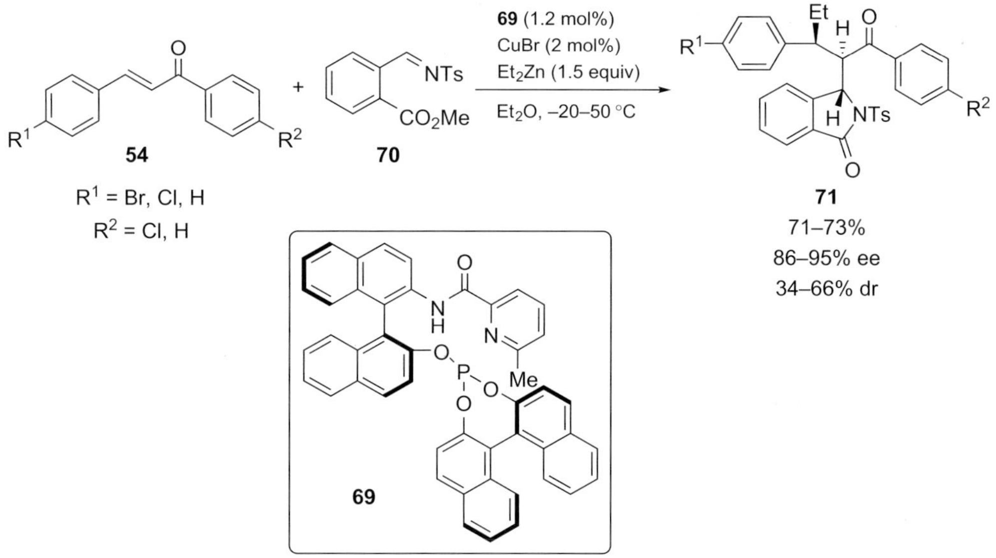

Scheme 11.17 Conjugate addition/Mannich reaction for the synthesis of isoindoline **71** [27].

Toward the development of an asymmetric conjugate domino reaction that constructs three contiguous stereocenters in acyclic systems, Huang [28] established an efficient method for assembling two distinct electrophiles into one molecule through the Cu-catalyzed dual Michael addition reaction. The initial Michael addition takes place from the *re* face of the enone **54a** through the control of the chiral ligand **69**. The nitro group of the nitroalkene is able to coordinate with the Cu center in the eight-membered cyclic transition state **73** on the second Michael addition and thus make the carbon–carbon double bond more electrophilic toward the enolate from its *re* face to facilitate a substrate-induced enantioselective Michael addition and give the γ-nitro ketone product **74** with high stereochemical control. This protocol using the chiral ligand **69** provides a reliable and rapid approach for the synthesis of chiral pyrrolidines **76**, which are five-membered *N*-heterocycles with multiple stereocenters (Scheme 11.18).

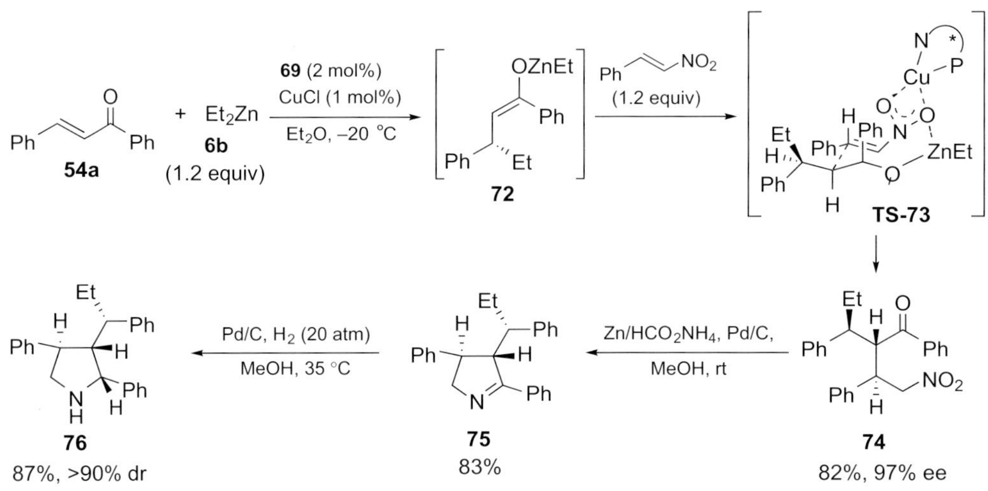

Scheme 11.18 Synthesis of chiral pyrrolidine derivatives **76** [28].

Chiral Pd complexes are one of the most widely investigated catalysts in asymmetric processes [4b]. So far, many chiral Pd-catalyzed Michael reactions have been reported [29]. Lu uncovered the diastereo- and enantioselective and highly efficient domino annulation of 2-formylarylboronic acids **78** and allenes **79** catalyzed by a chiral cationic Pd complex prepared from the ligand **77**. As shown in Scheme 11.19, after the transmetalation between the Pd complex **81** and arylboronic acid **78**, owing to the vacant coordination site on the cationic Pd complex **82**, allenes **79** could easily coordinate to the Pd center to give the intermediate **83**. Then, the η^1-allyl–Pd complex **84** was formed when allenes **79** were inserted into the Pd–carbon bond. The high Lewis acidity of the cationic Pd species **84** activates the carbonyl group by the coordination, and the high nucleophilic property of the cationic η^1-allyl–Pd complex **84** results in intramolecular 1,2-addition to produce the intermediate **85**. The subsequent hydrolysis of **85** would afford product **80** and complete the catalytic cycle. The success for the domino cyclization of allenes without the use of umpolung reagents showed the advantage using the cationic Pd complex as a catalyst for allenes [30].

The Henry reaction has been often associated with the Michael reaction in successful asymmetric domino sequences. Huang [31] reported a novel, highly enantio- and diastereoselective, Cu-catalyzed domino Michael/Michael/Henry reaction. The α-keto esters **87** and two nitroalkene molecules **88** produced the corresponding highly functionalized cyclohexanecarboxylates **89** with six stereogenic centers including one tetrasubstituted carbon in excellent yields and diastereo- and enantioselectivities when using $Cu(OAc)_2 \cdot H_2O$ and the chiral ligand **86** (Scheme 11.20).

The asymmetric Friedel–Crafts (FC) reaction is one of the most powerful methods to synthesize optically active aromatic compounds and has been included in various enantioselective domino reactions. Arai [32] reported the enantioselective

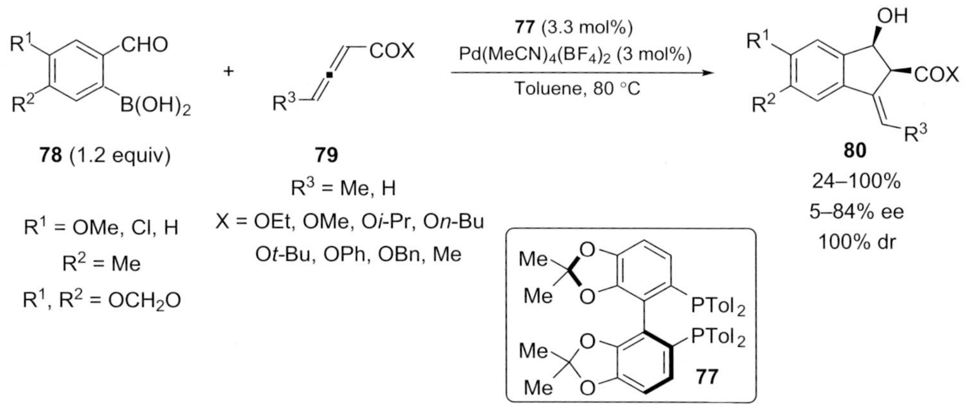

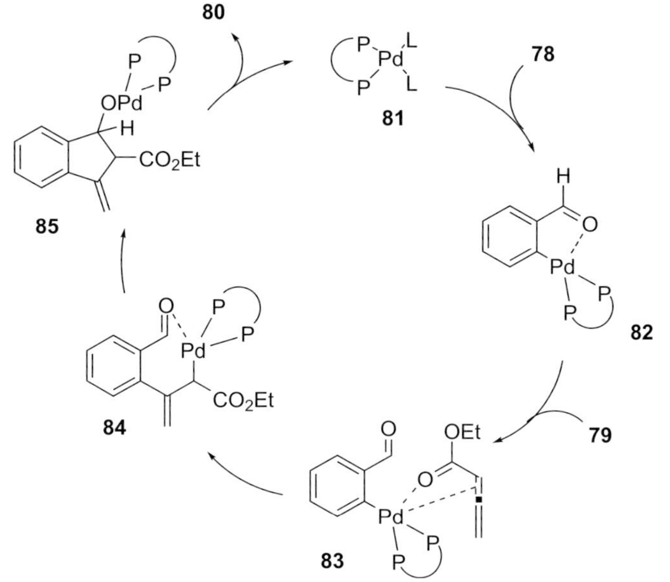

Scheme 11.19 Pd-mediated addition of boronic acid to the β-position of allenes/aldol sequence [30].

domino FC/Henry reaction of indoles, nitroalkenes, and aldehydes catalyzed by a combination of $Cu(OTf)_2$ with the chiral imidazoline-aminophenol ligand **90**. The corresponding domino products **93** bearing three contiguous stereocenters were obtained in high yields, remarkable enantioselectivities (up to 99% ee), and moderate to good diastereoselectivities (up to 90% dr). As shown in Scheme 11.21, in the first step, the nitroalkene **88** is activated by the Lewis acid catalyst **95** to start the enantioselective FC reaction. The diastereoselective Henry reaction is then promoted by the Cu nitronate functionality of the intermediate **97**, which results from the FC addition. The isolated FC product **94** did not react with the aldehyde **92** under the reaction conditions. Moreover, the enantiomeric excess value of

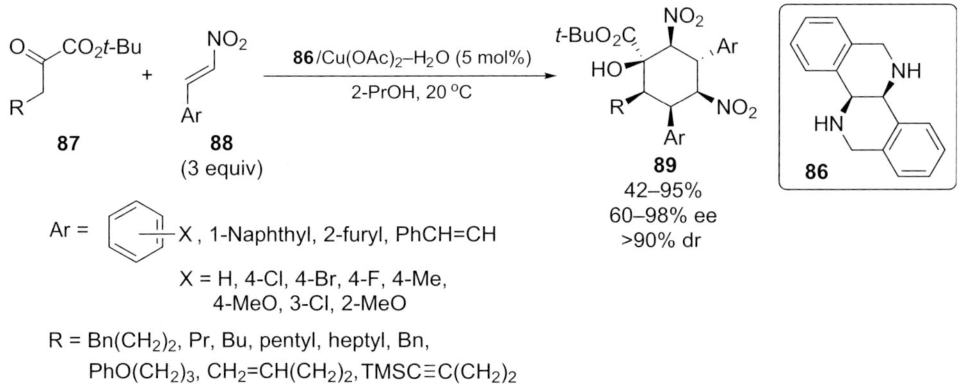

Scheme 11.20 Cu-catalyzed domino Michael/Michael/Henry reaction [31].

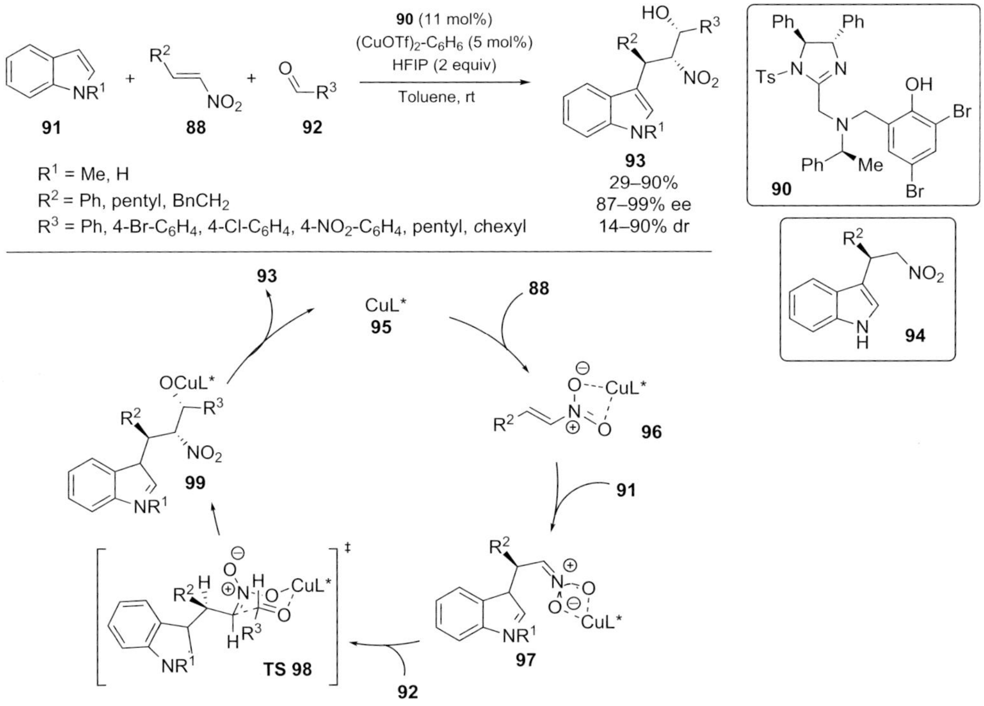

Scheme 11.21 Cu-catalyzed domino Friedel–Crafts/Henry reaction [32].

93a ($R^1 = H$, $R^2 = R^3 = Ph$) greatly improved to 99% from the 70% observed for the corresponding FC adduct **94**. Thus, the Henry reaction appears to proceed directly from the Cu nitronate intermediate **97** and is accompanied by kinetic resolution of the nitronate. The Cu-catalyzed FC reaction provided (*R*)-enriched **94**; therefore, the major (*R*)-Cu-nitronate intermediate **97** reacts with the aldehyde

92 to give the Cu alkoxide **99**. The formation of the (*R,R,R*) product **93** indicates that the diastereoselective Henry reaction proceeds in a syn-selective manner. This syn selectivity can be explained by considering the intermediate **98** with a Cu-containing six-membered ring. The relatively strong Lewis acidity of the CuOTf catalyst would favor the diastereoselective Henry reaction via this cyclic transition state. Finally, protonation of the Cu alkoxide **99** and aromatization furnishes the desired product **93** and regenerates the catalyst **95**. The additive 1,1,1,3,3,3-hexafluoro-2-propanol (HFIP) enhances the release of the products **93** from the catalyst. These products **93** could be readily envisaged as key starting materials in the synthesis of hydroxytryptamines [33].

Pyrrolidines are widely present in biologically significant compounds and thus have become a core skeleton of many drug candidates. Arai also investigated a catalytic asymmetric exo′-selective [3+2] cycloaddition of imino esters **99** with *trans*-nitroalkenes **88** to produce multifunctional pyrrolidines **100** via a Michael/Mannich domino sequence [34]. Usually, when a *trans*-nitroalkene **88** is used in the 1,3-dipolar cycloaddition, the stereo-conjunction between the 3- and 4-positions is fixed in a trans conformation, and four diastereomers are possible, classified as exo′, endo, exo, and endo′ isomers **100** [35]. Screenings of the metal salts to study the exo′ adduct ratio have found that nickel salts facilitated the selective production of the exo′ products **100**. Arai performed the exo′-selective reaction of imino esters **99** and *trans*-nitroalkenes **88** by using a combination of $Ni(OAc)_2$ with chiral imidazoline-aminophenol **90** as catalyst, which provided the corresponding pyrrolidines **100** in good yields and diastereoselectivities of up to 84% dr combined with high enantioselectivities for the major isomers. The present catalysis would be controlled thermodynamically in the stepwise Michael/Mannich cyclization reactions. As shown in Scheme 11.22, the nucleophilic addition at the C2 position of the nitroalkene **88** would be controlled by an interaction between the nitro functionality and the Ni center to direct the addition in anti-selective manner. However, after the anti-selective Michael addition via **103**, because the neutral Ni center cannot coordinate to both the nitro functionality and the imino ester at the same time, the Ni atom spontaneously flips to the nitronate for opening the strained cyclic intermediate **104**. This novel methodology represented the first general success in the catalytic asymmetric exo′-selective reaction of imino esters **99** and nitroalkenes **88**.

11.1.2
Domino Reaction Initiated by Cycloaddition

Enantioselective intermolecular [2+2+2] cycloadditions of 1,6-diynes with α,β-unsaturated carbonyl compounds are valuable methods for the synthesis of chiral cyclohexadienes in a single step [36]. The intramolecular Diels–Alder reaction (IMDA) is known to be a powerful strategy for the construction of complex multicyclic skeletons [37]. Tanaka reported an efficient enantioselective domino intermolecular [2+2+2]/IMDA cycloaddition reaction occurring between 1,6-diyne **106** and amide-linked 1,5-diene **107** bearing two sterically and/or electronically

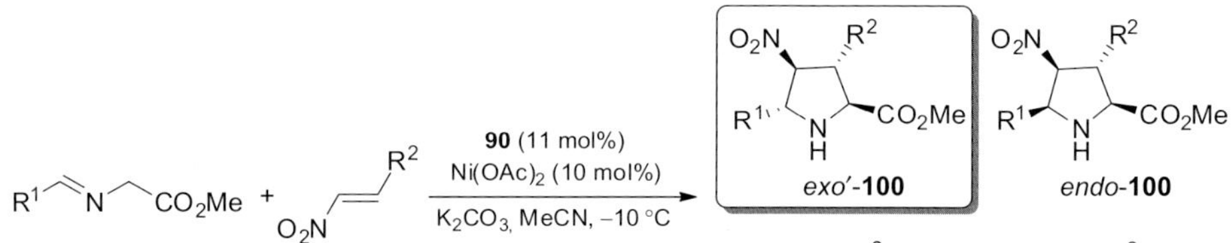

99 88

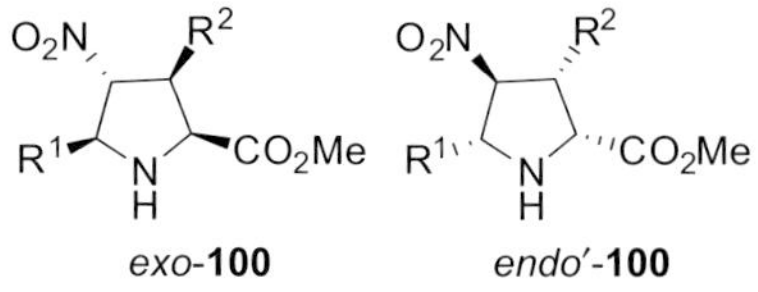

99a: R^1 = Ph
99b: R^1 = 4-MeO-C_6H_4
99c: R^1 = 3-MeO-C_6H_4
99d: R^1 = 4-Cl-C_6H_4
99e: R^1 = 3,4-$(OCH_2O)_4$-C_6H_3
99f: R^1 = 1-naphthyl
99g: R^1 = 2-naphthyl

88a: R^2 = Ph
88b: R^2 = 4-MeO-C_6H_4
88c: R^2 = 4-Br-C_6H_4
88d: R^2 = 2-Br-C_6H_4
88e: R^2 = 3-NO_2-C_6H_4
88f: R^2 = $CH_3(CH_2)_4$
88g: R^2 = $BnCH_2$

No.	Imino esters	Nitroalkenes	Yield (%)	Ratio of *exo′/endo/exo/endo′*	ee of exo′ (%)
1	**99a**	**88a**	99	82:16:1:1	97
2	**99b**	**88a**	79	90:10:0:0	93
3	**99c**	**88a**	67	89:11:0:0	96
4	**99d**	**88a**	84	87:9:0:4	97
5	**99e**	**88a**	70	88:10:0:2	91
6	**99f**	**88a**	67	85:8:0:7	91
7	**99g**	**88a**	70	92:7:0:1	99
8	**99a**	**88b**	93	86:13:0:1	96
9	**99a**	**88c**	85	85:14:0:1	97
10	**99a**	**88d**	94	80:17:0:3	96
11	**99a**	**88e**	78	81:15:1:3	97
12	**99a**	**88f**	68	68:28:0:4	92
13	**99a**	**88g**	64	79:21:0:0	94

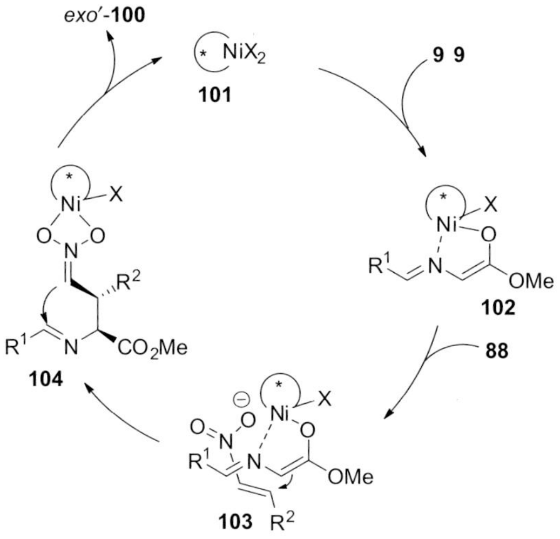

Scheme 11.22 Ni-catalyzed Michael/Henry sequence [34].

different alkene units [38]. This domino reaction was induced by a cationic Rh(I)/(*R*)-SEGPHOS ((+)-5,5′-Bis(diphenylphosphino)-4,4′-bi-1,3-benzodioxole, [4(*R*)-(4,4′-bi-1,3-benzodioxole)-5,5′-diyl]bis[diphenylphosphine]) (**105**) complex and provided the corresponding amides **109** in good yields and excellent enantioselectivities. The mechanism of the reaction involved a [2+2+2] cycloaddition between the two substrates, which was induced by the chiral cationic Rh(I) catalyst, to form the corresponding cyclohexadiene **108** containing a pendant alkene unit. A subsequent IMDA reaction of the cyclohexadiene furnished the final chiral bridged multicyclic product **109** (Scheme 11.23). Next, Tanaka investigated the enantioselective intermolecular [2+2+2] three-component co-cyclotrimerization/IMDA reaction cascade. After screening substrate combinations and catalysts, a cationic Rh(I)/(*R*)-BINAP complex was found to effectively catalyze the desired chemo-, regio-, and enantioselective cycloaddition cascade of dimethyl acetylenedicarboxylate (**110**), trimethylsilylacetylene (**111**), and 1,5-diene **112** at room temperature to yield the corresponding tricyclic amide **114** as a single regioisomer with excellent enantiomeric excess.

Enantioselective 1,3-dipolar cycloaddition of imino esters to electron-deficient alkenes is one of the most powerful and atom-economical C–C bond-forming reaction that facilitates the synthesis of a range of structurally and stereochemically rich pyrrolidines [35]. Wang developed the asymmetric 1,3-dipolar cycloaddition of naphthalene-1,4-dione (**116**) with imino esters **117** catalyzed by the Cu(I) complex of ferrocenyl ligand **115**, followed by silica-gel-promoted aromatization [39]. This one-pot process gave isoindolines **118** bearing a tetrasubstituted stereogenic center and a tertiary stereogenic center in high diastereoselectivity and excellent enantioselectivity (Scheme 11.24).

11.1.3 Domino Reaction Initiated by Carbometalation

Lautens reported a new enantioselective Rh-catalyzed domino reaction that gives access to fused heterocycles **122** by desymmetrization of alkyne-tethered cyclohexadienones **120** [40]. The syn addition of the Rh-aryl species **124** onto the alkyne led to two new C–C bonds and two stereocenters with good enantioselectivity (Scheme 11.25).

11.2 Domino Reaction Initiated by C–H Bond Formation

11.2.1 Domino Reaction Initiated by Conjugate Addition

The aldol reaction is a classical method for the creation of C–C bonds in organic synthesis [41]. Reductive aldol reaction of α,β-unsaturated carbonyl compounds with saturated carbonyl compounds promoted by catalytic amounts of various

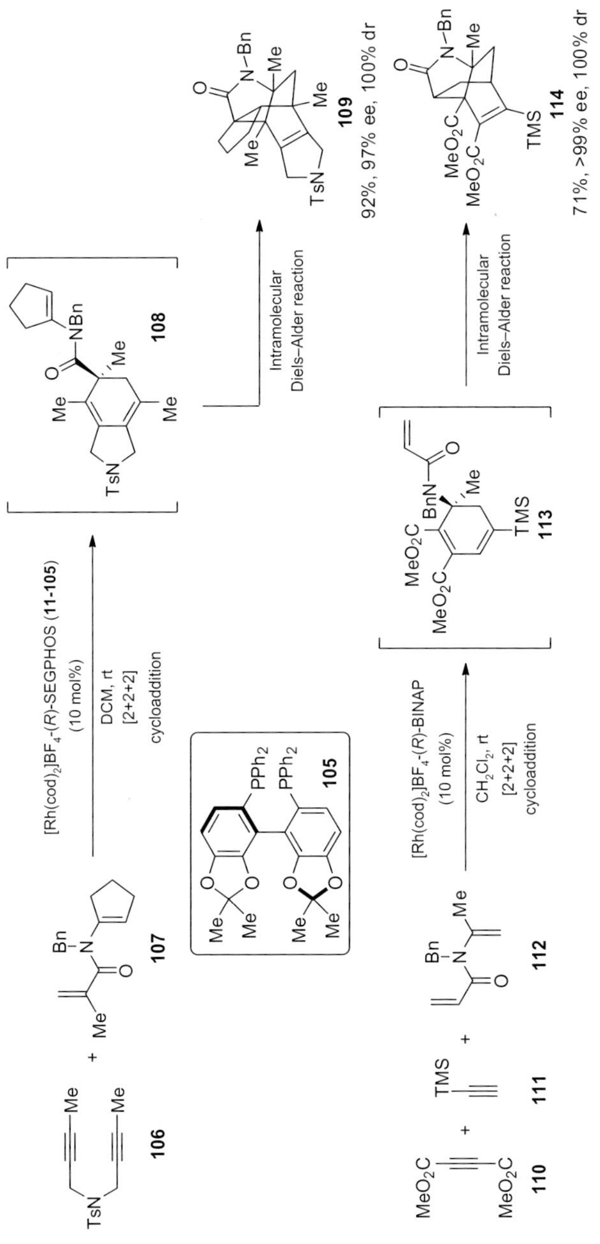

Scheme 11.23 Intermolecular [2+2+2] cycloaddition/intramolecular Diels–Alder (IMDA) cascade [38].

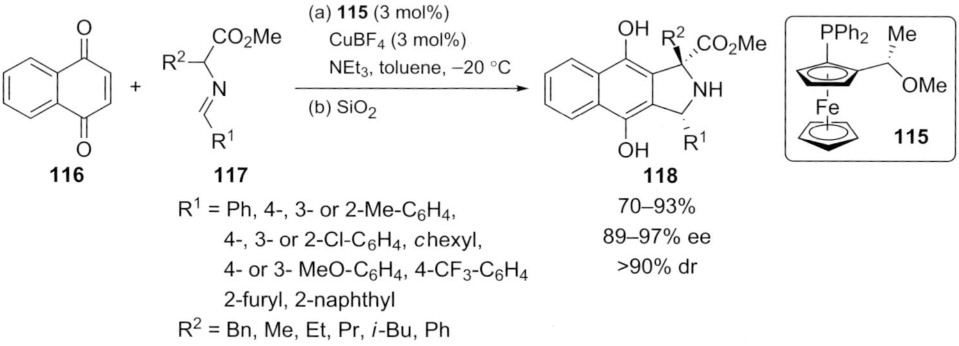

Scheme 11.24 1,3-Dipolar cycloaddition/aromatization [39].

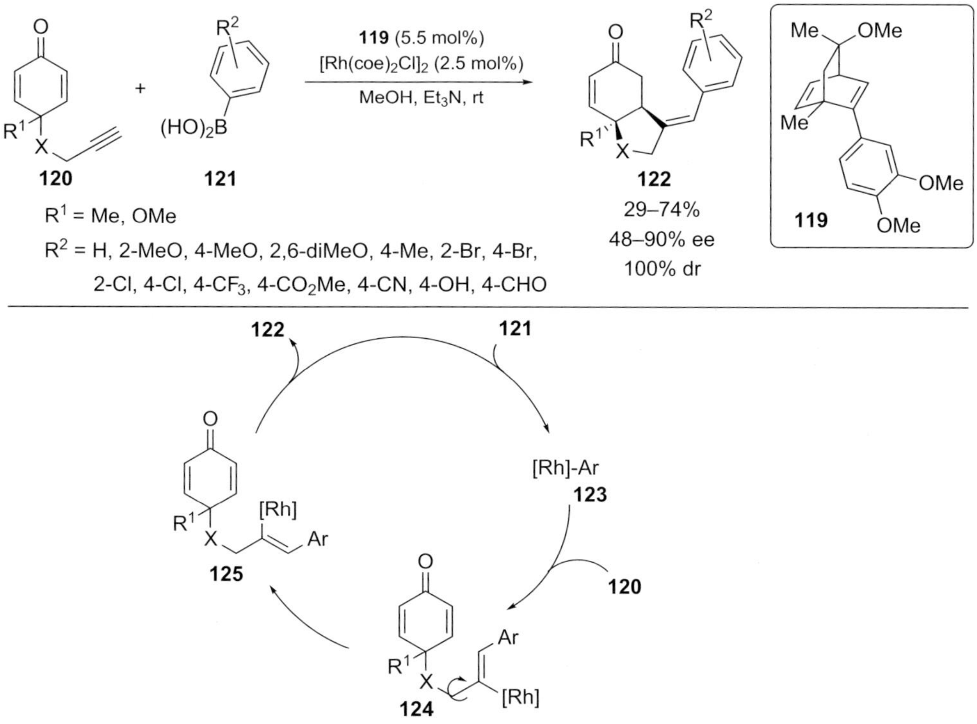

Scheme 11.25 Rh-catalyzed domino transformations of alkynylcyclohexadienones with organoborane [40].

transition-metal complexes and reducing agents is a powerful tool for the stereocontrol of C–C bond formation. In the reductive aldol reactions, enantioselectivity of the product depends on the second addition of the generating metal enolate with chiral ligands to electrophiles. So far, excellent progress has been achieved in the area of domino reductive aldol reactions for the construction of several contiguous stereocenters in a one-pot synthesis [42]. In 2006, Shibasaki and Kanai [43] described the first example of a catalytic, enantioselective, intermolecular reductive aldol reaction. Cu-(*R*)-Tol-BINAP-catalyzed reductive aldol reactions of acetophenone (**126a**) and methyl acrylate (**127**) were performed in the presence of triethoxysilane as reducing agent to produce the corresponding tertiary alcohol **128** quantitatively but in only 29% ee and 10% dr. Since then, Shibasaki and Kanai found that the Cu(I) complex of Taniaphos-based diphosphine ligand **129** promoted the intermolecular α-selective reductive aldol reaction of ketones **126** and allenoates **130** to give the corresponding tertiary alcohols **131** in high yields and good enantio- and diastereoselectivities (Scheme 11.26). The α- and γ-addition of allenoates **130** could be switched depending on the structure of chiral ligand. When employing the Cu(I) complex of (*R*)-DTBM-SEGPHOS, the reaction became γ-selective and provided the corresponding γ-adduct in high yields (70–97%) and excellent enantioselectivities (84–99%) [44].

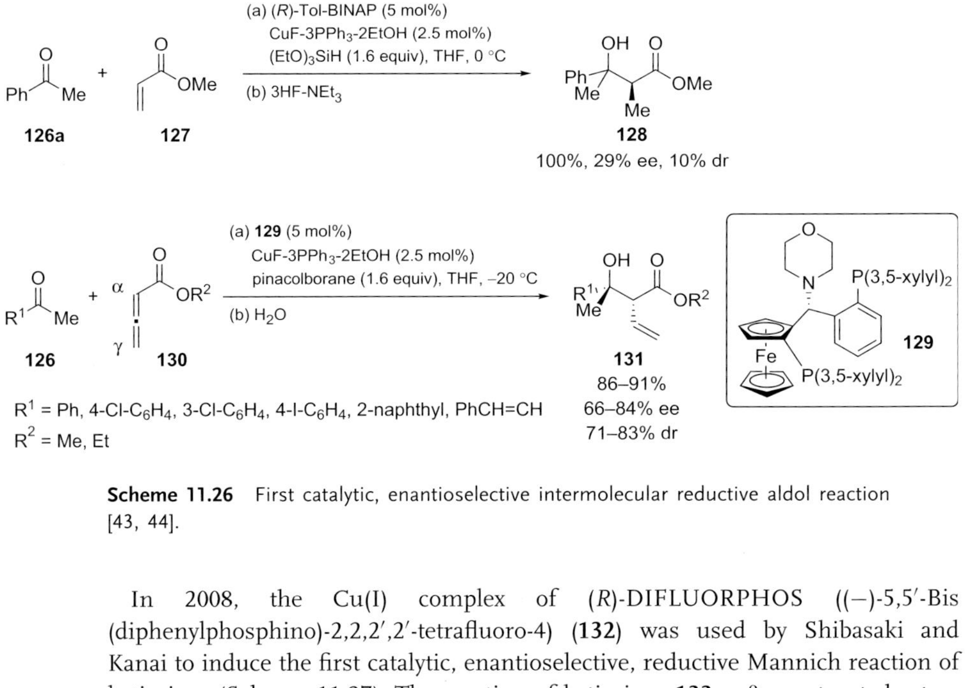

Scheme 11.26 First catalytic, enantioselective intermolecular reductive aldol reaction [43, 44].

In 2008, the Cu(I) complex of (*R*)-DIFLUORPHOS ((−)-5,5′-Bis (diphenylphosphino)-2,2,2′,2′-tetrafluoro-4) (**132**) was used by Shibasaki and Kanai to induce the first catalytic, enantioselective, reductive Mannich reaction of ketimines (Scheme 11.27). The reaction of ketimines **133**, α,β-unsaturated esters **134**, and triethoxysilane as a reducing agent provided the corresponding amines **135**, containing contiguous tetrasubstituted carbons, in high yields with good diastereoselectivities and high enantioselectivities. This methodology constituted the first entry to the catalytic asymmetric synthesis of $\beta^{2,3,3}$-amino acid derivatives [45].

Riant employed other chiral Taniaphos-based ligands to induce chirality in Cu(I)-catalyzed reductive aldol reactions of aldehydes **92** with methyl acrylate **127** using phenylsilane as a reducing agent. Although the reaction was highly chemoselective,

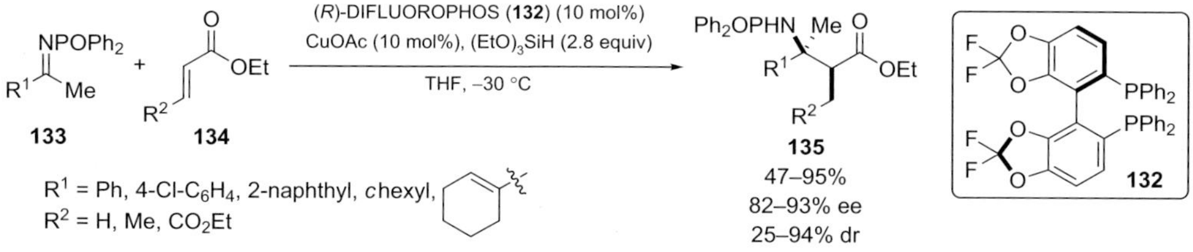

Scheme 11.27 First catalytic, enantioselective reductive Mannich reaction of ketimine [45].

it gave moderate syn diastereoselectivities. High enantioselectivities were obtained for the domino products arising from a wide range of cyclic aliphatic, aromatic, and heteroaromatic aldehydes **92** when the Taniaphos-type ligand **59** was employed [46]. The scope of this methodology was extended to the reductive aldol reactions of ketones under similar conditions. The domino reaction of ketones **126** with methyl acrylate **127** and phenylsilane as a reducing agent gave rise chemoselectively, by using chiral Taniaphos-type ligand **137**, to the corresponding tertiary alcohols **138** (Scheme 11.28) [47].

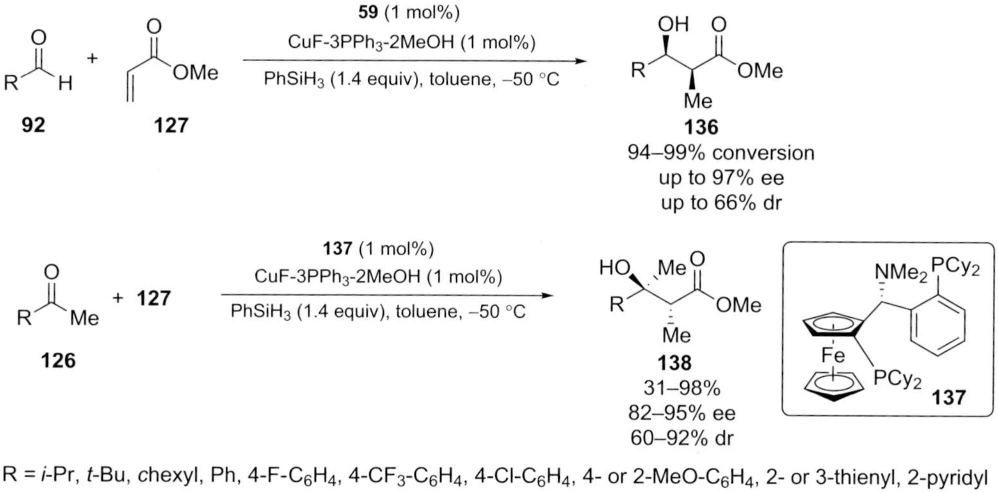

Scheme 11.28 Reductive aldol reaction catalyzed by Cu–Taniaphos complexes [46, 47].

In 2008, Krische reported the first enantioselective reductive aldol couplings of vinyl ketones, which were achieved through the design of a new class of TADDOL-derived phosphonite ligand **139** (Scheme 11.29). The Rh-catalyzed hydrogenative aldol coupling of vinyl ketones **140** with aldehydes **141**, performed in the presence of chiral ligand **139**, allowed the corresponding linear aldol adducts **142** to be obtained with high diastereo- and enantioselectivities [48].

In 2008, Lipshutz reported the first hydrometallative intramolecular reductive cycloaddition, leading to three new contiguous stereocenters in a one-pot process (Scheme 11.30). The enantioselective domino conjugate reduction/intramolecular

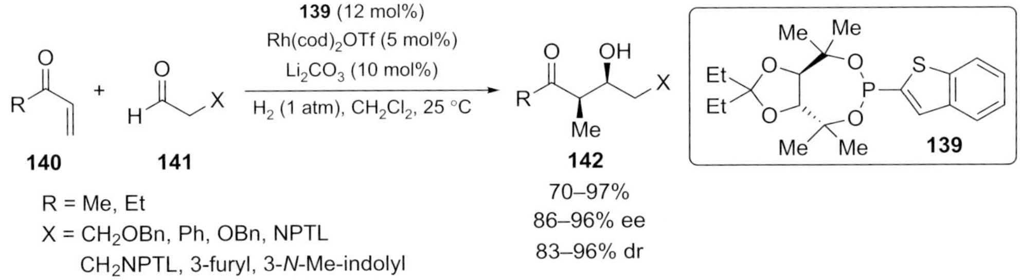

Scheme 11.29 First enantioselective, reductive aldol couplings of vinyl ketones with aldehydes [48].

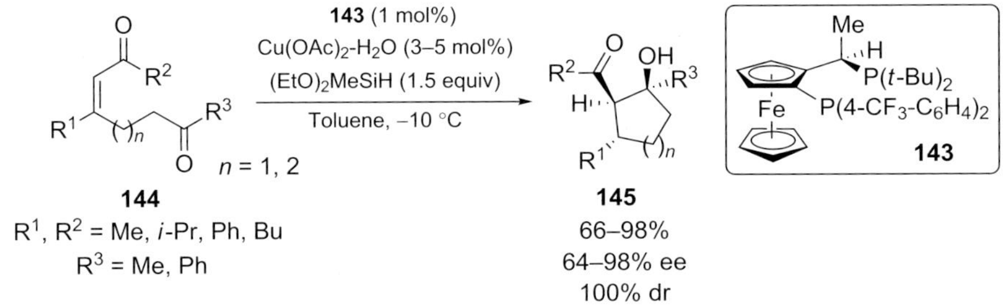

Scheme 11.30 First hydrometallative, intramolecular, reductive cycloaddition leading to three new contiguous stereocenters [49].

aldol reaction of β,β-disubstituted keto enones **144** provided the functionalized cyclic alkanols **145** as single diastereomers in high yields and enantioselectivities. The generation of the three contiguous stereocenters was achieved by using a combination of $Cu(OAc)_2$ with chiral bisphosphine **143** in the presence of diethoxymethylsilane as a reductant. In this process, the initial conjugate hydride addition generated an intermediary chiral Cu enolate through facial discrimination, which subsequently participated in an intramolecular aldol addition to the terminal ketones. Transmetalation of the resulting Cu alkoxide with a particular stoichiometric silane regenerated the chiral Cu hydride species [49].

Riant developed a versatile methodology for the diastereo- and enantioselective domino reductive aldol cyclization reaction of functionalized α,β-unsaturated esters **147** into the corresponding bicylic domino products **148** in high yields and enantioselectivities. The reaction was catalyzed by a Cu(I) complex of chiral bisphosphine **146** in the presence of phenylsilane as reductant. The product **148** with three contiguous stereocenters was formed with high diastereoselectivities. The stereoselectivity of the process increased with the steric hindrance of the ester moiety, since the best results were obtained with the *tert*-butyl esters (Scheme 11.31) [50].

Fukuzawa [51] designed click-ferrophos **149** and tested for the Cu(I)-catalyzed, enantioselective reductive aldol reaction of ketones with an acrylic ester in the

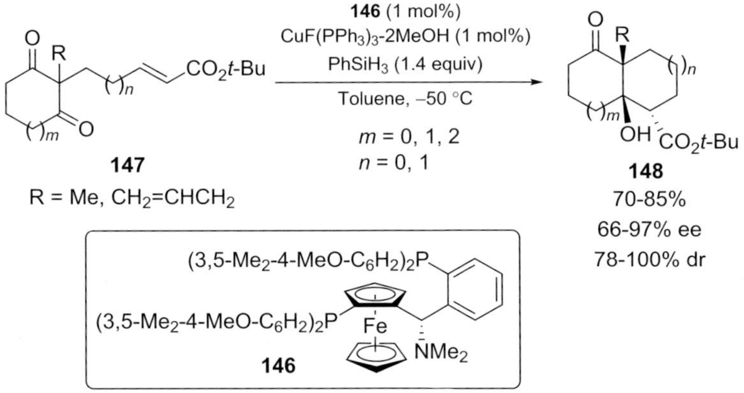

Scheme 11.31 Domino reductive aldol cyclization of functionalized α,β-unsaturated ester [50].

presence of Ph_3SiH. The Cu(I)–click-ferrophos complex was efficient for the reaction of acetophenone (**126a**) with methyl acrylate (**127**) to afford the erythro adduct **150** both in high diastereo- and enantioselectivity. The diastereomeric ratio of erythro/threo was improved when compared to Taniaphos-type ligand **59** (Scheme 11.32).

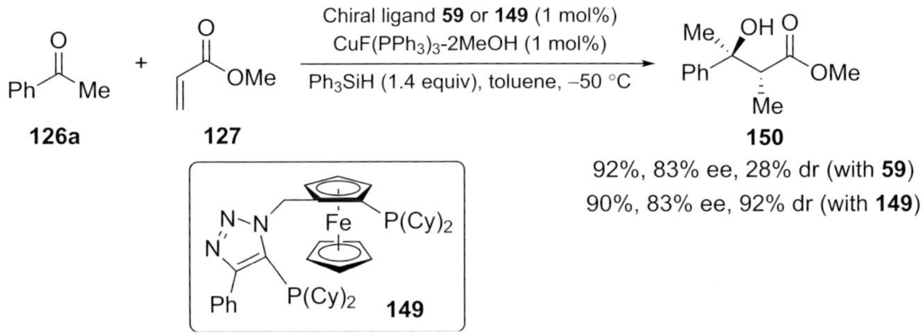

Scheme 11.32 Reductive aldol reaction of ketone with acrylic ester catalyzed by Cu(I)–click-ferrophos [51].

Chiu reported a range of enethioate derivatives of 1,3-dione **151** which underwent reductive aldol cyclization catalyzed by a chiral Cu hydride generated *in situ* from 5 mol% Taniaphos-type ligand **59**, 5 mol% $Cu(OAc)_2{\cdot}H_2O$, 5 mol% bipyridine, and 2.0 equiv of $PhSiH_3$, to furnish polycyclic β-hydroxythioester product **153** with >96% dr and in 93% yield and 93% ee (Scheme 11.33). The use of an amine such as bipyridine or 2,6-lutidine as an additive results in an increase of the overall reaction rate. The major bicyclic aldol product **153** has all substituents in cis form, which can be rationalized by a reductively generated (*Z*)-enolate reacting with the dione via the cyclic transition state **152** [52].

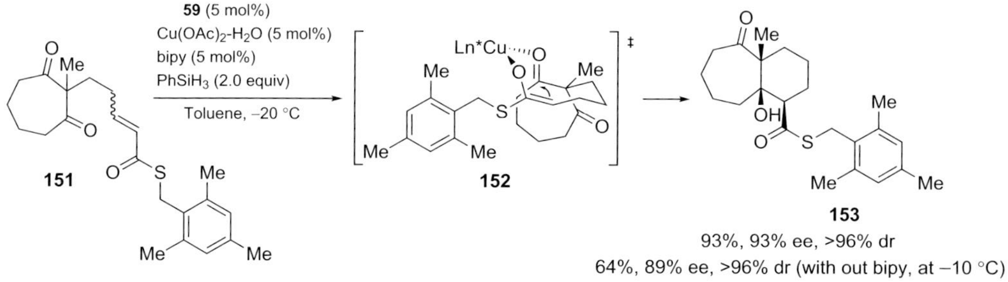

Scheme 11.33 Desymmetrical reductive aldol cyclization of enethioate of 1,3-dione [52].

By combining the utility of Cu hydride catalysis with the ability of C=N containing azaarenes to activate adjacent alkenes toward nucleophilic additions, the enantioselective reductive coupling of alkenyl azaarene **154** with ketone **155** has been developed by Lam (Scheme 11.34). This process is tolerant to a wide variety of azaarenes and ketones, and provides aromatic heterocycles **157** bearing tertiary-alcohol-containing side chains with high levels of diastereo- and enantioselection via a six-membered chair-like transition state **156** [53].

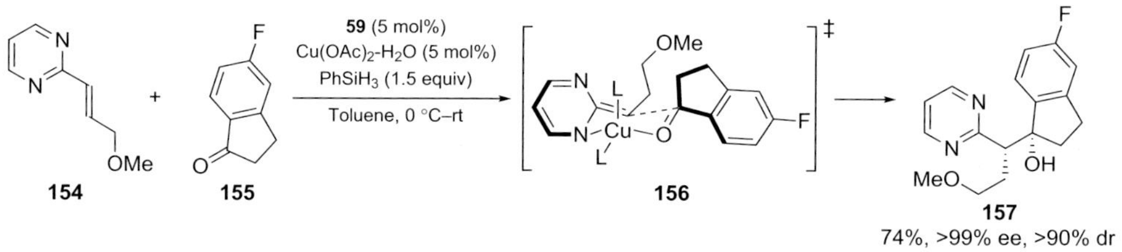

Scheme 11.34 Enantioselective, Cu-catalyzed reductive coupling of alkenyl azaarene with ketone [53].

11.3 Domino Reaction Initiated by C–N Bond Formation

11.3.1 Domino Reaction Initiated by Imine Formation

Sansano has reported a highly efficient silver-catalyzed enantioselective domino imine formation/1,3-dipolar cycloaddition reaction [54]. As shown in Scheme 11.35, benzaldehyde (**92a**) and *N*-methylmaleimide (**158**) or 3-pyridinecarbaldehyde (**92b**) and disulfone **161**, glycine methyl ester hydrochloride (**159**), triethylamine (1.05 equiv), and (*S*)-BINAP–$AgSbF_6$ (5 mol%) were mixed together in toluene, and the resulting mixture was allowed to react at room temperature. The outcomes

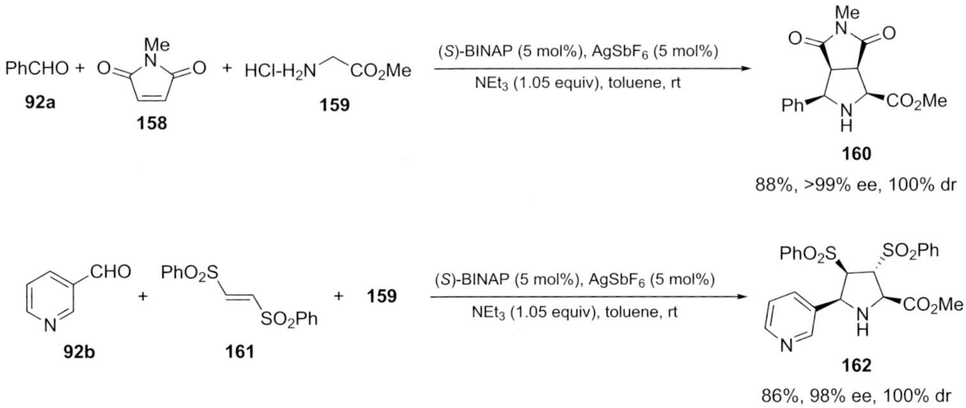

Scheme 11.35 Silver-catalyzed, enantioselective domino imine formation/1,3-dipolar cycloaddition reaction [54].

obtained for the compound *endo*-**160** or *endo*-**162** were impressive (88% yield, >99% ee; or 86% yield, 98% ee, respectively).

In 2011, Wulff described the first three-component catalytic asymmetric aziridination reaction of an aldehyde **92**, bis(dimethylanisyl)methylamine (**163**), and ethyl diazoacetate (**164**) to provide the corresponding chiral aziridine-2-carboxylic esters (**165**) [55]. When promoted by a chiral boroxinate catalyst *in situ* generated from $B(OPh)_3$ and chiral ligand (*S*)-VAPOL (2,2′-diphenyl-(4-biphenanthrol)), the reaction afforded products **165** with good yields and excellent diastereoselectivities and enantioselectivities (Scheme 11.36). This novel methodology furnished an effective solution to the problem of unstable imines derived from aliphatic aldehydes that cannot be purified.

11.3.2
Domino Reaction Based on Cycloaddition

Yang developed a Pd(II)-catalyzed enantioselective oxidative domino cyclization reaction with the chiral ligand **166** comprising quinoline and oxazoline under molecular oxygen as a green oxidant [56]. While the use of (−)-sparteine as a ligand provided good enantioselectivities in the case of oxidative domino cyclization of monosubstituted olefinic substrates **167** ($R^1 = R^2 = H$) [57], its use for the reaction of disubstituted olefinic substrates **167** (R^1 or $R^2 = H$) gave only 14% ee. The use of novel catalyst systems, such as the combination of $Pd(OAc)_2$ with chiral ligand **166**, allowed the formation of the corresponding tricyclic products **169** in good yields and excellent enantioselectivities and diastereoselectivities via a *re* face cyclization intermediate **168** as shown in Scheme 11.37.

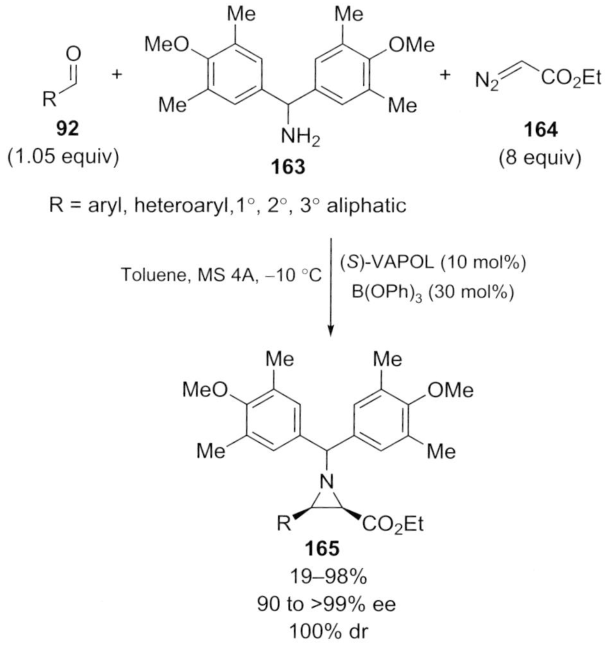

Scheme 11.36 First three-component catalytic, asymmetric, aziridination reaction [55].

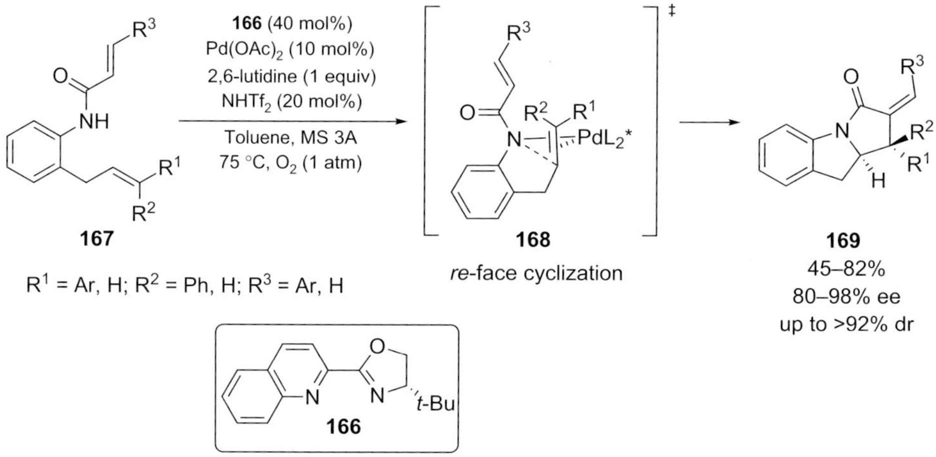

Scheme 11.37 Pd(II)-catalyzed enantioselective oxidative domino cyclization [56].

11.4
Domino Reaction Initiated by C–O Bond Formation

11.4.1
Domino Reaction Initiated by Carbonyl Ylide Formation

The synthetic potential of diazo compounds, in particular of α-diazoketones and α-diazoesters, is greatly widened by the ability of the derived carbene or metal–carbene intermediates to undergo inter- and intramolecular formation of N-, O-, S-, and other ylides. The benefit of these transformations, which are performed with Rh-, Cu-, Ni-, Ag-, or Yb-based catalysts, is given by subsequent rearrangement or addition reactions of the reactive ylides [2g,k, 35]. Among them, enantioselective catalytic domino carbonyl ylide formation/1,3-dipolar cycloaddition offers an elegant route to highly substituted oxygen-containing heterocycles. The development of a catalytic enantioselective version of this domino reaction has become a challenging objective. The primary work in the chiral Rh catalysis was reported by Hodgson *et al.* [58] in 1997. As a recent excellent example in this domino process, Suga developed 1,3-dipolar cycloaddition reactions of carbonyl ylides *in situ* generated from α-diazoacetophenone derivatives **171** with 3-crotonoyl-2-oxazolidinones **172** by using a combination of a chiral Pybox ligand **170** with $Yb(OTf)_3$ as the chiral Lewis acid catalyst. The reaction of *o*-(*p*-bromobenzyloxy)carbonyl-α-diazoacetophenone **171a** ($R^1 = 4\text{-Br-}C_6H_4CH_2$) and **172a** ($R^2 = Me$) afforded the corresponding endo cycloadduct **173a** as a sole product with >98% dr and with extremely high enantioselectivity of 96% ee [59a]. In 2007, Suga also reported the first successful example of reverse electron demand dipole–LUMO (lowest unoccupied molecular orbital)/dipolarophile–HOMO (highest occupied molecular orbital) controlled cycloaddition reactions between carbonyl ylides, which were *in situ* generated from *o*-methoxycarbonyl-α-diazoacetophenone derivatives **175** in the presence of $Rh_2(OAc)_4$, and vinyl ether derivatives activated by chiral Lewis acids [59b]. When cyclohexyl vinyl ether (**177**) was activated by the Ni(II) complex of ligand **174** as a chiral Lewis acid, it reacted with α,α′-dicarbonyl diazo compounds **175** to yield the corresponding endo cycloadducts **178** in high yields, almost complete diastereoselectivity, and high enantioselectivities (Scheme 11.38).

In 2007, Fu described the first examples of diastereo- and enantioselective Cu-catalyzed [4+1] cycloadditions of enones **180** with diazo compounds **181** using the planar chiral bipyridine ligand **179**. This new method furnishes synthetically useful, highly substituted 2,3-dihydrofuran derivatives **183** with good efficiency and stereoselection (Scheme 11.39) [60].

Hashimoto has reported the enantioselective domino carbonyl ylide formation/1,3-dipolar cycloaddition reaction of a range of α-diazo ketones **185** with various aromatic aldehydes **92**, promoted by the chiral Rh(II)–tetrakis[*N*-benzene-fused-phthaloyl-(*S*)-valinate] complex (**184**). The present reaction afforded the corresponding bicyclic cycloadducts **186** in good yields and enantioselectivities. The exo diastereomer **186** was diastereoselectively generated as a single product in almost all the cases (Scheme 11.40) [61].

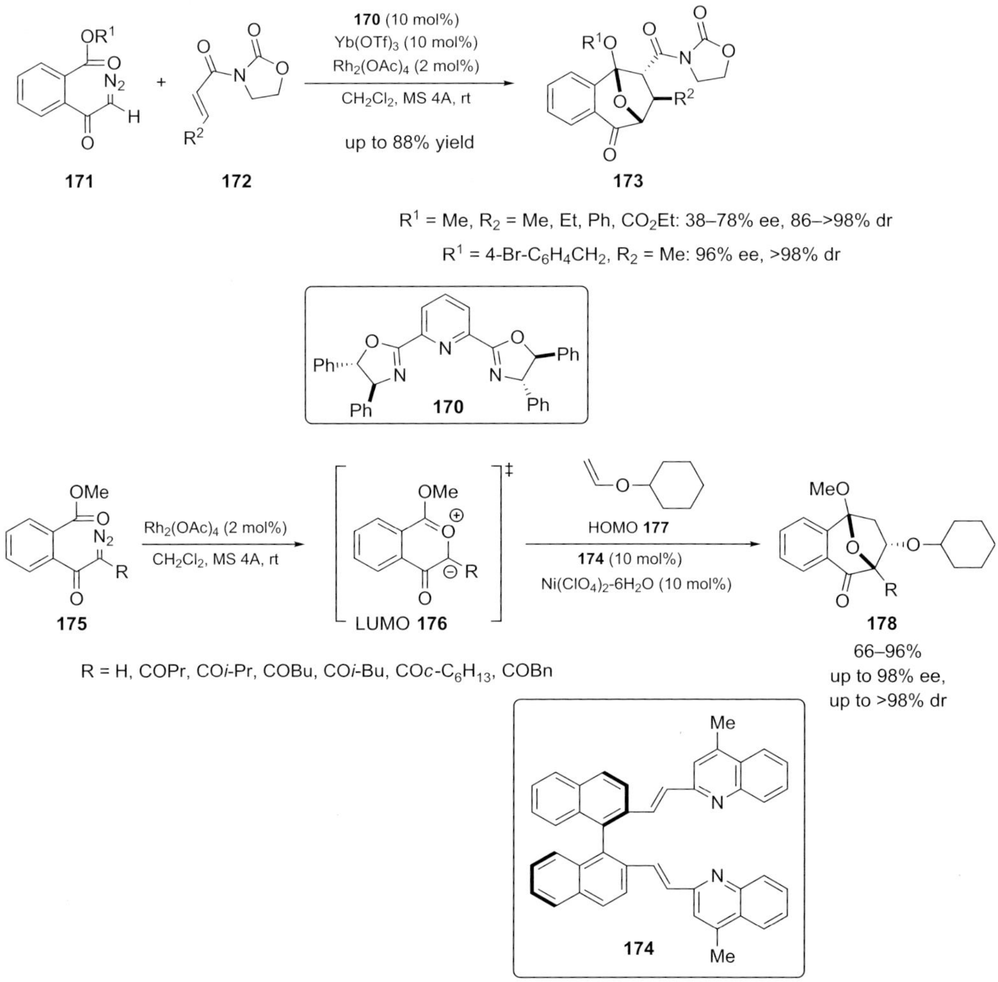

Scheme 11.38 Enantioselective domino cycloaddition developed by Suga [59].

In addition, Hashimoto [62a] found that the Rh_2(II)–tetrakis[*N*-tetrachlorophthaloyl-(*S*)-*tert*-leucinate] complex (**187**) was an exceptionally effective catalyst for the enantioselective domino carbonyl ylide formation/1,3-dipolar cycloaddition reaction of 2-diazo-3,6-diketo esters **188** with dipolarophiles such as acetylenes **189**, providing the corresponding cycloadducts **190** in good to high yields and excellent enantioselectivities and diastereoselectivities. This procedure was applied to the synthesis of the natural product **191** in nine steps starting from the domino cyclic product **190a** [62b]. Furthermore, the scope of this methodology could be extended to other dipolarophiles such as styrenes **192** and indoles **91** (Scheme 11.41) [62].

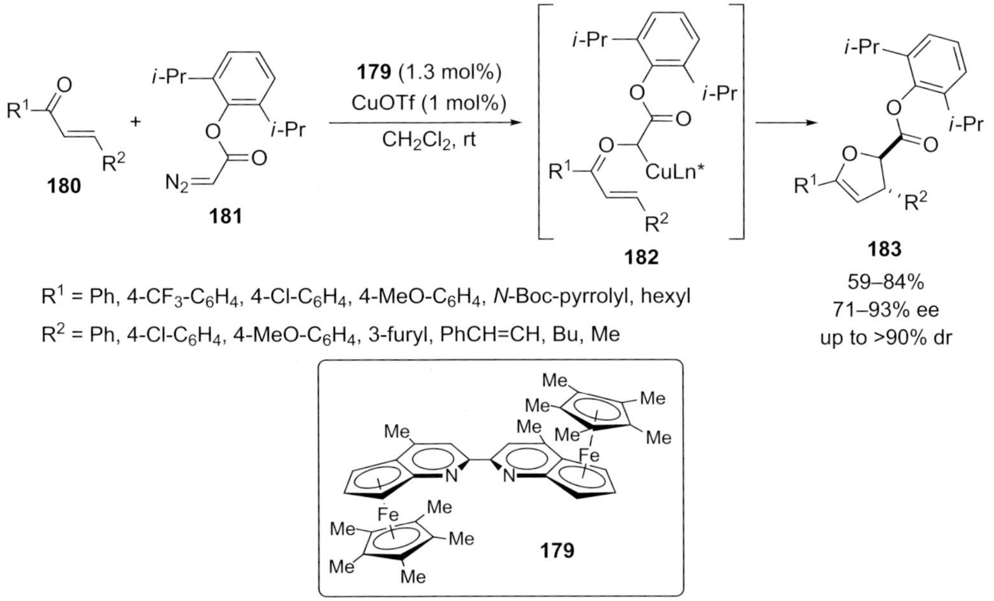

Scheme 11.39 Cu-catalyzed [4+1] cycloaddition of α,β-unsaturated ketone with diazoacetate [60].

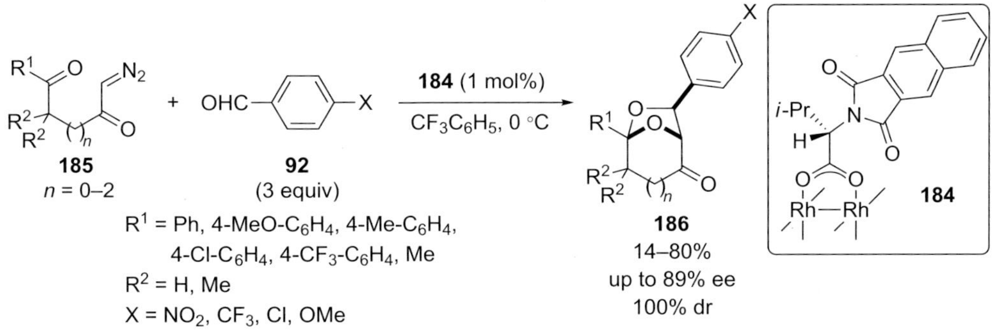

Scheme 11.40 Chiral Rh_2(II)-catalyzed domino carbonyl ylide formation/1,3-dipolar cycloaddition [61].

Hodgson [63] investigated the enantioselective intramolecular domino carbonyl ylide formation/1,3-dipolar cycloaddition reaction employing the sulfonyl functionality as a different type of electron-withdrawing group at the ylidic carbon, using unsaturated α-diazo-β,ϵ-diketo sulfones **193**. Among a series of chiral catalysts tested, the chiral Rh catalyst **184** provided the best, albeit moderate, enantioselectivity of 43% ee for the cycloadduct **194a** derived from phenyl sulfone **193a** along with 75% yield (Scheme 11.42).

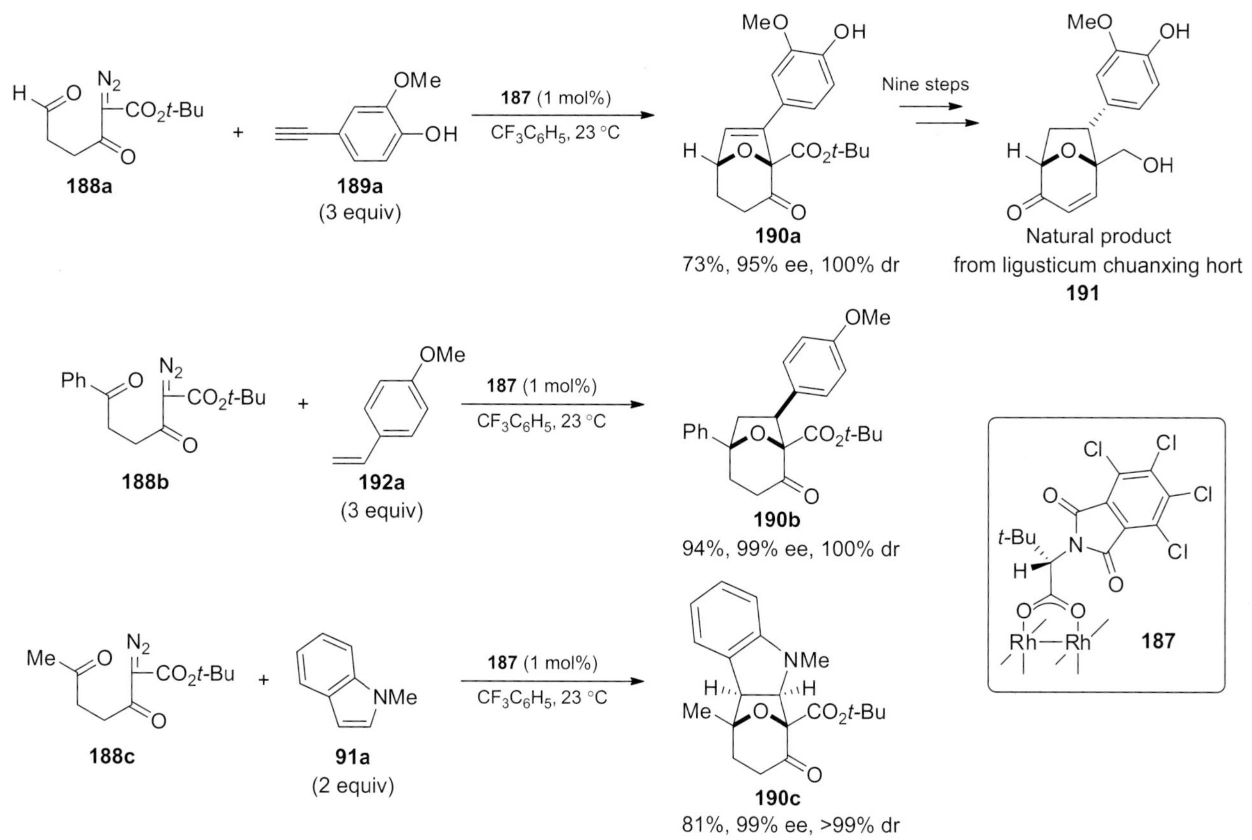

Scheme 11.41 Domino carbonyl ylide formation/1,3-dipolar cycloaddition reaction developed by Hashimoto [62].

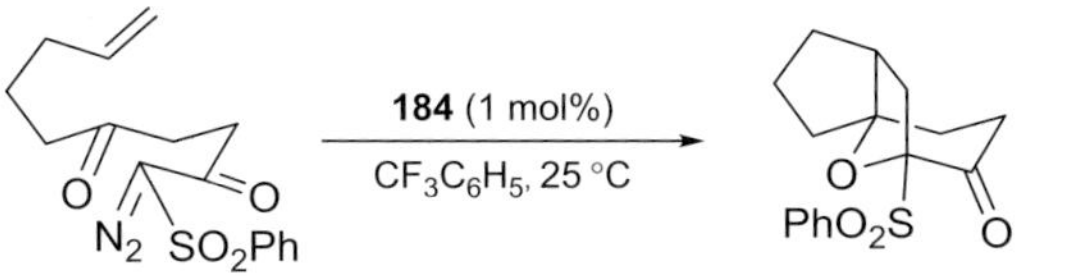

193a

194a
75%, 43% ee,100% dr

Scheme 11.42 Intramolecular domino carbonyl ylide formation/1,3-dipolar cycloaddition [63].

Davies [64] has developed a novel domino reaction initiated by the formation of an oxygen ylide, which involved vinyl diazoacetates **196** and allylic alcohols **197**. The enantioselective domino reaction generated highly functionalized chiral cyclopentanes **201** bearing four stereogenic centers as almost single diastereomers and moderate to high enantioselectivities (Scheme 11.43). The process was catalyzed by a combination of **195** and $Sc(OTf)_3$ and was supposed to evolve through four steps. The first step of the sequence was a Rh-catalyzed oxygen ylide formation,

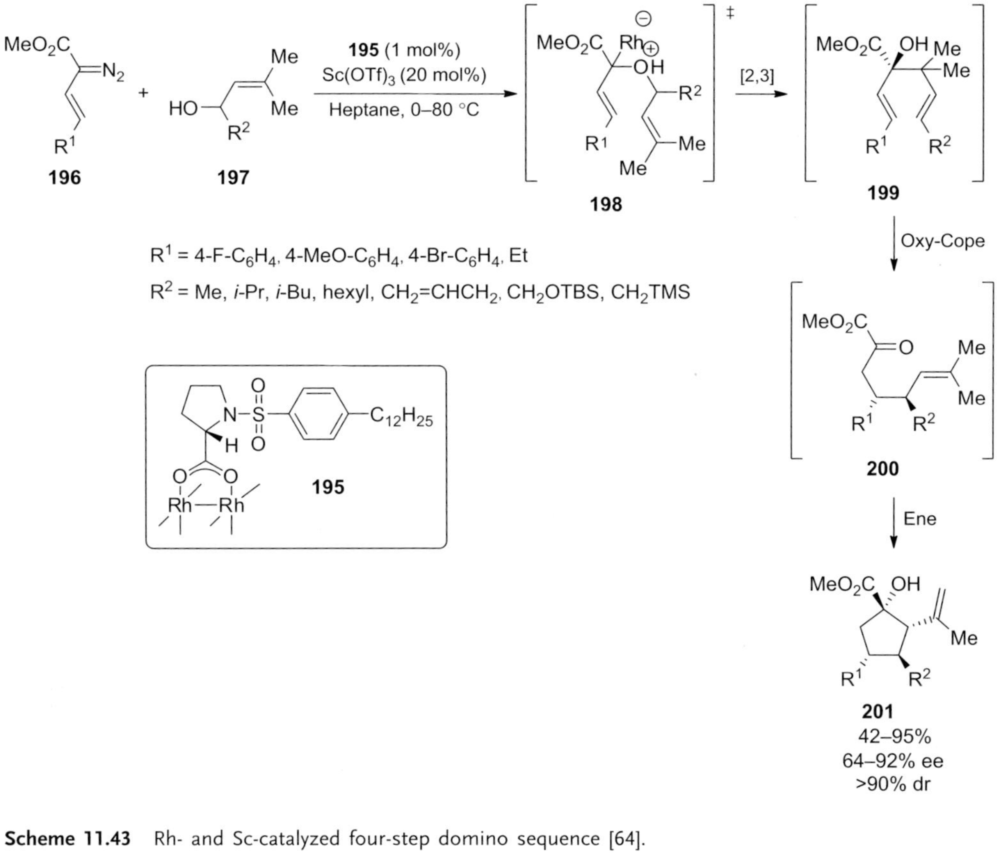

Scheme 11.43 Rh- and Sc-catalyzed four-step domino sequence [64].

which was followed by a [2,3]-sigmatropic rearrangement of **198**, an oxy-Cope rearrangement of **199**, and then finally a carbonyl ene reaction of **200**.

Somfai has described three-component 1,3-dipolar cycloadditions of *in situ* generated carbonyl ylides to aldimines induced by chiral Rh(II) carboxylate catalysts **202** [65]. The reaction of ethyl diazoacetate **164**, benzaldehyde (**92a**), and benzylidene benzylamine (**203**) gave rise to the corresponding *syn*-β-amino alcohol **206** in 62% yield and 88% dr with 24% ee. In spite of its low enantioselectivities, this process provided interesting building blocks for the synthesis of natural products and other biologically important products that are *syn*-β-amino alcohols (Scheme 11.44).

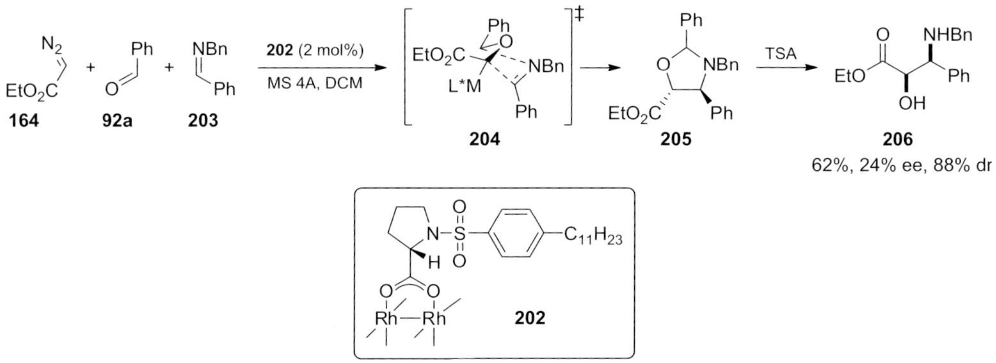

Scheme 11.44 Three-component domino carbonyl ylide formation/1,3-dipolar cycloaddition [65].

11.4.2
Domino Reaction Initiated by Oxonium Ylide Formation

Combination of $Rh(OAc)_4$ with a chiral zirconium complex of a 1,1′-bi-2-naphthol (BINOL) derivative was demonstrated by Hu to catalyze an efficient three-component, enantioselective, aldol-type reaction of aryl diazoacetates **207**, benzyl alcohol (**208**), and arylaldehydes [66a]. The process occurred by trapping a reactive alcoholic oxonium ylide **209**, which was formed *in situ* from the diazoacetate **207** and benzyl alcohol **208**, with aldehyde activated by chiral zirconium complex to give through an aldol-type addition the final products **210**. This three-component reaction provided a convenient and highly enantioselective route to the construction of an important class of compounds in medicinal chemistry, namely α,β-dihydroxy acid derivatives containing two contiguous tetrasubstituted carbon centers. The products **210** were obtained in moderate to good yields and erythro:threo ratios, combined with high enantioselectivities of up to 98% ee for the major erythro product [66b]. Hu has also shown that the cooperative catalysis of $[Rh(OAc)_4]_2$ with a chiral zinc complex of (*S*)-*t*-Bu-box ligand (**211**) applied to the reaction of diazo compounds **207** with H_2O (**212**) and α,β-unsaturated 2-acyl imidazole **213** afforded the corresponding γ-hydroxy ketone **214** with a chiral tetrasubstituted carbon center (Scheme 11.45). These

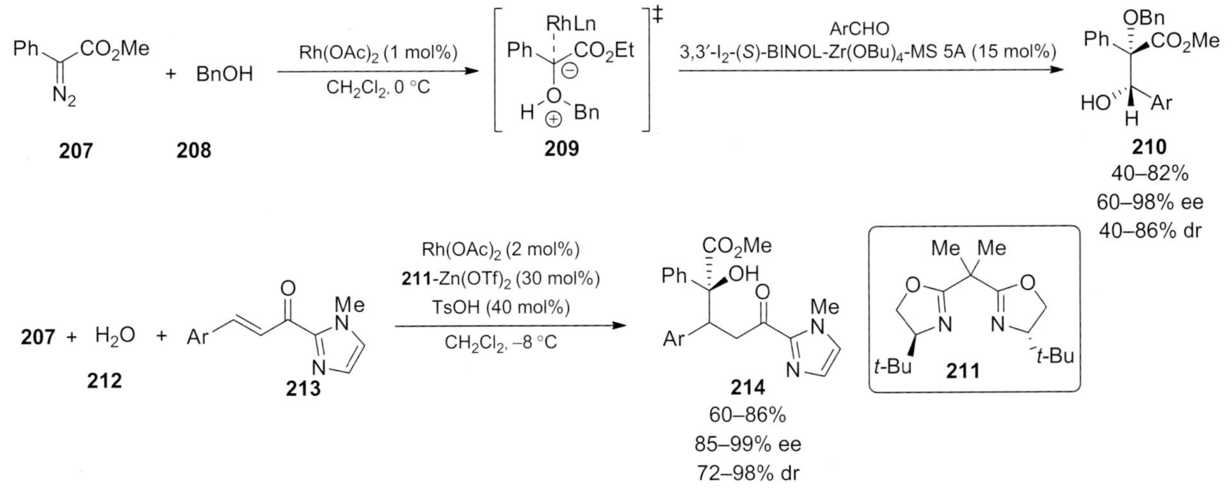

Scheme 11.45 $Rh(OAc)_2$ and chiral Lewis acid-catalyzed reaction to produce γ-hydroxyketone with a tetrasubstituted carbon center [66b].

highly functionalized products **214** were achieved in good yields and high diastereo- and enantioselectivities. The products **214** were readily converted into the corresponding γ-hydroxy carboxylic acid derivatives, which constitute key intermediates in the construction of natural products and pharmaceuticals.

11.4.3
Domino Reaction Based on Cycloaddition

Zhang reported a Au(I)-catalyzed diastereo- and enantioselective domino cyclization/[3+3] cycloaddition of 2-(1-alkynyl)-2-alken-1-one **216** with nitrone **217**, affording the corresponding chiral heterobicyclic furo[3,4-*d*][1,2]oxazine **218** [67]. This reaction is attractive by its easy scale-up possibility (gram scale) and the chemoselective functional group transformation in the product (Scheme 11.46).

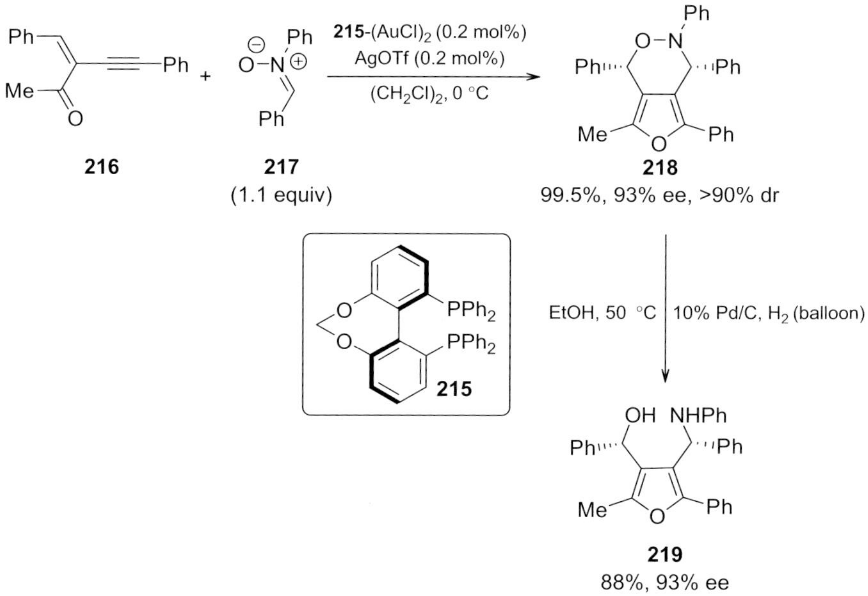

Scheme 11.46 Au(I)-catalyzed intermolecular domino cyclization/[3+3]cycloaddition [67].

Peters applied the cooperative activation by a soft bimetallic catalyst **220**, a hard Brønsted acid, and a hard Brønsted base to the formation of highly enantioenriched, diastereomerically pure masked α-amino acids **225** bearing adjacent tetrasubstituted and tertiary carbon stereocenters on the basis of a domino azlactone formation/Michael addition reaction starting from *N*-benzoylated amino acids **221** and α,β-unsaturated ketones **180** (Scheme 11.47). Since the activated catalyst was stable toward acetic anhydride, the *in situ* formation of azlactones **223** could be achieved through *O*-acylation with acetic anhydride of *N*-benzoylated amino acids

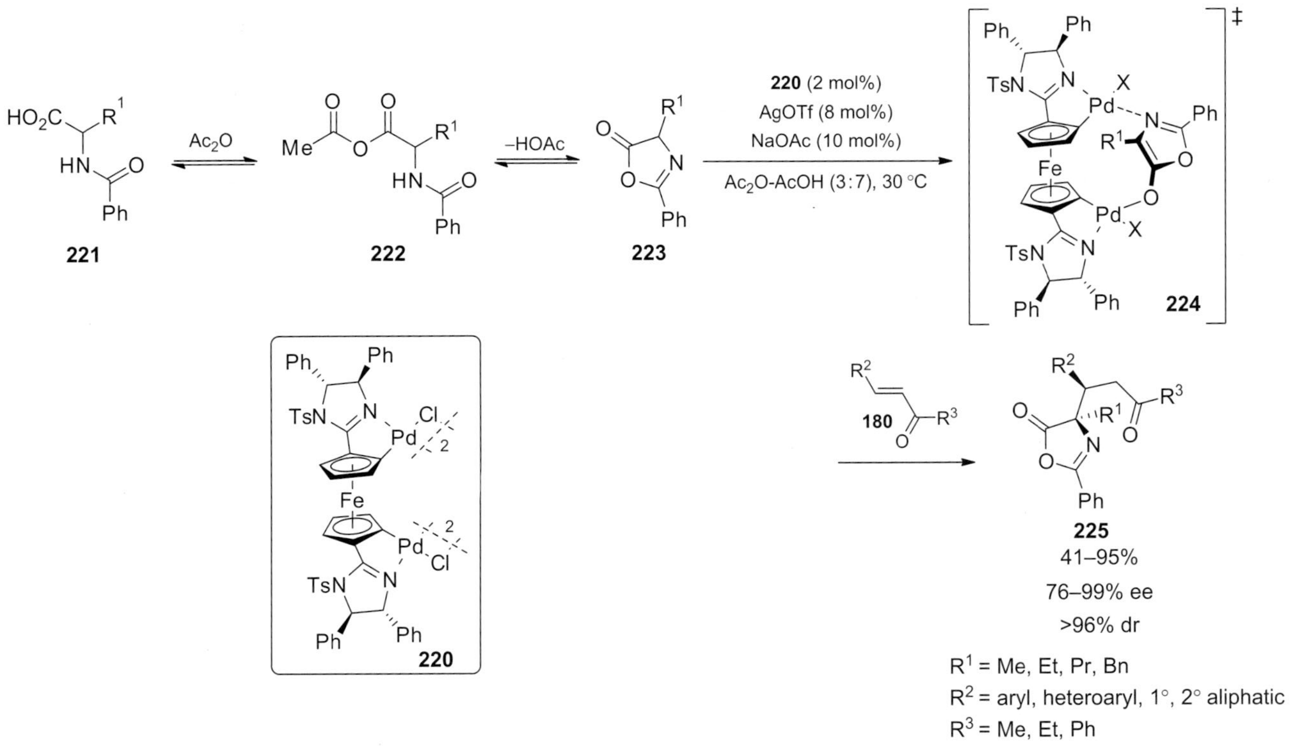

Scheme 11.47 Bispalladacycle-catalyzed Brønsted acid/base-promoted domino azlactone formation/Michael addition [68].

221, generating mixed anhydrides. The soft bimetallic catalyst **220** would allow a transition state **224** in which the azlactone **223** would be activated by enolization via coordination of the N and O atom. In a second stage, the activated azlactones added to α,β-unsaturated ketones **180** through a Michael addition to give the final products **225** in high yields combined with excellent diastereo- and enantioselectivities [68].

11.4.4 Domino Reaction Based on Pd(II)/Pd(IV) Catalysis

Recently, catalytic reactions via Pd(IV) intermediates generated from a Pd(II) precursor by the action of a powerful oxidant (e.g., a hypervalent iodine reagent) have been developed [69]. In 2007, Tse [69a] and Sanford [69b] independently reported a new achiral Pd-catalyzed oxidation reaction for the stereospecific conversion of enynes **227** into bicyclo[3.1.0]hexanes **228**. Unlike the related Pd(II)/Pd(0)-, Au-, and Pt-catalyzed cyclopropane-forming reactions, these transformations via Pd(II)/Pd(IV)^{IV} catalysis proceed with a net inversion of geometry with respect to the starting olefin. As a first example of enantioselective Pd(II)/Pd(IV) catalysis, in 2009 Sasai found that spiro bis(isoxazoline) ligands, abbreviated as SPRIXs (**226**) [70], could serve as effective chiral ligands in the Pd(II)/Pd(IV)-catalyzed enantioselective oxidative reaction. The high affinity of SPRIXs for Pd(II) and the remarkable stability of SPRIXs under oxidative conditions play an important role in promoting the enantioselective cyclization of enyne derivatives **227** to give chiral cyclopropyl ketones **228** in high yields and enantioselectivities [70a]. The obtained chiral lactones **228** are a versatile building block for biologically active molecules, such as an antiherpetic agent [71a], a protein kinase C-β inhibitor (JTT-010) [71b], and an anticonvulsant drug (pregabalin) [71c] (Scheme 11.48).

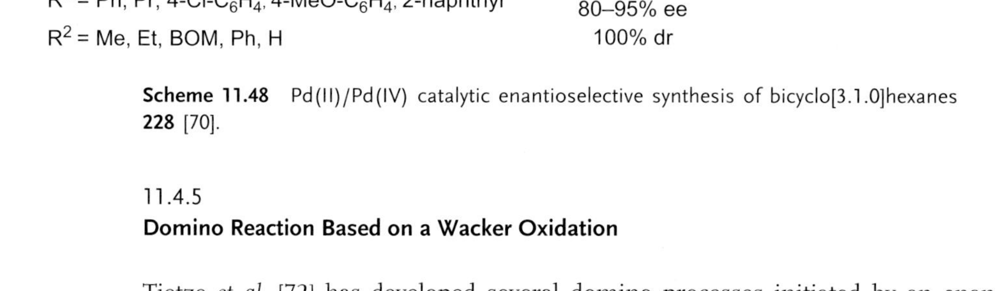

Scheme 11.48 Pd(II)/Pd(IV) catalytic enantioselective synthesis of bicyclo[3.1.0]hexanes **228** [70].

11.4.5 Domino Reaction Based on a Wacker Oxidation

Tietze *et al.* [72] has developed several domino processes initiated by an enantioselective Wacker oxidation, which is followed by a Heck reaction and a carbonylation/methoxylation, respectively, for the synthesis of natural products such as vitamin E (α-tocopherol) (**229**), diversonol (**230**), and blennolide A (**231**), as well as several heterocycles. The enantiomeric excess values range from 96 to >99%. The synthesis of the natural products is described in detail in Chapter 14 (Scheme 11.49).

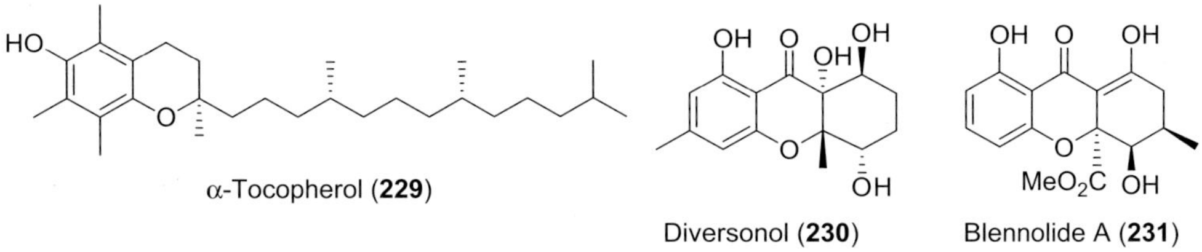

Scheme 11.49 Synthesis of natural products initiated by an enantioselective Wacker oxidation [72].

11.5 Domino Reaction Initiated by C–B and C–Si Bond Formation

11.5.1 Domino Reaction Initiated by Conjugate Addition

There has been significant interest in the development of synthetic methods for the preparation of chiral, optically enriched organoboronates and organosilicons as precursors of enantio-enriched organic compounds [73]. Shibasaki and Kanai [74] developed an enantioselective conjugate boration of trisubstituted enones **233** to produce enantiomerically enriched tertiary boronates, catalyzed by a Cu complex of QuinoxP* **232**. The racemic system was originally developed independently by Hosomi [75] and Miyaura [76]. The generating boron enolate **235** can react further with electrophiles. Shibasaki and Kanai successfully extended this catalytic reaction to an enantioselective three-component reaction (Scheme 11.50). The domino product **236**, derived through enantioselective conjugate boration, diastereoselective aldol reaction, and oxidation in a one-pot process, was obtained in 71% yield and 91% ee. The stereochemistries of three contiguous stereogenic centers involving a tetrasubstituted carbon were controlled to a high level, including their absolute configuration.

Lam [77] has described highly enantioselective Cu(I)-catalyzed borylative aldol cyclizations of enone diones **237** via a chair-like transition state **238**, which resulted in densely functionalized decalin-, hydrindane-, and diquinane-based products **239** containing four contiguous stereocenters, two of which are quaternary (Scheme 11.51).

1,4-Additions of silicon nucleophiles are similar to conjugate borylation. Hoveyda showed that Cu complexes with chiral carbene ligand **240** are excellent catalysts

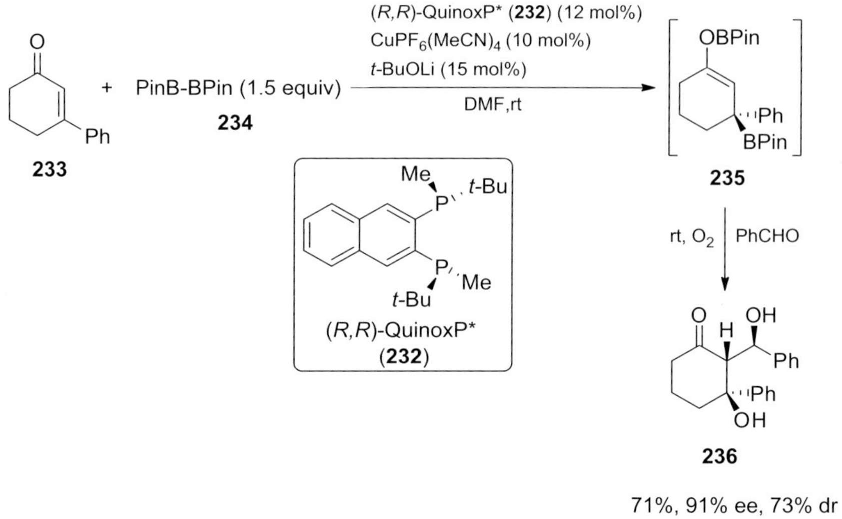

Scheme 11.50 Enantioselective conjugate boration, diastereoselective aldol reaction, and oxidation in one pot [74].

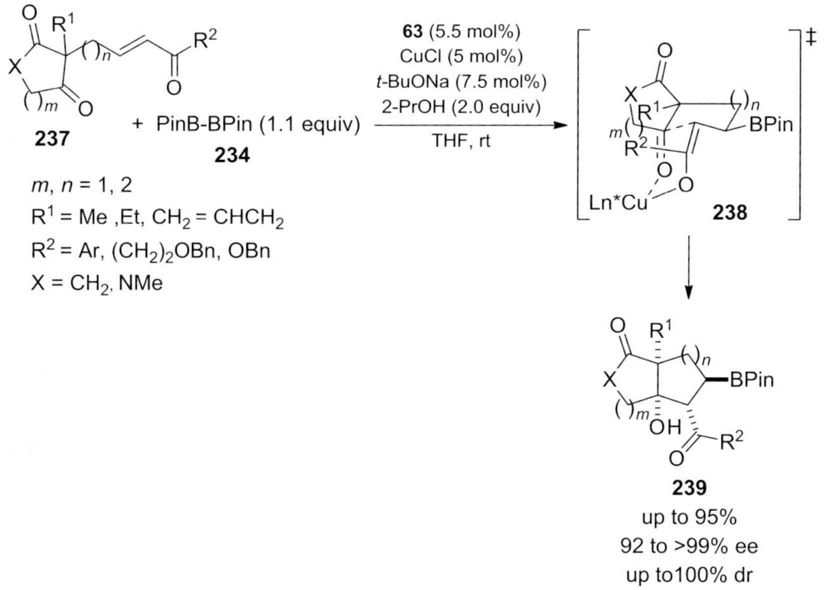

Scheme 11.51 Enantioselective Cu(I)-catalyzed borylative aldol cyclization [77]

for conjugate silylation (Scheme 11.52). The resulting boron enolate **242** can be trapped by treatment with methyl 2-bromoacetate. The domino product **243** was obtained by a one-pot procedure in 92% yield and 95% ee with >98% dr [78]. The racemic system was developed earlier by Tietze [79] with >99% dr.

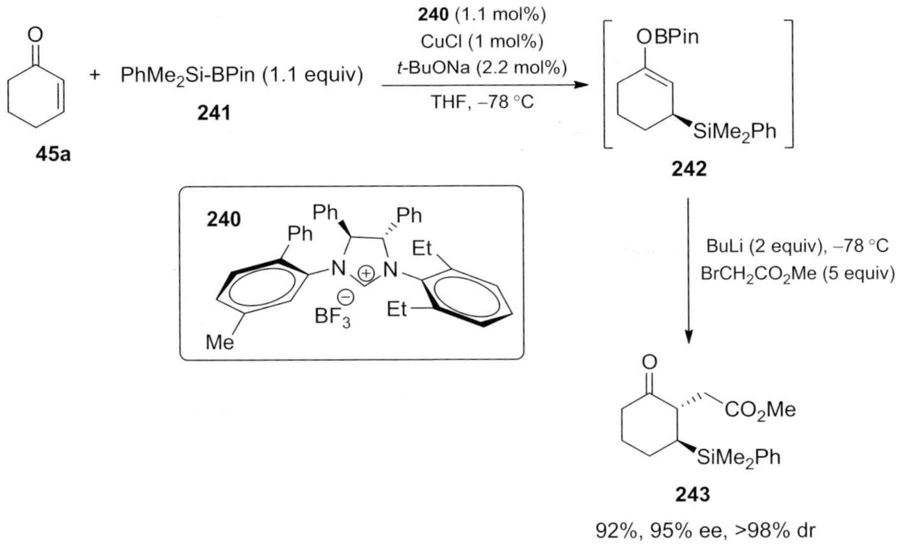

Scheme 11.52 One-pot synthesis of highly functionalized chiral organosilicon [78]

11.6 Conclusion and Outlook

In this chapter, we described recent advances in the catalytic, metal-mediated, enantioselective C–C bond-forming domino transformations from α,β-unsaturated carbonyl compounds to produce multiple stereogenic centers. The last 10 years have seen a rapid growth in the enantio- and diastereoselective domino reactions. The metal-mediated domino catalysis is the most powerful methodology to give facile access to very complex chiral molecules in a single operation and/or a one-pot process. Development of the efficient and elegant domino transforms will receive continued attention from chemists. Recently, the combination of metal catalysts, organocatalysts, and/or biocatalysts is set to provide new reaction protocols that furnish important building blocks in previously unknown ways [80]. Also, the studies on the mechanisms of the enantio- and diastereoselective domino reactions are highly desirable, as this will help accelerate the search to find new catalyst systems. Further ground-breaking discoveries can be expected with certainty in the near future.

References

1. (a) Tietze, L.F. and Beifuss, U. (1993) *Angew. Chem., Int. Ed. Engl.*, **32**, 131–163; (b) Tietze, L.F. (1996) *Chem. Rev.*, **96**, 115–136; (c) Tietze, L.F. and Modi, A. (2000) *Med. Res. Rev.*, **20**, 304–322; (d) Tietze, L.F. and Rackelmann, N. (2005) in *Multicomponent Reactions* (eds J. Zhu and H. Bienaymé), Wiley-VCH Verlag GmbH, Weinheim, pp. 121–168; (e) Tietze, L.F., Brasche, G., and Gericke, K. (2006) *Domino Reaction in Organic Syhthesis*, Wiley-VCH Verlag GmbH, Weinheim. (f) Tietze, L.F. and Levy, L. (2009) in *Domino Processes in The Mizoroki-Heck Reaction* (ed. M. Oestreich), Wiley-VCH Verlag GmbH, Chichester, pp. 281–344; (g) Tietze, L.F., Kinzel, T., and Brazel, C. (2009) *Acc. Chem. Res.*, **42**, 367–378; (h) Tietze, L.F. and Düfert, A. (2010) in *Catalytic Asymmetric Conjugate Reactions* (ed. A. Cordova), Wiley-VCH Verlag GmbH, Weinheim, pp. 321–350; (i) Tietze, L.F. and Düfert, A. (2010) *Pure Appl. Chem.*, **82**, 1375–1392; (j) Tietze, L.F., Düfert, M.A., and Schild, S.-C. (2012) in *Comprehensive Chirality*, Vol. **2** (eds E.M. Carreira and H. Yamamoto), Elsevier, Amsterdam, pp. 97–121; (k) Tietze, L.F., Stewart, S.G., and Düfert, A. (2012) in *Modern Tools for the Synthesis of Complex Bioactive Molecules* (eds J. Cossy and S. Arseniyadis), John Wiley & Sons, pp. 271–334.
2. Reviews on the asymmetric metal-catalyzed domino transformations, see: (a) Guo, H.-C. and Ma, J.-A. (2006) *Angew. Chem. Int. Ed.*, **45**, 354–366; (b) Pellissier, H. (2006) *Tetrahedron*, **62**, 1619–1665; (c) Pellissier, H. (2006) *Tetrahedron*, **62**, 2143–2173; (d) Nicolaou, K.C., Edmonds, D.J., and Bulger, P.G. (2006) *Angew. Chem. Int. Ed.*, **45**, 7134–7186; (e) Zhang, L., Sun, J., and Kozmin, S.A. (2006) *Adv. Synth. Catal.*, **348**, 2271–2296; (f) Chapman, C.J. and Frost, C.G. (2007) *Synthesis*, 1–21; (g) Padwa, A. and Bur, S.K. (2007) *Tetrahedron*, **63**, 5341–5378; (h) D'Souza, D.M. and Muller, T.J.J. (2007) *Chem. Soc. Rev.*, **36**, 1095–1108; (i) Michelet, V., Toullec, P.Y., and Genet, J.-P. (2008) *Angew. Chem. Int. Ed.*, **47**, 4268–4315; (j) Nicolaou, K.C. and Chen, J.S. (2009) *Chem. Soc. Rev.*, **38**, 2993–3009; (k) Toure, B.B. and Hall, D.G. (2009) *Chem. Rev.*, **109**, 4439–4486; (l) Padwa, A. (2009) *Chem. Soc. Rev.*, **38**, 3072–3081; (m) Biggs-Houck, J.E., Younai, A., and Shaw, J.T. (2010) *Curr. Opin. Chem. Biol.*, **14**, 371–382.
3. Reviews on the asymmetric organocatalyzed domino transformations, see: (a) Domling, A. (2006) *Chem. Rev.*, **106**, 17–89; (b) Enders, D., Gronal, C., and Huteel, M.R.M. (2007) *Angew. Chem. Int. Ed.*, **46**, 1570–1581; (c) Guillena, G., Ramon, D.J., and Yus, M. (2007) *Tetrahedron: Asymmetry*, **18**, 693–700; (d) Erkkila, A., Majander, I., and Pihko, P.M. (2007) *Chem. Rev.*, **107**, 5416–5470; (e) Yu, X. and Wang, W. (2008) *Org. Biomol. Chem.*, **6**, 2037–2046; (f) Dondoni, A. and Massi, A. (2008) *Angew. Chem. Int. Ed.*, **47**, 4638–4660; (g) Melchiorre, P., Marigo, M., Carlone, A., and Bartoli, G. (2008) *Angew. Chem. Int. Ed.*, **47**, 6138–6171; (h) Bertelsen, S. and Jorgensen, K.A. (2009) *Chem. Soc. Rev.*, **38**, 2178–2189; (i) Poulin, J., Grisé-Bard, C.M., and Barriault, L. (2009) *Chem. Soc. Rev.*, **38**, 3092–3101; (j) Grondal, C., Jeanty, M., and Enders, D. (2010) *Nat. Chem.*, **2**, 167–178; (k) Bonne, D., Coquerel, Y., Constantieux, T., and Rodriguez, J. (2010) *Tetrahedron: Asymmetry*, **21**, 1085–1109.
4. (a) Ruiz, M., Lopez-Alvarado, P., Giorgi, G., and Menendez, J.C. (2011) *Chem. Soc. Rev.*, **40**, 3445–3454; (b) Vlaar, T., Ruijter, E., and Orru, R.V.A. (2011) *Adv. Synth. Catal.*, **353**, 809–841; (c) Ruijter, E., Scheffelaar, R., and Orru, R.V.A. (2011) *Angew. Chem. Int. Ed.*, **50**, 6234–6246; (d) Galestokova, Z. and Sebesta, R. (2012) *Eur. J. Org. Chem.*, 6688–6695; (e) Clavier, H. and Pellissier, H. (2012) *Adv. Synth. Catal.*, **354**, 3347–3403; (f) Pellissier, H. (2012) *Adv. Synth. Catal.*, **354**, 237–294; (g) de Graaff, C., Ruijter, E., and Orru, R.V.A. (2012) *Chem. Soc.*

Rev., **41**, 3969–4009; (h) Grossmann, A. and Enders, D. (2012) *Angew. Chem. Int. Ed.*, **51**, 314–325; (i) Marson, C.M. (2012) *Chem. Soc. Rev.*, **41**, 7712–7722; (j) Pellissier, H. (2013) *Chem. Rev.*, **113**, 442–524.

5. (a) Lopez, F., Minnaard, A.J., and Feringa, B.L. (2007) *Acc. Chem. Res.*, **40**, 179–188; (b) Hartyunyan, S.R., den Hartog, T., Geurts, K., Minnaard, A.J., and Feringa, B.L. (2008) *Chem. Rev.*, **108**, 2824–2852; (c) Alexakis, A., Backvall, J., Krause, N., Pamies, O., and Dieguez, M. (2008) *Chem. Rev.*, **108**, 2796–2823; (d) Jerphagnon, T., Pizzuti, M.G., Minnaard, A.J., and Feringa, B.L. (2009) *Chem. Soc. Rev.*, **38**, 1039–1075; (e) Hawner, C. and Alexakis, A. (2010) *Chem. Commun.*, **46**, 7295–7306.
6. (a) Suzuki, M., Yanagisawa, A., and Noyori, R. (1988) *J. Am. Chem. Soc.*, **110**, 4718–4726; (b) Kitamura, M., Miki, T., Nakano, K., and Noyori, R. (1996) *Tetrahedron Lett.*, **37**, 5141–5144.
7. (a) Feringa, B.L., Pineschi, M., Arnold, L.A., Imbos, R., and de Vries, A.H.M. (1997) *Angew. Chem., Int. Ed. Engl.*, **36**, 2620–2623; (b) Keller, E., Maurer, J., Naasz, R., Schader, T., Meetsma, A., and Feringa, B.L. (1998) *Tetrahedron: Asymmetry*, **9**, 2409–2413; (c) Mandoli, A., Arnold, L.A., de Vries, A.H.M., Salvadori, P., and Feringa, B.L. (2001) *Tetrahedron: Asymmetry*, **12**, 1929–1937.
8. Li, K. and Alexakis, A. (2007) *Chem. Eur. J.*, **13**, 3765–3771.
9. Aikawa, K., Okamoto, T., and Mikami, K. (2012) *J. Am. Chem. Soc.*, **134**, 10329–10332.
10. (a) Sunazuka, T., Hirose, T., Shirahata, T., Harigaya, Y., Hayashi, M., Komiyama, K., Ōmura, S., and Smith, A.B. III, (2000) *J. Am. Chem. Soc.*, **122**, 2122–2123; (b) Hirose, T., Sunazuka, T., Yamamoto, D., Kojima, N., Shirahata, T., Harigaya, Y., Kuwajima, I., and Ōmura, S. (2005) *Tetrahedron*, **61**, 6015–6039.
11. Howell, G.P., Fletcher, S.P., Geurts, K., Horst, B., and Feringa, B.L. (2006) *J. Am. Chem. Soc.*, **128**, 14977–14985.
12. den Hartog, T., Rudolph, A., Maciá, B., Minnaard, A.J., and Feringa, B.L. (2010) *J. Am. Chem. Soc.*, **132**, 14349–14351.
13. Teichert, J.F. and Feringa, B.L. (2011) *Chem. Commun.*, **47**, 2679–2681.
14. Ngoc, D.T., Albicker, M., Schneider, L., and Cramer, N. (2010) *Org. Biomol. Chem.*, **8**, 1781–1784.
15. Tsuda, T., Satomi, H., Hayashi, T., and Saegusa, T. (1987) *J. Org. Chem.*, **52**, 439–443.
16. (a) Minagawa, K., Kouzuki, S., Yoshimito, J., Kawamura, Y., Tani, H., Iwata, T., Terui, Y., Nakai, H., Yagi, S., Hattori, N., Fujiwara, T., and Kamigauchi, T. (2002) *J. Antibiot.*, **55**, 155–164; (b) Minagawa, K., Kouzuki, S., and Kamigauchi, T. (2002) *J. Antibiot.*, **55**, 165–171; (c) Taishi, T., Takechi, S., and Mori, S. (1998) *Tetrahedron Lett.*, **39**, 4347–4350; (d) Yoshimoto, J., Kakui, M., Iwasaki, H., Fujiwara, T., Sugimoto, H., and Hattori, N. (1999) *Arch. Virol.*, **144**, 865–878; (e) Weber, H.A., Baenziger, N.C., and Gloer, J.B. (1988) *J. Org. Chem.*, **53**, 4567–4569; (f) Djura, P., Stierle, D.B., Sullivan, B., Faulkner, D.J., Arnold, E., and Clardy, J. (1980) *J. Org. Chem.*, **45**, 1435–1441; (g) Wright, A.E., Rueth, S.A., and Cross, S.S. (1991) *J. Nat. Prod.*, **54**, 1108–1111.
17. Šebesta, R., Pizzuti, M.G., Minnaard, A.J., and Feringa, B.L. (2007) *Adv. Synth. Catal.*, **349**, 1931–1937.
18. Welker, M., Woodward, S., and Alexakis, A. (2010) *Org. Lett.*, **12**, 576–579.
19. Degrado, S.J., Mizutani, H., and Hoveyda, A.H. (2001) *J. Am. Chem. Soc.*, **123**, 755–756.
20. Rathgeb, X., March, S., and Alexakis, A. (2006) *J. Org. Chem.*, **71**, 5737–5742.
21. (a) Martin, D., Kehrli, S., D'Augustin, M., Clavier, H., Mauduit, M., and Alexakis, A. (2006) *J. Am. Chem. Soc.*, **128**, 8416–8417; (b) Lee, K.-S., Brown, M.K., Hird, A.W., and Hoveyda, A.H. (2006) *J. Am. Chem. Soc.*, **128**, 7182–7184.
22. Kehrli, S., Martin, D., Rix, D., Mauduit, M., and Alexakis, A. (2010) *Chem. Eur. J.*, **16**, 9890–9904.
23. Xu, Y.-J., Liu, Q.-Z., and Dong, L. (2007) *Synlett*, 273–277.
24. (a) Gonzalez-Gomez, J.C., Foubelo, F., and Yus, M. (2008) *Tetrahedron Lett.*, **49**, 2343–2347; (b) Gonzalez-Gomez, J.C.,

Foubelo, F., and Yus, M. (2009) *J. Org. Chem.*, **74**, 2547–2553.

25. (a) Galestoková, Z. and Sebesta, R. (2011) *Eur. J. Org. Chem.*, 7092–7096; (b) Bilčík, F., Drusan, M., Marák, J., and Šebesta, R. (2012) *J. Org. Chem.*, **77**, 760–765.
26. Drusan, M., Lölsberg, W., Skvorcová, A., Schmalz, H.-G., and Sebesta, R. (2012) *Eur. J. Org. Chem.*, 6285–6290.
27. Guo, S., Xie, Y., Hu, X., Xia, C., and Huang, H. (2010) *Angew. Chem. Int. Ed.*, **49**, 2728–2731.
28. Guo, S., Xie, Y., Hu, X., and Huang, H. (2011) *Org. Lett.*, **13**, 5596–5599.
29. (a) Tietze, L.F., Ila, H., and Bell, H.P. (2004) *Chem. Rev.*, **104**, 3453–3516; (b) Hamashima, Y. (2006) *Chem. Pharm. Bull.*, **54**, 1351–1354; (c) Sodeoka, M. and Hamashima, Y. (2006) *Pure Appl. Chem.*, **78**, 477–494.
30. Yu, X. and Lu, X. (2009) *Org. Lett.*, **11**, 4366–4369.
31. Shi, D., Xie, Y., Zhou, H., Xia, C., and Huang, H. (2012) *Angew. Chem. Int. Ed.*, **51**, 1248–1251.
32. Arai, T. and Yokoyama, N. (2008) *Angew. Chem. Int. Ed.*, **47**, 4989–4992.
33. Arai, T., Wasai, M., and Yokoyama, N. (2011) *J. Org. Chem.*, **76**, 2909–2912.
34. Arai, T., Yokoyama, N., Mishiro, A., and Sato, H. (2010) *Angew. Chem. Int. Ed.*, **49**, 7895–7898.
35. Recent reviews on enantioselective [3+2] cycloaddition of imino esters with electron deficient alkenes, see: (a) Pellissier, H. (2007) *Tetrahedron*, **63**, 3235–3285; (b) Kanemasa, S. (2010) *Heterocycles*, **82**, 87–200; (c) Adrio, J. and Carretero, J.C. (2011) *Chem. Commun.*, **47**, 6784–6794.
36. Recent reviews on enantioselective [2+2+2] cycloadditions, see: (a) Tanaka, K. (2007) *Synlett*, 1977–1993; (b) Shibata, T. and Tsuchikama, K. (2008) *Org. Biomol. Chem.*, **6**, 1317–1323; (c) Tanaka, K. (2009) *Chem. Asian J*, **4**, 508–518; (d) Shibata, T., Uchiyama, T., Hirashima, H., and Endo, K. (2011) *Pure Appl. Chem.*, **83**, 597–605; (e) Shibata, Y. and Tanaka, K. (2012) *Synthesis*, **44**, 323. (f) Tanaka, K. (2012) *Heterocycles*, **85**, 1017–1043.
37. (a) Juhl, M. and Tanner, D. (2009) *Chem. Soc. Rev.*, **38**, 2983–2992; (b) Tadano, K. (2009) *Eur. J. Org. Chem.*, 4381–4394.
38. (a) Kobayashi, M., Suda, T., Noguchi, K., and Tanaka, K. (2011) *Angew. Chem. Int. Ed.*, **50**, 1664–1667; (b) Masutomi, K., Sakiyama, N., Noguchi, K., and Tanaka, K. (2012) *Angew. Chem. Int. Ed.*, **51**, 13031–13035.
39. He, Z., Liu, T., Tao, H., and Wang, C.-J. (2012) *Org. Lett.*, **14**, 6230–6233.
40. Keilitz, J., Newman, S.G., and Lautens, M. (2013) *Org. Lett.*, **15**, 1148–1151.
41. Recent reviews on asymmetric aldol reactions, see: (a) Heravi, M.M. and Asadi, S. (2012) *Tetrahedron: Asymmetry*, **23**, 1431–1465; (b) Dias, L.C., de Lucca, E.C. Jr., Ferreira, M.A.B., and Polo, E.C. (2012) *J. Braz. Chem. Soc.*, **23**, 2137–2158; (c) Miao, Z. and Chen, F. (2012) *Synthesis*, 2506–2514.
42. Reviews on asymmetric reductive-aldol reactions, see: (a) Han, S.B., Hassan, A., and Krische, M.J. (2008) *Synthesis*, 2669–2679; (b) Lipshutz, B.H. (2009) *Synlett*, 509–524.
43. Zhao, D., Oisaki, K., Kanai, M., and Shibasaki, M. (2006) *Tetrahedron Lett.*, **47**, 1403–1407.
44. Zhao, D., Oisaki, K., Kanai, M., and Shibasaki, M. (2006) *J. Am. Chem. Soc.*, **128**, 14440–14441.
45. Du, Y., Xu, L.-W., Oisaki, K., Kanai, M., and Shibasaki, M. (2008) *J. Am. Chem. Soc.*, **130**, 16146–16147.
46. Chuzel, O., Deschamp, J., Chausteur, C., and Riant, O. (2006) *Org. Lett.*, **8**, 5943–5946.
47. Deschamp, J., Chuzel, O., Hannedouche, J., and Riant, O. (2006) *Angew. Chem. Int. Ed.*, **45**, 1292–1297.
48. Bee, C., Han, S.B., Hassan, A., Iida, H., and Krische, M.J. (2008) *J. Am. Chem. Soc.*, **130**, 2746–2747.
49. Lipshutz, B.H., Amorelli, B., and Unger, J.B. (2008) *J. Am. Chem. Soc.*, **130**, 14378–14379.
50. (a) Deschamp, J. and Riant, O. (2009) *Org. Lett.*, **11**, 1217–1220; (b) Deschamp, J., Hermant, T., and Riant, O. (2012) *Tetrahedron*, **68**, 3457–3467.

51. Kato, M., Oki, H., Ogata, K., and Fukuzawa, S.-I. (2009) *Synlett*, 1299–1302.
52. (a) Ou, J., Wong, W.-T., and Chiu, P. (2012) *Org. Biomol. Chem.*, **10**, 5971–5978; (b) Ou, J., Wong, W.-T., and Chiu, P. (2012) *Tetrahedron*, **68**, 3450–3456.
53. Saxena, A., Choi, B., and Lam, H.W. (2012) *J. Am. Chem. Soc.*, **134**, 8428–8431.
54. Rodríguez, M.M., Nájera, C., Sansano, J.M., Costa, P.R.R., de Lima, E.C., and Dias, A.G. (2010) *Synlett*, 962–966.
55. Gupta, A.K., Mukherjee, M., and Wulff, W.D. (2011) *Org. Lett.*, **13**, 5866–5869.
56. He, W., Yip, K.-T., Zhu, N.-Y., and Yang, D. (2009) *Org. Lett.*, **11**, 5626–5628.
57. Yip, K.-T., Yang, M., Law, K.-L., Zhu, N.-Y., and Yang, D. (2006) *J. Am. Chem. Soc.*, **128**, 3130–3131.
58. Hodgson, D.M., Stupple, P.A., and Johnstone, C. (1997) *Tetrahedron Lett.*, **38**, 6471–6472.
59. (a) Suga, H., Suzuki, T., Inoue, K., and Kakehi, A. (2006) *Tetrahedron*, **62**, 9218–9225; (b) Suga, H., Ishimoto, D., Higuchi, S., Ohtsuka, M., Arikawa, T., Tsuchida, T., Kakehi, A., and Baba, T. (2007) *Org. Lett.*, **9**, 4359–4362; (c) Suga, H., Higuchi, S., Ohtsuka, M., Ishimoto, D., Arikawa, T., Hashimoto, Y., Misawa, S., Tsuchida, T., Kakehi, A., and Baba, T. (2010) *Tetrahedron*, **66**, 3070–3089.
60. Son, S. and Fu, G.C. (2007) *J. Am. Chem. Soc.*, **129**, 1046–1047.
61. (a) Tsutsui, H., Shimada, N., Abe, T., Anada, M., Nakajima, M., Nakamura, S., Nambu, H., and Hashimoto, S. (2007) *Adv. Synth. Catal.*, **349**, 521–526; (b) Kurosaki, Y., Shimada, N., Anada, M., Nambu, H., and Hashimoto, S. (2010) *Bull. Korean Chem. Soc.*, **31**, 694–696.
62. (a) Shimada, N., Anada, M., Nakamura, S., Nambu, H., Tsutsui, H., and Hashimoto, S. (2008) *Org. Lett.*, **10**, 3603–3606; (b) Shimada, N., Hanari, T., Kurosaki, Y., Takeda, K., Anada, M., Nambu, H., Shiro, M., and Hashimoto, S. (2010) *J. Org. Chem.*, **75**, 6039–6042; (c) Nambu, H., Hikime, M., Krishnamurthi, J., Kamiya, M., Shimada, N., and Hashimoto, S. (2009) *Tetrahedron Lett.*, **50**, 3675–3678; (d) Shimada, N., Hanari, T., Kurosaki, Y., Anada, M., Nambu, H., and Hashimoto, S. (2010) *Tetrahedron Lett.*, **51**, 6572–6575; (e) Shimada, N., Oohara, T., Krishnamurthi, J., Nambu, H., and Hashimoto, S. (2011) *Org. Lett.*, **13**, 6284–6287.
63. Hodgson, D.M., Glen, R., and Redgrave, A.J. (2009) *Tetrahedron: Asymmetry*, **20**, 754–757.
64. Parr, B.T., Li, Z., and Davies, H.M.L. (2011) *Chem. Sci.*, **2**, 2378–2382.
65. Torssell, S. and Somfai, P. (2006) *Adv. Synth. Catal.*, **348**, 2421–2430.
66. (a) Zhang, X., Huang, H., Guo, X., Guan, X., Yang, L., and Hu, W. (2008) *Angew. Chem. Int. Ed.*, **47**, 6647–6649; (b) Guan, X.-Y., Yang, L.-P., and Hu, W. (2010) *Angew. Chem. Int. Ed.*, **49**, 2190–2192.
67. Liu, F., Qian, D., Li, L., Zhao, X., and Zhang, J. (2010) *Angew. Chem. Int. Ed.*, **49**, 6669–6672.
68. Weber, M., Jautze, S., Frey, W., and Peters, R. (2010) *J. Am. Chem. Soc.*, **132**, 12222–12225.
69. (a) Tong, X., Beller, M., and Tse, M.K. (2007) *J. Am. Chem. Soc.*, **129**, 4906–4907; (b) Welbes, L.L., Lyons, T.W., Cychosz, K.A., and Sanford, M.S. (2007) *J. Am. Chem. Soc.*, **129**, 5836–5837; (c) Liu, H., Yu, J., Wang, L., and Tong, X. (2008) *Tetrahedron Lett.*, **49**, 6924–6928; For recent reviews on Pd(IV) catalysis, see: (d) Muñiz, K. (2009) *Angew. Chem. Int. Ed.*, **48**, 9412–9423; (e) Sehnal, P., Taylor, R.J.K., and Fairlamb, I.J.S. (2010) *Chem. Rev.*, **110**, 824–889; (f) Xu, L.M., Li, B.J., Yang, Z., and Shi, Z.J. (2010) *Chem. Soc. Rev.*, **39**, 712–733; (g) Neufeldt, S.R. and Sanford, M.S. (2012) *Acc. Chem. Res.*, **45**, 936–946; (h) Malacria, M. and Maestri, G. (2013) *J. Org. Chem.*, **78**, 1323–1328.
70. (a) Tsujihara, T., Takenaka, K., Onitsuka, K., Hatanaka, M., and Sasai, H. (2009) *J. Am. Chem. Soc.*, **131**, 3452–3453; (b) Bajracharya, G.B., Arai, M.A., Koranne, P.S., Suzuki, T., Takizawa, S., and Sasai, H. (2009) *Bull. Chem. Soc. Jpn.*, **82**, 285–302.

71. (a) Sekiyama, T., Hatsuya, S., Tanaka, Y., Uchiyama, M., Ono, N., Iwayama, S., Oikawa, M., Suzuki, K., Okunishi, M., and Tsuji, T. (1998) *J. Med. Chem.*, **41**, 1284–1298; (b) Tanaka, M., Ubukata, M., Matsuo, T., Yasue, K., Matsumoto, K., Kajimoto, Y., Ogo, T., and Inaba, T. (2007) *Org. Lett.*, **9**, 3331–3334; (c) Ok, T., Jeon, A., Lee, J., Lim, J.H., Hong, C.S., and Lee, H.-S. (2007) *J. Org. Chem.*, **72**, 7390–7393.
72. (a) Tietze, L.F., Sommer, K.M., Zinngrebe, J., and Stecker, F. (2005) *Angew. Chem. Int. Ed.*, **44**, 257–259; (b) Tietze, L.F., Stecker, F., Zinngrebe, J., and Sommer, K.M. (2006) *Chem. Eur. J.*, **12**, 8770–8776; (c) Tietze, L.F., Spiegl, D.A., Stecker, F., Major, J., Raith, C., and Grosse, C. (2008) *Chem. Eur. J.*, **14**, 8956–8963; (d) Tietze, L.F., Jackenkroll, S., Raith, C., Spiegl, D.A., Reiner, J.R., and Ochoa Campos, M.C. (2013) *Chem. Eur. J.*, **19**, 4876–4882; (e) Tietze, L.F., Ma, L., Reiner, J.R., Jackenkroll, S., and Heidemann, S. (2013) *Chem. Eur. J.*, early view.
73. (a) Mantilli, L. and Mazet, C. (2010) *ChemCatChem*, **2**, 501. (b) Hartmann, E., Vyas, D.J., and Oestreich, M. (2011) *Chem. Commun.*, **47**, 7917–7932.
74. Chen, I.-H., Yin, L., Itano, W., Kanai, M., and Shibasaki, M. (2009) *J. Am. Chem. Soc.*, **131**, 11664–11665.
75. Ito, H., Yamanaka, H., Tateiwa, J., and Hosomi, A. (2000) *Tetrahedron Lett.*, **41**, 6821–6825.
76. Takahashi, K., Ishiyama, T., and Miyaura, N. (2001) *J. Organomet. Chem.*, **625**, 47–53.
77. Burns, A.R., Gonzalez, J.S., and Lam, H.W. (2012) *Angew. Chem. Int. Ed.*, **51**, 10827–10831.
78. Lee, K.-S. and Hoveyda, A.H. (2010) *J. Am. Chem. Soc.*, **132**, 2898–2900.
79. Tietze, L.F., Tölle, N., Kratzert, D., and Stalke, D. (2009) *Org. Lett.*, **11**, 5230–5233.
80. (a) Shao, Z. and Zhang, H. (2009) *Chem. Soc. Rev.*, **38**, 2745–2755; (b) Rueping, M., Koenigs, R.M., and Atodiresei, I. (2010) *Chem. Eur. J.*, **16**, 9350–9365; (c) Zhong, C. and Shi, X. (2010) *Eur. J. Org. Chem.*, 2999–3025; (d) Zhou, J. (2010) *Chem. Asian J.*, **5**, 422–434; (e) Albrecht, L., Jiang, H., and Jorgensen, K.A. (2011) *Angew. Chem. Int. Ed.*, **50**, 8492–8509.

12
Domino Processes under Microwave Irradiation, High Pressure, and in Water

Bo Jiang, Shu-Jiang Tu, and Guigen Li

12.1
Introduction

It has been well known that domino reactions for the formation of two or more chemical bonds under an identical set of conditions can result in numerous chemical and bioactive molecules in an environmentally friendly and economically beneficial manner [1–3]. Those domino reactions conducted under microwave irradiation, high pressure, and in water are even greener in regard to shortened reaction periods, energy and manpower savings, and reduced consumption of starting materials and organic solvents. When domino reactions are performed intermolecularly, small fragments and units can be readily incorporated into larger functional structures; and when they are conducted intramolecularly, multiple domino cyclizations can be well controlled, leading to more diverse molecules [4, 5]. For the synthesis of complex molecules, domino reactions can improve not only the chemo- and regioselectivity but also the diastereo- and enantioselectivity [6–9]. A typical example of domino applications was shown by the combinatorial and automated medicinal screening where a larger number of new functional compounds are rapidly obtained for drug discovery and development [10]. Recently, it has been realized that many domino reactions can be run with less solvent, enabling the products to precipitate out of the reaction solution after reaching completion. Domino products can also be obtained by using the group-assisted purification (GAP) operation by simply washing the crude products with common solvents, thereby avoiding the traditional purification by column chromatography and recrystallization [11]. Obviously, the domino synthesis via the GAP workup can further reduce the use of materials such as silica gel and solvents as well as energy and manpower, and moreover can reduce waste generation. For aqueous domino reactions, the workup of products is simplified by effortless extractions. In this chapter, recent developments in domino reactions under microwave irradiation, high pressure, and in aqueous media are summarized by covering the work during the past 3 years.

Domino Reactions: Concepts for Efficient Organic Synthesis, First Edition. Edited by Lutz F. Tietze.

12.2
Microwave-Assisted Domino Reactions

Since 1986, when Gedye and Giguere published their pioneering work on microwave-irradiated reactions [12, 13], microwave chemistry has had a huge impact on organic synthesis [14–18]. Many reaction processes, such as the formation of heterocycles by condensation and cycloaddition reactions, have found widespread applications, particularly after accurately controllable microwave reactors were made commercially available. In fact, the high-density microwave irradiation has attained reliable and repeatable status for accelerating reaction processes. Direct microwave heating can shorten reaction times from hours to minutes or even seconds. Moreover, it can also inhibit side-product formation as well as enhance chemical yields and reproducibility. As a result, more and more academic and industrial teams are using microwave-assisted organic synthesis (MAOS) as a forefront technology for rapid system optimizations. It is believed that microwave heating will result in many more discoveries of organic reactions, particularly on cyclizations for forming new heterocycles.

12.2.1
Intramolecular Domino Reactions under Microwave Heating

The development of new Diels–Alder (DA) reactions has been actively pursued in the past several decades because they can readily provide an atom-economic access to a wide spectrum of biologically relevant compounds [19]. Moreover, the integration of DA reactions in a domino process and under microwave irradiation is a very fruitful procedure.

Dai and coworkers [20] studied the Wittig/intramolecular Diels–Alder (Wittig–IMDA)cycloaddition of ester-tethered 1,3,9-decatrienes **3** possessing a carbonyl substituent at C11 under controlled microwave heating in MeCN, at 180 °C (Scheme 12.1).

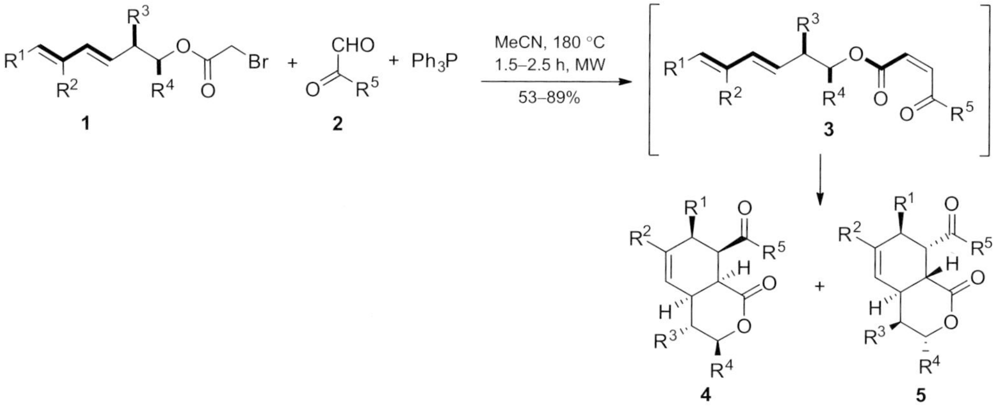

Scheme 12.1 Domino synthesis of isochromen-1-ones **4** and **5**.

The reaction afforded the cis and trans isomers of 3,4,4a,7,8,8a-hexahydroisochromen-1-ones **4** and **5** in 53–89% yields and in a 64 : 36 to 79 : 21 ratio. Gratifyingly, they found that by starting from α-bromoacetates of 3,5-hexadien-1-ols **1** and glyoxalate or phenylglyoxal hydrates **2** in the presence of PPh_3 and 2,6-lutidine under the same microwave heating conditions, the domino Wittig-IMDA cycloaddition could be achieved, leading to the formation of 3,4,4a,7,8,8a-hexahydroisochromen-1-one adducts in 73–91% yields in favor of the cis stereoisomer **4**. During this process, three consecutive C–C bonds were efficiently formed within a short time.

Microwave-assisted unimolecular domino reactions of propargyl vinyl ethers (PVEs) **6** in 1,2-dichloroethane (DCE) in the presence of molecular sieves of 4 Å to bind methanol directly led to the formation of the salicylaldehyde derivatives **7** in 27–89% yields (Scheme 12.2) [21]. This domino process is believed to involve a propargyl Claisen rearrangement/pseudo-pericyclic [1,3]-hydride shift/*E*/*Z* isomerization/1,5-hydrid shift/enolization/electrocyclization/aromatization. Later on, the same authors further investigated the reaction by running it in methanol to give the β,γ-unsaturated malonates **8** with complete regio- and stereoselectivity (Scheme 12.3). Interestingly, they found that the use of water as the medium resulted in the corresponding trisubstituted β,γ-unsaturated acids and esters **9**, again with high stereoselectivity (up to 19/1) [22].

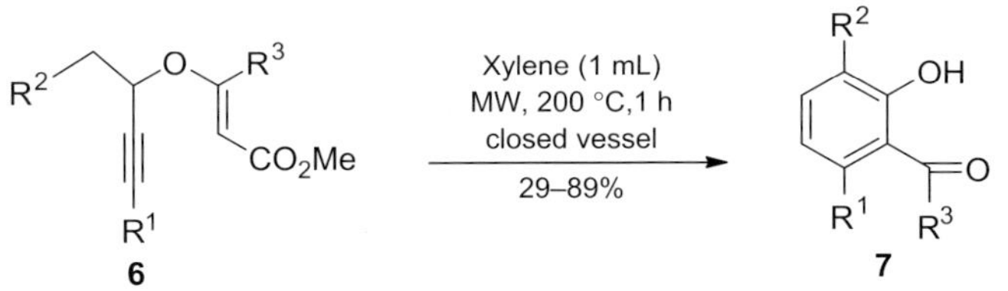

Scheme 12.2 Domino synthesis of salicylaldehyde derivatives **7**.

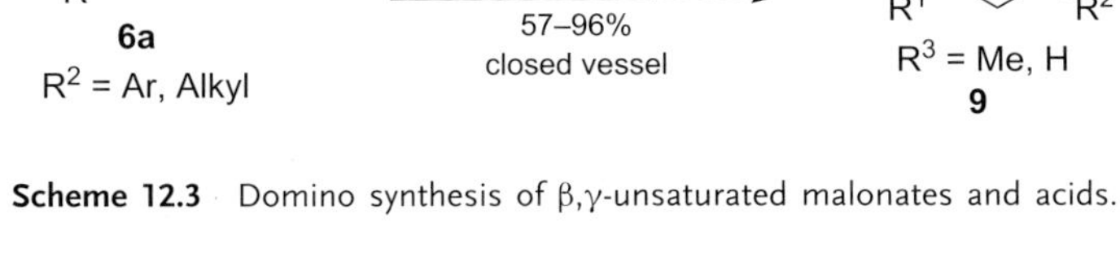

Scheme 12.3 Domino synthesis of β,γ-unsaturated malonates and acids.

12.2.2 Two-Component Domino Reaction under Microwave Heating

Inspired by the importance of nitrogen-containing heterocycles in studying bioactive substances, such as pharmaceuticals and agricultural chemicals, West and

Johnston [23] conducted a domino synthesis of iminooxazinones **12** in 42–93% yield through microwave heating of dioxinones **10** in the presence of carbodiimides **11** (Scheme 12.4).

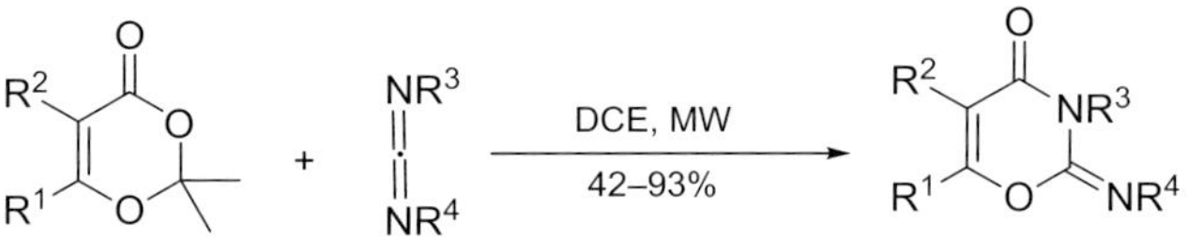

10 11 12

Scheme 12.4 Domino synthesis of iminooxazinones.

It was found that unsymmetrically substituted carbodiimides gave excellent regioselectivity, which allowed efficient assembly of the anticipated rings, with the nitrogen of lower nucleophilicity to be directed in the exocyclic imino group. High yields of 95–98% were achieved under microwave heating, along with high reaction rates.

The transformation of cinnamic acid hydrazide **13** and triethyl orthoesters **14** could be completed within 10 min to afford highly functionalized 5-substituted 2-styryl-1,3,4-oxadiazoles **15** (Scheme 12.5) [24].

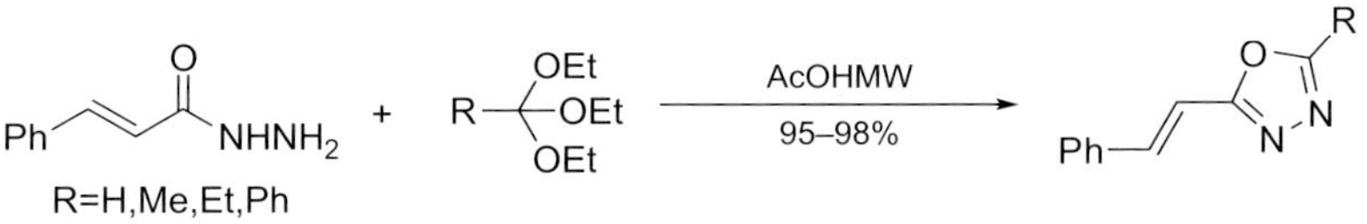

13 14 15

Scheme 12.5 Domino synthesis of 2-styryl-1,3,4-oxadiazoles.

The cyclic pyrido[3,2,1-*jk*]carbazoles **18** have been synthesized through microwave-assisted domino Fischer indole reaction/intramolecular cyclization of phenylhydrazine hydrochloride **16** and 2-(3-oxo-1,3-diarylpropyl)-1-cyclohexanones **17** [25]. After optimizing the process, the best results were obtained by running the reaction in water at 140 °C for 15 min (Scheme 12.6). The reaction showed a very good scope with 30 examples of pyrido[3,2,1-*jk*]carbazoles **18** in excellent chemical yields (90–96%).

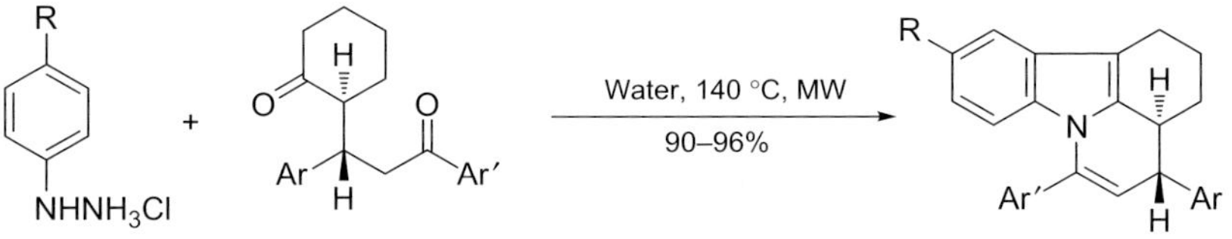

16 17 18

Scheme 12.6 Domino synthesis of pyrido[3,2,1-*jk*]carbazoles.

Prakash and coworkers reported a domino method of hydrolysis/dehydrohalogenation/Heck coupling to synthesize styrene sulfonate salts **21** in up to

89% yield using iodoarenes **19** and chloroethane sulfonyl chloride **20** as starting materials (Scheme 12.7). This methodology was found to be suitable for heterocyclic as well as disubstituted iodoarenes. However, the substrates 2-iodobenzoic acid and 3-iodophenol failed to give the desired products. It should be noted that this domino reaction was performed in aqueous media without the use of phosphine catalysts and additives [26].

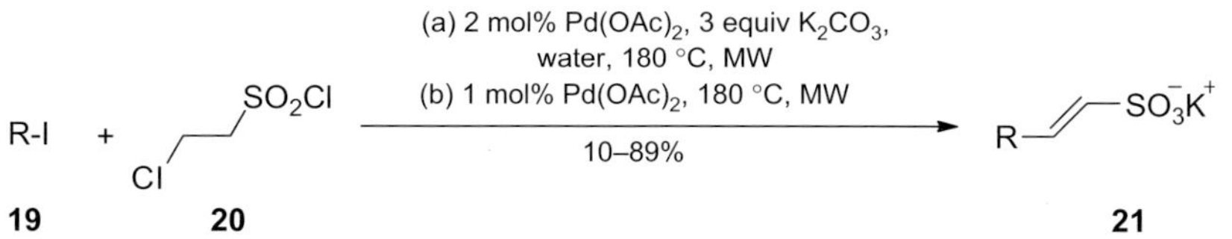

Scheme 12.7 Domino synthesis of styrene sulfonate salts.

Microwave-assisted DA reactions of substituted 3-acylamino-2H-pyran-2-ones **22** and vinyl-containing dienophiles **23** to give aniline derivatives **24** in 63–74% yield were reported by Kocevar *et al.* [27]. The reaction was carried out in the presence of 1,4-diazabicyclo[2.2.2]octane (DABCO) as an organocatalyst (Scheme 12.8). Alternatively, the reaction could also be performed under high pressure (13–15 kbar) at ambient temperatures in the absence of DABCO; as intermediates, 2-oxabicyclo[2.2.2]oct-5-ene **25** and alkoxycyclohexadiene have been isolated to support the proposed mechanism (Scheme 12.9).

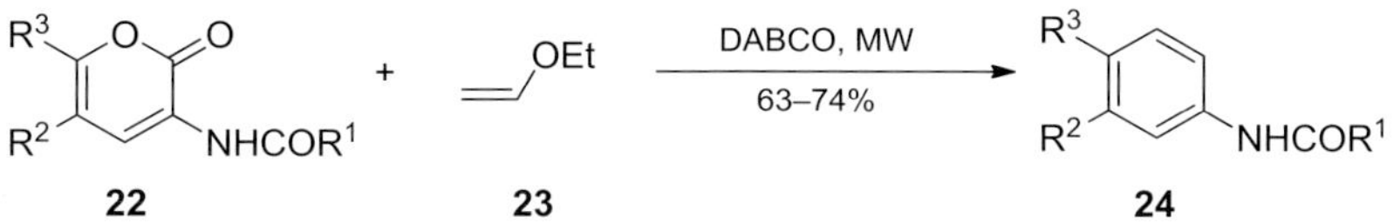

Scheme 12.8 Domino synthesis of aniline derivatives.

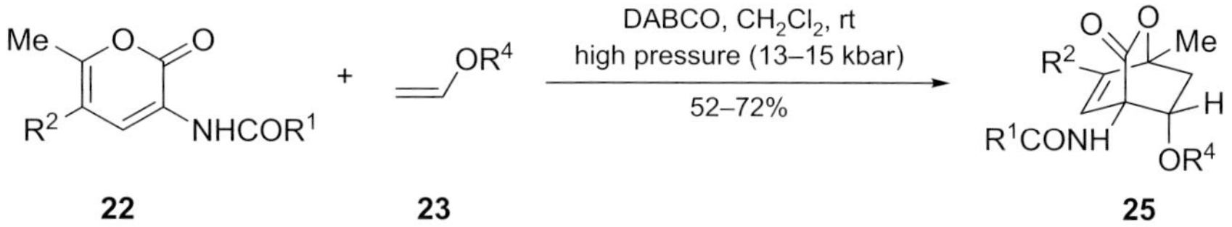

Scheme 12.9 2-Oxabicyclo[2.2.2]oct-5-enes as isolated intermediates in the domino reaction.

A domino synthesis of substituted phenanthrenes **28** was achieved by Dallemagne and coworkers [28] through reaction of 2-(2-bromophenyl)acetates **26** and cyanoboronic ester **27** (Scheme 12.10). The reaction resulted in the desired products in 60–87% yields through a Suzuki–Miyaura cross-coupling followed by a Dieckmann–Thorpe ring-closure reaction.

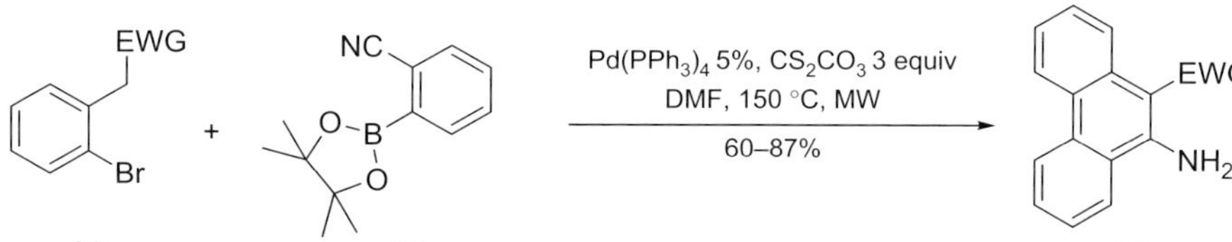

26 **27** **28**

Scheme 12.10 Domino synthesis of phenanthrenes.

Sadek and coworkers [29] reported two-component domino reactions of 1,1,3-tricyano-2-aminopropionitrile **29** with nitroolefins **30** in 1,4-dioxane under microwave irradiation, efficiently providing poly-substituted diaminobenzonitriles **31** in 70–73% yields (Scheme 12.11).

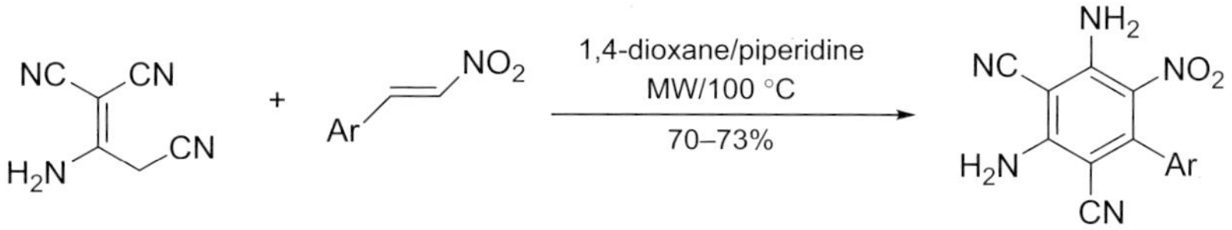

29 **30** **31**

Scheme 12.11 Domino synthesis of poly-substituted diaminobenzonitriles.

The fused pyrazolo[3,4-*b*]pyrazines **34** were prepared by Quiroga and coworkers by microwave-assisted cyclocondensation reaction of *ortho*-aminonitrosopyrazoles **32** with cyclic β-diketones **33**. Dimethylformamide (DMF) was proven to be the most suitable solvent for giving high yields under microwave irradiation, taking less than 18 min for completion. The synthesis showed a good substrate scope of 26 examples, affording the desired products in 40–85% yield (Scheme 12.12) [30].

32 **33** **34**

Scheme 12.12 Domino synthesis of pyrazolo[3,4-*b*]pyrazines.

Subsequently, Mane and coworkers [31] reported a domino synthesis of pyrazolo[3,4-*b*]quinolines **37** by carrying out the condensation of 2-chloro-3-formyl quinolines **35** with hydrazine hydrate/phenyl hydrazine **36** under microwave heating in water (Scheme 12.13). The reaction was accomplished in 2.5 h with yields of 93–97%. The same reaction took 7 h at 100 °C under conventional heating.

Carboxylic acids **38** were successfully converted into methylthiomethyl (MTM) esters **39** under microwave-assisted condition, as reported by Zimmerman *et al.*

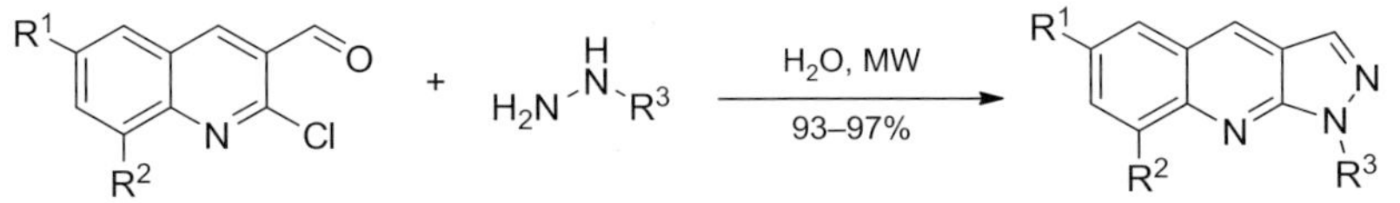

35 36 37

Scheme 12.13 Domino synthesis of pyrazolo[3,4-*b*]quinolines.

[32]. Chemical yields of 65–95% were achieved. This process allows the rapid introduction of an MTM ester protecting group to a variety of carboxylic acids including alkyl, electron-rich aromatic, and long-chain unsaturated carboxylic acids (Scheme 12.14). The products were isolated by a one-time extraction without using column chromatography and recrystallization, which belongs to GAP chemistry [11]. The procedure can be readily carried out on a large scale without deterioration of yields.

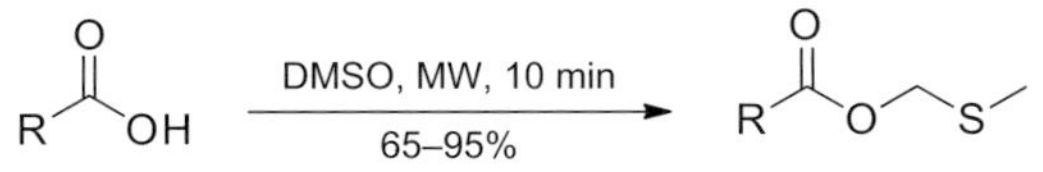

38 39

Scheme 12.14 Domino synthesis of methylthiomethyl (MTM) esters.

Dihydrobenzofurans belong to an important class of heterocycles, widely embedded in many biologically interesting natural products [33, 34] and have shown important biological activities [35]. Perumal and coworkers developed a facile stereoselective protocol for the synthesis of *trans*-2,3-dihydrobenzofurans **41** in moderate yields (Scheme 12.15). The domino reaction of cyclic 1,3-diketones **33** with 2,2′-sulfonylbis(1,3-diarylprop-2-en-1-ones) **40** in the presence of 1,8-diazabicyclo[5.4.0]undec-7-ene (DBU) in ethanol was performed under microwave irradiation as well as conventional thermal conditions. This transformation involved a domino Michael addition/proton exchange/annulation sequence [36].

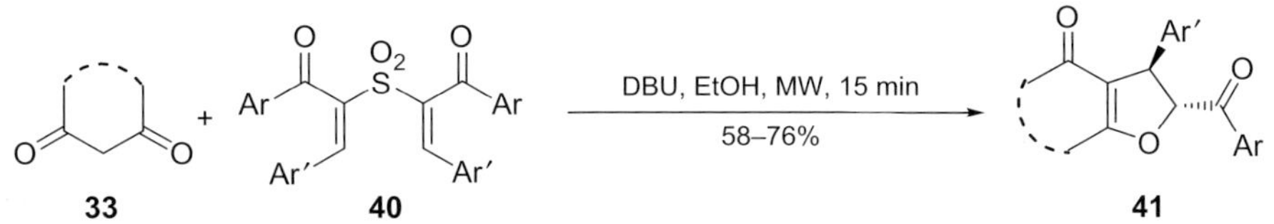

33 40 41

Scheme 12.15 Domino synthesis of *trans*-2,3-dihydrobenzofurans.

Spencer and coworkers reported the domino synthesis of 11 examples of 2-substituted-5-aminooxazole-4-carbonitriles **44** in 71–96% yields by microwave-mediated and flow chemistry (Scheme 12.16). Modifications around the oxazole motif could be made at positions 2, 4, and 5, affording a number of "rule of three" fragments. They obtained three oxazoles, which showed micromolar *Ki* values against cannabinoid (CB1/CB2) receptors [37].

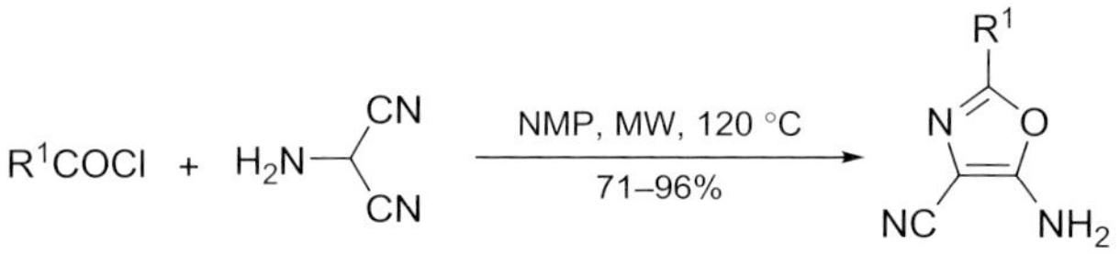

42 43 44

Scheme 12.16 Domino synthesis of 5-aminooxazole-4-carbonitriles.

Perlmutter and coworkers reported an improved reaction of amides **45** with solid-supported P_2S_5 reagent in tetrahydrofuran (THF) at 60 °C under microwave irradiation to generate 11 new thioamide derivatives **46** in 76–92% yields (Scheme 12.17). Compared to that with conventional heating, the reaction time is significantly reduced by microwave irradiation. Besides the shortened reaction periods, this thionation method also allowed more convenient workup procedure [38].

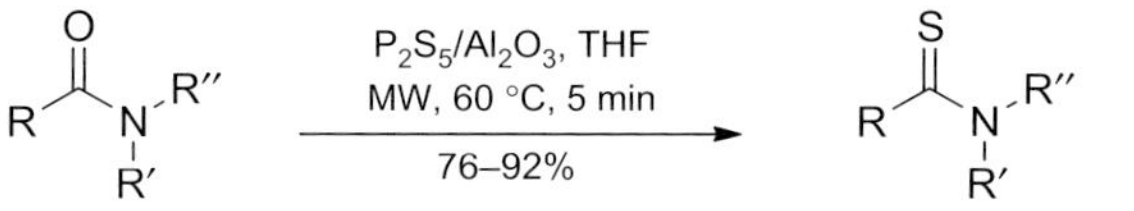

45 46

Scheme 12.17 Synthesis of thioamide derivatives.

Recently, an efficient preparation of *N*-acyl-*N*′-arylhexahydropyrimidines **48** with 66–95% yield was described by Orelli *et al.* [39] (Scheme 12.18). Microwave-assisted cyclodehydration of aminoamides **47** in water in the absence of any catalyst was achieved. The procedure also led to the formation of new *N*-acyl-*N*′-arylhexahydro-1,3-diazepines.

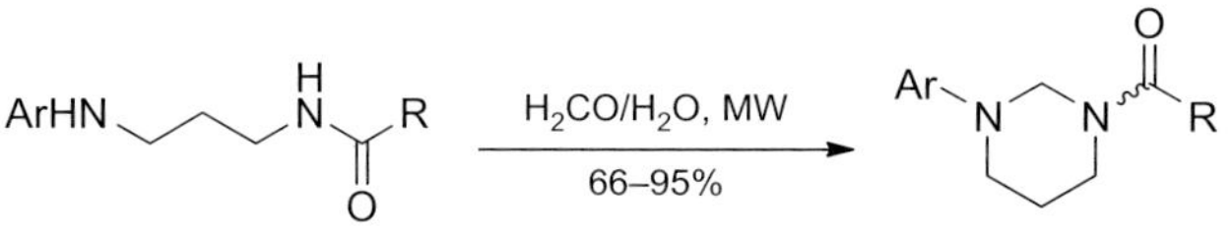

47 48

Scheme 12.18 Domino synthesis of *N*-acyl-*N*′-arylhexahydropyrimidines.

Microwave irradiation promotes a domino Sonogashira coupling/5-exo-dig cycloisomerization reaction of substituted 2-halobenzamides **49** with phenylacetylene **50** to give 3-(phenylmethylene) isoindolin-1-ones **51** in 42–94% yields. The best result was observed with the aid of DBU and $PdCl_2(MeCN)_2$/BINAP (2,2′-Bis(diphenylphosphino)-1,1′-binaphthalene) in DMF at 120 °C for 30 min (Scheme 12.19). Microwave irradiation not only accelerated the reaction rate but also increased the yield of the products. This domino method has been extended to different substrates of amidines and sulfonamides [40].

Poly-substituted 3-benzylquinoxalin-2(1*H*)-ones **54**, benzo[*d*]imidazoles **55**, and β-amino dipeptides **56** were all prepared using 4-arylidene-2-phenyloxazol-5-ones **52** and a large number of diverse *ortho*-diamines **53** in a highly selective manner.

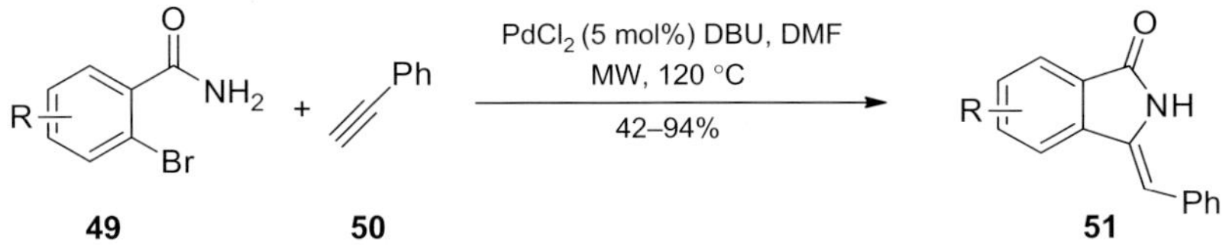

Scheme 12.19 Domino synthesis of isoindolin-1-ones.

The product selectivity solely depended on the used solvent and the strength of the acidic environment (Scheme 12.20). The reactions could be finished within 13–16 min to give 70–90% yield and high chemo- and regioselectivity without performing tedious workup and purification [41].

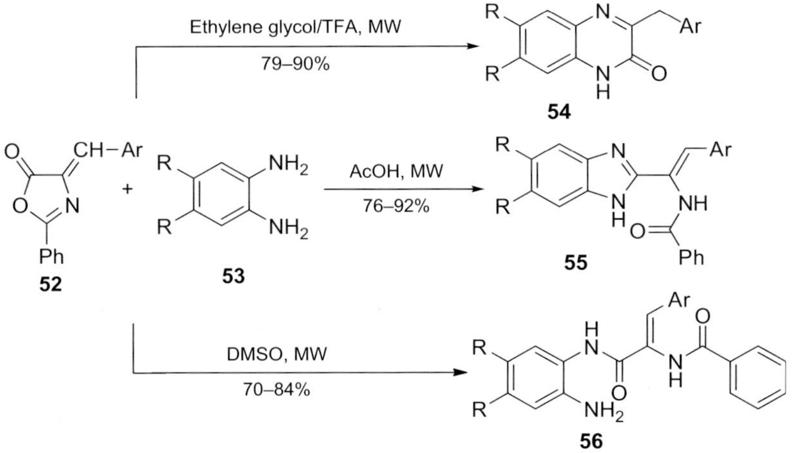

Scheme 12.20 Divergent reactions between oxazol-5-ones and *ortho*-diamines.

Efficient domino approaches to multifunctionalized tricyclic fused pyrroles **59** and dibenzo[*b*,*e*][1,4]diazepin-1-ones **60** have been established starting from enaminones **57** with different substitution patterns and arylglyoxal monohydrates **58**. The reaction pathway is controlled by the substituent at the N atom in **57**. The domino process can easily be performed by simply mixing **58** and *N*-amino acid enaminones or 3-(2-aminophenylamino)-5,5-dimethylcyclohex-2-enones **57** in the presence of AcOH under microwave irradiation (Scheme 12.21). The reaction occurs within 12–36 min, with water as the second product. A complete anti diastereoselectivity was achieved within the formation of tricyclic fused pyrrole derivatives [42].

A set of new two-carbon-tethered fused acridine/indoles **62** were synthesized via a AcOH-promoted domino reaction of indoline-2,3-dione **61** and C_2-tethered indol-3-yl enaminones **57a** (Scheme 12.22) [43]. The reaction was further expanded to prepare C-tethered fused acridine/pyridines pairs, N-substituted amino acids, *N*-cyclopropyl, and *N*-aryl-substituted fused acridine derivatives. In these domino processes, a fused acridine skeleton with concomitant formation of two new rings was readily achieved in good yields.

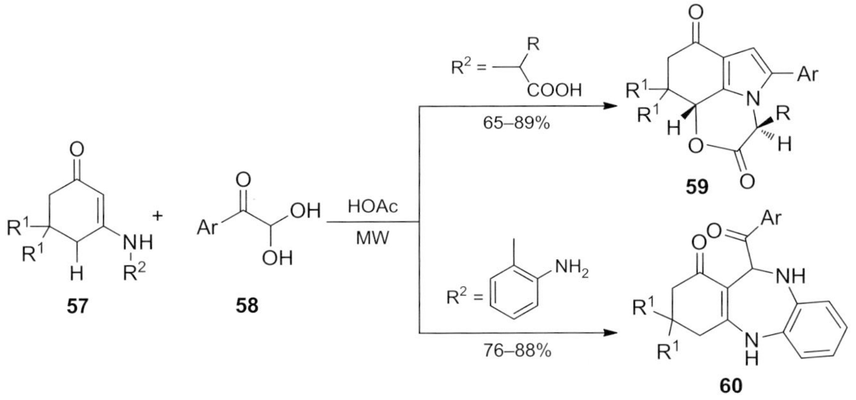

Scheme 12.21 Divergent reactions of arylglyoxal monohydrates with enaminones.

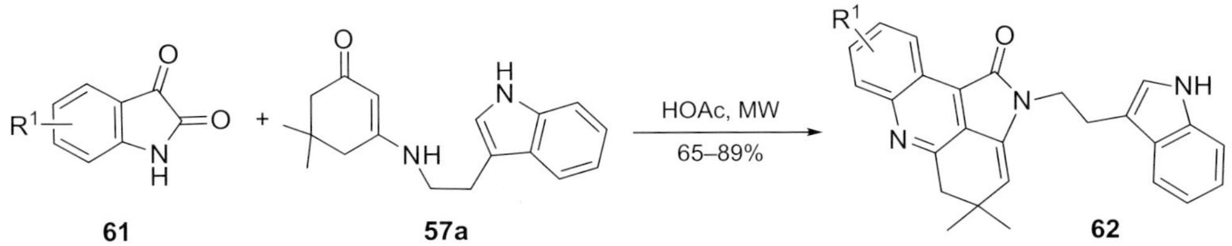

Scheme 12.22 Domino synthesis of fused acridine/indole pairs.

Domino [3+3] heterocyclization reactions have been established for the synthesis of *anti*-tetrahydropyrimidone derivatives **64** in 70–86% yields (Scheme 12.23) [44]. These reactions were conducted by reacting 4-arylidene-2-phenyloxazol-5(4*H*)-ones **52** with aryl amidines **63** under solvent-free and microwave irradiation conditions within 17–28 min. The selective construction of the *anti*-tetrahydropyrimidinone skeleton with one amide function was readily achieved under TEA (triethylamine) catalyzed condition by ring-opening of the oxazoles in a one-pot operation.

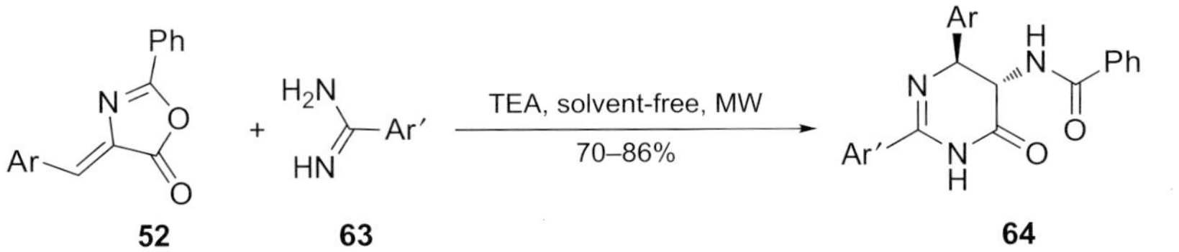

Scheme 12.23 Domino synthesis of *anti*-tetrahydropyrimidines.

12.2.3
Multicomponent Domino Reactions under Microwave Heating

A multicomponent reaction that employs an unsaturated carboxylic acid **65**, a 1,2- or 1,3-amino alcohol **66**, and gaseous CO and H_2 as starting materials has been

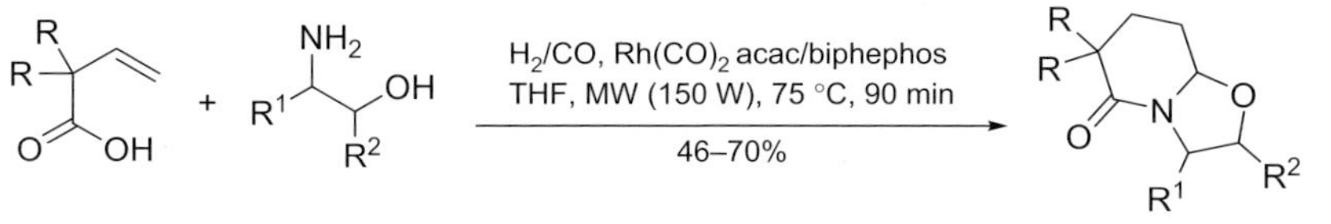

Scheme 12.24 Multicomponent domino reactions of unsaturated carboxylic acids.

reported by Taddei's group [45] (Scheme 12.24). The best yield was obtained when $Rh(CO)_2$acac/biphephos was used together with pyridinium *p*-toluenesulfonate (PPTS) at 90 °C under microwave heating. Under optimized conditions, a range of aminoalcohols were converted to the desired bicylic, tricyclic, tetracyclic, and spirocyclic oxazolopiperidines **67** in 46–70% yields through intramolecular lactamization.

Jeong and coworkers [46] accomplished a rapid synthesis of 2*H*-indazolo[2,1-*b*]phthalazine-triones **70** under solvent-free conditions by the reaction of phthalhydrazide **68**, 5,5-dimethylcyclohexane-1,3-dione **33a**, and aldehydes **69** (Scheme 12.25). After optimization of the reaction conditions, the best results were obtained at 80 °C under microwave irradiation using montmorillonite K-10 as catalyst. This approach offers many advantages, including excellent yields of 80–96%, short reaction times of only 5–15 min, easy isolation of the products, and environmentally benign conditions.

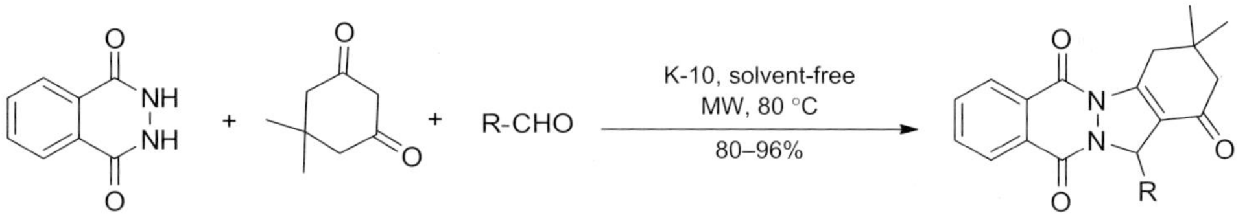

Scheme 12.25 Multicomponent synthesis of indazolo[2,1-*b*]phthalazine-triones.

A similar version of this domino process to give pyrazolo[1,2-*b*]phthalazine-5,10-diones **72** using **68**, **69**, and **71** as starting materials was reported by Singh and Raghuvanshi [47] (Scheme 12.26).

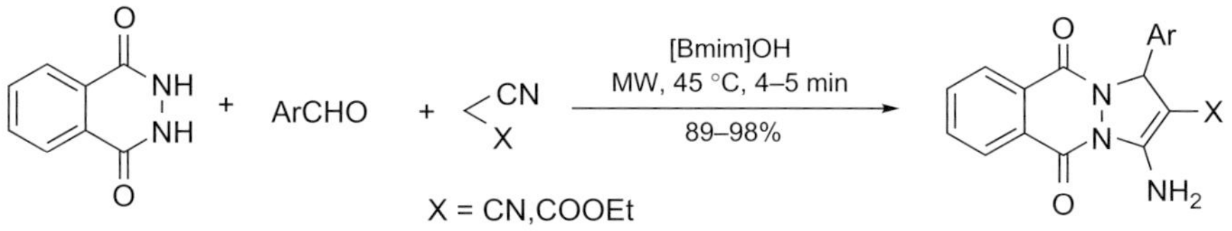

Scheme 12.26 Multicomponent synthesis of pyrazolo[1,2-*b*]phthalazine-5,10-diones.

Leadbeater presented a direct and rapid preparation of diaryl sulfides **75** from aryl iodides **73** using potassium thiocyanate **74** as the source of sulfur. The catalytic system consisted of a copper salt, tetrabutylammonium bromide as a phase-transfer agent, and water as the solvent. Microwave heating at 200 °C for 60 min allowed the conversion of a wide range of aryl iodides to give the corresponding diaryl sulfides in 36–75% yields (Scheme 12.27) [48].

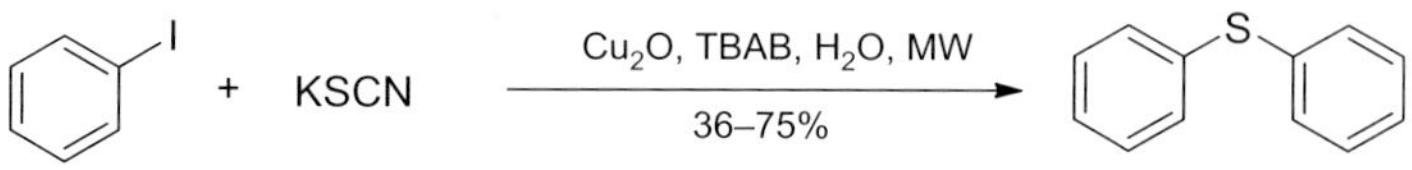

73 **74** **75**

Scheme 12.27 Multicomponent synthesis of diaryl sulfides.

The microwave-assisted synthesis of *N*-carbazolyldihydropyridines **78** via a four-component domino reaction of 3-amino-9-ethylcarbazole **76**, malononitrile **71a**, aromatic aldehydes **69**, and acetylenic esters **77** was reported by Prasad *et al.* [49] by using $InCl_3$ as catalyst (Scheme 12.28). The use of microwave heating allowed shortened reaction times and higher yields.

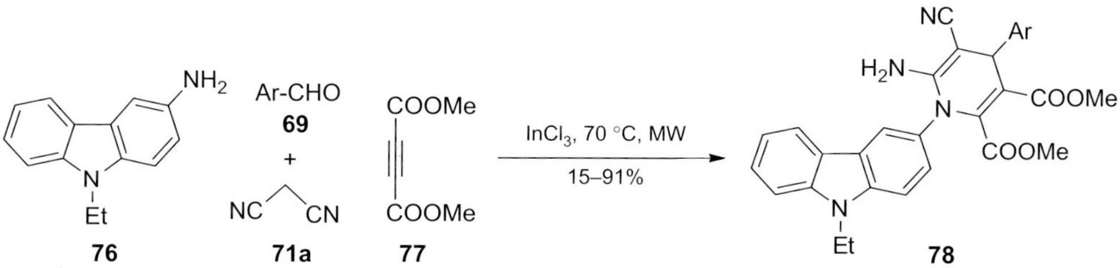

76 **71a** **77** **78**

Scheme 12.28 Multicomponent synthesis of fully substituted pyridines.

Functionalized macrocyclane-fused pyrazolo[3,4-*b*]pyridine derivatives **81** with aryl group residing in 2-position of the pyridine nucleus were assembled by a microwave-assisted three-component reaction of aldehydes **69**, 5-aminopyrazoles **80**, and macrocyclic ketones **79** in HOAc in the presence of 1.0 equiv of trifluoroacetic acid (TFA) as the promoter (Scheme 12.29) [50]. This method provided an efficient and promising synthetic strategy for the construction of 44 examples of differently substituted pyrazolopyridines **81**. To further explore the scope of the reaction, the same group investigated the domino reaction of aldehydes **69**, 5-aminopyrazoles **80**, and cyclopentanone **79a** under acidic and basic conditions. As expected, reaction in AcOH led to pyrazolo[3,4-*b*]pyridines **82** via a selective [4+2]

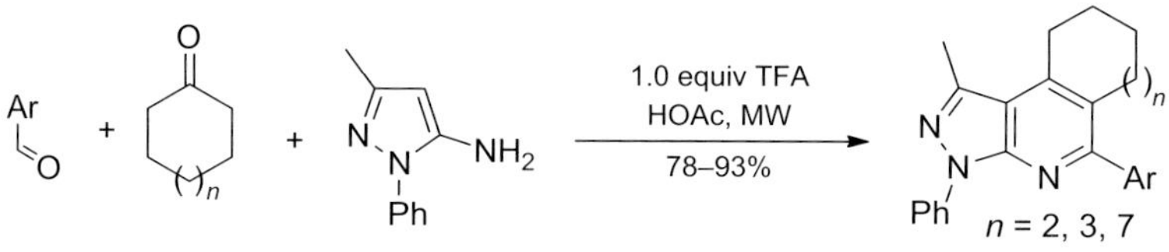

69 **79** **80** **81**

Scheme 12.29 Multicomponent synthesis of macrocyclane-fused pyrazolo[3,4-*b*]pyridines.

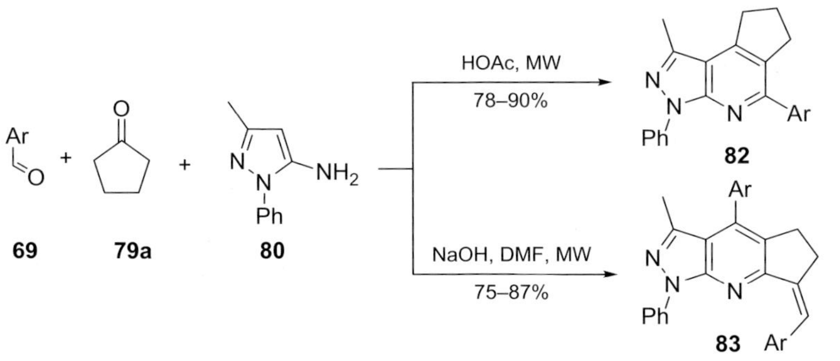

Scheme 12.30 Divergent synthesis of pyrazolo[3,4-*b*]pyridines.

heterocyclization in 78–93% yields, and to the fused pyrazolopyridine framework **83** under base conditions (Scheme 12.30) [51].

Subsequently, aryl amidines **63**, instead of 5-aminopyrazoles, reacted with aromatic aldehydes **69** and tetrahydrothiopyran-4-one (or tetrahydropyran-4-one) **79b**, respectively, to give the annulated pyrimidine derivatives **84** (Scheme 12.31) [52].

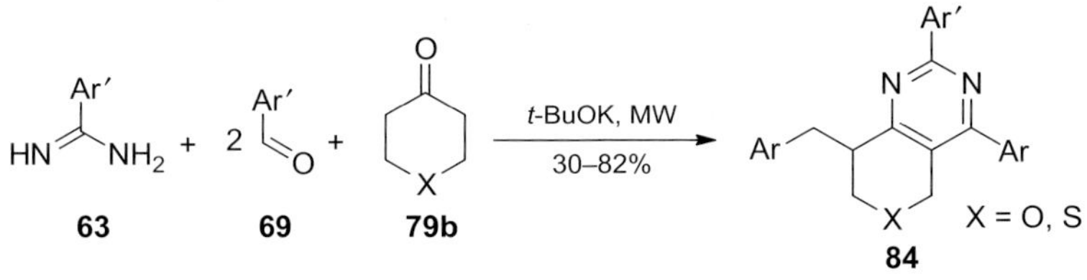

Scheme 12.31 Multicomponent synthesis of fused pyrimidine derivatives.

The same group also established a rapid and regioselective synthesis of benzo[*h*]pyrazolo[3,4-*b*]quinolines **85** (Scheme 12.32) [53]. The reaction could be conducted by using aldehydes **69**, 5-aminopyrazoles **80**, and 2-hydroxynaphthalene-1,4-dione **33b** as substrates under microwave irradiation within short a time of 10–26 min. Good to excellent chemical yields (61–91%) and complete regioselectivity were achieved for 22 examples.

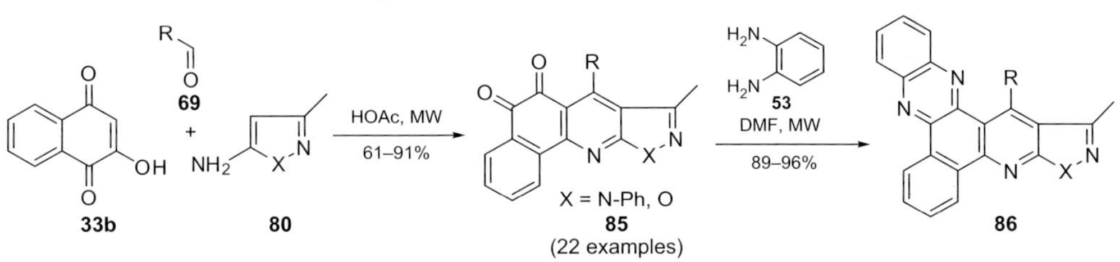

Scheme 12.32 Regioselective synthesis of benzo[*h*]pyrazolo[3,4-*b*]quinolines and their derivatives.

The resulting benzoquinolines **85** can readily be converted into quinoxaline-fused benzo[*h*]isoxazolo[5,4-*b*]quinoline analogs **86** by treatment with benzene-1,

2-diamine **53** under microwave irradiation (Scheme 12.32). The structural assignment has been unambiguously confirmed by X-ray analysis. A similar version to allow a rapid synthesis of highly functionalized benzo[*h*]naphtho[2,3-*a*]acridine-15,16(5*H*,14*H*)-diones **88** and their derivatives **89** was developed by Tu and coworkers (Scheme 12.33) [54].

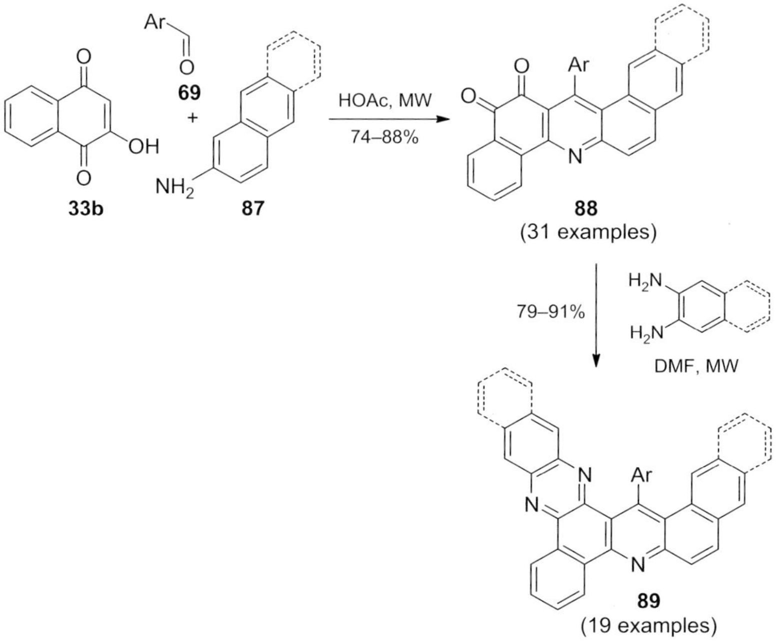

Scheme 12.33 Regioselective synthesis of benzo[*h*]naphtho[2,3-*a*]acridines and their derivatives.

A series of poly-functionalized dihydroquinoline derivatives **91** (26 examples) were synthesized by a three-component domino reaction of aldehydes **69** (1 equiv) and 1-arylethylidenemalononitriles **90** (2 equiv) in ethylene glycol using NaOH as a base promoter under microwave irradiation within 8–20 min (Scheme 12.34) [55]. In this process, the concomitant formation of two new cycles was readily achieved.

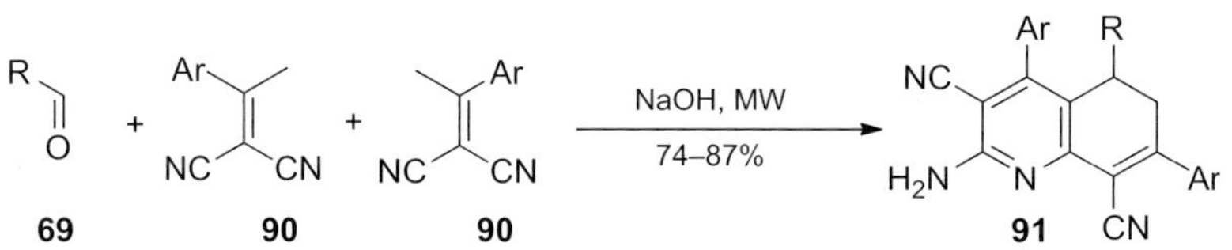

Scheme 12.34 Three-component synthesis of dihydroquinoline derivatives.

Tu and coworkers [56] synthesized new substituted 2,4-diaminopyridine-3,5-dicarbonitriles **92** under microwave irradiation and solvent-free conditions (Scheme 12.35). The multicomponent domino reaction of 2 mol of malononitrile

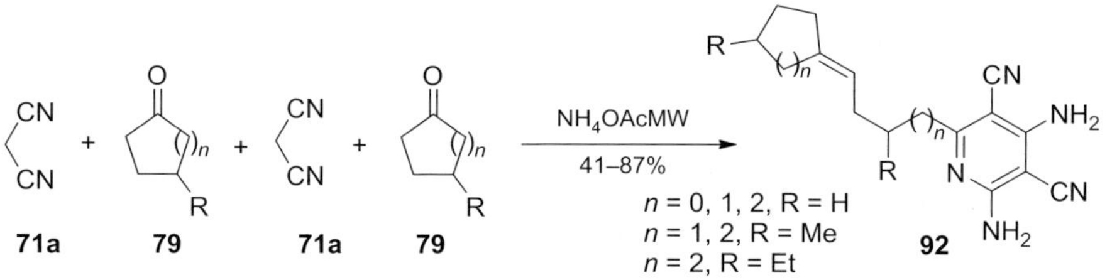

Scheme 12.35 Multicomponent synthesis of 2,4-diaminopyridine derivatives.

71a with 2 mol of cycloketones **79** and NH_4OAc allowed the efficient assembly of five new sigma bonds including two C–N bonds in a one-pot operation. A mechanism was proposed that involves a novel sequence consisting of deprotonation/imine formation/anionic carbonyl addition.

Later, Tu and coworkers [57] reported domino synthesis of furo[3,4-*b*][1,8]naphthyridines **93** through three-component reactions of aldehydes **69**, 2-aminoprop-1-ene-1,1,3-tricarbonitrile **29**, and N-substituted 4-aminofuran-2(5*H*)-ones **57** in EtOH using EtONa as a base (Scheme 12.36). N-substituted 3-aminocyclohex-2-enones were also suitable for this domino reaction, affording 21 examples of benzo[*b*][1,8]naphthyridines in 69–84% chemical yield, with the concomitant formation of two new pyridine rings.

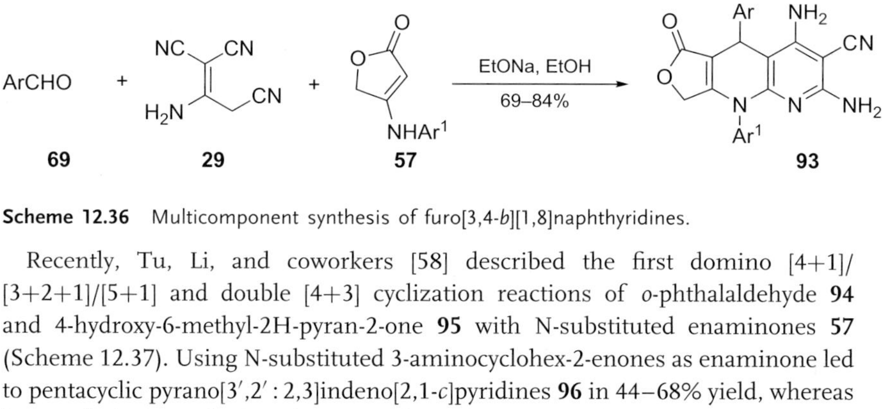

Scheme 12.36 Multicomponent synthesis of furo[3,4-*b*][1,8]naphthyridines.

Recently, Tu, Li, and coworkers [58] described the first domino [4+1]/[3+2+1]/[5+1] and double [4+3] cyclization reactions of *o*-phthalaldehyde **94** and 4-hydroxy-6-methyl-2H-pyran-2-one **95** with N-substituted enaminones **57** (Scheme 12.37). Using N-substituted 3-aminocyclohex-2-enones as enaminone led to pentacyclic pyrano[3′,2′:2,3]indeno[2,1-*c*]pyridines **96** in 44–68% yield, whereas by employing N-substituted 4-aminofuran-2(5*H*)-ones the reaction occurred in another direction to form multifunctionalized pentacyclic pyrano[4,3-*b*]oxepines **97**. This work provides an attractive strategy for the construction of structurally diverse pentacyclic oxa-azaspiro and oxa-azabridged skeletons.

Tu, Li, and coworkers also developed a new allylic functionalization without the use of metal catalysts [59–61]. Efficient annulation of enaminones **57** with arylglyoxal monohydrates **58** and the subsequent allylic functionalization with aliphatic carboxylic acids **38** as nucleophile reagents provided multifunctionalized indoles **98** with 75–89% yields (Scheme 12.38) [59]. To further explore the scope of this domino reaction, N-substituted 3-aminocyclohex-2-enones were employed. Interestingly, two molecules of arylglyoxal monohydrate and N-substituted

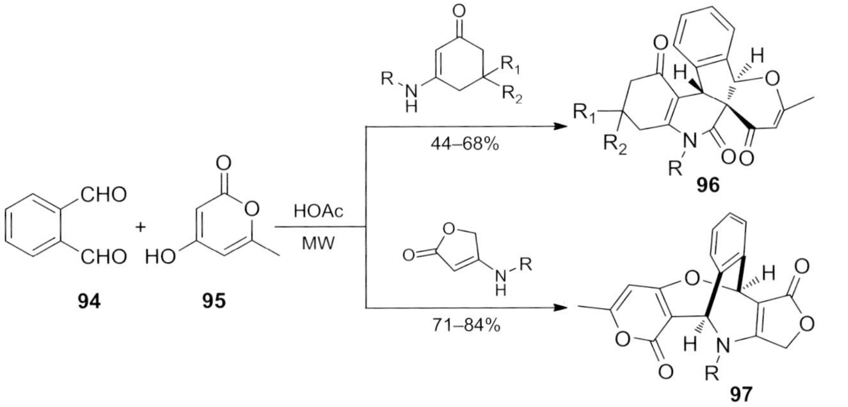

Scheme 12.37 Divergent reactions of *o*-phthalaldehyde with N-substituted enaminones and 4-hydroxy-6-methyl-2H-pyran-2-one.

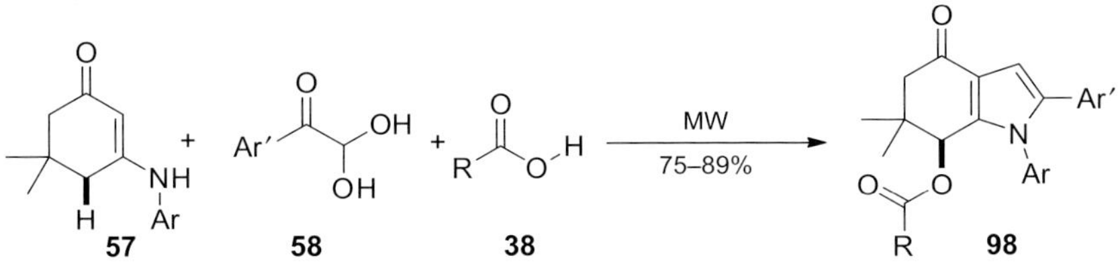

Scheme 12.38 Allylic esterification-based multicomponent domino reaction of enaminones.

3-aminocyclohex-2-enones reacted to give the final poly-substituted bis-indoles in a very short time of 26–30 min. When thiophenols **99** were introduced into this system, a domino [3+2] heterocyclization took place to give 3-arylthio-substituted fused pyrroles **100** in 70–89% yields. Contrarily, a two-step one-pot operation, wherein the thiophenol was added after the formation of the pyrrole derivative, gave the fused pyrrole framework **101** through an allylic thiolation with formation of a C(sp^3)–S bond. The authors have shown the feasibility of these two pathways with 44 examples (Scheme 12.39) [60].

Later, the same authors established a similar three-component domino reaction providing a divergent approach to 40 examples of fused pyrroles (Scheme 12.40) [61]. The reaction proceeded as a domino [3+2] heterocyclization to afford fused pyrroles **103** in 65–86% yields through an allylic amination. On the contrary, using **57** with R = carboxymethyl group, an intermolecular arylation took place with formation of **104**.

An efficient three-component domino strategy for the synthesis of multi-functionalized tetracyclic indeno[1,2-*b*]indole derivatives has again been reported by Tu, Li, and coworkers [62]. The domino reaction between enaminones **57** and 2,2-dihydroxyindene-1,3-dione **105** in acid anhydride generated the fused pyrazoles **106** (tetracyclic 6-5-5-6 skeleton), which involved a nucleophilic substitution/cyclization/methyl migration/aromatization/esterification. On the other hand,

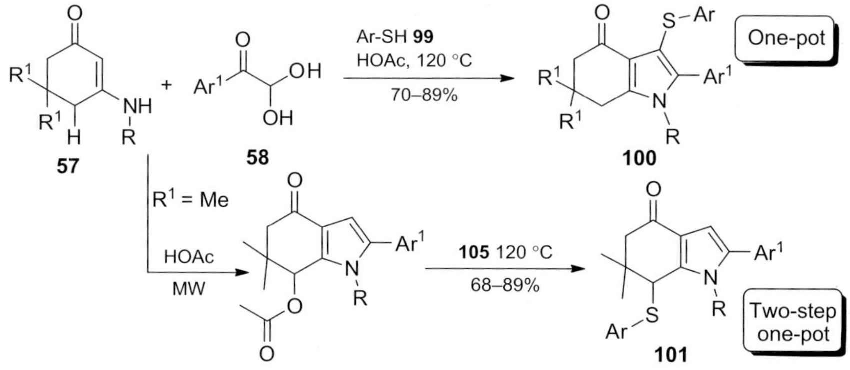

Scheme 12.39 Divergent synthesis of fused pyrroles with different substituted patterns.

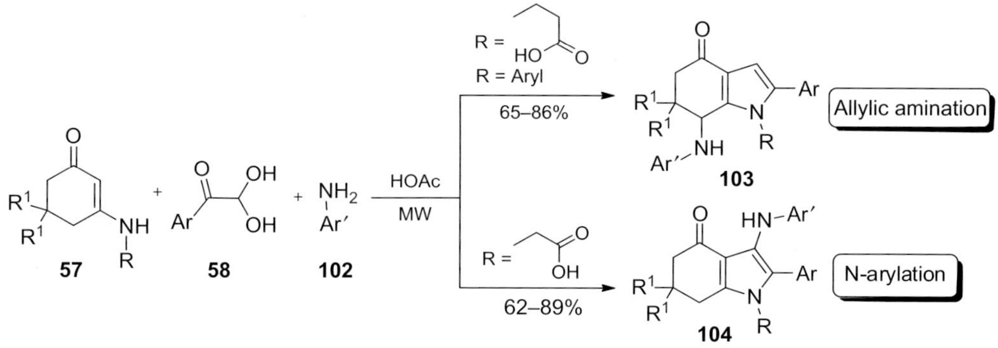

Scheme 12.40 Divergent synthesis of fused pyrroles with different substituted patterns.

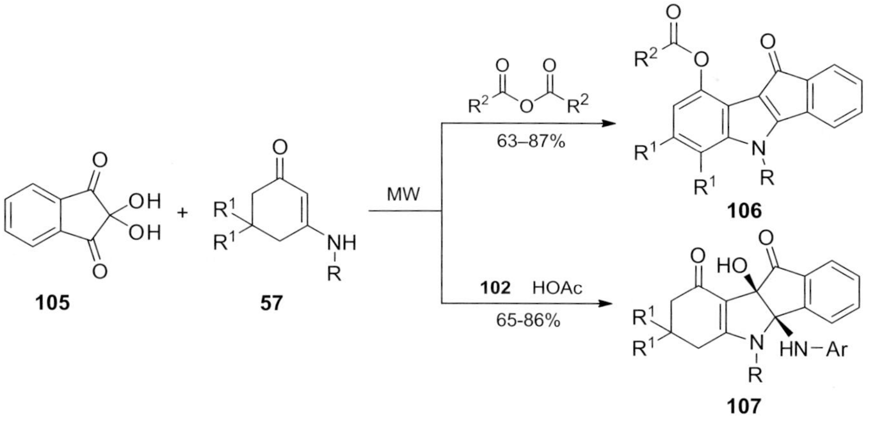

Scheme 12.41 Selective synthesis of tetracyclic indeno[1,2-*b*]indoles.

transformation of aromatic amines **102**, **57**, and **105** in HOAc led to compounds **107** with high syn diastereoselectivity (Scheme 12.41). Both reactions showed attractive features, including mild conditions, convenient one-pot operation, short reaction times of 15–32 min, and excellent regio- and/or stereoselectivity.

12.3 Aqueous Domino Reactions

12.3.1 Two-Component Domino Reactions in Water

The use of water as reaction medium has attracted widespread interest in current organic chemistry not only because of its environmental and economical characteristics but also because of the unique reactivity and selectivities that sometimes cannot be accomplished by using organic solvents. Many synthetic methodologies have been developed with great success by using water as the reaction medium, particularly those under microwave irradiation as discussed previously [63, 64]. The great success of using water as solvent in microwave-assisted transformations can be attributed to the unique capability of water for delivering the microwave energy to reactants in the reaction system [65]. Thus, water has been proven to be crucial for developing new multicomponent reactions by accelerating the reaction rates and by selective activation of intermediate species.

Wang and coworkers [66] developed a concise synthesis of substituted isocoumarins **109** by a domino reaction of *o*-allylbenzaldehydes **108** in water (Scheme 12.42). The substrate was found to undergo a 6-exo-trig cyclization in the presence of $PdCl_2$ and $CuCl_2$ in water at reflux to afford substituted isocoumarins **109** in 76–90% yields. This domino reaction consists of a 6-exo-trig cyclization, the addition of water, the elimination of PdHCl, the isomerization of a carbon–carbon double bond, the oxidation of a hemiacetal with the elimination of PdHCl, and the regeneration of $PdCl_2$.

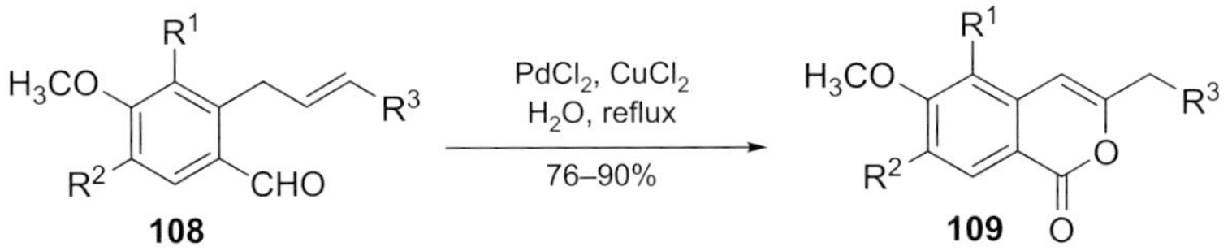

Scheme 12.42 Aqueous synthesis of isocoumarins.

The synthesis of spiro cyclohexadienone lactones **111** was accomplished by a domino reaction of 1-(*p*-hydroxyaryl)cyclobutanols **110** in water using hexafluoroisopropanol as cosolvent, being promoted by phenyl iodide(III) diacetate, in 54–76% yield (Scheme 12.43) [67].

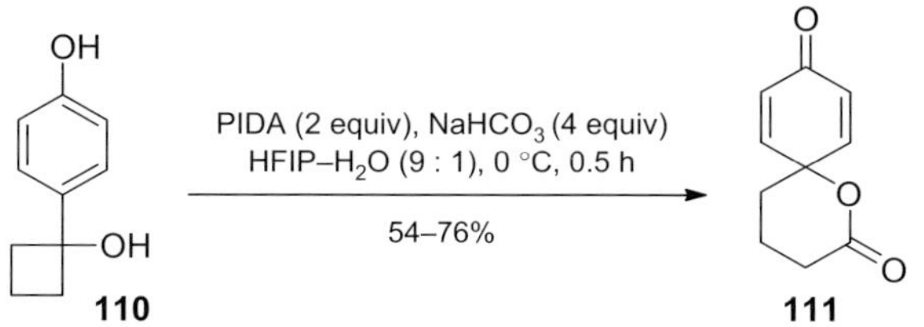

110 **111**

Scheme 12.43 Domino synthesis of spiro cyclohexadienone lactones.

Gu and coworkers reported an 2-iodoxybenzoic acid (IBX) induced oxidative coupling of *in situ* formed Baylis–Hillman adducts **114** with a nucleophile in water (Scheme 12.44) [68]. As nucleophiles, styrenes, β-dicarbonyl compounds, benzamides, and less reactive indoles can be used. This strategy offers an alternative path to the methylenylation of β-dicarbonyl compounds with formaldehyde for the formation of a methylene intermediate. In the reaction of **112** and the styrenes **113** (R = Ph), a 5,6-dihydropyran **114** is formed.

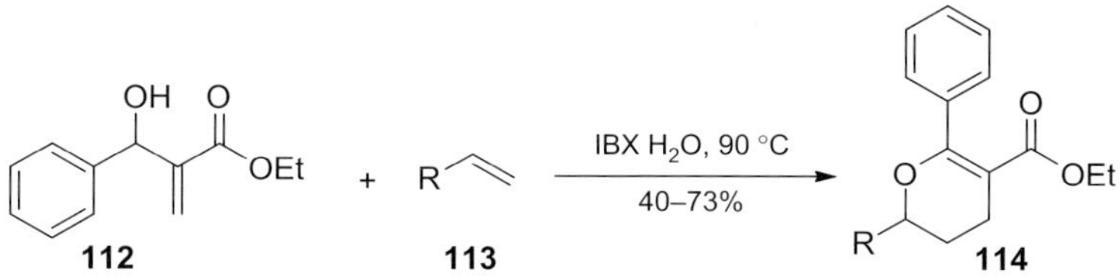

112 **113** **114**

Scheme 12.44 Domino synthesis of 5,6-dihydropyrans.

Lee and coworkers presented a facile synthesis of fused pyrans **116** in 65–90% yields in water by reacting a variety of cyclic 1,3-dicarbonyls **33** with α,β-unsaturated aldehydes **115** (Scheme 12.45). The key step is a formal [3+3] cycloaddition though a domino Knoevenagel/6π-electrocyclization. This methodology has been applied to the synthesis of biologically interesting pyranocoumarin, pyranoquinolinone, and pyranonaphthoquinone derivatives [69].

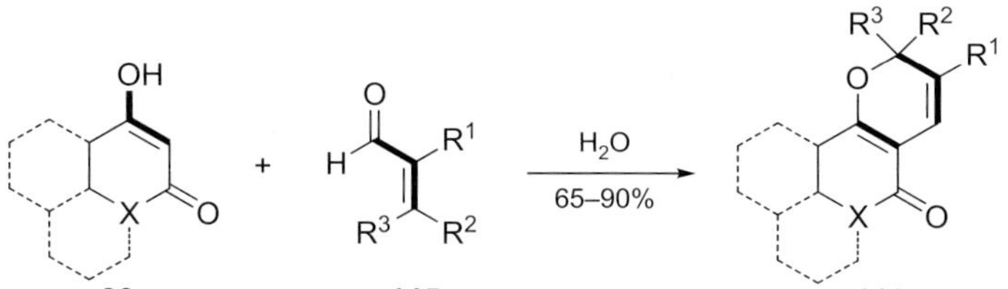

33 **115** **116**

Scheme 12.45 Domino synthesis of fused pyrans.

Perumal, Menéndez, and coworkers synthesized a set of *trans*-4,5-dihydrofuran-3-carbonitriles **119** in excellent diastereoselectivity and 83–95% yield; these can serve as valuable precursors for the syntheses of many natural products and

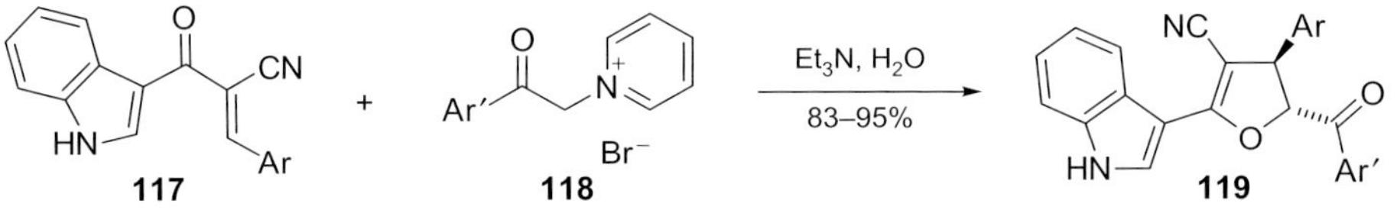

Scheme 12.46 Domino synthesis of *trans*-4,5-dihydrofuran-3-carbonitriles.

pharmaceuticals. The reaction was performed by reacting 2-(3-indolylcarbonyl)-3-aryl-2-propenenitriles **117** with (2-aryl-2-oxoethyl)pyridinium bromides **118** in water in the presence of TEA (Scheme 12.46) [70]. The resulting products could be isolated by simply filtering off the product from the aqueous reaction medium. In this domino reaction, one C–C and one C–O bond are formed; the sequence includes the formation of a pyridinium ylide, a Michael addition, and an annulation by an intramolecular nucleophilic substitution.

Jia and coworkers [71] reported a very nice palladium-catalyzed domino reaction involving a C–H activation process (Scheme 12.47). The regioselectivity of the attack of the nucleophile at the intermediately formed palladacyle **121** is solely controlled by the reaction conditions using the same substrate. In the Pd-catalyzed transformation of **120** in DMF in the absence of water, the aryl-substituted product **122** was obtained in 47–95% yield, and, in the presence of water, the alkyl-substituted product **123** in 60–95% yield. However, it should be noted that the reaction temperature has also some influence. The palladacycle intermediate **121** was successfully trapped by cyanation, Heck reaction, secondary C–H activation, and Suzuki coupling.

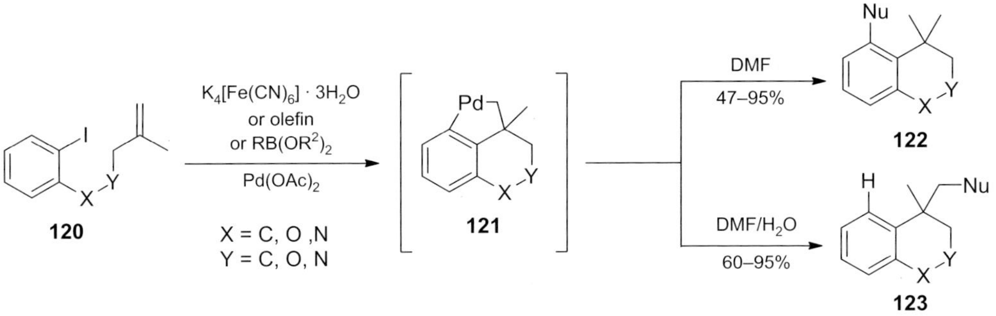

Scheme 12.47 Divergent reaction of o-substituted iodobenzene derivatives.

The enaminones **57a** and **b** were found to react with β-bromonitrostyrenes **124** in water, leading to the formation of a set of trisubstituted pyrroles **125** and **126**, respectively, in 72–90% yield (Scheme 12.48) [72].

Besides the use of **57** as substrates, also 1,3-dicarbonyl compounds **33** were subjected to the reaction with **124** to afford a set of functionalized bi- and tricyclic 2,3-dihydrofurans **126** and other tetrasubstituted 2,3-dihydrofurans with high diastereoselectivity (Scheme 12.49) [73].

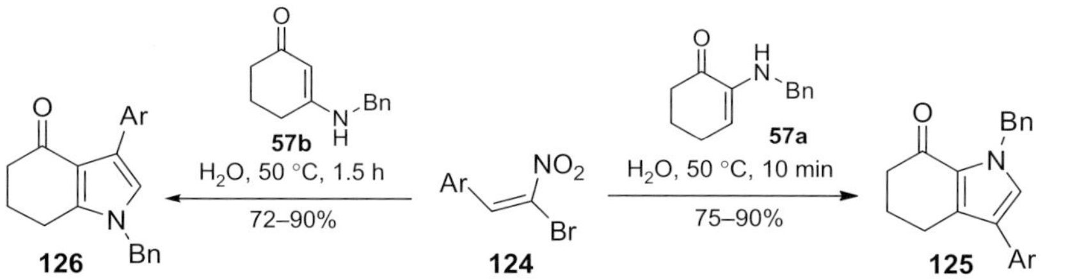

Scheme 12.48 Domino synthesis of trisubstituted pyrroles.

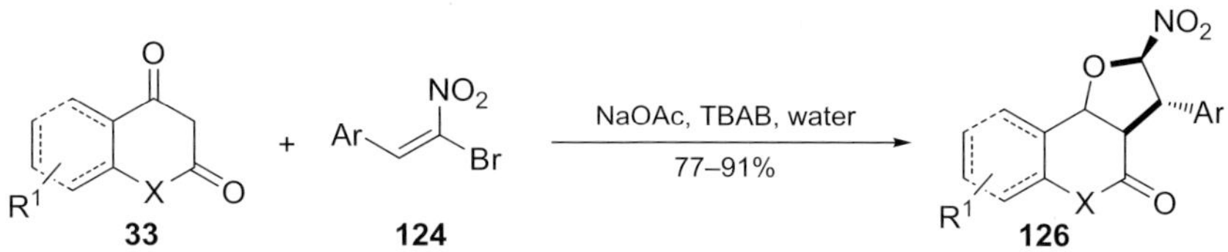

Scheme 12.49 Domino synthesis of 2,3-dihydrofurans.

Xie and coworkers [74] presented a similar synthesis of 3-phenylnaphtho[2,3-*b*]furan-4,9-diones **127** by reaction of **33b** with **124** (Scheme 12.50).

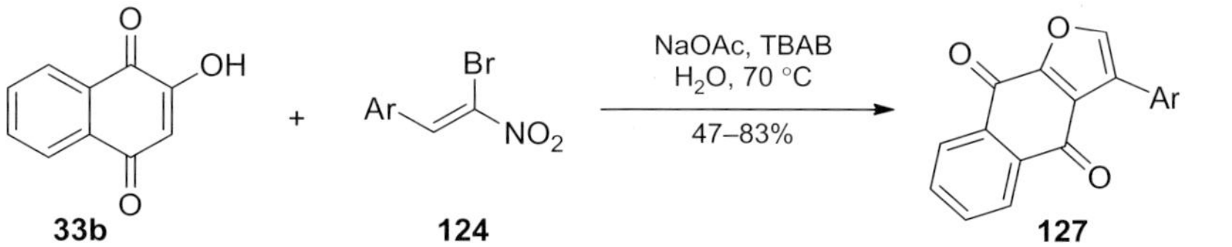

Scheme 12.50 Domino synthesis of naphtho[2,3-*b*]furan-4,9-diones.

Menendez and coworkers [75] developed a domino reaction of α-nitrocycloalkanones **128** with α-alkyl α,β-unsaturated aldehydes **129** in water, which led to functionalized bridged bicyclic lactones **130** containing 10-, 11-, 13-, and 15-membered rings with complete stereocontrol depending on the ring size (Scheme 12.51). The structures of the products were determined by X-ray diffraction analysis.

2,3-Dihydroisoxazoles are an important class of heterocycles and exist widely in a number of bioactive molecules. Xie and coworkers [76] developed a facile synthesis of trisubstituted 2,3-dihydroisoxazoles **132** by reacting hydroxylamine **131** with α,α-dicyanoolefins **90** under weak basic conditions (pH 9 and 10) in aqueous media (Scheme 12.52).

A domino Knoevenagel/hetero-DA reaction of 2-formyl-4-phenyl (*E*)-2-phenylethenesulfonates **133** with dimedone **33a** or *N,N*-dimethylbarbituric acid in water was reported by Ghandi and coworkers [77] (Scheme 12.53). The optimized conditions showed that performing the reaction in water without using any catalyst gave the best result. Subsequently, a series of hexahydrochromenes and

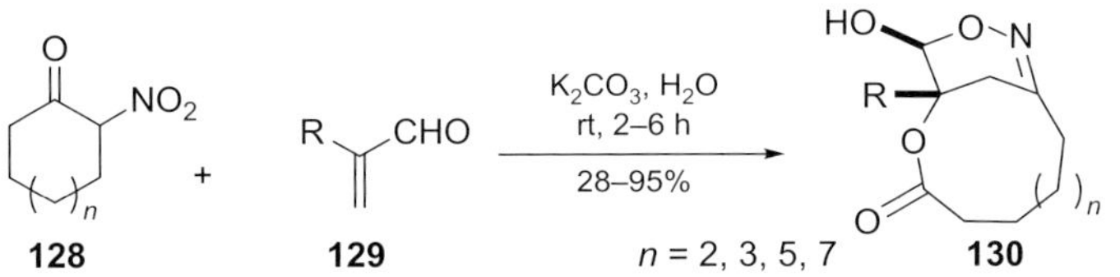

Scheme 12.51 Domino synthesis of bicyclic lactones.

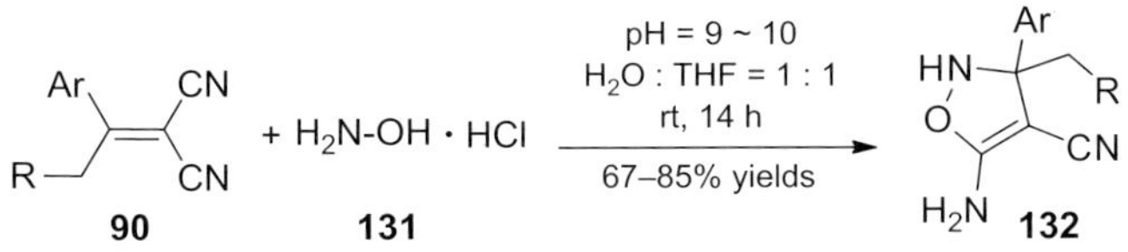

Scheme 12.52 Domino synthesis of 2,3-dihydroisoxazoles.

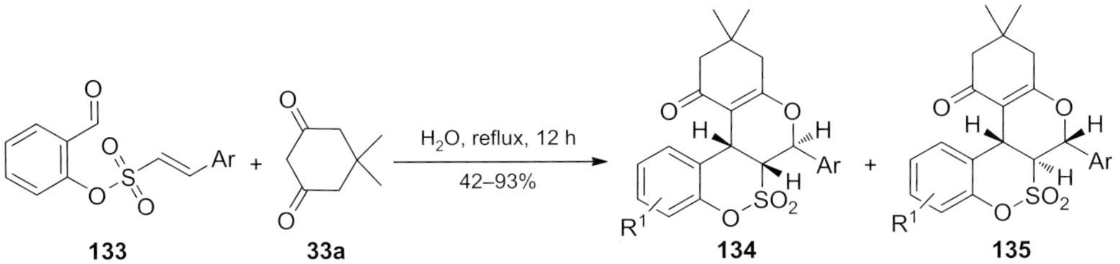

Scheme 12.53 Domino synthesis of fused benzo-δ-sultones.

pyrano[2,3-*d*]pyrimidine-annulated benzo-δ-sultones **134** and **135** were formed in 42–93% yields.

A similar aqueous domino Knoevenagel/hetero-DA reaction of *o*-propargylated salicylaldehyde with 1-methylindoline-2-thione in the absence of any Lewis acid has been developed by Majumdar and coworkers [78]. Seven examples of new indole-annulated pentacyclic heterocycles **138** were obtained in 72–80% yields (Scheme 12.54).

12.3.2
Multicomponent Domino Reaction in Water

Shinokubo and coworkers [79] performed a domino coupling reaction of aryl boronic acids **139** with internal alkynes **140** and acrylates **141** by using a rhodium catalyst in water to give nine examples of the 1,3-diene derivatives **142** in 37–81% yields (Scheme 12.55). The reaction indicates that the use of water offers the possibility to optimize known reactions, leading to novel transformations.

A phosphane-mediated domino reaction of suitable electron-poor dienes **143** with *N*-tosylimines **144** using water both as a solvent and a hydrogen source was reported by Marinetti and coworkers [80] (Scheme 12.56).

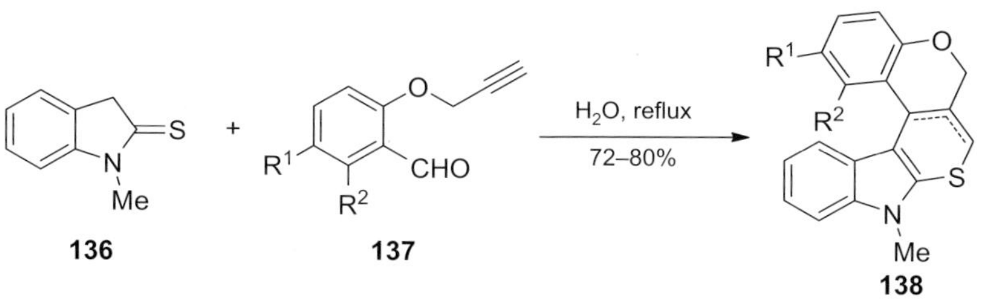

Scheme 12.54 Domino synthesis of indole-annulated heterocycles.

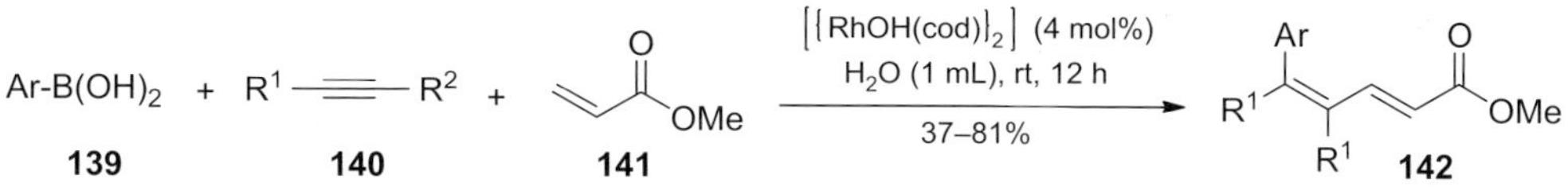

Scheme 12.55 Rhodium-catalyzed domino synthesis of 1,3-dienes.

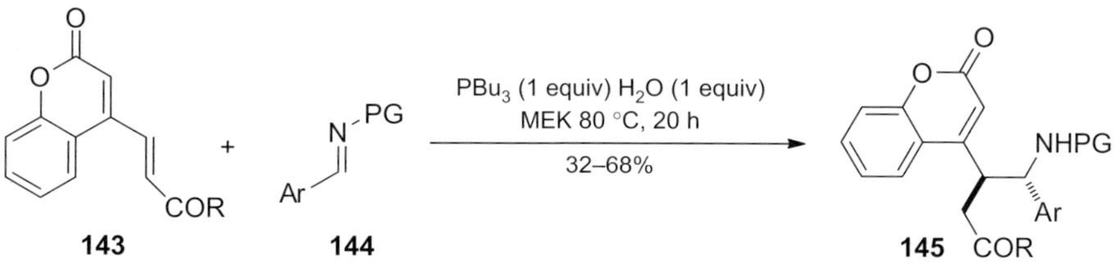

Scheme 12.56 Multicomponent domino synthesis of coumarin derivatives.

The domino process involves an aza-Morita–Baylis–Hillman/reduction sequence. A series of tosylamide functionalized coumarin derivatives **145** were assembled with high diastereoselectivity concerning the newly created contiguous carbon centers.

Jerome and coworkers [81] reported multicomponent domino reactions of formaldehyde **69a**, 1,3-dicarbonyl derivatives **146**, and styrenes **113** (or indole derivatives) in water to afford a wide range of dihydropyrans **147** and C3-substituted indole derivatives in 22–82% yields (Scheme 12.57). The described multicomponent domino reactions exhibit a good substrate scope and provide direct access to a wide range of small organic skeletons traditionally produced through multistep processes.

An efficient method for the synthesis of dihydrothiophene ureidoformamides **149** from commercially available materials in a four-component domino process was reported by Cai *et al.* [82] (Scheme 12.58). The reaction efficiently assembled aromatic aldehydes **69**, malodinitrile **71a**, 1,3-thiazolidinedione **148**, and anilines **102** with high yields by conducting the reaction in a cosolvent mixture of polyethylene glycol (PEG) 400/H_2O (1 : 1) at 80 °C. Under optimized conditions, a wide range of aldehydes and amines were converted into the desired heterocycles with yields ranging from 27% to 90%.

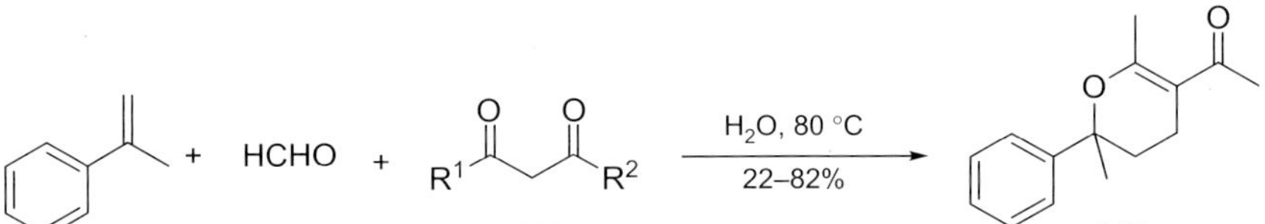

113 69a 146 147

Scheme 12.57 Multicomponent domino synthesis of dihydropyrans.

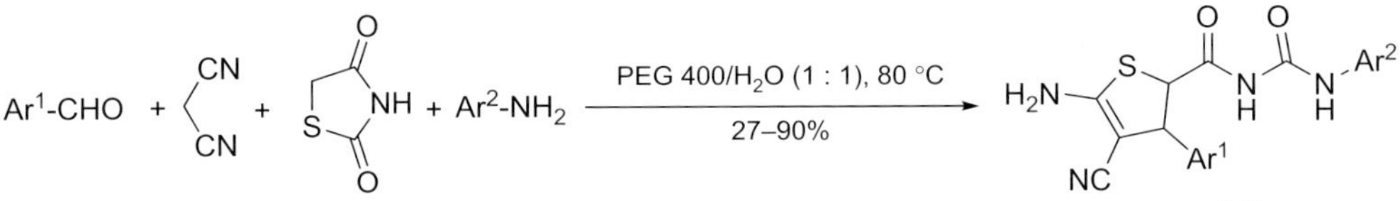

69 71a 148 102 149

Scheme 12.58 Multicomponent domino synthesis of ureidoformamides.

Majumdar and coworkers [83] synthesized fused thiopyran derivatives **150** in 83–98% yields through a three-component domino reaction of **71a**, **69**, and indoline-2-thione **136** or 4-hydroxy-2*H*-thiochromene-2-thione in water at 100 °C (Scheme 12.59). The resulting heterocyclic products could be conveniently separated from the reaction mixture without the use of any volatile organic solvent. In the reaction, three new bonds and one stereocenter were formed. When indoline-2-thione was replaced with sodium azide **151**, the three-component domino reaction provided 5-substituted-tetrazoles **152** in 67–88% yields through a Knoevenagel condensation/1,3 dipolar cycloaddition sequence (Scheme 12.60) [84].

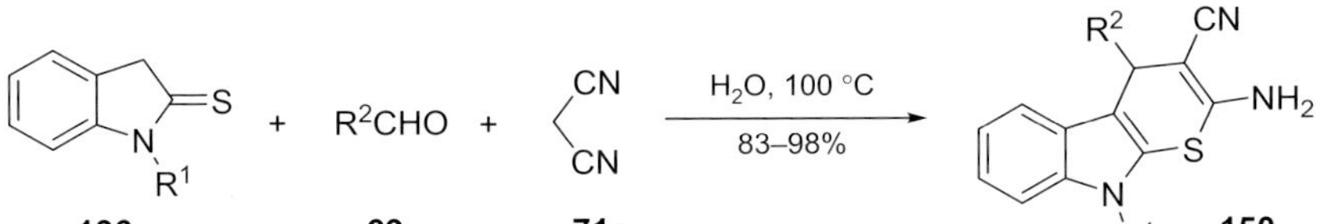

136 69 71a 150

Scheme 12.59 Multicomponent domino synthesis of fused thiopyrans.

An efficient approach to fused pyrimidine derivatives **154** was developed by Singh and coworkers [85]. This multicomponent reaction was performed with **69**, 6-amino-1,3-dimethyluracil **153**, and **33a**, and promoted by *p*-TsOH in water (Scheme 12.61). The protocol avoided the use of expensive catalysts, toxic solvents, and chromatographic separation. Both aromatic and aliphatic aldehydes could be utilized for this reaction.

Dandia and coworkers [86] reported a three-component reaction of isatin, α-cyanoacetic ester, and 5-amino-3-methylpyrazole to produce pyrazolopyridinyl spirooxindoles **155** in 82–93% yields by using sodium chloride as the inductor in water (Scheme 12.62).

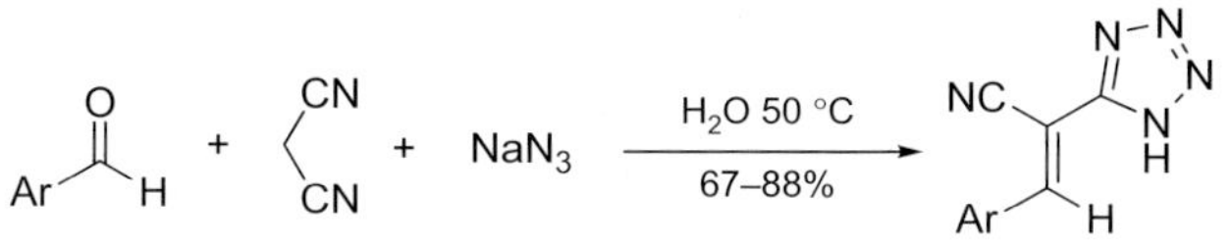

69 71a 151 152

Scheme 12.60 Multicomponent domino synthesis of 5-substituted-tetrazoles.

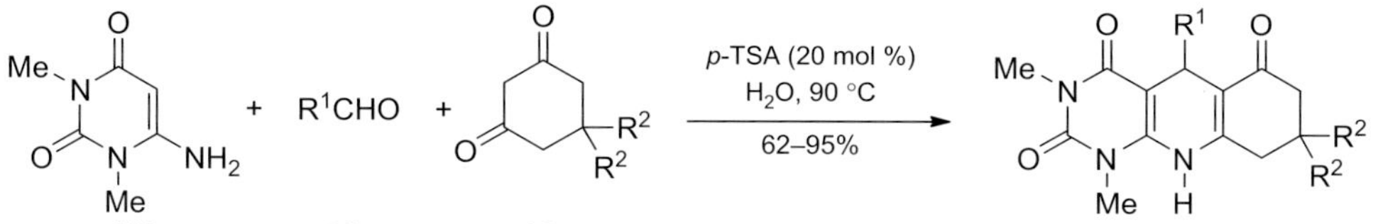

153 69 33a 154

Scheme 12.61 Multicomponent domino synthesis of fused pyrimidine derivatives.

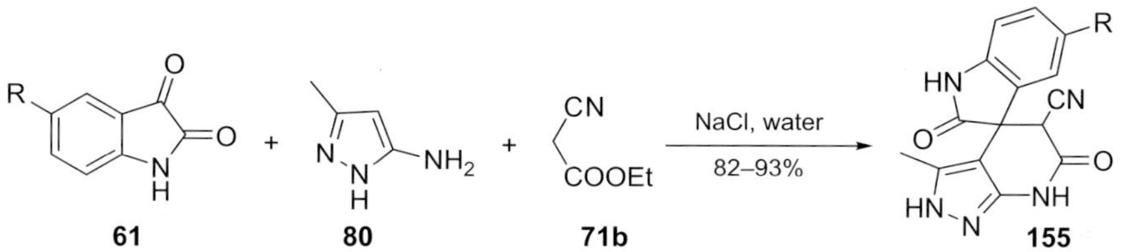

61 80 71b 155

Scheme 12.62 Multicomponent domino synthesis of spirooxindoles.

Tu and coworkers reported a new and highly stereoselective four-component protocol for the domino reaction of 2,6-diaminopyrimidine-4-one **153** with structurally diverse aryl aldehydes **69** and various barbituric acids **156**, resulting in 6-spiro-substituted pyrido[2,3-*d*]pyrimidines **157** and **158** with high diastereoselectivity (up to 99 : 1), in which the major diastereomer bears a cis relationship between the substituents at the 5- and 7-positions (Scheme 12.63) [87].

These reactions employ microwave heating and water as the reaction medium at 100 °C for 7–9 min. The mechanism for diastereoselectivity was confirmed by density functional theory (DFT) (B3LYP) calculations. Subsequently, new multicomponent domino reactions of Meldrum's acid **159**, aromatic aldehydes **69**, and electron-rich heteroaryl-amines **80** have been established for the synthesis of spiro{pyrazolo[1,3]dioxanes-pyridine}-4,6-diones and spiro{isoxazolo[1,3]dioxanes-pyridine}-4,6-diones **160** in aqueous solution under microwave irradiation (Scheme 12.64) [88]. A total of 26 examples were examined to show the broad substrate scope and high overall yields (76–93%). A new mechanism was proposed to explain the reaction process and the resulting chemo-, regio-, and stereoselectivity.

Moreover, Tu, Li, and coworkers [89] presented a regioselective [4+2+1] domino cyclization for the synthesis of spiro-substituted benzo[*b*]furo[3,4-*e*][1,4]diazepine

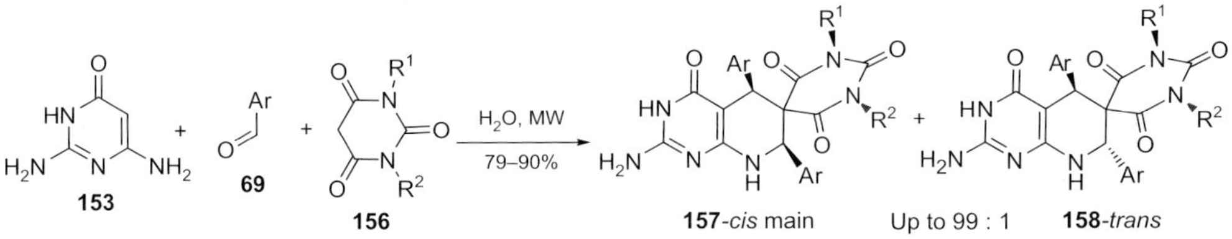

Scheme 12.63 Diastereoselective synthesis of 6-spiro-substituted pyrido[2,3-*d*]pyrimidines.

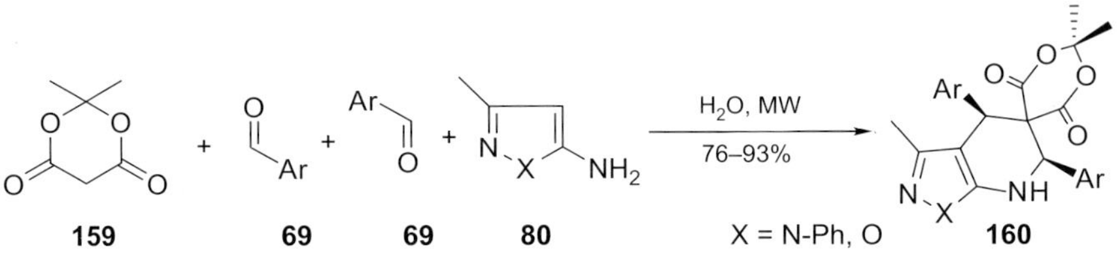

Scheme 12.64 Multicomponent synthesis of spiro{pyrazolo[1,3]dioxanes-pyridine}-4,6-diones.

derivatives **162** and **163** in 78–88% yields by varying the ratio of the starting materials (Scheme 12.65). The multicomponent domino reactions were conducted by reacting benzene-1,2-diamines **53** with tetronic acid **161** and 2,2-dihydroxy-2*H*-indene-1,3-dione **105** in aqueous solution under microwave irradiation for 18 min.

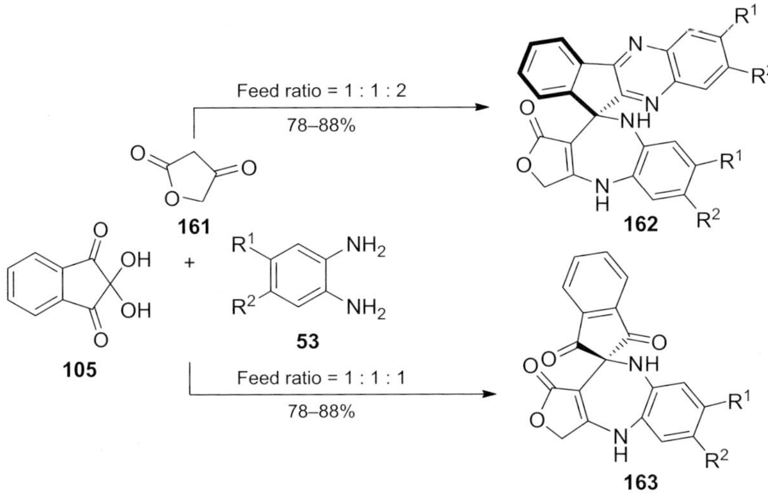

Scheme 12.65 Divergent synthesis of spiro-substituted benzo[*b*]furo[3,4-*e*][1,4]diazepines.

By replacing **105** with aromatic aldehydes **69** in this three-component domino reaction, a total of 38 examples of benzo[*f*]azulen-1-ones **164** in good overall yields (70–89%) have been synthesized with high regioselectivity (Scheme 12.66) [90].

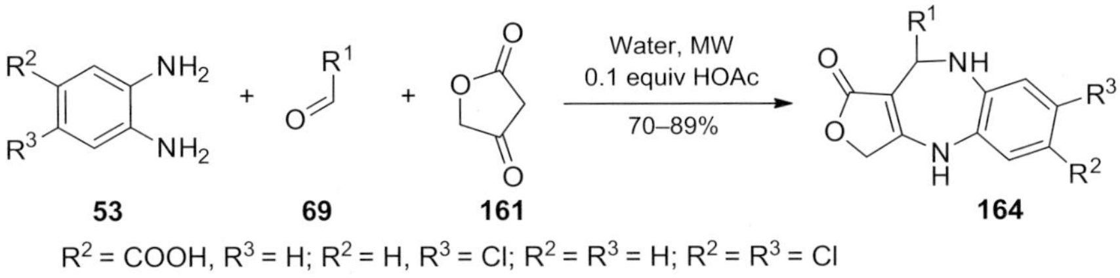

Scheme 12.66 Domino synthesis of benzo[*f*]azulen-1-ones.

The reaction was performed in aqueous media under microwave irradiation with a short reaction time of 7–19 min. Noteworthy, there was no need to add any strong acid or metal promoters.

In similar aqueous multicomponent reactions, a spectrum of pentacyclic isoindazole-fused benzo[*b,e*][1,4]diazepines **165** (Scheme 12.67) [91] and indoline-spiro benzofurodiazepines **166** (Scheme 12.68) [92] were accessible.

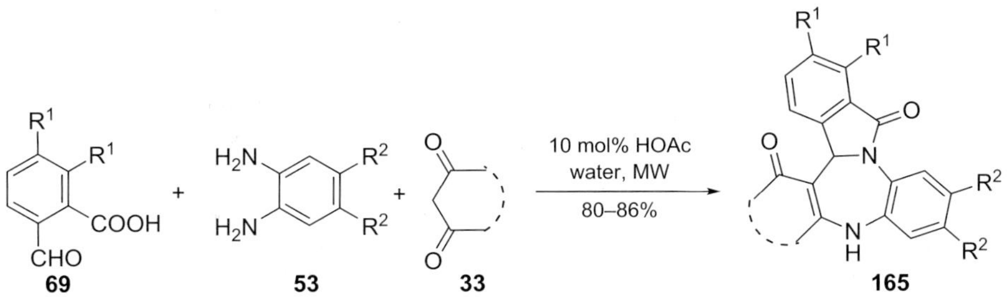

Scheme 12.67 Domino synthesis of isoindazole-fused benzo[*b,e*][1,4]diazepines.

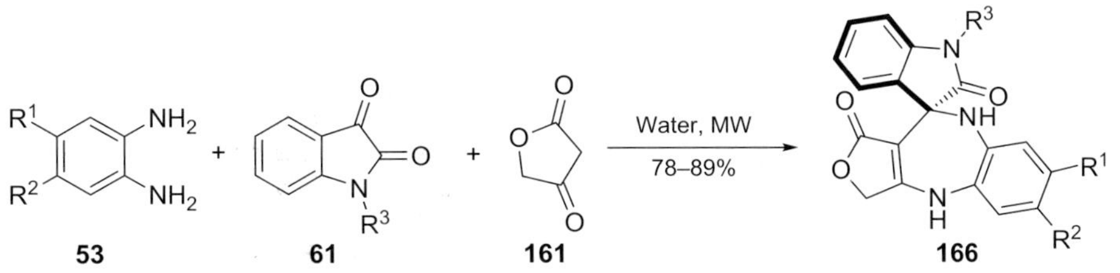

Scheme 12.68 Domino synthesis of indoline-spiro benzofurodiazepines.

12.4
High-Pressure-Promoted Domino Reactions

Similar to microwave and ultrasonic techniques, high-pressure reactions are a mild, nondestructive alternative tool to drive those reactions of high activation energy to occur. However, applying high pressure is useful only for reactions with a large negative volume of activation. A typical example for such a transformation is the DA reaction [93–95]. Under these circumstances, not only the reaction rate but also the regioselectivity and the diastereoselectivity can be improved [95].

Domino reactions applying high pressure have not been investigated extensively so far. On the other hand, several examples of DA reactions under high pressure are known. After the pioneering work of Tietze *et al.* [96] about the influence of high pressure on the endo/exo selectivity of intermolecular DA reactions, Chataigner, Piettre, and coworkers synthesized tricyclic benzofurans **169** through a DA reaction of electron-poor benzofurans **167** and all-carbon dienes **168** such as 2,3-dimethylbuta-1,3-diene or cyclohexadiene in THF under high pressures (up to 1.6 GPa) (Scheme 12.69) [97].

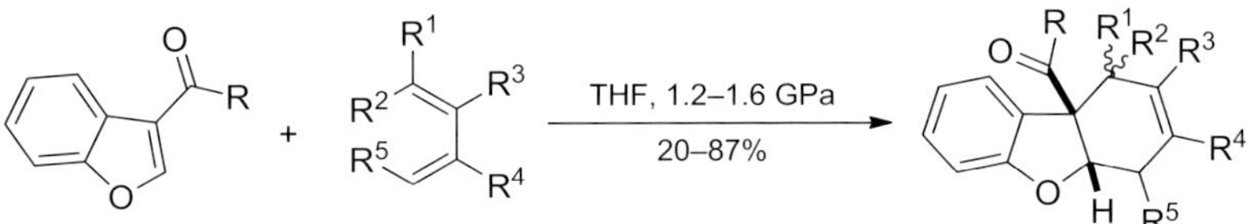

167 **168** **169**

Scheme 12.69 Domino synthesis of tricyclic benzofurans under high pressure.

Also, in this reaction, applying a high pressure resulted not only in higher yields but also better selectivity.

A real domino reaction under high pressure is the DA/retro-DA reaction of 4,6-diphenyl-[1,2]oxathiine-2,2-dioxide **170** with dimethyl but-2-ynedioate **77** to give *m*-terphenyl dicarboxy derivatives **171** in 67% yield (Scheme 12.70) [98].

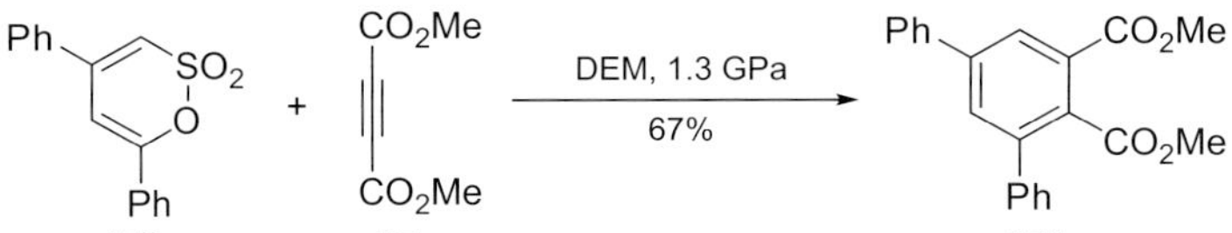

170 **77** **171**

Scheme 12.70 Domino synthesis of *m*-terphenyl dicarboxy derivatives under high pressure.

An excellent high-pressure domino reaction is shown by the four-component transformation of 2-methoxybuta-1,3-diene **172** with β-nitrostyrene **174** and 2 equiv of *N*-phenylmaleimide **173**, as reported by Scheeren and coworkers (Scheme 12.71) [99]. The reaction was completed within 42 h at 15 kbar to give a mixture of the two diastereomeric nitroso acetals *rac*-**175** and *rac*-**176** in yields of 84%. The domino process consists of a [4+2]/[4+2]/[3+2] cycloaddition and leads to the products **175**/**176** with the formation of six new bonds and eight stereogenic centers in good yield and high stereoselectivity just forming only the two diastereomers in a ratio of 7 : 3.

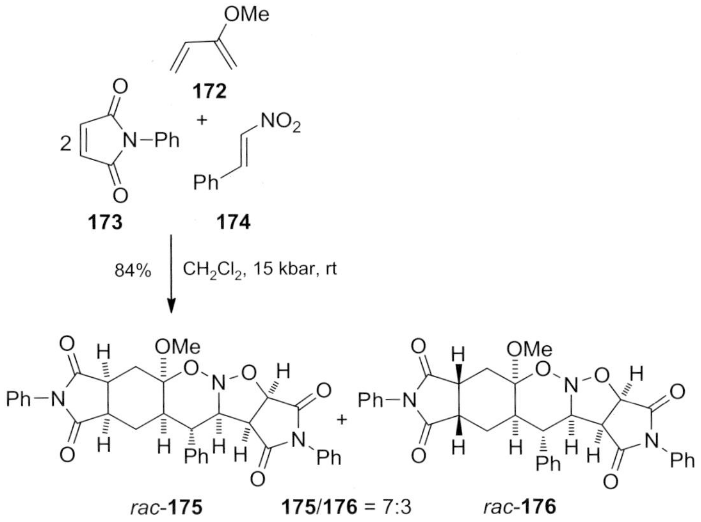

Scheme 12.71 Domino synthesis of nitroso acetals under high pressure.

Finally, Chataigner and Piettre [100] also developed a multicomponent domino [4+2]/[3+2] cycloaddition under high pressure of 3-nitroindole **177**, ethoxyethene **178**, and acrylates **141** (Scheme 12.72). The starting material was consumed in 24 h at room temperature, and the formed cycloadduct **179** was isolated in 83% yield and a diastereomeric ratio of 55 : 45. This domino reaction can be extended to the use of 3-nitropyrroles instead of nitroindoles at 1.6 GPa and 50 °C to afford diastereomeric nitrosoacetals.

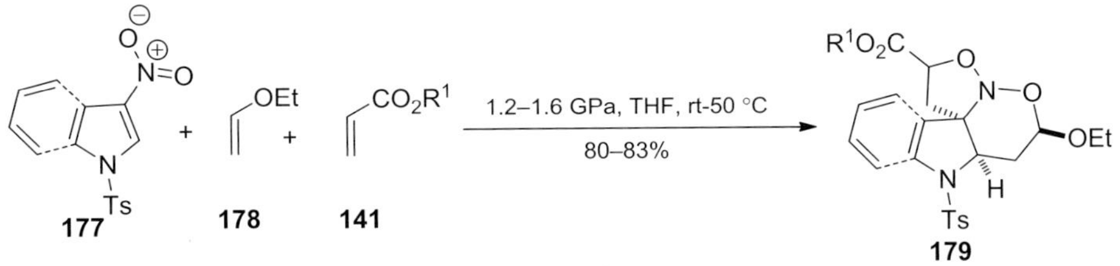

Scheme 12.72 High-pressure-promoted domino reaction of ethoxyethene.

12.5 Conclusion and Outlook

In conclusion, domino reactions under microwave irradiation, high pressure, and in water have become powerful and efficient tools in organic synthesis with attractive, greener characteristics. The combination of the advantages of the domino concept, microwave irradiation, and use of an aqueous medium allows the efficient assembly of important, useful, and interesting molecules of high complexity. Besides the bond-forming economy (multiple formations of carbon–carbon or carbon–heteroatom bonds), domino reactions in aqueous media and under microwave irradiation also show the following advantages: high reaction rates and simple optimization processes, high chemo and stereoselectivity, consecutive bond-forming pattern, and easy workup in environmentally benign environments. Recent advances have witnessed many new domino reactions under microwave irradiation, high pressure, and in water, which may not have been realized in traditional systems. It is anticipated that by taking the gains of domino reactions under microwave irradiation, high pressure, and in aqueous media, many new organic reactions with new mechanism will be established in the near future with the formation of products with diversified and complex structures, which will be of great advantage to chemical and biomedical libraries.

Acknowledgments

We are grateful to PAPD of Jiangsu Higher Education Institutions (China), Jiangsu Innovational Team Grant (China), and Robert A. Welch Foundation (D-1361, USA) for their generous support. We thank Dr. Guanhui An for his assistance.

References

1. Tietze, L.F. (1996) *Chem. Rev.*, **96**, 115–136.
2. Tietze, L.F. and Haunert, F. (2000) in *Stimulating Concepts in Chemistry* (eds F. Vogtle, J.F. Stoddart, and M. Shibasaki), Wiley-VCH Verlag GmbH, Weinheim, pp. 39–64.
3. Tietze, L.F., Brasche, G., and Gericke, K. (2006) *Domino Reactions in Organic Synthesis*, Wiley-VCH Verlag GmbH, Weinheim, p. 617.
4. Domling, A. (2006) *Chem. Rev.*, **106**, 17–89.
5. Zhu, J. and Bienaymé, H. (eds) (2005) *Multicomponent Reactions*, Wiley-VCH Verlag GmbH, Weinheim.
6. Tietze, L.F., Kinzel, T., and Brazel, C.C. (2009) *Acc. Chem. Res.*, **42**, 367–378.
7. Ikeda, S.-I. (2000) *Acc. Chem. Res.*, **33**, 511–519.
8. Domling, A. and Ugi, I. (2000) *Angew. Chem. Int. Ed.*, **39**, 3168–3210.
9. D'Souza, D.M. and Mueller, T.J.J. (2007) *Chem. Soc. Rev.*, **36**, 1095–1108.
10. Loupy, A. (2002) *Microwaves in Organic Synthesis*, Wiley-VCH Verlag GmbH Co. KGaA, Weinheim, p. 499.
11. For Group-Assisted Purification (GAP) chemistry, see: (a) Kattamuri, P.V., Ai, T., Pindi, S., Sun, Y., Gu, P., Shi, M., and Li, G. (2011) *J. Org. Chem.*, **76**, 2792–2797; (b) Kaur, P., Pindi, S., Wever, W., Rajale, T., and Li, G. (2010) *Chem. Commun.*, **46**, 4330–4332; (c) Kaur, P., Nguyen, T., and Li, G. (2009) *Eur. J. Org. Chem.*, **2009**, 912–916; (d) Kattuboina, A., Kaur, P., Ai, T., and Li, G. (2008) *Chem. Biol. Drug Des.*, **71**, 216–223.
12. Gedye, R., Smith, F., Westaway, K., Ali, H., Baldisera, L., Laberge, L., and Rousell, J. (1986) *Tetrahedron Lett.*, **27**, 279–282.
13. Giguere, R.J., Bray, T.L., Duncan, S.M., and Majetich, G. (1986) *Tetrahedron Lett.*, **27**, 4945–4948.
14. Kappe, C.O. (2004) *Angew. Chem. Int. Ed.*, **43**, 6250–6284.
15. Kappe, C.O. and Stadler, A. (2005) *Microwaves in Organic and Medicinal*

Chemistry, Wiley-VCH Verlag GmbH, Weinheim, p. 410.

16. Kappe, C.O. and Dallinger, D. (2009) *Mol. Diver.*, **13**, 71–193.

17. Kappe, C.O., Dallinger, D., Kim, Y.H., Lee, H., Kim, Y.J., Kim, B.T., and Heo, J.-N. (2008) *J. Org. Chem.*, **73**, 495–501.

18. Loupy, A. (2006) *Microwaves in Organic Synthesis*, 2nd, Completely Revised and Enlarged edn, Vol. 2, Wiley-VCH Verlag GmbH, Weinheim, p. 482.

19. (a) Takao, K.-I., Munakata, R., and Tadano, K.-I. (2005) *Chem. Rev.*, **105**, 4779–4807; (b) Kumar, A. (2001) *Chem. Rev.*, **101**, 1–20; (c) Winkler, J.D. (1996) *Chem. Rev.*, **96**, 167–176.

20. Wu, J., Jiang, X., Xu, J., and Dai, W.-M. (2011) *Tetrahedron*, **67**, 179–192.

21. Tejedor, D., Mendez-Abt, G., Cotos, L., Ramirez, M.A., and Garcia-Tellado, F. (2011) *Chem. Eur. J.*, **17**, 3318–3321.

22. Tejedor, D., Mendez-Abt, G., Cotos, L., and Garcia-Tellado, F. (2012) *Chem. Eur. J.*, **18**, 3468–3472.

23. Johnston, J.R. and West, F.G. (2012) *Tetrahedron Lett.*, **53**, 5479–5482.

24. Kudelko, A. and Zielinski, W. (2012) *Tetrahedron Lett.*, **53**, 76–77.

25. Chitra, S., Paul, N., Muthusubramanian, S., and Manisankar, P. (2011) *Green Chem.*, **13**, 2777–2785.

26. Prakash, G.K.S., Jog, P.V., Krishnan, H.S., and Olah, G.A. (2011) *J. Am. Chem. Soc.*, **133**, 2140–2143.

27. Juranovic, A., Kranjc, K., Perdih, F., Polanc, S., and Kocevar, M. (2011) *Tetrahedron*, **67**, 3490–3500.

28. Yougnia, R., Rochais, C., Santos, J.S.O., Dallemagne, P., and Rault, S. (2010) *Tetrahedron*, **66**, 2803–2808.

29. Sadek, K.U., Shaker, R.M., Elrady, M.A., and Elnagdi, M.H. (2010) *Tetrahedron Lett.*, **51**, 6319–6321.

30. Quiroga, J., Sánchez, N.E., Acosta, P., Insuasty, B., and Abonia, R. (2012) *Tetrahedron Lett.*, **53**, 3181–3187.

31. Mali, J.R., Pratap, U.R., Jawale, D.V., and Mane, R.A. (2010) *Tetrahedron Lett.*, **51**, 3980–3982.

32. McCarthy, A., Spatney, R., Manpadi, M., Myers, B.J., and Zimmerman, J.R. (2012) *Tetrahedron Lett.*, **53**, 4782–4784.

33. (a) Michael, J.P. (2000) *Nat. Prod. Rep.*, **17**, 603–620; (b) Michael, J.P. (1997) *Nat. Prod. Rep.*, **14**, 605–618; (c) Lipshutz, B.H. (1986) *Chem. Rev.*, **86**, 795–819; (d) Jacobi, P.A. and Selnick, H.G. (1990) *J. Org. Chem.*, **55**, 202–209; (e) Schuda, P.F. (1980) *Top. Curr. Chem.*, **91**, 75–111.

34. (a) Lee, J., Li, J.H., Oya, S., and Snyder, J.K. (1992) *J. Org. Chem.*, **57**, 5301–5312; (b) Kubo, I., Lee, Y.W., Balogh-Nair, V., Nakanishi, K., and Chapya, A. (1976) *Chem. Commun.*, 949–950; (c) Schulte, G., Scheuer, P.J., and McConnell, O.J. (1980) *Helv. Chim. Acta*, **63**, 2159–2167.

35. (a) MacRae, W.D. and Towers, G.H.N. (1984) *Phytochemistry*, **23**, 1207–1220; (b) Apers, S., Paper, D., Burgermeister, J., Baronikova, S., Van Dyck, S., Lemiere, G., Vlietinck, A., and Pieters, L.J. (2002) *Nat. Prod.*, **65**, 718–720; (c) Sun, L.Q., Takaki, K., Chen, J., Bertenshaw, S., Iben, L., Mahle, C.D., Ryan, E., Wu, D., Gao, Q., and Xu, C. (2005) *Bioorg. Med. Chem. Lett.*, **15**, 1345–1349; (d) Coy, E.D., Cuca, L.E., and Sefkow, M. (2009) *Bioorg. Med. Chem. Lett.*, **19**, 6922–6925.

36. Vinosha, B., Perumal, S., Renuga, S., and Almansour, A.I. (2012) *Tetrahedron Lett.*, **53**, 962–966.

37. Spencer, J., Patel, H., Amin, J., Callear, S.K., Coles, S.J., Deadman, J.J., Furman, C., Mansouri, R., Chavatte, P., and Millet, R. (2012) *Tetrahedron Lett.*, **53**, 1656–1659.

38. Lagiakos, H.R., Walker, A., Aguilar, M.-I., and Perlmutter, P. (2011) *Tetrahedron Lett.*, **52**, 5131–5132.

39. Bisceglia, J., Díaz, J.E., Torres, R.A., and Orelli, L.R. (2011) *Tetrahedron Lett.*, **52**, 5238–5240.

40. Hellal, M. and Cuny, G.D. (2011) *Tetrahedron Lett.*, **52**, 5508–5511.

41. Wang, S.-L., Jie, D., Jiang, B., Gao, Y., and Tu, S.-J. (2011) *ACS Comb. Sci.*, **13**, 572–577.

42. Jiang, B., Li, Q.-Y., Zhang, H., Tu, S.-J., Pindi, S., and Li, G. (2012) *Org. Lett.*, **14**, 700–703.

43. Jiang, B., Wang, X., Li, M.-Y., Wu, Q., Ye, Q., Xu, H.-W., and Tu, S.-J. (2012) *Org. Biomol. Chem.*, **10**, 8533–8538.

44. Wang, Q., Dai, A.X., Yi, M.S., Jiang, B., Tu, S.J., and Li, G. (2012) *Heterocycles*, **86**, 411–424.

45. Salvadori, J., Airiau, E., Girard, N., Mann, A., and Taddei, M. (2010) *Tetrahedron*, **66**, 3749–3753.
46. Reddy, M.V., Reddy, G.C.S., and Jeong, Y.T. (2012) *Tetrahedron*, **68**, 6820–6828.
47. Raghuvanshi, D.S. and Singh, K.N. (2011) *Tetrahedron Lett.*, **52**, 5702–5705.
48. Kelly, B.B., Lee, C., and Leadbeater, N.E. (2011) *Tetrahedron Lett.*, **52**, 4587–4589.
49. Yamuna, E., Zeller, M., and Prasad, K.J.R. (2011) *Tetrahedron Lett.*, **52**, 6805–6808.
50. Jiang, B., Liu, Y.-P., and Tu, S.-J. (2011) *Eur. J. Org. Chem.*, **16**, 3026–3035.
51. Wang, S.-L., Liu, Y.-P., Xu, B.-H., Wang, X.-H., Jiang, B., and Tu, S.-J. (2011) *Tetrahedron*, **67**, 9417–9425. doi: 10.1016/j.tet.2011.05.002#doilink
52. Jiang, B., Xue, L.-Y., Wang, X.-H., Tu, M.-S., Liu, Y.-P., and Tu, S.-J. (2012) *Tetrahedron Lett.*, **53**, 1261–1264.
53. Jiang, B., Zhang, G., Ma, N., Shi, F., Tu, S.-J., Kaur, P., and Li, G. (2011) *Org. Biomol. Chem.*, **9**, 3834–3838.
54. Wang, S.-L., Zhang, G., Ding, J., Jiang, B., Wang, X.-H., and Tu, S.-J. (2012) *Comb. Chem. High Throughput Screening*, **15**, 400–410.
55. Yu, Y., Tu, M.-S., Jiang, B., Wang, S.-L., and Tu, S.-J. (2012) *Tetrahedron Lett.*, **53**, 5071–5075.
56. Jiang, B., Wang, X., Shi, F., Tu, S.-J., and Li, G. (2011) *Org. Biomol. Chem.*, **9**, 4025–4028.
57. Li, J., Yu, Y., Tu, M.-S., Jiang, B., Wang, S.-L., and Tu, S.-J. (2012) *Org. Biomol. Chem.*, **10**, 5361–5365.
58. Jiang, B., Feng, B.-M., Wang, S.-L., Tu, S.-J., and Li, G. (2012) *Chem. Eur. J.*, **18**, 9823–9826.
59. Jiang, B., Yi, M.-S., Shi, F., Tu, S.-J., Pindi, S., McDowell, P., and Li, G. (2012) *Chem. Commun.*, 808–810.
60. Jiang, B., Yi, M.-S., Tu, M.-S., Wang, S.-L., and Tu, S.-J. (2012) *Adv. Synth. Catal.*, **354**, 2504–2510.
61. Jiang, B., Li, Y., Tu, M.-S., Wang, S.-L., Tu, S.-J., and Li, G. (2012) *J. Org. Chem.*, **77**, 7497–7505.
62. Jiang, B., Li, Q.-Y., Tu, S.-J., and Li, G. (2012) *Org. Lett.*, **14**, 5210–5213.
63. (a) Butler, R.N. and Coyne, A.G. (2010) *Chem. Rev.*, **110**, 6302–6337; (b) Raj, M. and Singh, V.K. (2009) *Chem. Commun.*, 6687–6703; (c) Morten, C.J., Byers, J.A., Van Dyke, A.R., Vilotijevic, I., and Jamison, T.F. (2009) *Chem. Soc. Rev.*, **38**, 3175–3192.
64. (a) Herrerias, C.I., Yao, X., Li, Z., and Li, C.J. (2007) *Chem. Rev.*, **107**, 2546–2562; (b) Li, C.J. and Chen, L. (2006) *Chem. Soc. Rev.*, **35**, 68–82; (c) Polshettiwar, V. and Varma, R.S. (2008) *Acc. Chem. Res.*, **41**, 629–639; (d) Polshettiwar, V. and Varma, R.S. (2008) *Chem. Soc. Rev.*, **37**, 1546–1557; (e) Dallinger, D. and Kappe, C.O. (2007) *Chem. Rev.*, **107**, 2563–2591.
65. (a) Lindstroem, U.M. (2002) *Chem. Rev.*, **102**, 2751–2771; (b) Simon, M.O. and Li, C.J. (2012) *Chem. Soc. Rev.*, **41**, 1415–1427.
66. Chen, P.Y., Huang, K.S., Tsai, C.C., Wang, T.P., and Wang, E.C. (2012) *Org. Lett.*, **14**, 4930–4933.
67. Fujioka, H., Komatsu, H., Nakamura, T., Miyoshi, A., Hata, K., Ganesh, J., Murai, K., and Kita, Y. (2010) *Chem. Commun.*, **46**, 4133–4135.
68. Tan, J.N., Lia, H., and Gu, Y. (2010) *Green Chem.*, **12**, 1772–1782.
69. Jung, E.J., Park, B.H., and Lee, Y.R. (2010) *Green Chem.*, **12**, 2003–2011.
70. Gunasekaran, P., Balamurugan, K., Sivakumar, S., Perumal, S., Menéndez, J.C., and Almansour, A.I. (2012) *Green Chem.*, **14**, 750–757.
71. Lu, Z., Hu, C., Guo, J., Li, J., Cui, Y., and Jia, Y. (2010) *Org. Lett.*, **12**, 480–483.
72. Rueping, M. and Parra, A. (2010) *Org. Lett.*, **12**, 5281–5283.
73. Xie, J.W., Li, P., Wang, T., and Zhou, F.T. (2011) *Tetrahedron Lett.*, **52**, 2379–2382.
74. Zhang, R., Xu, D., and Xie, J. (2012) *Chin. J. Chem.*, **30**, 1690–1694.
75. Giorgi, G., Miranda, S., Lopez-Alvarado, P., Avendano, C., Rodriguez, J., and Menendez, J.C. (2005) *Org. Lett.*, **7**, 2197–2200.
76. Li, P., Teng, B.T., Jin, F.G., Li, X.S., Zhu, W.D., and Xie, J.W. (2012) *Org. Biomol. Chem.*, **10**, 244–247.
77. Ghandi, M., Mohammadimehr, E., Sadeghzadeh, M., and Bozcheloei, A.H. (2011) *Tetrahedron*, **67**, 8484–8491.

78. Majumdar, K.C., Taher, A., and Ponra, S. (2010) *Tetrahedron Lett.*, **51**, 147–150.
79. Kurahashi, T., Shinokubo, H., and Osuka, A. (2006) *Angew. Chem. Int. Ed.*, **45**, 6336–6338.
80. Duvvuru, D., Retailleau, P., Betzer, J.F., and Marinetti, A. (2012) *Eur. J. Org. Chem.*, **5**, 897–900.
81. Gu, Y., De Sousa, R., Frapper, G., Bachmann, C., Barraulta, J., and Jerome, F. (2009) *Green Chem.*, **11**, 1968–1972.
82. Lu, G.P., Zeng, L.Y., and Cai, C. (2011) *Green Chem.*, **13**, 998–1003.
83. Majumdar, K.C., Ponra, S., and Ghosh, T. (2012) *RSC Adv.*, **2**, 1144–1152.
84. Tisseh, Z.N., Dabiri, M., Nobahar, M., Khavasi, H.R., and Bazgir, A. (2012) *Tetrahedron*, **68**, 1769–11770.
85. Verma, G.K., Raghuvanshi, K., Kumar, R., and Singh, M.S. (2012) *Tetrahedron Lett.*, **53**, 399–402.
86. Dandia, A., Laxkar, A.K., and Singh, R. (2012) *Tetrahedron Lett.*, **53**, 3012–3017.
87. Jiang, B., Cao, L.-J., Tu, S.-J., Zheng, W.R., and Yu, H.Z. (2009) *J. Comb. Chem.*, **11**, 612–616.
88. Ma, N., Jiang, B., Zhang, G., Tu, S.-J., Wever, W., and Li, G. (2010) *Green Chem.*, **12**, 1357–1361.
89. Cheng, C., Jiang, B., Tu, S.-J., and Li, G. (2011) *Green Chem.*, **13**(8), 2107–2115.
90. Wang, S.-L., Cheng, C., Wu, F.-Y., Jiang, B., Shi, F., Tu, S.-J., Rajale, T., and Li, G. (2011) *Tetrahedron*, **67**, 4485–4493. doi: 10.1016/j.tet.2011.05.002#doilink
91. Wang, S.-L., Cheng, C., Wu, F.-Y., Li, J., Jiang, B., and Tu, S.-J. (2011) *Chem. Lett.*, **40**, 834–836.
92. Wang, S.-L., Cheng, C., Gong, F., Wu, F.-Y., Jiang, B., Zhou, J.-F., and Tu, S.-J. (2011) *Chin. J. Chem.*, **29**, 2101–2108.
93. van Eldik, R., Asano, T., and le Noble, W.J. (1989) *Chem. Rev.*, **89**, 549–688.
94. For overviews see: (a) van Eldik, R. and Klärner, F.G. (2002) *High Pressure Chemistry*, Wiley-VCH Verlag GmbH, Weinheim; (b) Matsumoto, K., Kaneko, M., Hayashi, N., Uchida, T., and Acheson, R.M. (1998) *Heterocycles*, **47**, 1135–1177; (c) van Berkom, L.W.A., Kuster, G.J.T., and Scheeren, H.W. (2003) *Mol. Diver.*, **6**, 271–282; (d) Tietze, L.F. and Kettschau, G. (1997) *Top. Curr. Chem.*, **189**, 1–120. (e) Tietze, L.F. and Steck, P. (2002) in *High Presssure Chemistry* (eds R. van Eldik and F.G. Klärner), Wiley-VCH Verlag GmbH, Weinheim, pp. 239–283.
95. (a) Aben, R.W.M., Braverman, S., and Scheeren, H.W. (2003) *Eur. J. Org. Chem.*, **2003**, 894–897. (b) Misumi, Y. and Matsumoto, K. (2002) *Angew. Chem. Int. Ed.*, **41**, 1031–1033; (c) Malinowska, M., Salanski, P., Caille, J.-C., and Jurczak, J. (2002) *Synthesis*, 2707–2710; (d) Hayashi, Y., Okado, K., Ashimine, I., and Shoji, M. (2002) *Tetrahedron Lett.*, **43**, 8683–8686; (e) Hayashi, Y. and Nishimura, K. (2002) *Chem. Lett.*, 296–297; (f) Camara, C., Joseph, D., Dumas, F., d'Angelo, J., and Chiaroni, A. (2002) *Tetrahedron Lett.*, **43**, 1445–1448; (g) Has-Becker, S., Bodmann, K., Kreuder, R., Santoni, G., Rein, T., and Reiser, O. (2001) *Synlett*, 1395–1398.
96. Tietze, L.F., Hübsch, T., Voss, E., Buback, M., and Tost, W. (1988) *J. Am. Chem. Soc.*, **110**, 4065–4066.
97. Chopin, N., Gerard, H., Chataigner, I., and Piettre, S.R. (2009) *J. Org. Chem.*, **74**, 1237–1246.
98. Gaitzsch, J., Rogachev, V.O., Metz, P., Yusubov, M.S., Filimonov, V.D., and Kataeva, O. (2009) *J. Sulfur Chem.*, **30**, 4–9.
99. van Berkom, L.W.A., Kuster, G.J.T., Kalmoua, F., de Gelder, R., and Scheeren, H.W. (2003) *Tetrahedron Lett.*, **44**, 5091–5093.
100. Chataigner, I. and Piettre, S.R. (2007) *Org. Lett.*, **9**, 4159–4162.

13
Domino Reactions in Library Synthesis

Vincent Eschenbrenner-Lux, Herbert Waldmann, and Kamal Kumar

13.1
Introduction

The ability to *create* molecules of varying complexity is unique to and characteristic of organic synthesis. Despite the major developments made in this field of science, the synthesis of small molecules with the desired functional decorations remains a major challenge. Longer and multistep synthesis routes often become unavoidable when shorter and concise paths are too demanding in terms of chemo-, regio-, and stereoselectivity. In light of the unexpected underperformance of combinatorial compound collections in the discovery of hit-and-lead structures for drug discovery, scaffolds and structures targeted by organic chemists differ significantly from simple commercially available molecules. Domino reactions, wherein many reactions happen consecutively in one-pot conditions, are viable, reliable, and successful alternatives to multistep sequences to synthesize collections of compounds with desired levels of structural diversity and complexity [1, 2]. Therefore, these strategies to access complex structures are highly desired and sought after. Although domino reactions and one-pot multiple component reactions yielding various classes of molecules have developed rapidly, there is still a lot to be achieved in this field, for instance, in the construction of complex molecules with multiple stereogenic centers. The development of robust, scalable, economical, and stereoselective domino reactions that are also amenable to compound library synthesis is a major challenge. These libraries are highly useful sources of chemical probes for chemical biology research and novel lead structures for drug discovery [3, 4]. Moreover, unlike the multistep syntheses that often employ protection and deprotection steps, these reactions also address the challenge of atom economy in most of the cases [5].

In this chapter, we present an overview of domino reactions that have been used to construct various types of compound libraries targeting either natural product (NP) structures and privileged scaffolds or diverse ring systems.

Domino Reactions: Concepts for Efficient Organic Synthesis, First Edition. Edited by Lutz F. Tietze.

13.2
Domino Reactions in Natural-Product-Inspired Compound Collection Syntheses

NPs remain a major source of bioactive molecules and are of inspiration to generate compound libraries with structural fingerprints of their frameworks. Such libraries may be as good as or even better than the parent NPs in providing interesting bioactive molecules for various discovery processes. Many NPs show multiple activities against a spectrum of targets, and therefore the exploration of the chemical space around these inherently active molecules is highly desirable to find candidates with better activity and selectivity for a given target or in an assay.

In contrast to the total synthesis of an NP, which often is a long, multistep, and time-consuming endeavor, generating NP-like compounds should follow easy and efficient synthesis routes to provide a large number of compounds. Domino reaction sequences can play a major role in providing relatively complex and NP-based compound collections. Chirality of the product is another important factor that determines the biological effect of a molecule. Therefore, enantioselective and diastereoselective domino reactions are of particular interest in this area.

13.2.1
Coinage Metal-Catalyzed Domino Synthesis

Activation of acetylenes with coinage metals such as silver and gold can trigger a domino sequence on a well-designed substrate to provide a desired complex, NP-based framework and consequently a compound collection. Kumar, Waldmann, and coworkers [6] developed a domino synthesis of the fascaplysin-inspired alkaloids. In their approach, imine **3** (formed from aniline **1** incorporating a pendant nucleophile) and acetylenic benzaldehyde **2** undergo a cycloisomerization

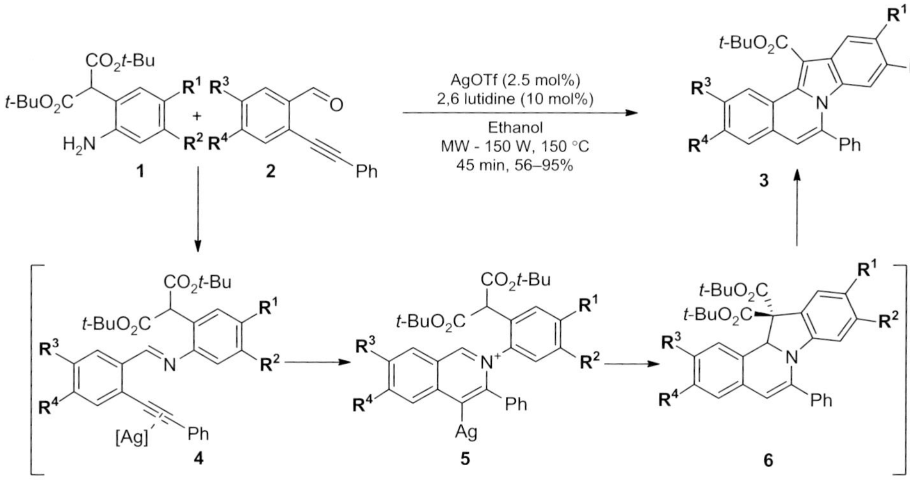

Scheme 13.1 Silver-catalyzed synthesis of indoloisoquinolines.

in the presence of silver to yield the isoquinolinium intermediate **5**. A further intramolecular nucleophilic attack of the pendant malonate nucleophile followed by decarboxylation of the tetracyclic intermediate **6** yields the targeted indoloisoquinolines **3** (Scheme 13.1).

This strategy led to the synthesis of a small compound library of indoloisoquinolines with functional group variations on the aromatic rings and with yields between 42 and 95%. Furthermore, a concise synthesis of the NPs themselves was achieved by employing this domino reaction sequence as a key transformation. Pentacyclic substrate **9** was obtained through the silver-catalyzed domino polycyclization between acetylenic indole-2-carbaldehyde **7** and aniline **8**. Pentacycle **9** was subsequently transformed into homofascaplysin C (**10**) in only one formylation step (overall yield of 54%), while two further transformations provided fascaplysin (**11**) in overall yield of 52% (Scheme 13.2).

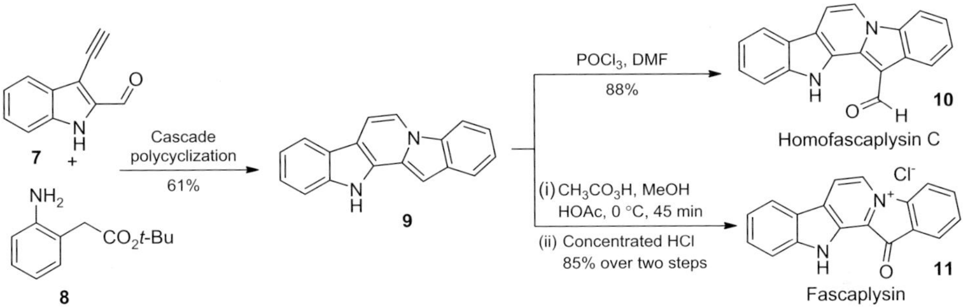

Scheme 13.2 Synthesis of natural products fascaplysin and homofascaplysin C.

A further example for a metal-initiated reaction sequence providing an indole-inspired NP-based complex scaffold was reported by Van der Eycken and coworkers [7]. In this case, a Ugi reaction provided the desired substrate for the sequence of cyclization reactions initiated by the activation of the acetylene with a gold catalyst. The Ugi four-component reaction employed the indole-carboxaldehyde **12**, the nitrile **13**, amine **14**, and carboxylic acid **15** and led to the formation of the intermediate **16**. In the presence of a gold(I) catalyst, instead of the expected indoloazepinone that could be formed via an endo-dig cyclization, a tetracyclic spiroindoline **17** was formed diastereoselectively (Scheme 13.3).

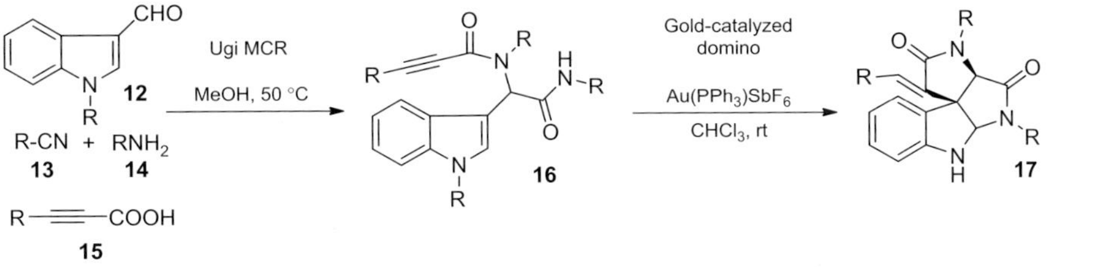

Scheme 13.3 Synthetic strategy to produce a spiroindole library.

Formation of **17** can be explained by a domino sequence that begins with an exo-dig cyclization followed by intramolecular trapping of the iminium indole cation by the amide NH to yield **17**. The nucleophilic attack of amide on the indole is possible only after the exo-dig addition of the indole to the alkyne from the front side, thus keeping the indole imine in close proximity to the amide. In case of a rear face exo-dig attack, subsequent nucleophilic attack on the iminium is sterically blocked, which pushes the equilibrium of the reaction toward ring opening of intermediate **19** back to the Ugi product **18** (Scheme 13.4). Overall, in this domino process, two new stereogenic centers are formed in a complex tetracyclic molecule.

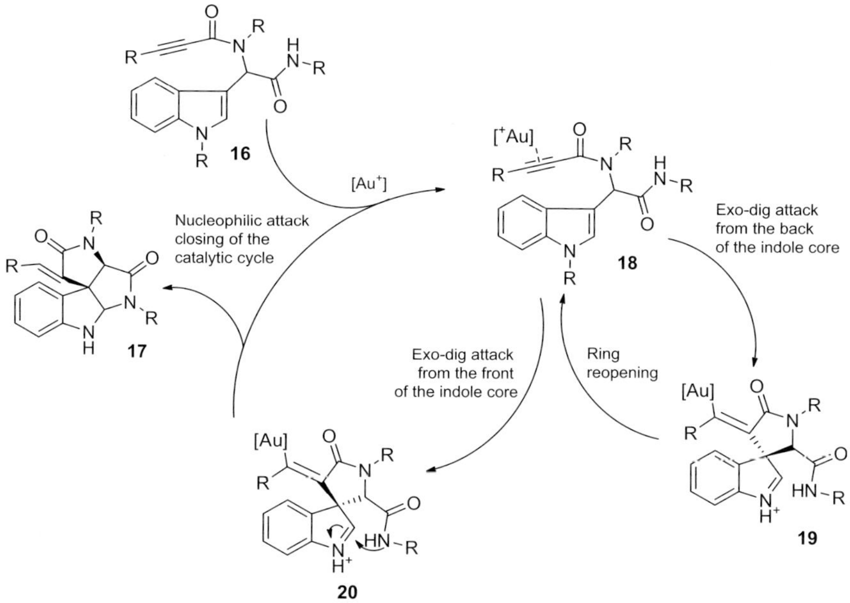

Scheme 13.4 Mechanism of the post-Ugi gold-catalyzed domino reaction.

13.2.2
Multicatalytic Domino Processes

Multicatalytic domino processes are synthetic strategies in which two or more catalytic domino reactions are sequentially used in the synthesis of a compound library [8–10]. The product obtained from the first domino reaction is the substrate for subsequent cascades, which yield products with either a different scaffold or modified functional group formation.

Ramachary and Reddy developed one of these sequences using an amino acid-catalyzed three-component reductive alkylation (TCRA) as the first reaction of their MCC. This multicomponent reaction involved an aldehyde **21**, a malonate **22**, and the Hantzsch ester **24** under amino acid catalysis starting with a Knoevenagel condensation, which was followed by hydrogenation of the formed olefin **23** to form the precursor **25** for the subsequent processes. This precursor could be isolated with yields of 78–98% (Scheme 13.5). Chiral aldehydes do not racemize during the domino Knoevenagel/hydrogenation process.

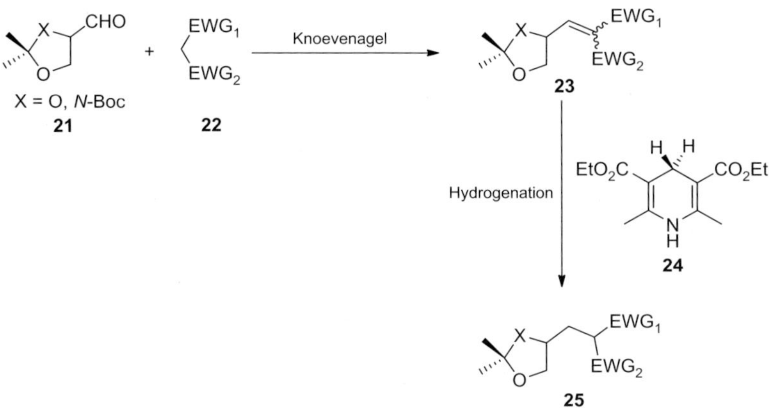

Scheme 13.5 Three-component reductive alkylation (TCRA) as first cascade of the multi-catalysis domino reaction.

When the same reaction was performed in the presence of an alcohol under acidic conditions, the TCRA product followed a second domino sequence in which the dioxolanes **26** were hydrolyzed. One of the free carboxylic acids formed the lactone **28** with one of the free diols, while the remaining carboxylic acid was esterified to yield compound **29** (Scheme 13.6). It could be used as a common precursor for the synthesis of various NPs or biologically active compounds, such as TAN 1364 A, the antibiotic agglomerins, hexanolides, brefeldin-A, or RK-682 (Scheme 13.7).

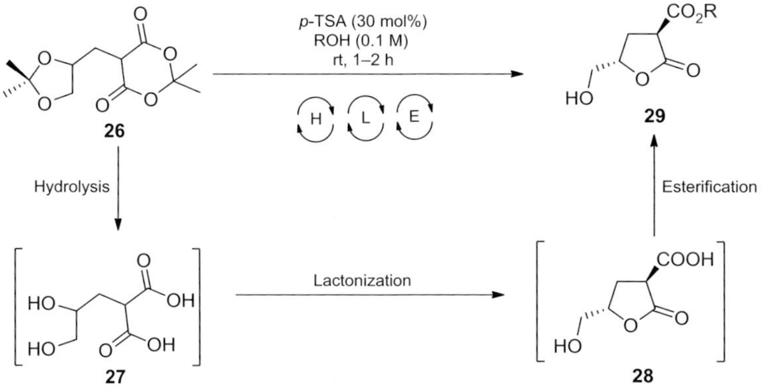

Scheme 13.6 Domino hydrolysis/lactonization/esterification reaction of the TCRA product.

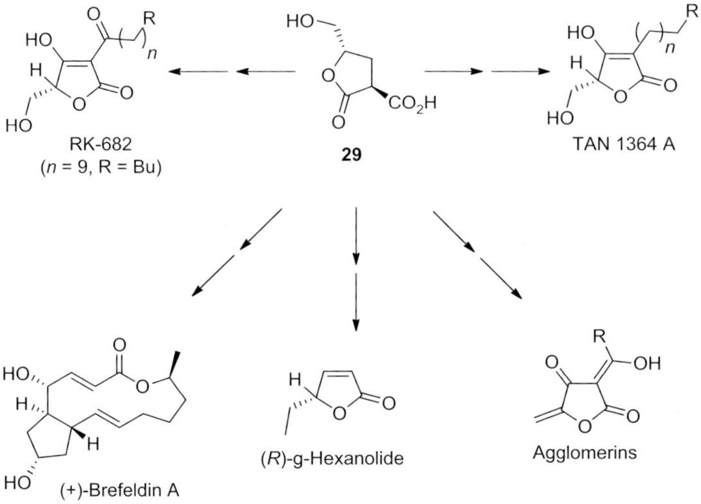

Scheme 13.7 Examples of natural products accessible from building block **29**.

A combination of different domino reactions was also possible by tuning the structure of the initial aldehyde and malononitrile and introducing various nucleophiles. In these cases, domino alkylation/ketenization/esterification (AKE), double TCRA-Robinson (2T-RA) annulation, or domino hydrolysis/oxa-Michael/dehydration processes were observed, leading to different classes of molecules (**30**, **32** and **33**, **35**) (Scheme 13.8).

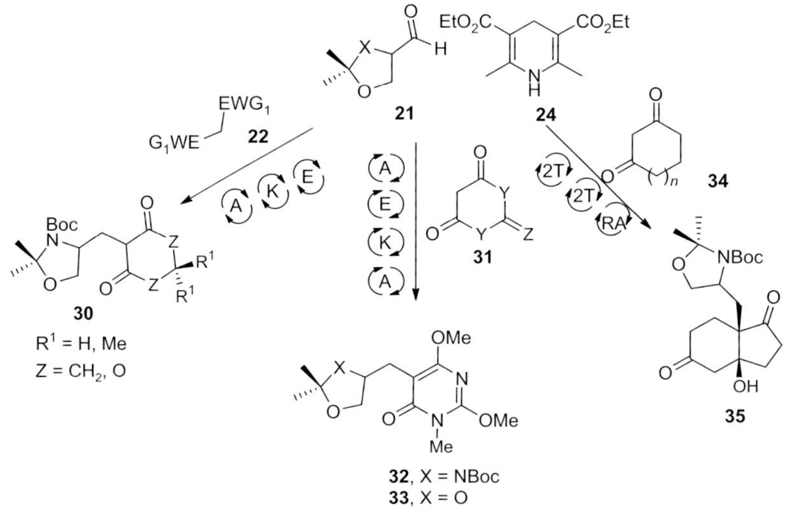

Scheme 13.8 Combination of domino processes leading to different compound classes.

13.2.3
Synthesis of Natural-Product-Inspired Centrocountins Using Domino Reactions

An unusually 12-step-long domino reaction sequence leading to an NP-inspired compound library of indoloquinolizines was recently reported by Kumar, Waldmann, and coworkers [11, 12]. The tetracyclic core of indoloquinolizine is an important scaffold that exists in many indole alkaloids such as yohimbine, an anticancer agent, or hydroxyangustine, a cytotoxic molecule also considered for anticancer treatment and therefore a very important core for a library synthesis. The straightforward synthesis of indoloquinolizines was performed by the reaction of 3-formylchromones with acetylene dicarboxylates in the presence of a phosphine followed by addition of tryptamine and trifluoroacetic acid (TFA) or camphor sulfonic acid (CSA) at 80 °C in toluene. Differently substituted formylchromones, acetylene dicarboxylates, and tryptamines were employed to obtain a collection of indoloquinolizines functionalized at eight different substitution sites (Scheme 13.9).

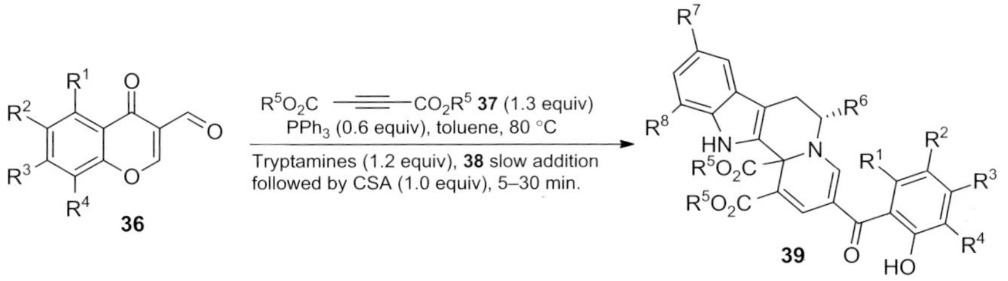

Scheme 13.9 Domino reaction to the indoloquinolizines.

The mechanism of this reaction was thoroughly investigated. The domino synthesis of the indoloquinolizines commences with the phosphine-catalyzed [4+2] annulation of 3-formylchromones **36** with alkynes **37** to yield tricyclic benzopyrones **42**. Benzopyrones **42** embody multiple electrophilic sites, inviting attack of nucleophiles such as tryptamines **38**. A conjugate addition of tryptamine to ring C of the benzopyrone accompanied by ring opening in which a phenol moiety serves as leaving group provides the intermediate **43**. Another conjugate addition of the generated phenol in **43** followed by pyran ring opening generates a ketoester in proximity to an enamine in **44a**. The triene (**44b**) thus generated may undergo 6π-electrocyclization to provide the cyclic aminal **45**. Another conjugate addition of phenol and elimination of water lead to the tricyclic dihydropyridines **46**. Acid-promoted opening of the chromone ring, with a phenol serving again as a good leaving group, and re-addition of the phenolate to the intermediary formed pyridinium salts **47** generate tricyclic dienes **48** an set the stage for a sigmatropic aza-Claisen rearrangement which yields the corresponding iminoesters **49**. The neighboring indole ring undergoes Pictet–Spengler cyclization with the activated imine to generate a newly formed ring incorporating a secondary amine **50**. The final steps of the sequence follow the conjugate aza-Michael addition of the

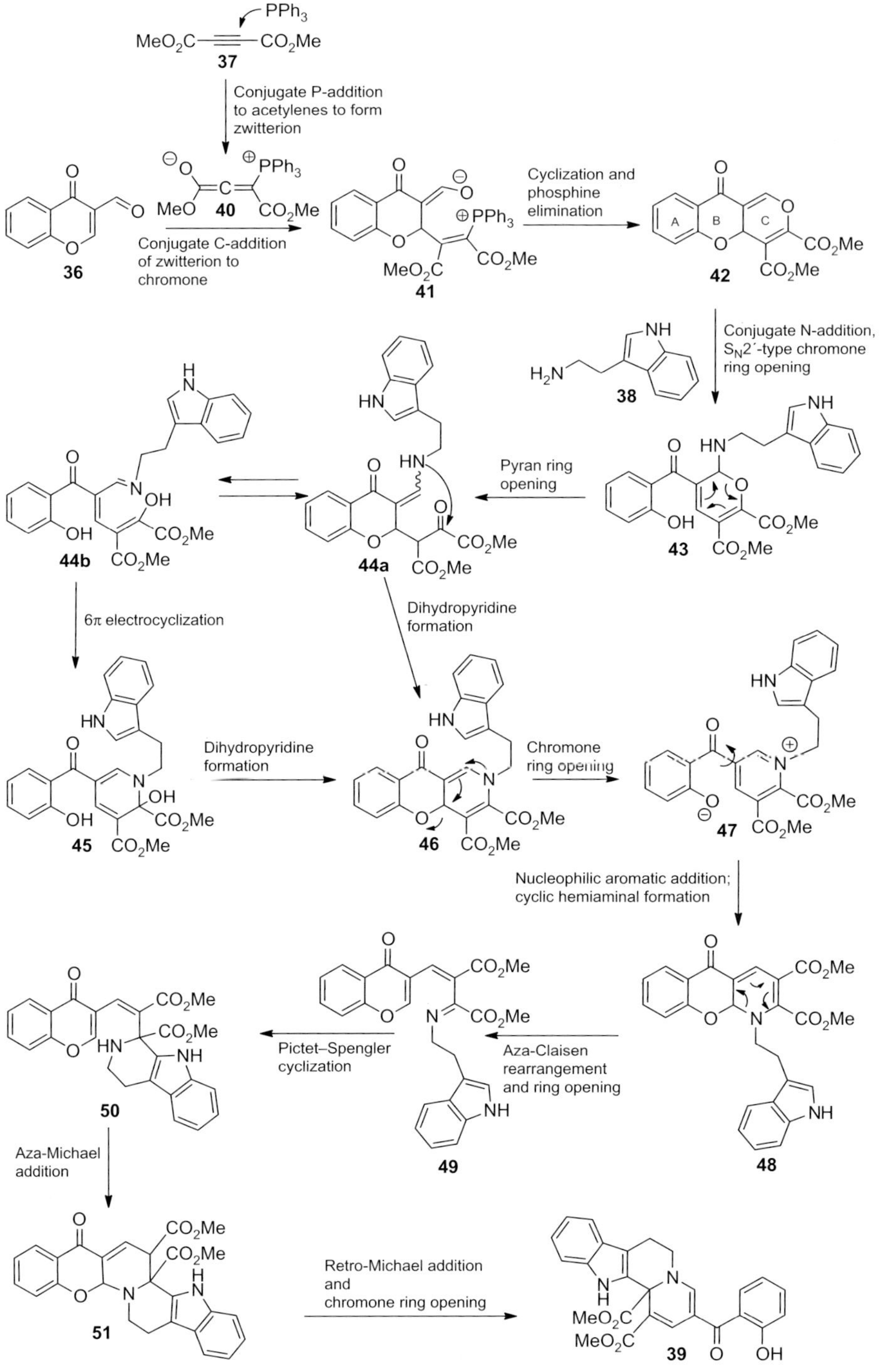

Scheme 13.10 Mechanism of the 12-step domino sequence yielding the indoloquinolizine (**39**).

secondary amines to the doubly vinylogous esters, leading to **51** and final acid-mediated pyran ring opening to culminate in the formation of indoloquinolizines **38** (Scheme 13.10). The intermediates (**42**, **45**, and **48**) were isolated and characterized by ^{1}H and ^{13}C NMR and single-crystal X-ray analysis.

This strategy yielded a small collection of indoloquinolizines with yields ranging from 20 to 91%. Low yields were observed in the case of halogen-substituted chromones, and the low-yielding step was found to be the phosphine-catalyzed [4+2] annulation of 3-formylchromones **36** with acetylene dicarboxylates **37**. Therefore, an alternative domino reaction was developed to bypass this phosphine-catalyzed reaction in order to increase the yield of these halogenated indoloquinolizines. It was achieved by reacting the tryptamine **38** with the formylchromone **36** to form the corresponding imine **52**, followed by an aza-Diels–Alder reaction with acetylene dicarboxylates **37** catalyzed by zinc chloride in dimethyl sulfoxide (DMSO) under dry conditions (molecular sieves). This reaction provided a direct access to the intermediate **51** of the 12-step domino process to yield the halogenated indoloquinolizines **39** in significantly higher yields (53–91%) (Scheme 13.11).

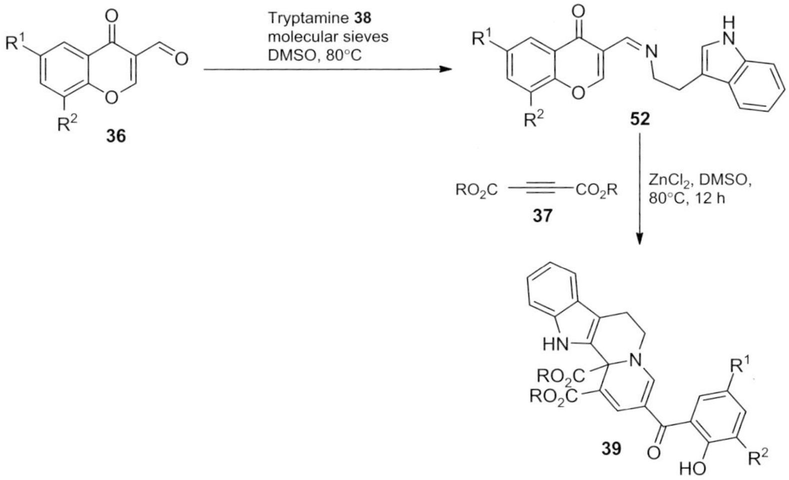

Scheme 13.11 An aza-Diels–Alder strategy to indoloquinolizines **39**.

A recent report by Antonchick, Waldmann, and others [13] describes a catalytic enantioselective domino synthesis of structurally and stereochemically complex molecules, which forms four carbon–carbon bonds and sets eight stereogenic centers with high regio-, diastereo-, and enantioselectivity. It can be programmed to yield different stereoisomers by varying only the order of combination of a common set of reagents and catalysts. In this tandem sequence, 1,4-benzoquinone **53** reacts with 2 equiv of azomethine ylide **54** in a double 1,3-dipolar [3+2] cycloaddition to form selectively the chiral compounds **56** or **57** (Scheme 13.12). Through this synthesis strategy, 1 out of 512 possible stereoisomers is formed with high selectivity. In principle, with this method a compound library of 240 600

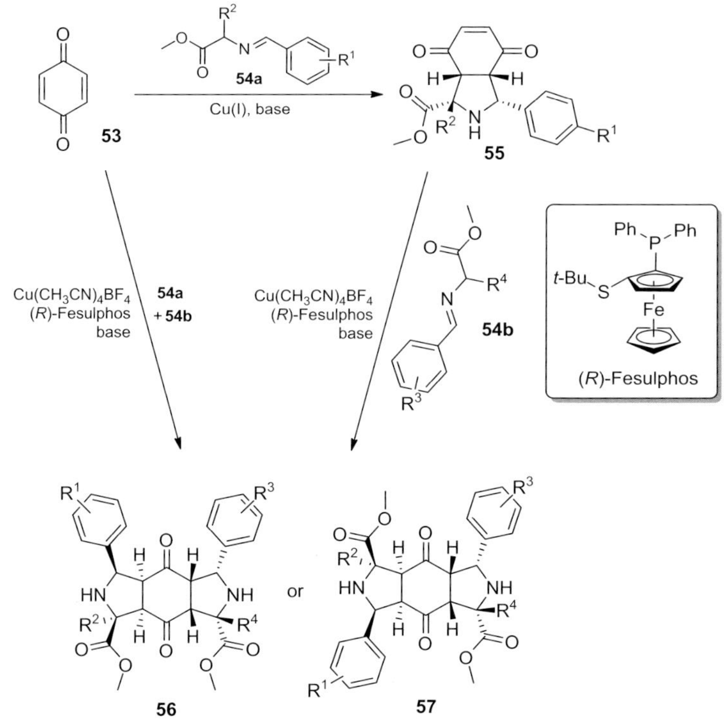

Scheme 13.12 Catalytic tandem double 1,3-dipolar cycloaddition of 1,4-benzoquinone **53**.

members encompassing more than 1.9 million stereocenters could be prepared by using only 20 amino acids, 20 aldehydes, 1,4-benzoquinone, and 1 chiral ligand.

13.3
Domino Approaches Targeting Scaffold Diversity

One of the major and important goals of diversity-oriented synthesis (DOS) is to provide many scaffold structures capable of generating small, diverse compound collections. These endeavors are meant to reach out to the unexplored chemical space with the least possible synthetic efforts. Domino reactions can play a significant role in developing these concise and efficient protocols. Different approaches are used in diversity synthesis to provide scaffold diversity [14]. In the substrate-based approach, also called the *folding pathway*, diversity is obtained by submitting different kinds of substrates to the same set of reagents, which result in products with diverse frameworks (Figure 13.1). In the reagent-based approach, termed *branching pathway*, a common substrate undergoes different reactions by changing the reagents (or also other synthesis parameters such as temperature or solvent) and forming different scaffold structures. Another strategy is the *build–couple–pair* approach to scaffold diversity (see below).

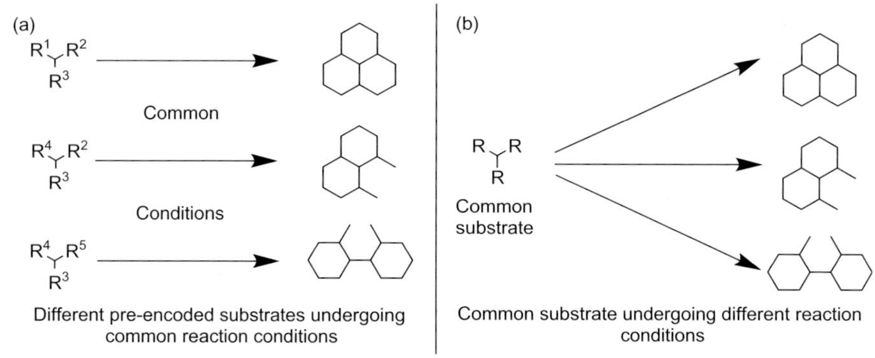

Figure 13.1 Diversity-oriented synthesis (DOS). (a) Substrate-based approach. (b) Reagent-based approach.

13.3.1
Substrate-Based Approach: the Metathesis/Metathesis Domino Process

The most often used reaction sequence among the substrate-based diversity syntheses is the domino metathesis/metathesis reaction, in which ROM (ring-opening metathesis), RCM (ring-closing metathesis), and CM (cross-metathesis) can occur in different orders throughout the synthesis to form various ring structures. Hanson and coworkers applied the metathesis domino strategy to obtain skeletally diverse sultams. The metathesis substrate sulfonamide **58** was obtained in two

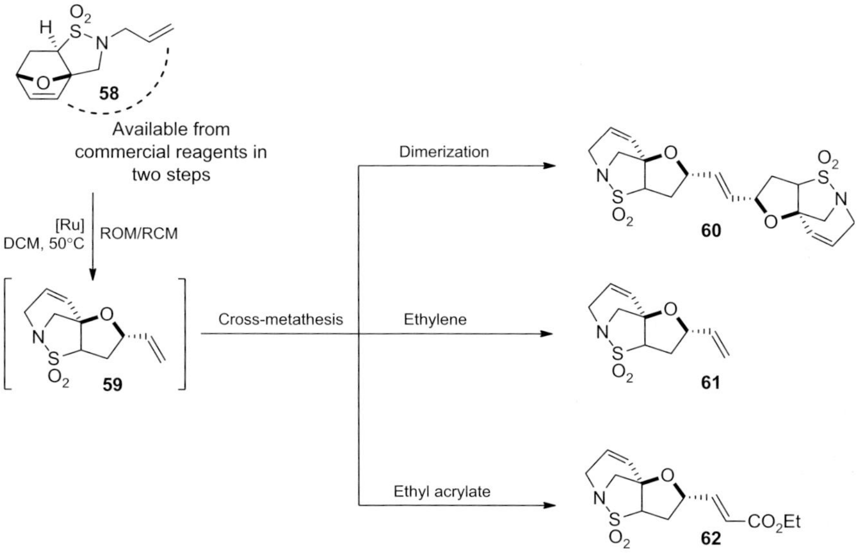

Scheme 13.13 Domino ROM/RCM/CM reaction to the sultams.

steps from commercially available materials, decorated on one side with an alkene-bearing furan ring and on the other side with a sulfonamide incorporating allyl or propargyl groups on nitrogen. First, only the ring-opening/ring-closing cascade was performed to determine the scope of the reaction in which the ring-fused furan in **58** was opened, and another cycle closed the terminal alkene to form the sultam **59**. To avoid dimerization, a further CM was introduced into the domino sequence with either ethylene or ethyl acrylate to form products **60–62** (Scheme 13.13).

Once the domino ROM/RCM/CM reaction was optimized, different types of terminal alkenes or alkynes could be used to expand the scope of the reaction and to obtain new scaffolds [15]. This implies, for example, blocking the RCM by introducing an alkyne (**63**) instead of the terminal alkene (**64**, Eq. (1), Scheme 13.14), but also modifying the position of the alkene (**65**) to influence the RCM and yield a tricyclic compound (**66**, Eq. (2), Scheme 13.14).

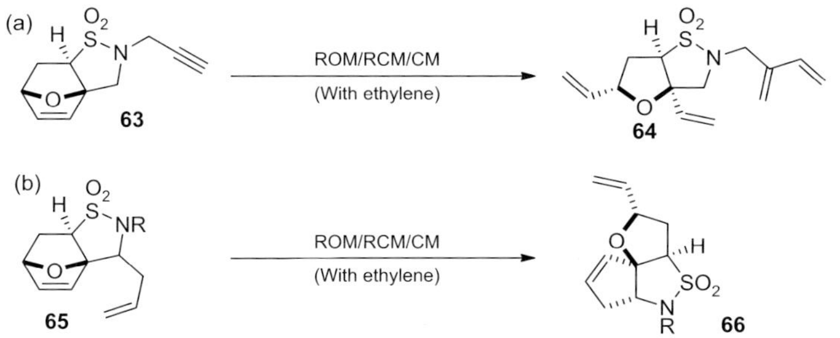

Scheme 13.14 Scaffold diversity introduced by changing the position and the nature of the terminal alkene.

The domino ROM/RCM/CM reaction can be further modulated to provide higher order structural diversity in the compounds by introducing the so-called capping and propagating groups and paving the path for diverse cyclization modes. Nelson and coworkers exploited this strategy extremely well in a combinatorial approach for the preparation of a highly complex and diverse compound collection [16, 17]. In their approach, simple building blocks suitable for a series of cascading RCM and ROM or ene–yne metathesis reactions were employed (Scheme 13.15). These core building blocks (**67**) consisted of three parts: a constant fluorinated tag, a propagating core with cyclic structure or appended alkynes, and a capping group. The fluorinated tag was employed for simplification of the purification using fluorous solid-phase extraction. The cyclic propagating core facilitated the propagation of the domino sequence through RCM and ROM, while the alkyne-containing propagating unit allowed the propagation of the reaction sequence of ring-closing ene–yne metathesis. These reaction sequences were initiated through the terminal alkene capping groups. This strategy rapidly provided more than 84 different molecules possessing unusual scaffolds (**68–75**). Similar approaches utilizing domino reactions are crucial for the exploration of underexploited parts of the chemical space.

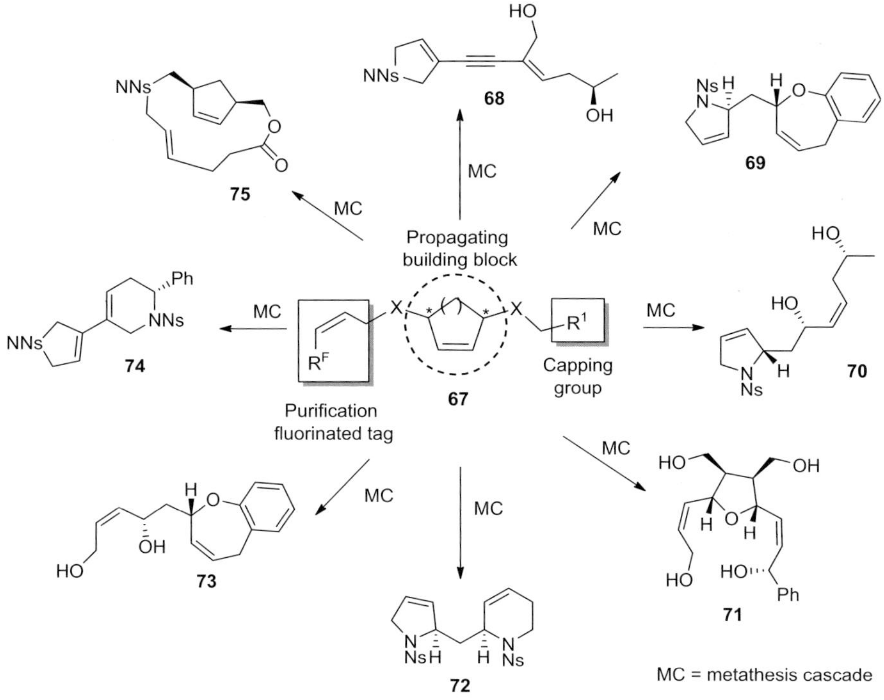

Scheme 13.15 Capping-propagation strategy to control domino metathesis/metathesis reactions.

Multicomponent reactions might be one of the most straightforward ways to introduce diversity and variability in functional groups in a compound library. In the work by Nelson and coworkers [18], two multicomponent reactions were combined to introduce skeletal diversity. Enamine **79** formed from aldehyde **77** and secondary amine **76** underwent an inverse electron demand Diels–Alder reaction with triazole **78** to form the intermediate **80**, which released nitrogen to form the diene **81**. Depending on the presence of an alkene at different positions on **81**, that is, whether on R^1 or R^3, it underwent a Diels–Alder reaction to form two different tricyclic scaffolds **82** and **83**. These were then subjected to a Jouillé–Ugi reaction with various carboxylic acids and nitriles to yield compounds **84–86**. Further skeletons were produced by manipulation of the side chains of the Ugi product, so that this strategy led to 26 different scaffolds, out of which 10 were formed directly by the folding-pathway approach (Scheme 13.16).

13.3.2 Reagent-Based Domino Approaches

The reagent-based approach or branching pathways in diversity synthesis exploits different functional or reactive sites of a common substrate to yield diverse scaffold

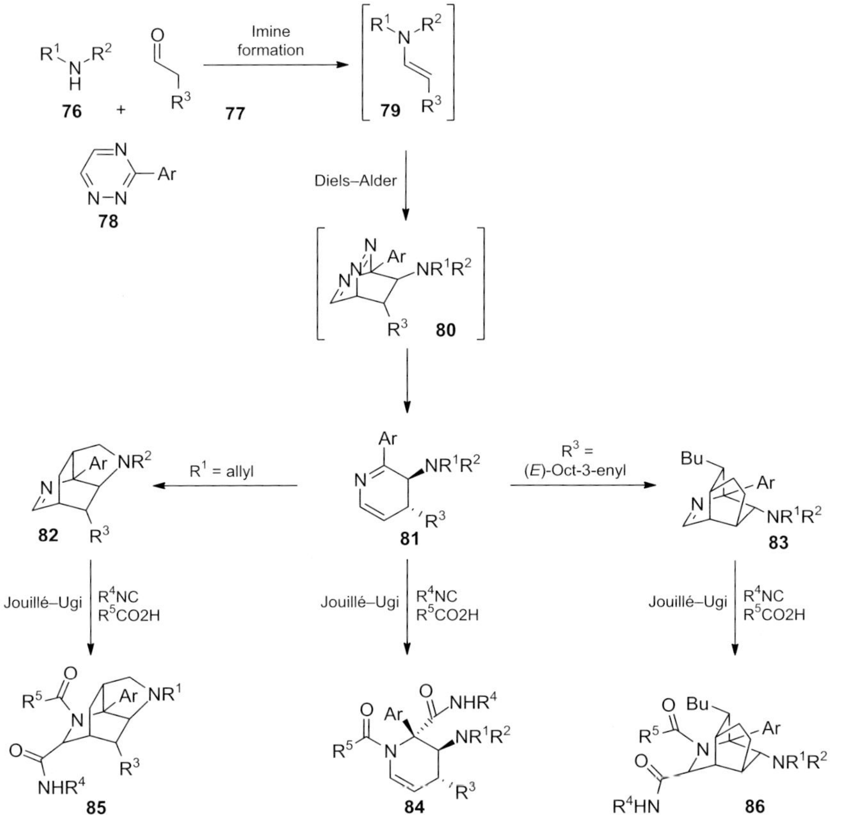

Scheme 13.16 Generation of skeletally diverse compounds by three-component reactions.

structures. On the basis of this approach, Kumar *et al.* [19] introduced the *branching domino* strategy wherein the common substrate was designed to allow different reaction sequences and each domino reaction might lead to a different molecular framework well decorated with functional groups for further modifications. A highly electrophilic ketoester **87** was used as common substrate to react with a set of diverse nucleophiles, which included mononucleophiles, bisnucleophiles, and also nucleophilic zwitterions (**88–95**). Each of these nucleophiles triggered a different domino reaction sequence, providing a new ring system. Thereby, a library of about 50 molecules embodying 12 different scaffolds was generated. Scheme 13.17 depicts some of the scaffolds (**96–103**) generated from the addition of mono- and bisnucleophiles to the common substrate **87**. Moreover, almost all of the electrophilic sites of the common substrate were utilized by at least one of the nucleophiles employed. In addition to NP-based and medicinal-chemistry-based scaffolds, the strategy also provided novel chemical entities for further chemical and biological investigations.

In another example, Kumar and coworkers [20] developed domino double annulations to provide ring diversity around the privileged chromone scaffold in a highly

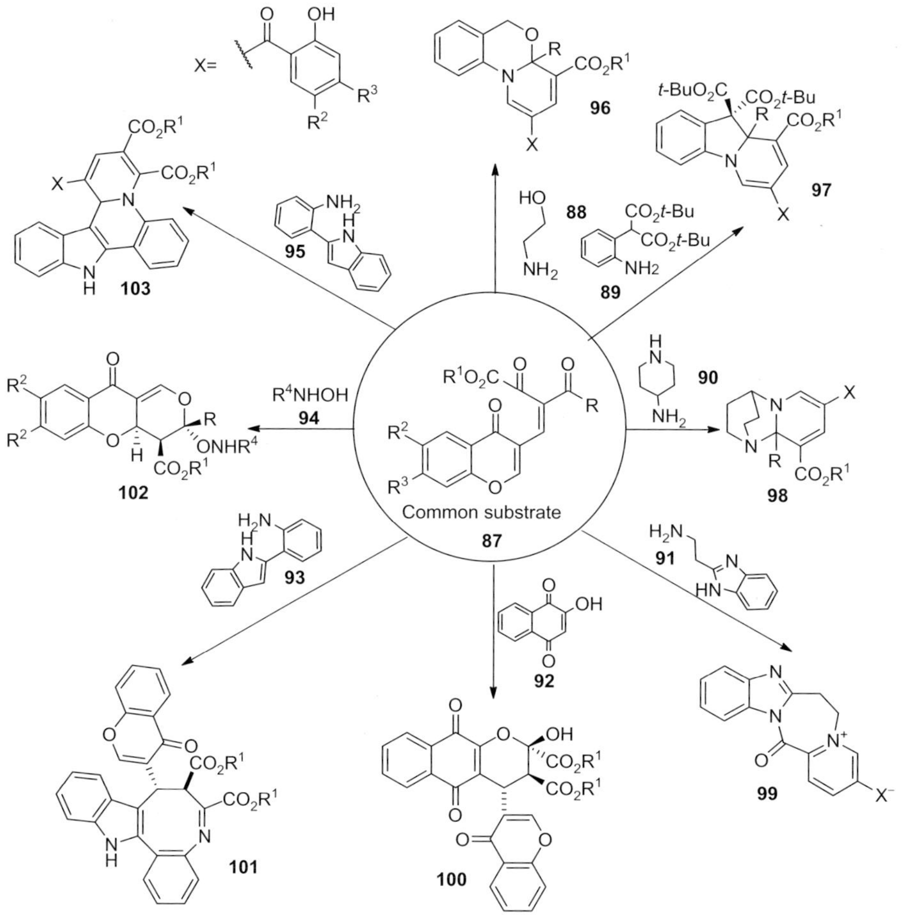

Scheme 13.17 Branching domino strategy to yield scaffold diversity.

diastereoselective manner. Different rings were incorporated by different zwitterions on to the chromone ring in a sequential manner, thereby providing tetracyclic benzopyrones decorated with three consecutive stereogenic centers including an all-carbon quaternary center. Diverse pairs of zwitterions were generated by the addition of a phosphine to electron-poor azadicarboxylates (**104**), acetylenedicarboxylates (**37**), and allenes (**106**). As shown in Scheme 13.18, the double annulation incorporated different rings, which included aza-, oxa-, and carbocycles, to the common benzopyrone and thus added great structural diversity to a small-sized compound collection (**109–112**). Importantly, these annulations transformed the relatively flat substrates **36** into highly sp^3-enriched new chemical entities, with high efficiency and excellent stereoselectivity. Easy accessibility of all substrates employed in the domino sequence along with the milder reaction conditions makes the reactions amenable to the synthesis of a compound collection.

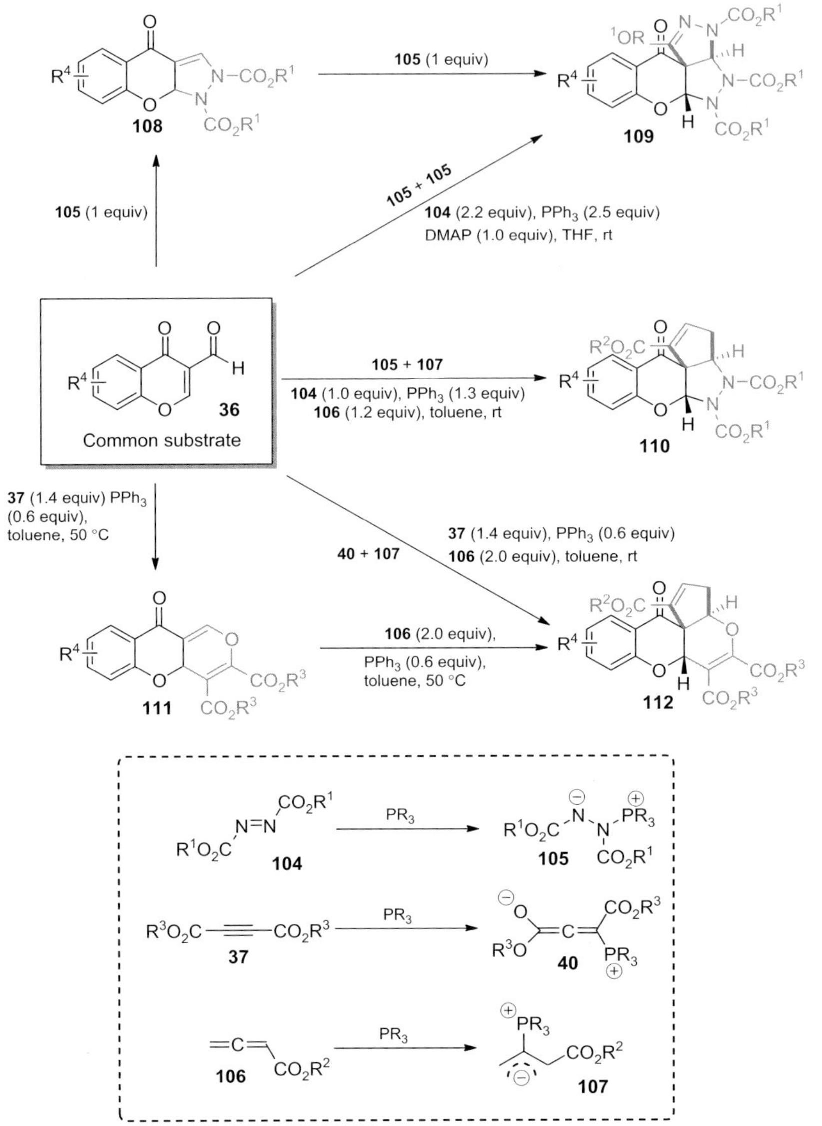

Scheme 13.18 Stereoselective domino double annulation to ring-fused benzopyrones.

Recently, the branching domino strategy was successfully exploited in a metal catalysis approach by Patil and coworkers [21]. In this strategy, alkynoic acids **113** were reacted with amine-bearing heterocyclic nucleophiles **114**, the so-called scaffold-building agents, to form an amide that is turned into ketoamide **116**. A gold catalyst (**115**) could further assist in an intramolecular nucleophilic attack of amide and nucleophilic aromatics to produce various heterocyclic scaffolds **117**. Using 30 different scaffold-building agents in combination with 8 alkynoic acids

(**118–125**), 65 compounds embodying 30 different polyheterocyclic scaffolds with five-, six-, and seven-membered rings were formed (selected examples: **126–133**) (Scheme 13.19).

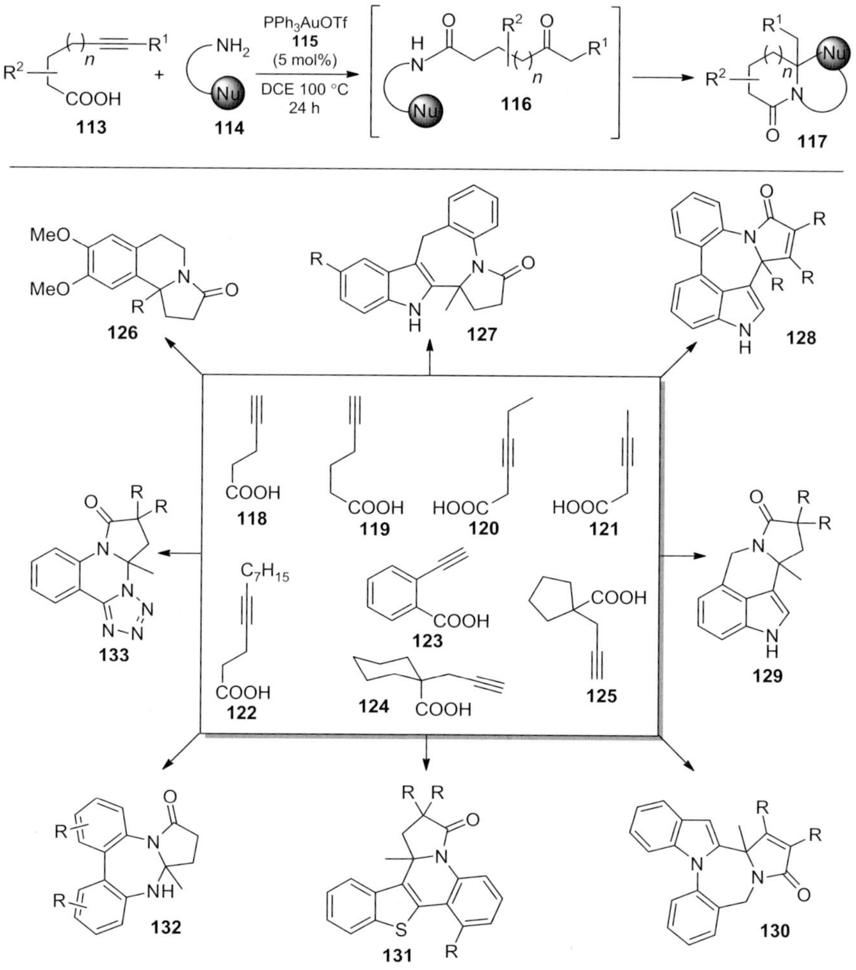

Scheme 13.19 Relay catalytic branching domino of alkynoic acids with heterocyclic amines (scaffold-building agents).

Two-directional synthesis can be used in a folding pathway approach, so that a symmetrical substrate can be used as a "molecular rope" that is "tied into knots" to get efficient formation of a range of linear and polycyclic architectures with NP-like features. This nice piece of work was reported by Stockman *et al.* [22], who used diethyl-7-oxotrideca-2,11-dienedioate **134** as the common substrate to yield 12 scaffolds (Scheme 13.20). Although simple reactions such as the Grignard reaction yielding the corresponding tertiary alcohol **135** did not induce domino reaction in this case, more elaborate domino reactions, such as a domino hydrazine formation/aza-Michael/prototropic shift/[3+2] cycloaddition/elimination that yielded compound **140** (41%) or a Hünig-base-induced domino condensation/azomethine ylide formation/[3+2] cycloaddition to yield octahydrocyclopenta[b]pyrrole **138** as a single diasteroisomer (71% yield), were advantageously used to construct complex molecular frameworks (Scheme 13.20).

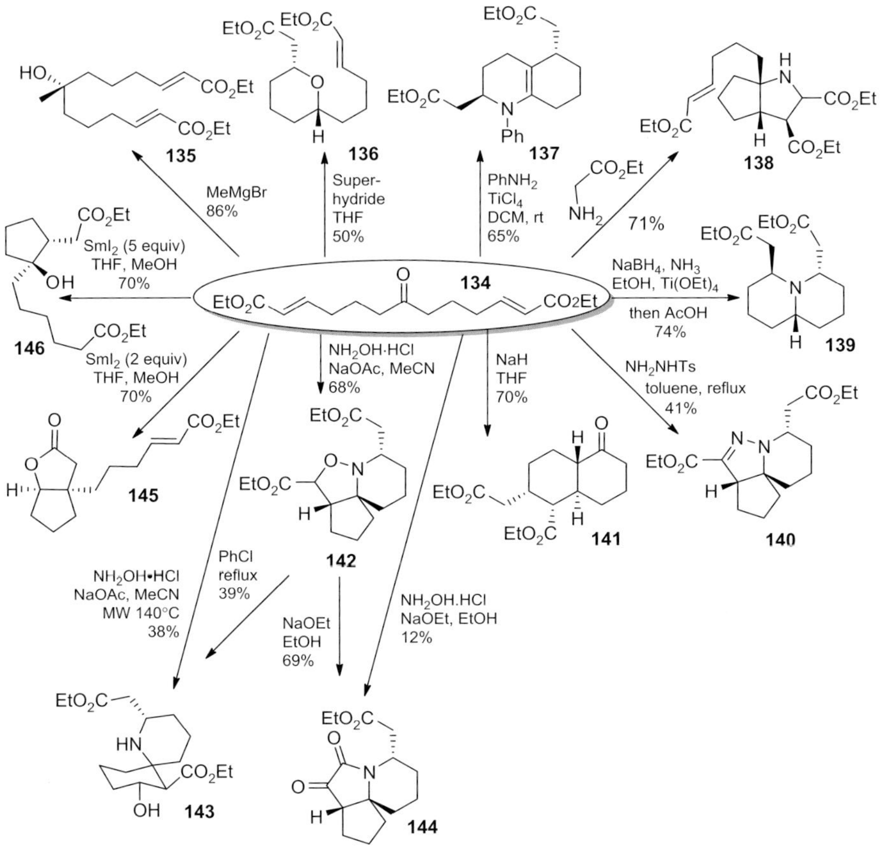

Scheme 13.20 Twelvefold branching synthesis using two-directional synthesis.

Reagent-based and substrate-based approaches can also be combined in a single two-directional synthesis in which different symmetrical substrates can be used under different reaction conditions to achieve predictable scaffold diversity [23]. In an example by Spring *et al.*, two domino reactions were used to synthesize polycyclic alkaloids from a common substrate **147**. A Boc-removal/bicyclization process using $AlCl_3$ as Lewis acid from **147** was used to produce 5-5, 5-6, and 6-6 bicyclic compounds (**148–153**), and a domino Michael addition/Dieckmann process/Michael addition reaction yielded 6-6-6, 6-6-5 tricyclic compounds (**154–156**). The 6-5-5 tricyclic scaffold **156** could not be synthesized by means of the same domino

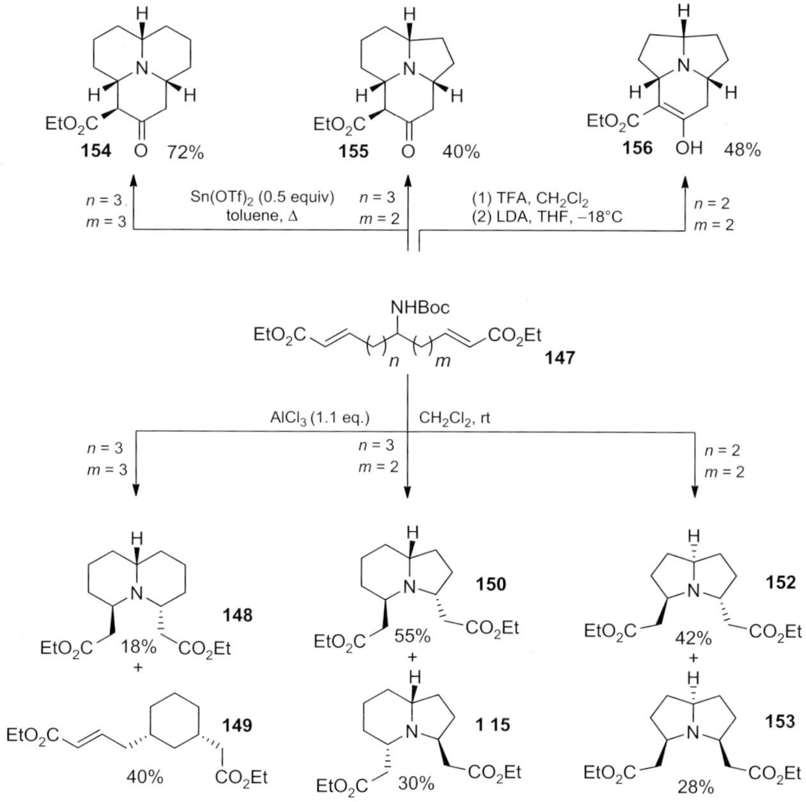

Scheme 13.21 Two-directional synthesis of alkaloid scaffolds using Lewis acid-catalyzed domino reaction.

reaction (Scheme 13.21). Domino products **148** and **154** are precursors for NPs, such as hippodamine and myrrhine, which are accessible after two to three steps.

13.3.3
Domino Reactions in the Build–Couple–Pair Approach for Library Synthesis

In the build–couple–pair approach to library synthesis, the building blocks are first synthesized from commercial material (build phase), which are then coupled to form a substrate (couple phase) that can further react under certain conditions and pair itself intramolecularly to form the cyclic products depicting the scaffold framework [24]. It is in the last two steps that the domino reaction can intervene to yield a more complex scaffold and with higher ring diversity.

Schreiber and Luo exploited domino sequences in both the coupling and pairing phases [25]. A gold-catalyzed cyclization of alkyne **157** was employed to obtain molecules that could subsequently be used in the pairing reaction. Activation of the alkyne by a gold(I) complex yielded the cyclized cationic species **160**,

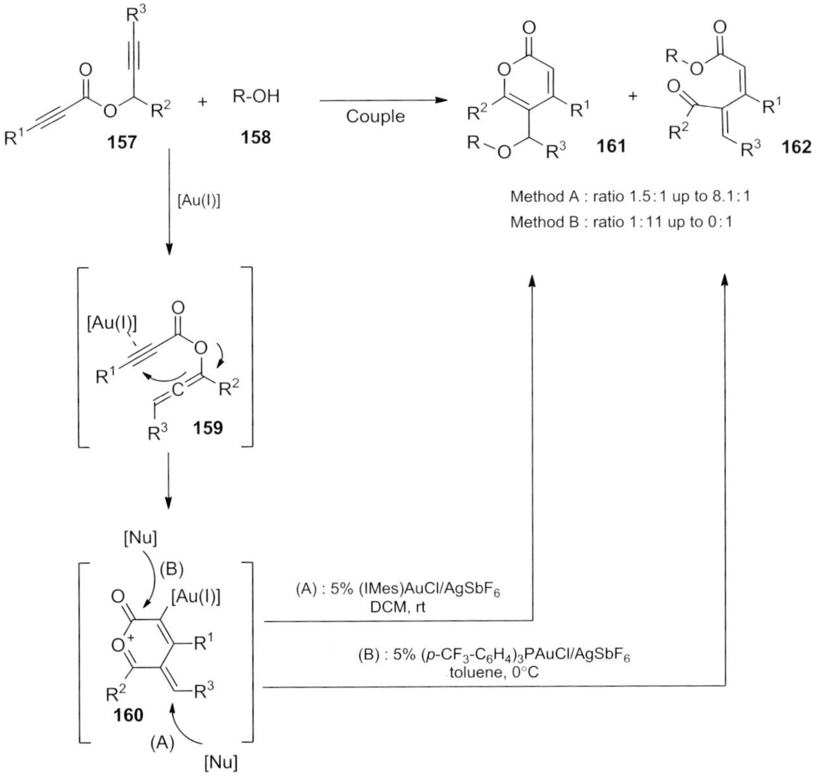

Scheme 13.22 Gold-catalyzed domino reaction used for the coupling phase.

which underwent a nucleophile addition either on the exocyclic olefin to yield the lactone **161** or on the carbonyl moiety to form the "open" pairing substrate **162** (Scheme 13.22).

Substrate **163** was built from commercial reagents and was coupled with azido phenyl propanol to form the pairing substrate **164**. The pairing process used in this build–couple–pair synthesis was a domino Staudinger/aza-Wittig/6π-disrotatory ring-closing reaction to furnish the new bicyclic ring system **166** (Scheme 13.23). The products obtained from this domino sequence were further transformed into the 2-pyridones **170** bearing two olefins moieties. An RCM thus provided the novel macrolactones **171** in good overall yields.

13.4 Solid-Phase Domino Syntheses of Compound Collections

An interesting example of a domino reaction used in a solid-phase strategy was reported by Meldal *et al.* [26], wherein the formation of fused (5-5), (5-6), (6-5), and (6-6) bicyclic ring systems was targeted using a common *N*-acyliminium

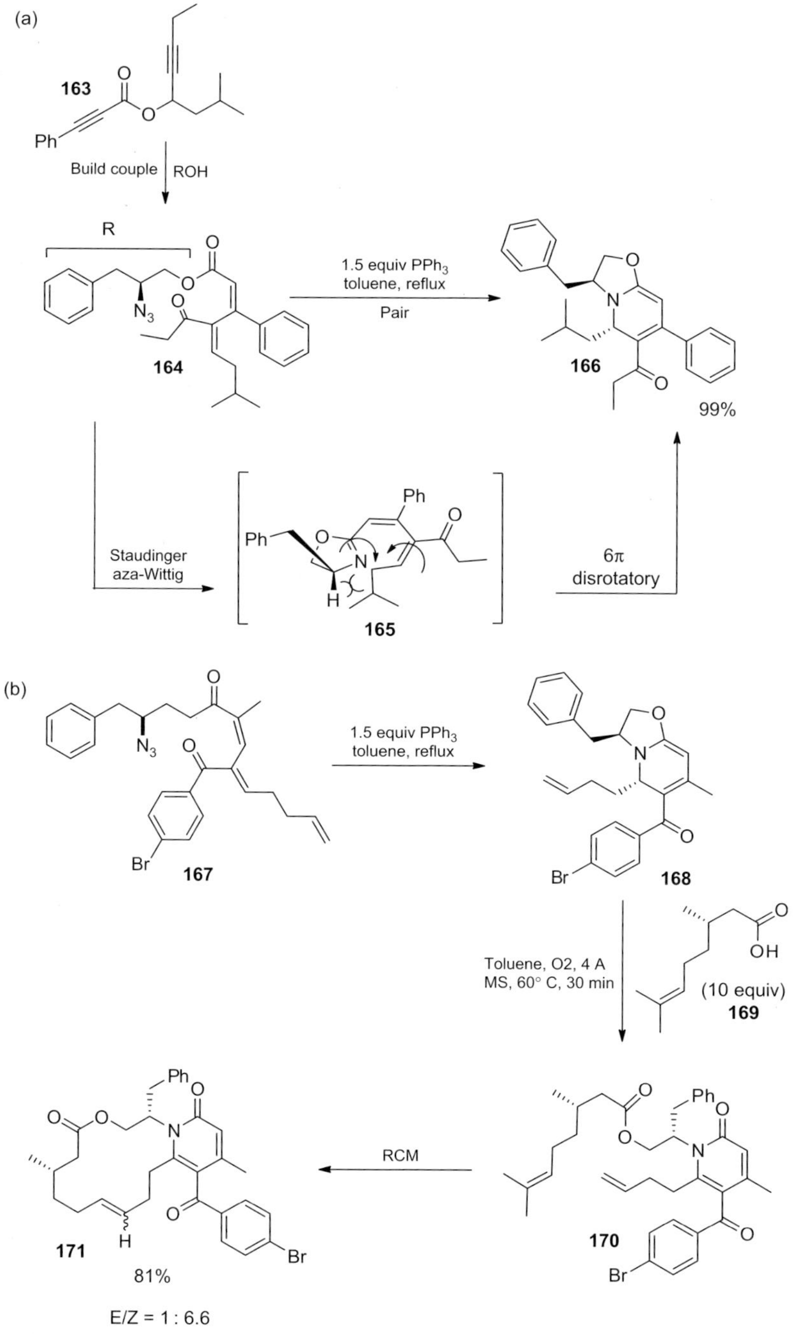

Scheme 13.23 (a, b) Domino Staudinger/aza-Wittig/6π-disrotatory ring-closing reaction used in the pairing phase.

intermediate. The synthesis strategy involved the coupling of masked aldehydes to the N-terminus of solid-supported peptides. After deprotection, the aldehydes **172** condensed with the amide backbone to generate hydroxylactams **173** of various ring sizes. Upon addition of acid, a domino reaction sequence followed, which involved shifting of the equilibrium toward *N*-acyliminium ion intermediates **174** and an intramolecular addition of nucleophiles (C- or heteroatom-based nucleophiles) that were conveniently positioned in the vicinity of the reactive center by the architecture of the molecular framework to form diverse ring structures. Overall, this domino reaction yielded the bicyclic structure **175** attached to a peptide sequence, which might be easily cleaved from the solid support (Scheme 13.24). Thus, different types of bi-, tri-, and tetracyclic scaffolds were produced with the acyliminium strategy.

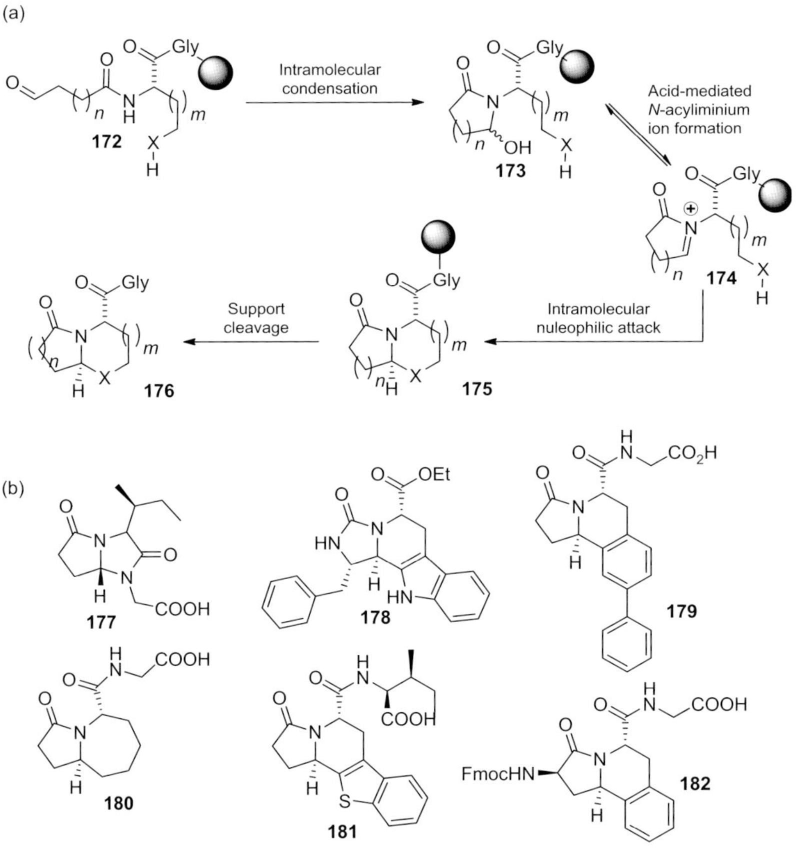

Scheme 13.24 (a) Intramolecular domino reactions of solid-supported cyclic *N*-acyliminium. (b) Examples of various heterocycles synthesized with this method.

In contrast to the example described above, where the domino reaction occurs as the final step of the synthesis, Bräse and Kapeller [27] used the domino reaction at the very beginning of the solid-phase synthesis of cannabinoids. In this synthesis, a salicylaldehyde motif bearing molecules **183** attached to a polystyrene backbone reacted with various Michael acceptors in a domino oxa-Michael/aldol (DOMA) condensation. A Wittig reaction of the resin-bound pyranoaldehydes **185** yielded dienes, which underwent a Diels–Alder reaction with enol ethers **186** to yield cannabinoid-based molecules **188** in only four steps. With one additional step, natural cannabinoids such as dronabinol, cannabidiol, or nabilone were synthesized with moderate to good overall yields (Scheme 13.25).

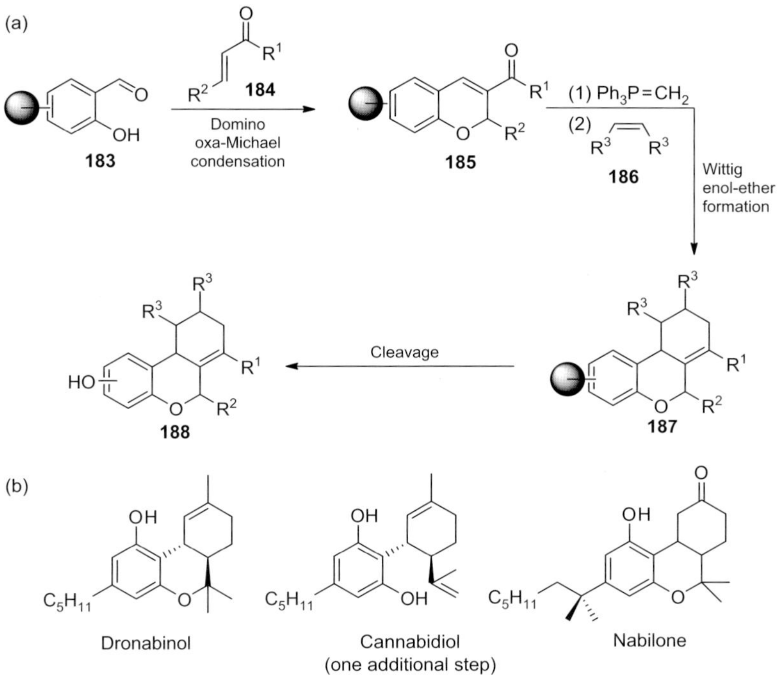

Scheme 13.25 Solid-supported domino synthesis of cannabinoids. (a) Synthetic strategy. (b) Examples of natural products synthesized through this method.

13.5 Conclusion

Domino reactions are well suited to create compound collections with high structural diversity and molecular complexity. The well-designed reaction sequences are reliable and efficient methods, giving synthetic access to functionalized small- to medium-sized molecules. These reactions are atom-economical and time-efficient, as they avoid multistep reactions, and reduce the number of tedious purification steps, which is a major advantage in the synthesis of large compound libraries. Recent developments in scaffold diversity synthesis utilizing common substrates for domino reactions to create a plethora of novel ring systems clearly demonstrate the power of these reactions in organic synthesis, in particular to generate diverse, small, focused compound libraries. Furthermore, domino reactions are amenable to most of the new techniques that are generally encountered in the synthesis of compound libraries, such as solid-phase synthesis, microwave irradiation [28], green chemistry [29] or the use of fluorescent-tagged reagents, and so on. In the present scenario, when new chemical entities are urgently required, for instance in drug discovery research, and chemists try to explore the novel chemical space of three-dimensionally complex small molecules, the development of new domino-reaction-based strategies that can be easily employed for compound library synthesis should be highly encouraged. In particular, there is still much to be achieved in the development of catalytic asymmetric domino reactions, and chemists can take a lot of inspiration from Mother Nature to reach these goals.

References

1. Tietze, L.F. (1996) *Chem. Rev.*, **96**, 115–136.
2. Ruijter, E., Scheffelaar, R., and Orru, R.V.A. (2011) *Angew. Chem. Int. Ed.*, **50**, 6234–6246.
3. Tietze, L.F. and Lieb, M.E. (1998) *Curr. Opin. Chem. Biol.*, **2**, 363–371.
4. Tietze, L.F. and Rackelmann, N. (2004) *Pure Appl. Chem.*, **76**, 1967–1983.
5. Huggins, D.J., Venkitaraman, A.R., and Spring, D.R. (2011) *ACS Chem. Biol.*, **6**, 208–217.
6. Waldmann, H., Eberhardt, L., Wittstein, K., and Kumar, K. (2010) *Chem. Commun.*, **46**, 4622–4624.
7. Modha, S.G., Kumar, A., Vachhani, D.D., Jacobs, J., Sharma, S.K., Parmar, V.S., Van Meervelt, L., and Van der Eycken, E.V. (2012) *Angew. Chem. Int. Ed.*, **51**, 9572–9575.
8. Ramachary, D.B. and Reddy, Y.V. (2010) *J. Org. Chem.*, **75**, 74–85.
9. Ramachary, D.B. and Kishor, M. (2007) *J. Org. Chem.*, **72**, 5056–5068.
10. Ramachary, D.B., Narayana, V.V., Prasad, M.S., and Ramakumar, K. (2009) *Org. Biomol. Chem.*, **7**, 2053–2062.
11. Dückert, H., Pries, V., Khedkar, V., Menninger, S., Bruss, H., Bird, A.W., Maliga, Z., Brockmeyer, A., Janning, P., Hyman, A., Grimme, S., Schürmann, M., Preut, H., Hübel, K., Ziegler, S., Kumar, K., and Waldmann, H. (2012) *Nat. Chem. Biol.*, **8**, 179–184.
12. Eschenbrenner-Lux, V., Dückert, H., Khedkar, V., Bruss, H., Waldmann, H., and Kumar, K. (2013) *Chem. Eur. J.*, **19**, 2294–2304.
13. Potowski, M., Schürmann, M., Preut, H., Antonchick, A.P., and Waldmann, H. (2012) *Nat. Chem. Biol.*, **8**, 428–430.
14. Galloway, W.R.J.D., Isidro-Llobet, A., and Spring, D.R. (2010) *Nat. Commun.*, **1**, 80–92.
15. Jeon, K.O., Rayabarapu, D., Rolfe, A., Volp, K., Omar, I., and Hanson, P.R. (2009) *Tetrahedron*, **65**, 4992–5000.
16. Morton, D., Leach, S., Cordier, C., Warriner, S., and Nelson, A. (2009) *Angew. Chem. Int. Ed.*, **48**, 104–109.
17. O'Leary-Steele, C., Pedersen, P.J., James, T., Lanyon-Hogg, T., Leach, S., Hayes, J., and Nelson, A. (2010) *Chem. Eur. J.*, **16**, 9563–9571.
18. Murrison, S., Maurya, S.K., Einzinger, C., McKeever-Abbas, B., Warriner, S., and Nelson, A. (2011) *Eur. J. Org. Chem.*, **2011**, 2354–2359.
19. Liu, W., Khedkar, V., Baskar, B., Schürmann, M., and Kumar, K. (2011) *Angew. Chem. Int. Ed.*, **50**, 6900–6905.
20. Baskar, B., Wittstein, K., Sankar, M.G., Khedkar, V., Schürmann, M., and Kumar, K. (2012) *Org. Lett.*, **14**, 5924–5927.
21. Patil, N.T., Shinde, V.S., and Sridhar, B. (2013) *Angew. Chem. Int. Ed.*, **52**, 2251–2255.

22. Robbins, D., Newton, A.F., Gignoux, C., Legeay, J.C., Sinclar, A., Rejzek, M., Laxon, C.A., Yalamanchili, S.K., O'Connell, M.A., and Stockman, R.A. (2011) *Chem. Sci.*, **2**, 2232–2235.
23. O'Connell, K.M.G., Diaz-Gavilàn, M., Galloway, W.R.J.D., and Spring, D.R. (2012) *Beilstein J. Org. Chem.*, **8**, 850–860.
24. Nielsen, T.E. and Schreiber, S.L. (2008) *Angew. Chem. Int. Ed.*, **47**, 48–56.
25. Luo, T. and Schreiber, S.L. (2009) *J. Am. Chem. Soc.*, **131**, 5667–5674.
26. Le Quement, S.T., Nielsen, T.E., and Meldal, M. (2007) *J. Comb. Chem.*, **9**, 1060–1072.
27. Kapeller, D.C. and Bräse, S. (2011) *ACS Comb. Sci.*, **13**, 554–561.
28. Santra, S. and Andreana, P.R. (2011) *Angew. Chem. Int. Ed.*, **50**, 9418–9422.
29. Gunasekaran, P., Balamurugan, K., Sivakumar, S., Perumal, S., Menéndez, J.C., and Almansour, A.I. (2012) *Green Chem.*, **14**, 750–757.

14
Domino Reactions in the Total Synthesis of Natural Products

Svenia-C. Düfert, Judith Hierold, and Lutz F. Tietze

While all the diverse fields and aspects of organic chemistry, including methodology, bioorganic chemistry, and material sciences, constantly evolve and grow in importance in both academic and industrial research, the total synthesis of natural products can still be considered the masterpiece of the area.

The requirements of modern total synthesis are conciseness and efficiency with respect to step count and yield, with a major focus now being placed on economic and ecological demands. The principles of domino reactions (see below) meet all of the aforementioned criteria. Transformations that include several bond-forming steps in one reaction have the potential to greatly enhance structural complexity and minimize waste production during a synthesis, thereby saving resources. In addition, the overall number of steps required for the total synthesis is reduced, decreasing the time required to access the target molecule.

During the last decades, the number of total syntheses featuring domino reactions has increased exponentially. Accordingly, numerous reviews and book chapters have been dedicated to the subject. Thus, we do not aim to be comprehensive in the following survey on total syntheses, but rather want to give an overview of the most recent elegant examples, which not only meet the definition of domino reactions but also provide a maximum increase in structural complexity.

As was the case in our earlier publications, the total syntheses described herein were classified according to the first step of the domino process they feature. Hence, the distinction has been made between cationic, anionic, radical, pericyclic, transition-metal-catalyzed, and reductive or oxidative domino processes.

14.1
Cationic Domino Reactions

Physalins are highly oxygen-functionalized steroidal compounds that occur in *Physalis* plants and exhibit a wide range of biological activities, including antitumor, antimicrobial, antiparasitic, and immunosuppressive behaviour [1]. In 2012, Sodeoka and coworkers [2, 3] reported a domino cyclization approach to the DEFGH ring scaffold common to type B physalins, such as physalin B (**7**) (Scheme 14.1).

Domino Reactions: Concepts for Efficient Organic Synthesis, First Edition. Edited by Lutz F. Tietze.

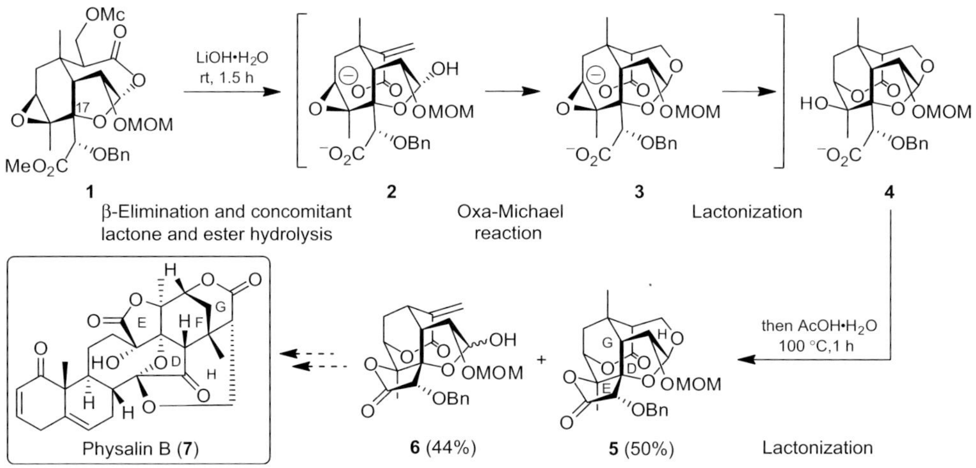

Scheme 14.1 Sodeoka's synthesis of the DEFGH ring scaffold of type B physalins by domino cyclization. Mc, monochloromethanesulfonyl.

The key step of Sodeoka's synthesis was the kinetically controlled preparation of the cage-shaped dilactone **5** from an advanced intermediate **1** using a domino reaction with subsequent lactonization.

Thus, treatment of α-hydroxyester **1** with lithium hydroxide initially furnished dicarboxylate **2** as a result of β-elimination and lactone hydrolysis with simultaneous hydrolysis of the methyl ester. Dicarboxylate **2** then likely underwent a 7-*endo*-oxa-Michael addition to form dicarboxylate **3**, which in turn cyclized to the G-ring lactone **4** upon opening of the epoxide (for nomenclature, see **7**). Consecutive treatment with acetic acid brought about a second lactonization, thus furnishing the final product **5**, constituting the DEFGH ring scaffold of type B physalins, in 50% isolated yield. The concomitant construction of the E ring was reported to be crucial for high yields since the cage-shaped dilactone **5**, with an architecture similar to that of the natural product family, was greatly stabilized by the functionality at C17, thereby shifting the equilibrium away from hemiacetal **6**, which was coisolated in 44% yield. While **5** was transformed into the completed DEFGH scaffold through a deprotection–oxidation sequence, the total synthesis of a natural product containing the prepared architecture, for example physalin B (**7**), has not yet been achieved using Sodeoka's approach.

In 2009, the group of Nicolaou [4] reported a strategy for the enantioselective total synthesis of hirsutellone B (**12**), a secondary metabolite isolated from the insect pathogenic fungus *Hirsutella nivea* BCC 2594 (Scheme 14.2). Members of the hirsutellone class of natural products exhibit promising biological properties, particularly activity against *Mycobacterium tuberculosis*, the causative pathogen of tuberculosis [5]. Nicolaou's approach featured a number of domino-based sequences, most prominently the stereoselective construction of the fused [6,5,6]-tricyclic core **11** from acyclic precursor **8** by a domino intramolecular epoxide opening/Diels–Alder reaction.

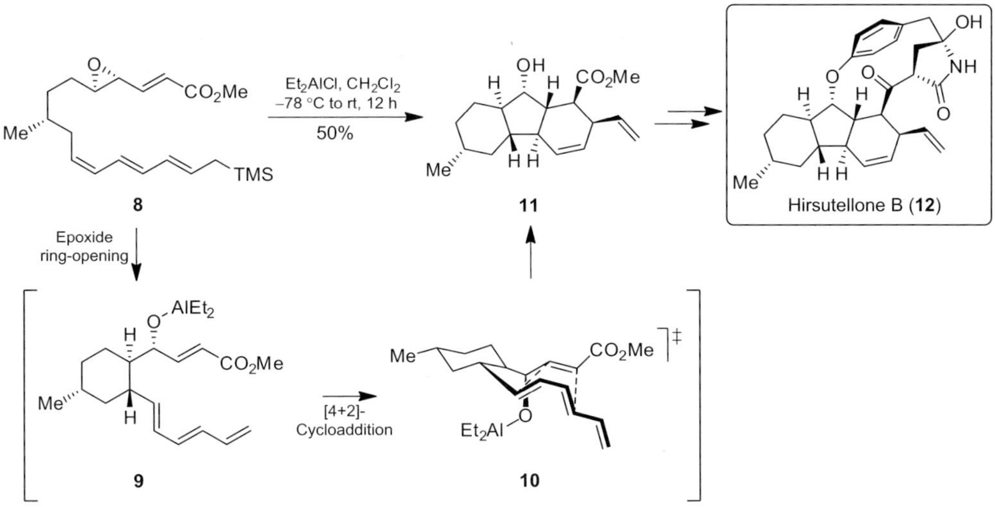

Scheme 14.2 Nicolaou's total synthesis of hirsutellone B (**12**) via domino epoxide opening/[4 + 2]-cycloaddition.

In the event, treatment of **8** with Lewis acidic Et_2AlCl afforded the tricyclic scaffold in 50% yield as a single diastereomer. Lewis acid activation of the trimethylsilyl (TMS)-epoxy tetraene brought about the first cyclization, which proceeded under opening of the activated epoxide through intramolecular nucleophilic substitution. The stereochemical outcome of this initial ring closure was proposed to stem from a chair-like transition state *en route* to cyclohexane **9**. A protonated form of intermediate **9** has been isolated, thus supporting the mechanistic proposal. Subsequent [4 + 2] cycloaddition led to the targeted tricyclic core **11**, likely through the endo transition state **10**. A series of further chemical transformations, including the formation of a 14-membered macrocyclic sulfone and a Ramberg–Bäcklund ring contraction, eventually resulted in the synthesis of the enantiopure natural product hirsutellone B (**12**).

A domino sequence initiated by Lewis acid-induced epoxide ring opening was featured in Katoh's enantioselective total synthesis of (+)-stachyflin (**18**) (Scheme 14.3) [6]. The sesquiterpenoidal alkaloid was isolated from a broth of *Stachybotrys* sp. RF-7260 and found to display extremely potent antiviral activity against the influenza H1N1 virus with a unique mechanism of action [7]. Katoh *et al.* [8] utilized a late-stage domino epoxide opening/rearrangement/cyclization sequence to access the completed pentacyclic ring skeleton of the natural product with cis-fused junctions between both the AB and the BC rings. Exposure of trans-fused decalin **13** (α-/β-epoxide 7 : 1) to an excess of $BF_3 \cdot Et_2O$ initially gave rise to carbocation **14**, which underwent migration of the C5 methyl group to the neighboring cationic center. A 1,2-hydride shift then furnished carbocation **16**, which was trapped by the inner phenolic hydroxy group to deliver the desired cyclized product **17**. The product was isolated as a mixture of diastereomers (α-/β-OH 66% : 9%), though

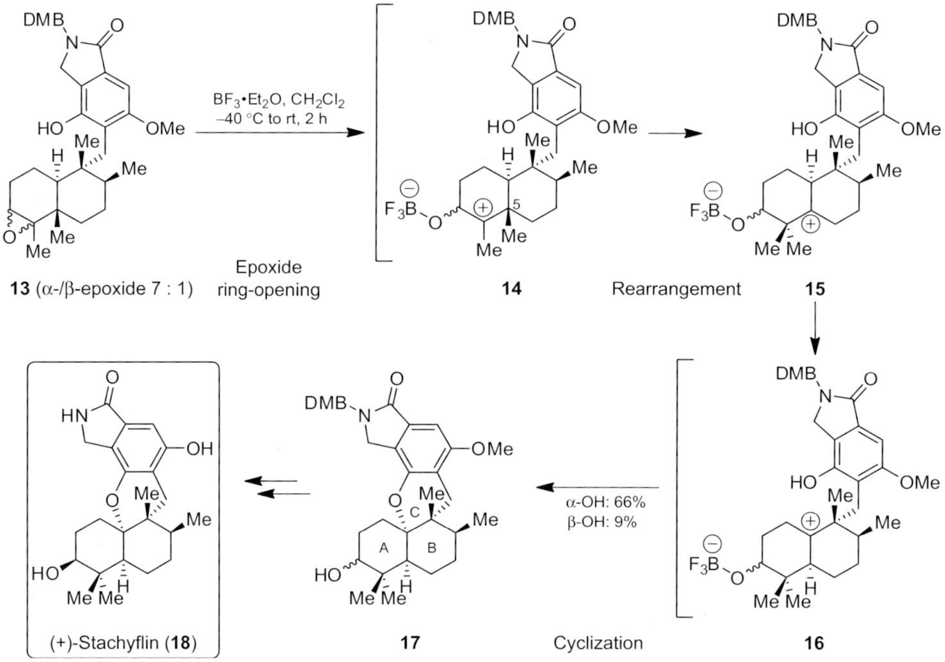

Scheme 14.3 Katoh's total synthesis of (+)-stachyflin (**18**) through domino epoxide opening/rearrangement/cyclization (DMB, 3,4-dimethoxybenzyl).

inversion of the configuration of the undesired α-OH stereoisomer was readily achieved at a later stage of the total synthesis.

(+)-Omaezakianol (**22**) is a member of the oxasqualenoid family of natural products, isolated from the red alga *Laurencia omaezakiana* [9]. The first total synthesis of the triterpene polyether was reported in 2009 by Morimoto, whose group accessed the natural product in 24 steps from farnesol [10]. One year later, Xiong *et al.* [11] published a remarkably short synthesis, generating **22** in only six steps from squalene using a biomimetic approach (Scheme 14.4). A key feature of the total synthesis was the three-step conversion of squalene-derived chlorohydrin **19** to (+)-omaezakianol (**22**), comprising Shi-epoxidation of **19** to pentaepoxide **20**, cationic polycyclization to pentakistetrahydofuran **21**, and subsequent reduction. The chlorohydrin group in **19** served as both a masking group for the terminal double bond of the natural product and the initiating group for the epoxide-opening domino sequence. The pivotal epoxidation/cationic polycyclization sequence proceeded in 21% yield over two steps.

In 2009, Wiemer and coworkers [12] reported the termination of an epoxide-initiated cationic cyclization by electrophilic aromatic substitution (Scheme 14.5). In this novel domino transformation, a MOM (methoxymethyl) "protected" phenol was cyclized with concomitant loss of the MOM group, which was then harnessed intramolecularly to forge a new C–C bond. The methodology was applied to the total synthesis of (+)-angelichalcone (**27**), a natural chalcone isolated from the

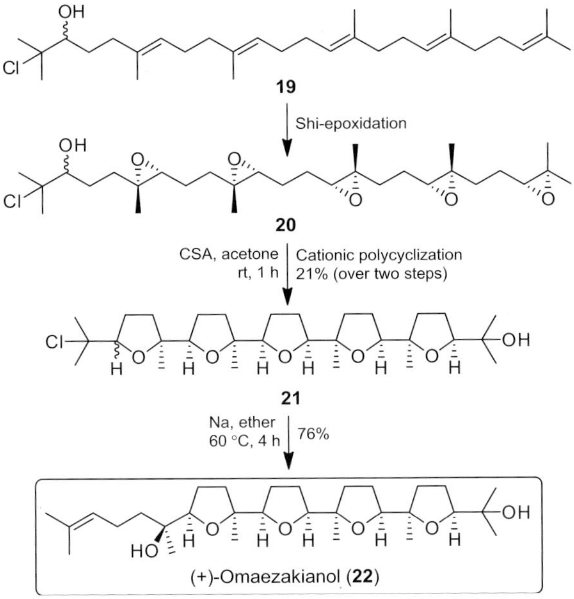

Scheme 14.4 Xiong's total synthesis of (+)-omaezakianol (**22**) through cationic polycyclization (CSA, camphorsulfonic acid).

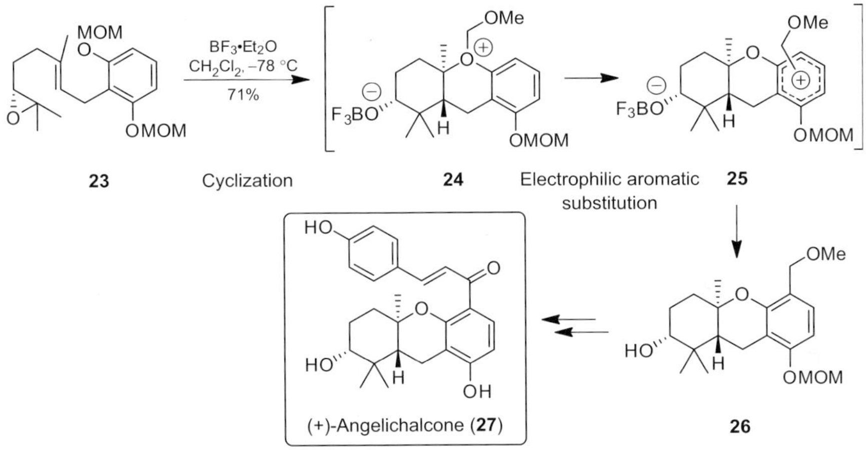

Scheme 14.5 Wiemer's total synthesis of (+)-angelichalcone (**27**) via domino cyclization/ electrophilic aromatic substitution.

Japanese herb *Angelica keiskei* [13]. Thus, treatment of resorcinol derivative **23** with $BF_3{\cdot}OEt_2$ initially led to oxonium ion **24**, which underwent loss of a $CH_3OCH_2^+$ cation. The adjacent Lewis basic arene rapidly trapped the electrophile, affording the π-complex **25** in an intramolecular process as supported by crossover experiments. The short-lived nature of the cationic species, i.e., $CH_3OCH_2^+$, resulted in exclusive substitution at the ortho carbon, with no substitution observed at the para position. Benzyl methyl ether **26**, the product of this domino cyclization/rearrangement process, was isolated in 71% yield and could readily be transformed into the corresponding benzaldehyde following 2,3-dichloro-5,6-dicyano-1,4-benzoquinone (DDQ)-induced oxidation. Further elaboration furnished the natural product (+)-angelichalcone in a total of seven steps and an overall yield of 21%. More recently, Wiemer applied this domino methodology to the first total synthesis of (+)-schweinfurthin A [14].

In 2008, the group of Williams [15] reported the first total synthesis of (±)-2-*O*-methylneovibsanin H (**33**) (Scheme 14.6). The diterpene, isolated from the leaves of *Viburnum awabuki* [16], was prepared by employing an acid-catalyzed domino sequence. Thus, the key transformation was the conversion of advanced intermediate **28** to cyclohexene **32** upon treatment with an excess of sulfuric acid in anhydrous methanol. Acid-mediated silyl deprotection first revealed alcohol **29**, which readily underwent an intramolecular oxa-Michael addition to yield tricyclic **30**. It was postulated that solvolysis and nucleophilic addition of methanol to the intermediary allyl cation then furnished acid **31**, which underwent Fischer esterification. The resultant highly functionalized cyclohexene **32** was isolated in 50% yield as a mixture of diastereomers at C2 (diastereomeric ratio (dr) = 85 : 15). The observed stereochemistry at the newly created stereocenters, i.e., at C2 and C5, was postulated to arise from the preexisting sterically congested stereocenters in the starting material (i.e., **28**). Cyclohexene **32** was eventually taken on to provide the natural

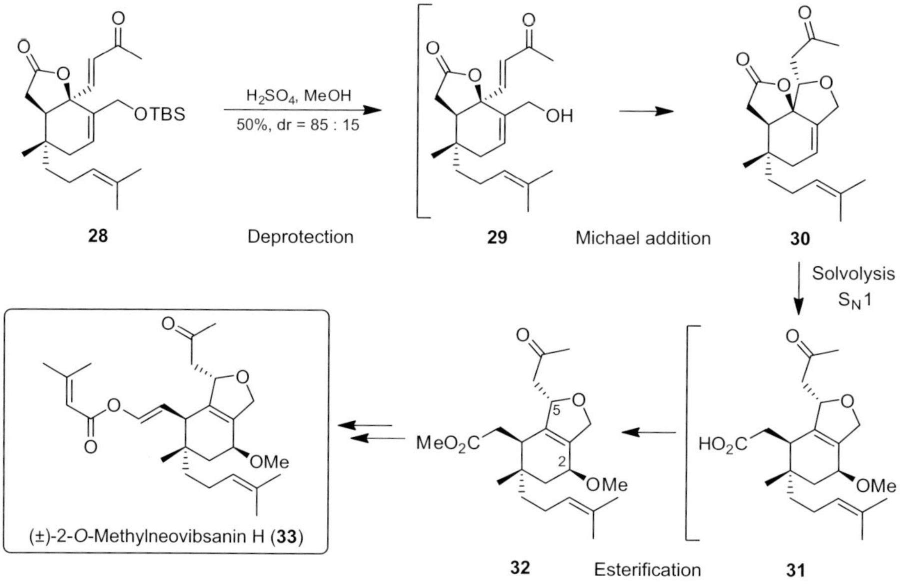

Scheme 14.6 Williams' total synthesis of (±)-2-*O*-methylneovibsanin H (**33**) via acid-induced domino sequence.

product in a total of 12 steps. Acid-induced domino sequences were also featured in the synthesis of a number of related vibsanins, as reported by Williams [17].

(−)-Cribrostatin 4 (**41**), a member of the tetrahydroisoquinoline family of natural products, was isolated in 2000 from the blue sponge *Cribrochalina* [18] and has been shown to display cytotoxic and antimicrobial activities at low micromolar concentrations [19]. Both the complex molecular architecture and biological profile of (−)-cribrostatin 4 have attracted the interest of the synthetic community, with a number of synthetic approaches towards the alkaloid reported to date [20]. In 2007, Chen and Zhu [21] detailed the total synthesis of the natural product using an acid-mediated domino β-elimination/cyclization reaction (Scheme 14.7). In Zhu's approach, the advanced intermediate **34** was subjected to Swern oxidation and tetra-*n*-butylammonium fluoride (TBAF)-mediated deprotection of the silyl ether, yielding a mixture of aldehyde **35** and hemiaminal **36**. Treatment of this mixture with 0.01% v/v methanesulfonic acid in dichloromethane initiated a process consisting of acyliminium formation, deprotonation, and domino β-elimination/cyclization. Thus, dehydration of hemiaminal **36** initially led to iminium ion **37**, which formed enamide **38** under loss of a proton. Elimination of the sulfur side chain then resulted in conjugated iminium ion **39**, which was set up to undergo a phenolic Mannich cyclization.

The targeted pentacycle **40** was thus isolated in 51% yield over three steps, along with an undesired regioisomer in 15% yield, resulting from minor nucleophilic attack at the 6-position of the tethered aromatic ring in the final domino step. Elaboration of **40** provided access to (−)-cribrostatin 4 (**41**) in a total of 21 steps and 4.3% overall yield. Concurrent with Zhu's report, the group of Williams [20b] independently developed a similar reaction sequence leading to the same natural product.

In 2010, Spring and coworkers [22] reported the diversity-oriented synthesis (DOS) of bicyclic and tricyclic scaffolds reminiscent of natural alkaloids by employing a Lewis acid-induced domino sequence. Linear aminoalkenes such as **42** were butoxycarbonyl (Boc)-deprotected when subjected to substoichiometric amounts of tin(II)triflate in refluxing toluene, thus allowing conjugate addition of the resultant amine **43** into the α,β-unsaturated ester groups. Concomitant Dieckmann condensation gave tricyclic amine **46** in 72% isolated yield. Mechanistic investigations indicated that the second conjugate addition might proceed after the Dieckmann condensation, supporting a deprotection/conjugate addition/Dieckmann condensation/conjugate addition domino sequence, as displayed in Scheme 14.8. Spring used the folding approach of linear aminoalkenes to generate a total of seven different polycyclic scaffolds, whose architectures were dependent on the length of the aminoalkene starting material and the nature and quantity of Lewis acid employed.

In 2010, Kim and Kim [23] reported a noteworthy domino approach to the first total synthesis of diptoindonesin G (**52**) (Scheme 14.9). The oligostilbenoid, possessing strong cytotoxic and immunosuppressive activity [24], was recently isolated from the tree bark of *Hopea mengarawan* [25]. In a domino cyclodehydration/Friedel–Crafts acylation/demethylation sequence, Kim

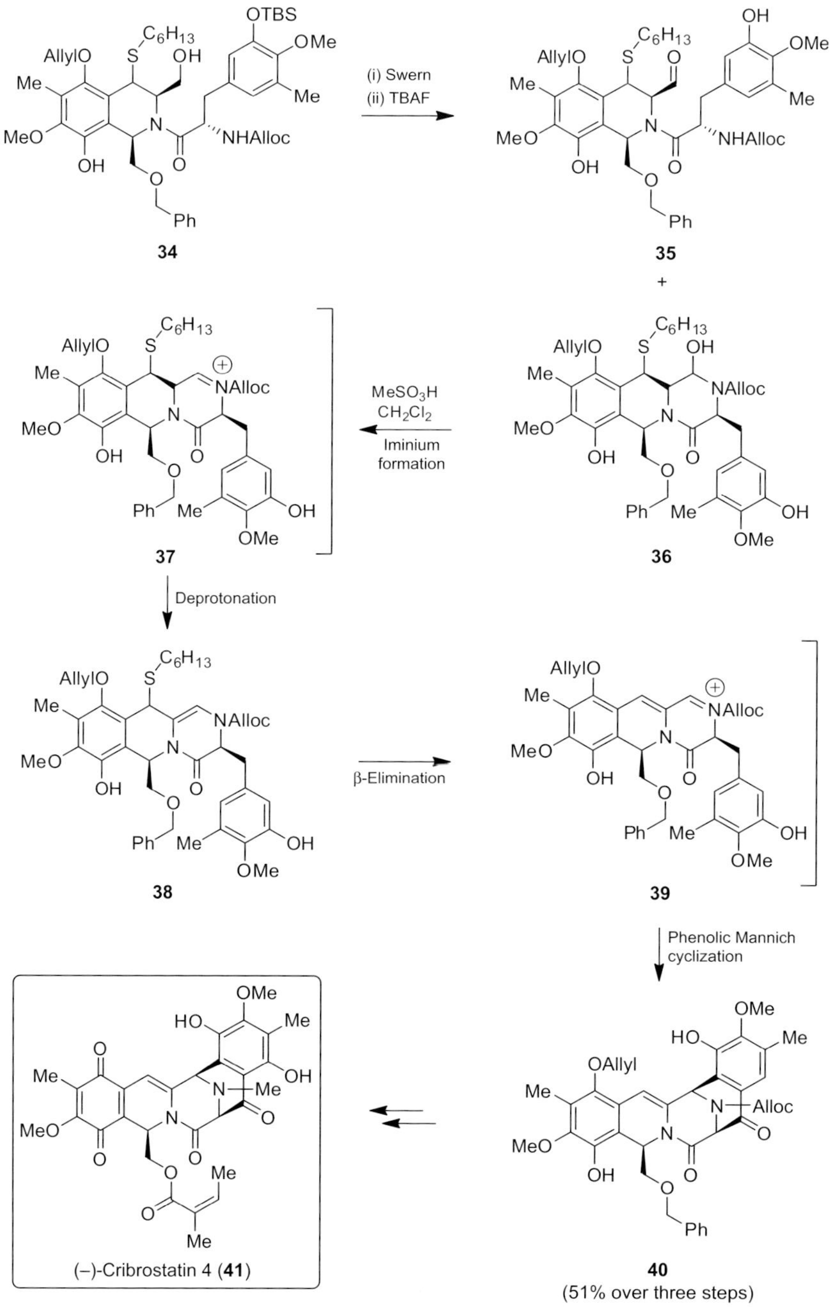

Scheme 14.7 Zhu's total synthesis of (−)-cribrostatin 4 (**41**) via domino β-elimination/cyclization.

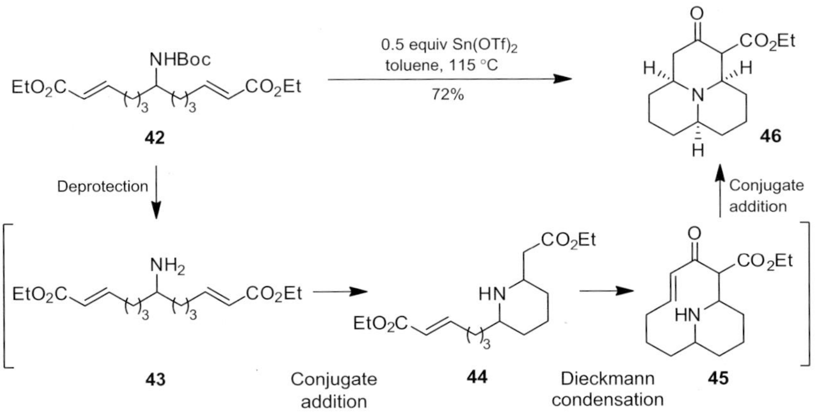

Scheme 14.8 Spring's diversity-oriented synthesis of polycyclic scaffolds via domino deprotection/double conjugate addition/Dieckmann condensation.

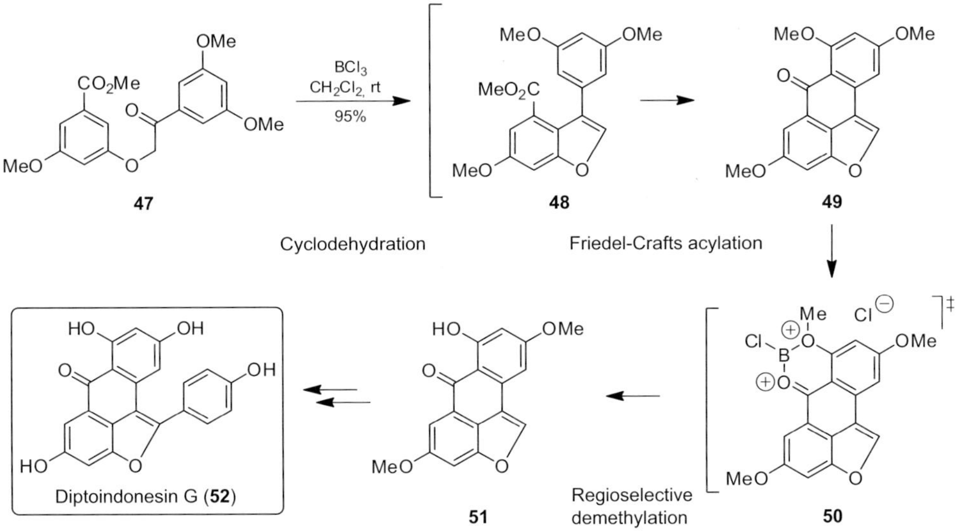

Scheme 14.9 Kim's total synthesis of diptoindonesin G (**52**) via BCl_3-facilitated domino double cyclization/demethylation.

constructed the tetracyclic 6*H*-anthra[1,9-*bc*]furan-6-one skeleton of the natural product directly from the aryloxyketone **47**. Mechanistically, treatment of **47** with BCl_3 initially triggered a cyclodehydration event to form benzofuran **48**. The Lewis acid then promoted an intramolecular Friedel–Crafts acylation to tetracyclic intermediate **49**, setting the stage for the regioselective demethylation of the ortho methoxy group. It was postulated that coordination of the boron to both the carbonyl group and the ortho methoxy oxygen accounted for the observed selectivity (see **50**). The resulting 6*H*-anthra[1,9-*bc*]furan-6-one **51** was isolated in an excellent yield of 95%, and led to the targeted natural product after further elaboration.

The erythrina alkaloids are an extensive class of natural products isolated from tropical and subtropical fabaceae plants of the *Erythrina* genus. They express a broad variety of bioactivities, including sedative, hypotensive, anticonvulsive, central nervous system (CNS) depressing, and curare-like properties [26]. In 2009, Tietze and coworkers [27] reported the formal synthesis of a member of this family, (+)-erysotramidine (**59**), in which the characteristic tetracyclic azaspiro skeleton of **59** was assembled using a Lewis acid-catalyzed domino process (Scheme 14.10). Thus, treatment of cyclohexanone **54** and primary arylethylamine **53** with trimethylaluminum in the presence of catalytic amounts of Lewis acidic indium triflate, followed by treatment with triflic acid, led to the targeted spirocyclic compound **55** in 92% yield (dr = 4 : 1). The proposed mechanism comprised the initial formation of an aluminum amide, which attacked the ester carbonyl of cyclohexanone **54** to form azaenolate **56**. The azaenolate in turn reacted with the ketone carbonyl, yielding enamine **57** as an inconsequential mixture of double-bond isomers. Upon treatment with triflic acid, the iminium ion **58** was formed, which underwent electrophilic aromatic substitution to afford the cis-annulated domino product **55** as the major isomer [28]. Further transformations, including a Tamao–Fleming oxidation, yielded an advanced intermediate previously utilized in the synthesis of **59** by Simpkins *et al.* [29], thus concluding the formal synthesis.

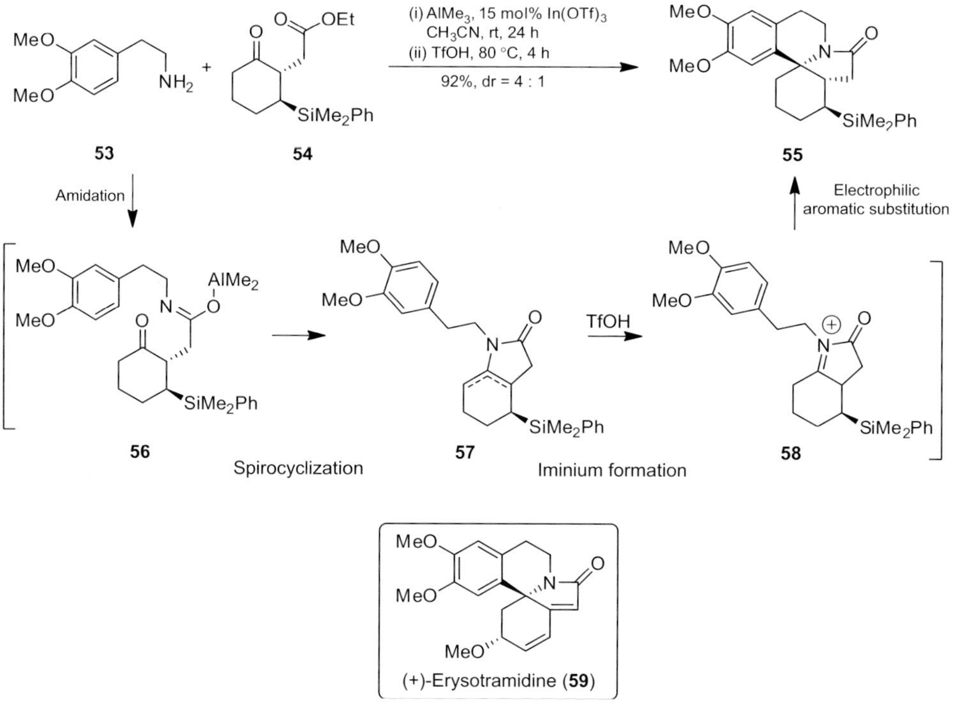

Scheme 14.10 Tietze's formal synthesis of (+)-erysotramidine (**59**) using a cationic domino process.

14.2
Anionic Domino Reactions

Anionic initiation of domino reactions is one of the most frequently encountered domino strategies applied in natural product synthesis. In 2013, Laws and Scheerer [30] reported the first enantioselective synthesis of the prenylated indole alkaloid (+)-malbrancheamide B (**65**), a novel calmodulin inhibitor (Scheme 14.11). Though a large number of routes to its [2.2.2]-diazabicyclic skeleton had already been developed [31], Scheerer's domino aldol condensation/isomerization/Diels–Alder cycloaddition approach rapidly generated molecular complexity while exhibiting good diastereofacial control [32]. For the domino synthesis of the pentacycle **64b**, indole carbaldehyde **61** was treated with diketopiperazine **60** under basic reaction conditions. Initially, aldol condensation product **62** was formed, which underwent alkene isomerization to endocyclic azadiene **63**, thus effecting subsequent intramolecular Diels–Alder cyclization to give **64**. The cycloadduct **64** was isolated in 85% yield as a 1 : 2 mixture of diastereomers (**64a**:**b**).

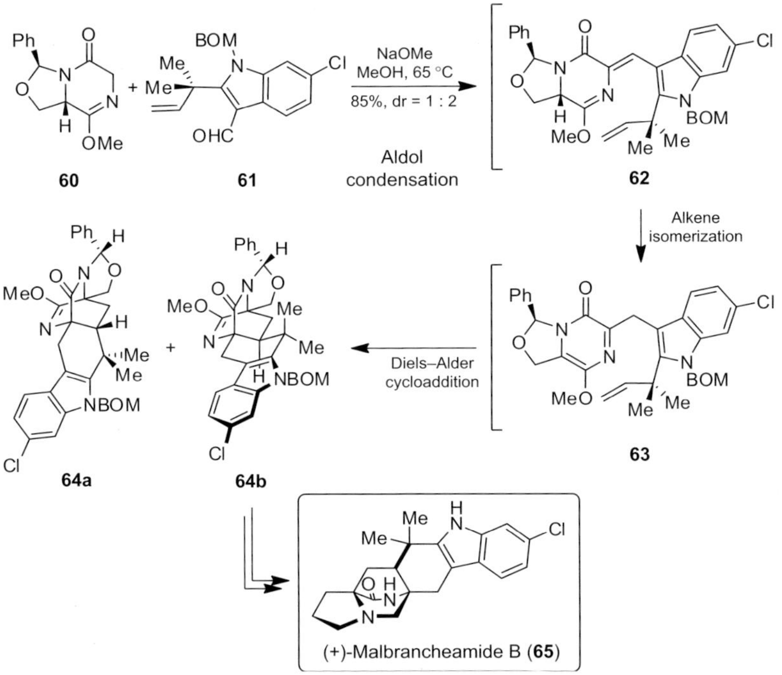

Scheme 14.11 Scheerer's synthesis of (+)-malbrancheamide B (**65**) via domino aldol condensation/isomerization/cycloaddition (BOM, benzyloxymethyl).

It was proposed that the biosynthetic pathway to **65** involved a similar intramolecu lar hetero-Diels–Alder cycloaddition [31, 33], but most biomimetic approaches so far have suffered from moderate stereoselectivity. Scheerer addressed this problem with the incorporation of a removable aminal auxiliary, which imparted a diastereofacial bias during the cycloaddition, guiding the terminal alkene to the azadiene face opposing the chiral phenyl amine. The correct diastereomer **64b** was further modified to provide (+)-malbrancheamide B (65) in 13 steps, as determined by the longest linear route.

An alternate approach to the synthesis of prenylated indole alkaloids was reported by Baran *et al.* [34, 35]. In 2005, his group targeted the [2.2.2]-diazabicycle-containing (+)- stephacidin A (**72**) using a beautiful reaction sequence, which, aside from utilizing two separate domino reactions, also included the first oxidative heterocoupling of two different types of enolates. The synthesis commenced with tryptophan **68**, which was accessed in gram-scale quantities through a domino condensation/palladium-catalyzed Heck-type coupling of *o*-iodoaniline **67** and pyroglutamate **66**. The addition of tetra-*n*-butylammonium iodide was found to be crucial for high yield by limiting the amount of deiodination observed for this transformation. Elaboration of **68** over several steps led to advanced intermediate **69**, which gave (+)- stephacidin A (**72**) in yields ranging from 28% to 45% after exposure to high temperatures (200 °C). This thermolytic domino sequence was initiated by Boc deprotection, followed by a formal ene reaction to generate the spirocyclic intermediate **71**. A ring-expanding 1,2-alkyl shift then afforded the natural product **72** (Scheme 14.12). In addition, Baran demonstrated the oxidative conversion of stephacidin A to natural (+)-avrainvillamide (**73**), which underwent spontaneous dimerization under various conditions to yield stephacidin B (**74**).

Dixon recently reported the total synthesis of the marine alkaloid manzamine A (**82**) using a highly convergent approach containing two domino reactions (Scheme 14.13) [36]. The piperidine ring of the natural product was assembled in a domino nitro-Mannich/lactamization process, induced by conversion of Michael adduct **75** with formaldehyde and hex-5-en-1-amine (**76**) in refluxing methanol. The intermediate β-nitro amine **77** spontaneously cyclized to give lactam **78**, which was further modified to pyrrolidinone **79** in four separate chemical transformations. Pyrrolidinone **79** was then converted to nitro amine **80** using a reductive nitro-Mannich reaction. Key to this second domino event was the chemoselective reduction of the pyrrolidinone carbonyl of **79** in the presence of the double bond as well as the acetal and the nitro groups, and without causing over-reduction to the pyrrolidine. This was achieved using titanium tetraisopropoxide and diphenylsilane, which facilitated carbonyl reduction to intermediate **80**, as well as subsequent intramolecular nitro-Mannich addition to deliver nitro amine **81** in 81% yield with high diastereoselectivity of 83 : 17 (epimeric at C7). The nitro-Mannich adduct **81** was then taken on to complete the synthesis of manzamine A in 18 steps, as determined by the longest linear sequence. In addition, Dixon applied his domino nitro-Mannich/lactamization strategy to the total syntheses of a number of related members of the manzamine family, including ircinol A (**83**) [36] and (−)-nakadomarin A (**84**) [37].

In 2007, Varseev and Maier [38] employed a domino Wittig–Horner/Diels–Alder cycloaddition reaction in the first total synthesis of the alkaloid (+)-neosymbioimine (**90**) (Scheme 14.14). The zwitterionic compound, comprising a tricyclic iminium

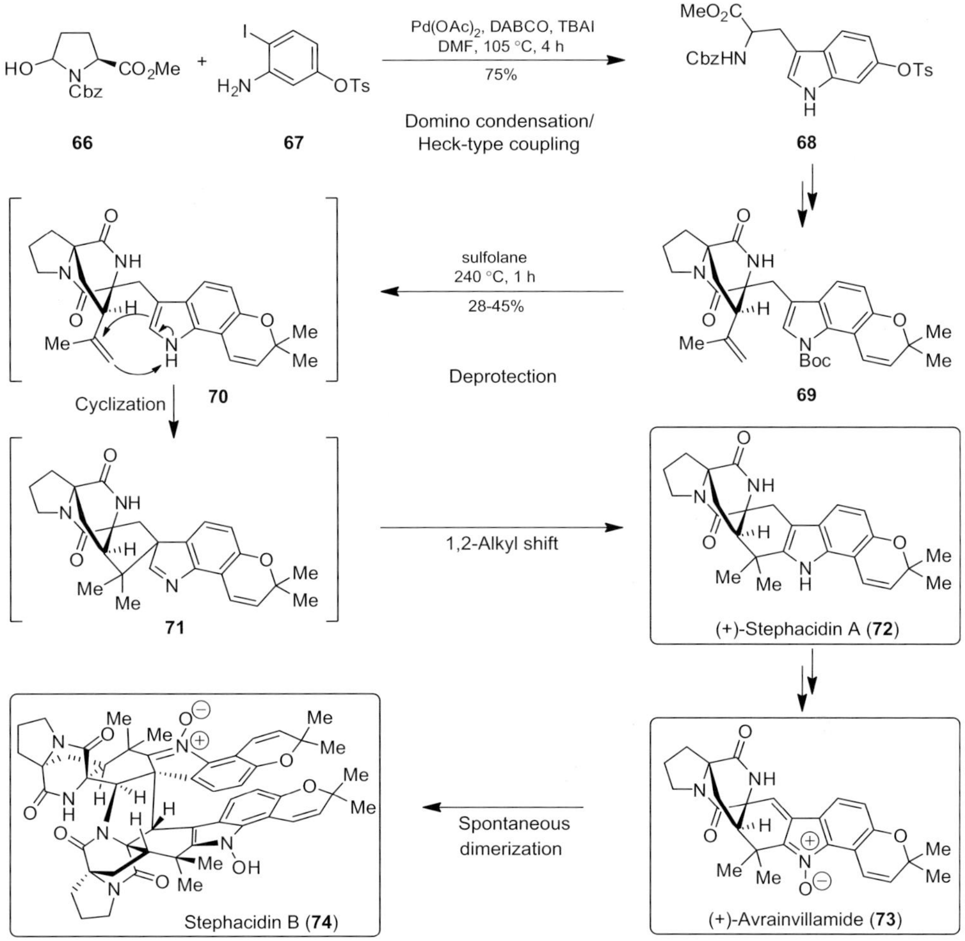

Scheme 14.12 Baran's syntheses of avrainvillamide (**73**) and the stephacidins (**72** and **74**) via domino deprotection/cyclization/rearrangement.

core and an anionic resorcinol unit, was constructed in 18 steps starting from *S*-citronellol (**85**). A series of Wittig transformations gave rise to key intermediate **86**, which was condensed with diethyl 3,5-dimethoxybenzylphosphonate (**87**) to give a 1 : 1 mixture of intermediate triene **88** and cyclized domino product **89**. This ratio could be improved to 5 : 95 by exposing the mixture to high temperatures, with an isolated yield for the domino product of 86%. The presence of a protected alcohol function at C5 (see **88**) was found to be essential for high endo/exo diastereoselectivity in the Diels–Alder step. Enantiopure cycloadduct **89** was further modified to provide (+)-neosymbioimine (**90**) in 10% overall yield.

Kawasaki and coworkers [39] developed an approach to natural products containing 3a-allylpyrrolo[2,3-*b*]indole by domino olefination/isomerization/Claisen (OIC) rearrangement (Scheme 14.15). The domino process exploited 3-indolinones

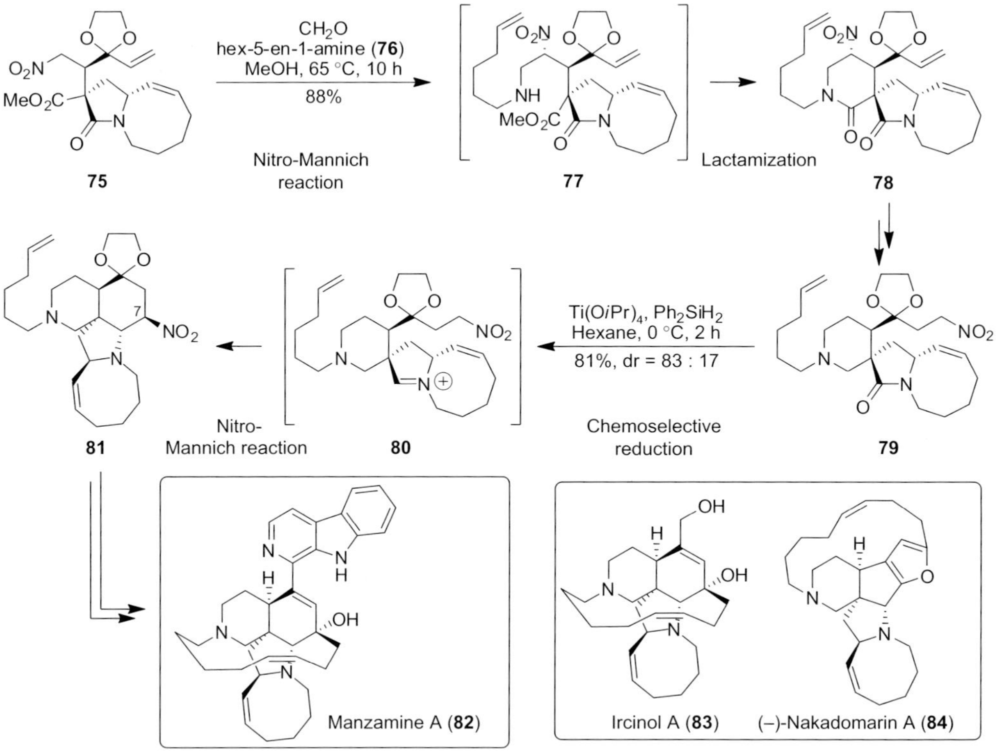

Scheme 14.13 Dixon's synthesis of manzamine A (**82**) via domino nitro-Mannich reaction/lactamization and reductive nitro-Mannich domino reaction.

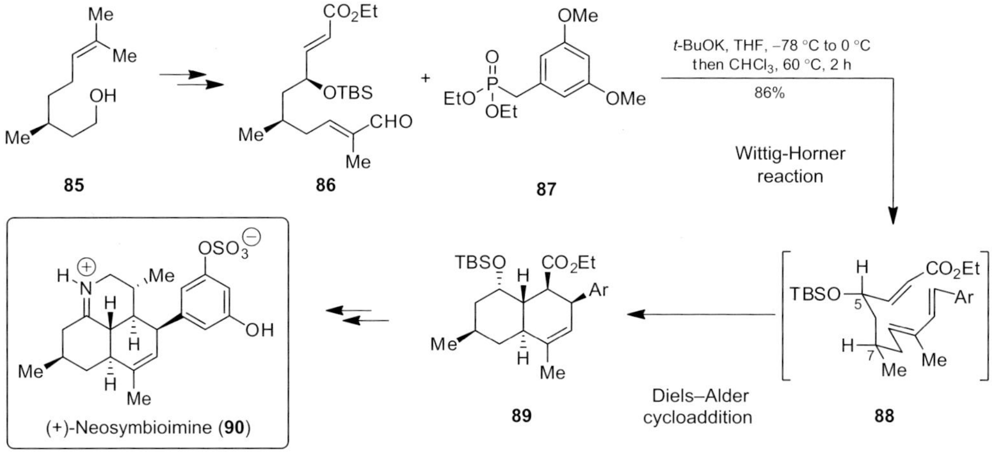

Scheme 14.14 Maier's synthesis of (+)-neosymbioimine (**90**) via domino Wittig–Horner/cycloaddition.

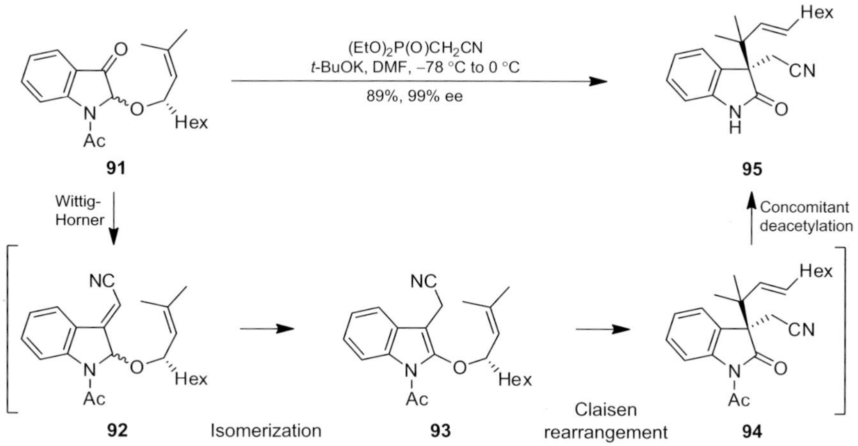

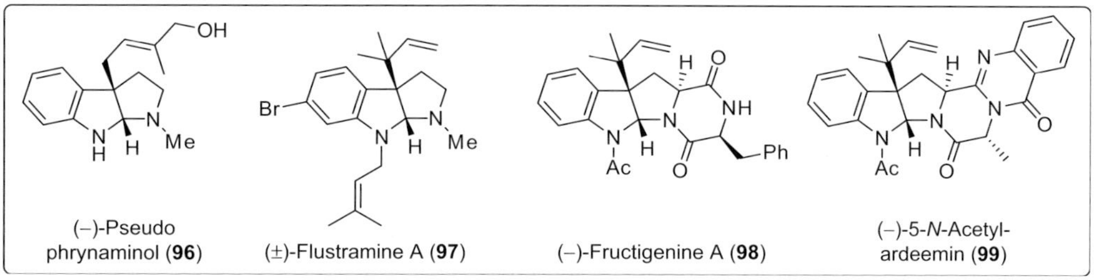

Scheme 14.15 Kawasaki's syntheses of pyrazino-pyrroloindole alkaloids via domino Wittig–Horner/isomerization/Claisen rearrangement.

such as **91**, which were converted with diethyl cyanomethylphosphonate under Wittig–Horner conditions. The resultant nitriles **92** underwent *in situ* isomerization of the newly formed double bond to give allyl vinyl ethers **93**, which are ideally suited substrates for a Claisen rearrangement. Concomitant deacetylation furnished the oxindoles **95**, which were obtained in high yield and enantiomeric excess (ee). The high stereoselectivity of the Claisen rearrangement has been attributed to the chair-like transition state of enantiopure allyl vinyl ether **93**. Domino products with varying substitutions were advanced to a number of naturally occurring alkaloids, including (−)-pseudophrynaminol (**96**) [40] and (−)-flustramine A (**97**) [41]. More recently, Kawasaki applied this strategy in combination with a Ugi reaction to the enantioselective syntheses of the pyrazino-pyrroloindoles (−)-fructigenine A (**98**) and (−)-5-*N*-acetylardeemin (**99**) [42].

Cannabinoids, a class of natural products isolated from the plant *Cannabis sativa*, are tricyclic terpenoid compounds with a long history of use both as medicinal agents and intoxicants [43]. In 2012, Nandaluru and Bodwell [44] reported the concise synthesis of cannabinol (**108**) using the multicomponent domino reaction of salicylaldehyde **100** with dimethyl glutaconate (**102**) and acetone (**101**) (Scheme 14.16). When reacted in the presence of pyrrolidine in refluxing

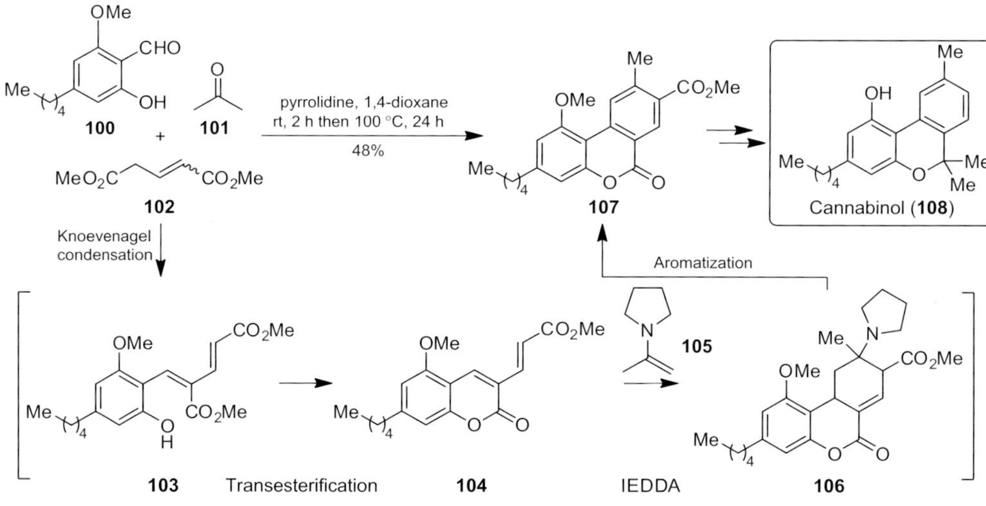

Scheme 14.16 Bodwell's synthesis of cannabinol (**108**) using a multicomponent domino reaction (IEDDA, inverse electron demand Diels–Alder).

1,4-dioxane for 24 h, the secondary amine mediated both the Knoevenagel condensation and transesterification between **100** and **102** and the formation of enamine **105** with acetone (**101**). The *in situ* generated diene **104** and dienophile **105** then underwent an inverse electron demand Diels–Alder reaction, which provided 6*H*-dibenzo[*b,d*]pyranone **107** in 48% yield following 1,2-elimination and transfer hydrogenation. Bodwell applied the multicomponent domino approach to a range of different substrates, furnishing dibenzopyranones of varying substitution in yields that were typically superior to those obtained using stepwise syntheses. To complete the total synthesis of cannabinol (**108**), **107** was subjected to four additional chemical transformations.

Deoxyaminosugars play an important role as glycosidic units in many natural products, such as aminoglycosides, macrolides, anthracyclines, or spinosyns, often being linked to the biological activity of these compounds. In 2009, Tietze *et al.* [45] reported the synthesis of the deoxyaminosugar (+)-D-forosamine (**114**), an integral part of the tetracyclic macrolactone spinosyn A (**115**) (Scheme 14.17), which is commercially used in a mixture with spinosyn D as a highly potent insecticide. The group employed a novel three-component domino Knoevenagel/hetero-Diels–Alder reaction in the synthesis of dihydropyran **113** from readily available starting materials *en route* to the carbohydrate **114**. Thus, treatment of nitroacetone (**109**) and formaldehyde (**110**) with an excess of ethyl vinyl ether (**111**) in dichloromethane at 80 °C gave the targeted dihydropyran in 37% yield. Initially, nitrobutenone **112** was formed, which underwent an inverse electron demand Diels–Alder reaction with ethyl vinyl ether (**111**). Further transformations, including an anti-selective hydrogenation of the C–C double bond of **113**, yielded the targeted carbohydrate (+)-D-forosamine

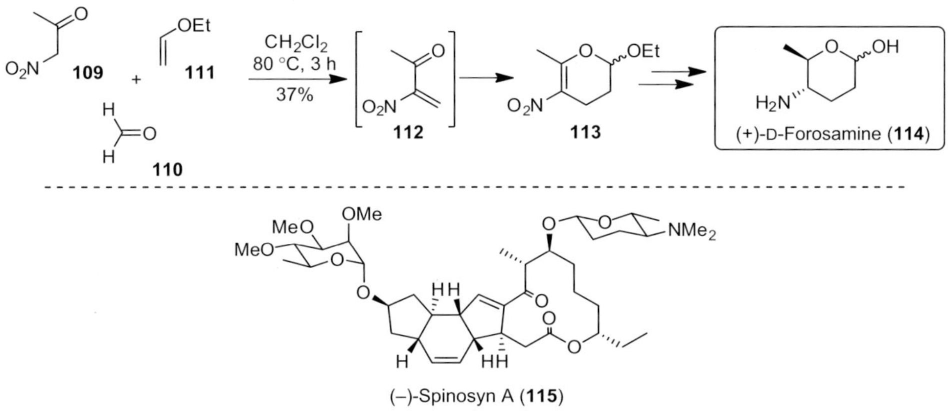

Scheme 14.17 Tietze's synthesis of deoxyaminosugar (+)-D-forosamine (**114**) using a domino Knoevenagel/hetero-Diels–Alder reaction.

(**114**). In related studies, the domino process was applied to the synthesis of 2-acetoxyforosamine and 2-acetoxyossamine by replacing ethyl vinyl ether with the 2-acetoxy derivative [46].

In 2011, Pandey and Kumara [47] reported an iminium-triggered domino reaction for the synthesis of the *Aspidosperma* alkaloid (+)-vincadifformine (**120**) (Scheme 14.18). This natural product exhibits cytotoxic activity against human cancer cell lines [48], and is a valuable precursor for vasodilators such as vincamine and vinpocetine [49]. Key step of Pandey's total synthesis was the reaction of 3,3-substituted tetrahydropyridine **116** with indole **117** in the presence of potassium iodide at elevated temperatures. Initially, the iminium ion **118** was formed by concomitant substitution of the chloride through the pyridine nitrogen and the

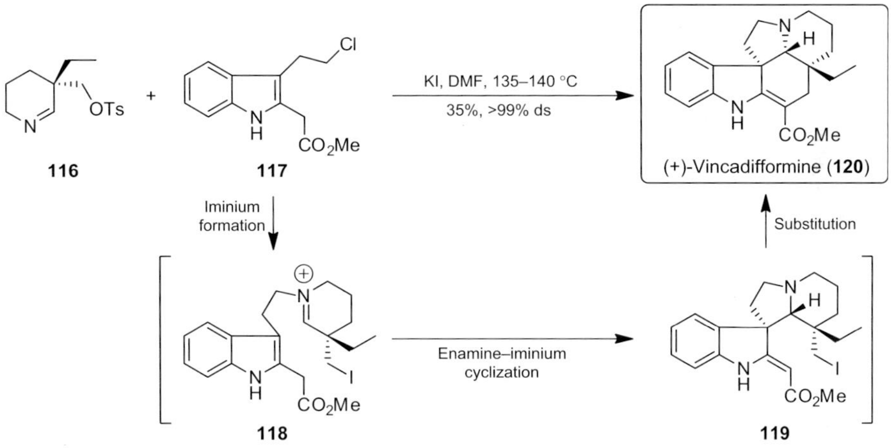

Scheme 14.18 Pandey's synthesis of (+)-vincadifformine (**120**) using an iminium-triggered domino reaction.

tosylate through the external iodide. Enamine–iminium cyclization followed by deprotonation and isomerization then led to enamine **119**, which was detectable as a mixture of diastereomers when the reaction was performed at a lower temperature (90 °C *vs.* 140 °C). Nucleophilic cyclization completed the domino process, which provided the complete pentacyclic skeleton of alkaloid **120** in 35% yield and high diastereoselectivity (>99% ds).

Another domino approach to *Aspidosperma* alkaloids was developed by Coldham *et al.* [50] (Scheme 14.19). In his domino cyclization/[3 + 2]-cycloaddition strategy, tricyclic amines were furnished in high yields and with high levels of regio- and stereocontrol from acyclic precursors. Thus, conversion of aldehyde **121** with glycine in the presence of catalytic amounts of camphorsulfonic acid gave amine **124** in 79% yield as a single diastereomer. Initial enamine formation induced *in situ* *N*-alkylation with concomitant decarboxylation, leading to the ylide **122**. The dipolar substrate then underwent a cycloaddition reaction with the tethered dipolarophile to yield the tricyclic target molecule **124**. The stereochemical outcome of the cycloaddition was proposed to result from an S-shaped transition state, in which the comparably long chain of the tethered enone allowed the preferential formation of the trans product. On heating with aqueous acid, amine **124** was deprotected and isomerized to provide ketone **125**, which was further modified to give the alkaloids (±)-aspidospermidine (**128**), (±)-aspidospermine (**127**), and (±)-quebrachamine (**126**). Recently, Boger *et al.* reported a domino hetero-Diels–Alder/cycloaddition approach to *Aspidosperma* alkaloids (−)-aspidospermine and (+)-spegazzinine [51], which will be elaborated in the appropriate subsection.

In 2012, Tietze *et al.* [52] reported the first total synthesis of (+)-(*R*)-pinnatolide (**134**), an isoprenoide isolated from the aerial parts of *Athanasia* species such as *A. pinnata* and *A. crithmifolia* [53] (Scheme 14.20). In an enantioselective approach, the group employed the facial-selective allylation (a multicomponent

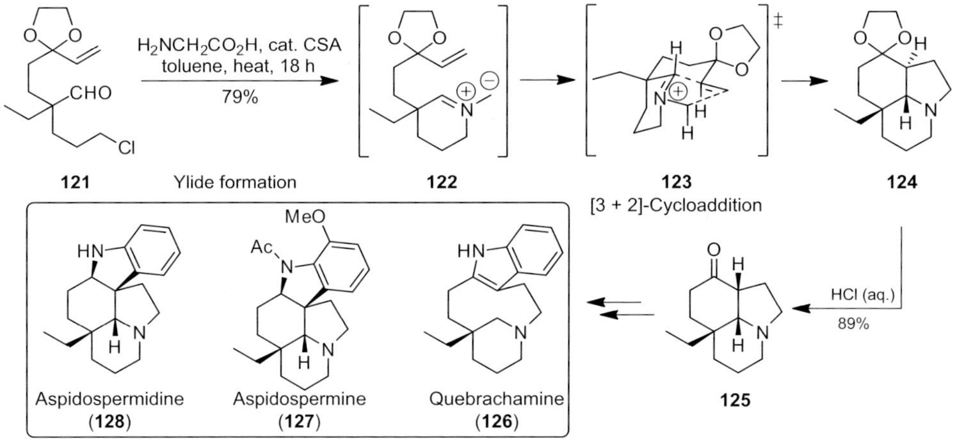

Scheme 14.19 Coldham's synthesis of *Aspidosperma* alkaloids via domino cyclization/[3 + 2]-cycloaddition (CSA, camphorsulfonic acid).

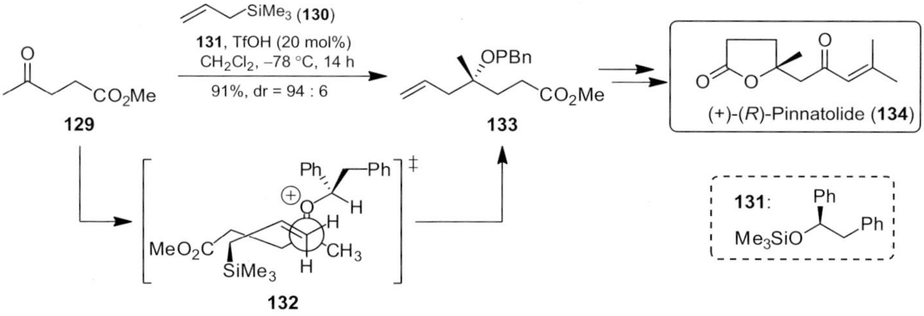

Scheme 14.20 Tietze's total synthesis of (+)-(*R*)-pinnatolide (**134**) via multicomponent domino allylation reaction.

reaction developed and thoroughly investigated within the Tietze group [54]) of a methyl ketone to furnish the stereocenter of the natural product. Thus, treatment of methyl levulinate (**129**) with allyltrimethylsilane (**130**) in the presence of the TMS ether of (*R*)-phenylbenzylcarbinol (**131**) and catalytic amounts of triflic acid in dichloromethane at $-78\,^{\circ}$C afforded the homoallylic ether **133** in 91% yield with a diastereoselectivity of 94 : 6. Carbinol **131** served as both chiral inductor and protecting group. The observed stereoselectivity was proposed to arise from the higher steric demand of the phenyl over the benzyl group, resulting in a *si*-face attack of the allylsilane on the intermediary formed oxeniumion (see the most favored synclinal transition state **132**). Homoallylic ether **133** was further elaborated to yield enantiopure (+)-(*R*)-pinnatolide (**134**) in five additional steps.

In 2001, Kobayashi and coworkers [55] reported the isolation of a novel class of *Daphniphyllum* alkaloids, daphnicyclidins A–H. In spite of their unique structural features, including a fused penta- or hexacyclic skeleton with three contiguous stereogenic centers, as well as their toxicity against both murine lymphoma L1210 and human epidermoid carcinoma KB cells, the first total synthesis of the daphnicyclidins is yet to be achieved [56]. In 2009, Iwabuchi *et al.* [57] reported an approach to the enantiocontrolled construction of the BCD ring system of daphnicyclidin A (**139**) using an acid-triggered domino sequence (Scheme 14.21). In this study, advanced intermediate **135** was converted with acetyl chloride in refluxing isopropyl alcohol. Under the reaction conditions, **135** first cyclized to acyliminium ion **136**, which underwent further cyclization to the Mannich product **137**/**138**. The tricyclic product was formed as a mixture of diastereomers (1 : 2), with the desired isomer **138** being disfavored due to the steric bulk of the pivaloyl protecting group. The domino cyclization of the unprotected alcohol with subsequent reprotection did consequently improve the ratio of desired **138** to undesired **137** to approximately 1 : 1. While conversion of **138** to the natural product is yet to be achieved, Iwabuchi's approach to the tricyclic core of daphnicyclidin A establishes all the stereogenic centers of **139**, thereby providing a useful entry to this class of alkaloids.

Kobayashi *et al.* [58] have applied a Ugi reaction to the formal synthesis of the proteasome inhibitor omuralide (**147**), which is derived from the natural product

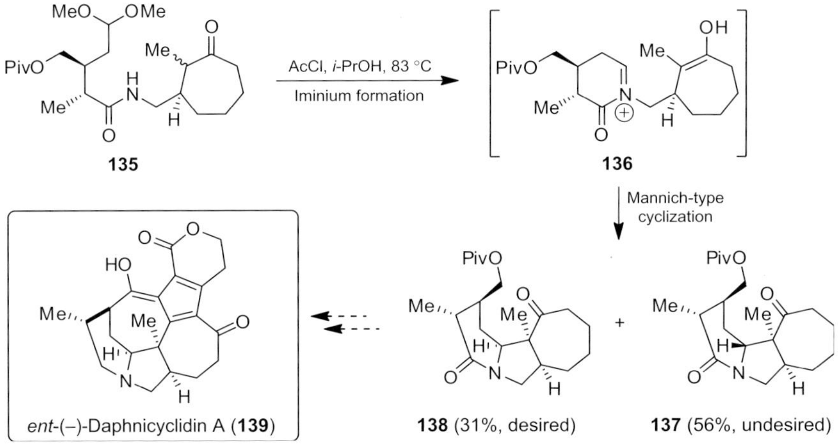

Scheme 14.21 Iwabuchi's synthesis of the BCD ring system of *ent*-(−)-daphnicyclidin A (**139**) via domino acyliminium cyclization/Mannich-type reaction.

lactacystin (Scheme 14.22). The novelty of his approach was the use of indole-isocyanide **141** as a convertible isocyanide in the multicomponent reaction, allowing the facile hydrolysis of the Ugi product to give the corresponding pyroglutamic acid. Thus, conversion of the ketoacid **140** with 4-methoxybenzylamine and isocyanide **141** in trifluoroethanol at ambient temperature furnished the expected Ugi product **144** in 78% yield as a single diastereomer. The stereoselectivity was proposed to arise from a preferred axial approach of the isocyanide nucleophile to intermediate imine **142**. The anilide **144** was amenable to cleavage under mild conditions, affording the methyl ester of pyroglutamic acid **145** via the corresponding *N*-acylindole. Further elaboration of **145** led to advanced intermediate **146**, at which

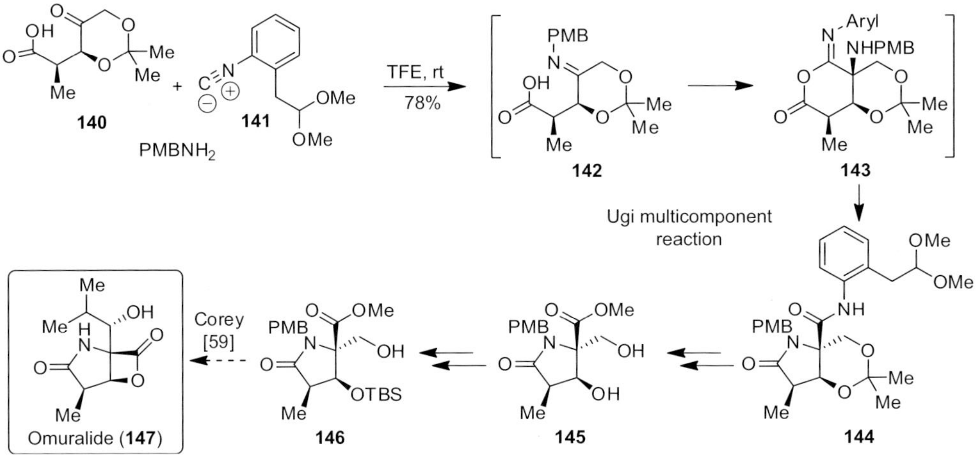

Scheme 14.22 Kobayashi's formal synthesis of omuralide (**147**) using a Ugi multicomponent reaction (TFE, trifluoroethanol).

point Kobayashi's approach to omuralide (**147**) intercepted the formerly reported total synthesis by Corey *et al.* [59].

A novel route toward guaianolide sesquiterpene lactones was reported by Rodriguez and Coquerel *et al.* [60] (Scheme 14.23). In 2008, the group developed a domino Michael/aldol/retro-Dieckmann (MARDi) approach to seven-membered rings [61], which was applied to access the bicyclo[5.3.0]decane scaffold common to guaianolides. Conversion of β-ketoester **148** with α,β-unsaturated aldehyde **149** under basic conditions (1,8-diazabicyclo[5.4.0]undec-7-ene (DBU) in methanol) initiated formation of Michael adduct **150**, which underwent subsequent regioselective aldol cyclization to β-hydroxyketone **151**. Retro-Dieckmann ring opening finally furnished domino product **152** in 49% yield as a mixture of at least six detectable diastereomers (dr = 20 : 7 : 3 : 3 : 3 : 1). Treatment of **152** with an excess of sodium borohydride resulted in the chemoselective reduction of the two major diastereomers only, thereby enabling ready separation from all minor isomers. The substrate-specific reduction was rationalized by an activating intramolecular hydrogen bond present only in the two reacting diastereomers. The diastereomeric diols **153** (dr = 2.9 : 1) were further reacted to provide lactone **154**, which closely resembles the guaianolide family of natural products.

The risk of a pandemic caused by influenza viruses such as the avian H5N1 virus has recently raised the profile of the anti-influenza neuraminidase inhibitor (−)-oseltamivir (**163**). Several novel synthetic methods to access the natural product have been proposed [62]. An outstanding approach is the enantioselective total synthesis reported by Hayashi and coworkers [63], in which the natural product was obtained after two one-pot sequences, featuring a domino reaction, in 60% overall yield (Scheme 14.24). Utilizing α-alkoxyaldehyde **155** and nitroalkene **156**, the first sequence commenced with an organocatalytic asymmetric Michael

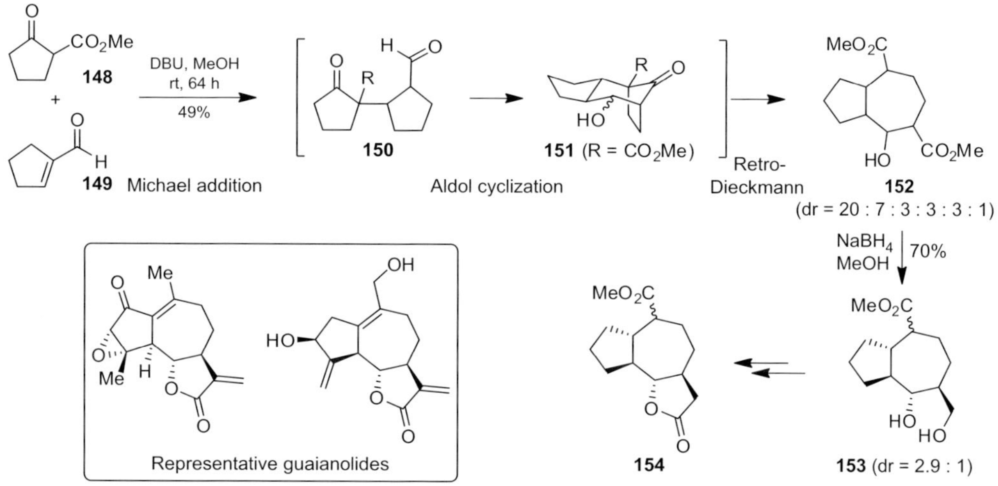

Scheme 14.23 Rodriguez and Coquerel's approach to guaianolide sesquiterpene lactones via MARDi reaction.

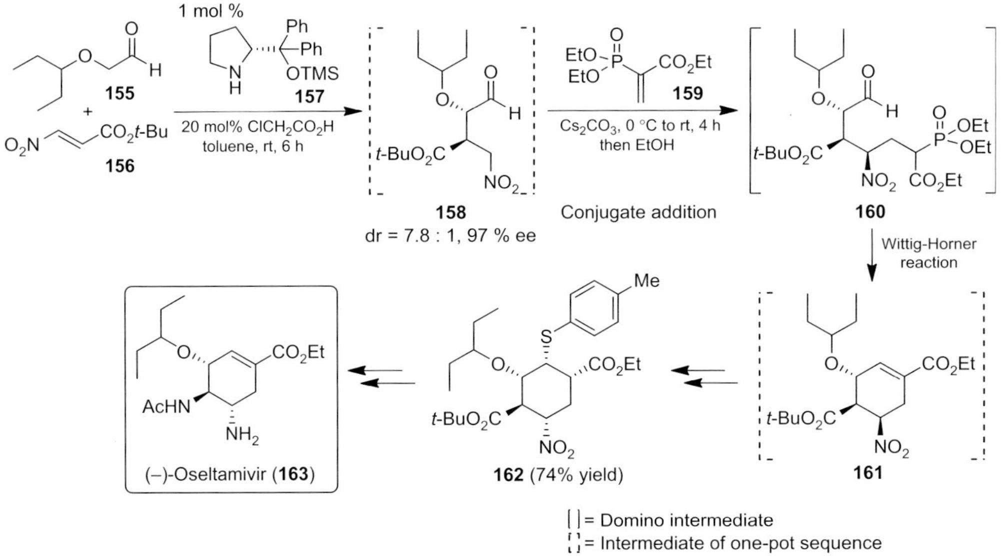

Scheme 14.24 Hayashi's synthesis of (−)-oseltamivir (**163**) using two one-pot sequences.

reaction to yield the adduct **158**, which subsequently underwent domino Michael addition/Wittig–Horner reaction upon exposure to the vinylphosphonate **159**. The resultant cyclohexene **161** was further manipulated to highly substituted cyclohexane **162**, at which stage the first purification of the total synthesis was performed by column chromatography, providing **162** in 74% yield. The second one-pot sequence, converting **162** to (−)-oseltamivir (**163**), was comprised of six separate stages, and proceeded in an overall yield of 81%.

A modification of Hayashi's (−)-oseltamivir synthesis was recently reported by Ma *et al.* [64]. The group was able to replace the Michael acceptor (*E*)-*tert*-butyl 3-nitroacrylate (**156**) with acetyl-protected 2-amino-1-nitroethene, thereby eliminating the three-step conversion of the ester to the acetylamino group required thus far.

In 2013, Bradshaw and Bonjoch *et al.* [65] reported the total synthesis of the *Lycopodium* alkaloid lycoposerramine Z (**170**) by means of a domino Robinson annulation/aza-Michael addition (Scheme 14.25). The phlegmarine alkaloid contains an unusual nitrone moiety which has been postulated to act as a radical trap, thus opening potential applications in the prevention of neurodegenerative diseases by disrupting harmful radical cascades [66]. A key feature of Bradshaw and Bonjoch's elegant approach was the rapid enantio- and diastereoselective assembly of the azabicyclic core of **170** from β-ketoester **164** and crotonaldehyde (**165**). Initially, an organocatalyzed Michael addition using the Jørgensen–Hayashi catalyst (**166**) was conducted, furnishing δ-ketoaldehyde **167**. Exposure to lithium hydroxide and water in isopropyl alcohol then promoted the aldol condensation of **167** to give cyclohexenone **168**, immediately followed by aza-Michael cyclization to *cis*-decahydroquinoline **169**, which was isolated in 72% yield and 85% ee.

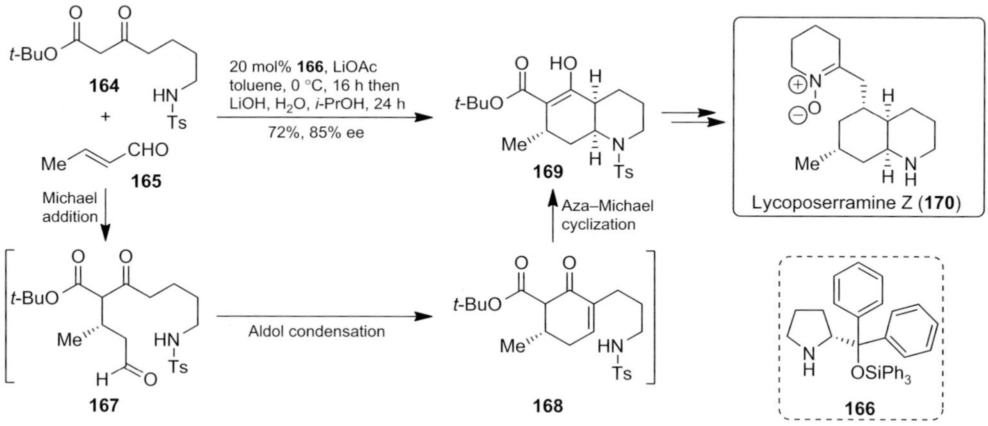

Scheme 14.25 Bradshaw and Bonjoch's synthesis of lycoposerramine Z (**170**) via domino Robinson annulation/aza-Michael addition.

A azabicycle **169** was further advanced to provide lycoposerramine Z (**170**) in only 10 steps with an overall yield of 20%.

The natural product α-tocopherol (**177**) is a prominent member of the vitamin E family, and is well known for its radical-chain-breaking antioxidant characteristics [67]. In 2008, the group of Woggon [68] reported one of the shortest syntheses of the chromanol to date using a domino aldol/oxa-Michael reaction (Scheme 14.26). In this approach, phytenal (**172**), itself obtained by oxidation of naturally occurring phytol, was reacted with *ortho*-hydroxyaldehyde **171** using 30 mol% of a diphenylprolinol-derived organocatalyst (**173**). Initially, enamine **174** was produced, which underwent stereoselective aldol reaction to give iminium intermediate **175**. Subsequent formation of the syn-substituted six-membered lactol **176** via oxa-Michael ring closure concluded the domino sequence, which proceeded in a yield of 58% and a diastereomeric excess of 97%. Lactol **176** was further advanced to provide α-tocopherol (**177**) in an overall yield of 29%. Woggon's approach constituted the first application of organocatalysis to the construction of chromanols. An alternate domino strategy toward **177**, which was reported by Tietze *et al.* in 2006 [69], will be elaborated in the context of transition-metal-catalyzed domino reactions.

Tetrahydroxanthones are a fast growing class of mycotoxins that possess interesting biological activity ranging from antibacterial and anti-HIV (human immunodeficiency virus) to cytostatic behavior [70]. While a number of approaches to tetrahydroxanthones have been reported [71], the first enantioselective total synthesis of diversonol (**185**), a prominent member of this natural product family, was only recently disclosed by Bräse and coworkers [72] using a domino vinylogous aldol reaction/oxa-Michael addition sequence (Scheme 14.27). In the event, salicylaldehyde **179** and prenal (**178**), when reacted in the presence of Jørgensen catalyst (**180**) and benzoic acid, underwent an asymmetric vinylogous aldol reaction to yield benzylic alcohol **182**. Subsequent ring closure through oxa-Michael addition

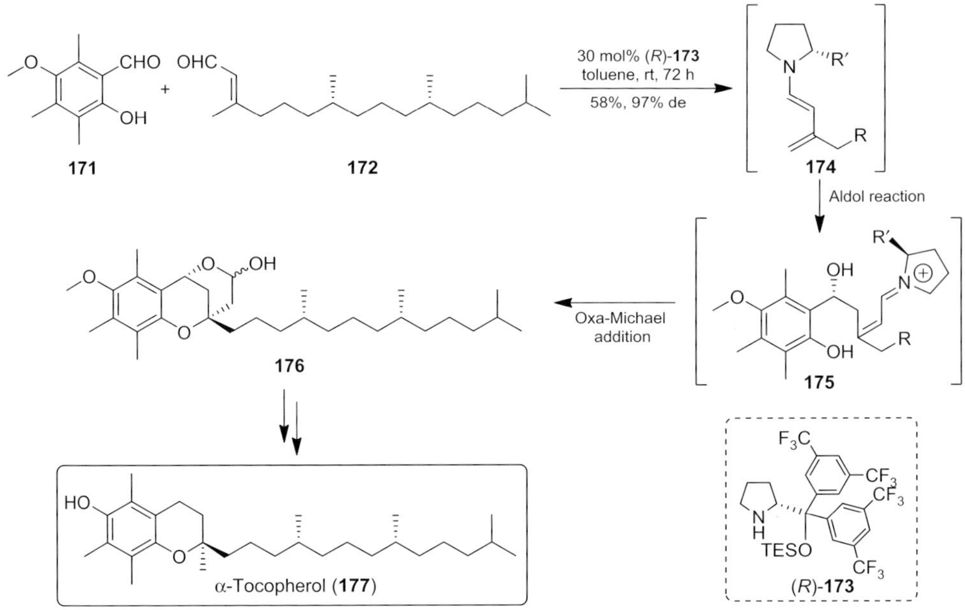

Scheme 14.26 Woggon's synthesis of α-tocopherol (**177**) via domino aldol/oxa-Michael reaction.

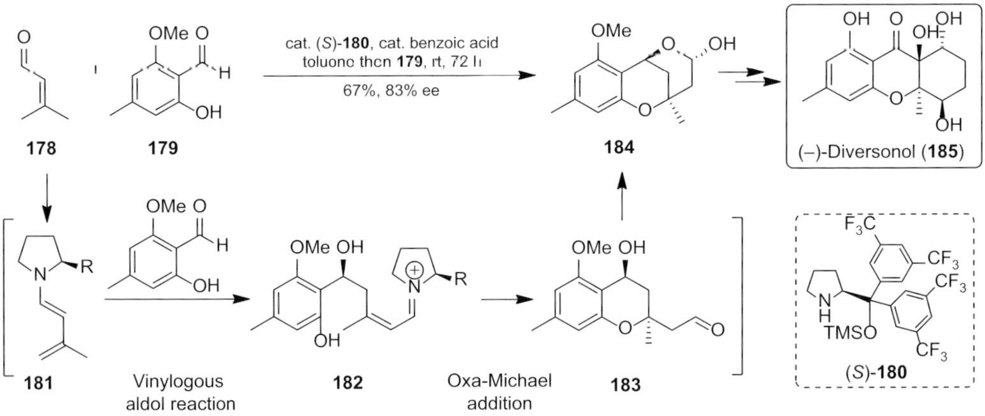

Scheme 14.27 Bräse's total synthesis of (−)-diversonol (**185**) via domino vinylogous aldol/oxa-Michael reaction.

then furnished chromanol **183**, which was isolated in its lactol form (**184**) in 67% yield and 83% ee. Further modification of lactol **184** yielded the (−)-enantiomer of the natural product in a total of 11 steps. A different but equally efficient domino approach to diversonol was recently reported by Tietze *et al.* [73] and will be elaborated in the context of transition-metal-catalyzed domino reactions.

The first total synthesis of the polycyclic natural products (±)-incarviditone (**188**) and (±)-incarvilleatone (**192**) has recently been accomplished by Lawrence *et al.* [74] employing a domino sequence of nucleophilic addition reactions (Scheme 14.28). The two compounds were isolated in 2009 by Zhang *et al.* [75] from *Incarvillea delavayi*, commonly known as the *Chinese trumpet flower*. In a biomimetic approach to (±)-incarviditone (**188**), Lawrence utilized the coisolated monomer (±)-rengyolone (**186**), which was subjected to catalytic amounts of potassium carbonate in dichloroethane. An oxa-Michael reaction between two stereochemically identical monomers gave dimer **187**, which then underwent Michael addition to yield the natural product **188** in 19% yield. Simultaneously, an oxa-Michael dimerization between two inverse enantiomers of rengyolone gave **189**. Subsequent Michael cyclization then afforded pentacyclic **190**, a diastereomer to incarviditone. In this case, the dimer was able to adopt the geometry shown in **191**, thereby enabling the aldol conversion of **190** to (±)-incarvilleatone (**192**), which was isolated in 23% alongside **188**.

In 2012, Greaney and coworkers [76] reported the formal synthesis of the sesquiterpene merrilactone A (**197**) via domino conjugate addition/aldol cyclization (Scheme 14.29). The neurotrophically active natural product exhibits an intriguing skeleton featuring five contiguous quaternary carbon centers, which attracted numerous previous efforts toward its total synthesis [77, 78]. The appeal of Greaney's domino approach is the rapid assembly of the highly congested central [3.3.0]bicyclooctane scaffold. Exposure of enone **193** to diethylaluminum cyanide and titanium tetrachloride in toluene at low temperatures resulted in the conjugate addition of the cyanide nucleophile to afford intermediate **194**. Aldol cyclization then furnished adduct **195** in 70% yield and a dr of 7 : 1, with the major isomer

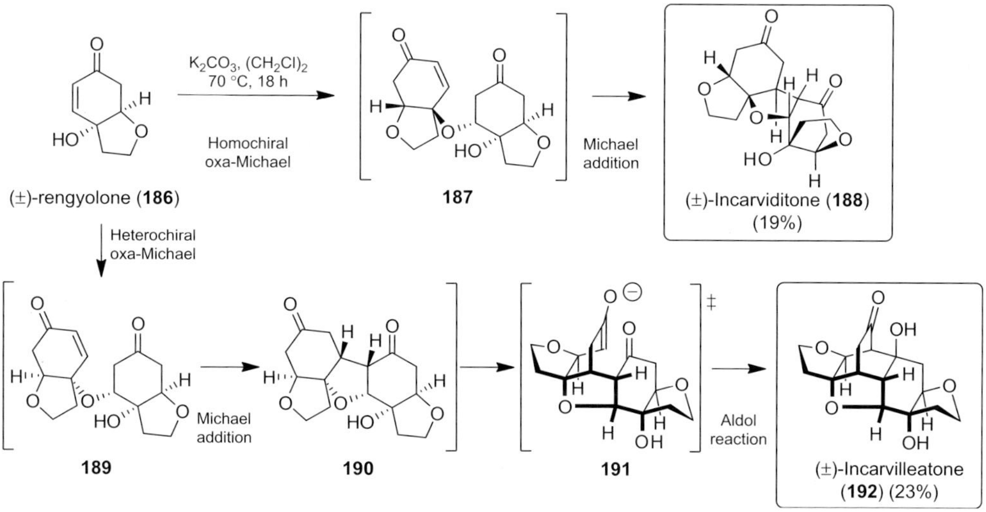

Scheme 14.28 Lawrence's syntheses of incarviditone (**188**) and incarvilleatone (**192**) using a domino sequence of nucleophilic addition reactions.

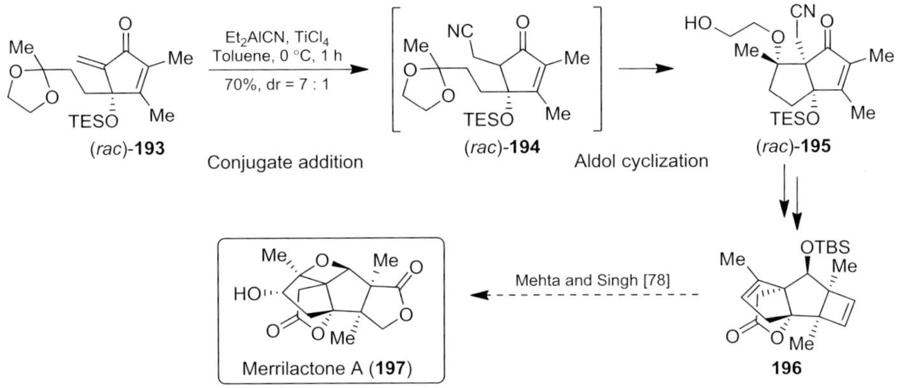

Scheme 14.29 Greaney's formal synthesis of merrilactone A (**197**) via domino conjugate addition/aldol cyclization.

exhibiting a cis relationship between the hydroxyethyl ether and the cyanomethyl group. Adduct **195** was taken on through a number of transformations to provide advanced intermediate **196**, at which point Greaney's synthesis intercepted the approach reported by Mehta and Singh [78].

In 2012, the longest domino sequence to date was reported by the groups of Waldmann and Kumar [79, 80] (Scheme 14.30). The researchers enabled the

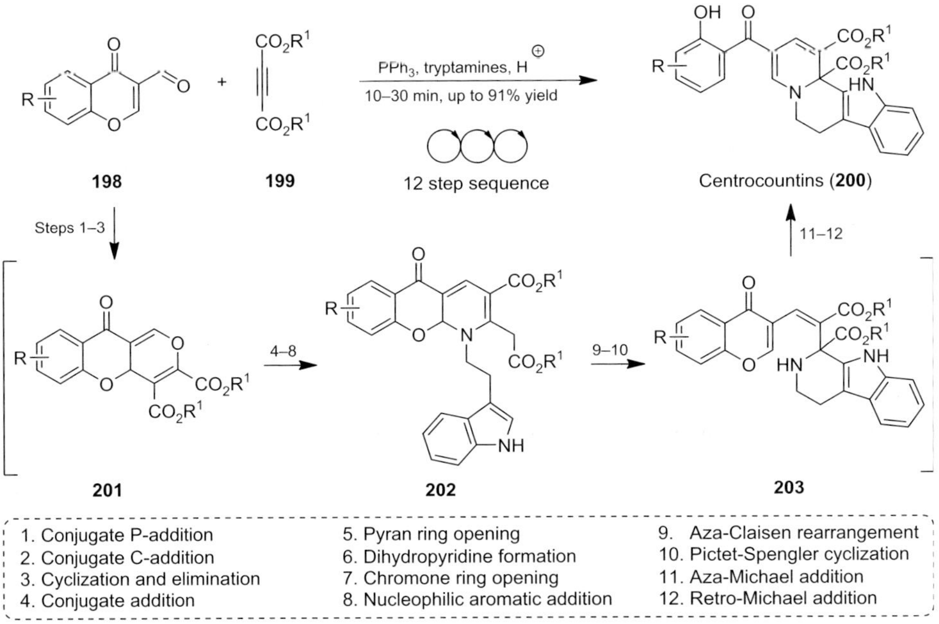

Scheme 14.30 Waldmann and Kumar's synthetic approach to centrocountins (**200**), showing key intermediates of the proposed 12-step domino sequence.

efficient synthesis of natural-product-inspired indoloquinolizines, termed *centrocountins* (**200**), from readily available starting materials following at least 12 consecutive domino steps. Centrocountins resemble the core tetracyclic scaffold of numerous polycyclic indole alkaloids and have been shown to induce delayed mitosis, chromosomal congressional defects and the formation of multipolar spindles, and target the centrosome-associated proteins nucleophosmin (NPM) and Crm1 [79]. Preparation of members of this compound class was achieved in 10–30 min by reaction between formylchromones **198** and alkynes **199** in the presence of triphenylphosphine, tryptamines and acid, affording centrocountins (**200**) in yields ranging from 20% to 91%. The following proposed mechanism of the reaction has been supported by isolation of key intermediates and various control experiments. The domino sequence commences with the phosphine-catalyzed stepwise [4 + 2]-annulation of formylchromone **198** and alkyne **199** to benzopyrone **201**. Nucleophilic attack of tryptamine on the sterically least hindered electrophilic site of **201**, accompanied by chromone ring opening, generates a phenol, which undergoes Michael addition to reform the chromone ring and open the pyran. Catalytic amounts of acid then promote condensation of the newly furnished enamine with the closeby α-ketoester, resulting in dihydropyridine formation. Acid-promoted chromone ring opening and readdition of the phenolate to the resultant pyridinium salt yielded key intermediate **202**. The tricyclic hemiaminal readily undergoes a sigmatropic aza-Claisen rearrangement to give an iminoester which acts as substrate for a Pictet–Spengler cyclization with the tethered indole ring to afford secondary amine **203**. The final steps of the sequence comprise the conjugate aza-Michael addition of the secondary amine to the chromone and an acid-mediated opening of the pyran ring, culminating in the formation of indoloquinolizines **200**. The exceptional length and efficiency of this synthesis is an elegant proof that domino processes in complex target synthesis are by no means limited to low number sequences, and continued success and advancement in the area can be expected.

14.3 Radical Domino Reactions

In 2012, the group of del Moral and Barrero [81] reported the asymmetric total synthesis of the natural triterpene (+)-*seco*-C-oleanane (**206**) using a stepwise radical cyclization (Scheme 14.31). The key domino process of their approach was mediated by the single electron transfer complex Cp_2TiCl, allowing the regio- and stereocontrolled cyclization of preoleanatetraene oxide (**205**) without the need for prefunctionalization. Thus epoxide **205**, itself prepared from the sesquiterpene farnesol (**204**), afforded the natural product in 44% yield upon treatment with catalytic amounts of Ti(III) generated from Cp_2TiCl_2 in the presence of excess manganese and trimethylsilyl chloride (TMSCl)–collidine, and subsequent TBAF-mediated desilylation. Interestingly, the radical polyene cyclization stopped at the bicyclic level. Prior to experimentation, this truncated cyclization had been predicted

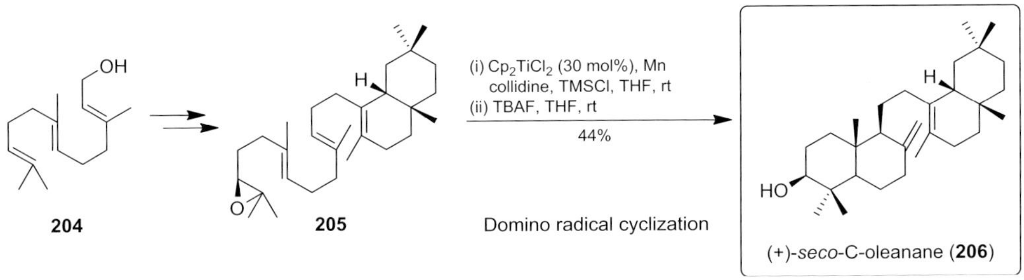

Scheme 14.31 Del Moral and Barrero's total synthesis of (+)-*seco*-C-oleanane (**206**) using a catalytic radical cyclization.

on the basis of the high activation energy calculated for the third ring closure, which was postulated to stem from steric crowding around the tetrasubstituted acceptor double bond. The more commonly employed carbocationic cyclization of polyprenes usually leads to tricyclic structures without the interception of either mono- or bicyclic carbocations. Accordingly, an attempted cationic variant of the cyclization using $SnCl_4$ failed to yield any *seco*-C-oleananes.

The bridged bicyclic peroxide yingzhaosu A (**214**) was isolated from *Artabotrys uncinatus*, extracts of which were used in traditional Chinese medicine as a remedy for malaria [82]. In 2005, Bachi and coworkers [83] reported a free-radical domino approach to access its characteristic 2,3-dioxabicyclo[3.3.1]nonane scaffold, the putative pharmacophore (Scheme 14.32). Treatment of the chiral pool material (*S*)-limonene (**207**) with oxygen, thiophenol, and radical initiator AiBN allowed access to β-sulfenyl endoperoxide **212** in gram-scale quantities. During the construction of the bridged bicyclic endoperoxide, five bonds were formed in a single operation. Initially, reaction of **207** with a thiophenol radical furnished tertiary radical **208**, which was trapped with molecular oxygen to afford peroxy radical **209**. Subsequent 6-exo-intramolecular addition resulted in the formation

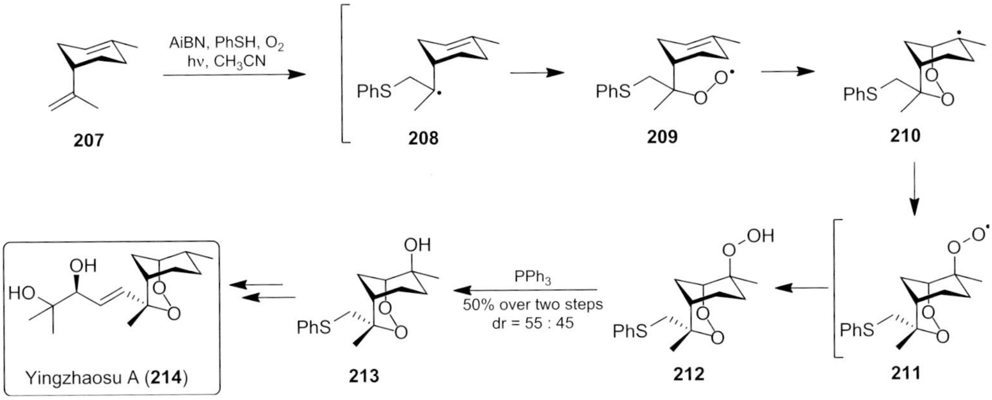

Scheme 14.32 Bachi's total synthesis of yingzhaosu A (**214**) using a free-radical domino reaction.

of tertiary carbon-centered radical **210**, which again was trapped with molecular oxygen. The resultant diperoxy radical **211** was quenched *in situ* with thiophenol, effecting the synthesis of both endoperoxide **212** and a thiophenol radical, thereby enabling a continuous radical cycle [84]. Reduction of the hydroperoxy group of **212** by treatment with triphenylphosphine in the same pot yielded endoperoxide **213** with low diastereoselectivity (dr = 55 : 45) in 50% yield over the two steps. While maintaining the integrity of the peroxide moiety, **213** was converted to the natural product yingzhaosu A (**214**), which was obtained in eight steps and an overall yield of 7.3%.

14.4
Pericyclic Domino Reactions

A prominent example of the *Aspidosperma* alkaloids is (+)-aspidospermine, which was first isolated in the late 1800s from the bark of *Aspidosperma quebracho* [85]. It shows a multitude of biological activities such as diuretic, hypertensive, and adrenergic blocking [86]. Even though the first racemic total synthesis was accomplished in 1963 by Storck and Dolfini [87], it still remains a favorite target for synthetic organic chemists. The group of Boger [51] achieved the total syntheses of the structurally closely related alkaloids (−)-aspidospermine (**127**) and (+)-spegazzinine (**217**) in 2012 (Scheme 14.33).

The domino reaction started with an intramolecular hetero-Diels–Alder reaction of **215** to give **218**, which was followed by elimination of molecular nitrogen to form the 1,3-dipole **219**. The final step in the sequence was an intramolecular [3 + 2]-cycloaddition to provide *rac*-**216** in 71% yield as a single diastereomer. Compound

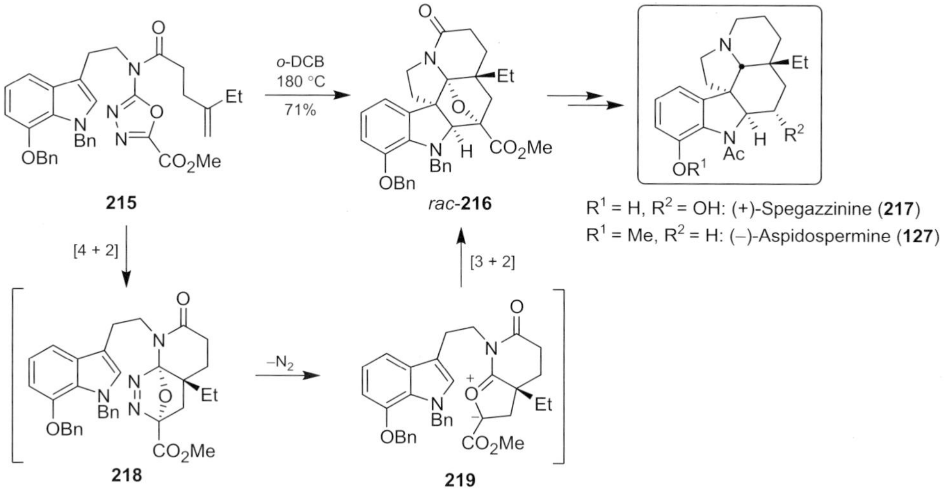

Scheme 14.33 Total syntheses of spegazzinine (**217**) and aspidospermine (**127**) by Boger and coworkers (*o*-DCB, *o*-dichlorobenzene).

216 was further manipulated to give both spegazzinine (**217**) and aspidospermine (**127**) as pure enantiomers using resolution by chiral phase chromatography. In the domino process presented here, four new C–C bonds and five stereocenters were formed; the approach had already been used for the total syntheses of the indole alkaloids vindoline and vindorosine [88]. Another synthesis of aspidospermine (**127**) by Coldham *et al.* [50] using an alternate domino approach is described in Section 14.2.

The natural product (+)-chamaecypanone C (**223**) was first isolated in 2007 by Kuo and coworkers [89] from the heartwood of *Chamaecyparis obtusa* var. *formosana* and shows strong anticancer potency. An enantioselective synthesis of this compound was described by Porco *et al.* in 2009 [90] using a retro-Diels–Alder/Diels–Alder reaction as key step (Scheme 14.34).

The retro-Diels–Alder reaction of dimer **220** led to monomer **221**, which underwent a [4 + 2]-cycloaddition with the cyclopentadienone formed *in situ* by dehydrogenation of **222**. The crude product was subsequently treated with BBr_3 to afford (+)-chamaecypanone C (**223**) in 53% yield over two steps. Starting material **220** was readily accessible in three steps and 99% ee from 2-hydroxy-5-isopropylbenzaldehyde. In advanced studies, a rhodium-catalyzed dehydrogenation has been applied to the synthesis of chamaecypanone C analogs [91].

Owing to their structural complexity, *Strychnos* alkaloids, especially strychnine itself, are a benchmark in synthetic organic chemistry and have been prepared numerous times [92]. Another alkaloid of the *Strychnos* family that enjoys increasing popularity nowadays is (+)-minfiensine. It was isolated from *Strychnos minfiensis* and features a 1,2,3,4-tetrahydro-9a,4a-(iminoethano)-9*H*-carbazole motif, and hence is closely related to echitamine and vincorine [93]. A total synthesis of this beautiful structure was accomplished by MacMillan and coworkers [94]. Their key step was a highly efficient enantioselective domino Diels–Alder/isomerization/amine cyclization reaction (Scheme 14.35). In this process, the chiral organocatalyst **226**·TBA (TBA, tribromoacetic acid) was used, which reacted first with propargyl aldehyde **225** to form iminium ion **229** as intermediate, which underwent an endo-selective Diels–Alder reaction with **224**

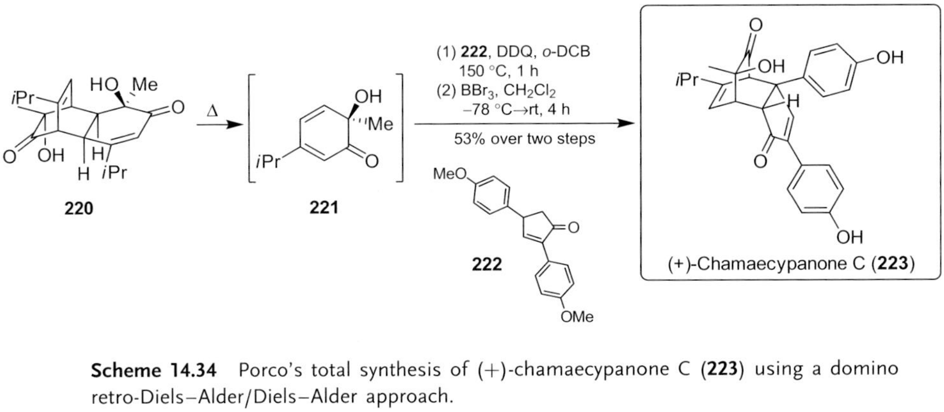

Scheme 14.34 Porco's total synthesis of (+)-chamaecypanone C (**223**) using a domino retro-Diels–Alder/Diels–Alder approach.

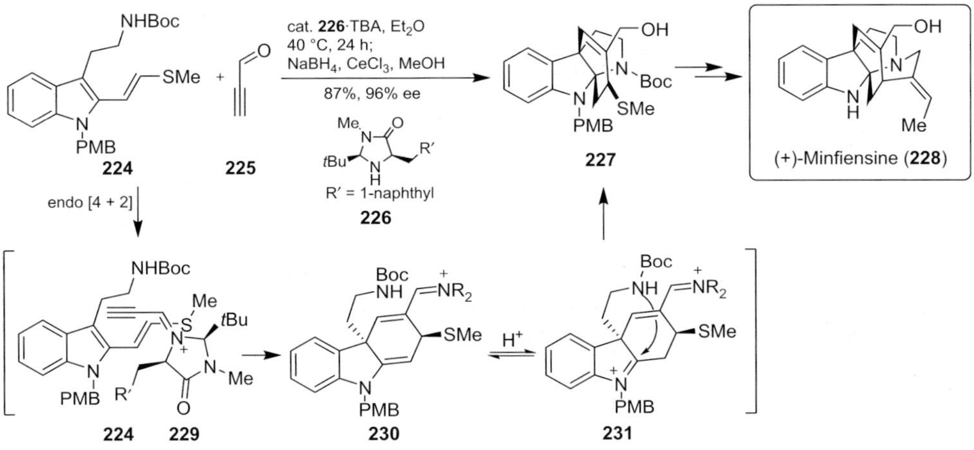

Scheme 14.35 Total synthesis of (+)-minfiensine (**228**) by MacMillan and coworkers ($NR_2 =$ **226**; TBA, tribromoacetic acid).

to give cycloadduct **230**. Enamine **230** then isomerized under acidic conditions to provide iminium ion **231**. The final step in this domino reaction was the intramolecular addition of the Boc-protected amino group to the iminium moiety, affording the tetracyclic **227** in 87% yield and very good ee. Altogether, this total synthesis consists of only nine steps, with an overall yield of 21%.

Minfiensine has also received attention from several other research groups. Because of the complexity of the molecule, it seems especially suitable for the application of domino processes, allowing an efficient and short access to the tetracyclic core. In 2008, Qin and coworkers [95] accomplished the total synthesis of the racemic natural product using a domino cyclopropanation/ring opening/amino cyclization reaction of a diazoketone to assemble the tetracycle. The group of Padwa synthesized the natural product using a domino Buchwald–Hartwig amidation/Diels–Alder approach, which will be discussed in Section 14.5. A domino Heck/isomerization/amine cyclization reaction was applied by Overman *et al.* [96] in the first enantioselective total synthesis of (+)-minfiensine (**228**).

In 2008, Aubé [97] reported the total syntheses of the *Stemona* alkaloids stenine (**240**) and neostenine (**236**) as well as the synthetic analog 13-*epi*-neostenine (**237**) using a domino Diels–Alder/Schmidt reaction as key step (Scheme 14.36). This class of alkaloids has been associated with interesting biological activities such as insecticidal, anthelmintic, antitussive and various neurochemical effects [98]. Mechanistically, the Diels–Alder reaction of diene **232** with cyclohexenone using $BF_3{\cdot}Et_2O$ allowed exclusive formation of the endo product **234** as a mixture of C-5 epimers. Interestingly, when using $SnCl_4$, the exo product **233** (endo/exo 1 : 3) was the major product. The carbonyl moiety was then attacked by the azide, and rearrangement under loss of nitrogen (via intermediate **238**) afforded the lactames **235** and **239**, respectively.

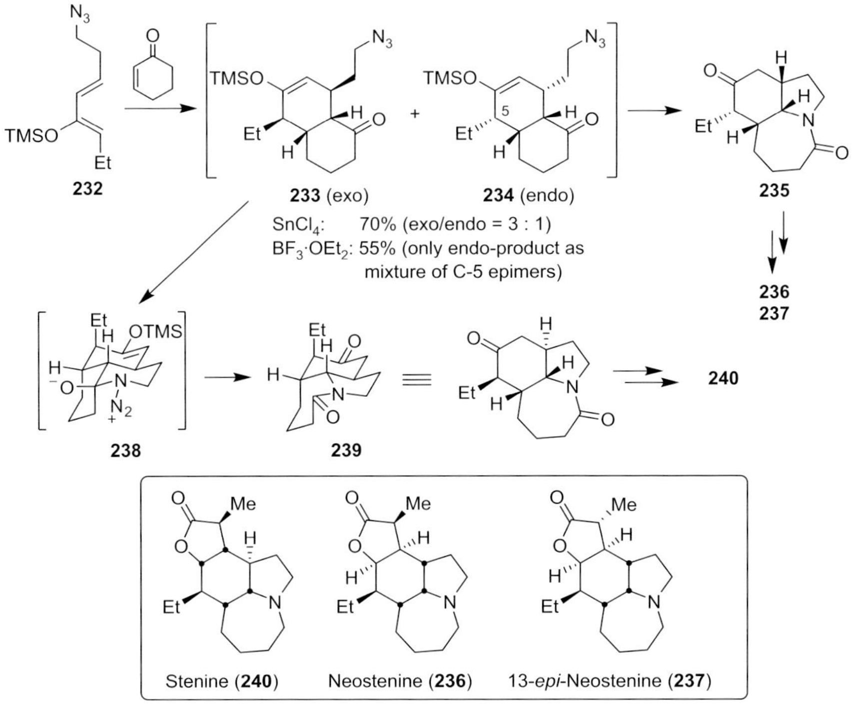

Scheme 14.36 Aubé's total syntheses of the natural products stenine (**240**) and neostenine (**236**) as well as the synthetic analog 13-*epi*-neostenine (**237**).

Further manipulation of **239** led to stenine (**240**), whereas **235** was converted to neostenine (**236**) and the synthetic analog 13-*epi*-neostenine (**237**). During the last 20 years, several syntheses of stenine have been reported, but this is the first total synthesis of neostenine.

14.5
Transition-Metal-Catalyzed Domino Reactions

Transition-metal-catalyzed reactions have become very popular in the total synthesis of natural products since they usually allow a late-stage assembly of functionalized structural fragments and, especially when used in a domino process, a rapid increase of complexity.

A nice example where both aspects are present is the synthesis of the ABCD core structure of active ingredients of the *Schisandra* genus by Parrain and Chouraqui [99] (Scheme 14.37). The initiating reaction of the domino process was a Sonogashira reaction of the racemic alkyne **241** and (*Z*)-3-bromoacrylic acid (**242**), resulting in formation of enyne **245**, which underwent a 5-*exo*-dig cyclization to afford lactone **246**. This was followed by a diastereoselective sigmatropic rearrangement,

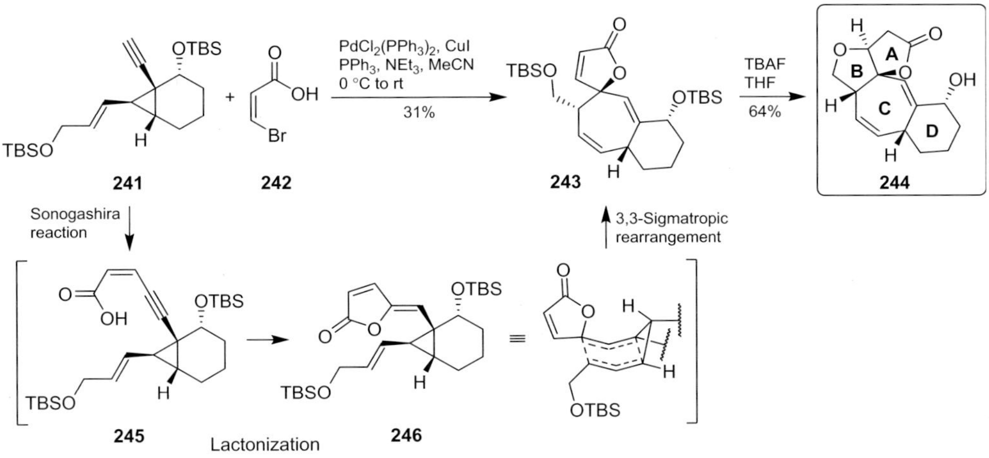

Scheme 14.37 Domino Sonogashira/lactonization/sigmatropic rearrangement sequence by Parrain and Chouraqui for the synthesis of the ABCD skeleton of natural products of the Schisandraceae family.

providing the racemic spirocycle **243** in 31% yield. *tert*-Butyldimethylsilyl (TBS) deprotection resulted in a hetero-Michael addition of the formed primary alcohol to the α,β-unsaturated lactone to give tetracycle **244** in 64%.

Alongside the synthesis of strychnos alkaloid minfiensine (**228**) by MacMillan *et al.* (see Section 14.4), Li and Padwa [100] have prepared this natural product using an alternate domino process (Scheme 14.38).

Key step was a copper-catalyzed domino Buchwald–Hartwig amidation/Diels–Alder reaction of amide **247** with 2-bromofuran **248** to give **249**. In the first

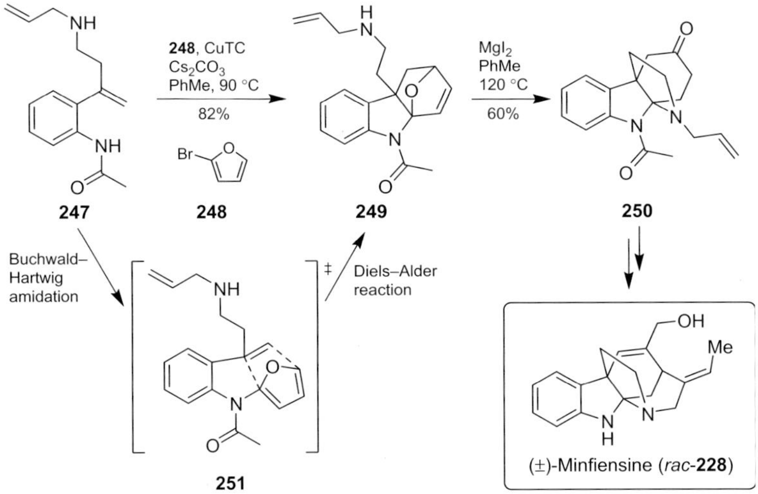

Scheme 14.38 Total synthesis of (±)-minfiensine (*rac*-**228**) by Padwa *et al.* (CuTC, copper(I)-thiophene-2-carboxylate).

step, furan **251** was formed, which underwent an intramolecular Diels–Alder reaction to afford **249** in 82% yield. This was further converted to tetracycle **250** in 60% yield employing a second domino process using magnesium iodide in toluene at 120 °C. Under these conditions, it is likely that a ring opening of the *N*,*O*-acetal occurs to give an iminium ion, which is attacked by the secondary amine. Moreover, the allylic alcohol formed in the ring opening isomerizes to give a carbonyl moiety. Further transformations of **250** finally afforded the racemic natural product minfiensine (*rac*-**228**).

The alkaloid (−)-galanthamine, isolated from Caucasian snow-drop (*Galanthus woronowii*) and other plant sources, is a competitive and reversible acetylcholinesterase inhibitor and an allosteric modulator of the neural nicotine receptors for acetylcholine. Galanthamine hydrobromide (commercially available as Razadyne) finds application in the treatment of symptoms of Alzheimer's disease [101]. Since the isolation of galanthamine (**256**) from natural sources is quite expensive, several total syntheses have been developed to date. The synthetic strategy to galanthamine (**256**) developed by Cho *et al.* [102] includes a domino Stille/Diels–Alder reaction as the key step (Scheme 14.39).

Palladium-catalyzed cross coupling of pyrone **252** with stannane **253** resulted in the intermediate **257**, which underwent a Diels–Alder reaction to give cycloadducts **254** and **255** in a combined yield of 45%, favoring the endo product **255** (endo/exo = 2.5 : 1). In this reaction, the addition of CuI was crucial, as without CuI only traces of the cycloadducts were obtained. Further transformations of **255** resulted in racemic galanthamine (**256**). In 2000, the group of Trost [103] reported an enantioselective synthesis of this alkaloid applying a Pd-catalyzed asymmetric allylic alkylation and an intramolecular Heck reaction to assemble the

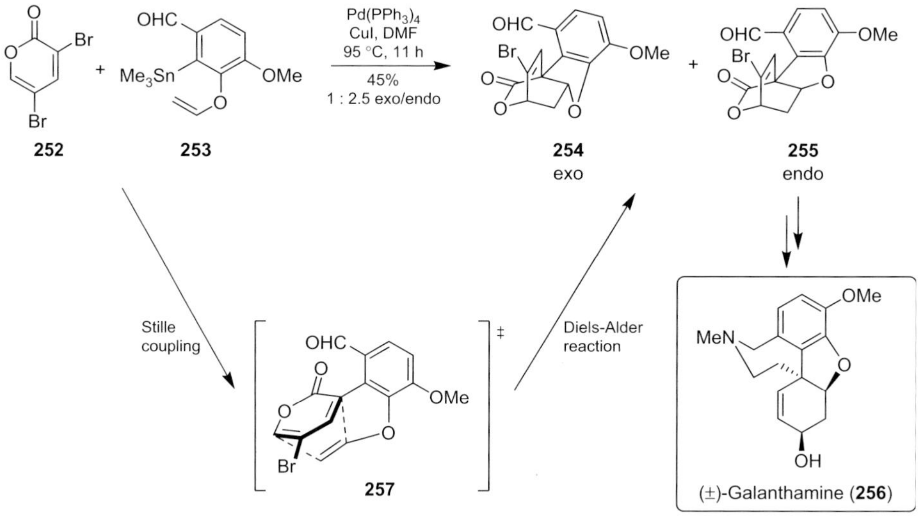

Scheme 14.39 Cho's total synthesis of racemic galanthamine (**256**) by a domino Stille/Diels–Alder reaction.

tricyclic core. Also noteworthy are the syntheses of galanthamine accomplished by Brown *et al.* in 2007 and Xie and Zhou in 2012 [104].

The shimalactones, isolated from the cultured marine fungus *Emericella variecolor*, are neuritogenic natural products which contain a bicyclo[4.2.0]octa-2,4-diene core structure in combination with an oxabicyclo[2.2.1]heptane moiety [105]. The complex structure of the shimalactones represents quite a synthetic challenge, which was overcome in 2008 by Trauner and coworkers [106] in a convergent synthetic approach using a domino Stille reaction/electrocyclization.

In this process, the Pd-catalyzed Stille cross-coupling reaction of enantiopure iodide **259** and stannane **258** afforded polyene **262**, which underwent a conrotatory 8π-electrocyclization resulting in the cyclooctatrienes **263** and **264**. These compounds reacted further in a disrotatory 6π-electrocyclization to a 5 : 1 mixture of the diastereomers shimalactone A (**260**) and B (**261**) in 55% and 11% yield, respectively (Scheme 14.40).

The biomimetic 8π/6π-electrocyclization process was also applied in several other syntheses of natural products containing a bicyclo[4.2.0]octadiene structure. Parker and Lim [107] used a domino Stille/8π/6π-electrocyclization reaction for the total synthesis of the immunosuppressants SNF4435 C (**267**) and D (**268**). Interestingly, the ratio of SNF4435 C and D found in nature (2.3 : 1) is close to that obtained in the synthetic approach (Scheme 14.41). This indicates that an

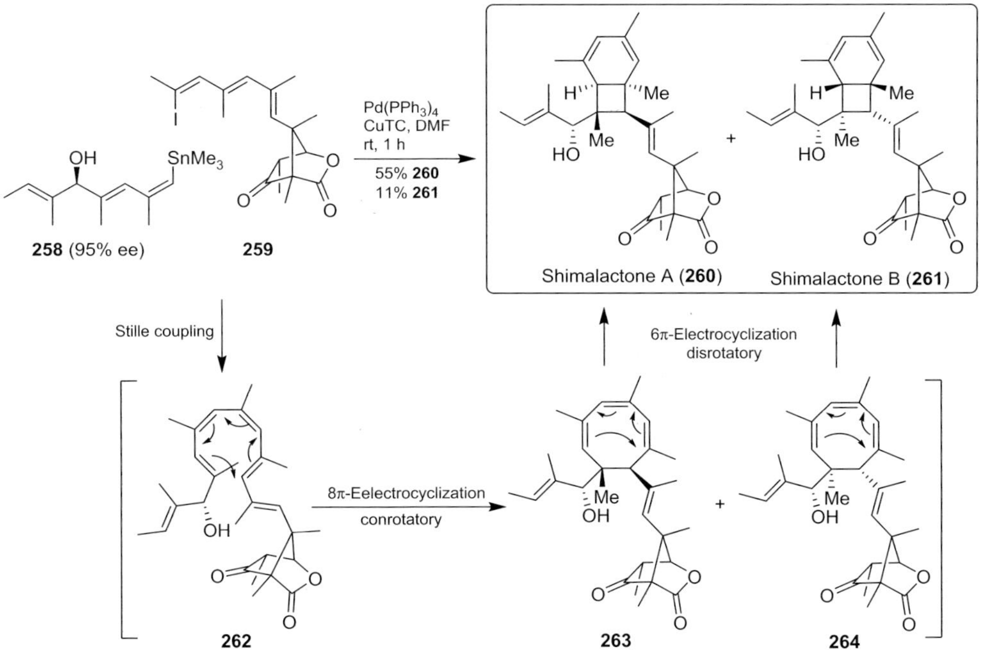

Scheme 14.40 Total synthesis of shimalactones A and B (**260** and **261**) by Trauner and coworkers applying a domino Stille/8π/6π-electrocyclization reaction.

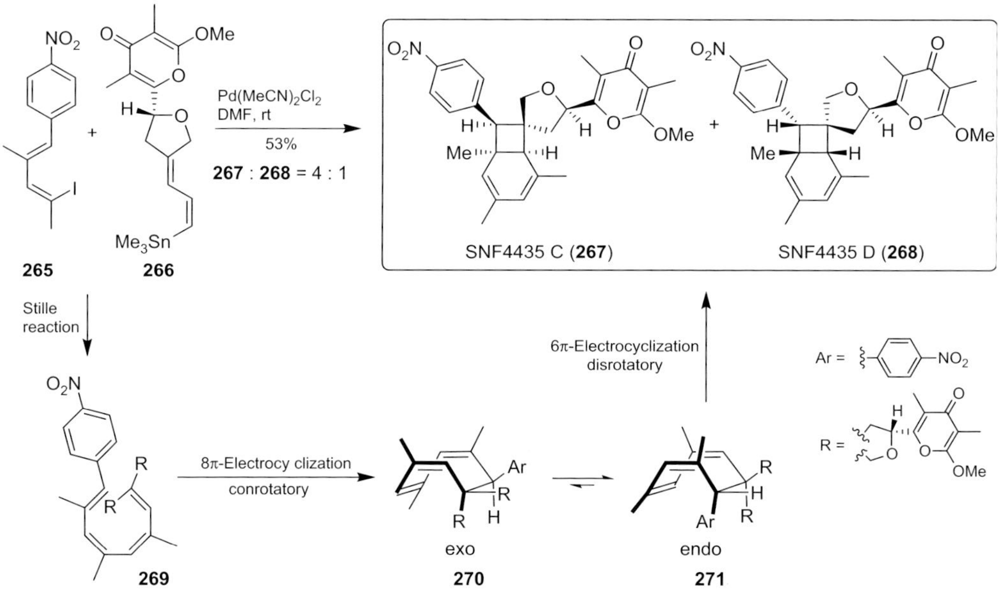

Scheme 14.41 Total synthesis of SNF4435 C and D (**267** and **268**) by Parker *et al.*

8π/6π-electrocyclization process might also be present in the biosynthesis of these compounds.

During the domino process, two new rings and four stereogenic centers were formed with good stereoselectivity, guided by the existing stereogenic center in starting material **258**. Thus, only 2 out of the possible 16 diastereomers were obtained. Moreover, the 6π-electrocyclization seems to be highly diastereoselective since none of the products has the aryl substituent on the exo face of the bicyclo[4.2.0]octadiene; the endo transition state is highly favored. The ratio of the obtained diastereomers therefore reflects the diastereoselectivity of the conrotatory 8π-electrocyclization [108].

Linoxepin (**275**) is a natural product that was first isolated from *Linum perenne* L. (Linaceae) by Schmidt and coworkers in 2007 [109]. It belongs to the aryldihydronaphthalene lignans and contains a novel benzonaphtho[1,8-*bc*]oxepine moiety. Even though no biological studies on linoxepin have been published, the structural resemblance to bioactive lignans such as the anticancer drugs etoposide and teniposide made it an attractive synthetic target. Recently, an efficient synthesis was published by the Tietze group [110] using a Pd-catalyzed domino carbopalladation/Heck reaction.

Treatment of aryl bromide **272** with Pd^0 resulted in an oxidative addition, which was followed by an intramolecular carbopalladation of the triple bond, resulting in with formation of the oxepine ring containing a vinyl-palladium species, i.e. **276** (Scheme 14.42). This functionality probably underwent a Heck-type reaction initiated by insertion into the allyl silane to give intermediate **277**. As shown by Tietze *et al.* [111] in earlier studies, the following β-elimination occurred in a

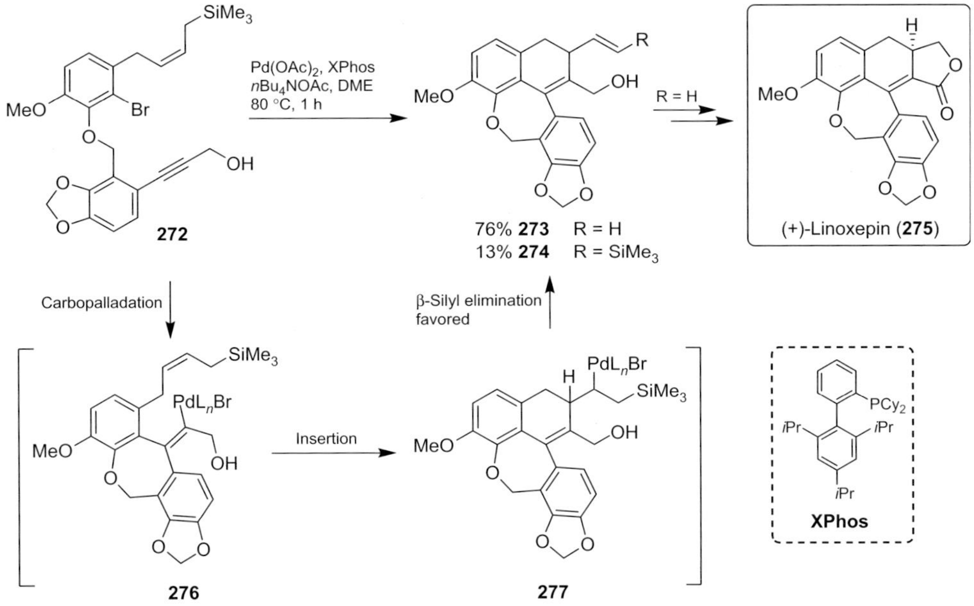

Scheme 14.42 Tietze's total synthesis of linoxepin (**275**) (DME, dimethoxyethane).

highly regioselective manner to give alkene **273** and vinyl silane **274** in 76% and 13% yield, respectively. Further transformation of **273** finally afforded the natural product. This concise total synthesis was performed in only 10 steps starting from commercially available substrates in almost 30% overall yield and without using any protecting groups.

The formation of functionalized seven-membered carbocycles is generally a demanding task. An efficient access to this class of compounds is provided by a domino cyclopropanation/Cope rearrangement [112]. Olson and Davies [113] used this enantioselective rhodium-catalyzed methodology in the formal total synthesis of frondosin B (**282**) to assemble its annulated seven-membered ring. Mechanistically, the initiating step was the formation of a rhodium-carbenoid species from **278** under loss of nitrogen (Scheme 14.43). This carbenoid reacted with diene **279** to give cyclopropane **284**, which underwent a Cope rearrangement followed by tautomerization to afford **280** in 57% yield and 97% ee. Ester **280** was further converted into ketone **281**, which is a key intermediate of Danishefsky's total synthesis of frondosin B (**282**) [114]. The domino key step, a formal [4 + 3]-cycloaddition, has been applied to the syntheses of several natural products, including (+)-barekoxide and (−)-barekol [115].

Since their discovery in 1966 [116], linear triquinanes such as hirsutene (**288**) remained a popular synthetic target for organic chemists, and many research groups have developed efficient and short syntheses [117]. In 2008, the Rh-catalyzed [(5 + 2) + 1] cycloaddition of Yu and coworkers [118] proved useful for the synthesis of the tricyclic skeleton of linear triquinanes. In detail, vinylcyclopropane

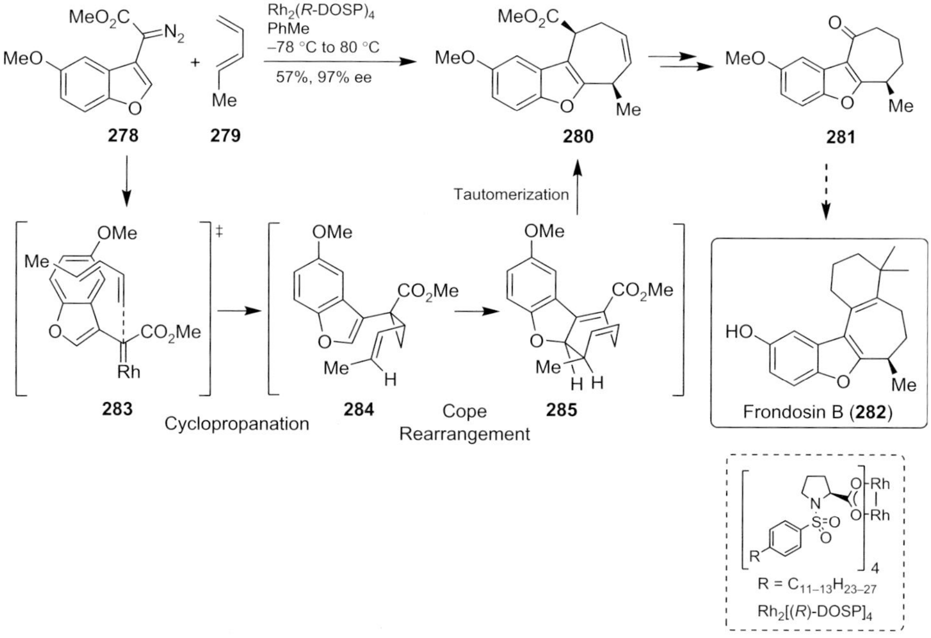

Scheme 14.43 Formal total synthesis of frondosin B (**282**) by Davies and coworkers.

286 was treated with a rhodium catalyst in the presence of carbon monoxide in dioxane at 80 °C. Under these conditions, *cis*-[(5 + 2) + 1]-cycloadduct **291** was formed presumably via the transition state **290** (Scheme 14.44).

It is assumed that the cycloaddition proceeds through cleavage of vinylcyclopropane, alkene insertion, CO insertion, and reductive elimination. The resultant

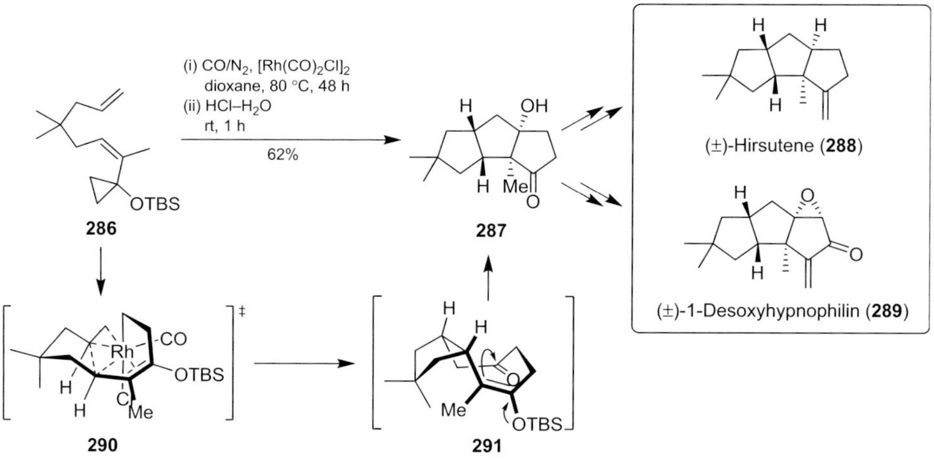

Scheme 14.44 Total synthesis of racemic hirsutene (**288**) and 1-desoxyhypnophilin (**289**) by Yu *et al.*

cyclooctane **291** undergoes a subsequent aldol reaction and affords the tricyclic compound **287** in 62% yield as a single diastereomer. Further transformations lead to either hirsutene (**288**) or 1-desoxyhypnophilin (**289**).

Cyanthiwigin U (**295**), first isolated in 2002 by Hamann and coworkers [119], is a diterpenoid and possesses a cyclohepta[*e*]indene ring system similar to that of allocyathin B_2, erinacine A, and sarcodonin G. Since members of this class show a broad spectrum of biological activity such as cytotoxicity and nerve-growth-factor stimulation, they have gained increased interest from synthetic organic chemists, and several total syntheses have been reported [120]. In 2005, the group of Phillips [121] described the total synthesis of (+)-cyanthiwigin U by means of a domino ring-opening/double ring-closing metathesis. For this purpose, dialdehyde **292** was converted into bis-enone **293**, which was treated with Grubbs II catalyst to afford the tricyclic bis-enone **294** in 43% yield. A likely reaction pathway, depicted in Scheme 14.45, comprises ring-opening metathesis of the bicyclo[2.2.2]octene **293** to give either **296** or **297**. Both compounds can then undergo two ring-closing metatheses leading to **294**, which was further manipulated to give (+)-cyanthiwigin U (**295**) in only 12 steps and with an overall yield of 17%.

Pfeiffer and Phillips [122] also reported a synthetic strategy for the conversion of **295** into cyanthiwigin W and Z in a few steps.

(+)-8-*epi*-Xanthatin (**301**), isolated from the aerial parts of various species of the genus *Xanthium*, is a sesquiterpene lactone which shows antimalarial and anticancer activity [123]. The first synthetic approach to this compound was reported by Martin *et al.* in 2005 [124]. Key step in the total synthesis of this natural product is a ruthenium-catalyzed domino ring-closing enyne/cross metathesis (Scheme 14.46).

When enyne **298** was treated with second-generation Hoveyda–Grubbs catalyst **299** in the presence of methyl vinyl ketone (**300**), the first step was a ring-closing

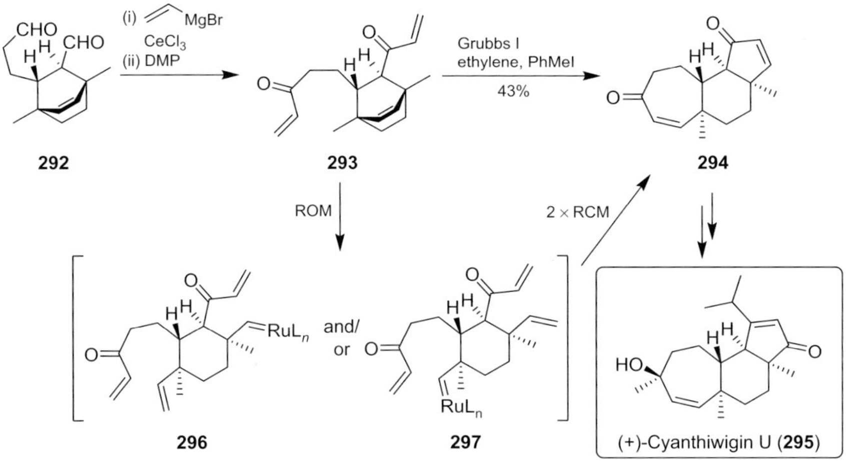

Scheme 14.45 Total synthesis of (+)-cyanthiwigin U (**295**) by Phillips and coworkers (DMP, Dess–Martin periodinane).

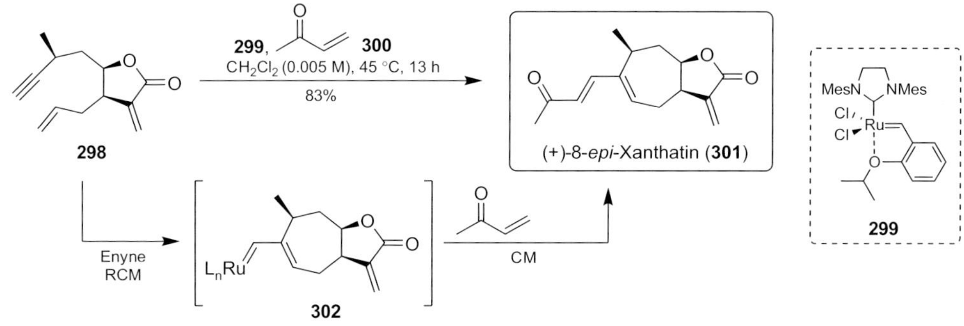

Scheme 14.46 Martin's total synthesis of (+)-8-*epi*-xanthatin (**301**).

metathesis presumably forming intermediate **302**, which underwent a subsequent cross metathesis with **300** to give the natural product **301** in 83% yield. The entire total synthesis required 14 steps in an overall yield of 5.5%.

The concept of enyne ring-closing metathesis was also used by Metz *et al.* [127] to give access to the enantiopure tetracyclic diterpenes kempene-2 (**305**), kempene-1 (**306**), and 3-*epi*-kempene-1 (**307**), which were isolated from the defense secretion of termite soldiers [125]. Due to the structural complexity of these compounds, their synthesis was a long-standing problem; thus, the only synthesis of racemic kempene-2 published so far was accomplished in 1991 by Dauben and coworkers [126]. The new domino enyne metathesis process is much more efficient, and Schubert and Metz [127] were able to form tetracyclic **304** in 92% yield from dienyne **303** using the Grubbs II catalyst (Scheme 14.47). Further transformations of **304** led to the desired natural products **305–307**. Comparison of the circular dichroism (CD) spectrum of synthetic (+)-kempene-2 (**305**) with that of the natural

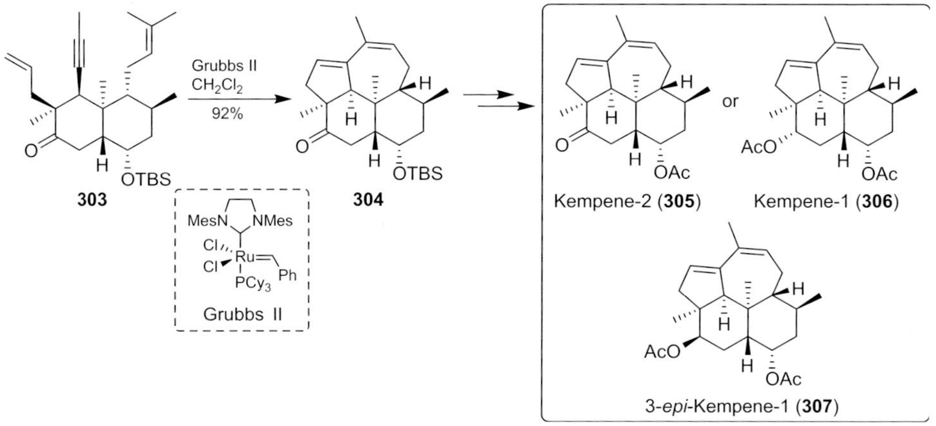

Scheme 14.47 Enantioselective synthesis of kempene-2 (**305**), kempene-1 (**306**), and 3-*epi*-kempene-1 (**307**) by Metz *et al.* using a domino metathesis reaction.

product showed conformity, indicating that the published absolute configuration was correct.

Several research groups are working toward the synthesis of various tetrahydroxanthones, which are a fast growing class of mycotoxins that possess interesting biological activity ranging from antibacterial and anti-HIV to cytostatic behavior. One prominent member of this class of natural products is diversonol (**185**), which was isolated by Turner in 1978 [128]. Its absolute configuration was determined by Krohn *et al.* [129] using CD spectroscopy and time dependent density functional theory calculations for electronic circular dichroism (TDDFT ECD) calculations. As discussed in Chapter 14.2, Bräse *et al.* [71, 72] published the first enantioselective synthesis of **185** in 2011 using an anionic domino process.

An alternate synthesis of enantiopure (−)-diversonol was recently reported by Tietze *et al.* using an enantioselective transition-metal-catalyzed domino process (Scheme 14.48). Key step in the synthesis is the formation of a chromane with concomitant introduction of the quaternary stereocenter at C-4a with 96% ee using a domino Wacker/carbonylation/methoxylation reaction in the presence of (*S,S*)-Bn-BOXAX (**310**) as ligand [73]. In this reaction phenol **308**, containing an alkene moiety, was treated with catalytic amounts of $Pd(tfa)_2$ and **310** in MeOH in the presence of CO and *p*-benzoquinone; the latter is necessary to reoxidize Pd^0 to Pd^{II}. One can assume that the first step is an enantiofacial coordination of the Pd^{II} species, which is attacked by the phenolic hydroxyl group to give intermediate **312**. Since a β-PdH elimination cannot take place, **312** reacts with CO to furnish **313**. The final step is the methoxylation of acyl-Pd species **313** to afford ester **309**, which is further converted into (−)-diversonol (**185**), the enantiomer of the natural product. However, use of (*R,R*)-Bn-BOXAX (*ent*-**310**), in the domino process allows for the preparation of (+)-diversonol (*ent*-**185**). Recently, a similar procedure has been applied to the first enantioselective total synthesis of blennolide A (**317**) starting from resorcinol (**314**) via the intermediates **315** and **316** (Scheme 14.49) [130].

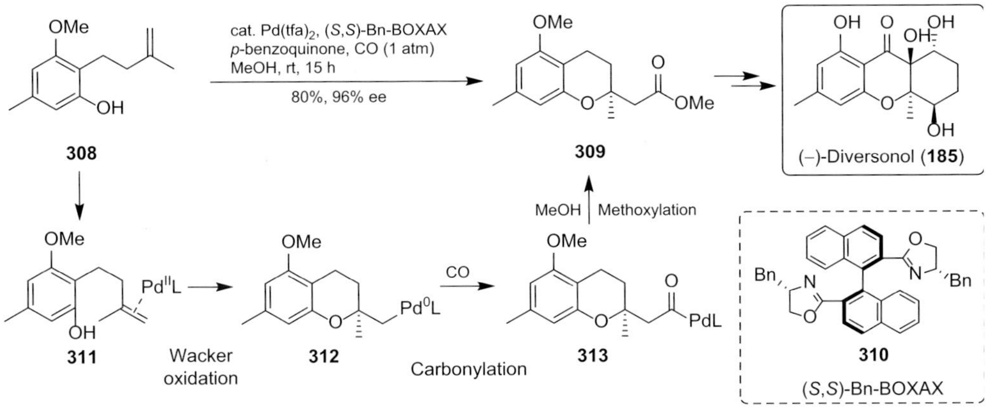

Scheme 14.48 Total synthesis of (−)-diversonol (**185**) by Tietze and coworkers applying a domino Wacker/carbonylation/methoxylation reaction.

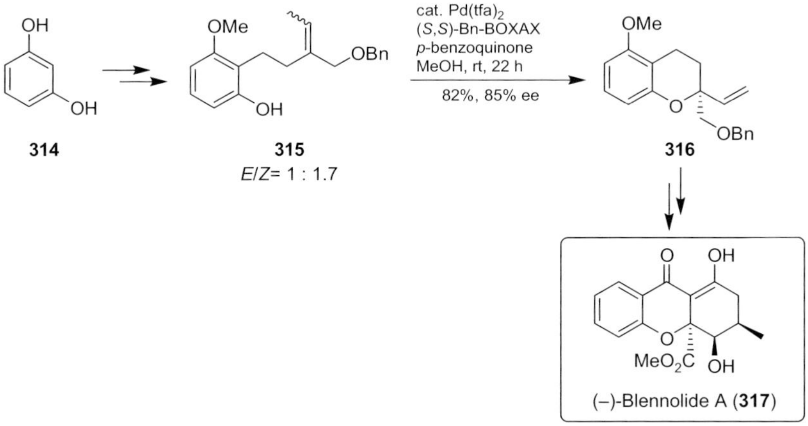

Scheme 14.49 Total synthesis of (−)-blennolide A (**185**) by Tietze and coworkers.

For the enantioselective total synthesis of α-tocopherol (**177**), which was also developed by Tietze *et al.* [69, 131], a domino Wacker/Heck reaction was used (Scheme 14.50).

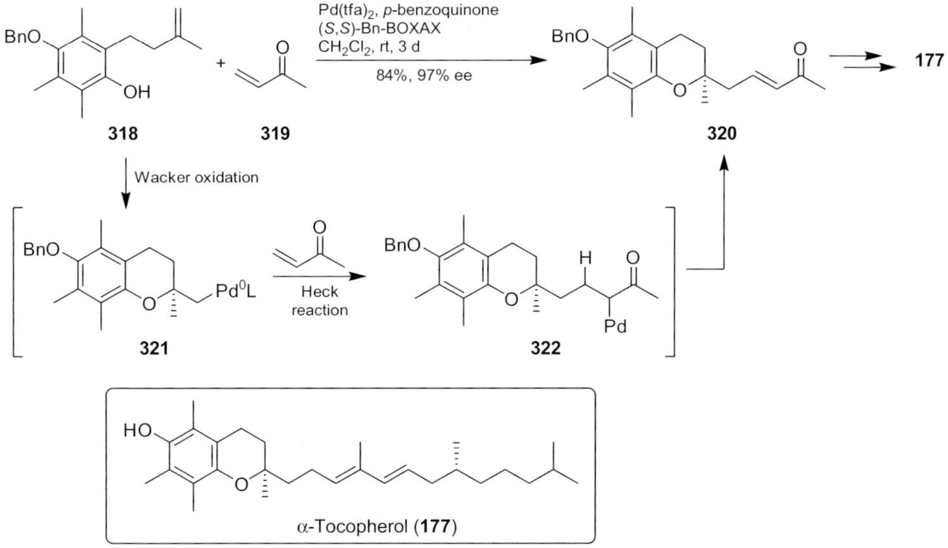

Scheme 14.50 Synthesis of α-tocopherol (**177**) by Tietze *et al.* applying an enantioselective domino Wacker/Heck reaction.

Thus, treatment of phenol **318** with catalytic amounts of Pd(tfa)$_2$ and (*S*,*S*)-Bn-BOXAX (**310**) led again to a Wacker oxidation, which was followed by an intermolecular Heck reaction with methyl vinyl ketone **319** to yield chromane **320**. This could then be converted into α-tocopherol (**177**). The conversion of **318** to **320** was one of the first reported enantioselective domino processes. Of course, many research groups shared interest in a synthetic approach to α-tocopherol (**177**) since it is the most active of the vitamin E family. Woggon and coworkers used a domino vinylogous aldol/oxa-Michael approach, as described in Section 14.2 [68], and the group of Trost [132] applied a Pd-catalyzed allylic alkylation for the formation of similar chromanes. Tietze *et al.* [131] have also used other enantioselective procedures for the synthesis of vitamin E.

In 2005, the structure elucidation and isolation of biyouyanagin A from *Hypericum* species *H. chinese* L. *var. salicifolium*, a plant that has been used as a folk medicine in Japan, was reported [133]. The proposed structure was revised by Nicolaou and coworkers in 2007 [134], proving that the natural product actually contained a cis, cis, trans, trans-fused instead of the proposed all-cis-fused four-membered ring. Biyouyanagin A (**326**) not only exhibits an impressive molecular architecture but also displays significant activity against HIV replication ($EC_{50} = 0.798\,\mu g\,ml^{-1}$) as well as inhibition of lipopolysaccharide-induced cytokine production and was therefore considered an interesting synthetic target [133]. Retrosynthetic considerations led to a [2 + 2]-cycloaddition for the formation of the highly substituted cyclobutane ring in **326**, rendering hyperolactone C (**325**) obvious key building block. For the synthesis of **325**, the precursor **324** was prepared by an interesting domino sequence starting from **323**. The Pd-catalyzed process transformed propargylic alcohol **323** and phenyl iodide into spirocyclic product **324** in 79% yield using $Pd(PPh_3)_4$ and, Et_3N under an atmosphere of CO and CO_2 at 200 psi (Scheme 14.51). The initiating step in the synthesis is probably a three-component Sonogashira-type reaction of phenyl iodide with CO and propargylic alcohol to acetylenic ketone **327** [135], which reacted with a molecule of CO_2 to form carbonate **328**. The carbonate is set up to undergo an intramolecular 5-*exo*-dig cyclization with the triple bond to furnish cyclic carbonate **329**, which, upon palladium catalysis, eliminates CO_2 resulting in π-allyl and σ-allyl Pd-complexes **330** and **331**, respectively. It was suggested that rearrangement of **331** afforded the palladacycle **332**, which underwent reductive elimination of Pd^0 to give **324** under retention of configuration. Further debenzylation, selenenylation, and oxidation/syn elimination afforded hyperolactone C (**325**), which was converted into biyouyanagin A (**326**).

In 2010, Stewart and Pfeffer reported a palladium-catalyzed access to tetrahydro-β-carbolines (also known as *tryptolines*) from 2-bromoindoles [136]. Alkaloids containing this type of core structure show significant biological activity, for example against cancer, and therefore represent interesting synthetic targets [137]. The present synthesis of several C-1-substituted tetrahydro-β-carbolines was achieved by a Pd-catalyzed domino reaction of aryl bromide **333** and several alkenes of the type **334**. After oxidative addition of the Pd catalyst into the carbon–bromine bond, a subsequent intermolecular Heck reaction with alkene **334** took place, followed by an intramolecular aza-Michael reaction of **336** to give substituted tryptolines **335** in a quick and efficient manner (64–83% yield) (Scheme 14.52).

The secodolastane diterpenoid (−)-indicol (**340**) was isolated from *Dictyota indica*, a brown alga from the Arabian Sea [138]. Secodolastanes possess a [5.4.0]undecane

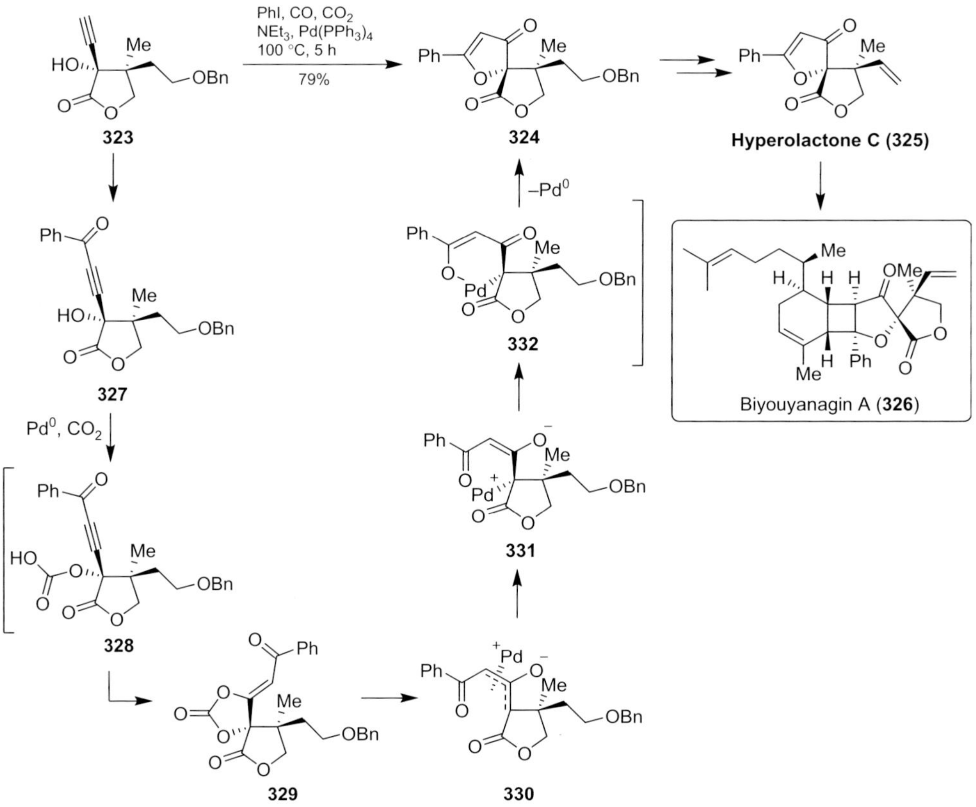

Scheme 14.51 Nicolaou's total synthesis of biyouyanagin A (**326**) by a Pd-catalyzed domino approach.

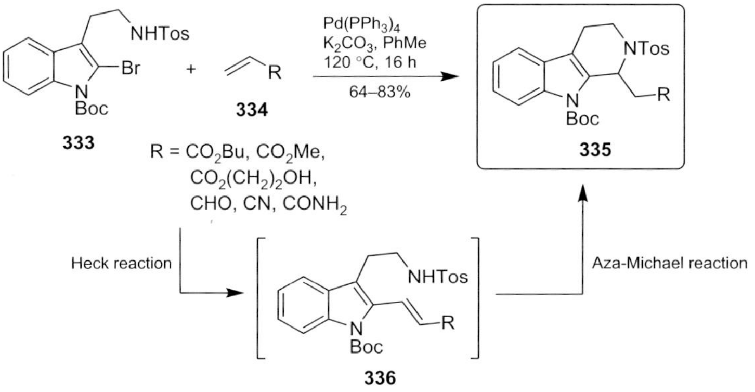

Scheme 14.52 Synthesis of tetrahydro-β-carbolines (**335**) by Stewart and Pfeffer *et al.* using a Pd-catalyzed domino Heck/aza-Michael reaction.

ring system bridged by a hemiacetal unit, and indicol in particular has four stereocenters of which two are quaternary.

The first total synthesis of this natural product was achieved by Chiu and Lam [139]. Key step of the synthesis is a rhodium-catalyzed domino cyclization/cycloaddition reaction to form the tricyclic core of the diterpenoid from linear α-diazoketone **337**. Concerning the mechanism of the reaction, it is likely that the rhodium catalyst, when reacted with **337** at 0 °C, formed a carbenoid species which immediately cyclized to **341** (Scheme 14.53). This 1,3-dipole then underwent an intramolecular cycloaddition with the alkene to give a mixture of two cycloadducts in 81% yield with **339** as the major product (dr = 1 : 3.1 **338** : **339**). The minor diastereomer **338** was probably formed via a less stable boat conformation of the tether in contrast to the chair conformation shown in **341**, leading to the desired product. Decreasing the temperature from 0 to −15 °C did not increase the dr but lowered the yield. It is also remarkable that the reaction afforded no more than 0.5 mol% of the rhodium(II)octanoate dimer ($[Rh_2(Oct)_4]$). Further transformation of **339** finally furnished (−)-indicol (**340**) in an overall yield of 10% over 21 steps.

In 2010, Overman [140] reported the first total synthesis of the alkaloid (+)-sieboldine A (**345**). The natural product was isolated from the Japanese club moss *Lycopodium sieboldii*, and shows potent inhibitory activity against acetylcholinesterase [141]. As part of his approach to the *cis*-hydrindanone core of the *Lycopodium* alkaloid, Overman utilized a mild Au(I)-promoted pinacol-terminated 1,6-enyne cyclization, a method originally introduced by Rhee and Kirsch *et al.* [142].

Starting from tetrahydrocyclopenta[*b*]furan-2-one **342**, enyne **343**, the substrate for the domino reaction, was prepared in 12 steps and with an overall yield of 45%. Exposure of **343** to the electron-rich gold(I) complex $(t\text{-Bu})_2P(o\text{-biphenyl})AuCl$ at room temperature afforded *cis*-hydrindanone **344** in 78% yield as a single stereoisomer (Scheme 14.54). The postulated mechanism involved Au(I) activation of the alkyne to initiate the cationic olefin cyclization of **346** to give carbocation **347**, which then underwent a pinacol rearrangement to the final product **344**. An originally attempted Lewis acid-catalyzed domino Prins/pinacol rearrangement of

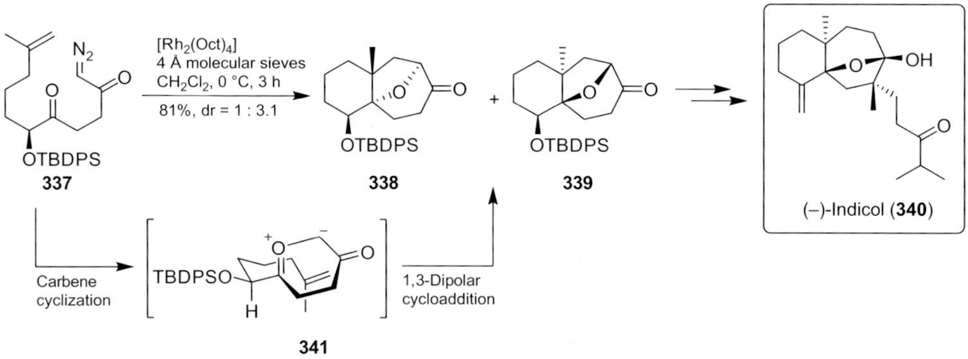

Scheme 14.53 Chiu's synthesis of (−)-indicol (**340**) by a rhodium-catalyzed domino carbene cyclization/cycloaddition reaction.

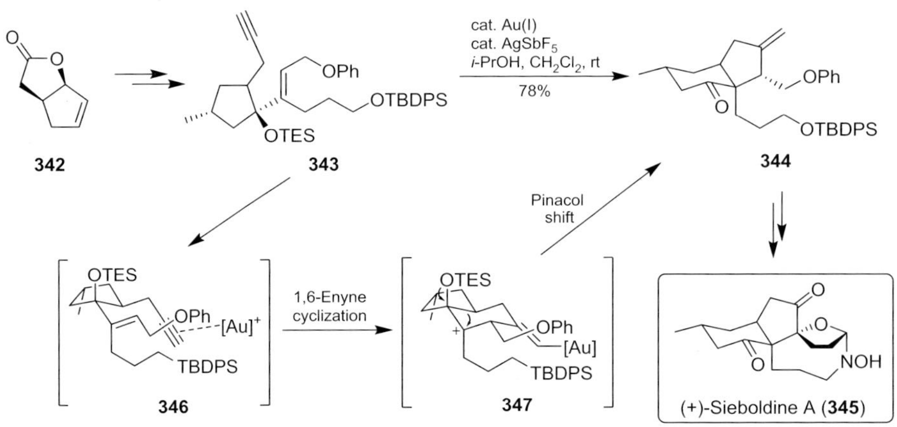

Scheme 14.54 Overman's total synthesis of (+)-sieboldine A (**345**) applying a domino 1,6-enyne cyclization/pinacol shift reaction.

the equivalent acetal had provided the domino product in unsatisfactory yields. With hydrindanone **344** in hand, the enantioselective total synthesis of (+)-sieboldine (**345**) was completed with a late-stage construction of the unprecedented *N*-hydroxyazacyclononane ring, yielding the natural product in a total of 20 steps.

14.6 Domino Reactions Initiated by Oxidation or Reduction

The natural product class of cortistatins is a compelling synthetic target because of their biological activity, their intriguing bridged 6/7/6/5-annulated ring skeleton, and the challenging stereochemistry. Cortistatin A (**350**) is the most potent member and inhibits the proliferation of human umbilical vein endothelial cells at nanomolar concentration, whereas it appears to have no adverse toxicological effects toward healthy cells [143]. Thus, this impressive activity profile has so far spawned at least four total syntheses and several formal total syntheses, as well as several approaches to the natural product core structure [144]. The group of Baran [145] reported the first total synthesis of cortistatin A (**350**) starting from the commercially available steroid prednisone, employing a domino oxidation/rearrangement/elimination reaction of advanced intermediate **348** (Scheme 14.55).

Single electron transfer to cyclopropyl bromide **348** by oxidation with SmI_2 led to the formation of **351**, which spontaneously rearranged under opening of the strained cyclopropyl ring. After elimination of bromine with generation of the trisubstituted double bond, the samarium enolate was captured by 2,4,4,6-tetrabromo-2,5-cyclohexadienone (TBHCB) as a bromine source, to finally lead to **349**. The obtained α-bromo ketone **349** was then further transformed into the natural product **350** in 6 more steps, thus completing the total synthesis in 15 steps

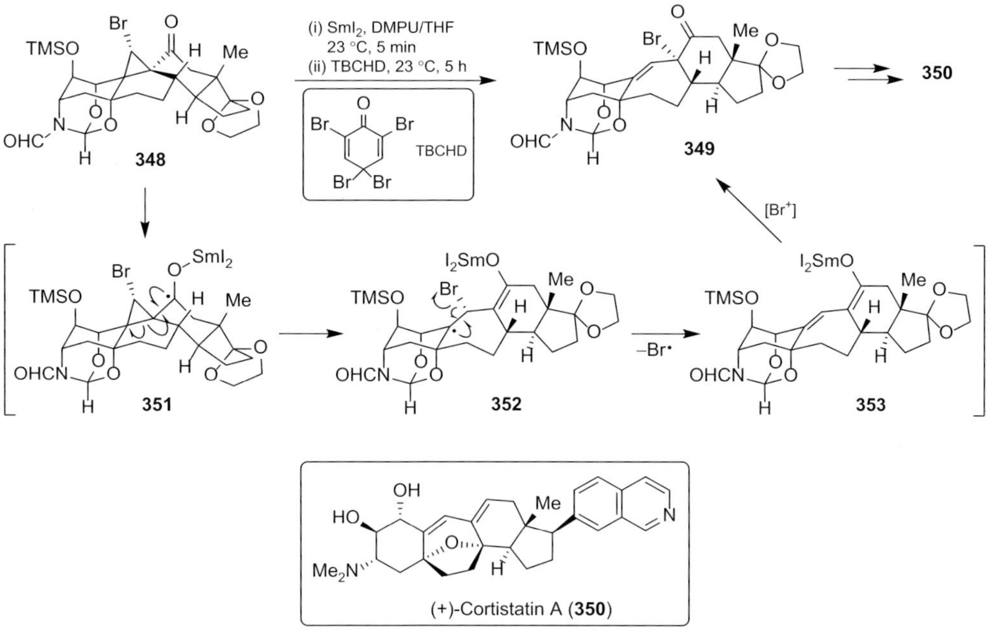

Scheme 14.55 Baran's total synthesis of (+)-cortistatin A (**350**).

and 3% overall yield. Recently, Baran disclosed an improved synthetic approach to cortistatin A utilizing the same starting material, prednisone [144b].

Apart from Baran's approach using a naturally abundant starting material with a partially assembled carbon skeleton, the group of Sorensen [146] opted for a fully synthetic access to the tetracyclic core structure of the cortistatin family. Their route relied on a domino cyclodearomatization/1,3-dipolar cycloaddition reaction that successively formed the highly oxygenated A and B rings in one process (Scheme 14.56).

Oxime **354** served as the starting material in the domino step, which, upon treatment with hypervalent iodine reagent $PhI(OAc)_2$, led to an oxidative dearomatization. The quinone congener **356** was formed by intramolecular nucleophilic attack of the free hydroxyl group. Concomitantly, the oxime functionality was oxidized to the nitrile oxide, which could then react with the α,β-unsaturated ketone in a 1,3-dipolar cycloaddition to form the tetracyclic cortistatin carbon skeleton **355**.

Another molecule that has received full attention of the synthetic community over decades is the indole alkaloid strychnine (**359**), being a popular target in total synthesis since Woodward's hallmark contribution in 1954 [147]. Since then, numerous approaches have been reported [92]. A short and elegant synthesis has recently been published by Beemelmanns and Reissig [148] (Scheme 14.57).

The synthetic endeavor to the synthesis of racemic **359** via the advanced intermediate previously employed in Rawal's approach (i.e. 358) [149] commenced using SmI_2 in a single electron transfer-induced domino sequence. After generation of

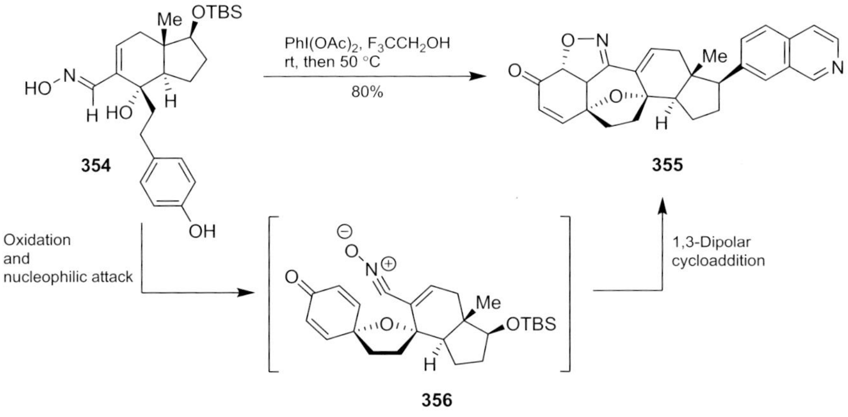

Scheme 14.56 Synthesis of the core structure of (+)-cortistatin A (**350**) by the group of Sorensen.

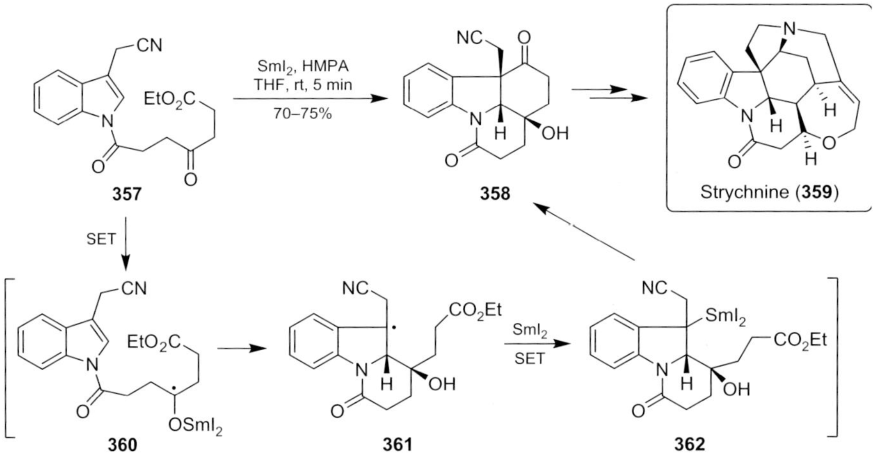

Scheme 14.57 Formal total synthesis of strychnine (**359**) by Reissig *et al.*

the carbon radical from the appendant carbonyl group in **357**, the first C–C bond was formed by attack at C-2 of the indole moiety. The formed stabilized benzyl radical **361** then underwent a second single electron transfer to give samarium nucleophile **362**. In the concluding step, 362 underwent closure of the second ring upon elimination of ethanol and installation of the ketone moiety, which later served as a handle to form the carbon–nitrogen bond of the second pyrrolidine ring in the natural product. Following Reissig's report, another route to racemic strychnine (**359**) was disclosed by Martin and Vanderwal [150], culminating in the shortest synthesis to date (six steps starting from tryptophyl bromide).

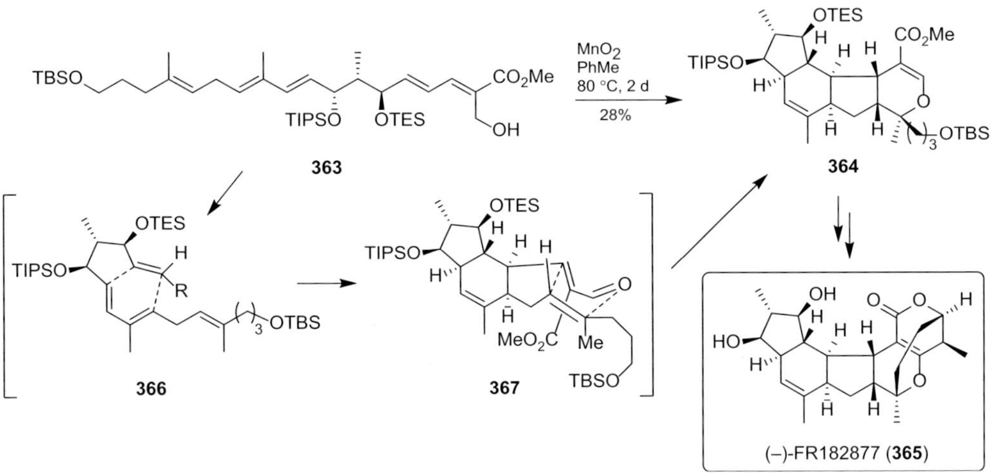

Scheme 14.58 Nakada's total synthesis of FR182877 (**365**).

The highly oxygenated natural product FR182877 (**365**) portrays a biological mode of action similar to that of paclitaxel; therefore, it features a strong anticancer activity through microtubule stabilization in a variety of tumor cell lines [151]. All three published total syntheses of FR182877 by Evans, Sorensen, and Nakada employed a domino reaction comprising two Diels–Alder variants as key step [152]. In Nakada's approach [152e], acyclic precursor **363** containing with a masked α,β-unsaturated aldehyde was used (Scheme 14.58).

Thus, oxidation of the allylic alcohol **363** to the corresponding aldehyde in the presence of a large excess of manganese dioxide initiated the domino oxidation/Diels–Alder/hetero-Diels–Alder sequence. After formation of **366** via an endo-*E*-syn transition state, the terminal aldehyde **367** then gave the tetracycle **364** in the concluding hetero-Diels–Alder step in 28% yield after 2 days. Intermediate **364** was then further transformed into the natural product **365** in an additional 14 steps, featuring an intramolecular Heck reaction and an iridium-catalyzed isomerization.

14.7
Conclusion

The immense potency of domino reactions cannot be presented in a better way as in their application to the total synthesis of complex natural products. Though nature has already shown that the domino concept is very useful in this field and several of the described natural product syntheses follow a biomimetic way, a multitude of other approaches have been developed over the last years, and we are convinced that in future most of the published total syntheses will be based on domino reactions.

References

1. (a) Antoun, M.D., Abramson, D., Tyson, R.L., Chang, C.-J., McLaughlin, J.L., Peck, G., and Cassady, J.M. (1981) *J. Nat. Prod.*, **44**, 579–585; (b) Magalhães, H.I.F., Veras, M.L., Torres, M.R., Alves, A.P.N.N., Pessoa, O.D.L., Silveira, E.R., Costa-Lotufo, L.V., Odorico de Moraes, M., and Pessoa, C. (2006) *J. Pharm. Pharmacol.*, **58**, 235–241; (c) Damu, A.G., Kuo, P.-C., Su, C.-R., Kuo, T.-H., Chen, T.-H., Bastow, K.F., Lee, K.-H., and Wu, T.-S. (2007) *J. Nat. Prod.*, **70**, 1146–1152.
2. Morita, M., Hirai, G., Ohkubo, M., Koshino, H., Hashizume, D., Maruoka, K., and Sodeoka, M. (2012) *Org. Lett.*, **14**, 3434–3437.
3. Ohkubo, M., Hirai, G., and Sodeoka, M. (2009) *Angew. Chem. Int. Ed.*, **48**, 3862–3866.
4. Nicolaou, K.C., Sarlah, D., Wu, T.R., and Zhan, W. (2009) *Angew. Chem. Int. Ed.*, **48**, 6870–6874.
5. For a recent review on this topic see:Sacchettini, J.C., Rubin, E.J., and Freundlich, J.S. (2008) *Nat. Rev. Microbiol.*, **6**, 41–52.
6. Watanabe, K., Sakurai, J., Abe, H., and Katoh, T. (2010) *Chem. Commun.*, **46**, 4055–4057.
7. Minagawa, K., Kouzuki, S., Yoshimoto, J., Kawamura, Y., Tani, H., Iwata, T., Terui, Y., Nakai, H., Yagi, S., Hattori, N., Fujiwara, T., and Kamigauchi, T. (2002) *J. Antibiot.*, **55**, 155–164.
8. Nakatani, M., Nakamura, M., Suzuki, A., Inoue, M., and Katoh, T. (2002) *Org. Lett.*, **4**, 4483–4486.
9. Matsuo, Y., Suzuki, M., Masuda, M., Iwai, T., and Morimoto, Y. (2008) *Helv. Chim. Acta*, **91**, 1261–1266.
10. Morimoto, Y., Okita, T., and Kambara, H. (2009) *Angew. Chem. Int. Ed.*, **48**, 2538–2541.
11. Xiong, Z., Busch, R., and Corey, E.J. (2010) *Org. Lett.*, **12**, 1512–1514.
12. Topczewski, J.J., Callahan, M.P., Neighbors, J.D., and Wiemer, D.F. (2009) *J. Am. Chem. Soc.*, **131**, 14630–14631.
13. (a) Akihisa, T., Tokuda, H., Ukiya, M., Iizuka, M., Schneider, S., Ogasawara, K., Mukainaka, T., Iwatsuki, K., Suzuki, T., and Nishino, H. (2003) *Cancer Lett.*, **201**, 133–137; (b) Aoki, N., Muko, M., Ohta, E., and Ohta, S. (2008) *J. Nat. Prod.*, **71**, 1308–1310.
14. Topczewski, J.J., Kodet, J.G., and Wiemer, D.F. (2011) *J. Org. Chem.*, **76**, 909–919.
15. Chen, A.P.-J. and Williams, C.M. (2008) *Org. Lett.*, **10**, 3441–3443.
16. Fukuyama, Y., Kubo, M., Minami, H., Yuasa, H., Matsuo, A., Fujii, T., Morisaki, M., and Harada, K. (2005) *Chem. Pharm. Bull.*, **53**, 72–80.
17. (a) Mak, J.Y.W. and Williams, C.M. (2012) *Chem. Commun.*, **48**, 287–289; (b) Mak, J.Y.W. and Williams, C.M. (2012) *Eur. J. Org. Chem.*, **2012**, 2001–2012; For a review see: (c) Mak, J.Y.W. and Williams, C.M. (2012) *Nat. Prod. Rep.*, **29**, 440–448.
18. Pettit, G.R., Knight, J.C., Collins, J.C., Herald, D.L., Pettit, R.K., Boyd, M.R., and Young, V.G. (2000) *J. Nat. Prod.*, **63**, 793 798.
19. For a review on the biological activity of tetrahydroisoquinolines see: Scott, J.D. and Williams, R.M. (2002) *Chem. Rev.*, **102**, 1669–1730.
20. (a) Chan, C., Heid, R., Zheng, S., Guo, J., Zhou, B., Furuuchi, T., and Danishefsky, S.J. (2005) *J. Am. Chem. Soc.*, **127**, 4596–4598; (b) Vincent, G. and Williams, R.M. (2007) *Angew. Chem. Int. Ed.*, **46**, 1517–1520; (c) Yokoya, M., Ito, H., and Saito, N. (2011) *Tetrahedron*, **67**, 9185–9192.
21. Chen, X. and Zhu, J. (2007) *Angew. Chem. Int. Ed.*, **46**, 3962–3965.
22. Díaz-Gavilán, M., Galloway, W.R.J.D., O'Connell, K.M.G., Hodkingson, J.T., and Spring, D.R. (2010) *Chem. Commun.*, **46**, 776–778.
23. Kim, K. and Kim, I. (2010) *Org. Lett.*, **12**, 5314–5317.
24. Ge, H.M., Yang, W.H., Shen, Y., Jiang, N., Guo, Z.K., Luo, Q., Xu, Q., Ma, J., and Tan, R.X. (2010) *Chem. Eur. J.*, **16**, 6338–6345.

25. Juliawaty, L.D., Sahidin, E.H.H., Achmad, S.A., Syah, Y.M., Latip, J., and Said, I.M. (2009) *Nat. Prod. Commun.*, **4**, 947–950.
26. Chawla, A.S. and Jackson, A.H. (1984) *Nat. Prod. Rep.*, **1**, 371–373.
27. Tietze, L.F., Tölle, N., Kratzert, D., and Stalke, D. (2009) *Org. Lett.*, **11**, 5230–5233.
28. Tietze, L.F., Tölle, N., and Noll, C. (2008) *Synlett*, 525–528.
29. Blake, A.J., Gill, C., Greenhalgh, D.A., Simpkins, N.S., and Zhang, F. (2005) *Synthesis*, **19**, 3287–3292.
30. Laws, S.W. and Scheerer, J.R. (2013) *J. Org. Chem.*, **78**, 2422–2429.
31. For a review see: Miller, K.A. and Williams, R.M. (2009) *Chem. Soc. Rev.*, **38**, 3160–3174.
32. Margrey, K.A., Chinn, A.J., Laws, S.W., Pike, R.D., and Scheerer, J.R. (2012) *Org. Lett.*, **14**, 2458–2461.
33. Porter, A.E.A. and Sammes, P.G. (1970) *J. Chem. Soc., Chem. Commun.*, 1103.
34. Baran, P.S., Guerrero, C.A., Hafensteiner, B.C., and Ambhaikar, N.B. (2005) *Angew. Chem. Int. Ed.*, **44**, 3892–3895.
35. Baran, P.S., Hafensteiner, B.D., Ambhaikar, N.B., Guerrero, C.A., and Gallagher, J.D. (2006) *J. Am. Chem. Soc.*, **128**, 8678–8693.
36. Jakubec, P., Hawkins, A., Felzmann, W., and Dixon, D.J. (2012) *J. Am. Chem. Soc.*, **134**, 17482–17485.
37. Jakubec, P., Cockfield, D.M., and Dixon, D.J. (2009) *J. Am. Chem. Soc.*, **131**, 16632–16633.
38. Varseev, G.N. and Maier, M.E. (2007) *Org. Lett.*, **9**, 1461–1464.
39. Kawasaki, T., Terashima, R., Sakaguchi, K.-E., Sekiguchi, H., and Sakamoto, M. (1996) *Tetrahedron Lett.*, **37**, 7525–7528.
40. Kawasaki, T., Ogawa, A., Takashima, Y., and Sakamoto, M. (2003) *Tetrahedron Lett.*, **44**, 1591–1593.
41. Kawasaki, T., Shinada, M., Kamimura, D., Ohzono, M., and Ogawa, A. (2006) *Chem. Commun.*, 420–422.
42. Takiguchi, S., Iizuka, T., Kumakura, Y.-S., Murasaki, K., Ban, N., Higuchi, K., and Kawasaki, T. (2010) *J. Org. Chem.*, **75**, 1126–1131.
43. ElSohly, M.A. and Slade, D. (2005) *Life Sci.*, **78**, 539–548.
44. Nandaluru, P.R. and Bodwell, G.J. (2012) *Org. Lett.*, **14**, 310–313.
45. Tietze, L.F., Böhnke, N., and Dietz, S. (2009) *Org. Lett.*, **11**, 2948–2950.
46. Tietze, L.F., Dietz, S., Böhnke, N., Düfert, M.A., Objartel, I., and Stalke, D. (2011) *Eur. J. Org. Chem.*, **2011**, 6574–6580.
47. Pandey, G. and Kumara, P. (2011) *Org. Lett.*, **13**, 4672–4675.
48. Lim, K.-H., Hiraku, O., Komiyama, K., Koyano, T., Hayashi, M., and Kam, T.-S. (2007) *J. Nat. Prod.*, **70**, 1302–1307.
49. (a) Calabi, L., Danieli, B., Lesma, G., and Palmisano, G. (1982) *J. Chem. Soc., Perkin Trans. 1*, 1371–1379; (b) Danieli, B., Lesma, G., Palmisano, G., and Gabetta, B. (1981) *J. Chem. Soc., Chem. Commun.*, 908–909; (c) Szántay, C. (1990) *Pure Appl. Chem.*, **62**, 1299–1302.
50. Coldham, I., Burrell, A.J.M., White, L.E., Adams, H., and Oram, N. (2007) *Angew. Chem. Int. Ed.*, **46**, 6159–6162.
51. Lajiness, J.P., Jiang, W., and Boger, D.L. (2012) *Org. Lett.*, **14**, 2078–2081.
52. Tietze, L.F., Wolfram, T., Holstein, J.J., and Dittrich, B. (2012) *Org. Lett.*, **14**, 4035–4037.
53. Bohlmann, F., Zdero, C., and Lehmann, L. (1991) *Phytochemistry*, **30**, 1161–1163.
54. (a) Tietze, L.F., Kinzel, T., and Wolfram, T. (2009) *Chem. Eur. J.*, **15**, 6199–6210; For a review see: (b) Tietze, L.F., Kinzel, T., and Brazel, C.C. (2009) *Acc. Chem. Res.*, **42**, 367–378.
55. (a) Kobayashi, J., Inaba, Y., Shiro, M., Yodashi, N., and Morita, H. (2001) *J. Am. Chem. Soc.*, **123**, 11402–11408; For a review on *Daphniphyllum* alkaloids see: (b) Kobayashi, J. and Kubota, T. (2009) *Nat. Prod. Rep.*, **26**, 936–962.
56. For the recent total synthesis of related *Daphniphyllum* alkaloid (+)-daphmanidin E see: (a) Weiss, M.E. and Carreira, E.M. (2011) *Angew. Chem. Int. Ed.*, **50**, 11501–11505; For selected

recent synthetic studies towards *Daphniphyllum* alkaloids see: (b) Dunn, T.B., Ellis, J.M., Kofink, C.C., Manning, J.R., and Overman, L.E. (2009) *Org. Lett.*, **11**, 5658–5661; (c) Sladojevich, F., Michaelides, I.N., Darses, B., Ward, J.W., and Dixon, D.J. (2011) *Org. Lett.*, **13**, 5132–5135; (d) Coldham, I., Burrell, A.J.M., Guerrand, H.D.S., and Oram, N. (2011) *Org. Lett.*, **13**, 1267–1269; (e) Yao, Y. and Liang, G. (2012) *Org. Lett.*, **14**, 5499–5501; (f) Darses, B., Michaelides, I.N., Sladojevich, F., Ward, J.W., Rzepa, P.R., and Dixon, D.J. (2012) *Org. Lett.*, **14**, 1684–1687; (g) Fang, B., Zheng, H., Zhao, C., Jing, P., Li, H., Xie, X., and She, X. (2012) *J. Org. Chem.*, **77**, 8367–8373.

57. Ikeda, S., Shibuya, M., Kanoh, N., and Iwabuchi, Y. (2009) *Org. Lett.*, **11**, 1833–1836.
58. Gilley, C.B., Buller, M.J., and Kobayashi, Y. (2007) *Org. Lett.*, **9**, 3631–3634.
59. Corey, E.J., Li, W.-D., and Nagamitsu, T. (1998) *Angew. Chem. Int. Ed.*, **37**, 1676–1679.
60. Reboul, I., Boddaert, T., Coquerel, Y., and Rodriguez, J. (2008) *Eur. J. Org. Chem.*, **2008**, 5379–5382.
61. Coquerel, Y., Filippini, M.-H., Bensa, D., and Rodriguez, J. (2008) *Chem. Eur. J.*, **14**, 3078–3092.
62. For a review see: Magano, J. (2009) *Chem. Rev.*, **109**, 4398–4438.
63. (a) Ishikawa, H., Suzuki, T., and Hayashi, Y. (2009) *Angew. Chem. Int. Ed.*, **48**, 1304–1307; (b) Ishikawa, H., Suzuki, T., Orita, H., Uchimaru, T., and Hayashi, Y. (2010) *Chem. Eur. J.*, **16**, 12616–12626.
64. Zhu, S., Yu, S., Wang, Y., and Ma, D. (2010) *Angew. Chem. Int. Ed.*, **49**, 4656–4660.
65. Bradshaw, B., Luque-Corredera, C., and Bonjoch, J. (2013) *Org. Lett.*, **15**, 326–329.
66. Sun, Y., Yu, P., Zhang, G., Wang, L., Zhong, H., Zhai, Z., Wang, L., and Wang, Y. (2012) *J. Neurosci. Res.*, **90**, 1662–1669.
67. For a comprehensive review on vitamin E see: Schneider, C. (2005) *Mol. Nutr. Food Res.*, **49**, 7–30.
68. Liu, K., Chougnet, A., and Woggon, W.-D. (2008) *Angew. Chem. Int. Ed.*, **47**, 5827–5829.
69. Tietze, L.F., Stecker, F., Zinngrebe, J., and Sommer, K.M. (2006) *Chem. Eur. J.*, **12**, 8770–8776.
70. For a review on mycotoxins see: Bräse, S., Encinas, A., Keck, J., and Nising, C.F. (2009) *Chem. Rev.*, **109**, 3903–3990.
71. For selected syntheses see: (a) Nising, C.F., Ohnemüller, U.K., Schmid, N., and Bräse, S. (2006) *Angew. Chem. Int. Ed.*, **45**, 307–309; (b) Nicolaou, K.C. and Li, A. (2008) *Angew. Chem. Int. Ed.*, **47**, 6579–6582; (c) Qin, T., Johnson, R.P., and Porco, J.A. Jr., (2011) *J. Am. Chem. Soc.*, **133**, 1714–1717; (d) Tietze, L.F., Spiegel, D.A., Stecker, F., Major, J., Raith, C., and Große, C. (2008) *Chem. Eur. J.*, **14**, 8956–8963; (e) Volz, N., Bröhmer, M.C., Nieger, M., and Bräse, S. (2009) *Synlett*, 550–553.
72. Bröhmer, M.C., Bourcet, E., Nieger, M., and Bräse, S. (2011) *Chem. Eur. J.*, **17**, 13706–13711.
73. Tietze, L.F., Jackenkroll, S., Raith, C., Spiegl, D.A., Reiner, J.R., and Campos, M.C.O. (2013) *Chem. Eur. J.*, **19**, 4876–4882.
74. Brown, P.D., Willis, A.C., Sherburn, M.S., and Lawrence, A.L. (2012) *Org. Lett.*, **14**, 4537–4539.
75. Chen, Y.Q., Shen, Y.H., Su, Y.Q., Kong, L.Y., and Zhang, W.D. (2009) *Chem. Biodivers.*, **6**, 779–783.
76. Nazef, N., Davies, R.D.M., and Greaney, M.F. (2012) *Org. Lett.*, **14**, 3720–3723.
77. For selected total syntheses see: (a) Meng, Z. and Danishefsky, S.J. (2005) *Angew. Chem. Int. Ed.*, **44**, 1511–1513; (b) Inoue, M., Sato, T., and Hirama, M. (2006) *Angew. Chem. Int. Ed.*, **45**, 4843–4848; (c) He, W., Huang, J., Sun, X., and Frontier, A.J. (2008) *J. Am. Chem. Soc.*, **130**, 300–308; (d) Shi, L., Meyer, K.M., and Greaney, M.F. (2010) *Angew. Chem. Int. Ed.*, **49**, 9250–9253; (e) Chen, J., Gao, P., Yu, F., Yang, Y., Zhu, S., and Zhai,

H. (2012) *Angew. Chem. Int. Ed.*, **51**, 5897–5899.

78. Mehta, G. and Singh, S.R. (2006) *Angew. Chem. Int. Ed.*, **45**, 953–955.

79. Dückert, H., Pries, V., Khedkar, V., Menninger, S., Bruss, H., Bird, A.W., Maliga, Z., Brockmeyer, A., Janning, P., Hyman, A., Grimme, S., Schürmann, M., Preut, H., Hübel, K., Ziegel, S., Kumar, K., and Waldmann, H. (2012) *Nat. Chem. Biol.*, **8**, 179–184.

80. Eschenbrenner-Lux, V., Dückert, H., Khedkar, V., Bruss, H., Waldmann, H., and Kumar, K. (2013) *Chem. Eur. J.*, **19**, 2294–2304.

81. Domingo, V., Arteaga, J.F., Pérez, J.L.L., Peláez, R., del Moral, J.F.Q., and Barrero, A.F. (2012) *J. Org. Chem.*, **77**, 341–350.

82. Liang, X.T., Yu, D.Q., Wu, W.L., and Deng, H.C. (1979) *Acta Chim. Sinica*, **37**, 215–230.

83. Szpilman, A.M., Korshin, E.E., Rozenberg, H., and Bachi, M.D. (2005) *J. Org. Chem.*, **70**, 3618–3632.

84. Korshin, E.E., Hoos, R., Szpilman, A.M., Konstantinovski, L., Posner, G.H., and Bachi, M.D. (2002) *Tetrahedron*, **58**, 2449–2469.

85. Fraude, G. (1878) *Ber. Dtsch. Chem. Ges.*, **11**, 2189–2191.

86. (a) Floriani, L. (1931) *An. Farm. Bioquim.*, **2**, 215. (b) Deutsch, H.F., Evenson, M.A., Drescher, P., Sparwasser, C., and Madsen, P.O. (1994) *J. Pharm. Biomed. Anal.*, **12**, 1283–1287.

87. (a) Storck, G. and Dolfini, J.E. (1963) *J. Am. Chem. Soc.*, **85**, 2872–2873; (b) Fukuda, Y., Shindo, M., and Shishido, K. (2003) *Org. Lett.*, **5**, 749–751; (c) Ban, Y., Sato, Y., Inoue, I., Nagai, M., Oishi, T., Terashima, M., Yonemitsu, O., and Kanaoka, Y. (1965) *Tetrahedron Lett.*, **6**, 2261–2268; (d) Pearson, A.J. (1981) *Tetrahedron Lett.*, **22**, 4033–4036.

88. (a) Kato, D., Sasaki, Y., and Boger, D.L. (2010) *J. Am. Chem. Soc.*, **132**, 3685–3687; (b) Elliott, G.J., Velcicky, J., Ishikawa, H., Li, Y., and Boger, D.L. (2006) *Angew. Chem. Int. Ed.*, **45**, 60–622; (c) Ishikawa, H., Elliott, G.I., Velcicky, J., Choi, Y., and Boger, D.L. (2006) *J. Am. Chem. Soc.*, **128**, 10596–10612.

89. (a) Chien, S.-C., Chang, J.-Y., Kuo, C.-C., Hsieh, C.-C., Yang, N.-S., and Kuo, Y.-H. (2007) *Tetrahedron Lett.*, **48**, 1567–1569; (b) Hsieh, C.-C., Kuo, Y.-H., Kuo, C.-C., Chen, L.-T., Cheung, C.-H.A., Chao, T.-Y., Lin, C.-H., Pan, W.-Y., Chang, C.-Y., Chien, S.-C., Chen, T.-W., Lung, C.-C., and Chang, J.-Y. (2010) *Biochem. Pharmacol.*, **79**, 1261–1271.

90. Dong, S., Hamel, E., Bai, R., Covell, D.G., Beutler, J.A., and Porco, J.A. Jr., (2009) *Angew. Chem. Int. Ed.*, **48**, 1494–1497.

91. Dong, S., Quin, T., Hamel, E., Beutler, J.A., and Porco, J.A. Jr., (2012) *J. Am. Chem. Soc.*, **134**, 19782–19787.

92. Cannon, J.S. and Overman, L.E. (2012) *Angew. Chem. Int. Ed.*, **51**, 4288–4311.

93. Massiot, G., Thépenier, P., Jacquier, M.-J., Le Men-Olivier, L., and Delaude, C. (1989) *Heterocycles*, **29**, 1435–1438.

94. Jones, S.B., Simmons, B., and MacMillan, D.W.C. (2009) *J. Am. Chem. Soc.*, **131**, 13606–13607.

95. Shen, L., Zhang, M., Wu, Y., and Qin, Y. (2008) *Angew. Chem. Int. Ed.*, **47**, 3618–3621.

96. Dounay, A.B., Humphreys, P.G., Overman, L.E., and Wrobleski, A.D. (2008) *J. Am. Chem. Soc.*, **130**, 5368–5377.

97. Frankowski, K.J., Golden, J.E., Zeng, Y., Lei, Y., and Aubé, J. (2008) *J. Am. Chem. Soc.*, **130**, 6018–6024.

98. (a) Götz, M. and Edwards, O.E. (1967) in *The Alkaloids*, Vol. 9 (ed. R.H.F. Manske), Academic Press, New York, pp. 545–551; (b) Götz, M. and Strunz, G.M. (1973) in *Alkaloids*, Vol. 9 (ed. K. Wiesner), Butterworth, London, pp. 143–160; (c) Lin, W.-H., Ye, Y., and Xu, R.S. (1992) *J. Nat. Prod.*, **55**, 571–576; (d) Xu, R.-S. (2000) *Stud. Nat. Prod. Chem.*, **21**, 729–772; (e) Pilli, R.A. and Ferreira de Oliveira, M.d.C. (2000) *Nat. Prod. Rep.*, **17**, 117–127; (f) Greger, H. (2006) *Planta Med.*, **72**, 99–113; (g) Xu, Y.T., Hon, P.-M., Jiang, R.-W., Cheng, L., Li, S.-H., Chan, Y.-P., Xu, H.-X., Shaw, P.-C.,

and But, P.P.-H. (2006) *J. Ethnopharmacol.*, **108**, 46–53.
99. Bartoli, A., Chouraqui, G., and Parrain, J.-tL. (2012) *Org. Lett.*, **14**, 122–125.
100. Li, G. and Padwa, A. (2011) *Org. Lett.*, **13**, 3767–3769.
101. (a) Marco-Contelles, J., Carreiras, M.C., Rodríguez, C., Villarroya, M., and García, A.G. (2006) *Chem. Rev.*, **106**, 116–133; (b) Popa, R.V., Pereira, E.F.R., Lopes, C., Maelicke, A., and Aluquerque, E.X.J. (2006) *J. Mol. Neurosci.*, **30**, 227–232; (c) Lilienfeld, S. (2002) *CNS Drug Rev.*, **8**, 159–176.
102. Chang, J.H., Kang, H.-U., Jung, I.-H., and Cho, C.-G. (2010) *Org. Lett.*, **12**, 2016–2018.
103. Trost, B.M. and Toste, F.D. (2000) *J. Am. Chem. Soc.*, **122**, 11262–11263.
104. (a) Satcharoen, V., McLean, N.J., Kemp, S.C., Camp, N.P., and Brown, R.C.D. (2007) *Org. Lett.*, **9**, 1867–1869; (b) Chen, J.-Q., Xie, J.-H., Bao, D.-H., Liu, S., and Zhou, Q.-L. (2012) *Org. Lett.*, **14**, 2714–2717.
105. (a) Wei, H., Itoh, T., Konishita, M., Kotoku, N., and Kobayashi, M. (2005) *Tetrahedron*, **61**, 8054–8058; (b) Wei, H., Itoh, T., Konishita, M., Kotoku, N., and Kobayashi, M. (2006) *Heterocycles*, **68**, 111–123.
106. Sofiyev, V., Navarro, G., and Trauner, D. (2008) *Org. Lett.*, **10**, 149–152.
107. Parker, K.A. and Lim, Y.-H. (2004) *J. Am. Chem. Soc.*, **126**, 15968–15969.
108. Beaudry, C.M. and Trauner, D. (2005) *Org. Lett.*, **7**, 4475–4477.
109. Schmidt, T.J., Vossing, S., Klaes, M., and Grimme, S. (2007) *Planta Med.*, **73**, 1574–1580.
110. Tietze, L.F., Duefert, S.-C., Clerc, J., Bischoff, M., Maaß, C., and Stalke, D. (2013) *Angew. Chem. Int. Ed.*, **52**, 3191–3194.
111. (a) Tietze, L.F., Kahle, K., and Raschke, T. (2002) *Chem. Eur. J.*, **8**, 401–407; (b) Tietze, L.F., Thede, K., Schimpf, R., and Sannicolo, F. (2000) *Chem. Commun.*, 583–584; (c) Tietze, L.F. and Modi, A. (2000) *Eur. J. Org. Chem.*, **2000**, 1959–1964.
112. (a) Reddy, R.P. and Davies, H.M.L. (2007) *J. Am. Chem. Soc.*, **129**, 10312–10313; (b) Davies, H.M.L., Matasi, J.J., and Ahmed, G. (1996) *J. Org. Chem.*, **61**, 2305–2013; (c) Davies, H.M.L., Calvo, R., and Ahmed, G. (1997) *Tetrahedron Lett.*, **38**, 1737–1740.
113. Olson, J.P. and Davies, H.M.L. (2008) *Org. Lett.*, **10**, 573–576.
114. (a) Inoue, M., Carson, M.W., Frontier, A.J., and Danishefsky, S.J. (2001) *J. Am. Chem. Soc.*, **123**, 1878–1889; (b) Inoue, M., Frontier, A.J., and Danishefsky, S.J. (2000) *Angew. Chem. Int. Ed.*, **39**, 761–764.
115. Lian, Y., Miller, L.C., Born, S., Sarpong, R., and Davies, H.M.L. (2010) *J. Am. Chem. Soc.*, **132**, 12422–12425.
116. Comer, F.W. and Trotter, J. (1966) *J. Chem. Soc. Phys. Org.*, **1**, 11–18.
117. (a) Chandler, C.L. and List, B. (2008) *J. Am. Chem. Soc.*, **130**, 6737–6739; (b) Wang, J.-C. and Krische, M.J. (2003) *Angew. Chem. Int. Ed.*, **42**, 5855–5857; (c) Toyota, M., Nishikawa, Y., Motoki, K., Yoshida, N., and Fukumoto, K. (1993) *Tetrahedron*, **49**, 11189–11204; (d) Cossy, J., Belotti, D., and Pete, J.P. (1990) *Tetrahedron*, **46**, 1859–1870; (e) Hua, D.H., Sinai-Zingde, G., and Venkataraman, S. (1985) *J. Am. Chem. Soc.*, **107**, 4088–4090; (f) Tatsuta, K., Akimoto, K., and Kinoshita, M. (1979) *J. Am. Chem. Soc.*, **101**, 6116–6118.
118. Jiao, L., Yuan, C., and Yu, Z.-X. (2008) *J. Am. Chem. Soc.*, **130**, 4421–4430.
119. Peng, J., Walsh, K., Weedman, V., Bergthold, J.D., Lynch, J., Lieu, K.L., Braude, I.A., Kelly, M., and Hamann, M.T. (2002) *Tetrahedron*, **58**, 7809–7819.
120. (a) Trost, B.M., Dong, L., and Schroeder, G.M. (2005) *J. Am. Chem. Soc.*, **127**, 2844–2845; (b) Takano, M., Umino, A., and Nakada, M. (2004) *Org. Lett.*, **6**, 4897–4900; (c) Piers, E., Gilbert, M., and Cook, K.L. (2000) *Org. Lett.*, **2**, 1407–1410; (d) Snider, B.B., Vo, N.H., O'Neil, S.V., and Foxman, B.M. (1996) *J. Am. Chem. Soc.*, **118**, 7644–7645.
121. Pfeiffer, M.W.B. and Phillips, A.J. (2005) *J. Am. Chem. Soc.*, **127**, 5334–5335.
122. Pfeiffer, M.W.B. and Phillips, A.J. (2008) *Tetrahedron Lett.*, **49**, 6860–6861.

123. (a) McMillan, C., Chavez, P.I., and Mabry, T.J. (1975) *Biochem. Syst. Ecol.*, **3**, 137–141; (b) Kawazu, K., Nakajima, S., and Ariwa, M. (1979) *Experientia*, **35**, 1294–1295; (c) Bohlmann, F., Singh, P., Joshi, K.C., and Singh, C.L. (1982) *Phytochemistry*, **21**, 1441–1443; (d) Ghazy, N.M., Omar, A.A., Elrashidy, E.M., and Metwally, A.M. (1988) *Egypt. J. Pharm. Sci.*, **29**, 39–42; (d) Ahmed, A.A., Jakupovic, J., Bohlmann, F., Regaila, H.A., and Ahmed, A.M. (1990) *Phytochemistry*, **29**, 2211–2215.

124. Kummer, D.A., Brenneman, J.B., and Martin, S.F. (2005) *Org. Lett.*, **7**, 4621–4623.

125. (a) Prestwich, G.D., Solheim, B.A., Clardy, J., Pilkiewicz, F.G., Miura, I., Tanis, S.P., and Nakanishi, K. (1977) *J. Am. Chem. Soc.*, **99**, 8082–8083; (b) Prestwich, G.D., Goh, S.H., and Tho, Y.P. (1981) *Experientia*, **37**, 11–13.

126. Dauben, W.G., Farkas, I., Bridon, D.P., Chuang, C.-P., and Henegar, K.E. (1991) *J. Am. Chem. Soc.*, **113**, 5883–5884.

127. Schubert, M. and Metz, P. (2011) *Angew. Chem. Int. Ed.*, **50**, 2954–2956.

128. Turner, W.B. (1978) *J. Chem. Soc., Perkin Trans. 1*, 1621.

129. Siddiqui, I.N., Zahoor, A., Hussain, H., Ahmed, I., Ahmad, V.U., Padula, D., Draeger, S., Schulz, B., Meier, K., Steinert, M., Kurtán, T., Flörke, U., Pescitelli, G., and Krohn, K. (2011) *J. Nat. Prod.*, **74**, 365–373.

130. Tietze, L.F., Ma, L., Reiner, J.R., and Jackenkroll, S. (2013) *Chem. Eur. J.*, **19**, 8610–8614.

131. (a) Tietze, L.F., Sommer, K.M., Zinngrebe, J., and Stecker, F. (2005) *Angew. Chem. Int. Ed.*, **44**, 257–259. (b) Tietze, L.F., Görlitzer, J., Schuffenhauer, A., and Hübner, M. (1999) *Eur. J. Org. Chem.*, **1999**, 1075–1084; (c) Tietze, L.F. and Görlitzer, J. (1998) *Synthesis*, 873–878.

132. Trost, B.M., Shen, H.C., Dong, L., Surivet, J.-P., and Sylvain, C. (2004) *J. Am. Chem. Soc.*, **126**, 11966–11983.

133. Tanaka, N., Okasaka, M., Ishimaru, Y., Yakaishi, Y., Sato, M., Okamoto, M., Oshikawa, T., Ahmed, S.U., Consentino, L.M., and Lee, K.-H. (2005) *Org. Lett.*, **7**, 2997–2999.

134. Nicolaou, K.C., Sarlah, D., and Shaw, D.M. (2007) *Angew. Chem. Int. Ed.*, **46**, 4708–4711.

135. Inoue, Y., Ohuchi, K., Yen, I.-F., and Imaizumi, S. (1989) *Bull. Chem. Soc. Jpn.*, **62**, 3518–3522.

136. Priebbenow, D.L., Henderson, L.C., Pfeffer, F.M., and Stewart, S.G. (2010) *J. Org. Chem.*, **75**, 1787–1790.

137. (a) Wang, H., Usui, T., Osada, H., and Ganesan, A. (2000) *J. Med. Chem.*, **43**, 1577–1585; (b) Jenkins, P.R., Wilson, J., Emmerson, D., Garcia, M.D., Smith, M.R., Gray, S.J., Britton, R.G., Mahale, S., and Chaudhuri, B. (2008) *Bioorg. Med. Chem.*, **16**, 7728–7739.

138. Bano, S., Parveen, S., and Ahmad, V.U. (1990) *J. Nat. Prod.*, **53**, 492–495.

139. Lam, S.K. and Chiu, P. (2007) *Chem. Eur. J.*, **13**, 9589–9599.

140. (a) Canham, S.M., France, D.J., and Overman, L.E. (2010) *J. Am. Chem. Soc.*, **132**, 7876–7877; (b) Canham, S.M., France, D.J., and Overman, L.E. (2013) *J. Org. Chem.*, **78**, 9–34.

141. Hirasawa, Y., Morita, H., Shiro, M., and Kobayashi, J. (2003) *Org. Lett.*, **5**, 3991–3993.

142. (a) Baskar, B., Bae, H.J., An, S.E., Cheong, J.Y., Rhee, Y.H., Duschek, A., and Kirsch, S.F. (2008) *Org. Lett.*, **10**, 2605–2607; (b) Menz, H., Binder, J.T., Crone, B., Duschek, A., Haug, T.T., Kirsch, S.F., Klahn, P., and Liébert, C. (2009) *Tetrahedron*, **65**, 1880–1888.

143. Aoki, S., Watanabe, Y., Sanagawa, M., Setiawan, A., Kotoku, N., and Kobayashi, M. (2006) *J. Am. Chem. Soc.*, **128**, 3148–3149.

144. (a) For a comprehensive account of published syntheses until 2010 see: Hardin-Narayan, A.R., Simmons, E.M., and Sapong, R. (2010) *Eur. J. Org. Chem.*, **2010**, 3553–3567; (b) Shi, J., Manolikakes, G., Yeh, C.-H., Guerrero, C.A., Shenvi, R.A., Shigehisa, H., and Baran, P.S. (2011) *J. Am. Chem. Soc.*, **133**, 8014–8027; (c) Yamashita, S., Iso, K., Kitajima, K., Himuro, M., and Hirama, M. (2011) *J. Org. Chem*, **76**, 2408–2425; (d) Flyer, A.N., Si, C., and Myers, A.G. (2010) *Nat.*

Chem, **2**, 886–892; (e) Simmons, E.M., Hardin-Narayan, A.R., Guo, X., and Sarpong, R. (2010) *Tetrahedron*, **66**, 4696–4700.

145. Shenvi, R.A., Guerrero, C.A., Shi, J., Li, C.-C., and Baran, P.S. (2008) *J. Am. Chem. Soc.*, **130**, 7241–7243.

146. Frie, J.L., Jeffrey, C.S., and Sorensen, E.J. (2009) *Org. Lett.*, **11**, 5394–5397.

147. Woodward, R.B., Cava, M.P., Ollis, W.D., Hunger, A., Daeniker, H.U., and Schenker, K. (1954) *J. Am. Chem. Soc.*, **76**, 4749–4751.

148. Beemelmanns, C. and Reissig, H.-U. (2010) *Angew. Chem. Int. Ed.*, **49**, 8021–8025.

149. Rawal, V.H. and Iwasa, S. (1994) *J. Org. Chem.*, **59**, 2685–2686.

150. Martin, D.B.C. and Vanderwal, C.D. (2011) *Chem. Sci.*, **2**, 649–651.

151. (a) Sato, B., Muramatsu, H., Miyauchi, M., Hori, Y., Takese, S., Hino, M., Hashimoto, S., and Terano, H. (2000) *J. Antibiot.*, **53**, 123–130; (b) Sato, B., Nakajima, H., Hori, Y., Hino, M., Hashimoto, S., and Terano, H. (2000) *J. Antibiot.*, **53**, 204–206; (c) Yoshimura, S., Sato, B., Kinoshita, T., Takase, S., and Terano, H. (2000) *J. Antibiot.*, **53**, 615–622; (d) Yoshimura, S., Sato, B., Kinoshita, T., Takase, S., and Terano, H. (2002) *J. Antibiot.*, **55**, C1. (e) Yoshimura, S., Sato, B., Takase, S., and Terano, H. (2004) *J. Antibiot.*, **57**, 429–435.

152. (a) Evans, D.A. and Starr, J.T. (2002) *Angew. Chem. Int. Ed.*, **41**, 1787–1790; (b) Evans, D.A. and Starr, J.T. (2003) *J. Am. Chem. Soc.*, **125**, 13531–13540; (c) Vosburg, D.A., Vanderwal, C.D., and Sorensen, E.J. (2002) *J. Am. Chem. Soc.*, **124**, 4552–4553; (d) Vanderwal, C.D., Vosburg, D.A., Weiler, S., and Sorensen, E.J. (2003) *J. Am. Chem. Soc.*, **125**, 5393–5407; (e) Tanaka, N., Suzuki, T., Matsumura, T., Hosoya, Y., and Nakada, M. (2009) *Angew. Chem. Int. Ed.*, **48**, 2580–2583.

15
Multicomponent Domino Process: Rational Design and Serendipity

Qian Wang and Jieping Zhu

15.1
Introduction

The domino process, as defined by Tietze, is a combination of two or more bond-forming reactions under identical conditions wherein the subsequent reactions result as a consequence of the functionality formed in the previous step. The quality of a domino reaction can be correlated to the number of chemical bonds formed, as well as to the increase of molecular complexity and its general applicability. The greater the number of bonds formed, which usually goes hand in hand with an increase of complexity of the product, the more useful might be the process [1]. A multicomponent reaction (MCR) is a process in which three or more reactants are combined in a single chemical operation to produce a compound that incorporates substantial portions of all starting materials [2]. It is a sequence of bimolecular events that proceeds according to the domino principle: that is, the subsequent transformations are a consequence of the functionalities produced in the previous one. Therefore, an MCR is a domino process by definition.

Domino process can be unimolecular (Eq. (1) [3], bimolecular (Eq. (2) [4], or multicomponent (Eq. (3), Scheme 15.1) [5]. A common feature of these processes is the high bond-forming efficiency that is essential for the creation of molecular complexity. However, MCR has following added bonus relative to uni and bimolecular domino processes: (i) high convergence, (ii) use of easily accessible starting materials, (iii) multiple variables and hence easy introduction of structural diversity, and (iv) large chemical space: the number of theoretically accessible compounds by MCRs is extremely high.

Contrary to the general perception, MCR occupies an important position in the development of modern organic chemistry. Indeed, many important named reactions such as the Strecker amino nitrile synthesis (1850) [6], the Hantzsch dihydropyridine synthesis (1882) [7], the Biginelli dihydropyrimidine synthesis (1891) [8], the Mannich reaction (1912) [9], the isocyanide-based Passerini reaction (1921) [10], and the Ugi (1959) reaction [11], among others, are all multicomponent processes. In spite of the significant contribution of MCRs to the state of the art of modern organic chemistry and its demonstrated potential in the synthesis of

Domino Reactions: Concepts for Efficient Organic Synthesis, First Edition. Edited by Lutz F. Tietze.

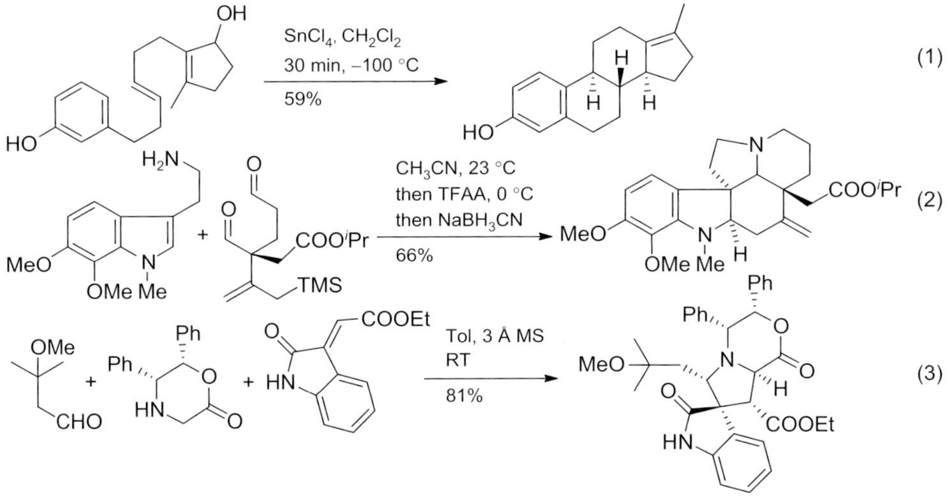

Scheme 15.1 Unimolecular, bimolecular, and multicomponent domino processes.

complex natural products, as exemplified in Robinson's landmark synthesis of tropinone in 1917 (Scheme 15.2) [12], little attention was paid to the development of novel MCRs during the period 1960–1990, with some important examples being listed in Scheme 15.3. However, from the mid-1990s, the research dedicated to the development and application of MCRs increased steadily probably because of the need for (i) rapid access to a large collection of compounds in parallel to the development of molecular biology and high-throughput biological screening facilities and (ii) a greener and more efficient chemistry. Indeed, MCR is an ideal reaction type to investigate in the quest for ideal chemical synthesis since this type of reaction is inherently (i) chemo- and regioselective; (ii) atom-economic [13] since most of them involve addition rather than substitution reactions; (iii) step-efficient and cost effective since it creates at least two chemical bonds in one operation; (iv) convergent and efficient in generating molecular complexity and diversity [14]; (v) operationally simple since most of the MCRs are performed under mild reaction conditions and in some cases even proceed spontaneously in the absence of an external reagent; and (vi) cost effective by decreasing the number of synthetic steps.

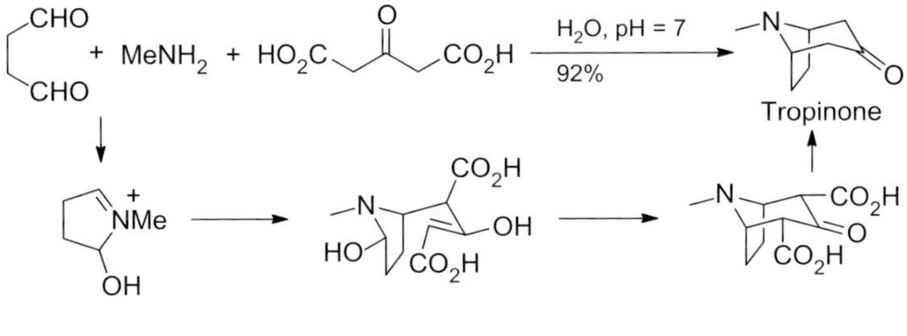

Scheme 15.2 Robinson's three-component synthesis of tropinone.

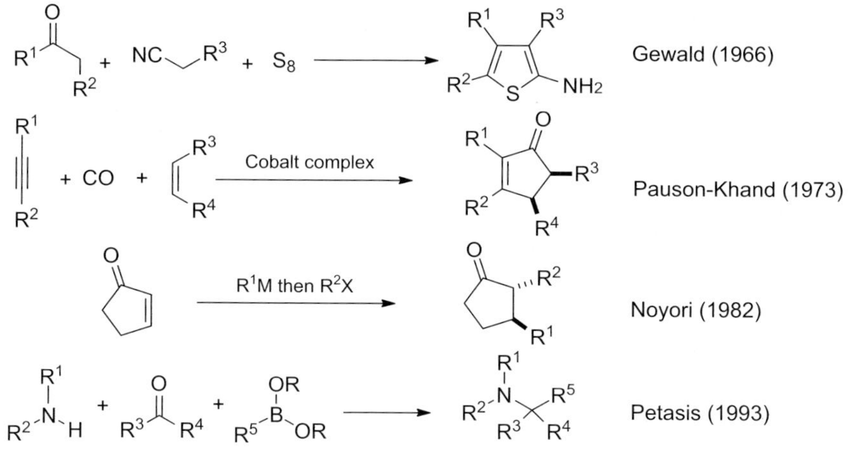

Scheme 15.3 Examples of some of the most recent named MCRs.

15.2 Basic Considerations of MCRs

MCR is composed of a series of elementary bimolecular reactions. Mixing three or more substrates, each bearing a reactive functional group, could provoke different reaction manifolds leading to different final products. In addition, the very domino nature of a MCR implies that any bimolecular reaction would generate a new intermediate function, which will in turn enter into the reaction process. Therefore, in devising any novel MCR, one has to carefully consider not only the reactivity match of the functional groups but also that of the incipient intermediate functionalities. An illustrative example is shown in Scheme 15.4. It is known that

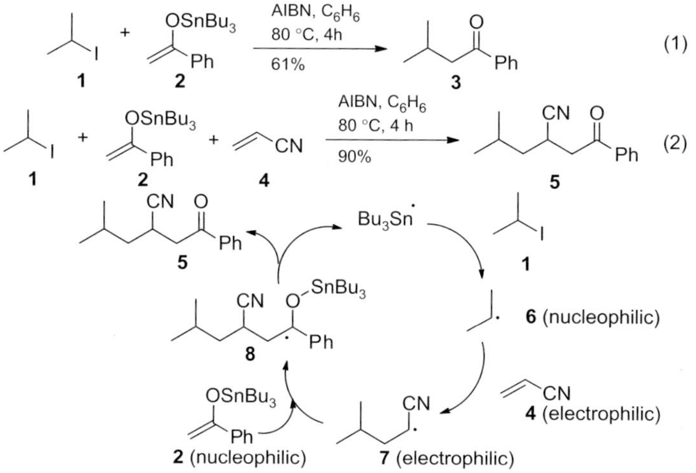

Scheme 15.4 Radical-based three-component reaction: example of reactivity match.

the alkyl radical generated from alkyl iodide **1** can undergo radical addition to tin enolate **2** to afford the homologated ketone **3** (Eq. (1), Scheme 15.4). However, in terms of reactivity, this is not a particularly matched case since both the alkyl radical and tin enolate are nucleophilic in nature. By adding an electron-deficient alkene into the above reaction mixture, an efficient three-component reaction was subsequently developed [15]. As shown in the catalytic cycle, the electron-rich radical **6** preferred to react with the electron-deficient acrylonitrile (**4**) to afford the radical **7**. The resulting radical **7**, with an electron-withdrawing group (EWG) at α position, was now electrophilic and added preferentially to the electron-rich alkene **2** to afford, after fragmentation, the three-component adduct **5** in excellent yield.

Most of the elementary reactions in the classic MCRs are equilibrium processes. Therefore, thermodynamic factors can significantly impact the reaction pathways in addition to the reaction kinetics. A classic example is the Strecker synthesis of α-amino nitrile **9** from aldehydes, amines, and cyanide (Scheme 15.5). The key step in this reaction is the nucleophilic addition of cyanide to the *in situ* formed iminium. However, condensation of a carbonyl compound with an amine leading to iminium is an equilibrium process, especially under aqueous conditions. Therefore, the desired addition reaction is in competition with direct addition of cyanide to the aldehyde, leading to cyanohydrin **10**. However, since the formation of both **9** and **10** were reversible, only the more stable adduct **9** was produced at the expense of cyanohydrin **10** under thermodynamically controlled conditions.

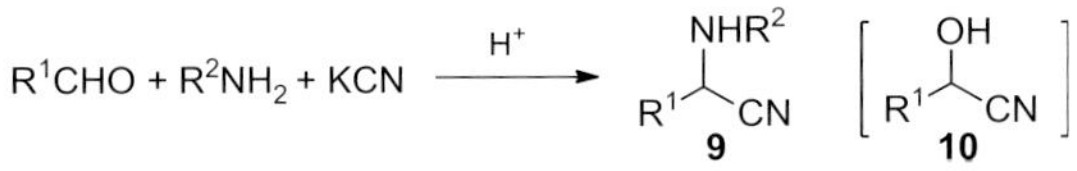

Scheme 15.5 The Strecker reaction.

A key point one can draw and should keep in mind from above analysis is that the presence of equilibrium steps is generally beneficial to MCR sequences. Otherwise, if an MCR was composed of a sequence of irreversible elementary reactions, then each step must be highly chemoselective in order to avoid the formation of side products, and examples belonging to this class are relative rare. Nevertheless, the Noyori three-component reaction involving the Michael addition of an organometallic species to enone followed by trapping of the resulting enolate by an alkyl halide could be categorized into this family.[1)]

As for any organic reactions, appropriate experimental conditions are of paramount importance. In Scheme 15.6 are depicted two most important isonitrile-based MCRs: the Passerini-3CR (P-3CR, (Eq. (1), Scheme 15.6) and the Ugi-4CR (U-4CR, Eq. (2), Scheme 15.6). While the P-3CR is generally performed in nonpolar aprotic solvents (e.g., CH_2Cl_2), the U-4CR prefers polar protic solvents (e.g., MeOH). Indeed, if one carries out the U-4CR in CH_2Cl_2, then the P-3CR

1) Michael addition could be reversible, depending on the nature of nucleophiles. However, when RM (R = alkyl, alkenyl, M = metals) was used as nucleophiles, the 1,4-addition became essentially irreversible.

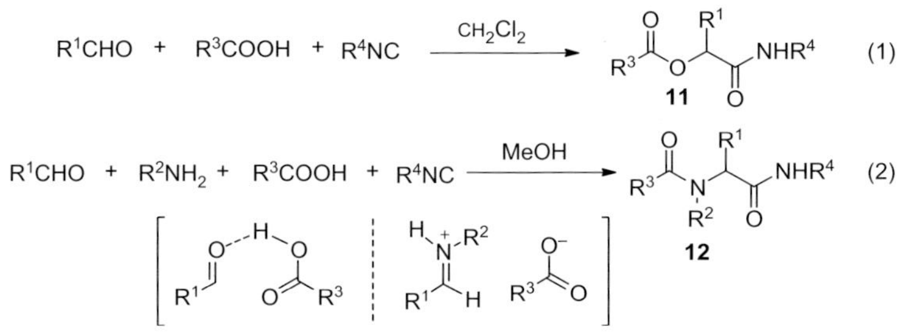

Scheme 15.6 The Passerini-3CR and the Ugi-4CR.

will become competitive, leading to α-acetoxy carboxamide **11** together with the desired α-acetamido carboxamide **12**. This intriguing solvent effect could in fact be explained by the fact that P-3CR is initiated by nucleophilic addition of isonitrile to the uncharged aldehyde function, while U-4CR involves an *in situ* formed positively charged iminium intermediate.

Many reviews dealing with the discovery and the application of MCRs in the drug development in the syntheses of heterocycles and bioactive natural products have appeared in recent years [16]. Instead of categorizing the reaction type, we will focus on the reaction development. As for any discovery process, there are many approaches, such as combinatorial method, union concept, rational design, and serendipity for discovering novel MCRs [17]. We will focus, in this chapter, on the rational "substrate design approach" for the development of MCRs and its interplay with serendipities. Indeed, as we have already stated, MCR is a combination of a series of two-component reactions where the product of the first reaction reacts with the third input to give a second product, which in turn may react with a fourth starting material, and so on. Therefore, by carefully considering the chemical reactivity of the individual starting material and relevant intermediates, it is possible to design novel MCRs through substrate design. If poly-functionalized substrates are designed and programmed in such a way that they will react in a highly ordered and productive manner to produce in high yield an interesting scaffold, then a novel MCR would be uncovered. Since every group has its own rational approach/analysis, we will illustrate the power of this approach using mainly examples of isocyanide-based MCRs developed in this group.

15.3
Substrate Design Approach in the Development of Novel MCRs

15.3.1
Chemistry of α-Isocyanoacetates

α-Isocyanoacetate (**13**) offers at least three reactive sites: (i) the isocyano group. The divalent carbon of isonitriles has a carbene-like reactivity undergoing readily α-addition to both nucleophiles and electrophiles. Note that it has such a

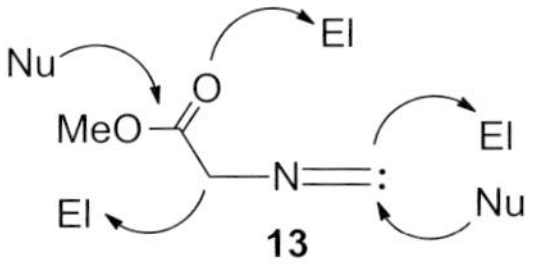

Figure 15.1 Reactivities of the α-isocyanoacetate.

pronounced nucleophilicity that most of the α-addition is initiated by nucleophilic addition of isocyano group to the polarized multiple bond; (ii) the ester function; and (iii) an acidic α-CH, hence a nucleophilic α-carbon upon deprotonation (Figure 15.1). Not surprisingly, α-isocyanoacetate (**13**), by virtue of its multiple functionalities, has become a much sought-after starting material in the development of MCRs [18].

The isocyano group, besides its carbene-like reactivity, is also an EWG. Therefore, the α-proton of α-isocyano acetate (**13**) is quite acidic because of the presence of two vicinal EWGs (ester and isocyano) and is readily deprotonated under even weakly basic conditions [19]. The nucleophilicity of the α-carbanion of the so-produced enolate is generally higher than that of the terminal divalent carbon of isonitrile. Therefore, its reaction with electrophiles, especially the polarized multiple bond, is generally initiated by the α-carbon of the α-isocyanoacetate. Matsumoto and coworkers [20] reported in 1978 that simply heating a solution of methyl α-isocyano acetate (**13**), 4-chlorobenzaldehyde (**14**), and piperidine (**15**) in MeOH led to the formation of amidine (**16**) in about 50% yield (Scheme 15.7). The Knoevenagel condensation followed by a formal α-addition of the secondary amine onto the isocyano group was proposed to account for the formation of **16**.

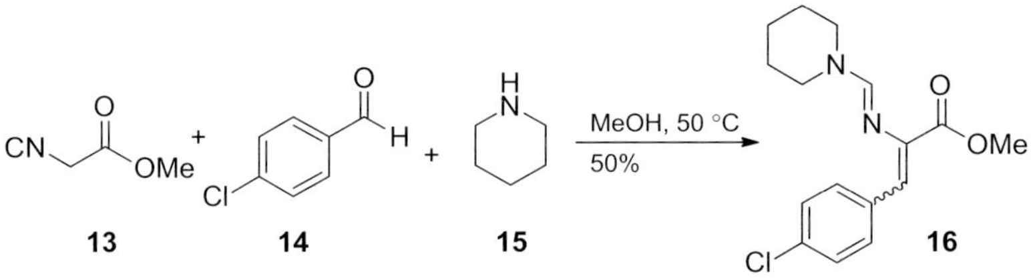

Scheme 15.7 Three-component synthesis of amidine from α-isocyanoacetate.

By carefully tuning the structure of the starting materials, Orru *et al.* [21] developed an efficient synthesis of imidazoline **18** by using essentially the same reaction partners (Scheme 15.8). The key to substrate design is the use of α-aryl-substituted α-isocyanoacetate (**17**) and the primary amine instead of the secondary amine as reaction partner of the aldehyde. The reaction was initiated by nucleophilic addition of the enolate anion of **17** onto the imine **19**, generated *in situ*. However, the lack of an additional acidic proton α to ester in the Mannich adduct **20** made the β-elimination impossible. On the other hand, the Mannich adduct having a secondary amine remained nucleophilic and could add intramolecularly to the divalent isocyano carbon, leading, after protonation, to the imidazoline **18** in

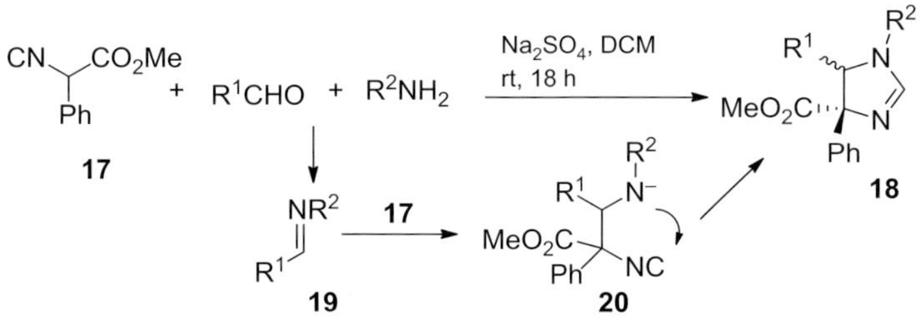

Scheme 15.8 Three-component synthesis of imidazoline.

excellent yield and diastereoselectivity.[2] The presence of an α-phenyl group was supposed to contribute further the acidity of the α-proton in **17**, allowing thus the deprotonation to occur under even mild conditions (CH_2Cl_2, room temperature (rt)).

These two examples illustrate how a subtle structural modification of the reactants can modify the reaction course, and hence the product of the MCR.

15.3.2
From α-Isocyanoacetates to α-Isocyanoacetamides

The divalent carbon of isocyanide has a pronounced nucleophilicity and undergoes readily α-addition to nucleophiles and electrophiles (Scheme 15.9). This chemical property is the basis of the Passerini three-component reaction (Eq. (1), Scheme 15.6) and the Ugi four-component reaction (Eq. (2), Scheme 15.6). However, the two aforementioned α-isocyanoacetate-based three-component reactions (cf. Scheme 15.7 and Scheme 15.8) were both initiated by the nucleophilicity of the α-carbon. The question one could ask is: can we initiate the reaction sequence by taking advantage of the nucleophilicity of the isocyano function using similar substrates? It is conceivable that, if the sequence can be initiated by the nucleophilicity of the isocyanide, a different reaction manifold will be induced leading to a different multicomponent adduct.

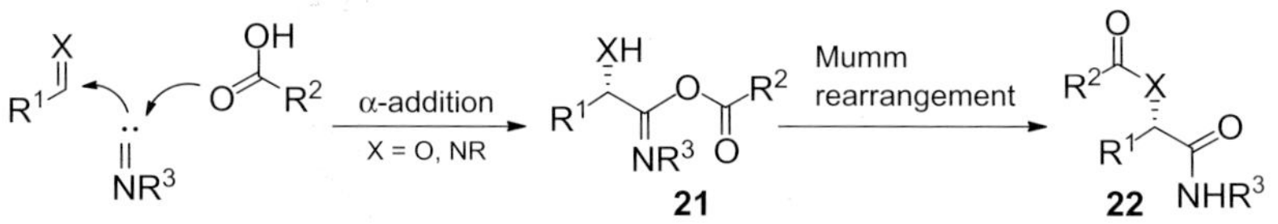

Scheme 15.9 α-Addition of isonitrile and subsequent Mumm rearrangement.

The isocyanide is intrinsically nucleophilic, whereas the α-carbon of α-isocyanoacetate becomes a nucleophile only upon deprotonation. It is therefore

2) Only strong nucleophiles such as organolithium, organomagnesium, and organozinc reagents can undergo nucleophilic addition to isonitrile. However, intramolecular nucleophilic addition to the isocyano group is much easier to proceed.

possible to channel the reactivity profile of α-isocyanoacetate by modulating the acidity of the α-CH. If α-isocyanoacetic acid derivatives have a less acidic α-proton and if the reaction conditions are sufficiently mild to prevent the α-deprotonation, then one can expect to have a reaction sequence initiated by the nucleophilicity of the isocyanide, leading consequently to a completely different product.

Toward this end, we set out to investigate the reactivity of the α-isocyanoacetamide **23** vis-à-vis aldehydes and imines. The design of **23** is based on the following considerations. First, the $pK_{a(\alpha CH)}$ of amide is known to be 3–5 units higher than that of ester. For example, the $pK_{a(\alpha CH)}$ of $NCCH_2CO_2Et$ is 13.1 (in dimethyl sulfoxide (DMSO)), while that of $NCCH_2CONEt_2$ is 17.1 (in DMSO). Consequently, the α-CH of amide should be less easily deprotonated. Note also that, in P-3CR and U-4CR, the nitrilium intermediate resulting from the nucleophilic addition of isocyano group to the polarized double bond is trapped by the carboxylic acid, leading to the α-adduct **21**, which then undergoes the irreversible Mumm rearrangement to afford the final multicomponent adduct **22** (Scheme 15.9). In the case of α-isocyanoacetamides, the pronounced Lewis basicity of the amide oxygen relative to the ester counterpart makes it a better nucleophile, which could potentially act as a nucleophile to trap the nitrilium intermediate, leading to a cyclic product.

This consideration led us to perform the same reaction as described by Matsumoto but using α-isocyanoacetamide **23** [22] instead of the α-isocyanoacetate as a reaction partner. As shown in Scheme 15.10, the reaction indeed proceeded in a completely different way to afford 5-aminooxazole **24** in excellent yield [23]. We hypothesized that, under these mild conditions, the deprotonation of α-proton of amide **23** did not occur. Consequently, the sequence is initiated by a nucleophilic addition of the isocyano carbon to the *in situ* generated imine **19** and led to the nitrilium intermediate **25**, which was in turn trapped by the amide oxygen to

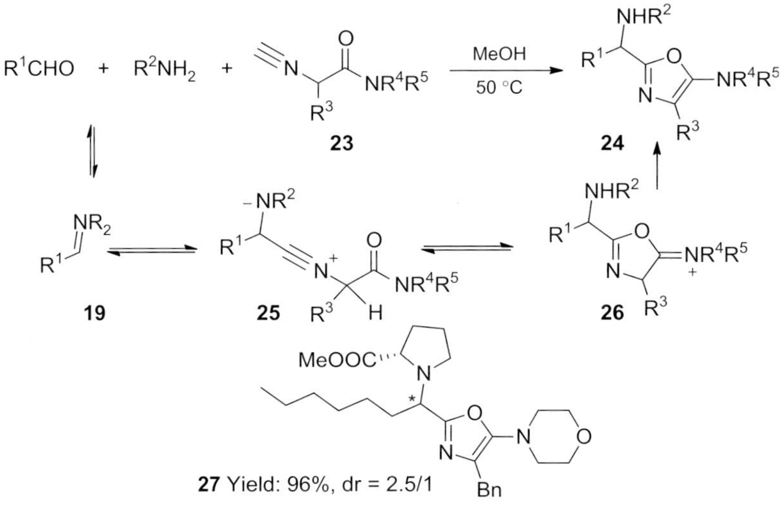

Scheme 15.10 Three-component synthesis of 5-aminooxazoles.

afford, after tautomerization, the oxazole **24**. Neither imidazoline nor amidine (if $R^3 = H$) resulting from the competitive Mannich and/or Knoevenagel type condensation was observed, which is the predominant pathway for α-isocyanoacetates. This example demonstrates how subtle structural modification of one input in a given MCR could modify the reaction sequence and diversify the structure of the resulting adducts.

The reaction turns out to be very general and can be performed with approximately equimolar amount of three components, simplifying thus the purification step. A variety of amines (primary, secondary, and aniline), aldehydes (aliphatic, aromatic), and isocyanides (α-unsubstituted, α-alkyl, and α-aryl substituted) participated in the reaction to give the corresponding oxazoles in good to excellent yields. It is worth noting that the reaction proceeded in the absence of any external reagent, constituting therefore an example of ideal "mix-and-go" protocol.

When enantiomerically pure isonitriles **23** ($R_3 = $ alkyl) were used as inputs, racemic oxazoles were obtained. On the other hand, using proline methyl ester as an amine input, a moderate asymmetric induction was observed, leading to two separable diastereomers **27** in the ratio 2.5 : 1.

As for the Ugi-4CR, methanol turned out to be the solvent of choice for the above three-component reaction. However, we found that the present 3CR worked equally well in toluene in the presence of a very weak Brønsted acid such as NH_4Cl. On the basis of this key observation, a catalytic enantioselective three-component synthesis of oxazole was subsequently developed. As shown in Scheme 15.11, simply mixing pivaldehyde (**28**), 4-trifluoromethylaniline (**29**), and α-phenyl-α-isocyanoacetamide (**30**) in the presence of a catalytic amount of chiral phosphoric acid **31**, the 5-amino oxazole **32** was isolated in 82% yield with 90% ee [24]. At the present stage of development, this catalytic enantioselective process is still limited in scope, as reasonable enantioselectivity was obtained only with sterically congested aldehydes. The use of aniline is also mandatory, as very low enatiomeric excess was obtained with aliphatic amines, most probably due to the problem of catalyst turnover in the later case.

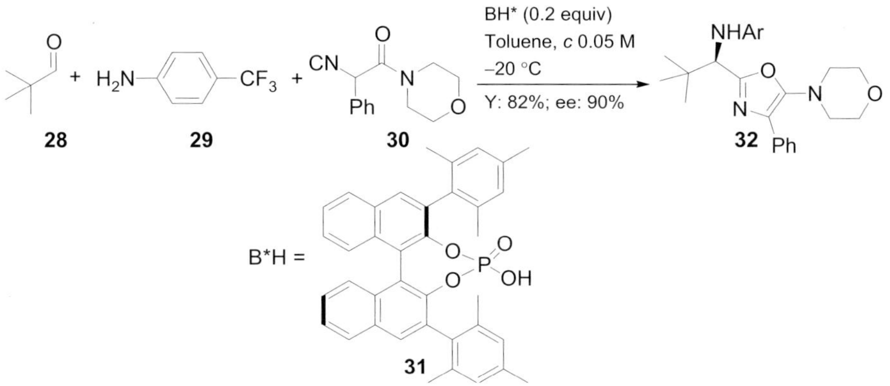

Scheme 15.11 Chiral phosphoric acid-catalyzed three-component synthesis of oxazoles.

Macrocycles by virtue of their widespread occurrence in nature and their intrinsic three-dimensional structure have found numerous applications in drug development, materials science, supramolecular chemistry, and so on. Not surprisingly, the challenging problem of macrocyclization has attracted the attention of synthetic chemists and provided impetus for the development of new technologies. Since macrocyclization has to be generally carried out at high dilution to avoid oligomerization/polymerization, whereas MCR is kinetically favored when it is performed at high concentration, the development of synthetically useful multicomponent synthesis of macrocycles is challenging [25]. Indeed, in contrast to the formidable development on the multicomponent synthesis of heterocycles, there were only sporadic reports on the multicomponent synthesis of macrocycles before 2003 [26]. Among different approaches, one option consisted of using two bifunctional building blocks to perform multiple MCRs involving one macrocyclization step (Eq. (1), Scheme 15.12). Here again, the proper design of substrate structures is

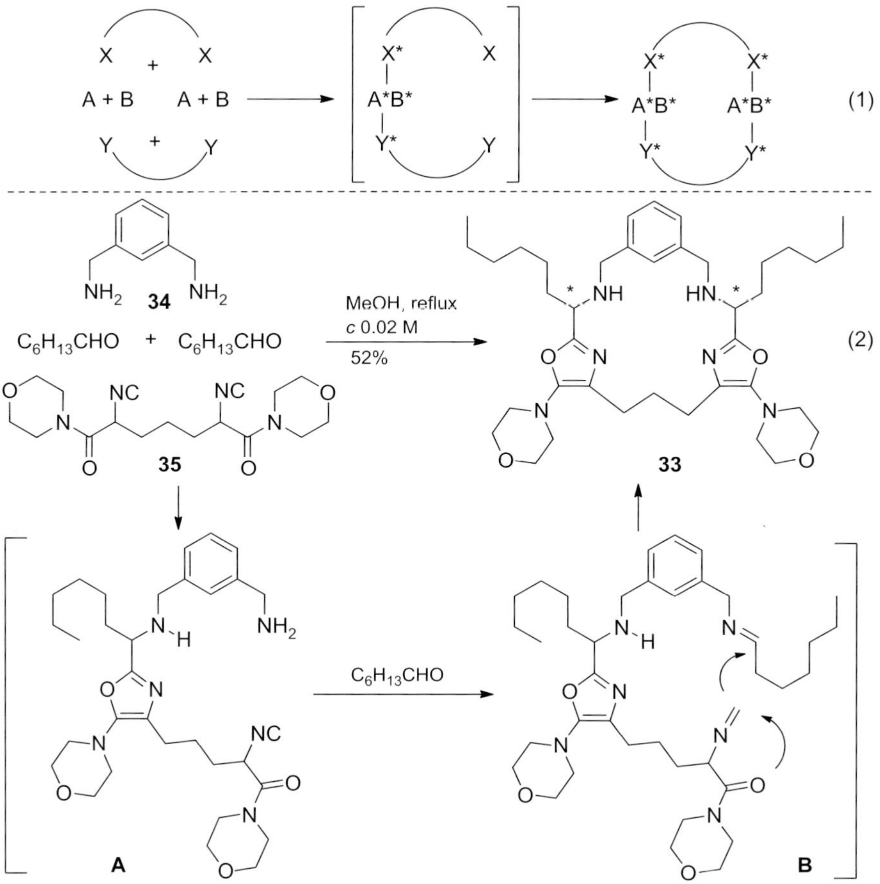

Scheme 15.12 Four-component (ABC_2) synthesis of *m*-cyclophane.

of paramount importance for the reaction to proceed in a single direction and to avoid competitive oligomerization/polymerization.

On the basis of this consideration, four-component (ABC_2) synthesis of *m*-cyclophane **33** was devised, implicating a twofold three-component synthesis of oxazole. Heating to reflux a solution of diamine **34**, a bis-α-isocyanoacetamide **35**, and 2 equiv of heptaldehyde afforded the macrocycle **33** in 52% yield (Eq. (2), Scheme 15.12) [27]. In this MCR, one macrocycle embedded with two heterocycles was produced in 52% yield via the creation of six chemical bonds, with water as the only by-product. The reaction proceeded through a three-component construction of oxazole **A** followed by its subsequent reaction with a second equivalent of aldehyde to provide the observed product via the imine intermediate **B**. Amazingly, the reaction performed at 0.1 M furnished cyclophane **33** in higher yield than when carried out at 0.01 M under otherwise identical conditions. That the multicomponent macrocyclization can be performed at such a high concentration is unique. Several factors could account for this observation: (i) the actual concentration of cyclization precursor **B**, generated by three-component reaction, is much lower than the concentration of the starting materials (0.1 M) and (ii) the *in situ* build-in oxazole ring in the cyclization precursor could potentially reduce the conformational mobility of the molecule, facilitating thus the desired head-to-tail cyclization [28].

15.3.3
From α-Isocyanoacetamides to α-Isocyanoacetic Acids

We subsequently hypothesized that α-isocyanoacetic acids **36** should display a similar reactivity profile as that of α-isocyanoacetamides. The reasoning was that the carboxylic acid function (COOH), being more acidic than any other protons in **36**, would be deprotonated first to the carboxylate. Since the formation of the dianion is highly impossible under weakly basic conditions, the reaction of **36** with the polarized double bond should again be initiated by the divalent carbon of the isocyano group. Indeed, reaction of an amine (2.0 equiv), an aldehyde (1.0 equiv), and the potassium salt of isocyanoacetic acid (**37**, 1.0 equiv),[3] under optimized conditions [NH_4Cl (1.5 equiv), toluene (*c* 0.12 M), rt], afforded the dipeptide **38** as a mixture of two diastereomers in good to excellent yield (Scheme 15.13) [29]. A reasonable mechanism would involve the nucleophilic addition of the isocyano group of **37** to the *in situ* generated iminium to afford the nitrilium intermediate **39**. Trapping the latter by an internal carboxylate oxygen would provide the oxazolone **40**, which might be in equilibrium with the 5-hydroxyoxazole **41**. An intermolecular nucleophilic attack of a second equivalent of amine on the oxazolone would then produce **38**. Overall, this MCR can be considered as a U-4CR (ABC_2)/five-center reaction.

A representative example is shown in Scheme 15.14. Simply stirring a toluene solution of sarcosine (**42**), 3-phenylpropanal (**43**), and potassium α-isocyanoacetate (**37**) in the presence of NH_4Cl at rt afforded the tetrapeptide **44** in 66% isolated

3) The salt is much more stable than the free acid.

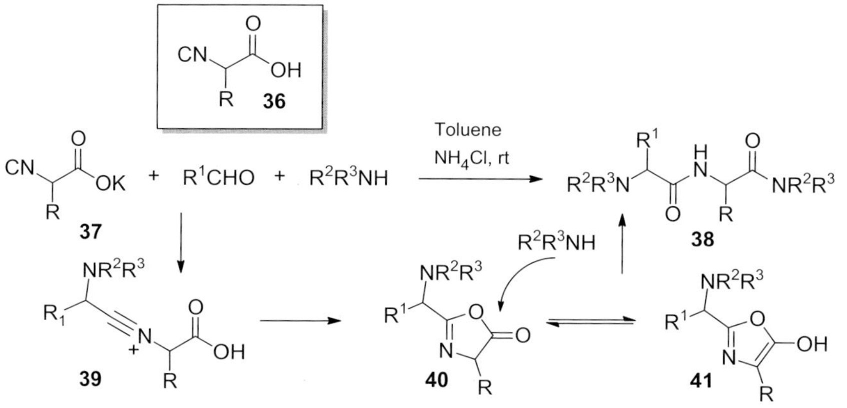

Scheme 15.13 Exploiting the reactivity of α-isocyanoacetic acids: three-component synthesis of peptides.

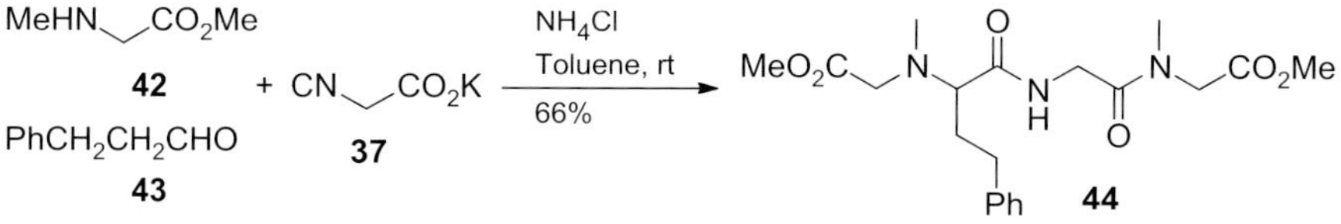

Scheme 15.14 Three-component synthesis of tripeptide **44**.

yield. In this reaction, one amino acid unit and two amide bonds were formed without the need for a coupling reagent.

15.3.4
Back to α-Isocyanoacetates

On going from α-isocyanoacetates to α-isocyanoacetamides, the acidity of the α-CH decreases because of the diminished electron-withdrawing power of the amide group. This subtle structural change impacted significantly the reactivity profile of these two substrates when engaged in the reaction with a polarized double bond. The next question we asked was, what will be the chemoselectivity by rendering the α-CH of α-isocyanoacetates more acidic? In terms of substrate design, this can be easily realized by introducing an additional EWG to the α-position of the α-isocyanoacetate or a heteroatom (X = N, for example) into the aromatic ring of the compound **17** (see **46** vs **17**). The principle that guided us in the design of **46** is completely different from that of α-isocyanoacetamides. The α-CH of **46** would be much more acidic than that of **17**, leading to its facile deprotonation even under very mild basic conditions. The natural consequence is that its conjugated base **48**, being highly stabilized, would display reduced nucleophilicity relative to **47**. Consequently, it is reasonable to assume that the reaction of **48** with a polarized double bond, such as imine, could be initiated by the nucleophilicity of the divalent carbon of isocyanide in spite of the concurrent existence of the carbanion. In other words, the α-(EWG-phenyl)-α-isocyanoacetates (**46**) would display isocyanoacetamide-type reactivity rather than that of the parent α-isocyanoacetate **17** (Figure 15.2).

The high acidity of the α-CH of **46a** was readily seen from the following control experiment (Scheme 15.15). Treating **17** with aqueous potassium hydroxide afforded the corresponding carboxylate **49**. On the contrary, compound **46a** was converted, under the identical conditions, exclusively to the enolate **48** instead of the potassium salt of the carboxylic acid (Scheme 15.15). These results clearly indicated (i) the high acidity of the α-CH of **46a** such that deprotonation of **46a** took place in preference to the nucleophilic attack of the hydroxide onto the ester group and (ii) the stability, and hence the low reactivity of the so-produced enolate, which persisted even in aqueous conditions.

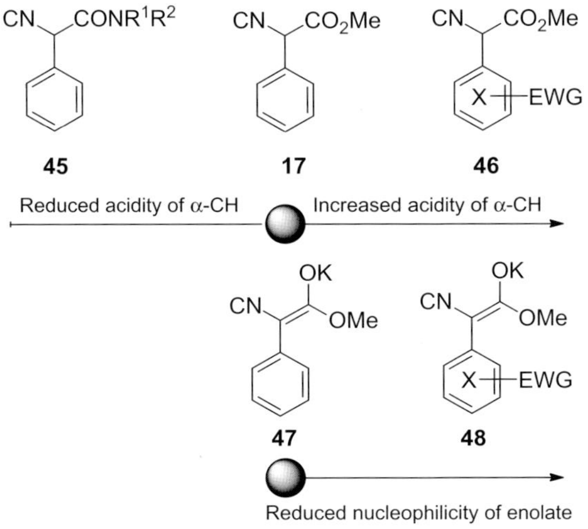

Figure 15.2 Dichotomy between Acidity of α-CH and nucleophilicity of the resulting enolate.

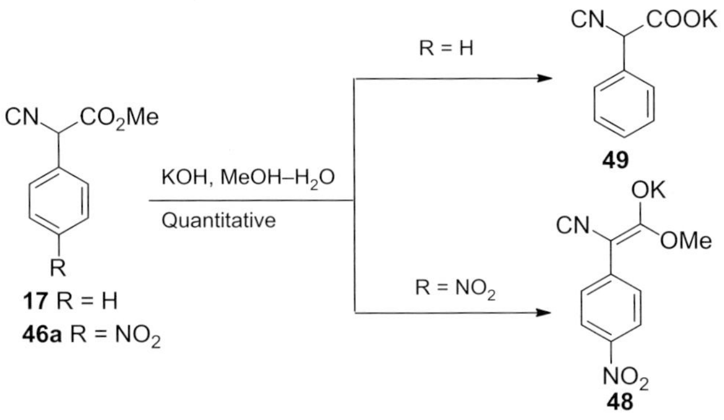

Scheme 15.15 Different behavior of **17** and **46a** under saponification conditions.

In line with our structure–reactivity analysis delineated above, the reaction of **46a**, imines, and aldehydes provided indeed 5-alkoxyoxazoles in good to excellent yields parallel to the chemistry of α-isocyanoacetamides (Scheme 15.16) [30]. The 3CR was performed with approximately equimolar quantities of the three components and reached completion between 4 and 16 h in toluene at rt or 60 °C. Both linear and α-branched aldehydes were found to be good substrates, as well as aromatic aldehydes and cinnamaldehyde. A wide range of aliphatic amines, including primary and cyclic secondary amines, were suitable partners to afford the corresponding 5-alkoxyoxazoles **50** in good to excellent yields. A concrete example is shown in Eq. (2) of Scheme 15.16.

α-Aryl-α-isocyanoacetates bearing an EWG at the ortho or para position of the aryl substituent (**46b–e**) and ethyl 2-isocyano-2-(pyridin-4-yl)acetate (**46f**, Figure 15.3) displayed the similar reactivity as **46a**. The reaction of these isocyanoacetates (**46a–f**) with aldehydes and amines afforded only the corresponding 5-alkoxyoxazole [31]. Imidazolines resulting from the competitive Mannich-type condensation were not observed in all these cases, validating thereby completely our working hypothesis.

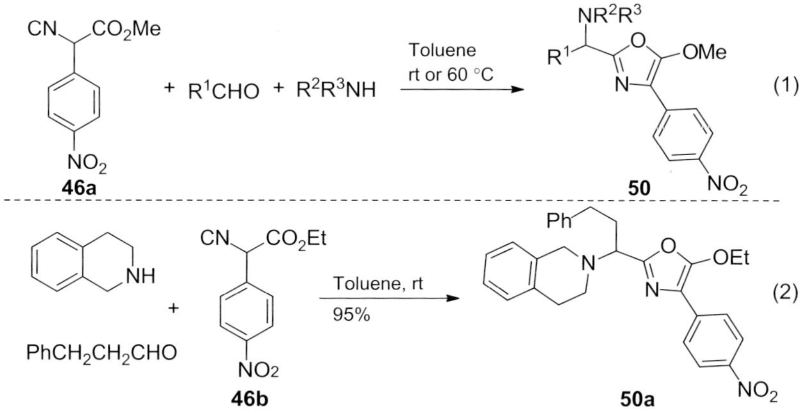

Scheme 15.16 Three-component synthesis of 5-alkoxyoxazoles.

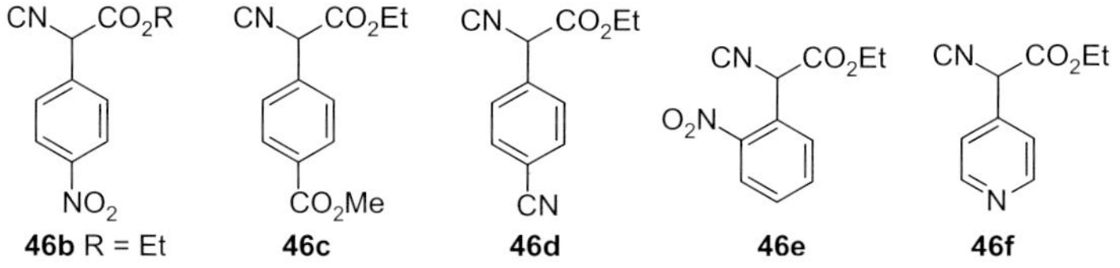

Figure 15.3 α-Aryl-α-isocyanoacetates.

15.3.5
Chemistry of Oxazoles

Stimulated by the efficient three-component synthesis of 5-aminooxazoles and aiming at developing novel MCRs using this heterocycle as a springboard, we became interested in exploiting the reactivity of this heterocycle. Kondrat'eva [32] had demonstrated that the aza-diene system of 5-alkoxyoxazoles reacted readily with the activated dienophiles to give [2 + 4] cycloadducts. Based on this reaction, a landmark synthesis of vitamin B6 (pyridoxine) had been developed by chemists at Merck in the 1960s [33]. Since intermolecular reaction of 5-aminooxazole with dienophile was known to be quite sensitive to the reaction conditions, and various pathways involving [2 + 4], [2 + 3], and even [2 + 2] cycloaddition could occur, we set out to examine the previously unexplored intramolecular cycloaddition of 5-aminooxazole in order to channel these different reaction pathways into a productive process [34]. In combination with the three-component synthesis of 5-aminooxazole, several possibilities existed:

1) A functionalized dienophile is introduced as an additional component.
2) The dienophile is tethered to one of the three inputs (amine, aldehyde, and α-isocyanoacetamide) leading directly, after formation of the transient oxazole, the Diels–Alder (DA) cycloadduct.

15.3.5.1 Dienophile as an Additional Component

Four-Component Synthesis of Pyrrolopyridines If a primary amine were used as an input, the three-component reaction shown in Scheme 15.10 would produce an oxazole bearing a secondary amine. In other words, the so-produced adduct **24** would contain both a diene (oxazole) and a nucleophilic site (amine). Therefore, a dienophile bearing an electrophilic center would pair perfectly with the dual reactivity of **24**. In practice, if an activated α,β-unsaturated carboxylic acid derivative were introduced after the three-component reaction, a sequence of *N*-acylation/intramolecular Diels–Alder reaction (IMDA) might occur to produce the four-component adducts.

Eventually, stirring a toluene solution of oxazole **24a** and acid chloride **51a** at 0 °C for 30 min followed by heating to reflux the reaction mixture provided the pyrrolopyridine **52a** in 65% yield. A possible reaction scenario involving a triple domino process is shown in Scheme 15.17. Thus, acylation of the secondary amine of oxazole **24a** by acyl chloride **51a** gave the amide **53**, which underwent IMDA reaction affording the bridged tricycle intermediate **54**. Base-promoted retro-Michael cycloreversion furnished then the pyrrolopyridine **52a**. The following facts were in accord with this reaction sequence: (i) In sharp contrast to Kondrat'eva's reports, reaction between oxazole **24a** and *N*-phenylmaleimide did not give the corresponding cycloadduct. This result supported the idea that acylation preceded cycloaddition. (ii) When the reaction was carried out in CH_2Cl_2 at rt, we were able to isolate the intermediate **54**. The coupling constant between H_a and H_b ($J_{H_a-H_b}$ = 4.1 *Hz*) indicated a gauche relationship (dihedral angle of 40 °C or so)

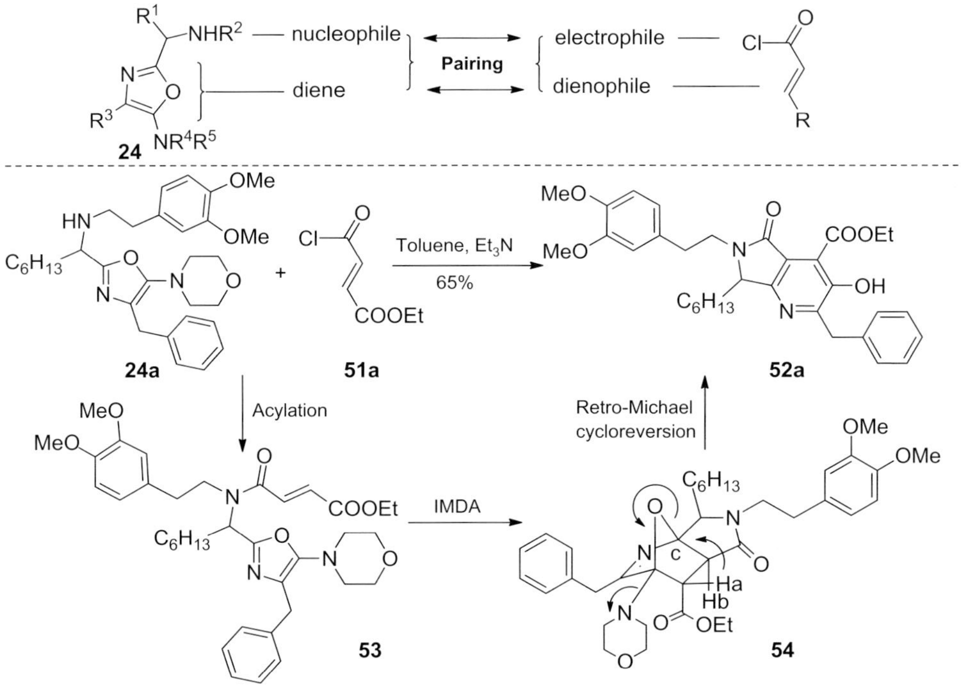

Scheme 15.17 Substrate-pairing in the development of a bimolecular domino process.

between these two protons. For the inherent ring strain imposed by the connecting bridge, only the lactam-*exo*-ester endo mode of cycloaddition was possible, leading to compound **54**. That the retro-Michael cycloreversion dominated over the alternative fragmentation assisted by the lone-pair electrons on the morpholine nitrogen could be explained on the basis of conformational analysis. In the intermediate **54**, the proton H_b is properly aligned with the C_c–O bond (Scheme 15.17), facilitating therefore the otherwise difficult 5-*endo*-trig reversal process.

With these conditions in hand, the development of a one-pot four-component synthesis of pyrrolopyridine was straightforward (Scheme 15.18). Thus, heating a toluene solution (60 °C) of an aldehyde, an amine, and an α-isocyanoacetamide (**23**) in the presence of ammonium chloride (1.5 equiv) for 4 h afforded the oxazole. After cooling the above reaction mixture to 0 °C, an α,β-unsaturated acyl chloride and triethylamine were introduced to it. Heating to reflux the resulting solution produced then pyrrolo[3,4-*b*]pyridin-5-one (**52**). In this one-pot four-component reaction, two C–N bonds and three C–C bonds were created with the formation of a bicyclic core.

Amino ester took part in this reaction effectively, leading to the functionalized pyrrolopyridine as a separable mixture of two diastereomers. The conditions were sufficiently mild so that no epimerization was observed.

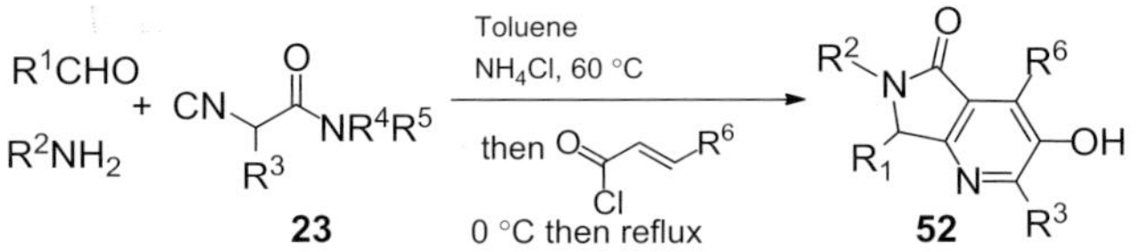

Scheme 15.18 Four-component reaction leading to pyrrolopyridines.

The above reaction has been successfully applied to a three-step synthesis of tetracyclic compounds, akin to the lennoxamine- and nuevamine-type natural products [35].

Five-Component Synthesis of Hexasubstituted Benzene Cycloaddition of oxazole with olefin afforded functionalized pyridine after fragmentation of the oxa-bridged cycloadduct (Eq. (1), Scheme 15.19). Logically, if an alkyne were used as a partner of oxazole, the cycloaddition would give an oxa-bridged cyclohexadiene that could undergo the retro-DA reaction to provide a substituted furan (Eq. (2), Scheme 15.19).

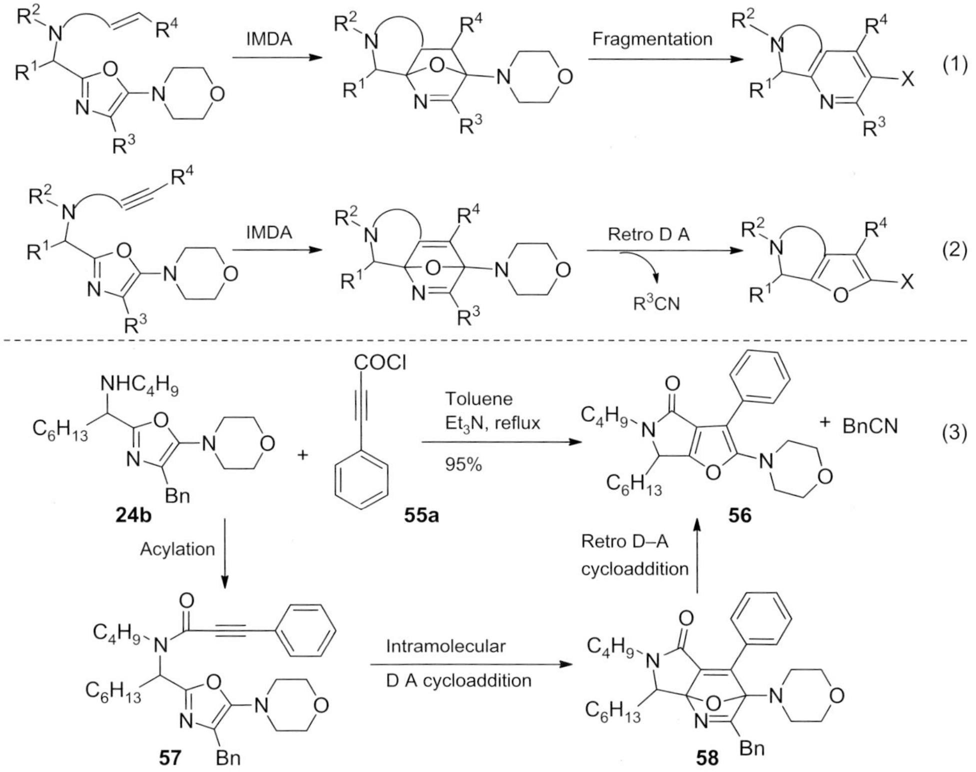

Scheme 15.19 A bimolecular domino acylation/intramolecular DA cycloaddition/retro DA sequence.

While the cycloaddition of oxazole with acetylene is a well-established method for furan synthesis [36], the corresponding reaction of 5-amino oxazole was unknown at the outset of this work. As shown in Eq. (3) of Scheme 15.19, the reaction of **24b** with 3-phenyl-2-propynoyl chloridephenylpropioloyl chloride (**55a**) proceeded smoothly to provide the 5,6-dihydro-furo[2,3-*c*]pyrrol-4-one (**56**) in greater than 95% yield. A triple domino sequence involving acylation/intramolecular DA cycloaddition/retro DA could explain the reaction outcome. That the domino process was initiated by acylation was evidenced by the fact that **24b** failed to react with dimethyl acetylenedicarboxylate (DMAD) under the identical conditions.

The power of the substrate design approach is therefore once again demonstrated from the above two examples. By simply changing the nature of the dienophile (double bond vs triple bond), we were able to synthesize completely different heterocycles starting from the same oxazole, which was in turn synthesized *in situ* by a three-component reaction.

Interestingly, the pyrrolofuran **56** obtained from the above bimolecular domino process is again a potential dienophile and is susceptible to reacting with another dienophile. Indeed, simply refluxing a toluene solution of **56** and *N*-phenylmaleimide (**59a**) afforded the polycycle **60a** having a hexasubstituted benzene ring in over 90% yield. If the reaction was heated at 70 °C, the intermediate **61** was isolated in 47% yield together with its fragmentation product **60a**. Maleic anhydride (**59b**) and naphthalene-1,4-dione (**59c**) are also effective dienophiles for this cycloaddition reaction (Scheme 15.20).

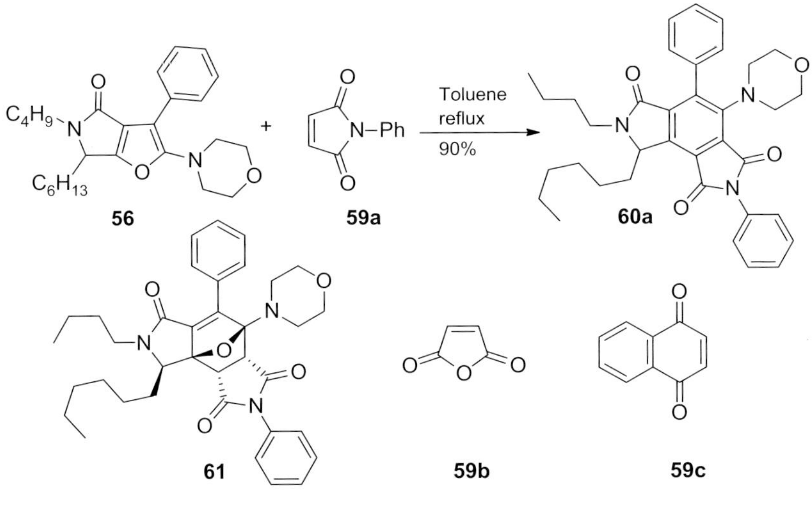

Scheme 15.20 Intermolecular [2 + 4] cycloaddition of pyrrolofuran with dienophiles.

By combing the three-component synthesis of oxazole with the above two bimolecular domino processes, a five-component reaction was subsequently developed [37]. The optimized conditions consisted of performing the reaction in toluene in the presence of a catalytic amount of camphorsulfonic acid (CSA, 0.1 equiv).

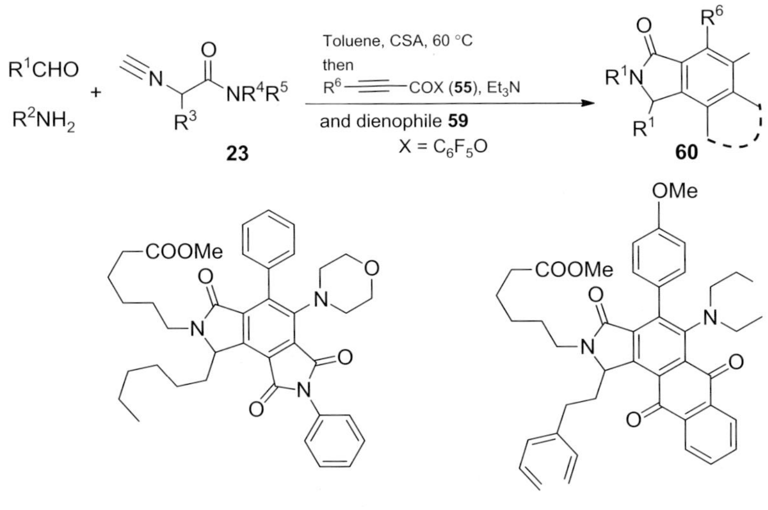

Scheme 15.21 Five-component synthesis of polycyclic compounds.

The generic reaction is depicted in Scheme 15.21. Because of the instability of the substituted propynoyl chloride, we preferred to use the pentafluorophenyl ester **55** ($X = C_6F_5O$), which is much more stable and can even be purified by flash chromatography.

In this one-pot transformation, *seven* functional groups reacted with each other in a highly ordered manner, leading to the creation of *seven* chemical bonds and a polyheterocyclic scaffold with a hexasubstituted benzene core. Not less than *nine* elementary reactions were involved in this experimentally simple MCR. Two dienes (oxazole and furan, both are isolable) were generated in the course of the reaction, which subsequently reacted with two different dienophiles. A catalytic amount of CSA (0.1 equiv) was the only reagent required to catalyze the entire reaction sequence.

15.3.5.2 **Using Dienophile-Containing Inputs**

If a reactive dienophile was incorporated into one of the following inputs, namely, amine, aldehyde, and α-isocyanoacetamide, then the resulting oxazoles **61–63** having both the diene and dienophile functions could undergo directly the IMDA reaction. Depending on the connectivity and the nature of the dienophile (double bond or triple bond), different products would then be produced after a sequence of three-component reaction/DA reaction/fragmentation (Figure 15.4).

On the basis of above consideration, a series of MCRs were developed. Two such examples using anilines having a tethered dienophile as a key reaction partner are shown in Scheme 15.22. Reaction of methyl *ortho*-amino cinnamate (**64**) with α-isocyanoacetamide and heptaldehyde in the presence of LiBr afforded the oxa-bridged polyheterocycles **66** in excellent yield (Eq. (1), Scheme 15.22) [38]. The initial three-component adduct **65** underwent *in situ* DA reaction with the tethered double bond to afford **66** as a mixture of two separable diastereomers in 94% yield.

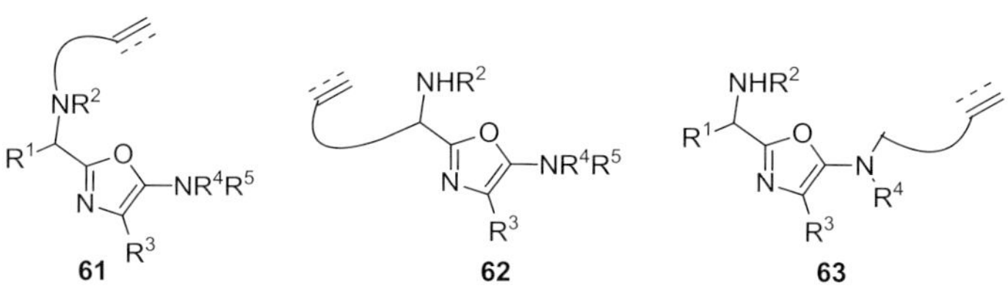

Figure 15.4 Dienophile-tethered oxazoles.

It is worth noting that one C–N, one C–O, and three C–C bonds were formed with the concomitant creation of five asymmetric centers in this one-pot process. The oxazole **65** could be isolated if the reaction was performed at rt. Significantly, the same reaction using *ortho*-aminophenyl propiolate **67** instead of **64** in the presence of ammonium chloride furnished furoquinoline **69** in 75% yield (Eq. (2), Scheme 15.22) [39]. The intramolecular DA cycloaddition of **68** followed by retro DA and oxidation accounted for the formation of **69**. These two examples demonstrated how subtle structural modification of one input (**64** vs **67**) in a given MCR could diversify the structure of the reaction product. Compound **66** is a product with up to five stereocenters, while **69** is a flat aromatic heterocycle.

By exploring exactly the same reaction sequence but changing the tethering position of dienophile, completely different products can again be prepared. Using *N*-(ω-alkenyl)-*N*-alkyl-α-isocyanoacetamide (**70**) as a polyfunctional substrate, a three-component synthesis of pyrrolo aza-indoline **71** was developed. By simply heating a toluene solution of **70**, a primary amine, and an aldehyde afforded the tricyclic compound **71** in good to excellent yield [40]. A possible reaction sequence

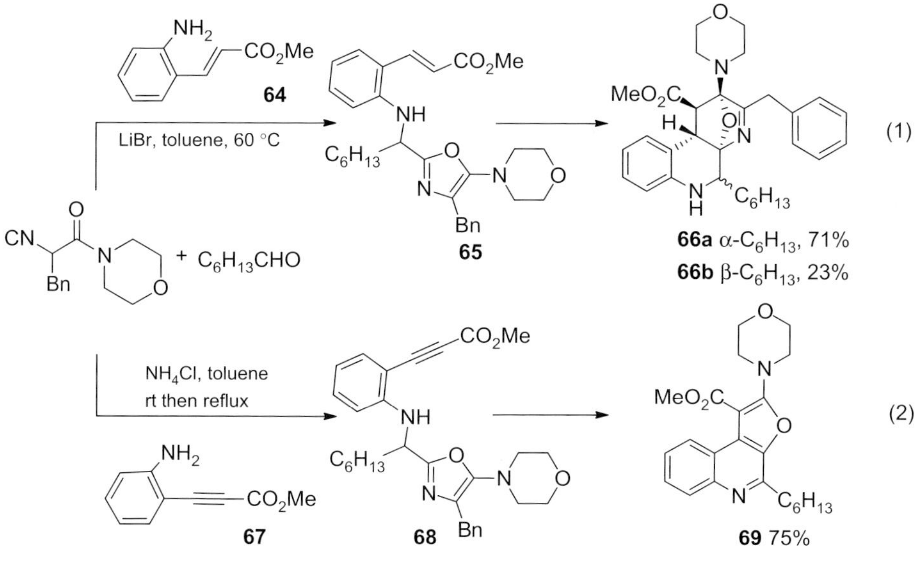

Scheme 15.22 Structure diversity by substrate design.

that accounted for the formation of **71** is depicted in Scheme 15.23. It involved (i) a three-component synthesis of oxazole **72**, (ii) intramolecular DA cycloaddition of **72** leading to **73**, (iii) fragmentation of bridged amino ether function assisted by the lone pair of nitrogen to 6-azaindoline **74**, and (iv) intramolecular transamidation to **71**. The designed substrate **70** contains four functionalities: an isocyano group, an amide, a double bond, and an ester. Initiated by nucleophilic addition of the isocyano carbon to an *in situ* generated iminium, all these reactive functionalities participated in the reaction sequence in a highly ordered manner to afford the observed product. No external reagent was required, and heating was the only external energy supply needed to promote this MCR leading to the creation of five chemical bonds with the concurrent formation of the tricyclic ring system. Water and methanol were the only by-products produced in this MCR. We noted that, under these conditions, Michael addition between amine and the enoate moiety of isocyanoacetamide (**70**) did not occur, which could otherwise have interrupted the entire reaction sequence. The condensation between amine and aldehyde is apparently a much faster process or is thermodynamically more favorable than the undesired Michael addition under this circumstance.

Introducing a functionality that does not participate in the multicomponent condensation but can react with the multicomponent adduct is the basic principle that guided our substrate design in the above examples. Another approach in designing novel multicomponent domino reaction consisted in appending two paired functional groups to the two starting materials which react intramolecularly after the multicomponent process was terminated. In this context, Paulvannan [41] reported an elegant synthesis of bridged tricyclic compounds by a combination of Ugi 4CR and IMDA reactions. Key to this process is the incorporation of a diene and a dienophile in two of the four components of the Ugi reaction. As shown in Scheme 15.24, stirring a methanol solution of furaldehyde (**75**), benzylamine (**76**), benzyl isonitrile (**77**), and ethyl fumarate (**78**) at rt for 36 h provided the cycloadduct **79** in 89% yield (dr = 92/8). The initially formed Ugi adduct **80**, which was isolable,

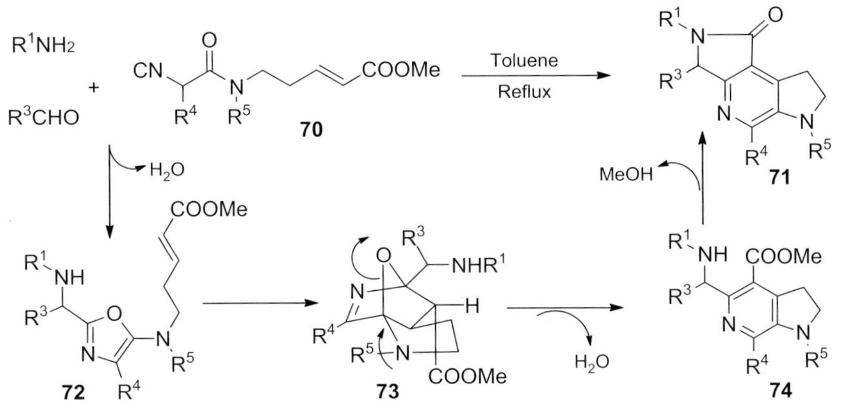

Scheme 15.23 Three-component synthesis of pyrrolo-azaindolines.

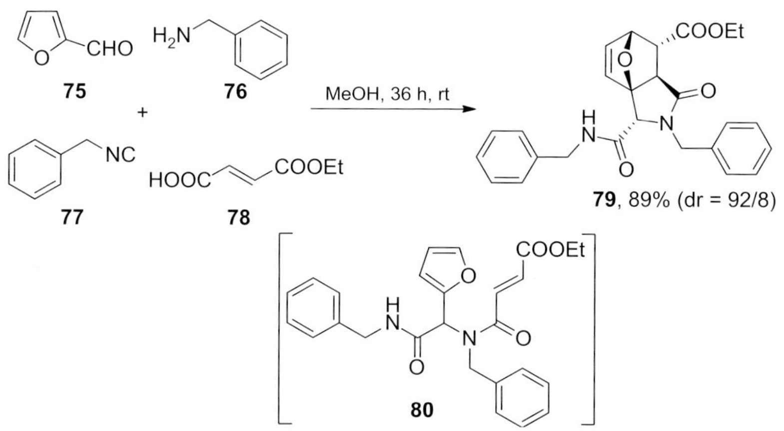

Scheme 15.24 Ugi-4CR/intramolecular Diels–Alder reaction to oxa-bridged heterocycle.

underwent intramolecular [4 + 2] cycloaddition to afford the observed heterocycle. The relative stereochemistry of the major isomer, undetermined at that time, was drawn in light of the X-ray structure of similar compounds obtained by Schreiber and coworkers [42].

On the basis of the same principle, we designed the azido amine **81** and α-isocyanoacetamide bearing an alkyne group **82**. Reaction of **81**, **82**, and heptaldehyde afforded effectively the macrocycle **83** in 45% yield (Scheme 15.25) [43]. The reaction involved a sequence of three-component synthesis of oxazole followed by an intramolecular [3 + 2] cycloaddition between the azide and alkyne. The azido and

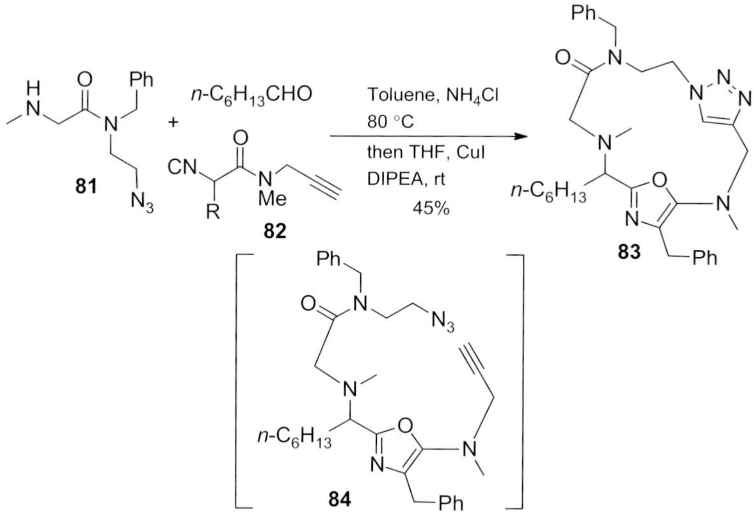

Scheme 15.25 Three-component synthesis of macrocycles by a sequence of 3CR-[3 + 2] cycloaddition.

alkyne functions were not directly involved in the three-component construction of oxazole, but reacted intramolecularly once the oxazole **84** was formed. This reaction created five chemical bonds with the concurrent formation of one macrocycle, one oxazole, and one triazole.

15.3.6 Serendipity

In chemistry, many important chemical transformations were discovered accidentally. Indeed, serendipity, which emanates from researchers with keen observation, open-mindedness, good instincts, and scientific intuition, has played a key role in science in general.

15.3.6.1 Groebke–Blackburn–Bienaymé Reaction

While investigating the U-4CR using 2-aminopyridine as amine input, three industrial research groups led by Bienaymé at Rhone-Poulenc [44], Groebke at Hoffmann-La Roche [45], and Blackburn at Millennium pharmaceuticals [46] discovered that the normal Ugi adduct was not produced. Instead, imidazo[1,2-*a*]pyridine was formed in excellent yields. The acetic acid used in these reactions did not participate in the Ugi reaction as was initially expected; instead, it promoted the formation of the imidazo[1,2-*a*]-heterocycles **86** (Scheme 15.26). The proposed

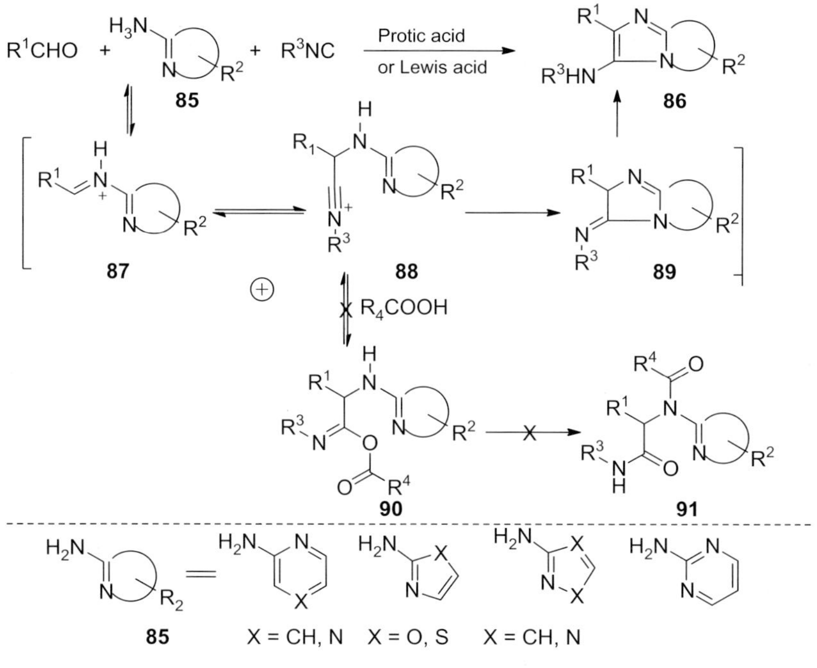

Scheme 15.26 The Groebke–Blackburn–Bienaymé reaction.

mechanism accounting for the formation of **86** involved the nucleophilic addition of isocyanide to iminium **87** leading to nitrilium **88**, which was then trapped intramolecularly by the neighboring nitrogen atom to afford a formal [4 + 1] cycloadduct **89**. A subsequent [1,3-*H*] prototropic shift gave then the final product **86**. The conversion of intermediate **88** and **89** is apparently kinetically faster (a 5-*exo*-dig cyclization) than the classic Ugi pathway involving the addition of carboxylic acid to the nitrilium which would furnish the normal Ugi adduct **91** via the imidate **90**. Other acids such as $HClO_4$ and $Sc(OTf)_3$ can also catalyze efficiently the above three-component reactions.

Needless to say that this unexpected result is much more interesting than if the classic Ugi-4CR adduct **91** were obtained. Indeed, the importance of imidazo[1,2-*a*]pyridine and the related bicyclic heterocycles has stimulated intensive studies on the scope and application of this novel three-component reaction. There seems to be no major limitation regarding the selection of aldehydes and isonitriles. For 2-aminoazenes, both heteroaromatic amidines and heteroaromatic guanidines participated in this novel 3CR. However, aliphatic amidines were found to be inactive. Many variants have since been reported from different research groups, both from academia and industry, because of the medicinal relevance of this family of heterocycles [47].

15.3.6.2 One-Carbon Oxidative Homologation of Aldehydes to Amides

We have shown in Scheme 15.27 that reaction of α-alkyl-substituted α-isocyanoacetic acids with amines (2.0 equiv) and aldehydes afford the corresponding dipeptides. Further exploring the scope of the reaction, we synthesized a potassium salt of α-phenyl-α-isocyanoacetic acid **94a** and submitted it to reaction with the hydrochloride salt of dimethyl amine (**92**), cyclohexanecarbaldehyde (**93**). While the expected dipeptide **38a** was indeed formed in about 5% yield, the *N*-acyl-α-iminoamide **95** was isolated as a major product. Compound **95** was isolable but could be directly converted into amide **97** and ketoamide **96a** upon

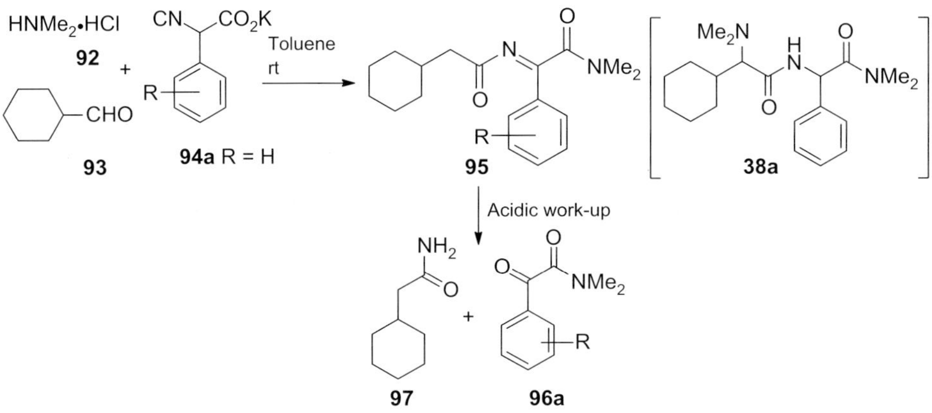

Scheme 15.27 Unexpected internal redox process.

acidic work-up (3 N HCl). If one focused on the structure of aldehyde **93** and amide **97**, one would then realize that this unexpected transformation represented in fact an interesting oxidative one-carbon homologation of aldehydes to amides.

Homologation of carbonyl compounds by one-carbon extension is a demanding transformation in organic synthesis [48]. While several methods exist for the one-carbon homologation of carboxylic acid derivatives including the classic Arndt–Eistert reaction [49], methods allowing the oxidative homologation of an aldehyde to the carboxylic acid derivatives are rare [50]. The mild reaction conditions and the intriguing mechanistic issue, in conjunction with the general lack of this type of homologation procedure, prompted us therefore to examine in detail this serendipity.

Systematically varying the structure of amines and α-isocyano acetic acids allowed us to draw the following conclusions: (i) Hydrochloride salt of dimethyl amine (**92**) was the best reaction partner. Using the free dimethyl amine (a tetrahydrofuran (THF) solution) led to no reaction, and the hydrochloride salt of other dialkyl amine produced the amide in much reduced yield. (ii) Using the potassium salt of α-aryl substituted-α-isocyano acetic acid **94** was a prerequisite to drive the reaction toward the formation of **97**, with **94b** (R = 4-OMe) being the best. The optimum conditions found consisted of treating the aldehyde with the potassium salt of α-(4-methoxyphenyl)-α-isocyano acetic acid (**94b**, R = 4-OMe, 1.0 equiv) and **92** (2.0 equiv) in toluene at rt followed by work-up with 3 N HCl. Under these conditions, the amide **97** was isolated in 81% yield. Both aliphatic and aromatic aldehydes including those with an EWG or electron-donating group were suitable substrates. Naphthaldehyde and pyridine carboxyaldehyde were also homologated. Chiral aldehydes such as (*S*)-(−)-citronellal have been transformed into the homologated amides without racemization. Cyclic ketone can be homologated, but aliphatic ketone failed to participate in the reaction [51].

A plausible mechanism for the formation of the compound **97** is shown in Scheme 15.28. The condensation of the ammonium salt **92** with aldehyde would give the iminium that would be trapped by nucleophilic addition of isocyano carbon to provide the putative nitrilium intermediate **98**. Intramolecular addition of carboxylate to the nitrilium would lead to oxazolone **99**, which would be in equilibrium with 5-hydroxyoxazole **100**. Depending on the nature of the substituent R^1, the subsequent reaction diverged. In the case of an alkyl group, the oxazolone **99** might exist predominantly over 5-hydroxyoxazole **100**. Nucleophilic attack of amine on **99** would then provide the dipeptide **38**. However, when R^1 was an aryl group, the equilibrium between **99** and **100** should shift toward the latter species because of the increased acidity of the proton α to the carbonyl function and additional stabilization offered by the conjugation with the aromatic ring. The 1,6-elimination of dimethylamine assisted by the 5-hydroxy group from **100** would lead to the pseudooxazolone **101**. This process was apparently favored over reprotonation, which was anyway degenerative since the oxazolone **99** would tautomerize back to the resonance-stabilized hydroxyoxazole **100**. Ring-opening of pseudooxazolone **101** by dimethyl amine would afford the *N*-acyl imino amide **95**. To validate this mechanistic proposal, authentic pseudooxazolone **101b** ($R = R^1 = Ph$) was

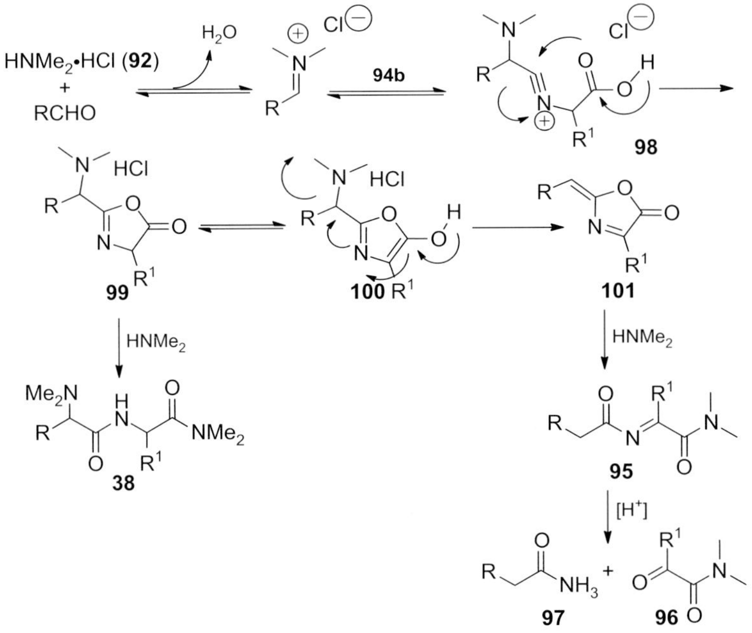

Scheme 15.28 Oxidative homologation of aldehydes to amides: mechanistic proposal.

synthesized from readily available mandelic acid and phenylglycine [52]. Simply mixing **101b** and **92** in toluene in the presence of triethylamine produced **97b** (R = Ph) in 90% yield. This control experiment revealed that the pseudooxazolone could well be the intermediate of the present homologation process.

Application of this methodology in natural product synthesis has recently been reported from the group of Reddy. Hydroxyaldehyde **102a**, in equilibrium with its cyclic hemiacetal form **102b**, was homologated under standard conditions to the amide **103**, which was then converted to (+)-palmyrolide A (**104**), an enantiomer of a neuroactive macrolide isolated from group of cyanobacteria comprised of *Leptolyngbya* and *Oscillatoria* species (Scheme 15.29) [53].

15.3.6.3 **One-Carbon Oxidative Homologation of Aldehydes to α-Ketoamides**

In connection with our project related to peptidomimetics, we had the occasion to examine the U-4CR of *N*-methyl-hydroxylamine **105a** (R = Me), heptanal, benzylisocyanide **106a**, and acetic acid, developed earlier by Guanti and coworkers [54]. Under the conditions used by the above authors (MeOH, rt), we did obtain the expected U-4CR adduct **107a** (R = Me) and the nonacylated adduct **108a** (R = Me) in yields of 45% and 18%, respectively (Scheme 15.30). However, *N*-benzyl-2-oxoheptanamide **109** was also produced unexpectedly in 10% isolated yield. The formation of **109** represented a formal oxidative coupling of aldehyde and isocyanide [55].

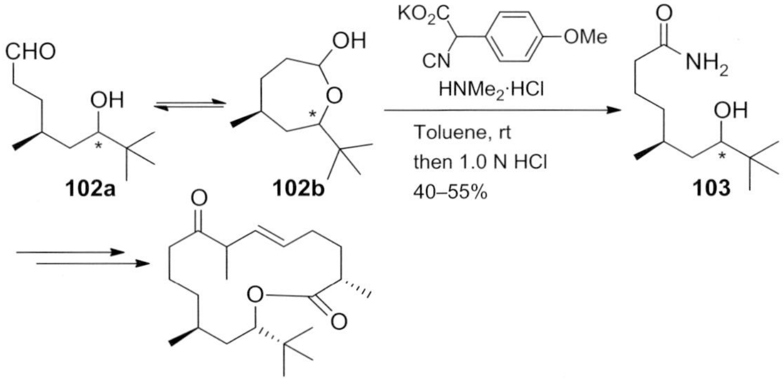

104 (+)-Palmyrolide A

Scheme 15.29 Reddy's synthesis of *ent*-(−)-palmyrolide A.

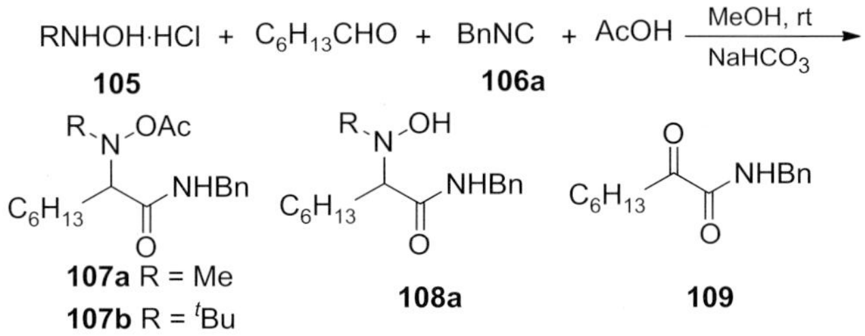

Scheme 15.30 An unexpected product from U-4CR of hydroxyl amine.

The α-ketoamides have been found in a number of biologically important natural products such as FK 506 and cyclotheonamide and in the designed protease inhibitors. Although many synthetic methodologies have been developed for the access to α-ketoamides, a general and efficient synthesis of α-ketoamides is still highly demanding. The fact that the direct coupling of aldehyde with isocyanide to α-ketoamide was unknown prompted us to examine in detail this unexpected result. After survey of a number of reaction parameters, the optimum conditions consisted in performing the reaction in MeOH using the hydrochloride salt of *N*-methylhydroxyamine **105a** (1.0 equiv) and an excess of AcOH (9.0 equiv) in the presence of 4 Å molecular sieves and $NaHCO_3$ (2.0 equiv) at rt. Under these conditions, the α-ketoamide **109** was isolated in 65% yield (Scheme 15.30).

Both aromatic and aliphatic isocyanides including α-isocyanoacetate provided α-ketoamides with moderate to good yields. However, the sterically hindered *tert*-butyl isocyanide provided the coupling product in low yield. Aliphatic aldehydes including functionalized ones are effective substrates. With α-amino aldehyde, *N*,*N*-double protection of the α-amino group was crucial for the reaction to occur properly. Chiral aldehyde **110** has been efficiently transformed into the corresponding α-ketoamide **111** in good yield without epimerization at the anomeric center (Scheme 15.31).

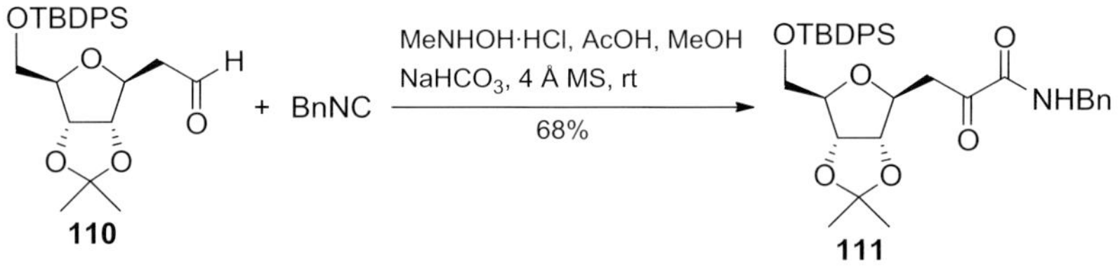

Scheme 15.31 From aldehyde to α-ketoamide: a selected example.

The so-optimized conditions failed to convert the aromatic aldehydes to the corresponding α-ketoamides. It was later found that zinc chloride could promote the coupling between aromatic aldehydes and isonitriles to afford the corresponding α-ketoamides in good yields (Scheme 15.32) [56].

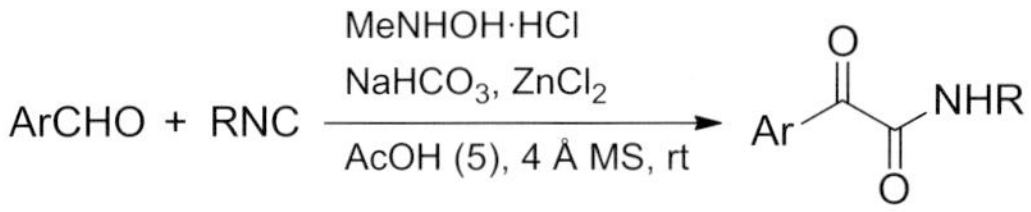

Scheme 15.32 $ZnCl_2$-promoted coupling of aromatic aldehydes with isonitriles to α-ketoamides.

A plausible reaction sequence that accounted for the formation of α-ketoamides is shown in Scheme 15.33. The condensation of an *N*-hydroxylmethylamine and an aldehyde gave the nitrone **112**, which reacted with an isocyanide to afford the nitrilium intermediate **113**. The classic Ugi reaction manifold would lead to the formation of α-acyloxyaminoamide **115** via the imidate intermediate **114**. β-Elimination of acetic acid under the optimized conditions would afford the α-iminoamide **116**, which would then be hydrolyzed to **117**.

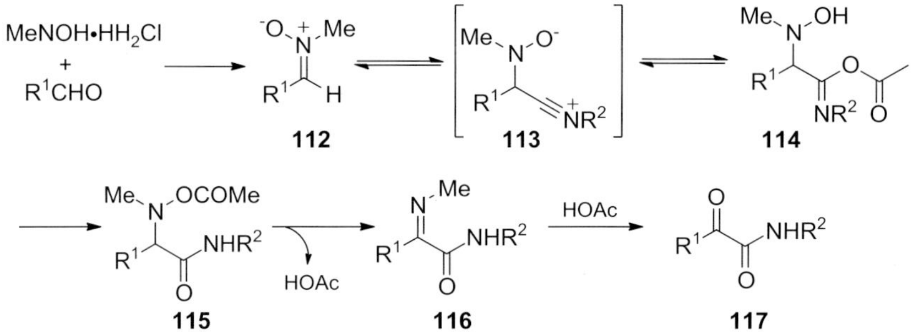

Scheme 15.33 Mechanism for the coupling of aldehydes with isonitriles to α-ketoamides.

15.4 Conclusion

In analyzing the biosynthesis of vitamin B12, Professor Eschenmoser stated: "These outwardly complex structural elements are found to 'self assemble' with surprising ease under structurally appropriate preconditions; the amount of 'external instruction' required for their formation turns out to be surprisingly small in view of the complexity and specificity of these structural elements [57]." While it is still difficult to match Nature's efficiency in contemplating the synthesis of complex molecules, multicomponent domino process is without doubt a powerful tool in the development of efficient and sustainable synthesis of targeted and designed molecules. We believe that the rational "substrate design approach" could be a highly rewarding approach toward the development of ideal MCRs. Indeed, with the critical mechanistic insight of various classic bimolecular reactions, development of new reactive chemical entities, and new activation methods of otherwise "inactive" functional groups, we could be optimistic that many new and synthetically useful MCRs will be developed on rational grounds.

However, the outcome of any organic reaction is highly sensitive to the stereoelectronic and steric properties of the given substrate as well as the subtle modification of the reaction conditions. This is particularly true in the field of MCRs because of the presence of multiple functionalized substrates. Therefore, whatever the soundness of one's working hypothesis, unexpected results always come up. Seizing the serendipity and critically analyzing the experimental feedback could therefore lead to the discovery of new reactions that are often more interesting and useful than the originally planned transformation. As in any scientific domain, the interplay between rational design and serendipity would provide a powerful driving force in the development of novel multicomponent domino reactions.

References

1. (a) Tietze, L.F., Brasche, G., and Gericke, K. (eds) (2006) *Domino Reactions in Organic Synthesis*, Wiley-VCH Verlag GmbH, Weinheim. (b) Tietze, L.F. (1996) *Chem. Rev.*, **96**, 115–136.
2. Tietze, L.F. and Rackelmann, N. (2005) in *Multicomponent Reactions* (eds J. Zhu and H. Bienaymé), Wiley-VCH Verlag GmbH, Weinheim, pp. 121–168.
3. Bartlett, P.A. and Johnson, W.S. (1975) *J. Am. Chem. Soc.*, **95**, 7501–7502.
4. He, F., Bo, Y., Altom, J.D., and Corey, E.J. (1999) *J. Am. Chem. Soc.*, **121**, 6771–6772.
5. Sebahar, P.R. and Williams, R.M. (2000) *J. Am. Chem. Soc.*, **122**, 5666–5667.
6. (a) Strecker, A. (1850) *Justus Liebigs Ann. Chem.*, **75**, 27. (b) Strecker, A. (1850) *Justus Liebigs Ann. Chem.*, **91**, 349. For reviews, see: (c) Enders, D. and Shilvock, J.P. (2000) *Chem. Soc. Rev.*, **29**, 359–373; (d) Gröger, H. (2003) *Chem. Rev.*, **103**, 2795–2827.
7. (a) Hantzsch, A. (1882) *Justus Liebigs Ann. Chem.*, **215**, 1. (b) Bossert, F., Meyer, H., and Wehinger, E. (1981) *Angew. Chem., Int. Ed. Engl.*, **20**, 762–769.
8. (a) Biginelli, P. (1891) *Ber. Dtsch. Chem. Ges.*, **24**, 2962. (b) Biginelli, P. (1893) *Ber. Dtsch. Chem. Ges.*, **26**, 447. (c) Kappe, C.O. (2000) *Acc. Chem. Res.*, **33**, 879–888.

9. (a) Mannich, C. and Krosche, W. (1912) *Arch. Pharm (Weinheim)*, **250**, 647. (b) Arend, M., Westermann, B., and Risch, N. (1998) *Angew. Chem., Int. Ed. Engl.*, **37**, 1045–1070.
10. (a) Passerini, M. (1922) *Gazz. Chim. Ital.*, **52**, 126–129; (b) Passerini, M. (1922) *Gazz. Chim. Ital.*, **52**, 181–189; (c) Banfi, L. and Riva, R. (2005) in *Organic Reactions*, Vol. **65** (ed A.B. Charette), John Wiley & Sons, Inc, pp. 1–140.
11. Dömling, A. and Ugi, I. (2000) *Angew. Chem. Int. Ed.*, **39**, 3168–3210.
12. Robinson, R. (1917) *J. Chem. Soc.*, 762–768.
13. Trost, B.M. (1995) *Angew. Chem., Int. Ed. Engl.*, **34**, 259–281.
14. Bienaymé, H., Hulme, C., Oddon, G., and Schmitt, P. (2000) *Chem. Eur. J.*, **6**, 3321–3329.
15. Miura, K., Fujisawa, N., Saito, H., Wang, D., and Hosomi, A. (2001) *Org. Lett.*, **3**, 2591–2594.
16. For reviews, see: (a) Posner, G.H. (1986) *Chem. Rev.*, **86**, 831–844; (b) Armstrong, R.W., Combs, A.P., Tempest, P.A., Brown, S.D., and Keating, T.A. (1996) *Acc. Chem. Res.*, **29**, 123–131; (c) Hulme, C. and Gore, V. (2003) *Curr. Med. Chem.*, **10**, 51–80; (d) Orru, R.V.A. and De Greef, M. (2003) *Synthesis*, **10**, 1471–1499; (e) von Wangelin, A.J., Neumann, H., Gördes, D., Klaus, S., Strübing, D., and Beller, M. (2003) *Chem. Eur. J.*, **9**, 4286–4294; (f) Balme, G., Bossharth, E., and Monteiro, N. (2003) *Eur. J. Org. Chem.*, **2003**, 4101–4111; (g) Simon, C., Constantieux, T., and Rodriguez, J. (2004) *Eur. J. Org. Chem.*, **2004**, 4957–4980; (h) Wessjohann, L.A. and Rüijter, E. (2005) *Top. Curr. Chem.*, **243**, 137–184; (i) Dömling, A. (2006) *Chem. Rev.*, **106**, 17–89; (j) D'Souza, D.M. and Müller, T.J.J. (2007) *Chem. Soc. Rev.*, **36**, 1095–1108; (k) Akritopoulou-Zane, I. and Djuric, S.W. (2007) *Heterocycles*, **73**, 125–147; (l) Isambert, N. and Lavilla, R. (2008) *Chem. Eur. J.*, **14**, 8444–8454; (m) El Kaim, L. and Grimaud, L. (2009) *Tetrahedron*, **65**, 2153–2171; (n) Ganem, B. (2009) *Acc. Chem. Res.*, **42**, 463–472; (o) Arndtsen, B.A. (2009) *Chem. Eur. J.*, **15**, 302–313; (p) Sunderhaus, J.D. and Martin, S.F. (2009) *Chem. Eur. J.*, **15**, 1300–1308; (q) Godineau, E. and Landais, Y. (2009) *Chem. Eur. J.*, **15**, 3044–3055; (r) Touré, B.B. and Hall, D.G. (2009) *Chem. Rev.*, **109**, 4439–4486; (s) Biggs-Houck, J.E., Younai, A., and Shaw, J.T. (2010) *Curr. Opin. Chem. Biol.*, **14**, 371–382; (t) van Berkel, S.S., Bögels, B.G.M., Wijdeven, M.A., Westermann, B., and Rutjes, F.P.J.T. (2012) *Eur. J. Org. Chem.*, **2012**, 3543–3559.
17. Ruijter, E., Scheffelaar, R., and Orru, R.V.A. (2011) *Angew. Chem. Int. Ed.*, **50**, 6234–6246.
18. (a) Marcaccini, S. and Torroba, T. (1993) *Org. Prep. Proced. Int.*, **25**, 141–208; (b) Gulevich, A.V., Zhdanko, A.G., Orru, R.V., and Nenajdenko, V.G. (2010) *Chem. Rev.*, **110**, 5235–5331.
19. (a) Hoppe, D. (1974) *Angew. Chem., Int. Ed. Engl.*, **13**, 789–804; (b) Schöllkopf, U. (1977) *Angew. Chem., Int. Ed. Engl.*, **16**, 339–348.
20. Suzuki, M., Nunami, K.I., Moriya, T., Matsumoto, K., and Yoneda, N. (1978) *J. Org. Chem.*, **26**, 4933–4935.
21. (a) Bon, R.S., Hong, C., Bouma, M.J., Schmitz, R.F., De Kanter, F.J.J., Lutz, M., Spek, A.L., and Orru, R.V.A. (2003) *Org. Lett.*, **5**, 3759–3762; (b) Bon, R.S., Van Vliet, B., Sprenkels, N.E., Schmitz, R.F., De Kanter, F.J.J., Stevens, C.V., Swart, M., Bickelhaupt, F.M., Groen, M.B., and Orru, R.V.A. (2005) *J. Org. Chem.*, **70**, 3542–3553.
22. The α-isocyanoacetamides are easily available, see: (a) Fayol, A., Housseman, C., Sun, X., Janvier, P., Bienaymé, H., and Zhu, J. (2005) *Synthesis*, 161–165. (b) Housseman, C. and Zhu, J. (2006) *Synlett*, 1777–1779; (c) Dömling, A., Beck, B., Fuchs, T., and Yazbak, A. (2006) *J. Comb. Chem.*, **8**, 872–880.
23. (a) Sun, X., Janvier, P., Zhao, G., Bienaymé, H., and Zhu, J. (2001) *Org. Lett.*, **3**, 877–880; (b) Janvier, P., Sun, X., Bienaymé, H., and Zhu, J. (2002) *J. Am. Chem. Soc.*, **124**, 2560–2567.
24. Yue, T., Wang, M.X., Wang, D.X., Masson, G., and Zhu, J. (2009) *Angew. Chem. Int. Ed.*, **48**, 6717–6721.

25. Wessjohann, L.A., Rivera, D.G., and Vercillo, O.E. (2009) *Chem. Rev.*, **109**, 796–814.
26. Failli, A., Immer, H., and Götz, M. (1979) *Can. J. Chem.*, **57**, 3257–3261.
27. Janvier, P., Bois-Choussy, M., Bienaymé, H., and Zhu, J. (2003) *Angew. Chem. Int. Ed.*, **42**, 811–814.
28. Conformation-directed macrocyclization, see:Blankenstein, J. and Zhu, J. (2005) *Eur. J. Org. Chem.*, **2005**, 1949–1964.
29. (a) Bonne, D., Dehkane, M., and Zhu, J. (2004) *Org. Lett.*, **6**, 4771–4774; Similar reaction performed in MeOH has very low substrate scope, see: (b) Bossio, R., Marcaccini, S., Paoli, P., and Pepino, R. (1994) *Synthesis*, 672–674.
30. Bonne, D., Dekhane, M., and Zhu, J. (2007) *Angew. Chem. Int. Ed.*, **46**, 2485–2488.
31. Lalli, C., Bouma, M.J., Bonne, D., Masson, G., and Zhu, J. (2011) *Chem. Eur. J.*, **17**, 880–889.
32. For a review, see: (a) Karpeiskii, M.Y. and Florent'ev, V.L. (1969) *Russ. Chem. Rev.*, **38**, 540–546.
33. (a) Harris, E.E., Firestone, R.A., Pfister, K. III,, Boettcher, R.R., Cross, F.J., Currie, R.B., Monaco, M., Peterson, E.R., and Reuter, W. (1962) *J. Org. Chem.*, **27**, 2705–2706; (b) Firestone, R.A., Harris, E.E., and Reuter, W. (1967) *Tetrahedron*, **23**, 943–955.
34. For related examples, see: (a) Levin, J.L. and Weinreb, S.M. (1984) *J. Org. Chem.*, **49**, 4325–4332; (b) Jung, M.E. and Street, L.J. (1984) *J. Am. Chem. Soc.*, **106**, 8327–8329; (c) Subramanyam, C., Noguchi, M., and Weinreb, S.M. (1989) *J. Org. Chem.*, **54**, 5580–5585; (d) Levin, J.L. (1989) *Tetrahedron Lett.*, **30**, 2355–2358; (e) Jung, M.E. and Dansereau, S.M.K. (1994) *Heterocycles*, **39**, 767–778; (f) Vedejs, E., Piotrowski, D.W., and Tuci, F.C. (2000) *J. Org. Chem.*, **65**, 5498–5505; (g) Padwa, A., Dimitroff, M., and Liu, B. (2000) *Org. Lett.*, **2**, 3233–3235; (h) Paquette, L.A. and Efrenov, I. (2001) *J. Am. Chem. Soc.*, **123**, 4492–4501.
35. (a) Gámez-Montaño, R. and Zhu, J. (2002) *Chem. Commun.*, 2448–2449; (b) Islas-Jácome, A., González-Zamora, E., and Gámez-Montaño, R. (2011) *Tetrahedron Lett.*, **52**, 5245–5248; (c) Zamudio, A., Garciá-González, M.C., Padilla, J., and González-Zamora, E. (2010) *Tetrahedron Lett.*, **51**, 4837–4839.
36. (a) For a review on intramolecular DA reaction of oxazole and acetylene, see:Jacobi, P.A. (1992) in *Advances in Heterocyclic Natural Product Synthesis*, Vol. **2** (ed W.H. Pearson), Jai Press Inc, Greenwich, CT, pp. 251–298.
37. Janvier, P., Bienaymé, H., and Zhu, J. (2002) *Angew. Chem. Int. Ed.*, **41**, 4291–4294.
38. (a) González-Zamora, E., Fayol, A., Bois-Choussy, M., Chiaroni, A., and Zhu, J. (2001) *J. Chem. Soc., Chem. Commun.*, 1684–1685; (b) Fayol, A., González-Zamora, E., Bois-choussy, M., and Zhu, J. (2007) *Heterocycles*, **73**, 729–746.
39. Fayol, A. and Zhu, J. (2002) *Angew. Chem. Int. Ed.*, **41**, 3633–3635.
40. Fayol, A. and Zhu, J. (2005) *Org. Lett.*, **7**, 239–242.
41. Paulvannan, K. (1999) *Tetrahedron Lett.*, **40**, 1851–1854.
42. Lee, D., Sello, J.K., and Schreiber, S.L. (2000) *Org. Lett.*, **2**, 709–712.
43. Pirali, T., Tron, G.C., and Zhu, J. (2006) *Org. Lett.*, **8**, 4145–4148.
44. Bienaymé, H. and Bouzid, K. (1998) *Angew. Chem., Int. Ed. Engl.*, **37**, 2234–2237.
45. Groebke, K., Weber, L., and Mehlin, F. (1998) *Synlett*, 661–663.
46. Blackburn, C., Guan, B., Fleming, P., Shiosaki, K., and Tsai, S. (1998) *Tetrahedron Lett.*, **39**, 3635–3638.
47. (a) Chen, C., Xiang, J., Wang, G.X., Fathi, R., Quan, J.M., and Yang, Z. (2007) *J. Comb. Chem.*, **9**, 982–989; (b) Hulme, C. and Lee, Y.S. (2008) *Mol. Diversity*, **12**, 1–15.
48. Stowell, J.C. (1984) *Chem. Rev.*, **84**, 409–435.
49. Ye, T. and Mckervey, M.A. (1994) *Chem. Rev.*, **94**, 1091–1160.
50. Via dibromoalkene, see: (a) Huh, D.H., Jeong, J.S., Lee, H.B., Ryu, H., and Kim, Y.G. (2002) *Tetrahedron*, **58**, 9925–9932; (b) Shen, W. and Kunzer, A. (2002) *Org. Lett.*, **4**, 1315–1317.

51. Bonne, D., Dehkane, M., and Zhu, J. (2005) *J. Am. Chem. Soc.*, **127**, 6926–6927.
52. Breitholle, E.G. and Stammer, C.H. (1974) *J. Org. Chem.*, **39**, 1311–1312.
53. Philkhana, S.C., Seetharamsingh, B., Dangat, Y.B., Vanka, K., and Reddy, D.S. (2013) *Chem. Commun.*, **49**, 3342–3344.
54. Basso, A., Banfi, L., Guanti, G., and Riva, R. (2005) *Tetrahedron Lett.*, **46**, 8003–8006.
55. Insertion of isocyanide to C–O bond of acetal, see: (a) Pellissier, H., Meou, A., and Gil, G. (1986) *Tetrahedron Lett.*, **27**, 2979–2980; (b) Tobisu, M., Kitajima, A., Yoshioka, S., Hyodo, I., Ospita, M., and Chatani, N. (2007) *J. Am. Chem. Soc.*, **129**, 11431–11437.
56. Bouma, M., Masson, G., and Zhu, J. (2010) *J. Org. Chem.*, **75**, 2748–2751.
57. Echenmoser, A. (1988) *Angew. Chem., Int. Ed. Engl.*, **27**, 5–39.

Index

Domino Reactions: Concepts for Efficient Organic Synthesis, First Edition. Edited by Lutz F. Tietze.

d

e